Neotropical Hard Ticks (Acari: Ixodida: Ixodidae)

Alberto A. Guglielmone • Santiago Nava •
Richard G. Robbins

Neotropical Hard Ticks (Acari: Ixodida: Ixodidae)

A Critical Analysis of Their Taxonomy, Distribution, and Host Relationships

Springer

Alberto A. Guglielmone
Inst Nacional de Tecnolog. Agropecuaria
Estacion Experimental Agropecuaria
Rafaela, Argentina

Santiago Nava
Inst Nacional de Tecnolog. Agropecuaria
Estación Experimental Agropecuaria
Rafaela, Argentina

Richard G. Robbins
Dept Entomology, Smithsonian Institution
Walter Reed Biosystematics Unit
Suitland, MD, USA

ISBN 978-3-030-72352-1 ISBN 978-3-030-72353-8 (eBook)
https://doi.org/10.1007/978-3-030-72353-8

This Springer imprint is published by the registered company Springer Nature Switzerland AG.
The registered company address is: Gewerbestrasse 11, 6330 Cham, Switzerland

This work is dedicated to the memory of
Carl Ludwig KOCH (1778–1857)
For his founding contributions to tick (Ixodida) classification, originally proposed in 1844 and basically unchanged to this day.

Preface

The family Ixodidae, whose members are commonly known as hard ticks, currently comprises 137 Neotropical species, most of which are endemic to this region. Here we examine each Neotropical ixodid, providing information on original descriptions and redescriptions, geographic distribution, hosts, and human parasitism, discussed in both general and specific terms, together with a review of any related problems, such as tick invasiveness or the confusion that can arise when incorrect names are applied to tick taxa. A critical analysis of the scientific literature indicates that there is a pressing need to reevaluate the taxonomic status of several Neotropical tick species, research that in turn will impact our understanding of tick systematics, ecology, and evolution while also casting new light on the importance of these arthropods as vectors of human and animal diseases. We believe that this synopsis will be welcomed by biologists, veterinarians, public health professionals, or anyone interested in ticks as fascinating organisms in their own right, whether working in the Neotropics or elsewhere. The literature search for all tick species treated here concluded on October 31, 2020.

Rafaela, Argentina — Alberto A. Guglielmone
Rafaela, Argentina — Santiago Nava
Suitland, MD, USA — Richard G. Robbins

Acknowledgments

We warmly recognize the enthusiastic collaboration of the following colleagues who located publications, provided critically important information, or discussed issues relevant to this study: *Dmitry A. Apanaskevich*, United States National Tick Collection, Georgia Southern University, Statesboro, USA; *Darci M. Barros-Battesti*, Laboratório de Parasitología, Instituto Butantan, São Paulo, Brazil; *Lorenza Beati*, United States National Tick Collection, Georgia Southern University, Statesboro, USA; *Sergio E. Bermúdez*, Departamento de Investigación en Entomología Médica, Instituto Conmemorativo Gorgas de Estudio de la Salud, Panama; *Lance A. Durden*, Department of Biology, Georgia Southern University, Statesboro, USA; *Fernando S. Flores*, Facultad de Ciencias Veterinarias, Universidad Nacional del Litoral, Esperanza, Argentina; *Agustín Estrada-Peña*, Facultad de Veterinaria, Universidad de Zaragoza, Zaragoza, Spain; *Carmen Guzmán-Cornejo*, Departamento de Biología, Facultad de Ciencias, Universidad Autónoma de México, Distrito Federal, México; *Marcelo B. Labruna*, Departamento de Medicina Veterinária Preventiva e Saúde Animal, Faculdade de Medicina Veterinária e Zootecnia, Universidade de São Paulo, Brazil; *Thiago F. Martins*, Departamento de Medicina Veterinária Preventiva e Saúde Animal, Faculdade de Medicina Veterinária e Zootecnia, Universidade de São Paulo, Brazil; *Mariano Mastropaolo*, Facultad de Ciencias Veterinarias, Universidad Nacional del Litoral, Esperanza, Argentina; *Gourdaz Moalei*, Center for Vector Biology and Zoonotic Diseases, Connecticut Agricultural Experimental Station and Yale School of Public Health, Connecticut, USA; *Trevor N. Petney*, Department of Ecology and Parasitology, Institute of Zoology, Karlsruhe Institute of Technology, Karlsruhe, Germany; *Marta E. Sánchez*, Estación Experimental Agropecuaria Rafaela, Instituto Nacional de Tecnología Agropecuaria, Rafaela, Argentina; *Matias J. P. Szabó*, Faculdade de Medicina Veterinária, Universidade Federal de Uberlândia, Uberlândia, Brazil; *Evelina L. Tarragona*, Estación Experimental Agropecuaria Rafaela, Instituto Nacional de Tecnología Agropecuaria, Rafaela, Argentina; and *José M. Venzal*, Facultad de Veterinaria, Universidad de la República, Salto, Uruguay.

In Argentina, the Instituto Nacional de Tecnología Agropecuaria and the Asociación Cooperadora of the Estación Experimental Agropecuaria provided support to Alberto A. Guglielmone and Santiago Nava. In Washington, DC, Richard G. Robbins was ably assisted by the staff of the Smithsonian Institution Libraries, especially Richard Greene, Amanda Landis, Carrie Smith, Gil Taylor, Wanda West, and Bonnie White, who patiently sought and supplied many of the papers cited here. All material in this book has been reviewed by the Walter Reed Army Institute of Research. There is no objection to its presentation and/or publication. The opinions or assertions contained herein are the private views of the authors and are not to be construed as official or reflecting the true views of the U.S. Department of the Army or the Department of Defense.

Introduction

Hard ticks (Acari: Ixodidae) are blood-feeding ectoparasites of amphibians, reptiles, birds, and mammals worldwide. Ticks may injure their hosts by causing dermatoses, anemia, paralysis, and other debilitating conditions, but their chief importance lies in their capacity to transmit pathogens, some of which may be fatal to domestic and wild animals as well as humans. The Neotropical Zoogeographic Region covers a vast territory extending from the eastern and western flanks of the Mexican Plateau southward to southern Argentina, Chile, and their adjacent islands plus the Greater and Lesser Antilles and the Galápagos Islands (Cox and More 2005), a definition accepted earlier by Guglielmone et al. (2014) in their analysis of the Ixodidae of the world. This definition is problematic only insofar as its northern limit is concerned. Guglielmone et al. (2003a) stated that in the east the Neotropics includes the Mexican state of Veracruz, while in the west it comprises western Nayarit south of 22° N through western Jalisco, all of Colima State, western Michoacán, and western Guerrero southward through Central America. However, the western Neotropical limit has since been reevaluated and currently includes most of the states of Michoacán and Guerrero, while Morelos is now considered as belonging *in toto* to the Neotropical Region and is herein referred to as "southern Mexico." As a consequence, two species of Ixodidae formerly listed as occurring only in the Nearctic Region (Guglielmone et al. 2003a, 2014) are now treated as Neotropical taxa, and one species heretofore regarded as Nearctic is currently treated as found in both the Nearctic and Neotropical Zoogeographic Regions.

The Neotropical Ixodidae were earlier reviewed by Guglielmone et al. (2003a). Later in-depth studies focused on Neotropical species of the genus *Amblyomma* (Voltzit 2007), compared the Neotropical tick fauna with those of other zoogeographic regions (Guglielmone et al. 2014), addressed the ticks of particular Neotropical countries (e.g., González-Acuña and Guglielmone 2005; Mastropaolo et al. 2014; Dantas-Torres et al. 2019b), or examined the ecology and taxonomy of ticks from several southern South American countries (Nava et al. 2017). More recently, Estrada-Peña et al. (2020) analyzed host–tick interactions of several Neotropical species belonging to the genera *Amblyomma*, *Haemaphysalis,* and *Ixodes*, while

Guglielmone et al. (2020) compiled descriptions and redescriptions of all species of Ixodidae known by the end 2019, an exercise that underscored some significant problems in the morphological identification of several Neotropical taxa.

The initial review of Neotropical Ixodidae by Guglielmone et al. (2003a) provided information on type depositories, known tick life history stages, geographical distribution by country, principal hosts, and literature for identifying each species, together with a general discussion of the impact of ticks on human and animal health. Apart from the appearance of several new tick species descriptions, a vast amount of biological information has been published on Neotropical ticks over the last 16 years, prompting the present attempt to update our knowledge of the Neotropical tick fauna. Further, it is now possible to address the host profiles of Neotropical ixodids in some detail—these were very brief in Guglielmone et al. (2003a) and extended only to the family level in Guglielmone et al. (2014). Nava and Guglielmone (2013) analyzed host specificity in adults of 42 species of Neotropical tick as well as the immature stages of 22 Neotropical tick taxa. Estrada-Peña et al. (2020) grouped the parasitic stages of Neotropical species of *Amblyomma*, *Haemaphysalis,* and *Ixodes* into clusters according to their host affinities, yielding conclusions that sometimes disagree with those presented here. However, a full analysis of tick–host specificity in the Neotropical Region remains a work in progress. Previously, no attempt had been made to list every species of host for each stage of every tick species. Here we include such information, but with the proviso that a thorough understanding of Neotropical ixodid–host relationships remains a long-term objective.

As noted above, Guglielmone et al. (2003a) summarized the medical and veterinary importance of Neotropical ticks, but this topic is outside the scope of this study. Emerging and reemerging tick-borne pathogens or potential pathogens are receiving increasing attention by many workers, and such studies have to date resulted in a myriad of publications that would easily justify a separate text. With that in mind, we have focused on updating the basic biological information for novel and previously described tick species and stages, with an emphasis on their geographical distribution, and principal and exceptional host relationships. Information on invasive tick species is also provided, as is a guide to the numerous incorrect names that have historically appeared in the Neotropical tick literature. While compiling the data for this study, several new nomenclatural and taxonomic problems emerged, and these are discussed together with some as yet unresolved historical problems (e.g., the *Amblyomma maculatum–Amblyomma triste* complex, and the *Ixodes auritulus* species group). We hope that this single-volume summary of the Neotropical tick fauna will prove helpful to the many biologists, epidemiologists, veterinarians, sanitary workers, and other public health professionals whose work brings them into regular contact with Neotropical ticks and tick-borne diseases.

Methodology

The family Ixodidae comprises 758 species (Guglielmone et al. 2015 and updates), a significant increase from the 702 species listed in Guglielmone et al. (2010b). Historically, the Ixodidae has been divided into two groups: the Prostriata, containing ticks in which the anal groove borders the anus anteriorly, and the Metastriata, in which the anal groove, if present, curves posterior to the anus. The Prostriata group consists solely of the genus *Ixodes* and currently contains 265 species, while the Metastriata consists of 492 taxa in 16 genera as follows: *Africaniella* (2 species), *Amblyomma* (136 species), *Anomalohimalaya* (3), *Archaeocroton* (1), *Bothriocroton* (7), *Compluriscutula* (fossil, 1 species), *Cornupalpatum* (fossil, 1 species), *Cosmiomma* (1), *Dermacentor* (42), *Haemaphysalis* (176), *Hyalomma* (27), *Margaropus* (3), *Nosomma* (2), *Rhipicentor* (2), *Rhipicephalus* (87), and *Robertsicus* (1 species), plus one taxon wrongly assigned to the genus *Haemaphysalis* in Chitimia-Dobler et al. (2018) as discussed in Guglielmone et al. (2020). The Prostriata are well represented in the Neotropical Region, as is the metastriate genus *Amblyomma* and, to a much lesser extent, the genera *Dermacentor*, *Haemaphysalis,* and *Rhipicephalus*. The relative representation of Neotropical ixodid genera has not significantly changed in decades.

Information Presented for Species of Ixodidae Each species of hard tick discussed here is introduced by its currently accepted scientific name, date of authorship, and author name(s), together with the paper in which the original description can be found, and any necessary comments on the taxonomic status of that name. This information is followed by:

Descriptions and redescriptions: this section provides authors and citations for the first description of each active life history stage (female, male, nymph, larva), together with a non-exhaustive list of studies that contain redescriptions of these stages.

Geographic distribution: this section first outlines each species' overall geographic range but is then followed by an alphabetical list of the countries or territories and their administrative divisions in which the species is known to occur, based on published literature.

Hosts: a general statement describing each tick species' host specificity is followed by the principal hosts of particular life history stages or combinations of stages, together with notes on provisional or potential principal hosts, all documented by literature references. An accompanying table lists the particular Neotropical host species of Amphibia, Reptilia, Aves, and Mammalia from which each tick stage has been collected. Where a tick species also occurs in zoogeographic regions outside the Neotropics, the families of hosts in those regions are provided.

Human parasitism: this section recapitulates data from Guglielmone and Robbins (2018) and updates that work with subsequently published records.

Comments: in those cases where reports conflicting with any of the information in the above sections are known to exist, they are discussed in depth here. Such commentary is considered important because it may point to the need for further research on the taxonomy, ecology, or evolution of particular tick species and may have a bearing on studies of tick-borne diseases.

Additional Comments on Geographical Distribution of Ixodidae In this work, the ranges of hard ticks are discussed on a country basis, but also with respect to cross-national subdivisions that highlight unique characteristics of the ixodid fauna. Some hard tick species have entered the Neotropical Region by a natural process of range expansion, as a consequence of the absence of significant geographic barriers between the southern Nearctic Region and the northern Neotropics. Additionally, long-term geological phenomena, such as the rise of the Isthmus of Panama *circa* three million years ago (late Cenozoic), ended nearly 50 million years of South American isolation. By connecting Central and South America, the Isthmus facilitated the southward movement of northern vertebrates and their ticks, and, conversely but to a lesser extent, the northward movement of South American vertebrates and their parasites, with the result that some Neotropical hard ticks eventually reached the Nearctic Zoogeographic Region. With the arrival of Europeans, tick species long associated with livestock or domestic animals, such as *Amblyomma variegatum*, *Rhipicephalus microplus,* and *Rhipicephalus sanguineus sensu stricto* and *sensu lato*, successfully invaded the Neotropical Region and the latter two species are now widespread. Improvements in transport have led to the interception of several additional exotic tick species, such as those associated with the worldwide pet trade, but their frequent detection should serve as a warning that such species may become invasive, especially in tropical and subtropical regions. Accordingly, a section of this book is devoted to introduced tick species, evidencing the need to develop protocols to mitigate or prevent the establishment of foreign tick species in Neotropical countries.

Extensive research on the Neotropical tick fauna, and the proliferation of scientific papers generated by that research, has often led to the citation of tick names that are taxonomically invalid or that have been erroneously or doubtfully ascribed to the Neotropical Region. As an aid to investigators who may be unfamiliar with the intricacies of zoological nomenclature, or for those workers who are specifically interested in taxonomic problems, we have provided a list of these confusing names, with explanatory remarks.

Additional Comments on Host–Tick Relationships The information presented here on hosts of Neotropical ticks is based on examination of more than 2500 published scientific papers; with a few exceptions, data presented during professional meetings are not included in this study. Additionally, papers that do not specifically indicate the tick stages found on particular host species were excluded from our analysis of host–tick associations.

Ixodid hosts have been grouped into four traditional classes: Amphibia, Reptilia, Aves, and Mammalia, and are treated here in that order. Some workers regard Aves as a subgroup of Reptilia or Sauropsida, but we have retained Aves as a matter of convenience. We follow the classificatory scheme of Clements et al. (2018) for

Aves; the International Union for Conservation of Nature (IUCN) (2018) for mammals; and the IUCN with supplementary sources for Amphibia and Reptilia, including the reptile list in Guglielmone and Nava (2010). Several older records of tick parasitism were published using host names that are currently considered invalid; in these cases, an effort has been made to trace the synonymies of such names, but where it has not been possible to associate older names with their current valid equivalents, the records involved have also been excluded from our host–parasite analyses. There are also some significant differences between the host classifications adopted here and those appearing in relatively recent discussions of tick–host specificity, such as Guglielmone et al. (2014), Nava et al. (2017), and Guglielmone and Robbins (2018). One example is the order Artiodactyla, which is sometimes included in the order Cetartiodactyla, merging Cetacea and Artiodactyla, as seen in the IUCN (2018), but some authors argue that Cetacea is a clade within Artiodactyla, a view accepted here. Obviously, host names, like those of their parasites, are subject to continuous change as our knowledge of their phylogenetics improves with advances in the methods employed by systematic biologists.

Our analysis of host–parasite relationships for the genera of Neotropical Ixodidae has necessarily been non-statistical, since records often do not exist for one or more tick parasitic stages and collection records vary from hundreds to sometimes only one (a not infrequent case). When an obvious host preference can be inferred, principal hosts are defined at the level of class and order. Alternatively, where only one or a few collection records exist, the host class and order are considered potential or provisional. For each ixodid genus, we further categorize combinations of classes and orders of hosts, with the goal of revealing tendencies in host utilization that may serve as a tentative basis for further studies of tick–host specificity when additional data are available. Again, where there is uncertainty in the interpretation of historical data, because of doubts concerning the true identity of either the hosts or ticks involved, such questionable associations appear as comments under individual tick species and in the last chapter of this study.

Human parasitism by Neotropical hard tick species, also a topic of concern here, has been accorded separate analysis and discussion, followed immediately by a list of all known hosts of Neotropical Ixodidae together with the stages that have been reported from these hosts.

The above material, for which the literature review ended on October 31, 2020, is presented in five chapters as follows:

Chapter 1: contains summaries of the species in the group Prostriata (genus *Ixodes*) together with general comments on prostriate ticks.

Chapter 2: contains summaries of the species in the metastriate genus *Amblyomma*, which numerically is the most important genus in the Neotropical Region, as well as general comments on this genus.

Chapter 3: contains summaries of the species in metastriate genera other than *Amblyomma*, which in the Neotropical Region includes the genera *Dermacentor*, *Haemaphysalis,* and *Rhipicephalus*, plus general comments pertaining to ticks in these genera.

Chapter 4: in this chapter, the geographic distribution of hard ticks is discussed on a country basis and by territorial subdivisions of the Neotropical Region, highlighting differences in the ixodid fauna throughout the region. A list of invasive species of Ixodidae detected in the Neotropics that apparently failed to become established is included in this chapter, along with a substantial but non-exhaustive list of names incorrectly applied to ixodids found in Neotropical territories.

Chapter 5: this chapter addresses host–tick associations. Within a given class of hosts (Amphibia, Reptilia, Aves, and Mammalia), it lists the orders, families, genera, and species parasitized by Neotropical Ixodidae. There follows a synopsis of tick parasitism within that class. This leads into a discussion of host utilization by the different genera of Ixodidae in the Neotropical Region, with remarks on some peculiarities of host–tick relationships and general comments related to their hypothetical evolution. The chapter concludes with a summary of human parasitism by Neotropical hard ticks. The appendix of this book, which is also related to this chapter, contains a full list of all known hosts of Neotropical Ixodidae together with the stages that have been reported from these hosts.

Contents

Chapter 1
Prostriata Group: Genus *Ixodes*

There are 265 species of *Ixodes* worldwide. This genus is most speciose in the Afrotropical Zoogeographic Region, where 70 species (26% of the total) occur, 67 of which are found nowhere else, followed by the Palearctic Zoogeographic Region with 61 species (23%), 44 of which are found exclusively in this region. A total of 55 species (21%) are known from the Neotropics, 43 of them endemic to this region.

1. *Ixodes abrocomae* Lahille, 1916

Lahille, F. (1916) Descripción de un nuevo ixódido chileno. *Revista Chilena de Historia Natural*, 20, 107–108.

Descriptions and Redescriptions: The male was described but not figured by Lahille (1916), while the female of *Ixodes abrocomae* was described almost one hundred years later by Guglielmone et al. (2010a).

Guglielmone et al. (2010a) redescribed the male, and both sexes of *Ixodes abrocomae* were redescribed by Nava et al. (2017).

The larva and nymph of *Ixodes abrocomae* remain unknown.

Geographical Distribution: A southern South American species that has only been found in **Chile**: Atacama and Coquimbo (Guglielmone et al. 2010a).

Hosts: As detailed in Table 1.1, all hosts of *Ixodes abrocomae* are from the order Rodentia: Abrocomidae and Cricetidae (Lahille 1916; Guglielmone et al. 2010a). *Ixodes abrocomae* is known from four males and two females, and the four host species below are treated as provisional principal hosts of this tick.

Human Parasitism: No.

Comments: Prior to the description of the female by Guglielmone et al. (2010a), *Ixodes abrocomae* had only been known from the male holotype, which Camicas et al. (1998) suggested was the male of *Ixodes sigelos*, but this synonymy was

A. A. Guglielmone et al., *Neotropical Hard Ticks (Acari: Ixodida: Ixodidae)*,
https://doi.org/10.1007/978-3-030-72353-8_1

Table 1.1 Hosts for males (M) and females (F) of *Ixodes abrocomae*

MAMMALIA			
RODENTIA: Abrocomidae		*Abrothrix olivaceous*	M
Abrocoma bennettii	M	*Phyllotis xanthopygus*	F
RODENTIA: Cricetidae			
Abrothrix longipilis	M		

rejected by Guglielmone et al. (2003a). Thereafter, Guglielmone et al. (2010a) presented morphological and molecular evidence to support the validity of *Ixodes abrocomae*, and to separate if from *Ixodes sigelos*; nevertheless, the hypothesis of conspecificity between *Ixodes abrocomae* and *Ixodes sigelos* has again been proposed by Muñoz-Leal et al. (2019a), who found three different populations of ticks morphologically close to *Ixodes abrocomae* and *Ixodes sigelos* but with dissimilar molecular sequences. Even so, Muñoz-Leal et al. (2019b) appear to recognize the validity of *Ixodes abrocomae*.

2. *Ixodes affinis* Neumann, 1899

Neumann, L.G. (1899) Révision de la famille des ixodidés (3e mémoire). *Mémoires de la Société Zoologique de France*, 12, 107–294.

Descriptions and Redescriptions: Neumann (1899) described the female of *Ixodes affinis* in a brief text lacking illustrations; the male was described with figures in Cooley and Kohls (1945), and the larva and nymph were described by Oliver et al. (1987).

Cooley and Kohls (1945) redescribed the female, while Keirans and Clifford (1978) provided diagnostic characters and illustrations of both sexes of *Ixodes affinis*, and more recently, Guzmán-Cornejo and Robbins (2010 without figures), Bermúdez et al. (2018a) and Saracho-Bottero et al. (2020) redescribed the male and the female.

There are no redescriptions of the nymph and larva of *Ixodes affinis*.

Geographical Distribution: *Ixodes affinis* has been recorded in several Neotropical countries from southern Mexico to Argentina, and it is also found in the Nearctic Region. However, problems have arisen in the identification of this species, and several records should be considered tentative. It is provisionally thought that *Ixodes affinis* and/or a species close to it are established in **Argentina**: Chaco, Corrientes, and Formosa (Saracho-Bottero et al. 2020; Sebastian et al. 2020); **Belize**: Belize, Cayo, and Stann Creek (Varma 1973; Lopes et al. 2016); **Colombia**: Antioquía, and Chocó (Mattar and López Valencia 1998); **Costa Rica**: Guanacaste, and Limón (Cooley and Kohls 1945; Carreno et al. 2001); **Ecuador** (Keirans et al. 1999; Saracho-Bottero et al. 2020); **Guatemala**: Petén (Cooley and Kohls 1945); **southern Mexico**: Chiapas, and Yucatán (Guzmán-Cornejo et al. 2007; Rodríguez-Vivas et al. 2016);

Panama: Bocas del Toro, Chiriquí, Colón, Darién, Guna Yala, Panamá, and Panamá Oeste (Fairchild et al. 1966); and **Peru**: Arequipa (provisional) (Fonseca 1960).

Hosts: Males and females of *Ixodes affinis* have frequently been collected from Artiodactyla: Bovidae and Cervidae, and Carnivora: Felidae and Canidae in the Neotropics; however, of 263 adult specimen records from the Neotropics, 139 (53%) are known to have been recovered from wild felids, chiefly *Leopardus pardalis* and *Panthera onca.* Consequently, these two species of Felidae are provisionally recognized as principal hosts of male and female *Ixodes affinis* in the Neotropical Region. Immature stages of this tick have been identified based on specimens collected from mammalian, avian, and squamatan hosts in the Nearctic but without a preferred host. The few Neotropical records of the nymph of *Ixodes affinis* are from hosts parasitized in the Nearctic, while records of larvae are from mammalian and squamatan hosts in Panama (Esser et al. 2016a) and avian hosts in Argentina (Flores et al. 2020; Esteban et al. 2020). Host preferences of the immature stages of Neotropical *Ixodes affinis* are regarded as basically undetermined. The host profile of *Ixodes affinis* is shown in Table 1.2, based on data provided by Cooley and Kohls (1945), Fonseca

Table 1.2 Hosts for males (M), females (F), nymphs (N), and larvae (L) of *Ixodes affinis*

MAMMALIA			
ARTIODACTYLA: Bovidae		*Puma concolor*	MF
Cattle*	MF	**CARNIVORA: Procyonidae**	
ARTIODACTYLA: Cervidae		*Nasua narica***	MF
*Mazama gouazoubira***	MF	**DIDELPHIMORPHIA: Didelphidae**	
Mazama temama	MF	*Didelphis marsupialis*	MF
*Odocoileus virginianus**	MF	*Marmosa robinsoni*	L
CARNIVORA: Canidae		**PERISSODACTYLA: Equidae**	
Domestic dog*	MF	Horse*	MF
CARNIVORA: Felidae		**RODENTIA: Cricetidae**	
Domestic cat*	MF	*Sigmodon peruanus*	F
Leopardus pardalis	MF	*Transandinomys bolivaris*	N
Panthera onca	MF		
REPTILIA			
SQUAMATA: Teiidae			
Ameiva sp.	NL		
AVES			
CATHARTIFORMES: Cathartidae		**PASSERIFORMES: Thraupidae**	
*Coragyps atratus**	F	*Tachyphonus rufus**	N
PASSERIFORMES: Furnariidae		*Thlypopsis sordida**	N
*Campylorhamphus trochilirostris**	NL	**PASSERIFORMES: Troglodytidae**	
PASSERIFORMES: Parulidae		*Troglodytes aedon*	N
*Basileuterus culicivorus**	L	**PASSERIFORMES: Turdidae**	
*Myiothlypis leucoblephara**	L	*Turdus amaurochalinus**	NL
PASSERIFORMES: Passerellidae		*Turdus rufiventris**	NL
*Arremon flavirostris**	L		

* Some specimens classified as "near *Ixodes affinis*"
** Specimens classified as "near *Ixodes affinis*"

(1960), Fairchild et al. (1966), Guzmán-Cornejo et al. (2007), Solís-Hernández et al. (2015), Esser et al. (2016a, b), Lopes et al. (2016), Peniche-Lara et al. (2018), Flores et al. (2020), Saracho-Bottero et al. (2020) and Sebastian et al. (2020). Several other hosts of *Ixodes affinis* have been recorded in the Nearctic Region, where adults and nymphs have been found on Carnivora: Procyonidae; adults alone have been collected from Carnivora: Ursidae; and immature stages have been recovered from Lagomorpha: Leporidae, Rodentia: Cricetidae, Muridae, and Sciuridae, Eulipotyphla: Soricidae, and Passeriformes: several families (Guglielmone et al. 2014; Heller et al. 2016; Scott et al. 2016; Nadolny and Gaff 2018).

Human Parasitism: Harrison et al. (2010) found *Ixodes affinis* on people but none of the ticks were attached. Guglielmone and Robbins (2018) discussed alleged records of human parasitism by this tick, concluding that they need confirmation. Consequently, *Ixodes affinis* was not listed as a tick causing human parasitism.

Comments: Clifford et al. (1973) list *Ixodes affinis* as a Nearctic species, but this is in error because Neumann (1899) described this tick from specimens collected in Costa Rica. A Palearctic record of *Ixodes affinis* from Japan in Kishida (1930) is treated as a diagnostic error in Yamaguti et al. (1971), Guglielmone et al. (2014), and here.

Ixodes affinis is morphologically close to *Ixodes fuscipes* and *Ixodes pararicinus.* Some authors treat *Ixodes aragaoi*, currently a synonym of *Ixodes fuscipes*, as a synonym of *Ixodes affinis* (Cooley and Kohls 1945; Fairchild et al. 1966; Keirans and Clifford 1978; Barros-Battesti and Knysak 1992; Durden and Keirans 1996; Guzmán-Cornejo and Robbins 2010; and probably others), ignoring the comparison of the females of *Ixodes fuscipes* (as *Ixodes aragaoi*) and *Ixodes affinis* in Aragão and Fonseca (1952). Keirans and Clifford (1978) state that the first description of the male of *Ixodes affinis* is in Fonseca (1935a) under the name *Ixodes ricinus aragaoi*, another synonym of *Ixodes fuscipes*, but this opinion could no longer be sustained after the redescription of *Ixodes fuscipes* (as *Ixodes aragaoi*) by Onofrio et al. (2009, 2014), as discussed in Labruna et al. (2020a).

Records of *Ixodes affinis* from Brazil and Uruguay (Homsher et al. 1991; Barros-Battesti and Knysak 1992) were considered doubtful in Guglielmone et al. (2003a). Nevertheless, recent developments have revealed problems attending the morphological and molecular identification of *Ixodes affinis.* Rodríguez-Vivas et al. (2016) in Mexico and Polsomboon et al. (2017) in Belize found *Ixodes* near *affinis*, implying that more than one species may be represented under the name *Ixodes affinis*. Saracho-Bottero et al. (2020) reclassified *Ixodes pararicinus* from northeastern Argentina as *Ixodes affinis* or a species close to it, adding that three alleged populations of *Ixodes* sp. cf. *I. affinis*, one from Argentina, another from Colombia and Ecuador, and a third population from Belize and USA, differ from one another at the molecular level, and are probably also unlike *Ixodes affinis sensu stricto.* Saracho-Bottero et al. (2020) stress that specimens from all populations of *Ixodes* near *affinis* and *bona fide Ixodes affinis* are morphologically very similar. It would therefore appear that more than one species exists under the name *Ixodes affinis.* Flores et al. (2020) and Sebastian et al. (2020) also identify as *Ixodes* sp. cf. *I. affinis* ticks collected in northeastern Argentina. Those records of *Ixodes* sp. cf. *I. affinis* are treated in Table 1.2 as "near *Ixodes affinis*." Clearly, the taxonomic statements above concerning this species should be considered provisional.

Fonseca (1960) recorded larvae and nymphs of *Ixodes affinis* in Peru and gave a succinct description of the nymph from specimens collected in nature that does not match the description of laboratory-reared material in Oliver et al. (1987). For that reason, these records of immature stages are not included in the host list above, and the record of a female collected from Cricetidae is regarded as provisionally valid. Esser et al. (2016a) found larvae and nymphs of *Ixodes affinis* on Squamata: Iguanidae, apparently the only record of an *Ixodes* from this type of host in the Neotropical Region. However, for the purposes of our analysis, Squamata are not considered significant hosts of *Ixodes affinis*.

Guglielmone et al. (2003a) listed "various" as the principal hosts of *Ixodes affinis* but most adult specimens of this tick have been collected from wild Felidae. Because wild felids are high on the food chain, they are not naturally abundant, and making collections from them can be difficult. However, the numerous collection records of this tick from wild felids strongly indicate that they are the preferred hosts of *Ixodes affinis*.

3. *Ixodes amarali* Fonseca, 1935

Fonseca, F. (1935b) Notas de acareologia. XVI. *Ixodes amarali*, sp. n. (Acarina: Ixodidae). *Memórias do Instituto Butantan*, 9, 137–143.

Descriptions and Redescriptions: Fonseca (1935b) described and illustrated the female of *Ixodes amarali*, and Martins et al. (2019a) described the male. Fonseca (1958a) described the nymph and larva, though without figures.

Onofrio et al. (2006a, 2009) redescribed the female, and Sénevet and Ripert (1967) redescribed the larva but, again, no figures of this stage were provided.

Geographical Distribution: *Ixodes amarali* is a South American species that has only been found in **Brazil**: Alagoas, Ceará, Federal District, Goiás, Mato Grosso do Sul, Minas Gerais, Pernambuco, Pará, Paraíba, Pernambuco, and Rio de Janeiro (Aragão 1936; Fonseca 1958a; Yoshizawa et al. 1996; Barros-Battesti and Knysak 1999; Faccini et al. 1999; Sponchiado et al. 2015; Martins et al. 2019a).

Hosts: Females, nymphs, and larvae of *Ixodes amarali* are usually collected from Didelphimorphia: Didelphidae, chiefly *Monodelphis domestica*, a principal host of this tick, but the geographic distribution of this host does not cover the entire range of *Ixodes amarali*. Immature stages have also frequently been recovered from Rodentia: Cricetidae, mostly *Cerradomys subflavus* and *Oligoryzomys nigripes*, which are provisionally regarded as the principal hosts of the nymph and larva of *Ixodes amarali*. The host profile of *Ixodes amarali* is shown in Table 1.3, constructed using data provided by Fonseca and Trindade (1957), Fonseca

Table 1.3 Hosts for females (F), nymphs (N), and larvae (L) of *Ixodes amarali*

MAMMALIA			
DIDELPHIMORPHIA: Didelphidae		*Necromys lasiurus*	FN
Didelphis albiventris	F	*Oligoryzomys eliurus*	N
Monodelphis brevicaudata	F	*Oligoryzomys microtis*	N
Monodelphis domestica	FNL	*Oligoryzomys nigripes*	NL
Monodelphis glirina	F	*Rhipidomys mastacalis*	L
Monodelphis kunsi	FNL	*Thaptomys nigrita*	L
RODENTIA: Cricetidae		**RODENTIA: Echimyidae**	
Akodon montensis	N	*Trinomys setosus*	L
Cerradomys subflavus	FNL	**RODENTIA: Muridae**	
Hylaeamys megacephalus	L	*Rattus rattus*	N

(1958a), Linardi et al. (1984), Barros-Battesti and Knysak (1999), Saraiva et al. (2012), and Martins et al. (2019b).

Human Parasitism: No.

Comments: Guglielmone et al. (2014) state that the larva of *Ixodes amarali* is undescribed, but there is a succinct description in Fonseca (1958a).

There is a doubtful record of unknown stages of *Ixodes amarali* collected from Sciuridae in Oliveira et al. (2014) that is excluded from the host list above.

4. *Ixodes andinus* Kohls, 1956

Kohls, G.M. (1956a) Eight new species of *Ixodes* from Central and South America (Acarina: Ixodidae). *Journal of Parasitology*, 42, 636–649.

Descriptions and Redescriptions: Kohls (1956a) described the female, nymph, and larva but illustrated only the female of *Ixodes andinus.*

Sénevet and Ripert (1967) redescribed but did not figure the larva of *Ixodes andinus*.

The male of *Ixodes andinus* remains unknown.

Geographic Distribution: *Ixodes andinus* is a South American species that has been recorded only from **Peru**: Ancash, the specimens having been collected in February 1954 (Kohls 1956a). There are no other records for this species.

Hosts: The known hosts of *Ixodes andinus* are Rodentia: Cricetidae; females, nymphs, and larvae have been collected from *Phyllotis darwini*, while larvae and nymphs have been recovered from *Calomys sorellus*. However, it would appear that *Phyllotis darwini* cannot be a host of *Ixodes andinus* because Wilson and Reeder (2005) and the International Union for Conservation of Nature (2018) state that this rodent is found only in Chile. Therefore, *Phyllotis darwini* is here treated as an unconfirmed host of *Ixodes andinus*, and the principal host of this tick is provisionally considered to be *Phyllotis* sp.

Human Parasitism: No.

Comments: No controversial issues related to *Ixodes andinus* have been found.

5. *Ixodes auritulus* Neumann, 1904 but previously identified as *Ixodes thoracicus* Koch, 1844 in Neumann (1899), who later realized that this species differs from the taxon described by Koch (1844) and accordingly renamed this tick *Ixodes auritulus*

Neumann, L.G. (1904) Notes sur les ixodidés. II. *Archives de Parasitologie*, 8, 444–464.
Koch, C.L. (1844) Systematische Übersicht über die Ordnung der Zecken. *Archiv für Naturgeschichte*, 10, 217–239.
Neumann, L.G. (1899) Révision de la famille des ixodidés (3e mémoire). *Mémoires de la Société Zoologique de France*, 12, 107–294.

Descriptions and Redescriptions: Neumann (1899) described and figured the prosoma and scutum of the female of *Ixodes auritulus* under the name *Ixodes thoracicus*; the male was described by Dumbleton (1953), and the larva and nymph were described by Nuttall (1916).

In keeping with this tick's wide geographic distribution, there are numerous published redescriptions of *Ixodes auritulus*, suggestive of a species group, and in fact the name *Ixodes auritulus* represents a species complex, as discussed below. The male, female, nymph, and larva were redescribed in Furman and Loomis (1984) and Lindquist et al. (2016); the male, female, and nymph were redescribed in Gregson (1956), Arthur (1960b), and Nava et al. (2017); the female, nymph, and larva were redescribed in Dumbleton (1953); the male and female were redescribed in Dumbleton (1961), Wilson (1967), Keirans and Clifford (1978) and Onofrio et al. (2006a, 2009); the female and nymph were redescribed in Ault (1943), Cooley and Kohls (1945) and Roberts (1970); the male alone was redescribed in Kohls and Clifford (1966); the female alone was redescribed in Neumann (1911a), Nuttall and Warburton (1911), Nuttall (1916), Boero (1957), Barros-Battesti et al. (2003) and Bermúdez et al. (2018a); the nymph alone was redescribed in Aragão (1938) and Durden and Keirans (1996), and diagnostic characters for the larva of *Ixodes auritulus* were provided in Webb et al. (1990).

Geographic Distribution: *Ixodes auritulus* has been found in the Australasian, Nearctic, and Neotropical Zoogeographic Regions. In the Neotropical Region, *Ixodes auritulus* has been found in **Argentina**: Buenos Aires, Ciudad Autónoma de Buenos Aires, Santa Cruz, and Tierra del Fuego, Antártida e Islas del Atlántico Sur (Ault 1943; Flores et al. 2014; González-Acuña et al. 2005b; Cicuttin et al. 2017); **Brazil**: Paraná, Rio de Janeiro, Rio Grande do Sul, and São Paulo (Barros-Battesti and Knysak 1999; Arzua et al. 2005; Michel et al. 2017); **Chile**: Libertador Bermardo O'Higgins, Los Ríos, Magallanes y Antártica Chilena, and Tarapacá

(González-Acuña et al. 2004a, 2005b, 2009; Llanos-Soto et al. 2017); **Colombia**: Cauca, and Cundinamarca (González-Acuña et al. 2005b); **Costa Rica**: Alajuela, Cartago, Guanacaste, and San José (Tonn et al. 1963; Arnold 1970; González-Acuña et al. 2005b); **Ecuador** (Santos Dias 1958a); **Guatemala**: Petén, and San Marcos (Arthur 1960b); **Panama**: Chiriquí (Bermúdez et al. 2015c); **Peru**: Cusco, Huánuco, Junín, Lima, and Puno (Need et al. 1991; González-Acuña et al. 2005b); **Uruguay**: Cerro Largo, Lavalleja, and Rocha (Venzal et al. 2003c; Carvalho et al. 2020) and **Venezuela**: Mérida, and Trujillo (Jones et al. 1972).

Hosts: Females, nymphs, and larvae of Neotropical *Ixodes auritulus* are parasites of Aves of several orders, with a few records from Rodentia: Cricetidae. Most Neotropical avian records of *Ixodes auritulus* are from Passeriformes, but within this order, the family Turdidae accounts for most collections of females, nymphs, and larvae, as shown by records from *Turdus albicollis*, *Turdus amaurochalinus*, *Turdus falcklandii,* and *Turdus rufiventris*, which are considered the principal hosts of this tick, but only within the South American distribution of *Ixodes auritulus*; other Neotropical birds are important hosts for nymphs and larvae of this tick in southern Mexico and Central America. The Neotropical host profile of *Ixodes auritulus* is shown in Table 1.4, constructed using data provided by Nuttall (1916), Ault (1943), Cooley and Kohls (1945), Arthur (1960b), Tonn et al. (1963), Jones et al. (1972), Arzua and Barros-Battesti (1999), Barros-Battesti and Knysak (1999), Robbins et al. (2001b), Arzua et al. (2003, 2005), Venzal et al. (2003c), González-Acuña et al. (2004a, 2005b, 2009), Nava et al. (2011), Flores et al. (2014), Bermúdez et al. (2015c, 2018a), Cicuttin et al. (2019b), Muñoz-Leal et al. (2019a) and Carvalho et al. (2020). Records of *Ixodes auritulus* outside the Neotropics encompass many avian orders, with a few specimens collected from Carnivora: Canidae, and Rodentia: Sciuridae (Guglielmone et al. 2014; Durden et al. 2016). All records are considered to represent the *Ixodes auritulus* group.

Human Parasitism: No.

Comments: The female from Punta Arenas (southern Chile) described by Neumann (1899) is the only account of *Ixodes auritulus sensu stricto* currently known. Arthur (1960b), Roberts (1970), Jones et al. (1972) and González-Acuña et al. (2005b) consider *Ixodes auritulus* a species group name. This tick has a wide distribution, and there is little doubt that more than one taxon exists under the name *Ixodes auritulus.* Roberts (1970) described morphological differences between Australian and New Zealand populations of the *Ixodes auritulus* complex and also emphasized that avian host utilization by this tick varies in different regions of the world where *Ixodes auritulus* is established, while Guglielmone et al. (2014) found differences in the morphological redescriptions of *Ixodes auritulus* by several authors. Dumbleton (1961) named New Zealand specimens as *Ixodes auritulus zealandicus*; most probably, *zealandicus* should become a name for a separate, valid taxon rather than a subspecies of *Ixodes auritulus.* Additionally, González-Acuña et al. (2009) found important molecular differences in specimens of this species from northern Chile and southern Argentina. Research comparing

Table 1.4 Hosts for females (F), nymphs (N), and larvae (L) of the *Ixodes auritulus* species group

	AVES		
COLUMBIFORMES: Columbidae		*Anisognathus lacrymosus*	L
Columbina talpacoti	F	*Diglossa albilatera*	L
FALCONIFORMES: Falconidae		*Diglossa plumbea*	N
Phalcoboenus australis	FNL	*Haplospiza unicolor*	F
GALLIFORMES: Cracidae		*Iridosornis rufivertex*	L
Oreophasis derbianus	F	*Microspingus cabanisi*	NL
Penelope sp.	NL	*Microspingus lateralis*	L
Penelope superciliaris	N	*Nephelornis oneilli*	L
GALLIFORMES: Odontophoridae		*Phrygilus fruticeti*	NL
Odontophorus guttatus	F	*Pipraeidea bonariensis*	FN
PASSERIFORMES: Conopophagidae		*Poospiza nigrorufa*	NL
Conopophaga lineata	NL	*Saltator aurantiirostris*	NL
PASSERIFORMES: Furnariidae		*Sicalis olivascens*	FN
Aphrastura spinicauda	NL	*Sporophila caerulescens*	L
Asthenes fuliginosa	N	*Stephanophorus diadematus*	NL
Cinclodes antarcticus	F	*Tangara vassorii*	L
Cinclodes fuscus	FN	*Thraupis cyanocephala*	F
Cinclodes patagonicus	N	*Trichothraupis melanops*	N
Clibanornis dendrocolaptoides	FNL	*Xenospingus concolor*	L
Furnarius rufus	F	**PASSERIFORMES: Troglodytidae**	
Lochmias nematura	F	*Cantorchilus nigricapillus*	L
Phacellodomus striaticollis	N	*Pheugopedius genibarbis*	L
Pygarrhichas albogularis	L	*Thryorchilus browni*	FNL
Synallaxis ruficapilla	NL	*Troglodytes* sp.	FN
Synallaxis spixi	NL	*Troglodytes aedon*	FNL
Syndactyla rufosuperciliata	N	**PASSERIFORMES: Turdidae**	
PASSERIFORMES: Icteridae		*Catharus gracilirostris*	L
Curaeus curaeus	F	*Turdus albicollis*	FNL
Leistes loyca	F	*Turdus amaurochalinus*	FNL
PASSERIFORMES: Parulidae		*Turdus falcklandii*	FNL
Basileuterus culicivorus	NL	*Turdus nigrescens*	FL
Geothlypis aequinoctialis	L	*Turdus nigriceps*	FL
Myiothlypis leucoblephara	L	*Turdus plebejus*	FL
Seiurus aurocapilla	L	*Turdus rufiventris*	FNL
PASSERIFORMES: Passerellidae		**PASSERIFORMES: Tyrannidae**	
Atlapetes pallidinucha	NL	*Elaenia albiceps*	L
Atlapetes schistaceus	L	*Knipolegus nigerrimus*	F
Junco vulcani	N	*Phylloscartes ventralis*	N
Zonotrichia capensis	NL	**PELECANIFORMES: Ardeidae**	
PASSERIFORMES: Rhinocryptidae		*Nycticorax nycticorax*	FN
Scytalopus sp.	L	**PICIFORMES: Picidae**	
PASSERIFORMES: Thamnophilidae		*Colaptes rupicola*	F
Thamnophilus caerulescens	FN	**PROCELLARIIFORMES: Procellariidae**	
Thamnophilus ruficapillus	FNL	*Pelecanoides magellani*	FNL
PASSERIFORMES: Thraupidae		**TINAMIFORMES: Tinamidae**	
Anisognathus igniventris	F	*Nothoprocta pentlandii*	FN
	MAMMALIA		
RODENTIA: Cricetidae		*Microryzomys minutus*	L
Aepeomys lugens	L	*Nephelomys albigularis*	L

populations of *Ixodes auritulus* from different parts of the world is needed to resolve the many questions concerning this name. Keirans and Clifford (1978) state that the first description of the male of *Ixodes auritulus* is in Kohls and Clifford (1966), but there were other descriptions prior to the study of Kohls and Clifford (1966), although none from the type locality in southern Chile; consequently, there are no descriptions of the male of *Ixodes auritulus sensu stricto.* In fact, the only parasitic stage of *Ixodes auritulus sensu stricto* known with certainty is the female because the larva and nymph were described by Nuttall (1916) from material collected in Canada and Peru, and no redescriptions have been made from species collected in southern Chile.

The Afrotropical Region was included within the range of *Ixodes auritulus* by Camicas et al. (1998, under the name *Scaphixodes auritulus*) and Guglielmone et al. (2014) but further study failed to reveal *bona fide* records of this tick from that region. Therefore, the Afrotropical Region is excluded from the geographical distribution of *Ixodes auritulus*. Castrezana (2010) states that *Ixodes auritulus* is present in the State of Sonora in the Nearctic portion of Mexico and, perhaps, occurs throughout the country. Nevertheless, *Ixodes auritulus* is not listed as a Mexican tick in Guzmán-Cornejo et al. (2007) and Guzmán-Cornejo and Robbins (2010), and Mexico is provisionally excluded from the range of *Ixodes auritulus.*

A number of authors believe that all parasitic stages of *Ixodes auritulus* accept several orders of Aves as hosts, although with an overall tendency to feed on Passeriformes. We consider this statement true; however, an analysis of host data for this tick indicates that birds of the family Turdidae are most commonly infested by all parasitic stages of *Ixodes auritulus* (González-Acuña et al. 2005b; Cicuttin et al. 2019b; Carvalho et al. 2020) in comparison with other families of Passeriformes found in South America. Accordingly, the family Turdidae is provisionally treated here as containing the principal hosts of *Ixodes auritulus*, although this statement has not been confirmed for tick populations outside South America.

6. *Ixodes bequaerti* Cooley and Kohls, 1945

Cooley, R.A. & Kohls, G.M. (1945) The genus *Ixodes* in North America. *National Institute of Health Bulletin*, (184) 246 pp.

Descriptions and Redescriptions: Cooley and Kohls (1945) described and illustrated the female of *Ixodes bequaerti.*

The female was redescribed without figures in Guzmán-Cornejo and Robbins (2010) and with figures in Bermúdez et al. (2018a).

The male, nymph, and larva of *Ixodes bequaerti* remain unknown.

Geographical Distribution: *Ixodes bequaerti* is a Neotropical species that has been found in southern North America and Central America in **Honduras**: Yoro (Cooley and Kohls 1945); **southern Mexico**: Chiapas (Cooley and Kohls 1945); **Nicaragua**: Estelí, and Matagalpa (Kohls and Clifford 1966); and **Panama**: Chiriquí (Bermúdez et al. 2015c).

Table 1.5 Hosts for the females (F) of *Ixodes bequaerti*

AVES			
PASSERIFORMES: Thraupidae		**TROGONIFORMES: Trogonidae**	
Volatinia jacarina	F	*Pharomachrus mocinno*	F
PASSERIFORMES: Turdidae		*Trogon collaris*	F
Catharus sp.	F		

Hosts: The only known hosts of the female of *Ixodes bequaerti* are Aves. The host profile of *Ixodes bequaerti* is shown in Table 1.5, constructed with information provided by Cooley and Clifford (1945), Kohls and Clifford (1966), and Bermúdez et al. (2015c, 2018a). The females of *Ixodes bequaerti* are known from 11 specimens plus an unknown number of females collected from *Catharus* sp. (Turdidae). Of the 11 ticks, ten were found on *Pharomachrus mocinno* and *Trogon collaris* (Trogonidae); consequently, this family of birds is provisionally thought to include the principal hosts of *Ixodes bequaerti.*

Human parasitism: No.

Comments: Clifford et al. (1973) and Camicas et al. (1998) state, erroneously, that *Ixodes bequaerti* is a Nearctic species, but it is a Neotropical taxon, as previously discussed in Guglielmone et al. (2003a, 2014).

7. *Ixodes bocatorensis* Apanaskevich and Bermúdez, 2017

Apanaskevich, D.A. & Bermúdez, S.E. (2017) Description of a new species of *Ixodes* Latreille, 1795 (Acari: Ixodidae) and redescription of *I. lasallei* Méndez & Ortiz, 1958, parasites of agoutis and pacas (Rodentia: Dasyproctidae, Cuniculidae) in Central and South America. *Systematic Parasitology*, 94, 463–475.

Descriptions and Redescriptions: The male of *Ixodes bocatorensis* was described in Kohls and Clifford (1966) under the name *Ixodes lasallei*, as explained in Apanaskevich and Bermúdez (2017), who also described the female of *Ixodes bocatorensis* from specimens previously identified as *Ixodes lasallei.*

Bermúdez et al. (2018a) and Onofrio et al. (2020, supplementary material) redescribed the male and female of *Ixodes bocatorensis.*

The nymph and larva of *Ixodes bocatorensis* remain unknown.

Geographic Distribution: *Ixodes bocatorensis* is a Neotropical tick species that has been found in southern Central America and northern South America as follows: **Brazil**: Amazonas (Onofrio et al. 2020); **Colombia**: Antioquía (Apanaskevich and Bermúdez 2017); **Panama**: Bocas del Toro, and Los Santos (Apanaskevich and Bermúdez 2017), and **Venezuela**: Bolívar, Carabobo, and Falcón (Apanaskevich and Bermúdez 2017).

Table 1.6 Hosts for the males (M) and females (F) of *Ixodes bocatorensis*

MAMMALIA			
PERISSODACTYLA: Tapiridae		**PILOSA: Myrmecophagidae**	
Tapirus sp.	F	*Tamandua tetradactyla**	F
PILOSA: Bradypodidae		**RODENTIA: Cuniculidae**	
*Bradypus tridactylus**	F	*Cuniculus paca*	MF
PILOSA: Cyclopedidae		**RODENTIA: Dasyproctidae**	
*Cyclopes didactylus**	F	*Dasyprocta leporina**	MF
		Dasyprocta punctata	F

* Some records of *Ixodes bocatorensis* from these hosts have previously been identified as *Ixodes cf I. fuscipes*

Hosts: The limited records of *Ixodes bocatorensis* in Apanaskevich and Bermúdez (2017) appear to indicate that females of this tick are prone to infest *Cuniculus paca* (Rodentia: Cuniculidae) and *Dasyprocta punctata* (Rodentia: Dasyproctidae), because 15 of the 16 specimens were female ticks, and 13 of the females had been found on these two rodents. However, in Brazil, Onofrio et al. (2020) later found many females and a few males on *Dasyprocta leporina*, another principal host of this tick. The host profile of *Ixodes bocatorensis* is shown in Table 1.6, constructed using data provided by Apanaskevich and Bermúdez (2017) and Onofrio et al. (2020).

Human Parasitism: No.

Comments: The male of *Ixodes bocatorensis* is known from a few specimens, one collected with several female ticks from an individual *Cuniculus paca*, and three found on *Dasyprocta leporina*. All the specimens classified as *Ixodes bocatorensis* in Apanaskevich and Bermúdez (2017) were identified as *Ixodes lasallei* by Kohls and Clifford (1966). It would be of interest to search collections of *Ixodes lasallei* to determine whether there are additional specimens of *Ixodes bocatorensis* classified under that name. Several adult specimens of *Ixodes* cf. *I. fuscipes* collected from Brazilian *Dasyprocta leporina* in Gianizella et al. (2018b) were reclassified as *Ixodes bocatorensis* by Onofrio et al. (2020); additionally, records of females of *Ixodes* cf. *I. fuscipes* from *Bradypus tridactylus*, *Cyclopes didactylus,* and *Myrmecophaga tetradactyla* in Gianizella et al. (2018b) are provisionally treated here as *Ixodes bocatorensis*.

8. ***Ixodes boliviensis*** Neumann, 1904

Neumann, L.G. (1904) Notes sur les ixodidés. II. *Archives de Parasitologie*, 8, 444–464.

Descriptions and Redescriptions: Neumann (1904) described the male, female, and nymph of *Ixodes boliviensis*, but without figures.

The male, female, and nymph were redescribed and figured in Nuttall and Warburton (1911), while Cooley and Kohls (1945, under the name *Ixodes bicornis*), Guzmán-Cornejo and Robbins (2010, no figures), and Bermúdez et al. (2018a) redescribed both sexes of *Ixodes boliviensis*. The female alone was redescribed under the name *Ixodes bicornis* in Neumann (1906).

The larva of *Ixodes boliviensis* has not been formally described but was provisionally recognized by Fairchild et al. (1966) from field-collected material.

Geographical Distribution: *Ixodes boliviensis* has a wide range in the mainland Neotropics from southern North America to central South America as follows: **Belize**: Cayo (Redell and Veni 1996); **Bolivia**: Cochabamba (Neumann 1904); **Colombia**: Boyacá (provisional), and Nariño (Osorno Mesa 1942; Oniki-Willis and Willis 2018); **Costa Rica**: Cartago, Guanacaste, Heredia, Puntarenas, and San José (Álvarez et al. 2005; Troyo et al. 2014; Moreira-Soto et al. 2016); **Ecuador**: Napo (Pesquera et al. 2015); **Guatemala**: Alta Verapaz, Petén, and Sololá (Bequaert 1938; Cooley and Kohls 1945; Instituto Interamericano de Cooperación para la Agricultura 1988); **Honduras**: Atlántida or Yoro (Cooley and Kohls 1945); **southern Mexico**: Chiapas, Guerrero, Nayarit, Oaxaca, Quintana Roo, and Veracruz (Neumann 1906; Cooley and Kohls 1945; Hoffmann 1962; Guzmán-Cornejo et al. 2007); **Nicaragua**: Jinotega, and Nueva Segovia (Bermúdez et al. 2015b); **Panama**: Bocas del Toro, Chiriquí, and Panamá Oeste (Fairchild et al. 1966; Bermúdez and Miranda 2011b; Gittens et al. 2020) and **Peru**: Piura (Blair et al. 2004). Many records of *Ixodes boliviensis* have been published under the name *Ixodes bicornis*.

Hosts: Males and females of *Ixodes boliviensis* have been recovered from several orders of mammals, but the great majority of specimens have been collected from Carnivora. However, of 350 adult ticks from all hosts, 223 specimens (63%) were taken from dogs, which are consequently treated as the principal hosts for *Ixodes boliviensis*. There is just one record for the larva of this tick and only a few records of nymphs, so it is impossible to determine the principal hosts for the immature stages of *Ixodes boliviensis*. The host profile of *Ixodes boliviensis* is shown in Table 1.7, constructed with data provided by Neumann (1904), Bequaert (1938), Osorno Mesa (1942), Fairchild et al. (1966), Guzmán-Cornejo et al. (2007), Esser et al. (2016a) and Oniki-Willis and Willis (2018).

Human Parasitism: Guglielmone and Robbins (2018) listed *Ixodes boliviensis* as a rare parasite of humans that has been found on people in Belize (Cayo), Mexico (Guerrero, and Veracruz), and Panama (Chiriquí). Most records of human parasitism by *Ixodes boliviensis* have resulted from bites by female ticks, with a single record of a male tick.

Comments: Camicas et al. (1998) do not treat *Ixodes boliviensis* as valid and consider this name to be a probable synonym of *Ixodes diversifossus*. The latter species is treated as a Nearctic tick in Guglielmone et al. (2014) but is listed as present in the Nearctic and Neotropical Regions in Camicas et al. (1998). Keirans and Clifford (1978) and Guglielmone et al. (2009), among others, state that *Ixodes boliviensis* is very close to the Nearctic *Ixodes diversifossus* and advise type comparisons to determine whether they are synonyms or valid species. However, if

Table 1.7 Hosts for males (M), females (F), nymphs (N), and larvae* (L) of *Ixodes boliviensis***

MAMMALIA			
ARTIODACTYLA: Bovidae		**CARNIVORA: Ursidae**	
Cattle	MF	*Tremarctos ornatus*	F
ARTIODACTYLA: Cervidae		**DIDELPHIMORPHIA: Didelphidae**	
Odocoileus virginianus	F	*Didelphis marsupialis*	F(N***)
CARNIVORA: Canidae		**PERISSODACTYLA: Equidae**	
Domestic dog	MFN	Horse	F
Speothos venaticus	MFN	Mule	F
CARNIVORA: Felidae		**PERISSODACTYLA: Tapiridae**	
Domestic cat	F	*Tapirus bairdii*	F
Leopardus pardalis	MF	**RODENTIA: Erethizontidae**	
Panthera onca	MF	*Coendou mexicanus*	F
CARNIVORA: Procyonidae		**RODENTIA: Sciuridae**	
Procyon lotor	MFNL**	*Sciurus* sp.	F
AVES			
GALLIFORMES: Cracidae		**GALLIFORMES: Phasianidae**	
Crax rubra	FN	Pheasant	F

* Provisional diagnosis because the larva of *Ixodes boliviensis* remains undescribed
** Several records of *Ixodes boliviensis* have been published under the name *Ixodes bicornis*
*** Nymph recorded as near *boliviensis* in Osorno Mesa (1942)

Ixodes bicornis is validated, it too should be part of any such comparison (see below).

As its name indicates, the type specimens of *Ixodes boliviensis* were collected in Bolivia (Cochabamba Department) and described in Neumann (1904). Allegedly this species was later found in Costa Rica (Nuttall 1916). The name *Ixodes boliviensis* was not cited again until Osorno Mesa (1942) named a nymph found on a marsupial in Colombia as *Ixodes* near *boliviensis.* Afterward, Kohls (1956c) synonymized *Ixodes bicornis* with *Ixodes boliviensis*; therefore, records under the former name are now treated as corresponding to the latter. Kohls' proposal (1956c) was based on examination of types and criticism of the figures in Nuttall and Warburton (1911) that supported the validity of both taxa. However, Kohls (1956c) provided no figure to support his claim, which until recently was undisputed. The synonymization of *Ixodes bicornis* with *Ixodes boliviensis* should be reexamined in light of the geographic distribution and hosts of these taxa, including human parasitism. We consider *Ixodes boliviensis* provisionally valid, pending a comparison of types of *Ixodes diversifossus* and *Ixodes bicornis*.

Bermúdez et al. (2015b) found Nicaraguan specimens of *Ixodes boliviensis* previously identified as *Ixodes scapularis*. Bermúdez and Miranda (2011b) did not find *Ixodes boliviensis* on dogs in the Panama Province of Boca del Toro, where this tick was found on dogs and other hosts by Fairchild et al. (1966). Pinto (1938) erroneously included Argentina within the range of *Ixodes boliviensis* (Guglielmone et al. 2003a). Hoffmann (1962) cited the presence of *Ixodes boliviensis* in the Mexican states of Colima, Jalisco, and Michoacán, but these states are ignored in

Guzmán-Cornejo et al. (2007). We exclude Argentina and the cited Mexican states from the geographic distribution of this tick.

Guglielmone and Robbins (2018) listed a record from a human in the Mexican state of Chiapas based on Guzmán-Cornejo et al. (2007), but this record is actually from the State of Veracruz.

9. *Ixodes capromydis* Černý, 1966

Černý, V. (1966a) Nueva especie de garrapata del género *Ixodes* Latreille (Ixodoidea, Ixodidae) en la jutía conga de la isla de Pinos. *Poeyana Serie A*, (24) 9 pp.

Descriptions and Redescriptions: Černý (1966a) described and figured the male, female, nymph and larva of *Ixodes capromydis*.

Černý et al. (1975) redescribed the male, female, and nymph, while Barros-Battesti et al. (2009) redescribed both sexes of *Ixodes capromydis*.

There are no redescriptions of the larva of *Ixodes capromydis*.

Geographic Distribution: *Ixodes capromydis* is a Neotropical Caribbean species that has been found only in **Cuba**: Isla de la Juventud (Barros-Battesti et al. 2009).

Hosts: The only known host of all parasitic stages of *Ixodes capromydis* is *Capromys pilorides* (Rodentia: Capromyidae) (Černý 1966a).

Human Parasitism: No.

Comments: No controversial issues related to *Ixodes capromydis* have been found.

10. *Ixodes catarinensis* Onofrio and Labruna, 2020 in Onofrio et al. (2020)

Onofrio, V.C., Guglielmone, A.A., Barros-Battesti, D.M., Gianizella, S.L., Marcili, A., Quadros, R.M., Marques, S. & Labruna, M.B. (2020) Description of a new species of *Ixodes* (Acari: Ixodidae) and first report of *Ixodes lasallei* and *Ixodes bocatorensis* in Brazil. *Ticks and Tick-borne Diseases*, 11 (4) (article 101423) 9 pp.

Descriptions and Redescriptions: Onofrio et al. (2020) described and figured the female of *Ixodes catarinensis*.

The male, nymph, and larva of *Ixodes catarinensis* remain unknown.

There are no redescriptions of the female of *Ixodes catarinensis*.

Geographic Distribution: *Ixodes catarinensis* is a Neotropical species that has been found only in **Brazil**: Santa Catarina and São Paulo (Onofrio et al. 2020).

Hosts: *Ixodes catarinensis* is known from six females, three of them found on cats (Carnivora: Felidae); two females were collected from *Marmosa demerarae* (Didelphimorphia: Didelphidae), and one was taken from *Sooretamys angouya*

(Rodentia: Cricetidae). The absence of relatedness between these hosts and the small number of specimens indicates that more information is needed on the host preferences of *Ixodes catarinensis*. Nevertheless, we consider it unlikely that a domestic mammal is the main host of female *Ixodes catarinensis*, and we hypothetically propose that rodents and didelphimorphs are the more likely principal hosts.

Human Parasitism: No.

Comments: *Ixodes catarinensis* was confused with *Ixodes fuscipes* prior to its description by Onofrio et al. (2020).

11. *Ixodes chilensis* Kohls, 1956

Kohls, G.M. (1956a) Eight new species of *Ixodes* from Central and South America (Acarina: Ixodidae). *Journal of Parasitology*, 42, 636–649.

Descriptions and Redescriptions: Kohls (1956a) described and figured the female of *Ixodes chilensis* from a single specimen with a broken hypostome, while the nymph was described in González-Acuña et al. (2018).

Nava et al. (2017) and González-Acuña et al. (2018) redescribed the female, but the redescription by the latter authors is more detailed.

The male and larva of *Ixodes chilensis* remain unknown.

Geographical Distribution: *Ixodes chilensis* is a Neotropical species that has only been found in southern South America in **Chile**: Araucanía, and Los Ríos (Kohls 1956a; González-Acuña et al. 2018).

Hosts: The female and nymph of *Ixodes chilensis* have been found on *Myotis chiloensis* (Chiroptera: Vespertilionidae) (González-Acuña et al. 2018).

Human Parasitism: No.

Comments: No hosts for *Ixodes chilensis* were reported in the description of Kohls (1956a), but Tagle (1971) speculated that this tick was collected on a horse, although he presented no evidence to support his statement, and González-Acuña et al. (2018) later found this species on bats.

12. *Ixodes cookei* Packard, 1869

Packard, A.S. (1869) Report of the curator of Articulata. *First Annual Report of the Trustees of the Peabody Academy of Sciences*, pp. 56–69

Descriptions and Redescriptions: Packard (1869) described the female of *Ixodes cookei* without providing figures. The male, nymph, and larva were described and figured by Banks (1908).

Gregson (1956) redescribed the male, female, and nymph of *Ixodes cookei*; Sonenshine (1979) redescribed the female, nymph, and larva; and Lindquist et al. (2016) redescribed all stages. The male and female were redescribed by Keirans and

Clifford (1978) and Keirans and Litwak (1989); the female was redescribed by Banks (1908) and Lancaster (1973). The nymph was redescribed by Durden and Keirans (1996), while the larva was redescribed by Clifford et al. (1961), Sénevet and Ripert (1967), Ko (1973) and Coley (2015).

Geographic Distribution: *Ixodes cookei* is basically a Nearctic species with a Neotropical record from **southern Mexico**: Veracruz (Montiel-Parra et al. 2007).

Hosts: The Neotropical record of *Ixodes cookei* is based on a male tick collected from the feces of *Urocyon cinereoargentatus* (Carnivora: Canidae) by Montiel-Parra et al. (2007). All stages of this species have been collected on several orders of mammals (mostly small and medium-sized), but larvae and nymphs have occasionally been taken on Aves (Passeriformes and Strigiformes) in the Nearctic Region (Guglielmone et al. 2014).

Human Parasitism: *Ixodes cookei* has been categorized as a very frequent parasite of people, with many records from Canada and the USA (Guglielmone and Robbins 2018), but there is no record of human parasitism in the Neotropics.

Comments: Montiel-Parra et al. (2007) provided no taxonomic support for their record of *Ixodes cookei* in southern Mexico. This record is treated as valid in Guzmán-Cornejo and Robbins (2010) and provisionally valid here.

A record in Nuttall (1916) of a nymph of *Ixodes hexagonus cookei* (a synonym of *Ixodes cookei* erroneously named *Ixodes angustus cookei* in Guglielmone et al. 2003a) from *Mustela frenata* in Costa Rica is treated as doubtful in Guglielmone et al. (2003a) and here.

13. *Ixodes cooleyi* Aragão and Fonseca, 1951

Aragão, H.B. & Fonseca, F. (1951b) Notas de ixodologia. I. Duas novas espécies do género *Ixodes* e um novo nomen para *Haemaphysalis kochi* Aragão, 1908 (Acari: Ixodidae). *Memórias do Instituto Oswaldo Cruz*, 49, 567–574.

Descriptions and Redescriptions: *Ixodes cooleyi* is known only from two females described and figured in Aragão and Fonseca (1951b).

There are no redescriptions of the female of *Ixodes cooleyi.*

The male, nymph, and larva of *Ixodes cooleyi* remain unknown.

Geographic Distribution: The only record of *Ixodes cooleyi* is from an unknown locality in **Bolivia,** according to Aragão and Fonseca (1951b), but Barros-Battesti and Knysak (1999) situated the record in the Department of Oruro. There are no other records for this species.

Hosts: The two female specimens of *Ixodes cooleyi* were collected from *Ctenomys* sp. (Rodentia: Ctenomyidae) (Aragão and Fonseca 1951b).

Human Parasitism: No.

Comments: Clifford et al. (1973) treat *Ixodes cooleyi* as a doubtful species, but no reasons have been found to support that statement.

Aragão and Fonseca (1961a) list *Ixodes cooleyi* as part of the Brazilian tick fauna, probably due to a typographical error.

14. *Ixodes copei* Wilson, 1980

Wilson, N. (1980) Records of ticks from Jamaica with the description of *Ixodes* (*Ixodes*) *copei* n. sp. from the mockingbird, *Mimus polyglottos* (Acari: Ixodidae). *International Journal of Acarology*, 6, 157–162.

Descriptions and Redescriptions: *Ixodes copei* is known from one female and three larvae, described and figured in Wilson (1980).

There are no redescriptions of the female of *Ixodes copei.*

The male and nymph of *Ixodes copei* remain unknown.

Geographical Distribution: *Ixodes copei* is a Neotropical Caribbean species that has only been found in **Jamaica**: Saint Thomas Parish (Wilson 1980).

Hosts: The female of *Ixodes copei* was collected from *Mimus polyglottos* (Passeriformes: Mimidae), and the larvae were taken from *Turdus jamaicensis* (Passeriformes: Turdidae), birds that are provisionally treated as the principal hosts for this tick.

Human Parasitism: No.

Comments: No controversial issues related to *Ixodes copei* have been found.

15. *Ixodes cornuae* Arthur, 1960

Arthur, D.R. (1960b) A review of some ticks (Acarina: Ixodidae) of sea birds. Part II. The taxonomic problems associated with the *Ixodes auritulus – percavatus* group of species. *Parasitology*, 50, 199–226.

Descriptions and Redescriptions: *Ixodes cornuae* was originally identified as *Ixodes auritulus* and given its current status by Arthur (1960b), who described and illustrated the nymph and the female.

Nava et al. (2017) redescribed the female and the nymph, but without figures.

The male and larva of *Ixodes cornuae* remain unknown.

Geographic Distribution: *Ixodes cornuae* is a Neotropical South American species that has been found in **Chile**: Magallanes y de la Antárctica Chilena, and **Ecuador** (Arthur 1960b).

Hosts: Information on the hosts of *Ixodes cornuae* is imprecise. The hosts for the nymph and female ticks found in Ecuador are considered to be Galliformes: Phasianidae (Guglielmone et al. 2014; Nava et al. 2017) based on the fact that the ticks were collected from a kind of "caille," but this bird may also be a member of the family Odontophoridae; therefore, the Phasianidae-*Ixodes cornuae* relationship should be regarded as provisionally valid. The host for the Chilean female of *Ixodes cornuae* is unknown.

Human Parasitism: No.

Comments: Arthur (1960b) cautions that the supposed nymph of *Ixodes cornuae* may not belong to this species, casting doubt on the validity of the species itself. *Ixodes cornuae* is a member of the *Ixodes percavatus* group, which contains species that remain poorly known.

16. *Ixodes cuernavacensis* Kohls and Clifford, 1966

Kohls, G.M. & Clifford, C.M. (1966) Three new species of *Ixodes* from Mexico and description of the male of *I. auritulus auritulus* Neumann, *I. conepati* Cooley and Kohls, and *I. lasallei* Méndez and Ortiz (Acarina: Ixodidae). *Journal of Parasitology*, 52, 810–820.

Descriptions and Redescriptions: Kohls and Clifford (1966) described and figured the female of *Ixodes cuernavacensis*.

Guzmán-Cornejo and Robbins (2010) redescribed the female but without figures. The male, nymph, and larva of *Ixodes cuernavacensis* remain unknown.

Geographic Distribution: *Ixodes cuernavacensis* is a Neotropical Mexican species. It has only been found in **southern Mexico**: Morelos (Kohls and Clifford 1966).

Hosts: The only known host for the female of *Ixodes cuernavacensis* is *Streptoprocne semicollaris* (Caprimulgiformes: Apodidae), which is provisionally treated as the principal host for this tick.

Human Parasitism: No.

Comments: Guglielmone et al. (2003a, 2014) state that *Ixodes cuernavacensis* is a Nearctic species, contrary to Clifford et al. (1973) and Camicas et al. (1998), who listed this tick as a Neotropical taxon. Reassessment of the type locality where the only two known specimens of *Ixodes cuernavacensis* were collected indicates that it lies within the boundaries of the Neotropical Region.

17. *Ixodes dicei* Keirans and Ajohda, 2003

Keirans, J.E. & Ajohda, S. (2003) *Ixodes* (*Ixodes*) *dicei*, a new tick species (Acari: Ixodida: Ixodidae) from a lagomorph in a restricted high-altitude habitat in Talamanca, Costa Rica. *International Journal of Acarology*, 29, 63–67.

Descriptions and Redescriptions: The male and female of *Ixodes dicei* were described and illustrated by Keirans and Ajohda (2003).

There are no redescriptions of the male and female of *Ixodes dicei*.

The larva and nymph of *Ixodes dicei* remain unknown.

Geographic Distribution: *Ixodes dicei* is a Central American species that has only been found in **Costa Rica**: San José (Keirans and Ajohda 2003).

Hosts: The only known host for the male and female of *Ixodes dicei* is *Sylvilagus dicei* (Lagomorpha: Leporidae) (Keirans and Ajohda 2003) and this lagomorph is treated here as potentially the principal host for this tick.

Human Parasitism: No.

Comments: No controversial issues related to *Ixodes dicei* have been found.

18. *Ixodes downsi* Kohls, 1957

Kohls, G.M. (1957b) *Ixodes downsi*, a new species of tick from a cave in Trinidad, British West Indies (Acarina – Ixodidae). *Proceedings of the Entomological Society of Washington*, 59, 257–264.

Descriptions and Redescriptions: The male, female, nymph, and larva of *Ixodes downsi* were described by Kohls (1957b).

There are no redescriptions of the male, female, nymph, and larva of *Ixodes downsi*.

Geographic Distribution: *Ixodes downsi* is a Neotropical South American species that has been found in **Peru**: Huánuco (Wilson and Baker 1988); **Trinidad and Tobago**: Tunapuna-Piarco (Kohls 1957b), and **Venezuela**: Monagas (González-Acuña et al. 2008).

Hosts: A female of *Ixodes downsi* was collected from a *Steatornis caripensis* (Caprimulgiformes: Steatornithidae) (Wilson and Baker 1988), and a larva was found on *Anoura geoffroyi* (Chiroptera: Phyllostomidae) (Kohls 1957b); these are provisionally treated as the principal hosts for this tick. Nymphs, males, and other females of *Ixodes downsi* have been recovered from a cave wall and bat guano (Kohls 1957b; González-Acuña et al. 2008).

Human Parasitism: No.

Comments: Kolonin (2008) states that adults of *Ixodes downsi* feed solely on Aves, which is strictly true only of the female tick for which we have a host record, but this tick was described from a cave shared by birds and bats, and the larva of *Ixodes downsi* was collected from the latter host; therefore, more data are needed before conclusions can be drawn concerning this tick's host specificity, as noted in Wilson and Baker (1988) and González-Acuña et al. (2008). Consequently, Aves are treated here as provisional principal hosts for females of *Ixodes downsi*, and *Anoura* *geoffroyi* is considered a potential principal host for the larva, pending the appearance of additional host records. Recent records of *Ixodes downsi* from *Rhinella marina* (Anura: Bufonidae) in Trinidad and Tobago by Kamilar et al. (2020) are treated as diagnostic errors.

19. *Ixodes fossulatus* Neumann, 1899

Neumann, L.G. (1899) Révision de la famille des ixodidés (3e mémoire). *Mémoires de la Société Zoologique de France*, 12, 107–294.

Descriptions and Redescriptions: *Ixodes fossulatus* is known only from a female specimen that was described by Neumann (1899) without figures.

Nuttall and Warburton (1911) redescribed the female type specimen of *Ixodes fossulatus*, stressing differences between their account and the morphological characters described in Neumann (1899).

The male, nymph, and larva of *Ixodes fossulatus* remain unknown.

Geographic Distribution: The female of *Ixodes fossulatus* was collected in **Ecuador** (Neumann 1899).

Host: Unknown (Neumann 1899).

Human Parasitism: No.

Comments: Clifford et al. (1973) treated *Ixodes fossulatus* as a doubtful species, while Kishida (1930) misidentified a Japanese tick (Palearctic Region) as *Ixodes fossulatus*, an error explained in Yamaguti et al. (1971).

Due to typographical errors, Estrada-Peña et al. (2020) listed adults and larvae of *Ixodes fossulatus* among a cluster of species prone to feed on Didelphimorphia and Rodentia, and the nymph of this tick was included as a parasite of Aves, but, as stated above, no hosts are known for *Ixodes fossulatus*.

20. *Ixodes fuscipes* Koch, 1844

Koch, C.L. (1844) Systematische Übersicht über die Ordnung der Zecken. *Archiv für Naturgeschichte*, 10, 217–239.

Descriptions and Redescriptions: Koch (1844) briefly described the female of *Ixodes fuscipes* but without figures, and the male was described, also without figures, by Fonseca (1935a) under the name *Ixodes ricinus aragaoi*, the original name of *Ixodes aragaoi*, which is currently considered a synonym of *Ixodes fuscipes*.

The male and female were redescribed by Fonseca (1937) as *Ixodes ricinus aragaoi*, by Calzada (1936) under the name *Ixodes ricinus rochensis*, and by Onofrio et al. (2006a, 2009, 2014), Nava et al. (2017) and Saracho-Bottero et al.

(2020) under the name *Ixodes aragaoi.* Koch (1847) redescribed and figured the female, while Fonseca (1935a) redescribed the female under the name *Ixodes ricinus aragaoi.* Labruna et al. (2020a) redescribed the holotype of *Ixodes fuscipes* along with a female from the syntype series of *Ixodes aragaoi* to demonstrate that the latter name is a synonym of the former.

The nymph and larva of *Ixodes fuscipes* have been identified by some workers from field-collected material under the names *Ixodes aragaoi* or *Ixodes pararicinus*, but there are no *bona fide* descriptions of these stages.

Geographic Distribution: *Ixodes fuscipes* is a Neotropical South American species found in **Brazil**: Paraná, Rio Grande do Sul, and São Paulo (Labruna et al. 2020a), and **Uruguay**: Cerro Largo, Florida, Lavalleja, Maldonado, Rocha, and Tacuarembó (Keirans et al. 1985; Venzal et al. 2005b, both under the name *Ixodes pararicinus*, and Labruna et al. 2020a).

Hosts: Males and females of *Ixodes fuscipes* have usually been found on cattle (Artiodactyla: Bovidae) and *Mazama gouazoubira* (Artiodactyla: Cervidae), with one record each from a dog and a horse. The undescribed nymph and larva have been found on mammals and birds, under the names *Ixodes aragaoi* and *Ixodes pararicinus*; the specimens were taken from several species of Didelphimorphia: Didelphidae, Rodentia: Cricetidae, and three families of Passeriformes, but it is unknown whether any of these vertebrates are principal hosts for the immature stages of this tick. The host profile of *Ixodes fuscipes* is shown in Table 1.8a, constructed using the information provided by Arzua et al. (2005), Venzal et al.

Table 1.8a Hosts for females (F), nymphs* (N), and larvae* (L) of *Ixodes fuscipes***

MAMMALIA			
ARTIODACTYLA: Bovidae		**PERISSODACTYLA: Equidae**	
Cattle	MF	Horse	M
ARTIODACTYLA: Cervidae		**RODENTIA: Cricetidae**	
Mazama gouazoubira	MF	*Akodon* sp.	L
Mazama rufina	M	*Akodon azarae*	N
CARNIVORA: Canidae		*Bruceipattersonius* sp.	L
Dog	F	*Juliomys pictipes*	L
DIDELPHIMORPHIA: Didelphidae		*Oligoryzomys delticola*	N
Monodelphis americana	NL	*Oligoryzomys nigripes*	NL
Monodelphis dimidiata	N		
AVES			
PASSERIFORMES: Furnariidae		**PASSERIFORMES: Turdidae**	
Phacellodomus striaticollis	NL	*Turdus albicollis*	NL
Syndactyla rufosuperciliata	NL	*Turdus rufiventris*	NL
PASSERIFORMES: Parulidae			
Myiothlypis leucoblephara	L		

* Provisional diagnosis because the nymph and larva of *Ixodes fuscipes* remain undescribed

** Most records of *Ixodes fuscipes* have been published under the names *Ixodes aragaoi* and *Ixodes pararicinus*

(2005a, b), Pinter (2007), Onofrio et al. (2014), Blanco et al. (2017), Michel et al. (2017) and Labruna et al. (2020a).

Human Parasitism: No.

Comments: *Ixodes fuscipes* has been considered a valid species with the exception of Cooley and Kohls (1945), who treated this species as doubtful, but this opinion has not been followed by most tick workers. Nonetheless, the history of the name *Ixodes fuscipes* is confused, and information concerning this species in Guglielmone et al. (2014) and others prior to the study of Labruna et al. (2020a) should be regarded as incorrect.

Koch (1844) described *Ixodes fuscipes* from a female tick collected from an unknown host in southern Brazil (Labruna et al. 2020a), and Neumann (1899) described *Ixodes spinosus* from females, nymphs, and larvae taken from *Dasyprocta aguti* (= *Dasyprocta* sp.) at an unknown Brazilian locality. However, Neumann (1901) reduced *Ixodes spinosus* to a synonym of *Ixodes fuscipes*, an opinion that was undisputed for more than a century. Nuttall and Warburton (1911) supported the synonymy proposed by Neumann (1901) and redescribed and produced the first figures of the female of *Ixodes fuscipes* using a female of the syntype series of *Ixodes spinosus*, which was taken to represent *bona fide Ixodes fuscipes* thereafter. Recently, Labruna et al. (2020a) compared the type of *Ixodes fuscipes* with the same female of the syntype series of *Ixodes spinosus* used by Nuttall and Warburton (1911) to redescribe the alleged female of *Ixodes fuscipes* but found that the two type specimens differ morphologically and reinstated *Ixodes spinosus* as a valid species. Additionally, Labruna et al. (2020a) compared the type of *Ixodes fuscipes* with females of the syntype series of *Ixodes aragaoi* (originally named *Ixodes ricinus aragaoi*) and found these specimens to be morphologically indistinguishable. Consequently, Labruna et al. (2020a) downgraded *Ixodes aragaoi* to a synonym of *Ixodes fuscipes*. These taxonomic decisions drastically alter our previous understanding of *Ixodes fuscipes*, as explained below.

Ixodes fuscipes is close to *Ixodes affinis* and *Ixodes pararicinus*. As already noted under *Ixodes affinis*, *Ixodes fuscipes* (under the name *Ixodes aragaoi*) has been treated as a synonym of *Ixodes affinis* by some authors, but this cannot be justified, as shown in Onofrio et al. (2014) and Saracho-Bottero et al. (2020), and as anticipated in Aragão and Fonseca (1952). All records of *Ixodes aragaoi* are treated here as *Ixodes fuscipes*, but this tick has also been partially confused with *Ixodes pararicinus* in Uruguay. *Ixodes pararicinus* is geographically limited to northwestern Argentina (Saracho-Bottero et al. 2020), but this tick was previously considered present in Uruguay. All Uruguayan records have since been changed first to *Ixodes aragaoi* and now to *Ixodes fuscipes*.

The geographic distribution of *Ixodes fuscipes* prior to the study of Labruna et al. (2020a) encompassed ten Brazilian states and two Peruvian regions, but its range is now thought to be restricted to just three states in southern Brazil and six Uruguayan departments, as shown above.

Labruna et al. (2020a) examined specimens classified as *Ixodes aragaoi* in several tick collections and found that all of them were *Ixodes fuscipes*, with the exceptions of a male tick collected from *Puma concolor* in Brazil (Labruna et al. 2005a) that belongs to an unidentified species of *Ixodes*, and a female tick found on a passeriform host in southern Brazil by Barros-Battesti and Knysak (1999) that is, in fact, *Ixodes auritulus*.

The actual host profile of *Ixodes fuscipes* differs markedly from that in Guglielmone et al. (2014). Several records of *Ixodes fuscipes* have been reclassified as *Ixodes spinosus*, as explained here under that species, but many other records are from specimens that are not considered to be either *Ixodes fuscipes* or *Ixodes spinosus*. Thus, two female ticks collected from *Dasyprocta variegata* in Brazil were identified as *Ixodes fuscipes* in Martins et al. (2014a) and Onofrio et al. (2014) but reclassified as *Ixodes lasallei* in Onofrio et al. (2020). The list of confusing specimens includes males, females, nymphs, and larvae of *Ixodes* cf *I. fuscipes* in Gianizella et al. (2018b) who collected ticks in the Brazilian state of Amazonas. Adults of *Ixodes* cf. *I. fuscipes* in Gianizella et al. (2018b) are considered to be *Ixodes bocatorensis*, in part following Onofrio et al. (2020), but the larvae and nymphs are treated here as *Ixodes* sp. Several other authors have identified as *Ixodes fuscipes* ticks that are now treated as *Ixodes* sp. The hosts allegedly infested with *Ixodes fuscipes* but now treated as infested with *Ixodes* sp. are listed in Table 1.8b, constructed using the information provided by Barros-Battesti and Knysak (1999), Arzua et al. (2005), Labruna et al. (2005d), Martins et al. (2014a), Blanco et al. (2017), Michel et al. (2017) and Gianizella et al. (2018b). The hosts of *Ixodes* sp. included in Table 1.8b are not included in the list of hosts of Neotropical Ixodidae provided in the appendix.

Table 1.8b Hosts for females (F), nymphs (N), and larvae (L) of *Ixodes* sp., previously identified as *Ixodes fuscipes* or *Ixodes* cf. *I. fuscipes*

MAMMALIA			
DIDELPHIMORPHIA: Didelphidae		*Oligoryzomys nigripes*	F
Lutreolina crassicaudata	N	**RODENTIA: Sciuridae**	
PILOSA: Myrmecophagidae		*Sciurus aestuans*	NL
*Tamandua tetradactyla**	N	**RODENTIA: Dasyproctidae**	
RODENTIA: Cricetidae		*Dasyprocta leporina**	NL
Akodon paranaensis	N		
AVES			
PASSERIFORMES: Furnariidae		**PASSERIFORMES: Thamonophilidae**	
Campylorhamphus falcularius	N	*Drymophila malura*	N
Syndactyla rufosuperciliata	N	**PASSERIFORMES: Thraupidae**	
Xiphorhynchus fuscus	N	*Tachyphonus coronatus*	N
PASSERIFORMES: Parulidae		*Trichothraupis melanops*	N
Basileuterus culicivorus	N		

* identified as *Ixodes cf. I. fuscipes*

21. *Ixodes galapagoensis* Clifford and Hoogstraal, 1980

Clifford, C.M. & Hoogstraal, H. (1980) A new species of *Ixodes* parasitizing the rice rat in the Galapagos (Ixodoidea: Ixodidae). *Proceedings of the Entomological Society of Washington*, 82, 378–383.

Descriptions and Redescriptions: The male, female, and nymph of *Ixodes galapagoensis* were described in Clifford and Hoogstraal (1980).

There are no redescriptions of the male, female, and nymph of *Ixodes galapagoensis*. The larva of *Ixodes galapagoensis* remains unknown.

Geographic Distribution: *Ixodes galapagoensis* is known from a single collection made on an island belonging to **Ecuador**: Galápagos (Clifford and Hoogstraal 1980).

Hosts: The female and nymph of *Ixodes galapagoensis* are known only as parasites of *Aegialomys galapagoensis* (Rodentia: Cricetidae); the only known male was found *in copula* (Clifford and Hoogstraal 1980).

Human Parasitism: No.

Comments: Clifford and Hoogstraal (1980) state that *Ixodes galapagoensis* is the only species of *Ixodes* found in the Galápagos Islands, and to date, no other member of this genus has been found there.

22. *Ixodes guatemalensis* Kohls, 1956

Kohls, G.M. (1956a) Eight new species of *Ixodes* from Central and South America (Acarina: Ixodidae). *Journal of Parasitology*, 42, 636–649.

Descriptions and Redescriptions: The female of *Ixodes guatemalensis* was described and figured in Kohls (1956a), and the male was described in Keirans and Eckerlin (2005).

Guzmán-Cornejo and Robbins (2010) redescribed the male and female but without figures, while Keirans and Eckerlin (2005) redescribed the female of *Ixodes guatemalensis*.

The larva and nymph of *Ixodes guatemalensis* remain unknown.

Geographic Distribution: *Ixodes guatemalensis* has been found in southern North America and Central America as follows: **Guatemala**: El Progreso, San Marcos, and Zacapa (Kohls 1956a; Keirans and Eckerlin 2005); and **southern Mexico**: Guerrero, and Veracruz (Guzmán-Cornejo et al. 2007).

Table 1.9 Hosts for males (M) and females (F) of *Ixodes guatemalensis*

MAMMALIA			
CARNIVORA: Mustelidae		**RODENTIA: Cricetidae**	
Mustela frenata	M*F	*Peromyscus grandis*	F
EULIPOTYPHLA: Soricidae		**RODENTIA: Sciuridae**	
Cryptotis goodwini	F	*Sciurus aureogaster*	F

* A male found *in copula* with a female feeding on *Mustela fenata*

Hosts: The female of *Ixodes guatemalensis* has been found on three orders of mammals, but the great majority of specimens have been collected from *Sciurus aureogaster* (Rodentia: Sciuridae), the principal host for this tick. The host profile of *Ixodes guatemalensis* is shown in Table 1.9, constructed with information provided by Kohls (1956a), Keirans and Eckerlin (2005), and Guzmán-Cornejo et al. (2007).

Human Parasitism: No.

Comments: No controversial issues related to *Ixodes guatemalensis* have been found.

23. *Ixodes jonesae* Kohls, Sonenshine and Clifford, 1969

Kohls, G.M., Sonenshine, D.E. & Clifford, C.M. (1969) *Ixodes* (*Exopalpiger*) *jonesae* sp. n. (Acarina: Ixodidae), a parasite of rodents in Venezuela. *Journal of Parasitology*, 55, 447–452.

Descriptions and Redescriptions: The male, female, nymph, and larva of *Ixodes jonesae* were described and figured in Kohls et al. (1969).

There are no redescriptions of the male, female, nymph, and larva of *Ixodes jonesae*.

Geographic Distribution: *Ixodes jonesae* is a Neotropical species from northern South America that has only been found in **Venezuela**: Mérida and Táchira (Kohls et al. 1969).

Hosts: Most records of *Ixodes jonesae* are from rodents (Cricetidae), as stated in Kohls et al. (1969), and *Thomasomys laniger* is thought to be the principal host for males, females, nymphs, and larvae of this tick. The host profile of *Ixodes jonesae* is shown in Table 1.10, based on information provided by Kohls et al. (1969) and Jones et al. (1972).

Human Parasitism: No.

Comments: No controversial issues related to *Ixodes jonesae* have been found.

Table 1.10 Hosts for males (M), females (F), nymphs (N), and larvae (L) of *Ixodes jonesae*

MAMMALIA			
PAUCITUBERCULATA: Caenolestidae		**RODENTIA: Cricetidae**	
Caenolestes fuliginosus	L	*Microryzomys minutus*	FL
		Thomasomys laniger	MFNL

24. *Ixodes lasallei* Méndez Arocha and Ortiz, 1958

Méndez Arocha, M. & Ortiz, I. (1958) Revisión de las garrapatas venezolanas del género *Ixodes* Latreille, 1795 y estudio de un nuevo *Amblyomma* (Acarina: Ixodidae). *Memorias de la Sociedad de Ciencias Naturales La Salle*, 18, 196–208.

Descriptions and Redescriptions: Méndez Arocha and Ortiz (1958) described and figured the female of *Ixodes lasallei.* The alleged male of *Ixodes lasallei* was described in Kohls and Clifford (1966) but is, in fact, the male of *Ixodes bocatorensis*, as discussed in Apanaskevich and Bermúdez (2017); consequently, the description of the male of *Ixodes lasallei* is to be found in Apanaskevich and Bermúdez (2017).

Apanaskevich and Bermúdez (2017) and Onofrio et al. (2020, supplementary material) redescribed the female.

The larva and nymph of *Ixodes lasallei* remain unknown.

Geographic Distribution: *Ixodes lasallei* is a South and Central American species that has been found in **Brazil**: Rondônia (Onofrio et al. 2020); **Colombia**: Arauca (Apanaskevich and Bermúdez 2017); **Peru**: Loreto (Apanaskevich and Bermúdez 2017); and **Venezuela**: Amazonas, and Apure (Apanaskevich and Bermúdez 2017).

Hosts: The principal host for females of *Ixodes lasallei* is provisionally considered to be *Cuniculus paca* (Rodentia: Cuniculidae) because 16 of 22 females of this tick were collected on this host, but adult ticks have also been collected from Rodentia: Dasyproctidae, Carnivora, and Didelphimorphia. The host profile of this species appears in Table 1.11, constructed with information provided by Guerrero (1996), Apanaskevich and Bermúdez (2017), and Onofrio et al. (2020).

Human Parasitism: No.

Comments: All records of adults of *Ixodes lasallei* from Panama in Kohls and Clifford (1966) and Fairchild et al. (1966) were identified as *Ixodes bocatorensis* by Apanaskevich and Bermúdez (2017); consequently, Panama is regarded as outside the range of *Ixodes lasallei.*

Guglielmone et al. (2014) treated as provisionally valid records of larvae and nymphs of *Ixodes lasallei* in Fairchild et al. (1966), Jones et al. (1972), and Guerrero (1996). However, the description of *Ixodes bocatorensis*, a close relative of *Ixodes lasallei*, by Apanaskevich and Bermúdez (2017) indicates that the taxonomic status

Table 1.11 Hosts for males (M) and females (F) of *Ixodes lasallei*

MAMMALIA			
CARNIVORA: Felidae		**RODENTIA:Dasyproctidae**	
Leopardus pardalis	F	*Dasyprocta fuliginosa*	F
DIDELPHIMORPHIA: Didelphidae		*Dasyprocta leporina*	F
Marmosa demerarae	MF*	*Dasyprocta variegata*	F
RODENTIA: Cuniculidae			
Cuniculus paca	MF		

* The diagnosis of these specimens of *Ixodes lasallei* is treated as provisionally valid

of those larvae and nymphs is uncertain because the two tick species have ranges that partially overlap. Venezuelan records of adults of *Ixodes lasallei* from *Marmosa demerarae* in Guerrero (1996) are treated as provisionally valid, while a male "close to *Ixodes lasallei*" from *Nephelomys albigularis* in Jones et al. (1972) is not included in the host list above. The female "close to *Ixodes lasallei*" found on a *Dasyprocta leporina* in Venezuela by Jones et al. (1972) was identified as *Ixodes bocatorensis* by Apanaskevich and Bermúdez (2017). Two female ticks collected from *Dasyprocta variegata* in Brazil were identified as *Ixodes fuscipes* by Martins et al. (2014a) and Onofrio et al. (2014), but both specimens were reidentified as *Ixodes lasallei* in Onofrio et al. (2020).

25. ***Ixodes longiscutatus*** Boero, 1944, as *Ixodes longsicutatum*, and amended to *Ixodes longiscutatus* in Camicas et al. (1998)

Boero, J.J. (1944a) Notas ixodológicas. I). *Ixodes longiscutatum* [sic], nueva especie. *Revista de la Asociación Médica Argentina*, 58, 353–354.

Camicas, J.L., Hervy, J.P., Adam, F. & Morel, P.C. (1998) *Les tiques du monde. Nomenclature, stades décrits, hôtes, répartition (Acarida, Ixodida)*. Orstom, Paris, 233 pp.

Descriptions and Redescriptions: The female of *Ixodes longiscutatus*, with a broken hypostome, was described and illustrated in Boero (1944a, under the name *Ixodes longsicutatus*); the larva and nymph were described by Kohls and Clifford (1967) under the name *Ixodes uruguayensis*, a synonym of *Ixodes longiscutatus*, as discussed in Guglielmone and Nava (2014).

The female of *Ixodes longiscutatus* was redescribed in Boero (1957, under the name *Ixodes longiscutatus*), and Nava et al. (2017) redescribed the female and nymph.

There are no redescriptions of the larva of *Ixodes longiscutatus* but Venzal et al. (2008) present figures of the basis capituli and marginal setae that are of diagnostic value.

The male of *Ixodes longsicutatus* remains unknown.

Table 1.12 Hosts for females (F), nymphs (N), and larvae (L) of *Ixodes longiscutatus*

MAMMALIA			
ARTIODACTYLA: Bovidae		**RODENTIA: Cricetidae**	
Cattle	F	*Akodon lutescens**	L
PERISSODACTYLA: Equidae		*Necromys obscurus*	L
Horse	F	*Oligoryzomys flavescens*	L
RODENTIA: Caviidae		*Oligoryzomys nigripes*	N
Cavia aperea	FNL	*Oxymycterus nasutus*	NL
Cavia tschudii	FNL	*Scapteromys tumidus*	NL

* It is uncertain whether this host of *Ixodes longiscutatus* is *Akodon lutescens* or a related species

Geographic Distribution: *Ixodes longiscutatus* is a Neotropical southern South American species that has been found in **Argentina**: Buenos Aires, Entre Ríos, and Salta (Venzal et al. 2008); **Brazil**: Rio Grande do Sul (Michel et al. 2017); and **Uruguay**: Cerro Largo, Florida, Maldonado, Rocha, San José, and Soriano (Venzal et al. 2008).

Hosts: Most records of larvae and nymphs of *Ixodes longiscutatus* are from *Oxymycterus nasutus* and *Scapteromys tumidus* (Rodentia: Cricetidae), with a few specimens from other Cricetidae and Rodentia: Caviidae. Female ticks are known from seven specimens, two collected from cattle, one specimen found on a horse, and four ticks recovered from Rodentia: Caviidae. The principal hosts of *Ixodes longiscutatus* females are hypothetically thought to be *Cavia aperea* and *Cavia tschudii* (Rodentia: Caviidae). The host profile of this species is shown in Table 1.12, constructed with information from Venzal et al. (2008), Dall'Agnol et al. (2017), and Michel et al. (2017).

Comments: Camicas et al. (1998) erroneously list the male of *Ixodes longiscutatus* as described.

The palps of the immature stages of *Ixodes longiscutatus* are unusual because they resemble ticks of the genus *Haemaphysalis.* Records of a nymph and a female of *Ixodes longiscutatus* in Ivancovich and Luciani (1992) published prior to the study of Venzal et al. (2008) that found that the nymph of *Ixodes longiscutatus* is identical to that of *Ixodes uruguayensis* are treated as doubtful and are excluded from the host list above.

Venzal et al. (2008) recorded nymphs of *Ixodes longiscutatus* feeding on *Akodon lutescens*, but according to the International Union for Conservation of Nature (2018) Argentina is outside the range of this rodent, although other authors have listed it as present in northwestern Argentina. Therefore, it is uncertain whether the host named in Venzal et al. (2008) is *Akodon lutescens* or a morphologically similar species.

26. *Ixodes loricatus* Neumann, 1899

Neumann, L.G. (1899) Révision de la famille des ixodidés (3e mémoire). *Mémoires de la Société Zoologique de France*, 12, 107–294.

Descriptions and Redescriptions: Neumann (1899) described the male, female, nymph, and larva of *Ixodes loricatus*, figuring both sexes.

Nuttall and Warburton (1911) and Marques et al. (2004) redescribed the male, female, nymph, and larva of *Ixodes loricatus*, while Cooley and Kohls (1945) and Nava et al. (2017) redescribed the male, female, and nymph. Neumann (1911a), Aragão and Fonseca (1961a), and Onofrio et al. (2006a, 2009) redescribed the male and female of *Ixodes loricatus*, and Boero (1957) redescribed the female alone. Sénevet and Ripert (1967) redescribed the larva of *Ixodes loricatus*.

Geographical Distribution: *Ixodes loricatus* is a Neotropical South American species that has been found in **Argentina**: Buenos Aires, Chaco, Córdoba, Corrientes, Formosa, Misiones, and Santa Fe (Guglielmone et al. 2011); **Brazil**: Alagoas, Bahia, Ceará, Goiás, Mato Grosso do Sul, Minas Gerais, Paraíba, Paraná, Pernambuco, Río de Janeiro, Río Grande do Sul, and São Paulo (Guglielmone et al. 2011; Fonseca et al. 2020); **Paraguay**: Cordillera, and Presidente Hayes (Guglielmone et al. 2011); and **Uruguay**: Canelones, Maldonado, and Montevideo (Guglielmone et al. 2011).

Hosts: *Ixodes loricatus* has been found on several types of mammalian hosts, but males, females, nymphs, and larvae are all prone to feed on Didelphimorphia: Didelphidae, especially *Didelphis albiventris*, *Didelphis aurita*, *Didelphis marsupialis,* and *Lutreolina crassicaudata*, although larvae and nymphs are also common on several species of Rodentia: Cricetidae named below. The host profile of *Ixodes loricatus* is shown in Table 1.13, constructed with information provided by Cooley and Kohls (1945), Fonseca and Aragão (1952a), Guglielmone and Nava (2011), Guglielmone et al. (2011), Dantas-Torres et al. (2012), Colombo et al. (2013), Coelho et al. (2016), Nava et al. (2017) and Luz et al. (2018a).

Human Parasitism: No.

Comments: *Ixodes loricatus* has been thought to occur from the southern tip of South America to southern Mexico, but the study of Guglielmone et al. (2011) shows that it is established in east-central and northeastern South America. The earlier range was a consequence of mislabeling or false information, as well as incorrect identifications, especially for early records of *Ixodes loricatus* published before the closely related *Ixodes luciae* was described (Guglielmone et al. 2011), but *Ixodes loricatus* was also confused with *Ixodes angustus* (see this name in Chap. 4). Recent records of *Ixodes loricatus* from northwestern Argentina, as in Nava et al. (2004), were reidentified as either *Ixodes luciae* or ticks that could not be determined to species, while ticks from Buenos Aires Province, Argentina, classified by Ivancovich and Luciani (1992) as *Ixodes luciae*, were later identified as *Ixodes loricatus* (Guglielmone et al. 2011).

Table 1.13 Hosts for adult ticks of undetermined sex (A), males (M), females (F), nymphs (N), and larvae (L) of *Ixodes loricatus*

MAMMALIA			
CARNIVORA: Procyonidae		**RODENTIA: Cricetidae**	
Nasua nasua	F	*Akodon azarae*	NL
CINGULATA: Dasypodidae		*Calomys* sp.	NL
Dasypus novemcinctus	F	*Cerradomys subflavus*	F
DIDELPHIMORPHIA: Didelphidae		*Euryoryzomys russatus*	N
Chironectes minimus	F	*Holochilus brasiliensis*	NL
Didelphis albiventris	MFNL	*Hylaeamys megacephalus*	NL
Didelphis aurita	MFNL	*Oligoryzomys flavescens*	NL
Didelphis marsupialis	MFL	*Oligoryzomys nigripes*	NL
Gracilinanus agilis	N	*Oxymycterus nasutus*	L
Lutreolina crassicaudata	MFNL	*Oxymycterus rufus*	NL
Marmosa sp.	FN	*Rhipidomys cariri*	NL
Marmosa demerarae	MFNL	*Scapteromys aquaticus*	NL
Marmosa murina	NL	*Scapteromys tumidus*	NL
Metachirus nudicaudatus	A	*Zygodontomys* sp.	F
Monodelphis americana	L	**RODENTIA: Cuniculidae**	
Monodelphis dimidiata	NL	*Cuniculus paca*	F
Monodelphis scallops	F	**RODENTIA: Muridae**	
Philander frenatus	FNL	*Mus* sp.	MF
Philander opossum	MF	*Rattus norvegicus*	MFL
RODENTIA: Caviidae		*Rattus rattus*	N
Cavia sp.	F		

Acevedo-Gutiérrez et al. (2020) stated that *Ixodes loricatus* had been listed as a Colombian tick, but these authors also stressed the need to confirm such diagnoses because of the morphological similarity between *Ixodes loricatus* and *Ixodes luciae.* We do not include Colombia within the range of *Ixodes loricatus*. The inclusion of *Ixodes loricatus* in identification keys for *Ixodes* found in Panama (Fairchild et al. 1966) and Mexico (Guzmán-Cornejo and Robbins 2010) is unjustified because this tick is not present in those countries. Several Brazilian records of this tick have been named as *Ixodes didelphidis*, a synonym of *Ixodes loricatus*, as demonstrated by Labruna et al. (2002a).

The record of adults of *Ixodes loricatus* parasitizing cricetid rodents in Venezuela (Vásquez et al. 2011) requires confirmation. Minoprio et al. (1994) in Argentina and Serra-Freire et al. (2011) in Brazil report human infestations by *Ixodes loricatus*, while Onofre-Barragán (2015) allegedly found this tick on cattle in southern Mexico (State of Guerrero), but all these records are here considered errors in identification.

27. *Ixodes luciae* Sénevet, 1940, but previously identified as *Ixodes loricatus spinosus* by Nuttall (1910) and replaced by *Ixodes loricatus vogelsangi* in Santos Dias (1954) because the name *spinosus* was preoccupied with *Ixodes spinosus* in Neumann (1899). The name *Ixodes luciae* prevails over the others

Sénevet, G. (1940) Quelques ixodidés de la Guyane française. Espèces nouvelles d'*Ixodes* et d'*Amblyomma. 6º Congreso Internacional de Entomología,* Madrid, España, setiembre 1935, pp. 891–898.

Neumann, L.G. (1899) Révision de la famille des ixodidés (3e mémoire). *Mémoires de la Société Zoologique de France*, 12, 107–294.
Nuttall, G.H.F. (1910) New species of ticks (*Ixodes*, *Amblyomma*, *Rhipicephalus*). *Parasitology*, 3, 408–416.
Santos Dias, J.A.T. (1954) Um novo nome para o *Ixodes loricatus spinosus* Nuttall, 1910 (*nomen bis lectum*). *Moçambique*, (79), 79–81.

Descriptions and Redescriptions: The female of *Ixodes luciae* was described and figured in Nuttall (1910), under the name *Ixodes loricatus spinosus*, a senior synonym of *Ixodes luciae*, as discussed in Guglielmone and Nava (2014). The male was described in Cooley and Kohls (1945); the larva and nymph of *Ixodes luciae* were described in Onofrio et al. (2010b).

The female of *Ixodes luciae* was redescribed by Sénevet (1940) as a new species. Nava et al. (2017) redescribed the male, female, and nymph. Floch and Fauran (1958), Aragão and Fonseca (1961a), Onofrio et al. (2006a, 2009, 2010b) and Bermúdez et al. (2018a) redescribed the male and female of *Ixodes luciae*. Cooley and Kohls (1945) redescribed the female.

There are no redescriptions of the larva of *Ixodes luciae*.

Geographic Distribution: *Ixodes luciae* is a Neotropical species characterized by a vast range from southern South America to southern North America as follows: **Argentina**: Jujuy, Salta, and Tucumán (Guglielmone et al. 2011; Ortiz et al. 2011), **Belize**: Stann Creek (Cooley and Kohls 1945); **Bolivia**: Santa Cruz (Guglielmone et al. 2011); **Brazil**: Acre, Amazonas, Maranhão, Mato Grosso, Mato Grosso do Sul, Pará, and Rondônia (Witter et al. 2016; Costa et al. 2020); **Colombia**: Antioquia, Boyacá, Cundinamarca, Meta, and Valle del Cauca (Guglielmone et al. 2011; Faccini-Martínez et al. 2017; Acevedo-Gutiérrez et al. 2020); **Costa Rica** (Guglielmone et al. 2011); **Ecuador** (Guglielmone et al. 2011); **French Guiana**: Cayena (Binetruy et al. 2019); **Guatemala**: Izabal (Monroy Lefebre and Cajas González 1988); **Honduras** (Guglielmone et al. 2011); **southern Mexico**: Chiapas, Colima, Oaxaca, Tabasco, and Veracruz (Guzmán-Cornejo et al. 2012); **Nicaragua** (Jones et al. 1972); **Panama**: Bocas del Toro, Colón, Darién, Kuna Yala, and Panamá (Guglielmone et al. 2011); **Peru**: Cajamarca, Cusco, Madre de Dios, Loreto, Piura, Puno, and Tumbes (Need et al. 1991; Norris et al. 1999; Díaz et al. 2007); **Suriname** (Keirans 1985b); **Trinidad and Tobago** (Aitken et al. 1968); and **Venezuela**: Amazonas, Araguá, Barinas, Carabobo, Distrito Federal, Falcón, Guárico, Lara, Miranda, Trujillo, Yaracuy, and Zulia (Vásquez et al. 2011).

Hosts: *Ixodes luciae* has been collected from several mammalian hosts, but males, females, nymphs, and larvae are usually found on Didelphimorphia: Didelphidae, such as *Didelphis albiventris*, *Didelphis aurita,* and *Philander opossum*, although Jones et al. (1972) found *Monodelphis brevicaudata* to be an important host for larvae and nymphs but not for adults of *Ixodes luciae*. Larvae and especially nymphs have also been frequently found on Rodentia: Cricetidae. Díaz et al. (2007) demonstrated the importance of *Hylaeamys perenensis* and *Hylaeamys yunganus* as hosts for the nymph in Peruvian populations of *Ixodes luciae*, but outside Peru it was not

Table 1.14 Hosts for adult ticks of undetermined sex (A), males (M), females (F), nymphs (N), and larvae (L) of *Ixodes luciae*

MAMMALIA			
CARNIVORA: Canidae		*Thylamys venustus*	N
Domestic dog	MF	*Tlacuatzin canescens*	N
DIDELPHIMORPHIA: Didelphidae		**RODENTIA: Cricetidae**	
Caluromys lanatus	F	*Calomys callosus*	N
Caluromys philander	L	*Hylaeamys megacephalus*	N
Didelphis albiventris	MFNL	*Hylaeamys perenensis*	N
Didelphis aurita	MF	*Hylaemys yunganus*	N
Didelphis marsupialis	MFNL	*Ichthyomys pittieri*	F
Lutreolina crassicaudata	F	*Nyctomys sumichrasti*	N
Marmosa constantiae	N	*Oecomys* sp.	A
Marmosa demerarae	FN	*Oecomys auyantepui*	N
Marmosa mexicana	N	*Oecomys bicolor*	L
Marmosa murina	FNL	*Oecomys concolor*	NL
Marmosa robinsoni	MFNL	*Oligoryzomys microtis*	N
Marmosops impavidus	F	*Oryzomys* sp.	N
Metachirus nudicaudatus	F	*Sigmodon* sp.	NL
Monodelphis sp.	A	*Zygodontomys brevicauda*	N
Monodelphis brevicaudata	NL	**RODENTIA: Cuniculidae**	
Philander andersoni	F	*Cuniculus paca*	F
Philander opossum	MFNL	**RODENTIA: Muridae**	
Thylamys cinderella	N	"Rat"	MF

possible to determine whether there is a cricetid species more prone to infestation with *Ixodes luciae* than other taxa in this family. The host profile of *Ixodes luciae* is shown in Table 1.14, constructed using data from Méndez Arocha and Ortiz (1958), Fonseca (1959), Fairchild et al. (1966), Aitken et al. (1968), Jones et al. (1972), Guerrero (1996), Labruna et al. (2005d), Autino et al. (2006), Díaz et al. (2007), Guzmán-Cornejo et al. (2007, 2012), Bermúdez et al. (2010a), Guglielmone et al. (2011), Luz et al. (2013b), Nava et al. (2017) and Binetruy et al. (2019).

Human Parasitism: No.

Comments: The name *Ixodes luciae* was interpreted as a new name for *Ixodes loricatus spinosus* Nuttall, 1910, preoccupied by *Ixodes spinosus* Neumann, 1899 in Keirans and Hillyard (2001); however, Sénevet (1940) described *Ixodes luciae* without mentioning either of those names, and he probably overlooked the description of *Ixodes loricatus spinosus*.

Specimens of alleged *Ixodes loricatus* collected in northwestern Argentina in Nava et al. (2004) have been reidentified as *Ixodes luciae* or *Ixodes* sp. Records of *Ixodes luciae* from Sciuridae collected in Rio de Janeiro State in Brazil by Oliveira et al. (2014) are treated as doubtful in Guglielmone et al. (2014); consequently, this state is not included within the range of this tick. Onofrio et al. (2010b) and Luz et al. (2013b) both identified as adults specimens of *Ixodes luciae* collected at two localities in the Brazilian state of Pará, but Luz et al. (2013b) listed "rats" as hosts for these ticks, while no hosts are provided by Onofrio et al. (2010b). "Rats" were not included as hosts of *Ixodes luciae* in Guglielmone et al. (2014) but are included as provisionally valid here. Serra-Freire et al. (1996) allegedly found nymphs of this tick on cervids in Brazil based on a key where the nymph of *Ixodes luciae* was not included because it was still undescribed at that time; this record was accepted as valid in Nava et al. (2017) and by others but is rejected here. Recent records of *Ixodes luciae* from *Rhinella marina* (Anura: Bufonidae) in Trinidad and Tobago by Kamilar et al. (2020) are treated as diagnostic errors.

The record of human parasitism by *Ixodes luciae* in Ivancovich and Luciani (1992) was treated as valid in Guglielmone et al. (2003a). However, this record is regarded as a diagnostic error in Guglielmone et al. (2011) because the supposed human-feeding specimen was not found, and all *Ixodes luciae* identified in Ivancovich and Luciani (1992) from the same locality where the alleged human parasitism took place were in fact *Ixodes loricatus* or *Ixodes auritulus*. *Ixodes luciae* is not listed as a parasite of humans in Guglielmone and Robbins (2018).

28. *Ixodes mexicanus* Cooley and Kohls, 1942

Cooley, R.A. & Kohls, G.M. (1942b) *Ixodes mexicanus* n. sp. et *Ixodes tancitarius* n. sp. dos nuevas garrapatas mexicanas. *Revista de la Sociedad Mexicana de Historia Natural*, 3, 149–154.

Descriptions and Redescriptions: Cooley and Kohls (1942b) described and figured the female of *Ixodes mexicanus*.

Cooley and Kohls (1945) and Guzmán-Cornejo and Robbins (2010) redescribed the female, but the latter authors provided no figures.

The male, nymph, and larva of *Ixodes mexicanus* remain unknown.

Geographic Distribution: *Ixodes mexicanus* is a Neotropical species that has only been found in **southern Mexico**: Michoacán (Cooley and Kohls 1942b; Guzmán-Cornejo et al. 2007).

Hosts: The female of *Ixodes mexicanus* is known from only two specimens, one collected from *Campylorhynchus gularis* (Passeriformes: Troglodytidae) and the other recovered from *Junco phaeonotus* (Passeriformes: Passerellidae) (Cooley and Kohls 1942b).

Human Parasitism: No.

Comments: Guglielmone et al. (2003a, 2014) state that *Ixodes mexicanus* is a Nearctic species, contrary to Clifford et al. (1973) and Camicas et al. (1998), who listed it as a Neotropical taxon. Reassessment of the type locality where the only specimens of *Ixodes mexicanus* have been collected indicates that the site lies within the boundaries of the Neotropical Region.

29. *Ixodes minor* Neumann, 1902

Neumann, L.G. (1902) Notas sur les ixodidés. *Archives de Parasitologie*, 6, 109–128.

Descriptions and Redescriptions: Neumann (1902) described the male and female of *Ixodes minor* but only figured the male. The nymph was described by Smith and Gouck (1947) under the name *Ixodes bishoppi*, a synonym of *Ixodes minor*, as stated in Keirans and Clifford (1978), while the larva of *Ixodes minor* was described in Clifford et al. (1961).

The male and female of *Ixodes minor* were redescribed in Nuttall and Warburton (1911), Keirans and Clifford (1978), and Keirans and Litwak (1989); the female was redescribed by Smith and Gouck (1947) under the name *Ixodes bishoppi*. Durden and Keirans (1996) and Sénevet and Ripert (1967) described the nymph and larva, respectively, providing diagnostic characters for these stages.

Geographic Distribution: *Ixodes minor* is found in the Nearctic and Neotropical Zoogeographic Regions; in the latter region, this species occurs in Central America as follows: **Costa Rica**: Limón (Ogrzewalska et al. 2015); and **Guatemala**: Chinaltenango (Kohls 1953).

Hosts: The Neotropical hosts for *Ixodes minor* are few, as shown in Table 1.15, constructed using information contained in Kohls (1953) and Ogrzewalska et al. (2015). Hosts for this tick are more numerous in the Nearctic than in the Neotropics, where adults, nymphs, and larvae have been found on Rodentia: Cricetidae, Muridae, and Sciuridae, while immature stages have been collected from Carnivora: Mephitidae, and several families of Passeriformes (Guglielmone et al. 2014). With that in mind, the Neotropical host profile of *Ixodes minor* should be regarded as preliminary, although the limited records conform with data from the Nearctic Region, and the hosts below are therefore treated as provisional principal hosts for Neotropical populations of *Ixodes minor*.

Human Parasitism: No.

Comments: *Ixodes minor* was treated as a doubtful species by Cooley and Kohls (1945), but subsequent studies proved its validity.

Table 1.15 Hosts for males (M), females (F), and larvae (L) of *Ixodes minor**

MAMMALIA			
RODENTIA: Muridae			
Rat	MF		
AVES			
PASSERIFORMES: Troglodytidae		**PASSERIFORMES: Turdidae**	
Troglodytes aedon	L	*Catharus frantzii*	L

* Neumann (1902) described M and F of *Ixodes minor* from specimens collected from *Hesperomys* sp.? [sic]

Smith and Gouck (1947) allegedly described the larva of *Ixodes minor* under the name *Ixodes bishoppi*; however, this description differs significantly from the account of *Ixodes minor* in Clifford et al. (1961), which is considered its first description in Keirans and Clifford (1978), a position also accepted here.

There is a Panamanian record of *Ixodes minor* in Fairchild (1943) that is treated as an error in Fairchild et al. (1966), and Panama is therefore excluded from the range of this tick.

30. *Ixodes montoyanus* Cooley, 1944

Cooley, R.A. (1944a) *Ixodes montoyanus* (Ixodidae) a new tick from Colombia. *Boletín de la Oficina Sanitaria Pan-Americana*, 23, 804–806.

Descriptions and Redescriptions: The female of *Ixodes montoyanus* was described and figured by Cooley (1944a), while the male, nymph, and larva were described and illustrated by Keirans (1973).

Keirans (1973) redescribed the female.

There are no redescriptions of the male, nymph and larva of *Ixodes montoyanus.*

Geographical Distribution: *Ixodes montoyanus* has been found in northern South America as follows: **Colombia**: Distrito Capital, and Valle del Cauca (Cooley 1944a; Keirans 1973); and **Venezuela**: Mérida (Keirans 1973).

Hosts: The principal hosts for females, nymphs, and larvae of *Ixodes montoyanus* are Artiodactyla: Cervidae, with one record of adult ticks from Carnivora, but Keirans (1973) states that *Pudu mephistophiles* is the chief host for all parasitic stages of this tick, an opinion also accepted here. The host profile of *Ixodes montoyanus* is shown in Table 1.16, constructed using information provided by Cooley (1944a) and Keirans (1973).

Human Parasitism: No.

Comments: The male of *Ixodes montoyanus* is known from one specimen, which Keirans (1973) used to describe it.

Acevedo-Gutiérrez et al. (2020) include the Colombian department of Cundinamarca within the range of *Ixodes montoyanus*, but no records were provided to support the presence of this tick there.

Keirans (1973) considers that most if not all records of *Ixodes montoyanus* from *Pudu* sp. correspond to *Pudu mephistophiles.*

Table 1.16 Hosts for males (M), females (F), nymphs (N), and larvae (L) of *Ixodes montoyanus*

MAMMALIA			
ARTIODACTYLA: Cervidae		*Pudu mephistophiles*	FNL
Mazama sp.	F	**CARNIVORA: Procyonidae**	
Pudu sp.	MFNL	*Nasuella olivacea*	F

31. *Ixodes nectomys* Kohls, 1956

Kohls, G.M. (1956a) Eight new species of *Ixodes* from Central and South America (Acarina: Ixodidae). *Journal of Parasitology*, 42, 636–649.

Descriptions and Redescriptions: *Ixodes nectomys* is known from a male that was described and illustrated in Kohls (1956a).

There are no redescriptions of the male of *Ixodes nectomys.*

The female, nymph, and larva of *Ixodes nectomys* remain unknown.

Geographic Distribution: *Ixodes nectomys* is a Neotropical species that has only been found in **Peru**: Cusco (Kohls 1956a).

Host: The only known host for the male of *Ixodes nectomys* is *Nectomys apicalis* (Rodentia: Cricetidae), named as *Nectomys squamipes garleppi* in Kohls (1956a).

Human Parasitism: No.

Comments: No controversial issues related to *Ixodes nectomys* have been found.

32. *Ixodes neuquenensis* Ringuelet, 1947

Ringuelet, R. (1947) la supuesta presencia de *Ixodes brunneus* Koch en la Argentina y descripción de una nueva garrapata *Ixodes neuquenensis* nov. sp. *Notas del Museo de la Plata*, 12, 207–216.

Descriptions and Redescriptions: The female of *Ixodes neuquenensis* was first described and figured by Boero (1945) under the name *Ixodes brunneus* (currently a Nearctic species), a diagnostic error, while the nymph and larva were described in Guglielmone et al. (2004).

Ringuelet (1947) redescribed the female and provided the current status of *Ixodes neuquenensis.* The female and nymph were redescribed in Nava et al. (2017), and Guglielmone et al. (2004) also redescribed the female of *Ixodes neuquenensis.*

There are no redescriptions of the larva.

The male of *Ixodes neuquenensis* remains unknown.

Geographic Distribution: *Ixodes neuquenensis* is a Neotropical species that has only been found in southern South America as follows: **Argentina**: Neuquén, and Río Negro (Ringuelet 1947; Guglielmone et al. 2004); and **Chile**: Región de los Lagos (Marín-Vial et al. 2007).

Hosts: The female, nymph, and larva of *Ixodes neuquenensis* have only been collected from *Dromiciops gliroides* (Microbiotheria: Microbiotheriidae).

Human Parasitism: No.

Comments: *Ixodes neuquenensis* has had a confusing history, with Boero (1945) identifying its female as *Ixodes brunneus* (Nearctic). The nymph of *Ixodes neuquenensis* described in Ringuelet (1947) belongs to another as yet undetermined species of *Ixodes* (Guglielmone et al. 2004), but Camicas et al. (1998) state, erroneously, that the only known stage of *Ixodes neuquenensis* is the nymph.

Daciuk et al. (1981) allegedly found the male of *Ixodes neuquenensis* on a fox in southern Argentina, but the tick was later identified as *Amblyomma tigrinum*, as discussed in Guglielmone and Nava (2005). Sebastian et al. (2016) identified as *Ixodes* cf *I. neuquenensis* nymphs collected from *Phyllotes xanthopygus* in Argentinean Patagonia, implying that a species close to *Ixodes neuquenensis* exists there.

Guglielmone and Nava (2011) hypothesized that *Ixodes neuquenensis* and its ancient marsupial hosts constitute a rare association of a precursor of Australian marsupials parasitized by a relatively new tick species, derived from the colonization of southern South America by ticks of the genus *Ixodes* after the Great American Interchange.

33. *Ixodes nuttalli* Lahille, 1913

Lahille, F. (1913) Sobre dos *Ixodes* de la República Argentina y la medición de las garrapatas. *Boletín del Ministerio de Agricultura, República Argentina*, 16, 278–289.

Descriptions and Redescriptions: Lahille (1913) described and illustrated the male of *Ixodes nuttalli*; the female, nymph, and larva were described and figured by Nuttall (1916).

Nava et al. (2017) provided a text redescription of the male, female, and nymph. Nuttall (1916) redescribed the male, while Sénevet and Ripert (1967) redescribed the larva of *Ixodes nuttalli*.

Geographic Distribution: *Ixodes nuttalli* is a South American species that has been found in **Argentina**: Mendoza (Lahille 1913), and **Peru**: "Ninahuanchi?" = Mina Huasi?: Lima Region? (Nuttall 1916).

Hosts: Males, females, nymphs, and larvae of *Ixodes nuttalli* have been collected from *Lagidium viscacia* (Rodentia: Chinchillidae).

Human Parasitism: No.

Comments: No controversial issues related to *Ixodes nuttalli* have been found apart from the indefinite Peruvian locality where this species was collected.

34. *Ixodes paranaensis* Barros-Battesti, Arzua, Pichorim, and Keirans, 2003

Barros-Battesti, D.M., Arzua, M., Pichorim, M. & Keirans, J.E. (2003) *Ixodes (Multidentatus) paranaensis* n. sp. (Acari: Ixodidae) a parasite of *Streptoprocne biscutata* (Sclater, 1865) (Apodiformes: Apodidae) birds in Brazil. *Memórias do Instituto Oswaldo Cruz*, 98, 93–102.

Descriptions and Redescriptions: The male, female, nymph, and larva of *Ixodes paranaensis* were described and figured in Barros-Battesti et al. (2003).

Table 1.17 Hosts for females (F), nymphs (N), and larvae (L) of *Ixodes paranaensis*

AVES			
CAPRIMULGIFORMES: Apodidae		*Streptoprocne biscutata*	FNL
Cypseloides fumigatus	F	*Streptoprocne zonaris*	FN
Cypseloides senex	FN		

The male and female were redescribed in Onofrio et al. (2006a, 2009).

There are no redescriptions of the nymph and larva of *Ixodes paranaensis.*

Geographic Distribution: *Ixodes paranaensis* is a Neotropical species that has only been found in **Brazil**: Minas Gerais, Paraná, and Rio de Janeiro (Barros-Battesti et al. 2003; Tolesano-Pascoli et al. 2014; Luz et al. 2016b).

Hosts: Females, nymphs, and larvae have been found on birds of the family Apodidae but most records are from *Streptoprocne biscutata*, the principal host of *Ixodes paranaensis*. The host profile of *Ixodes paranaensis* is shown in Table 1.17, constructed with information provided by Barros-Battesti et al. (2003), Tolesano-Pascoli et al. (2014), and Luz et al. (2016b). The male specimen of *Ixodes paranaensis* that was used to describe this species was collected from a bird's nest; subsequently, males and other stages of *Ixodes paranaensis* have been found inside caves used by the principal host (Dantas-Torres et al. 2009b; Muñoz-Leal et al. 2020).

Human Parasitism: No.

Comments: No controversial issues related to *Ixodes paranaensis* have been found.

35. *Ixodes pararicinus* Keirans and Clifford, 1985

Keirans J.E., Clifford, C.M., Guglielmone, A.A. & Mangold, A.J. (1985) *Ixodes (Ixodes) pararicinus* n. sp. (Acari: Ixodoidea: Ixodidae), a South American cattle tick long confused with *Ixodes ricinus*. *Journal of Medical Entomology*, 22, 401–407.

Descriptions and Redescriptions: The male and female of *Ixodes pararicinus* were described and illustrated in Keirans et al. (1985).

Nava et al. (2017) and Saracho-Bottero et al. (2020) redescribed the male and female of *Ixodes pararicinus*.

Venzal et al. (2005b) allegedly described the nymph and larva of *Ixodes pararicinus* but it is uncertain whether the descriptions are based on *bona fide* specimens of this tick. Therefore, the nymph and larva of *Ixodes pararicinus* are treated as undescribed, but some authors have been able to identify these stages from field-collected material.

Geographic Distribution: *Ixodes pararicinus* is a southern South American species with *bona fide* specimens known only from **Argentina**: Catamarca, Jujuy, Salta, and Tucumán (Keirans et al. 1985; Mastropaolo et al. 2008; Ortiz et al. 2011; Saracho-Bottero et al. 2020).

Hosts: Adults of *Ixodes pararicinus* have mostly been collected from cattle (Artiodactyla: Bovidae). The undescribed nymph and larva have usually been taken from Rodentia: Cricetidae and several families of Passeriformes, but Beldoménico et al. (2003a) found that the rodent *Calomys callosus* carried more nymphs and larvae than other members of its family (Cricetidae), while Flores et al. (2014) and Saracho-Bottero et al. (2017) found that *Turdus rufiventris* (Passeriformes: Turdidae) is the most common host for these stages among birds. Therefore, *Calomys callosus* and *Turdus rufiventris* are provisionally treated as principal hosts for the larvae and nymphs of *Ixodes pararicinus*. The host profile of *Ixodes pararicinus* is shown in Table 1.18, constructed using information provided by Keirans et al. (1985), Beldoménico et al. (2003a, b), Autino et al. (2006), Ortiz et al. (2011), Flores et al. (2014), Nava et al. (2017) and Saracho-Bottero et al. (2017).

Human Parasitism: There is an Argentinean record of a nymph of *Ixodes pararicinus* feeding on a human in the Province of Jujuy (Saracho-Bottero et al. 2018).

Comments: *Ixodes pararicinus*, *Ixodes affinis,* and *Ixodes fuscipes* are closely related species. Guglielmone et al. (2003a) speculated that the range of *Ixodes*

Table 1.18 Hosts for adult ticks of undetermined sex (A), males (M), females (F), nymphs (N)* and larvae (L)* of *Ixodes pararicinus*

	MAMMALIA		
ARTIODACTYLA: Bovidae		**PERISSODACTYLA: Equidae**	
Cattle	MF	Horse	MF
CARNIVORA: Canidae		**RODENTIA: Cricetidae**	
Domestic dog	N	*Calomys callosus*	NL
CHIROPTERA: Phyllostomidae		*Calomys fecundus*	NL
Desmodus rotundus	A	*Oligoryzomys destructor*	L
DIDELPHIMORPHIA: Didelphidae			
Thylamys venustus	L		
	AVES		
PASSERIFORMES: Corvidae		*Atlapetes citrinellus*	L
Cyanocorax chrysops	NL	**Thraupidae**	
PASSERIFORMES: Furnariidae		*Coryphospingus cucullatus*	L
Syndactyla rufosuperciliata	NL	**Troglodytidae**	
Parulidae		*Troglodytes aedon*	L
Myioborus brunniceps	L	**Turdidae**	
Myiothlypis bivittata	L	*Turdus amaurochalinus*	NL
Passerellidae		*Turdus nigriceps*	N
Arremon flavirostris	L	*Turdus rufiventris*	NL
Arremon torquatus	L		

* Records of nymphs and larvae of *Ixodes pararicinus* are provisionally valid because these stages remain undescribed

pararicinus may include several South American countries, but Guglielmone and Nava (2005) proposed that northeastern Argentinean populations named as *Ixodes pararicinus* were in fact *Ixodes fuscipes* (named as *Ixodes aragaoi*). This situation changed after confirmation that the species present in Uruguay is *Ixodes fuscipes*, not *Ixodes pararicinus*, and the emergence of evidence leading to the conclusion that ticks from northeastern Argentina classified as *Ixodes pararicinus* were actually *Ixodes affinis* or a species close to it (Saracho-Bottero et al. 2020). *Bona fide* populations of *Ixodes pararicinus* are now thought to be confined to northwestern Argentina in the Yungas phytogeographic province of the Amazonian domain, partly supporting the above hypothesis of Guglielmone and Nava (2005). Consequently, the list of hosts excludes several mammals recorded as parasitized by *Ixodes pararicinus* in Nava et al. (2017).

Nava et al. (2017) and Guglielmone and Robbins (2018) include Colombia and Peru within the range of *Ixodes pararicinus*, and Acevedo-Gutiérrez et al. (2020) list Colombian records of *Ixodes pararicinus*, but new information summarized above indicates that its presence in those countries requires verification. Therefore, the geographic distribution of this tick is provisionally limited to Argentina. An additional problem concerns the alleged descriptions of the larva and nymph of *Ixodes pararicinus* in Venzal et al. (2005b), which are based on specimens from Argentina and Uruguay, although *Ixodes pararicinus* is not established in the latter country. Therefore, records of the immature stages of *Ixodes pararicinus* are somewhat uncertain and should be treated as provisionally valid.

Estrada-Peña et al. (2020) include the nymph of *Ixodes pararicinus* in a cluster of species that feed on Rodentia and/or Didelphimorphia, and the larva in a cluster of ticks that feed on birds, partially disagreeing with the above statement that recognizes rodents and birds as principal hosts for nymphs and larvae of this tick.

36. *Ixodes pomerantzi* Kohls, 1956

Kohls, G.M. (1956a) Eight new species of *Ixodes* from Central and South America (Acarina: Ixodidae). *Journal of Parasitology*, 42, 636–649.

Descriptions and Redescriptions: The male and female of *Ixodes pomerantzi* were briefly described and illustrated in Kohls (1956a).

Bermúdez et al. (2018a) redescribed the male and female of *Ixodes pomerantzi*.

The nymph and larva of *Ixodes pomerantzi* remain undescribed, but some authors have been able to identify the nymph of this species from field-collected material.

Geographical Distribution: *Ixodes pomerantzi* is a Neotropical species found in Central and northeastern South America as follows: **Ecuador**: Pichincha (Keirans 1985b); **Guatemala**: Sacatépequez (Kohls 1956a); **Panama**: Chiriqui (Fairchild et al. 1966); and **Peru**: Piura (Kohls 1956a).

Hosts: Males, females, and nymphs of *Ixodes pomerantzi* are known to parasitize Lagomorpha: Leporidae, especially *Sylvilagus brasiliensis*. The host profile of

Table 1.19 Hosts for males (M), females (F), and nymphs (N)* of *Ixodes pomerantzi*

MAMMALIA			
LAGOMORPHA: Leporidae		*Sylvilagus dicei*	MFN
Sylvilagus brasiliensis	MFN**	*Sylvilagus floridanus*	M

* There is no formal description of the N of *Ixodes pomerantzi*
** Host not universally accepted

Ixodes pomerantzi is shown in Table 1.19, constructed with information provided by Kohls (1956a) and Bermúdez et al. (2018a).

Human Parasitism: No.

Comments: Camicas et al. (1998) and Kolonin (2009) state that the nymph of *Ixodes pomerantzi* is described, but it is not; nevertheless, Fairchild et al. (1966) and Clifford et al. (1973) were able to identify this stage of *Ixodes pomerantzi.*

Specimens of *Ixodes pomerantzi* from Ecuador were confused with *Ixodes dentatus*, as explained in Keirans (1985b).

Sylvilagus brasiliensis is regarded here as the principal host of *Ixodes pomerantzi*. In some studies, this mammal is treated as having a limited range in Brazil, while other investigators regard *Sylvilagus brasiliensis* as broadly distributed in the Neotropics. The latter view has been adopted here, but workers such as Bermúdez et al. (2018a, b) listed *Sylvilagus dicei* rather than *Sylvilagus brasiliensis* as the host of *Ixodes pomerantzi* in Panama. This issue remains unresolved, and *Sylvilagus brasiliensis* should be provisionally regarded as the principal host of *Ixodes pomerantzi*.

37. *Ixodes rubidus* Neumann, 1901

Neumann, L.G. (1901) Révision de la famille des ixodidés (4e mémoire). *Mémoires de la Société Zoologique de France*, 14, 249–372.

Descriptions and Redescriptions: Neumann (1901) described but did not figure the female of *Ixodes rubidus*, while Nuttall and Warburton (1911) described and figured the nymph.

Cooley and Kohls (1945) redescribed the female and nymph. Nuttall and Warburton (1911) redescribed the female and provided the first figures of this stage; Guzmán-Cornejo and Robbins (2010, without figures) and Bermúdez et al. (2018a) also redescribed the female of *Ixodes rubidus.*

The male of *Ixodes rubidus* remains unknown, while the larva is undescribed, but some authors have been able to identify this stage from field-collected material.

Geographical Distribution: *Ixodes rubidus* has been mostly recorded from the Neotropical Zoogeographic Region (southern North America and Central America), with a few records from the Nearctic portion of Mexico. Keirans (1985b) classified as *Ixodes rubidus* two female ticks collected from *Myocastor coypus* in South

America. In the Neotropics *Ixodes rubidus* has been found in **Guatemala**: Chimaltenango (Fairchild et al. 1966); **southern Mexico**: Guerrero (Guzmán-Cornejo et al. 2007); and **Panama**: Chiriquí (Fairchild et al. 1966), with a vague **South American** record in Keirans (1985b).

Hosts: Females of *Ixodes rubidus* are known from nine Neotropical records, seven of them involving specimens collected from several families of Carnivora. Female ticks have also been found on marsupials and rodents, with one record for each of these hosts. Nymphs have been recorded on four occasions, parasitizing Carnivora in all cases. There is only one record of *Ixodes rubidus* larvae, and the specimens were obtained from Carnivora: Procyonidae, but it is uncertain whether such carnivores are the preferred host for this stage. Additionally, Neumann (1901) listed one female and three nymphs collected from *Bassariscus astutus* (Carnivora: Procyonidae) in Guanajuato (Nearctic Mexico). The principal hosts for females and nymphs of this tick are here considered to be several families of Carnivora. The host profile of *Ixodes rubidus*, compiled from specimens collected in the Neotropical Region and based on information provided by Fairchild et al. (1966), Keirans (1985b), and Guzmán-Cornejo et al. (2007), is shown in Table 1.20.

Human Parasitism: No.

Comments: Clifford et al. (1973) claim that *Ixodes rubidus* is a Nearctic species (under the name *Ixodes rubidis*, a *lapsus*), while Camicas et al. (1998), under the name *Pholeoixodes rubidus*, state that it occurs exclusively in the Neotropical Region. As stated above, *Ixodes rubidus* is established in both zoogeographic regions (Guglielmone et al. 2003a). The locality record of the female ticks collected from *Myocastor coypus* is "South America" (Keirans 1985b), which is treated here as a valid but uncertain location.

Camicas et al. (1998) listed the larva of *Ixodes rubidus* as described, but no formal description of this stage has been found; therefore, the record of the undescribed larva in Fairchild et al. (1966) should be considered provisional.

Table 1.20 Hosts for females (F), nymphs (N), and larvae (L)* of *Ixodes rubidus*

MAMMALIA			
CARNIVORA: Canidae		*Nasua narica*	F
Urocyon cinereoargenteus	FN	*Procyon lotor*	FNL
CARNIVORA: Mephitidae		**DIDELPHIMORPHIA**	
Conepatus semistriatus	FN	**Didelphidae**	
CARNIVORA: Mustelidae		*Didelphis* sp.	F
Eira barbara	F	**RODENTIA**	
Mustela frenata	F	**Myocastoridae**	
CARNIVORA: Procyonidae		*Myocastor coypus*	F**
Bassaricyon gabbii	FN		

* Provisional identification because there is no formal description of the larva of *Ixodes rubidus*
** Allegedly collected in South America

38. *Ixodes scapularis* Say, 1821

Say, T. (1821) An account of the arachnides of the United States. *Journal of the Academy of Natural Sciences of Philadelphia*, 2, 59–82.

Descriptions and Redescriptions: Say (1821) described the female of *Ixodes scapularis*, but without figures; Banks (1908) briefly described the male, presenting a figure of the stigmal plate, while the larva and nymph were described with accompanying figures in Hooker et al. (1912).

The female, male, nymph, and larva were redescribed in Keirans et al. (1996a) and Lindquist et al. (2016); Cooley and Kohls (1945) redescribed the male, female, and nymph, while Sonenshine (1979) did the same with the female, nymph, and larva. Hooker et al. (1912), Keirans and Clifford (1978), and Keirans and Litwak (1989) redescribed the male and female, and Banks (1908) redescribed the female while also providing the first illustrations of the female of *Ixodes scapularis*. Dubie et al. (2017) redescribed the nymph and larva. Durden and Keirans (1996) redescribed the nymph alone, and Clifford et al. (1961), Sénevet and Ripert (1967), and Coley (2015) redescribed the larva of *Ixodes scapularis*.

Geographic Distribution: *Ixodes scapularis* is basically a Nearctic species, with several unconfirmed Neotropical records; therefore, Guglielmone et al. (2003a) stated that this tick is not present in the Neotropics, but recently *Ixodes scapularis* has been treated as occurring in **southern Mexico**: Chiapas, Jalisco, and Oaxaca (Guzmán-Cornejo et al. 2007; Guzmán-Cornejo and Robbins 2010), records that are regarded here as provisionally valid.

Hosts: Cattle and deer have been listed as hosts of *Ixodes scapularis* in southern Mexico (Guzmán-Cornejo et al. 2007; Guzmán-Cornejo and Robbins 2010), but tick stages found on these hosts were not provided; therefore, the Neotropical host profile for *Ixodes scapularis* is essentially undetermined. The host profile of this species in the Nearctic Region includes several orders of mammals for all parasitic stages, with larvae and nymphs being found on Passeriformes: several families, Galliformes: Odontophoridae, and Squamata: Anguidae and Scincidae, plus infrequent records from other types of hosts (Guglielmone et al. 2014).

Human Parasitism: *Ixodes scapularis* is a common parasite of humans in the Nearctic Region. Guglielmone and Robbins (2018) list many cases of human parasitism in Canada and the USA but none from Mexico.

Comments: *Ixodes scapularis* is a species of great medical and veterinary importance. Therefore, it is important to definitively confirm its presence in the Neotropics. In the absence of further evidence, records of *Ixodes scapularis* from Bolivia (Payno Balazans 1978), Costa Rica (Hooker et al. 1912), and Guatemala (Monroy and Solís 1988) are treated here as doubtful. Recently, Bermúdez et al. (2015b) found that Nicaraguan specimens of supposed *Ixodes scapularis* deposited in a tick collection in that country were, in fact, *Ixodes boliviensis*. Guzmán-Cornejo et al. (2007) and Guzmán-Cornejo and Robbins (2010) accept the presence of *Ixodes scapularis* in southern Mexico, a view also accepted here, but further confirmation of natural established populations of this tick and its hosts should be sought.

39. *Ixodes schulzei* Aragão and Fonseca, 1951

Aragão, H.B. & Fonseca, F. (1951b) Notas de ixodologia. I. Duas novas espécies do género *Ixodes* e um novo nomen para *Haemaphysalis kochi* Aragão, 1908 (Acari: Ixodidae). *Memórias do Instituto Oswaldo Cruz*, 49, 567–574.

Descriptions and Redescriptions: Aragão and Fonseca (1951b) described and figured the female of *Ixodes schulzei*, while Barros-Battesti et al. (2007b) described the larva and nymph.

Nava et al. (2017) redescribed the female and nymph, and Barros-Battesti et al. (2007b) and Onofrio et al. (2006a, 2009) redescribed the female.

There are no redescriptions of the larva of *Ixodes schulzei.*

The male of *Ixodes schulzei* remains unknown.

Geographic Distribution: *Ixodes schulzei* is a South American tick that has been found in **Argentina**: Misiones (Lamattina et al. 2016); and **Brazil**: Minas Gerais, Paraná, Rio de Janeiro, Rondônia, Santa Catarina, and São Paulo (Aragão and Fonseca 1951b; Labruna et al. 2003; Arzua et al. 2005; Onofrio et al. 2013).

Hosts: *Ixodes schulzei* has been collected from Rodentia: Cricetidae, but all females are known only from *Nectomys squamipes*. The host profile of *Ixodes schulzei* is shown in Table 1.21, constructed using the data of Aragão and Fonseca (1951b), Onofrio et al. (2013), and Martins et al. (2016a). Records of larvae and nymphs from *Nectomys squamipes* were inferred from the study of Onofrio et al. (2013), who collected 52 "immature stages" from this host, which is treated as the principal host for females of this tick and the provisional principal host for larvae and nymphs.

Human Parasitism: No.

Comments: Clifford et al. (1973) regarded *Ixodes schulzei* as a doubtful species, but there are no reasons to question its validity.

Table 1.21 Hosts for females (F), nymphs (N), and larvae (L) of *Ixodes schulzei*

MAMMALIA			
RODENTIA: Cricetidae		*Nectomys squamipes*	FN*L*
Akodon montensis	NL	*Oligoryzomys flavescens*	N
Euryoryzomys russatus	N	*Oligoryzomys nigripes*	N

* Records of nymphs and larvae from *Nectomys squamipes* are inferred

40. *Ixodes sigelos* Keirans, Clifford and Corwin, 1976

Keirans J.E., Clifford, C.M. & Corwin, D. (1976) *Ixodes sigelos*, n. sp. (Acarina: Ixodidae), a parasite of rodents in Chile, with a method for preparing ticks for examination by scanning electron microscopy. *Acarologia*, 18, 217–225.

Descriptions and Redescriptions: The female, nymph, and larva of *Ixodes sigelos* were described and illustrated in Keirans et al. (1976).

Nava et al. (2017) redescribed the female and nymph, while Guglielmone et al. (2005b) redescribed the nymph.

There are no redescriptions of the larva of *Ixodes sigelos*.

The male of *Ixodes sigelos* remains unknown.

Geographic Distribution: *Ixodes sigelos* is a southern South American species that has been found in **Argentina**: Catamarca, Chubut, Río Negro, Santa Cruz, and Tucumán (Guglielmone et al. 2005b; Sánchez et al. 2010; Sebastian et al. 2016); and **Chile**: Araucanía, Aysén, Bío Bío, Coquimbo, Metropolitana, and Ñuble (González-Acuña et al. 2004a; Guglielmone et al. 2005b; González-Acuña and Guglielmone 2005).

Hosts: All records of females, nymphs, and larvae of *Ixodes sigelos* are from Rodentia, with most hosts belonging to the family Cricetidae. A total of 22 females, 92 nymphs, and 107 larvae of *Ixodes sigelos* have been recognized in studies of this tick; 55%, 81%, and 87% of these stages, respectively, have been collected from *Abrothrix olivaceus* and *Phyllotis xanthopygus*, which are considered the principal hosts of *Ixodes sigelos*. The host profile of *Ixodes sigelos* is shown in Table 1.22, based on information in Keirans et al. (1976), González-Acuña et al. (2004a), Guglielmone et al. (2005b), Sánchez et al. (2010) and Mastropaolo et al. (2013).

Human Parasitism: No.

Comments: Camicas et al. (1998) suggested that *Ixodes sigelos* is a synonym of *Ixodes abrocomae*, but Guglielmone et al. (2010a) provided evidence for the validity of both species. Nevertheless, the hypothesis of conspecificity between *Ixodes abrocomae* and *Ixodes sigelos* has again been proposed by Muñoz-Leal et al.

Table 1.22 Hosts for females (F), nymphs (N), and larvae (L) of *Ixodes sigelos*

MAMMALIA			
RODENTIA: Abrocomidae		*Euneomys chinchilloides*	L
Abrocoma bennettii	N	*Loxodontomys micropus*	N
RODENTIA: Cricetidae		*Oligoryzomys longicaudatus*	FL
Abrothrix longipilis	NL	*Phyllotis xanthopygus*	FNL
Abrothrix olivaceus	FNL	*Reithrodon auritus*	F
Akodon spegazzinii	N	**RODENTIA: Muridae**	
Calomys musculinus	N	*Rattus norvegicus*	N
Chelemys macronyx	L	**RODENTIA: Octodontidae**	
Eligmodontia morgani	L	*Aconaemys fuscus*	FNL
Euneomys sp.	N	*Octodon degus*	L

(2019a), although later Muñoz-Leal et al. (2019b) appear to recognize the validity of *Ixodes abrocomae*.

Sánchez et al. (2010) provide molecular evidence to argue that the name *Ixodes sigelos* may represent more than one taxon, while Muñoz-Leal et al. (2019a) found three different populations of ticks on Chilean rodents that are morphologically close to *Ixodes abrocomae* and *Ixodes sigelos* but with dissimilar molecular sequences, and a fourth population was added in Muñoz-Leal et al. (2019b). The latter authors treat these populations as cryptic species belonging to the *Ixodes sigelos* group; host information on these taxa has not been included in this analysis, pending formal descriptions of the new species that should clarify the complex structure of this species group.

41. *Ixodes silvanus* Saracho-Bottero, Beati, Venzal, Guglielmone and Nava, 2021 in Saracho Bottero et al. (2021)

Saracho Bottero' M.N., Beati, L., Venzal, J.M., Claps, L.G., Thompson, C.S., Mangold, A.J. Guglielmone, A.A. & Nava, S. (2021) *Ixodes silvanus* n. sp. (Acari: Ixodidae), a new member of the subgenus *Trichotoixodes* Reznik, 1961, from northwestern Argentina. *Ticks and Tick-borne Diseases*, 12 (1) (article 101572) 11 pp.

Descriptions and Redescriptions: Boero (1957) first described and figured the female of *Ixodes silvanus*, under the name *Ixodes brunneus*. Saracho Bottero et al. (2021) described the nymph and larva of *Ixodes silvanus*

The female of *Ixodes silvanus* was redescribed and given its current status in Saracho Bottero et al. (2021).

There are no redescriptions of the nymph and larva of *Ixodes silvanus*.

The male of *Ixodes silvanus* remains unknown.

Geographic Distribution: *Ixodes silvanus* is a Neotropical tick from southern South America found in **Argentina**: Jujuy, Salta, and Tucumán (Saracho Bottero et al. 2021).

Hosts: All known parasitic stages of *Ixodes silvanus* have been collected from several families of Passeriformes. However, most females and nymphs have been found on *Turdus nigriceps* and *Turdus rufiventris* (Turdidae), and five of the seven larvae whose hosts are known were collected from *Turdus rufiventris*, and these species of *Turdus* are treated here as provisional principal hosts of this tick. The host profile of *Ixodes silvanus* is shown in Table 1.23, constructed using information provided by Boero (1957) and Saracho Bottero et al. (2021).

Human Parasitism: No.

Comments: Boero (1955) mentioned the presence of *Ixodes brunneus* parasitizing *Turdus nigriceps* in northwestern Argentina, and later Boero (1957) described and figured the alleged *Ixodes brunneus* found in northwestern Argentina. However,

Table 1.23 Hosts for females (F), nymphs (N), and larvae (L) of *Ixodes silvanus*

AVES			
PASSERIFORMES: Furnariidae		**PASSERIFORMES: Thraupidae**	
Synallaxis frontalis	L	*Thlypopsis sordida*	N
Syndactyla rufosuperciliata	N	**PASSERIFORMES: Tityridae**	
PASSERIFORMES: Parulidae		*Pachyramphus validus*	F
Myiothlypis bivittata	L	**PASSERIFORMES: Turdidae**	
PASSERIFORMES: Passerellidae		*Turdus nigriceps*	FN
Atlapetes citrinellus	N	*Turdus rufiventris*	FNL
Zonotrichia capensis	N		

Fairchild et al. (1966) suggested that the tick illustrated by Boero (1957) represented a new species, different from *Ixodes brunneus*. The opinion of Fairchild et al. (1966) was confirmed more than 50 years later by Saracho Bottero et al. (2021), and the tick named *Ixodes brunneus* by Boero (1955, 1957) was found to be the new species *Ixodes silvanus*.

42. *Ixodes sinaloa* Kohls and Clifford, 1966

Kohls, G.M. & Clifford, C.M. (1966) Three new species of *Ixodes* from Mexico and description of the male of *I. auritulus auritulus* Neumann, *I. conepati* Cooley and Kohls, and *I. lasallei* Méndez and Ortiz (Acarina: Ixodidae). *Journal of Parasitology*, 52, 810–820.

Descriptions and Redescriptions: The female of *Ixodes sinaloa* was described and figured in Kohls and Clifford (1966), while Keirans and Jones (1972) described and illustrated the nymph and larva.

The female was redescribed in Guzmán-Cornejo and Robbins (2010), who provide morphological diagnostic characters but no figures.

There are no redescriptions of the nymph and larva.

The male of *Ixodes sinaloa* remains unknown.

Geographic Distribution: *Ixodes sinaloa* is mostly a Neotropical species that has been found in southern North America and Central America, with some records fron the Nearctic area of Mexico. *Ixodes sinaloa* occurs in **Costa Rica**: Alajuela or Puntarenas (McCain et al. 2007); **southern Mexico**: Colima, Jalisco, Nayarit, and Oaxaca (Keirans and Jones 1972; Guzmán-Cornejo et al. 2012; Light et al. 2020); and **Nicaragua**: Matagalpa (Keirans and Jones 1972).

Hosts: Most females of *Ixodes sinaloa* have been found on *Heteromys pictus* (Rodentia: Heteromyidae), while the majority of nymphs and larvae have been collected from *Handleyomys alfaroi* (Rodentia: Cricetidae); these rodents are provisionally regarded as principal hosts of females and immature stages of *Ixodes sinaloa*, respectively. A female tick was taken from Cricetidae, and single nymphs

Table 1.24 Hosts for females (F), nymphs (N), and larvae (L) of *Ixodes sinaloa*

MAMMALIA			
DIDELPHIMORPHIA: Didelphidae		*Reithrodontomys sumichrasti*	N
Tlacuatzin canescens	N	*Tanyuromys aphrastus*	NL
RODENTIA: Cricetidae		**RODENTIA: Heteromyidae**	
Handleyomys alfaroi	NL	*Heteromys pictus*	F
Oryzomys couesi	F	*Heteromys salvini*	FN
Peromyscus mexicanus	NL		

were recovered from Heteromyidae and Didelphimorphia: Didelphidae. The host profile of *Ixodes sinaloa* is shown in Table 1.24, based on data provided by Kohls and Clifford (1966), Keirans and Jones (1972), McCain et al. (2007), and Guzmán-Cornejo et al. (2012). All records of *Ixodes sinaloa* from the Nearctic area of Mexico are from *Heteromys pictus* (Heteromyidae).

Human Parasitism: No.

Comments: Camicas et al. (1998) list the male as described, but no description has been found for this stage of *Ixodes sinaloa.*

Camicas et al. (1998) also state that *Ixodes sinaloa* is a Neotropical species, but as noted above, it has also been found in the Nearctic area of Mexico (Kohls and Clifford 1966).

43. *Ixodes spinipalpis* Hadwen and Nuttall, 1916 in Nuttall (1916), under the name *Ixodes dentatus spinipalpis*, and given its current status by Hubbard (1941)

Nuttall, G.H.F. (1916) Notes on ticks. IV. Relating to the genus *Ixodes* and including description of three new species and two varieties. *Parasitology*, 8, 294–337.

Hubbard, C.A. (1941) Ectoparasites of western Lagomorpha. Part I. The ectoparasites of 100 western conies (*Ochotonae*) Part II. The ectoparasites of western rabbits (*Lepus-Sylvilagus-Brachylagus*). *Pacific University Bulletin*, 37, 1–8.

Descriptions and Redescriptions: Keirans and Clifford (1978) explained that the male and female of *Ixodes spinipalpis* were first described and illustrated in Bishopp (1912) under the name *Ixodes diversifossus*; Hadwen and Nuttall in Nuttall (1916) described the nymph and larva under the name *Ixodes dentatus spinipalpis.* This tick's current specific status was established by Hubbard (1941).

Furman and Loomis (1984) and Lindquist et al. (2016) redescribed the male, female, nymph, and larva, while Cooley and Kohls (1945) and Gregson (1956) redescribed the male, female, and nymph. Cooley (1944b), under the name *Ixodes neotomae*, and Keirans and Clifford (1978) redescribed the male and female; Hadwen and Nuttall in Nuttall (1916) redescribed the female under the name *Ixodes dentatus spinipalpis* and were the first to use the epithet *spinipalpis*. Durden and Keirans (1996) redescribed the nymph, while Allred et al. (1960), Sénevet and Ripert

(1967), Webb et al. (1990), and Coley (2015) redescribed the larva of *Ixodes spinipalpis.*

Geographic Distribution: *Ixodes spinipalpis* is largely a Nearctic species, with a few Neotropical records from **southern Mexico**: Guerrero, Morelos, and Veracruz (Guzmán-Cornejo et al. 2007).

Hosts: The records of *Ixodes spinipalpis* in the Neotropical Region are from several nymphs found on *Peromyscus* sp. (Rodentia: Cricetidae), one nymphal record from *Sciurus* sp. (Rodentia: Sciuridae), and one female taken from *Romerolagus diazi* (Lagomorpha: Leporidae) (Guzmán-Cornejo et al. 2007). The predominance of Cricetidae as hosts for the nymphs of Neotropical *Ixodes spinipalpis* is an indication that rodents of this family are the principal hosts for the nymph. In the Nearctic Region, all parasitic stages of *Ixodes spinipalpis* are usually found on Rodentia: Cricetidae, but adults are also common on Lagomorpha: Leporidae; adults and immature stages have also been collected from Carnivora: Procyonidae, Lagomorpha: Ochotonidae, and Rodentia: Sciuridae; adults alone have been retrieved from Carnivora: Canidae, while immature stages have been recovered from Aves: Galliformes and Passeriformes, and a few mammals (Guglielmone et al. 2014; Guglielmone and Robbins 2018). It is provisionally proposed that females and nymphs of *Ixodes spinipalpis* utilize *Romerolagus diazi* and *Peromyscus* sp., respectively, as principal hosts in the Neotropical Region.

Human Parasitism: Guglielmone and Robbins (2018) list *Ixodes spinipalpis* as a sporadic parasite of humans, with several records from Canada and the USA. Xu et al. (2019) added new cases of people bitten by this tick in the USA, but there is no record of *Ixodes spinipalpis* feeding on people in the Neotropics.

Comments: As stated above, Keirans and Clifford (1978) assert that the first descriptions of the male and female of *Ixodes spinipalpis* are in Bishopp (1912) under the name *Ixodes diversifossus*, and that opinion is considered valid here. However, authors such as Gregson (1956), Keirans and Clifford (1978), and Camicas et al. (1998) claim that the female of *Ixodes diversifossus* redescribed in Neumann (1911a) is, in fact, *Ixodes spinipalpis.* This viewpoint is controversial because these authors treat as valid the name *Ixodes diversifossus* in Neumann (1899), who described this tick from two mutilated females, and no further specimens have been collected enabling the redescription of this tick. Additionally, Neumann (1911a) provided a summary of his original description of *Ixodes diversifossus* in Neumann (1899), and it is difficult to understand how the name *Ixodes diversifossus*, as understood in Neumann (1911a), could be considered identical to *Ixodes spinipalpis.* Clearly, the claim of Keirans and Clifford (1978), that the description of the female of *Ixodes diversifossus* in Neumann (1911a) is in fact a description of the female of *Ixodes spinipalpis*, conflicts with their view that the first description of the male and female of *Ixodes spinipalpis* is in Bishopp (1912) under the name *Ixodes diversifossus*, because, in that case, the first description of the female of *Ixodes spinipalpis* should be in Neumann (1911a).

The taxonomic situation described above can be analyzed alternatively as follows: Neumann (1899) described *Ixodes diversifossus* and redescribed this species in Neumann (1911a). If, as stated by several authors named above, Neumann (1911a), in fact, described *Ixodes spinipalpis*, the same can be said about the tick described as *Ixodes diversifossus* by Neumann (1899), leaving open the possibility that *Ixodes spinipalpis* is a synonym of *Ixodes diversifossus*.Many records of *Ixodes spinipalpis* have appeared under the name *Ixodes neotomae*, a name proven to be a synonym of *Ixodes spinipalpis* by Norris et al. (1997), but Camicas et al. (1998) and Barker and Murrell (2004, 2008) treat both names as valid.

There is a record of presumed adults of *Ixodes spinipalpis* collected from Aves in Costa Rica by Hernández et al. (2013) that needs confirmation; consequently, it was not included in the list of hosts of this tick.

44. *Ixodes spinosus* Neumann, 1899

Neumann, L.G. (1899) Révision de la famille des ixodidés (3e mémoire). *Mémoires de la Société Zoologique de France*, 12, 107–294.

Descriptions and Redescriptions: Neumann (1899) described but did not figure the female, nymph, and larva of *Ixodes spinosus*. Nuttall and Warburton (1911) redescribed the female (with figures) and translated the description of the nymph and larva of *Ixodes spinosus* in Neumann (1899), under the name *Ixodes fuscipes*.

Onofrio et al. (2006a, 2009) redescribed and illustrated the female, again under the name *Ixodes fuscipes*. Labruna et al. (2020a) also redescribed the female of *Ixodes spinosus*, providing figures of the only remaining type specimen in the original syntype series, along with figures of a specimen later collected in Brazil and reinstated *Ixodes spinosus*. Sénevet and Ripert (1967) repeated the description of the larva of *Ixodes spinosus* in Nuttall and Warburton (1911) as *Ixodes fuscipes*.

The male of *Ixodes spinosus* remains unknown.

Geographical Distribution: *Ixodes spinosus* is a Neotropical South American species found in **Brazil**: Acre, Rio de Janeiro, and Rondônia (Labruna et al. 2020a), and **Peru**: Madre de Dios or Puno (Labruna et al. 2020a).

Hosts: Most records of *Ixodes spinosus* females and the only record of nymphs and larvae are from *Dasyprocta* sp. (Rodentia: Dasyproctidae), provisionally treated as the principal host of this tick. All records of *Ixodes spinosus* are from Rodentia, as shown in Table 1.25, constructed with information provided by Labruna et al.

Table 1.25 Hosts for females (F), nymphs (N), and larvae (L) of *Ixodes spinosus**

MAMMALIA			
RODENTIA: Cuniculidae		**RODENTIA: Dasyproctidae**	
Cuniculus paca	F	*Dasyprocta* sp.	FNL
		Myoprocta pratti	F

* Most records of *Ixodes spinosus* have been published under the name *Ixodes fuscipes*

(2020a). Most records of *Ixodes spinosus* have been reported under the name *Ixodes fuscipes.*

Human Parasitism: No.

Comments: *Ixodes spinosus* was described by Neumann (1899) but relegated to the synonymy of *Ixodes fuscipes* by Neumann (1901). This view remained undisputed for more than a century, and redescriptions of the alleged female of *Ixodes fuscipes*, such as those above, were in fact redescriptions of the female of *Ixodes spinosus.* Labruna et al. (2020a) examined the types of both *Ixodes fuscipes* and *Ixodes spinosus*, determining that they represent different species and that *Ixodes fuscipes* is a senior synonym of *Ixodes aragaoi.*

Labruna et al. (2020a) listed a female of *Ixodes spinosus* found in Peru, based on Fairchild et al. (1966) and Need et al. (1991), but the former authors provided a locality in Puno Region for this record, while the latter placed the locality in Madre de Dios Region.

Neumann (1899) named *Dasyprocta aguti* as the host of *Ixodes spinosus*, but it has proved difficult to determine the taxonomic status of this host among the *Dasyprocta* species found in Brazil; therefore, *Dasyprocta* sp. is adopted here as the original host of this tick.

45. *Ixodes stilesi* Neumann, 1911, but previously named as *Ixodes elegans* by Neumann (1910), a name preoccupied by *Ixodes elegans* Guérin-Méneville, 1844 and consequently renamed *Ixodes stilesi*

Neumann, L.G. (1910) Sur quelques espèces d'Ixodidae nouvelles ou insuffisamment connues. *Annales des Sciences Naturelles, Zoologie, Série 9*, 12, 161–176.

Neumann, L.G. (1911b) Note rectificative à propos de deux espèces d'Ixodinae. *Archives de Parasitologie*, 14, 415.

Guérin-Méneville, F.E. (1844) Iconographie du règne animal de G. Cuvier (1829–1843), ou représentation d'après nature de l'une des espèces les plus remarquables et souvent non encore figurées de chaque genre d'animaux, (4), *Arachnides*, Paris, 20 pp.

Descriptions and Redescriptions: Neumann (1910) described and illustrated the female of *Ixodes stilesi* under the name *Ixodes elegans* (preoccupied); the male and nymph are described in Guglielmone et al. (2006b).

Nava et al. (2017) redescribed the male, female, and nymph, while Nuttall and Warburton (1911) redescribed the female of *Ixodes stilesi.* Guglielmone et al. (2006b) also redescribed the female but with some errors amended in Guglielmone et al. (2007b).

The larva of *Ixodes stilesi* remains undescribed, but Ivanova et al. (2014) were able to identify this stage from field-collected material.

Geographic Distribution: *Ixodes stilesi* is a southern South American species that has only been found in **Chile**: Bío Bío, Los Ríos, and Ñuble (Osorio 2001; González-Acuña and Guglielmone 2005; Guglielmone et al. 2007b).

Hosts: Males, females, and nymphs of *Ixodes stilesi* have usually been collected from *Pudu puda* (Artiodactyla: Cervidae), and one larva of *Ixodes stilesi* was found on the same host species and identified molecularly by Ivanova et al. (2014). One nymph was found on *Oligoryzomys longicaudatus* (Rodentia: Cricetidae) (Guglielmone et al. 2007b).

Human Parasitism: No.

Comments: Clifford et al. (1973) questioned the validity of *Ixodes stilesi,* but it is a valid species. Estrada-Peña et al. (2020) include the larva of *Ixodes stilesi* in a cluster formed by Neotropical species feeding mostly on Artiodactyla, but there is insufficient information to confirm this host-tick association.

46. *Ixodes taglei* Kohls, 1969

Kohls, G.M. (1969a) *Ixodes taglei* n. sp. (Acarina: Ixodidae) a parasite of the deer, *Pudu pudu* [sic] (Wol.), in Chile. *Journal of Medical Entomology*, 6, 280–283.

Descriptions and Redescriptions: Kohls (1969a) described and figured the male and female of *Ixodes taglei.*

Nava et al. (2017) redescribed the male and female.

The larva and nymph of *Ixodes taglei* remain unknown.

Geographical Distribution: *Ixodes taglei* is a southern South American species that has only been found in **Chile**: Bío Bío, and Los Ríos (Kohls 1969a; Verdugo et al. 2017; Muñoz-Leal et al. 2019a).

Hosts: *Pudu puda* (Artiodactyla: Cervidae) is the only known host for the male and female of *Ixodes taglei* (Kohls 1969a; Verdugo et al. 2017; Muñoz-Leal et al. 2019a).

Human Parasitism: No.

Comments: No controversial issues related to *Ixodes taglei* have been found.

47. *Ixodes tancitarius* Cooley and Kohls, 1942

Cooley, R.A. & Kohls, G.M. (1942b) *Ixodes mexicanus* n. sp. et *Ixodes tancitarius* n. sp. dos nuevas garrapatas mexicanas. *Revista de la Sociedad Mexicana de Historia Natural*, 3, 149–154.

Descriptions and Redescriptions: Cooley and Kohls (1942b) described and figured the female of *Ixodes tancitarius.*

Cooley and Kohls (1945) and Guzmán-Cornejo and Robbins (2010) redescribed the female, but the latter authors provided no figures.

The male, nymph, and larva of *Ixodes tancitarius* remain unknown.

Geographic Distribution: *Ixodes tancitarius* is a Nearctic and Neotropical Mexican species. It has been found in **southern Mexico**: Michoacán and Veracruz (Cooley and Kohls 1942b; Guzmán-Cornejo et al. 2007) with a Nearctic record from the State of México (Hoffmann 1969).

Hosts: The female of *Ixodes tancitarius* has been collected from *Peromyscus* sp. and *Reithrodontomys* sp. (Rodentia: Cricetidae).

Human Parasitism: The only Nearctic record for *Ixodes tancitarius* is a female tick collected from a human in Mexico (Hoffmann 1969) and treated as provisionally valid by Guglielmone and Robbins (2018).

Comments: Guglielmone et al. (2003a) state that *Ixodes tancitarius* is a Nearctic species, contrary to Camicas et al. (1998), who listed it as a Neotropical taxon. In fact, the first record of *Ixodes tancitarius* (Cerro Tancítaro, Michoacán State) lies within the limits of the Neotropical Region. Guglielmone et al. (2014) maintain, erroneously, that the first Neotropical record of *Ixodes tancitarius* is from Veracruz, as listed in Guzmán-Cornejo et al. (2007).

48. ***Ixodes tapirus*** Kohls, 1956

Kohls, G.M. (1956a) Eight new species of *Ixodes* from Central and South America (Acarina: Ixodidae). *Journal of Parasitology*, 42, 636–649.

Descriptions and Redescriptions: The female of *Ixodes tapirus* was described and figured in Kohls (1956a), and the male was described and illustrated in Apanaskevich et al. (2017).

The male and female were redescribed in Bermúdez et al. (2018a), and the female of *Ixodes tapirus* was also redescribed by Apanaskevich et al. (2017).

The nymph and larva of *Ixodes tapirus* remain unknown.

Geographic Distribution: *Ixodes tapirus* is a Central American and northern South American species that has been found in **Colombia**: Huila (Kohls 1956a); **Costa Rica**: San José, and **Panama**: Bocas del Toro, and Chiriquí (Apanaskevich et al. 2017).

Hosts: Males and females of *Ixodes tapirus* have been collected from *Tapirus bairdii*, while a female tick (the holotype for the species) was found on *Tapirus pinchaque* (Perissodactyla: Tapiridae) (Kohls 1956a; Fairchild, et al. 1966; Apanaskevich et al. 2017). *Tapirus bairdii* is provisionally treated here as the principal host for the male and female of *Ixodes tapirus*.

Human Parasitism: No.

Comments: No controversial issues related to *Ixodes tapirus* have been found.

49. *Ixodes tecpanensis* Kohls, 1956

Kohls, G.M. (1956a) Eight new species of *Ixodes* from Central and South America (Acarina: Ixodidae). *Journal of Parasitology*, 42, 636–649.

Descriptions and Redescriptions: The female of *Ixodes tecpanensis* was described and figured in Kohls (1956a).

There are no redescriptions of the female of *Ixodes tecpanensis*.

The male, nymph, and larva of *Ixodes tecpanensis* remain unknown.

Geographic Distribution: *Ixodes tecpanensis* is a Central American species that has only been found in **Guatemala**: Chimaltenango (Kohls 1956a).

Hosts: *Ixodes tecpanensis* is known from two females collected from *Orthogeomys grandis* (Rodentia: Geomyidae).

Human Parasitism: No.

Comments: No controversial issues related to *Ixodes tecpanensis* have been found.

50. *Ixodes texanus* Banks, 1909

Banks N. (1909) Three new ticks from the United States. *Proceedings of the Entomological Society of Washington*, 10, 170–173.

Descriptions and Redescriptions: Banks (1909) described and figured the female of *Ixodes texanus*, while the male and nymph were described in Cooley and Kohls (1945), and the larva of *Ixodes texanus* was described in Allred et al. (1960).

Redescriptions of the male, female, nymph, and larva of *Ixodes texanus* are provided in Furman and Loomis (1984) and Lindquist et al. (2016). The male, nymph, and larva were redescribed in Gregson (1956), while Sonenshine (1979) redescribed the female, nymph, and larva. Keirans and Clifford (1978) and Keirans and Litwak (1989) redescribed the male and female. Nuttall and Warburton (1911), Cooley and Kohls (1945), Lancaster (1973), and Guzmán-Cornejo and Robbins (2010) redescribed the female. Durden and Keirans (1996) redescribed the nymph alone. Clifford et al. (1961), Sénevet and Ripert (1967), Webb et al. (1990), Kleinjan and Lane (2008), and Coley (2015) redescribed the larva of *Ixodes texanus*.

Geographic Distribution: *Ixodes texanus* is a largely Nearctic species with a Neotropical record from **southern Mexico**: Guerrero (Guzmán-Cornejo et al. 2007).

Hosts: The southern Mexican record of *Ixodes texanus* in Guzmán-Cornejo et al. (2007) is for a female and a nymph collected from *Bassariscus* sp. (Carnivora: Procyonidae), a hypothetical Neotropical principal host for this tick. Nearctic records of adults and immature stages of *Ixodes texanus* are mostly from Carnivora: Mustelidae and Procyonidae, but all parasitic stages have also been recorded on several orders of small to medium-sized Mammalia (Guglielmone et al. 2014).

Human Parasitism: Guglielmone and Robbins (2018) listed a few cases of human parasitism in Canada and the USA, but none from the Neotropics.

Comments: Ondrejicka et al. (2016) present molecular evidence suggesting that more than one species may exist under the name *Ixodes texanus.*

51. ***Ixodes tiptoni*** Kohls and Clifford, 1962

Kohls, G.M. & Clifford, C.M. (1962) *Ixodes tiptoni*, a new species of tick from Panama (Acarina – Ixodidae). *Journal of Parasitology*, 48, 182–184.

Descriptions and Redescriptions: Kohls and Clifford (1962) described the female, nymph, and larva of *Ixodes tiptoni* but figures are only provided for the female and larva; the male was described without figures in a key to the *Ixodes* of Panama constructed by Fairchild et al. (1966).

Bermúdez et al. (2018a) redescribed the female, and Sénevet and Ripert (1967) redescribed the larva of *Ixodes tiptoni.*

There are no redescriptions of the male and nymph of *Ixodes tiptoni.*

Geographic Distribution: *Ixodes tiptoni* is a Neotropical Central American species that has only been found in **Panama**: Bocas del Toro and Chiriquí (Kohls and Clifford 1962; Fairchild et al. 1966).

Hosts: Females, nymphs, larvae, and the only known male of *Ixodes tiptoni* have been collected from *Sciurus granatensis* (Rodentia: Sciuridae) (Kohls and Clifford 1962; Fairchild et al. 1966).

Human Parasitism: No.

Comments: The male of *Ixodes tiptoni* is considered undescribed in Camicas et al. (1998) and Guglielmone et al. (2014), but there is a description, albeit without figures, of the only known male specimen in Fairchild et al. (1966), who constructed a key to identify the *Ixodes* of Panama.

52. ***Ixodes tropicalis*** Kohls, 1956

Kohls, G.M. (1956a) Eight new species of *Ixodes* from Central and South America (Acarina: Ixodidae). *Journal of Parasitology*, 42, 636–649.

Descriptions and Redescriptions: The female of *Ixodes tropicalis* was described and figured in Kohls (1956a).

Table 1.26 Hosts for females (F), nymphs* (N), and larvae* (L) of *Ixodes tropicalis*

MAMMALIA			
DIDELPHIMORPHIA: Didelphidae		*Thomasomys aureus*	F
Didelphis marsupialis	FNL	**RODENTIA: Echimyidae**	
RODENTIA: Cricetidae		*Dactylomys boliviensis*	F
Nephelomys albigularis	NL		

* Provisional diagnosis because the nymph and larva of *Ixodes tropicalis* remain undescribed

There are no redescriptions of the female of *Ixodes tropicalis.*

The male of *Ixodes tropicalis* is unknown, and the nymph and larva remain undescribed, but Trapido and Sanmartín (1971) have been able to identify the immature stages from field-collected material.

Geographic Distribution: *Ixodes tropicalis* is a South American species that has been found in **Colombia**: Antioquia and Valle del Cauca (Kohls 1956a; Trapido and Sanmartín 1971); and **Peru**: Puno (Kohls 1956a).

Hosts: Females, nymphs, and larvae of *Ixodes tropicalis* have been found on Didelphimorphia: Didelphidae, and Rodentia: Cricetidae, while females alone have been recorded from Rodentia: Echimyidae. Trapido and Sanmartín (1971) collected many *Ixodes tropicalis* during field studies in Colombia and stated that females were usually found on *Didelphis marsupialis* (Didelphidae), while larvae and nymphs were commonly found on *Nephelomys albigularis* (Cricetidae); these mammals are treated here as provisional principal hosts of *Ixodes tropicalis.* The host profile of this species is shown in Table 1.26, constructed with the information provided by Kohls (1956a) and Trapido and Sanmartín (1971).

Human Parasitism: No.

Comments: The records of *Ixodes tropicalis* from Didelphidae were not included in Guglielmone et al. (2014), while the diagnoses of the undescribed larva and nymph of *Ixodes tropicalis* from Colombia in Trapido and Sanmartín (1971) are treated as provisionally valid. Trapido and Sanmartín (1971) listed *Didelphis albiventris* as the principal host for adults of this tick, but Colombia is outside the range of that marsupial, so their specimens were probably collected from *Didelphis marsupialis*, which was also included in the study of Trapido and Sanmartín (1971) and is treated here as the provisional principal host for *Ixodes tropicalis.* Acevedo-Gutiérrez et al. (2020) listed *Hylaemys laticeps* as a host of this tick, but this statement was not confirmed with the references provided by these authors.

53. *Ixodes uriae* White, 1852

White, A. (1852) Insects and Aptera. *In*: *Journal of the voyage in Baffin's Bay and Barrow Straits, in the years 1850 – 1851*, 2 (appendix), 208–211.

Descriptions and Redescriptions: White (1852) described the female of *Ixodes uriae* and included a figure. The male was described without figures in Kramer and Neuman (1883) under the name *Ixodes fimbriatus*, a synonym of *Ixodes uriae*, as discussed in Guglielmone and Nava (2014). The nymph was described and figured in Pickard-Cambridge (1876) under the name *Hyalomma puta*, another synonym of *Ixodes uriae*, and the larva of *Ixodes uriae* was described and illustrated in Nuttall (1912) under the name *Ixodes putus* (= *Ixodes uriae*).

The male, female, nymph, and larva were redescribed in Arthur (1963), Wilson (1970), Filippova (1977), Furman and Loomis (1984), Lindquist et al. (2016), and Estrada-Peña et al. (2017), while the male, female, and nymph of *Ixodes uriae* were redescribed in Cooley and Kohls (1945), Roberts (1960, 1970) and Nava et al. (2017). Wilson (1967) redescribed the male, female, and larva; Keirans and Clifford (1978) redescribed the male and female; Durden and Keirans (1996) redescribed the nymph, while Sénevet and Ripert (1967), Roberts (1969), and Webb et al. (1990) redescribed the larva of *Ixodes uriae*.

Geographic Distribution: Circumpolar because *Ixodes uriae* has been found around the poles and adjacent regions of the Afrotropical, Australasian, Nearctic, Neotropical, and Palearctic Zoogeographic Regions as well as various islands (Guglielmone et al. 2014). Neotropical records of *Ixodes uriae* are from southern South America in **Argentina**: Tierra del Fuego, Antártida e Islas del Atlántico Sur; and **Chile**: Magallanes y de la Antárctica Chilena (Nava et al. 2017).

Hosts: All parasitic stages of *Ixodes uriae* worldwide have been found on Aves of several orders, with rare records from Carnivora: Mustelidae and Rodentia: Muridae. Most Neotropical specimens of males, females, nymphs, and larvae have been recorded from five species of Sphenisciformes: Spheniscidae, which are treated here as the principal hosts for all parasitic stages of *Ixodes uriae*. The Neotropical host profile of *Ixodes uriae* is shown in Table 1.27, constructed using information provided by Muñoz-Leal and González-Acuña (2015).

Human Parasitism: Guglielmone and Robbins (2018) listed *Ixodes uriae* as a frequent parasite of humans, with records from Canada, Denmark, Great Britain, Iceland, Norway, Russia, and the USA, but none from Argentina and Chile.

Comments: Several authors regard *Ixodes uriae* as a name that may include more than one taxon, but this situation remains unresolved.

Some of the records above are, in fact, closer to the Antarctic than the South American mainland. There is a record in Cooley and Kohls (1945) of a female of *Ixodes uriae* from Buenos Aires Province (Argentina), in a locality far from southern Argentina that requires confirmation. The male of *Ixodes uriae* is sometimes cited as

Table 1.27 Hosts for females (F), nymphs (N), and larvae (L) of *Ixodes uriae*

AVES			
CHARADRIIFORMES: Haematopodidae		**SPHENISCIFORMES: Spheniscidae**	
Haematopus ater	N	*Eudyptes chrysocome*	FNL
CHARADRIIFORMES: Laridae		*Pygoscelis adeliae*	FNL
Leucophaeus scoresbii	F	*Pygoscelis antarcticus*	FNL
CHARADRIIFORMES: Stercorariidae		*Pygoscelis papua*	FNL
Stercorarius antarcticus	F	*Spheniscus magellanicus*	FNL
PROCELLARIIFORMES: Procellariidae		**SULIFORMES: Phalacrocoracidae**	
Daption capense	N	*Phalacrocorax atriceps*	NL

a parasite of birds, as in Guglielmone et al. (2014) and Muñoz-Leal and González-Acuña (2015), among others, but the males of this tick do not feed.

Guglielmone and Nava (2017) consider *Ixodes uriae* a species of evolutionary importance, speculating that ancestors of this tick evolved with feathered dinosaurs, eventually transferring to modern aquatic birds that are currently the hosts of *Ixodes uriae*.

54. *Ixodes venezuelensis* Kohls, 1953

Kohls, G.M. (1953) *Ixodes venezuelensis*, a new species of tick from Venezuela, with notes on *Ixodes minor* Neumann, 1902 (Acarina: Ixodidae). *Journal of Parasitology*, 39, 300–303.

Descriptions and Redescriptions: The female and nymph of *Ixodes venezuelensis* were described and figured in Kohls (1953), and the larva was described in Durden and Keirans (1994).

The female and nymph were redescribed in Durden and Keirans (1994), while Bermúdez et al. (2018a) redescribed the female of *Ixodes venezuelensis*.

There are no redescriptions of the larva.

The male of *Ixodes venezuelensis* remains unknown.

Geographic Distribution: *Ixodes venezuelensis* is a Central American and northern South American species that has been found in **Colombia**: Antioquía (Kohls 1953); **Costa Rica**: Alajuela, and Heredia (Durden and Keirans 1994); **Panama**: Darién, and Panamá (Durden and Keirans 1994); and **Venezuela**: Amazonas, Apure, Aragua, Barinas, Bolívar, Carabobo, Distrito Federal, Falcón, Mérida, Miranda Táchira, and Trujillo (Kohls 1953; Jones et al. 1972; Durden and Keirans 1994).

Hosts: Females, nymphs, and larvae of *Ixodes venezuelensis* have been found on Didelphimorphia: Didelphidae, and Rodentia: Cricetidae; however, most females of this tick have been collected from *Heteromys anomalus* and *Heteromys desmarestianus* (Rodentia: Heteromyidae), while the majority of larvae and nymphs have been taken from *Monodelphis brevicaudata* (Didelphimorphia: Didelphidae),

Table 1.28 Hosts for females (F), nymphs (N), and larvae (L) of *Ixodes venezuelensis*

MAMMALIA			
DIDELPHIMORPHIA: Didelphidae		*Tylomys panamensis*	F
Didelphis marsupialis	NL	*Zygodontomys brevicauda*	F
Marmosa sp.	L	**RODENTIA: Dasyproctidae**	
Marmosa mexicana	N	*Dasyprocta leporina*	NL
Monodelphis adusta	F	*Myoprocta acouchy*	N
Monodelphis brevicaudata	FNL	*Myoprocta pratti*	N
Philander opossum	NL	**RODENTIA: Echimyidae**	
RODENTIA: Cricetidae		*Proechimys* sp.	NL
Ichthyomys pittieri	F	*Proechimys guyannensis*	N
Melanomys caliginosus	F	*Proechimys semispinosus*	N
Microryzomys minutus	NL	**RODENTIA: Heteromyidae**	
Nephelomys albigularis	F	*Heteromys anomalus*	FN
Oryzomys sp.	FL	*Heteromys desmarestianus*	F
Peromyscus mexicanus	N	**RODENTIA: Muridae**	
Rhipidomys fulviventer	L	*Mus musculus*	F
Sigmodontomys alfari	F	**RODENTIA: Sciuridae**	
Transandinomys talamancae	F	*Sciurus igniventris*	N

and these mammals are hypothetically regarded as the principal hosts of *Ixodes venezuelensis.* Female ticks have also been collected from two species of *Monodelphis.* The host profile of this tick is shown in Table 1.28, constructed using information provided by Kohls (1953), Jones et al. (1972), Guerrero (1996), and Durden and Keirans (1994).

Human Parasitism: No.

Comments: Vogelsang and Santos Dias (1953b) redescribed the female and nymph of *Ixodes venezuelensis*, but Méndez Arocha and Ortiz (1958) stated that several morphological characters in that female redescription do not agree with the original description in Kohls (1953), and the redescriptions Vogelsang and Santos Dias (1953b) are therefore excluded from the above list.

There are two records of "*Ixodes* probably *venezuelensis*" in Jones et al. (1972) that are not included in the host list of *Ixodes venezuelensis.*

Durden and Keirans (1994) believe that *Ixodes venezuelensis* is parthenogenetic or the male is not parasitic.

55. ***Ixodes woodi*** Bishopp, 1911, under the name *Ixodes angustus woodi* and given its current status in Cooley and Kohls (1945)

Bishopp, F.C. (1911) Some new North American Ixodidae with notes on other species. *Proceedings of the Biological Society of Washington*, 24, 197–208.

Cooley, R.A. & Kohls, G.M. (1945) The genus Ixodes in North America. *National Institute of Health Bulletin*, (184) 246 pp.

Descriptions and Redescriptions: The female of *Ixodes woodi* was described and figured in Bishopp (1911) under the name *Ixodes angustus woodi*, the male was described and illustrated in Kohls (1950), the nymph was described and figured in Cooley and Kohls (1945), and the larva was described and illustrated in Robbins and Keirans (1987).

The male, female, nymph, and larva of *Ixodes woodi* were redescribed in Robbins and Keirans (1992), while Furman and Loomis (1984) redescribed the male, female, and the nymph. Keirans and Clifford (1978) and Keirans and Litwak (1989) redescribed the male and female, and Cooley and Kohls (1945) redescribed the female. The nymph of *Ixodes woodi* was redescribed in Robbins and Keirans (1987) and Durden and Keirans (1996), and the larva was redescribed in Webb et al. (1990) and Kleinjan and Lane (2008).

Geographic Distribution: *Ixodes woodi* is essentially a Nearctic species with just one Neotropical record in **southern Mexico**: Morelos (Guzmán-Cornejo et al. 2007).

Hosts: A female of *Ixodes woodi* was collected on a *Neotoma* sp. (Rodentia: Cricetidae) in the Neotropics (Guzmán-Cornejo et al. 2007). All parasitic stages of this tick are usually collected on Rodentia: Cricetidae in the Nearctic Region, although they have also been recovered from Eulipotyphla: Soricidae; adult ticks alone have been taken from Carnivora: Procyonidae, and Rodentia: Sciuridae; immature stages have been found on Carnivora: Mephitidae, and Rodentia: Geomyidae and Heteromyidae (Guglielmone and Robbins 2018).

Human Parasitism: *Ixodes woodi* is a very rare parasite of humans in the Nearctic Region (Guglielmone and Robbins 2018).

Comments: *Ixodes woodi* was considered to be a Nearctic species, but this situation changed after the inclusion of the Mexican state of Morelos as part of the Neotropical Region (see Introduction). Serra-Freire et al. (2015) allegedly found a nymph of *Ixodes woodi* introduced on a person into Brazil (Neotropical Region) from Europe (Palearctic Region). This record of *Ixodes woodi* is doubtful.

56. The Special Neotropical Records of *Ixodes percavatus sensu lato*

According to Guglielmone et al. (2014), *Ixodes percavatus*, described in Neumann (1906), is found on islands in the Atlantic and Indian Oceans, where females, nymphs, and larvae have been collected from Procellariiformes: Diomedeidae, with a record of a single larva from Passeriformes: Turdidae, and unknown stages have allegedly been taken from other marine birds. Wilson (1970) treated records of *Ixodes percavatus* from islands in the Indian Ocean as misidentifications of *Ixodes kerguelenensis*, having inspected one female collected on the Kerguelen Islands that was previously identified as *Ixodes percavatus*. However, Paulian (1953) recorded several birds infested with *Ixodes percavatus* on those same islands. Therefore, the presence of *Ixodes percavatus* on islands in the Indian Ocean is treated here as provisionally valid.

Labruna et al. (2020b) documented records of 16 nymphs and 21 larvae collected from the pelagic bird *Thalassarche chlororhynchos* (Procellariiformes: Diomedeidae), either rescued or captured on a Brazilian beach in the State of Espírito Santo or offshore of Rio de Janeiro State. All these ticks were classified as *Ixodes percavatus sensu lato* because of difficulties in morphologically separating them from *Ixodes kerguelenensis*.

Breeding localities of *Thalassarche chlororhynchos* are restricted to the Tristan da Cunha Islands and Cough Island, all located in the central South Atlantic Ocean (Cuthbert et al. 2003). From these islands the birds depart to obtain food at sea, where they often enter the Neotropical Zoogeographic Region. However, neither *Ixodes percavatus* nor *Ixodes kerguelenensis* is part of the Neotropical ixodid fauna, and *Ixodes percavatus sensu lato* is treated as naturally invasive in this region.

Comments on *Ixodes*

Table 1.29 summarizes the known stages, zoogeographic distribution, and principal hosts for the 55 species of *Ixodes* found in the Neotropics. A total of 43 species are endemic to the region, but 12 species are also found in other zoogeographic regions. Thus one species (*Ixodes auritulus*, which in fact represents a species group) is an Australasian, Nearctic, and Neotropical species, while another tick (*Ixodes uriae*) has a range encompassing territories close to the poles, and the remaining ten species are found in both the Neotropical and Nearctic Regions. The total number of species of *Ixodes* established in the Neotropics as of 2020 represents an increase of 22% in relation to 2003 (Guglielmone et al. 2003a), when 45 species were found there, 39 endemic to this region and six taxa also found elsewhere. Six new species are now known to be found in both the Nearctic and Neotropical Regions: *Ixodes cookei* (provisional), *Ixodes scapularis*, *Ixodes spinipalpis*, *Ixodes tancitarius*, *Ixodes texanus,* and *Ixodes woodi.* Two species, *Ixodes cuernavacensis* and *Ixodes mexicanus*, were considered to be Nearctic species in Guglielmone et al. (2003a) but have been reevaluated and are currently treated as Neotropical ticks, while three new species, *Ixodes bocatorensis*, *Ixodes spinosus,* and *Ixodes silvanus***,** have been added to the list of Neotropical species compiled in 2003.

A total of 27 species (49% of the total) have one or two undescribed parasitic stages. Taxonomic research focusing on these undescribed stages is needed in order to better understand tick ecology and life histories, particularly as these relate to the transmission of tick-borne pathogens.

Host-tick associations show that Amphibia and Reptilia are irrelevant as food sources for Neotropical *Ixodes.* On the contrary, 73% (40 species) are known to feed only on Mammalia, 16% (nine species) have solely been collected from Aves, 5% (three species) have been found on Aves and Mammalia, and 5% (three species) have unknown hosts. Among mammalian hosts, rodents are central because 29 (57%) of the 51 species of Neotropical *Ixodes* whose hosts are known utilize rodents as principal (or potential principal) hosts for at least one parasitic stage, while 19 (37%) of them are known only from Rodentia. This situation is even more evident in the case of ticks endemic to the South American mainland, where 15 of the

Table 1.29 Known stages, zoogeographic distribution, and orders of principal hosts (or potential principal hosts) of Neotropical species of *Ixodes* and their parasitic stages. Hosts listed are for parasitic associations determined in Neotropical localities. M = male, F = female, N = nymph, L = larva. "Unknown" refers to tick stages with unidentified natural hosts. "Undetermined" refers to tick stages with at least one host identified, but principal hosts are uncertain

Species	Known stages	Zoogeographic Region	Principal hosts for adults	Principal hosts for nymphs	Principal hosts for larvae
1. *I. abrocomae*	MF	Neotropical	Rodentia	Unknown	Unknown
2. *I. affinis*	MFNL	Nearctic, Neotropical	Carnivora	Undetermined	Undetermined
3. *I. amarali*	FNL	Neotropical	Didelphimorphia	Didelphimorphia, Rodentia	Didelphimorphia, Rodentia
4. *I. andinus*	FNL	Neotropical	Rodentia	Rodentia	Rodentia
5. *I. auritulus*	MFNL	Australasian, Nearctic, Neotropical	Passeriformes	Passeriformes	Passeriformes
6. *I. bequaerti*	F	Neotropical	Trogoniformes	Unknown	Unknown
7. *I. bocatorensis*	MF	Neotropical	Rodentia	Unknown	Unknown
8. *I. boliviensis*	MFN	Neotropical	Carnivora	Undetermined	Undetermined
9. *I. capromydis*	MFNL	Neotropical	Rodentia	Rodentia	Rodentia
10. *I. catarinensis*	F	Neotropical	Didelphimorphia, Rodentia	Unknown	Unknown
11. *I. chilensis*	FN	Neotropical	Chiroptera	Chiroptera	Unknown
12. *I. cookei*	MFNL	Nearctic, Neotropical	Unknown	Unknown	Unknown
13. *I. cooleyi*	F	Neotropical	Rodentia	Unknown	Unknown
14. *I. copei*	FL	Neotropical	Passeriformes	Unknown	Passeriformes
15. *I. cornuae*	FN	Neotropical	Galliformes	Galliformes	Unknown
16. *I. cuernavacensis*	F	Neotropical	Caprimulgiformes	Unknown	Unknown
17. *I. dicei*	MF	Neotropical	Lagomorpha	Unknown	Unknown
18. *I. downsi*	MFNL	Neotropical	Caprimulgiformes	Unknown	Chiroptera
19. *I. fossulatus*	F	Neotropical	Unknown	Unknown	Unknown
20. *I. fuscipes*	MF	Neotropical	Artiodactyla	Undetermined	Undetermined
21. *I. galapagoensis*	MFN	Neotropical	Rodentia	Rodentia	Unknown
22. *I. guatemalensis*	MF	Neotropical	Rodentia	Unknown	Unknown
23. *I. jonesae*	MFNL	Neotropical	Rodentia	Rodentia	Rodentia
24. *I. lasallei*	MF	Neotropical	Rodentia	Unknown	Unknown
25. *I. longiscutatus*	FNL	Neotropical	Rodentia	Rodentia	Rodentia
26. *I. loricatus*	MFNL	Neotropical	Didelphimorphia	Didelphimorphia, Rodentia	Didelphimorphia, Rodentia
27. *I. luciae*	MFNL	Neotropical	Didelphimorphia	Didelphimorphia, Rodentia	Didelphimorphia, Rodentia
28. *I. mexicanus*	F	Neotropical	Passeriformes	Unknown	Unknown
29. *I. minor*	MFNL	Nearctic, Neotropical	Rodentia	Unknown	Passeriformes
30. *I. montoyanus*	MFNL	Neotropical	Artiodactyla	Artiodactyla	Artiodactyla
31. *I. nectomys*	M	Neotropical	Rodentia	Unknown	Unknown
32. *I. neuquenensis*	FNL	Neotropical	Microbiotheria	Microbiotheria	Microbiotheria
33. *I. nuttalli*	MFNL	Neotropical	Rodentia	Rodentia	Rodentia
34. *I. paranaensis*	MFNL	Neotropical	Caprimulgiformes	Caprimulgiformes	Caprimulgiformes

(continued)

Table 1.29 (continued)

35. *I. pararicinus*	MF	Neotropical	Artiodactyla	Rodentia, Passeriformes	Rodentia, Passeriformes
36. *I. pomerantzi*	MF	Neotropical	Lagomorpha	Lagomorpha*	Unknown
37. *I. rubidus*	FN	Nearctic, Neotropical	Carnivora	Carnivora	Undetermined
38. *I. scapularis*	MFNL	Nearctic, Neotropical	Unknown	Unknown	Unknown
39. *I. schulzei*	FNL	Neotropical	Rodentia	Rodentia	Rodentia
40. *I. sigelos*	FNL	Neotropical	Rodentia	Rodentia	Rodentia
41. *I. silvanus*	FNL	Neotropical	Passeriformes	Passeriformes	Passeriformes
42. *I. sinaloa*	FNL	Nearctic, Neotropical	Rodentia	Rodentia	Rodentia
43. *I. spinipalpis*	MFNL	Nearctic, Neotropical	Lagomorpha	Rodentia	Unknown
44. *I. spinosus*	FNL	Neotropical	Rodentia	Rodentia	Rodentia
45. *I. stilesi*	MFN	Neotropical	Artiodactyla	Artiodactyla	Undetermined
46. *I. taglei*	MF	Neotropical	Artiodactyla	Unknown	Unknown
47. *I. tancitarius*	F	Nearctic, Neotropical	Rodentia	Unknown	Unknown
48. *I. tapirus*	MF	Neotropical	Perissodactyla	Unknown	Unknown
49. *I. tecpanensis*	F	Neotropical	Rodentia	Unknown	Unknown
50. *I. texanus*	MFNL	Nearctic, Neotropical	Carnivora	Carnivora	Unknown
51. *I. tiptoni*	MFNL	Neotropical	Rodentia	Rodentia	Rodentia
52. *I. tropicalis*	F	Neotropical	Didelphimorphia, Rodentia	Didelphimorphia, Rodentia*	Didelphimorphia, Rodentia*
53. *I. uriae*	MFNL	Circumpolar	Sphenisciformes	Sphenisciformes	Sphenisciformes
54. *I. venezuelensis*	FNL	Neotropical	Didelphimorphia, Rodentia	Didelphimorphia, Rodentia	Didelphimorphia, Rodentia
55. *I. woodi*	MFNL	Nearctic, Neotropical	Rodentia	Unknown	Unknown

* Provisional result because the tick stage remains undescribed

23 taxa established there have been found parasiting rodents, largely species belonging to the family Cricetidae, which invaded South America during the Pliocene Epoch.

Aves are hosts of an important proportion of Neotropical *Ixodes* because 12 species (22% of the total whose hosts are known) have at least one stage that feeds on birds, and nine taxa (17%) have been found only on avian hosts. The order Passeriformes surpasses all other avian host orders, with six species (11%) feeding on passerine birds alone or in combination with mammals or birds of other orders. These associations are considered to be rather new in paleontological terms, occurring during the Cenozoic Era along with mammal and bird radiations.

Records of human parasitism by *Ixodes* in the Neotropics are few, with just three species known to rarely feed on people: *Ixodes boliviensis*, *Ixodes pararicinus,* and, provisionally, *Ixodes tancitarius.* This, however, represents an increase from Guglielmone et al. (2003a), when only *Ixodes boliviensis* and *Ixodes luciae* were reported feeding on humans (human parasitism by *Ixodes luciae* was subsequently not confirmed—see *Ixodes luciae* in this chapter).

Additional information on tick geographical distribution and host associations is provided in Chaps. 4 and 5, and the appendix.

Chapter 2
Metastriata Group: Genus *Amblyomma*

There are 493 species in the Metastriata group; 492 of these are distributed among 16 genera, and one species is of uncertain generic status. Only four metastriate genera are established in the Neotropics—*Amblyomma*, *Dermacentor*, *Haemaphysalis,* and *Rhipicephalus*—and these are treated in alphabetical order.

Genus *Amblyomma*

Worldwide there are 136 species in the genus *Amblyomma.* The largest group of species is found in the Neotropical Region, with 67 taxa (49% of the total), 54 of which are known only from this region. Therefore, *Amblyomma* may be regarded as the paradigmatic Neotropical metastriate genus. The Afrotropical Region is second in number of *Amblyomma* species, with 28 taxa (21% of the total), 24 of which are found only there.

1. *Amblyomma albopictum* Neumann, 1899, but originally named *Ixodes variegatus* by Lucas (1852), a name preoccupied by *Ixodes variegatus* (Fabricius, 1798) in Fabricius (1805), a synonym of *Amblyomma variegatum* and consequently renamed under the genus *Amblyomma* by Neumann (1899)

Neumann, L.G. (1899) Révision de la famille des ixodidés (3e mémoire). *Mémoires de la Société Zoologique de France*, 12, 107–294.

Fabricius, J.C. (1798) *Supplementum Entomologiae Systematicae*. Hafniae, 572 pp.

Fabricius, J.C. (1805) *Systema Antliatorum secundum ordines, genera, species adiectis synonymis, locis, observationibus, descriptionibus*. Carolum Reichard, Brunsvigae, 372 pp.

Lucas, H. (1852) *Ixodes variegatus*, Lucas. *Bulletin de la Société Entomologique de France, Série* 2, 7, 88–89.

Descriptions and Redescriptions: The male was described but not figured by Lucas (1852) under the name *Ixodes variegatus* (preoccupied, see above), while

A. A. Guglielmone et al., *Neotropical Hard Ticks (Acari: Ixodida: Ixodidae)*,
https://doi.org/10.1007/978-3-030-72353-8_2

the female was described, also without figures, by Vigueras (1934) and the larva was described by Černý (1969b).

The male and female of *Amblyomma albopictum* were redescribed by Clifford and Kohls (1962), Voltzit (2007), and Barros-Battesti et al. (2009), while the male was redescribed by Neumann (1899) and Robinson (1926).

There are no redescriptions of the larva of *Amblyomma albopictum.*

The nymph of *Amblyomma albopictum* remains undescribed, but some authors have been able to identify this stage from field-collected material.

Geographic Distribution: *Amblyomma albopictum* is a Neotropical species that has been found in the Caribbean Islands and Central America as follows: **Bahamas**: Black Point, and Central Andros (Durden et al. 2015); **Costa Rica** (Guglielmone et al. 2003a); **Cuba**: Artemisa, Camagüey, Isla de la Juventud, La Habana and Mayabeque (Černý 1969b; Álvarez et al. 2020); **Dominica** (Rawlins et al. 1993); **Dominican Republic**: Pedernales, and Santo Domingo (Clifford and Kohls 1962; Voltzit 2007); and **Honduras**: Islas de la Bahía (Morel 1967).

Hosts: Males, females, nymphs, and larvae of *Amblyomma albopictum* have been found on *Cyclura cychlura* and *Cyclura lewisi* (Squamata: Iguanidae), which have been regarded as the principal hosts of this tick; however, the range of *Amblyomma albopictum* is broader than the ranges of these iguanids, and other squamatan hosts surely play an important role as hosts. The host profile of *Amblyomma albopictum* is shown in Table 2.1, which is based on data from Pérez Vigueras (1934), Černý (1969b), Voltzit (2007), and Durden et al. (2015).

Human Parasitism: No.

Comments: Problems attend the identification of *Amblyomma albopictum*, particularly the female, because Vigueras (1934) described this stage without providing illustrations, and the first redescription of the female in Clifford and Kohls (1962) states that coxa I possesses a blunt rounded internal spur, whereas the corresponding figure shows coxa I with only an external spur, in agreement with the text description of Vigueras (1934). Černý (1969b) listed morphological differences among authors who defined the female of *Amblyomma albopictum*, while Clifford and Kohls (1962) found differences between females from Cuba and the Dominican Republic, and

Table 2.1 Hosts for males (M), females (F), nymphs (N), and larvae (L) of *Amblyomma albopictum**

REPTILIA			
SQUAMATA: Boidae		*Cyclura lewisi*	MFNL
Chilabothrus angulifer	MFN	*Cyclura nubila*	MF
SQUAMATA: Iguanidae		**SQUAMATA: Tropiduridae**	
Cyclura cornuta	MF	*Leiocephalus carinatus*	NL
Cyclura cychlura	MFNL		

* Provisional diagnosis due to differences in the definition of *Amblyomma albopictum* among authors

Kohls (1969c) collected specimens close to this species in Dominica that were later described as *Amblyomma antillorum*. Guglielmone et al. (2003a) speculated that more than one taxon might exist under the name *Amblyomma albopictum*. The male and female of *Amblyomma albopictum* were later redescribed by Voltzit (2007) and Barros-Battesti et al. (2009), but their descriptions of the female differ from that in Clifford and Kohls (1962), while the male and female in Voltzit (2007) differ morphologically from the corresponding descriptions in Barros-Battesti et al. (2009). Clearly, the morphology of *Amblyomma albopictum* should be reexamined throughout its range; meanwhile, the records above should be treated as provisionally valid.

Camicas et al. (1998) list the larva of *Amblyomma albopictum* as undescribed, probably because they consider the description in Černý (1969b) invalid.

The presence of *Amblyomma albopictum* in Haiti is doubtful because Schulze (1941a, under the name *Amblyomma haitianum*, a synonym of *Amblyomma albopictum*) cited this country but the locality provided is located in the Dominican Republic; therefore, Haiti is provisionally excluded from the geographic distribution of this tick. Floch and Fauran (1959a) state that Neumann (1899) redescribed the male of *Amblyomma albopictum* from a tick found in French Guiana, but Neumann (1899) specified that the specimen was found in Cuba; nevertheless, Burridge (2011) included French Guiana within the geographic distribution of this tick. *Amblyomma albopictum* was treated as a tick found in Brazil by Neumann (1899), Floch and Fauran (1958), and Guglielmone et al. (2003a), but the presence of this species there is doubtful (Dantas-Torres et al. 2009a), and Brazil is no longer included in the range of this tick. Álvarez et al. (2020) exclude the provinces of La Habana and Pinar del Río from the range of *Amblyomma albopictum* in Cuba, but Černý (1966b, 1969b) listed this tick from localities that are within the limits of those provinces according to the most recent arrangement of Cuban administrative divisions.

The Brazilian record of a male *Amblyomma albopictum* taken from *Coendou spinosus* (Rodentia: Erethizontidae) in Neumann (1899), a record repeated by several workers thereafter, is doubtful and may be due to mislabeling, while Cuban records of this tick in Barros-Battesti et al. (2009) from such squamatan hosts as *Alsophis* and *Leoicephalus*, and from *Coendou* are the result of typographical errors.

2. *Amblyomma antillorum* Kohls, 1969

Kohls, G.M. (1969b) A new species of *Amblyomma* from iguanas in the Caribbean (Acarina: Ixodidae). *Journal of Medical Entomology*, 6, 439–442.

Descriptions and Redescriptions: The male and female of *Amblyomma antillorum* were described and figured in Kohls (1969b), while the larva and nymph were described in Keirans (1985a).

Voltzit (2007) redescribed the male and female of *Amblyomma antillorum*.

There are no redescriptions of the larva and nymph of *Amblyomma antillorum*.

Geographical Distribution: *Amblyomma antillorum* is a Neotropical species that has been found in the Caribbean Region as follows: **Dominica**: Saint Joseph (Kohls 1969b); **Turks and Caicos Islands**: Caicos Islands (Keirans 1985a); and **Virgin Islands (United Kingdom)**: Anegada (Kohls 1969b).

Hosts: Males, females, nymphs, and larvae of *Amblyomma antillorum* have been found on *Cyclura pinguis* and *Iguana delicatissima* (Squamata: Iguanidae), which are treated here as principal hosts for this tick, but these iguanids are absent in the Turks and Caicos Islands, where other reptiles serve as hosts. The host profile of *Amblyomma antillorum* is shown in Table 2.2, which is based on data from Keirans (1985a) and Durden et al. (2015).

Human Parasitism: No.

Comments: Male specimens of *Amblyomma antillorum* were identified as *Amblyomma* near *albopictum* in Kohls (1969c).

Keirans (1985a) states that males of *Amblyomma antillorum* were collected from "iguana" in the Turks and Caicos Islands on July 25, 1930, and that the host was *Cyclura carinata* (named as *Iguana carinata*), but this is impossible to confirm.

3. ***Amblyomma argentinae*** Neumann, 1905, but originally named *Ixodes testudinis* by Conil (1877), a name preoccupied by *Ixodes testudinis* in Leydig (1855), and replaced by *Amblyomma argentinae* of Neumann (1905), who described this tick while unaware of its synonymy with *Ixodes testudinis* of Conil (1877)

Neumann, L.G. (1905) Notes sur les ixodidés. III. *Archives de Parasitologie*, 9, 225–241.

Leydig, F. (1855) Zum feineren Bau der Arthropoden. *In:* Müller, J. (Editor), *Archiv für Anatomie, Physiologie und Wissenschaftliche Medicin*, Veit et Comp, pp. 376–480.

Conil, P.A. (1877) Description d'une nouvelle espèce d'ixode, *Ixodes testudinis*. *Acta de la Academia Nacional de Ciencias Exactas (Argentina)*, 3, 25–37.

Descriptions and Redescriptions: The male and female of *Amblyomma argentinae* were described and figured in Conil (1877) under the name *Ixodes testudinis*, a senior synonym of *Amblyomma argentinae*, as discussed in Guglielmone and Nava (2014). The nymph was described without figures by Neumann (1905) under the name *Amblyomma argentinae*, while the larva was described in Estrada-Peña et al. (1993b) under the name *Amblyomma testudinis*.

Table 2.2 Hosts for males (M), females (F), nymphs (N), and larvae (L) of *Amblyomma antillorum*

REPTILIA			
SQUAMATA: Boidae		**SQUAMATA: Iguanidae**	
Boa nebulosa	M	*Cyclura pinguis*	MFNL
SQUAMATA: Dipsadidae		*Iguana delicatissima*	MFNL
Alsophis sibonius	L	**SQUAMATA: Teiidae**	
		Pholidoscelis fuscatus	L

Nava et al. (2017) redescribed the male, female, and nymph. Lahille (1905), Robinson (1926), and Boero (1957) redescribed the male and female under the name *Amblyomma testudinis*, and both sexes were redescribed under the name *Amblyomma argentinae* by Voltzit (2007). The nymph alone was redescribed in Martins et al. (2014c).

There are no redescriptions of the larva of *Amblyomma argentinae*.

Geographical Distribution: *Amblyomma argentinae* is a southern South American species that has only been found in **Argentina**: Catamarca, Chaco, Córdoba, Formosa, Mendoza, Salta, Santa Fe, Santiago del Estero, and Tucumán (Guglielmone and Nava 2006).

Hosts: Males, females, nymphs, and larvae of *Amblyomma argentinae* are usually found on *Chelonoidis chilensis* (Testudines: Testudinidae). The host profile of *Amblyomma argentinae* is shown in Table 2.3, which is based on data from Debárbora et al. (2015) and Nava et al. (2017).

Human Parasitism: No.

Comments: Barbará and Dios (1918) realized that Neumann's description of *Amblyomma argentinae* matched that of Conil (1877) of *Ixodes testudinis* and decided that the former name was a synonym of *Amblyomma testudinis*, although the epithet "*testudinis*" was preoccupied by *Ixodes testudinis* of Leydig (1855). Nevertheless, the opinion of Barbará and Dios (1918) prevailed for a long time, although the correct name of this species is *Amblyomma argentinae*, as discussed in Guglielmone et al. (2001), who supported Camicas et al. (1998) in unambiguously synonymizing *Ixodes testudinis* as described in Conil (1877) (preoccupied) with *Amblyomma argentinae*. Santos Dias (1958b, c) treated *Amblyomma cubanum* and *Aponomma thumbi*, both from Cuba, as synonyms of *Amblyomma argentinae* (under the name *Amblyomma testudinis*), but Santos Dias (1993) later reported *Aponomma thumbi* as a synonym of *Amblyomma dissimile*; and *Amblyomma cubanum* is also a synonym of *Amblyomma dissimile*, as discussed in Guglielmone and Nava (2010, 2014).

Table 2.3 Hosts for males (M), females (F), nymphs (N), and larvae (L) of *Amblyomma argentinae*

AMPHIBIA			
ANURA: Bufonidae			
Rhinella sp.	MF		
REPTILIA			
SQUAMATA: Boidae		*Tropidurus spinulosus*	N
Boa constrictor	MFN	**SQUAMATA: Viperidae**	
Epicrates cenchria	F	*Bothrops neuwiedi*	M
Eunectes notaeus	MF	*Crotalus durissus*	MFN
SQUAMATA: Teiidae		**TESTUDINES: Chelidae**	
Teius teyou	NL	*Phrynops hilarii*	F
SQUAMATA: Tropiduridae		**TESTUDINES: Testudinidae**	
Tropidurus etheridgei	N	*Chelonoidis chilensis*	MFNL

There are records of *Amblyomma argentinae* from Brazil, Chile, Cuba, Peru, Suriname, Uruguay, and Venezuela, but as stated in Nava et al. (2017), these records are either misidentifications or resulted from parasitized tortoises being introduced from Argentina, since there is no evidence that natural populations of *Amblyomma argentinae* are established in those countries. There is an odd record of two males of *Amblyomma argentinae* that were collected from a bird in Chile (González-Acuña et al. 2003b), and another unexpected record of a male tick found on a mammal in southern Argentina (Ivancovich and Luciani 1992), plus some specimens recovered from captive vertebrates, but none of these records are included in Table 2.3.

The first report of a fossil tick in amber from the Americas was of an *Amblyomma* male from the Dominican Republic (Lane and Poinar 1986) that was identified as being morphologically close to *Amblyomma argentinae* (under the name *Amblyomma testudinis*), based on the key of Jones et al. (1972) for the *Amblyomma* of the Western Hemisphere.

4. *Amblyomma aureolatum* (Pallas, 1772) under the name *Acarus aureolatus* and given its current status by Koch (1844)

Pallas, P.S. (1772) *Spicilegia Zoologica*, volume 40, fascicule 9, 87 pp.
Koch, C.L. (1844) Systematische Übersicht über die Ordnung der Zecken. *Archiv für Naturgeschichte*, 10, 217–239.

Descriptions and Redescriptions: Pallas (1772) described the female accompanied by a figure of no diagnostic value, under the name *Acarus aureolatus*, and Koch (1844) described but did not figure the male of *Amblyomma aureolatum*. The nymph was described in Martins et al. (2010).

Nava et al. (2017) redescribed the male, female, and nymph of *Amblyomma aureolatum*. The male and female were redescribed by Schulze (1941b), Aragão and Fonseca (1961b), Onofrio et al. (2006b, keys and figures), Voltzit (2007), and Dantas-Torres et al. (2019b, keys and figures). The nymph alone was redescribed in Martins et al. (2014c, key and figures).

The larva of *Amblyomma aureolatum* remains undescribed, but several authors have been able to identify this stage from field-collected material.

Geographical Distribution: *Amblyomma aureolatum* is a Neotropical species that has been found in South America as follows: **Argentina**: Buenos Aires, Chaco, Ciudad Autónoma de Buenos Aires, Entre Ríos, Misiones, and Santa Fe (Guglielmone and Nava 2006; Tarragona et al. 2012; Cicuttin et al. 2013; Colombo et al. 2016b); **Brazil**: Bahia, Ceará, Espírito Santo, Minas Gerais, Pará, Paraná, Pernambuco, Piauí, Rio de Janeiro, Rio Grande do Sul, Santa Catarina, São Paulo, and Sergipe (Neumann 1899; Aragão 1936; Aragão and Fonseca 1961b; Guimarães et al. 2001; Acosta et al. 2016; Moerbeck et al. 2016); **French Guiana**: Cayena, and Saint Laurent du Maroni (Floch and Fauran 1958); **Paraguay**: Itapuá, and San Pedro (Keirans 1985b; Ogrzewalska et al. 2014); and **Uruguay**: Artigas, Colonia,

Maldonado, Paysandú, Río Negro, Rocha, Salto, and Tacuarembó (Martins et al. 2014d; Rivero et al. 2017).

Hosts: Usual hosts of males and females of *Amblyomma aureolatum* are Carnivora: Canidae, Felidae, and *Procyon cancrivorus* in the family Procyonidae, while several families of Passeriformes are hosts for the nymph and the undescribed larva of this tick, and there are a few records of nymphs from Accipitriformes and Strigiformes. However, an analysis of 282 nymphs and 568 larvae of *Amblyomma aureolatum* listed in the literature shows that 78% of the nymphs were collected from Turdidae, and the great majority of nymphal specimens were taken from *Turdus rufiventris*, while 92% of the larvae were also found on Turdidae, mainly *Turdus rufiventris*, and, to a lesser extent, *Troglodytes aedon* (Troglodytidae). Therefore, *Turdus rufiventris* is treated here as a provisional principal host for the nymph and larva of *Amblyomma aureolatum.* Nevertheless, the geographical distribution of *Amblyomma aureolatum* is broader than the range of *Turdus rufiventris*, so other undetermined hosts must be of importance for the nymphal and larval stages of this tick. The extensive host profile of *Amblyomma aureolatum* is shown in Table 2.4, which is based on data from Guglielmone et al. (2003c), Salvador et al. (2007), Cicuttin et al. (2013, 2019b), Luz et al. (2014, 2017a, b, 2018a), Lavina et al. (2015), Nava et al. (2017), Ramírez et al.(2020) and Teixeira et al. (2020).

Human Parasitism: Guglielmone and Robbins (2018) list males and females of *Amblyomma aureolatum* as sporadic parasites of humans, providing records from Argentina, Brazil (Paraná, Rio de Janeiro, and São Paulo), French Guiana (Cayena), and Paraguay (San Pedro). Thereafter, Reck et al. (2018) reported cases of human parasitism at ten localities in Rio Grande do Sul (Brazil) that were caused by males, females, and nymphs of *Amblyomma aureolatum*, while Savani et al. (2019) described a case of human parasitism by a female tick in the Brazilian state of São Paulo, and Valente et al. (2020) recorded three males, three females and a nymph collected from people at five localities in the State of Paraná (Brazil). Guglielmone and Robbins (2018) state that the female of *Amblyomma aureolatum* is the stage most often found feeding on humans.

Comments: *Amblyomma aureolatum* and *Amblyomma ovale* are morphologically similar species that have been confused with each other. Records of *Amblyomma aureolatum* (frequently under the name *Amblyomma striatum*, a synonym of *Amblyomma aureolatum*) published prior to the study of Aragão and Fonseca (1961b) should be treated cautiously because several such records are the result of confusion with *Amblyomma ovale*, frequently under the name *Amblyomma fossum*, a synonym of *Amblyomma ovale.* Other workers, such as Vogelsang and Santos Dias (1953b), treat *Amblyomma aureolatum* (under the name *Amblyomma striatum*) as a synonym of *Amblyomma ovale*, rejecting the opinion of Tonelli Rondelli (1937), who described and depicted the male of *Amblyomma aureolatum* (under the name *Amblyomma striatum*) as morphologically different from *Amblyomma ovale.* However, time has proven Tonelli Rondelli (1937) correct, and it is now accepted that both *Amblyomma aureolatum* and *Amblyomma ovale* are valid species.

Table 2.4 Hosts for adult ticks of undetermined sex (A), males (M), females (F), nymphs (N), and larvae* (L) of *Amblyomma aureolatum*

	MAMMALIA		
ARTIODACTYLA: Bovidae		*Procyon cancrivorus*	MF
Cattle	MF	**DIDELPHIMORPHIA: Didelphidae**	
ARTIODACTYLA: Cervidae		*Didelphis albiventris*	M
Mazama gouazoubira	MF	*Didelphis aurita*	MFN
CARNIVORA: Canidae		*Lutreolina crassicaudata*	M
Cerdocyon thous	MFN	**PILOSA: Bradypodidae**	
Chrysocyon brachyurus	MF	*Bradypus tridactylus*	N
Domestic dog	MFN	*Bradypus variegatus*	F
Lycalopex gymnocercus	MF	**PILOSA: Myrmecophagidae**	
CARNIVORA: Felidae		*Tamandua tetradactyla*	F
Domestic cat	MF	**PRIMATES: Atelidae**	
Herpailurus yagouaroundi	A	*Alouatta guariba*	MF
Leopardus geoffroyi	MF	**RODENTIA: Caviidae**	
Leopardus guttulus	F	*Hydrochoerus hydrochaeris*	F
Leopardus pardalis	MF	**RODENTIA: Ctenomyidae**	
Leopardus tigrinus	A	*Ctenomys* sp.	MFL
Leopardus wiedii	MF	**RODENTIA: Echimyidae**	
Puma concolor	MFN	*Euryzygomatomys spinosus*	MFNL
CARNIVORA: Mustelidae		**RODENTIA: Erethizontidae**	
Galictis cuja	M	*Coendou insidiosus*	F
Galictis vittata	A	**RODENTIA: Sciuridae**	
CARNIVORA: Procyonidae		*Sciurus aestuans*	F
Nasua nasua	M		
	AVES		
ACCIPITRIFORMES: Accipitridae		*Thamnophilus ruficapillus*	L
Rupornis magnirostris	N	**PASSERIFORMES: Thraupidae**	
PASSERIFORMES: Conopophagidae		*Haplospiza unicolor*	N
Conopophaga lineata	NL	*Microspingus lateralis*	N
PASSERIFORMES: Furnariidae		*Poospiza nigrorufa*	N
Cranioleuca obsoleta	NL	*Saltator similis*	NL
Cranioleuca pallida	L	*Sporophila caerulescens*	N
Furnarius rufus	NL	*Tachyphonus coronatus*	L
Synallaxis cinerascens	NL	*Trichothraupis melanops*	NL
Synallaxis ruficapilla	N	**PASSERIFORMES: Troglodytidae**	
PASSERIFORMES: Parulidae		*Troglodytes aedon*	NL
Myiothlypis leucoblephara	NL	**PASSERIFORMES: Turdidae**	
PASSERIFORMES: Passerellidae		*Turdus albicollis*	NL
Arremon semitorquatus	N	*Turdus amaurochalinus*	NL
Zonotrichia capensis	N	*Turdus leucomelas*	N
PASSERIFORMES: Thamnophilidae		*Turdus rufiventris*	NL
Dysithamnus mentalis	N	**PASSERIFORMES: Tyrannidae**	
Myrmoderus squamosus	N	*Platyrinchus mystaceus*	L
Pyriglena leucoptera	NL	**STRIGIFORMES: Tytonidae**	
Thamnophilus caerulescens	NL	*Tyto alba*	N

* The larva of *Amblyomma aureolatum* remains undescribed

Voltzit (2007) depicts a female of *Amblyomma aureolatum* with a broken hypostome and figures the male with an unusual spur on coxa IV, casting doubt on whether these figures represent *bona fide Amblyomma aureolatum*.

The range of *Amblyomma aureolatum* is not well defined. The presence of *Amblyomma aureolatum* in Bolivia (Squire 1972), Colombia (Osorno Mesa 1942), Guatemala (Bequaert 1938), Panama (Neumann 1899), and Venezuela (Fiasson 1949) requires confirmation, and these countries are provisionally excluded from the range of *Amblyomma aureolatum*. Suriname was included within the range of

Amblyomma aureolatum by Guglielmone et al. (2003a, c) and Nava et al. (2017), but the only Surinamian record, in Reyne (1923), also requires confirmation, and this country is not included within the geographic distribution of this species. Floch and Fauran (1958) listed several records of *Amblyomma aureolatum* (under the name *Amblyomma striatum*) from French Guiana and clearly separated this species from *Amblyomma ovale* (named as *Amblyomma fossum*) in their text descriptions, but the figures used to support the morphological differences between these species are of poor quality. Moreover, Binetruy et al. (2019) were unable to confirm the presence of *Amblyomma aureolatum* in French Guiana, which is provisionally included within the range of this species. Guimarães et al. (2001) exclude the Brazilian states of Bahia, Minas Gerais, Pará, Pernambuco, Piauí, Sergipe (Aragão 1936), Pará (Neumann 1899), and Rio de Janeiro (Aragão and Fonseca 1961b) from the geographic distribution of this tick. However, Neumann (1899), Aragão (1936), and Aragão and Fonseca (1961b) listed records of *Amblyomma aureolatum* (under the name *Amblyomma striatum*) from those Brazilian states and separate this species from *Amblyomma ovale* (named as *Amblyomma fossum*).

Records of *Amblyomma aureolatum* collected from "snakes" in Floch and Fauran (1958, as *Amblyomma striatum*) and on Anura (Lavina et al. 2015) are not included in Table 2.4.

5. *Amblyomma auricularium* (Conil, 1878) under the name *Ixodes auricularius* and given its current status in Lahille (1905, using the name *Amblyomma auriculare*) and not in Aragão (1938), as cited in Guglielmone and Nava (2014), although Aragão (1938) used the proper name, *Amblyomma auricularium*

Conil, P.A. (1878) Description d'une nouvelle espèce d'ixode, *Ixodes auricularius*. *Acta de la Academia Nacional de Ciencias Exactas (Argentina)*, 3, 99–110.
Lahille, F. (1905) Contribution à l'étude des ixodidés de la République Argentine. *Anales del Ministerio de Agricultura, República Argentina*, 2, 166 pp.
Aragão, H.B. (1938) Nota sobre os ixodideos da Republica Argentina. *Memórias do Instituto Oswaldo Cruz*, 33, 319–327.
Guglielmone, A.A. & Nava, S. (2014) Names for Ixodidae (Acari: Ixodoidea): valid, synonyms, *incertae sedis*, *nomina dubia*, *nomina nuda*, *lapsus*, incorrect and suppressed names – with notes on confusions and misidentifications. *Zootaxa*, 3767, 1–256.

Descriptions and Redescriptions: The male and female of *Amblyomma auricularium* were described and figured by Conil (1878) under the name *Ixodes auricularius*, but the figure legends do not accord with the sex depicted. The nymph was described by Martins et al. (2010), and the larva of *Amblyomma auricularium* was described by Amorim and Serra-Freire (2000b).

The male, female, and nymph of *Amblyomma auricularium* were redescribed in Nava et al. (2017); the male and female were also redescribed in Robinson (1926, as *Amblyomma concolor*, a synonym of *Amblyomma auricularium*), Boero (1957, under the name *Amblyomma auriculare*), Onofrio et al. (2006b, keys and figures), Voltzit (2007), Guzmán-Cornejo et al. (2011, keys and figures), Bermúdez et al. (2018a) and Dantas-Torres et al. (2019b, keys and figures). The male alone was redescribed by Lahille (1905). The nymph alone was redescribed in Martins et al. (2014c, key and figures).

There are no redescriptions of the larva of *Amblyomma auricularium*.

Geographical Distribution: *Amblyomma auricularium* is basically a Neotropical species but it has also been found in the Nearctic Zoogeographic Region in northern Mexico and southern USA. *Amblyomma auricularium* has a wide geographic distribution in the Neotropics, including most of the mainland territory and one Caribbean country as follows: **Argentina**: Buenos Aires, Catamarca, Chaco, Chubut, Córdoba, Corrientes, Entre Ríos, Formosa, Jujuy, La Pampa, La Rioja, Salta, San Juan, San Luis, Santa Fe, Santiago del Estero, and Tucumán (Guglielmone and Nava 2006); **Belize** (Varma 1973); **Bolivia**: Beni (Robbins and Deem 2002); **Brazil**: Bahia, Ceará, Goiás, Maranhão, Mato Grosso, Mato Grosso do Sul, Minas Gerais, Pará, Paraiba, Pernambuco, Piauí, Rio Grande do Norte, Rio Grande do Sul, Rio de Janeiro, Rondônia, Santa Catarina, Sergipe, and Tocantins (Robinson 1926; Aragão 1936; Fonseca 1958b; Freire 1972; Keirans 1985b; Amorim and Serra-Freire 2000b; Linardi et al. 1991; Guimarães et al. 2001; Martins et al. 2014a; Lugarini et al. 2015; Witter et al. 2016); **Colombia**: Meta, Sucre, and Valle del Cauca (Osorno Mesa 1942; Acevedo-Gutiérrez et al. 2020); **Costa Rica**: Guanacaste, and Limón (Álvarez et al. 2005); **El Salvador** (Navarrete Abarca et al. 2014); **French Guiana**: Cayena (Floch and Fauran 1958); **Guyana** (Rawlins et al. 1993); **Honduras** (Guglielmone et al. 2003b); **southern Mexico**: Chiapas, Michoacán, Nayarit, Tabasco, and Yucatán (Hoffmann and López-Campos 2000; Guzmán-Cornejo et al. 2011; Rodríguez-Vivas et al. 2016); **Nicaragua** (Guglielmone et al. 2003b); **Panama** (Fairchild et al. 1966); **Paraguay**: Caa-guazú, Ñeembucú, and Presidente Hayes (Massi Pallarés and Benítez Usher 1982); **Suriname** (Rawlins et al. 1993); **Trinidad and Tobago** (Guglielmone et al. 2003b); **Uruguay**: Artigas (Venzal et al. 2002); **Venezuela**: Apure, Bolívar, Carabobo, Falcón, Lara, Miranda, Monagas, and Trujillo (Jones et al. 1972).

Hosts: Usual hosts for males, females, nymphs, and larvae of *Amblyomma auricularium* are Cingulata: Chlamyphoridae and Dasypodidae, but they have also been found on six orders of mammals, while several species of Passeriformes are hosts for the immature stages, and nymphs have been collected from one species of Strigiformes. There is an unusual record of a female of *Amblyomma auricularium* collected from a reptile and discussed in Guzmán-Cornejo et al. (2011). *Dasypus novemcinctus* (Cingulata: Dasypodidae), *Euphractus sexcinctus,* and *Tolypeutes matacus* (Cingulata: Chlamyphoridae) are provisionally treated here as principal hosts for the nymph and larva of this tick, but birds and other mammals may be more important hosts for *Amblyomma auricularium* than currently recognized. The host profile of *Amblyomma auricularium* in the Neotropics is extensive, as shown in Table 2.5, which is based on data from Boero (1944b), Guglielmone et al. (2003b), Esser et al. (2016a, b), Nava et al. (2017), Pereira et al. (2017), Oliveira et al. (2017),

Table 2.5 Hosts for adult ticks of undetermined sex (A), males (M), females (F), nymphs (N), and larvae (L) of *Amblyomma auricularium**

MAMMALIA			
ARTIODACTYLA: Tayassuidae		*Dasypus septemcinctus*	MFN
Pecari tajacu	N	**DIDELPHIMORPHIA: Didelphidae**	
CARNIVORA: Canidae		*Didelphis albiventris*	MFNL
Cerdocyon thous	M	*Didelphis marsupialis*	MFNL
Chrysocyon brachyurus	MF	*Gracilinanus agilis*	NL
Domestic dog	MNL	*Monodelphis domestica*	NL
Lycalopex vetulus	N	*Philander opossum*	MFNL
Urocyon cinereoargenteus	N	**PERISSODACTYLA: Equidae**	
CARNIVORA: Felidae		Horse	F
Domestic cat	N	**PILOSA: Myrmecophagidae**	
CARNIVORA: Mephitidae		*Myrmecophaga tridactyla*	MN
Conepatus semistriatus	MFNL	*Tamandua mexicana*	MF
CARNIVORA: Mustelidae		*Tamandua tetradactyla*	MFNL
Galictis cuja	MF	**RODENTIA: Caviidae**	
Galictis vittata	M	*Dolichotis salinicola*	FN
CARNIVORA: Procyonidae		*Galea spixii*	N
Nasua narica	M	*Hydrochoerus isthmius*	M
Nasua nasua	N	**RODENTIA: Chinchillidae**	
Procyon lotor	MN	*Lagostomus maximus*	MFN
CINGULATA: Chlamyphoridae		**RODENTIA: Cricetidae**	
Cabassous centralis	MF	*Oxymycterus delator*	N
Cabassous tatouay	MF	*Sigmodon* sp.	N
Chaetophractus vellerosus	MFN	**RODENTIA: Dasyproctidae**	
Chaetophractus villosus	MFN	*Dasyprocta azarae*	L
Euphractus sexcinctus	MFNL	**RODENTIA: Echimyidae**	
Tolypeutes matacus	MFNL	*Clyomys laticeps*	N
Tolypeutes tricinctus	MFN	*Hoplomys gymnurus*	NL
Zaedyus pichiy	MN	*Proechimys semispinosus*	NL
CINGULATA: Dasypodidae		*Thrichomys apereoides*	NL
Dasypus hybridus	M	*Thrichomys inermis*	NL
Dasypus kappleri	A	*Thrichomys laurentius*	FN
Dasypus novemcinctus	MFNL	**RODENTIA: Erethizontidae**	
Dasypus sabanicola	MF	*Coendou mexicanus*	N
AVES			
PASSERIFORMES: Cardinalidae		*Saltator similis*	N
Cyanoloxia brissonii	N	**PASSERIFORMES: Troglodytidae**	
PASSERIFORMES: Furnariidae		*Cantorchilus longirostris*	L
Furnarius leucopus	L	*Troglodytes aedon*	L
Megaxenops parnaguae	N	**PASSERIFORMES: Turdidae**	
PASSERIFORMES: Passerellidae		*Turdus amaurochalinus*	NL
Zonotrichia capensis	N	*Turdus grayi*	N
PASSERIFORMES: Pipridae		*Turdus rufiventris*	N
Neopelma pallescens	N	**PASSERIFORMES: Tyrannidae**	
PASSERIFORMES: Thamnophilidae		*Cnemotriccus fuscatus*	L
Gymnopithys leucaspis	N	*Stigmatura napensis*	N
Myrmorchilus strigilatus	N	**PASSERIFORMES: Vireonidae**	
Sakesphorus cristatus	N	*Cyclarhis gujanensis*	L
Thamnophilus doliatus	N	**STRIGIFORMES: Strigidae**	
PASSERIFORMES: Thraupidae		*Athene cunicularia*	N
Coryphospingus pileatus	L		
REPTILIA			
SQUAMATA: Iguanidae			
Iguana sp.	F		

*There is uncertainty concerning the descriptions and redescriptions of *Amblyomma auricularium*; therefore, the above records should be considered provisionally valid

Sousa et al. (2017), Bermúdez et al. (2018a), Lopes et al. (2018, 2019), Maia et al. (2018), Barbieri et al. (2019), Szabó et al. (2019), Martins et al. (2020b), Oliveira et al. (2020) and Teixeira et al. (2020). If the Nearctic Region is included, this host profile becomes even broader because adults and nymphs of this tick have also been found on Artiodactyla: Suidae, and larvae on Artiodactyla: Cervidae, and Accipitriformes: Accipitridae in the southern Nearctic, where larvae and nymphs are also commonly recovered from Didelphimorphia: Didelphidae (Mertins et al. 2017).

Human Parasitism: Guglielmone and Robbins (2018) discussed alleged records of human parasitism by *Amblyomma auricularium*, concluding that there are no *bona fide* records of this species feeding on humans. Nonetheless, Lopes et al. (2018) found a nymph attached to a human in Natal City, State of Rio Grande do Norte, Brazil, and Szabó et al. (2020) collected three nymphs from humans in the Brazilian state of Minas Gerais.

Comments: Lahille (1905) placed *Ixodes auricularius* in the genus *Amblyomma*, but this taxon was considered invalid by Nuttall and Warburton (1911) and Robinson (1926). The name *Amblyomma auricularium* was used again by Aragão (1935, as *Amblyomma auriculare*) and has continued to be used by many workers up to the present with no evident dispute concerning its diagnosis. However, data re-analysis demonstrates that *Amblyomma auricularium* can be difficult to recognize with certainty.

Amblyomma auricularium is by definition a tick species whose males and females possess inornate scuta. Although Conil (1878) confused the male with the female tick and *vice versa*, both sexes were consistently treated and figured as inornate, a morphological character confirmed by Lahille (1905) for the male of *Amblyomma auricularium*. Nevertheless, Vogelsang and Santos Dias (1953a) consider this species ornate. Onofrio et al. (2006b) constructed a key for the morphological diagnosis of *Amblyomma*, where the male and female of *Amblyomma auricularium* are treated as inornate, but despite the key's text, these authors highlighted the supposed scutal ornamentation in the corresponding figures of both sexes of *Amblyomma auricularium*. Dantas-Torres et al. (2019b) appear to indicate that some males and females of Brazilian *Amblyomma auricularium* are ornate, but others are not, although only figures of ornate specimens appear in their study. By contrast, Jones et al. (1972), in their keys to the species of *Amblyomma* from the Western Hemisphere, Guzmán-Cornejo et al. (2011) in their analysis of the Mexican *Amblyomma*, Nava et al. (2017), in their review of the ticks of the Southern Cone of South America, and Bermúdez et al. (2018a), in their study of Panamanian Ixodidae, depicted males and females of *Amblyomma auricularium* as having inornate scuta. It is unclear whether Jones et al. (1972), Guzmán-Cornejo et al. (2011), Nava et al. (2017), and Bermúdez et al. (2018a, b) were describing the same species, but it appears that those studies are of species other than the *Amblyomma auricularium* described in Onofrio et al. (2006b) and Dantas-Torres et al. (2019b). This taxonomic problem is difficult to resolve because the types used by Conil (1878) to describe

Amblyomma auricularium are not available for comparison and the Argentinean type locality is unknown.

The figures of males and females of Brazilian *Amblyomma auricularium* with ornate scuta in Onofrio et al. (2006b) and Dantas-Torres et al. (2019b) reveal minimal morphological differences from the figures of *Amblyomma pseudoconcolor*, also a common parasite of armadillos. Notably, Camicas et al. (1998) treat *Amblyomma pseudoconcolor* as a synonym of *Amblyomma auricularium*. Guglielmone and Nava (2006) state that the ornamentation of the male of *Amblyomma pseudoconcolor* is indistinct and reclassified several specimens of *Amblyomma auricularium* as *Amblyomma pseudoconcolor* after examining the specimens to confirm the presence of ornamentation. In this context, it is important to emphasize that *Amblyomma auricularium* was described from specimens collected in Argentina, but from 1979 to 2020, the tick collection of the Instituto Nacional de Tecnología Agropecuaria in Rafaela, Santa Fe Province, accessioned 20 samples of *Amblyomma pseudoconcolor* and none of *Amblyomma auricularium*. This collection contains four samples of *Amblyomma auricularium* (accession numbers 383, 384, 996, 1247) bearing the legend "under revision," a hint that doubts exist concerning the usefulness of this material in any attempt to redescribe *Amblyomma auricularium*. However, Costa et al. (2020) examined Brazilian ticks from Maranhão listed as *Amblyomma pseudoconcolor* in Arzua et al. (2005) and redetermined them as *Amblyomma auricularium*, a decision that demands confirmation, given current uncertainties over the diagnosis of true *Amblyomma auricularium*.

Additionally, Floch and Fauran (1958) stated that descriptions of *Amblyomma auricularium* in Floch and Abonnenc (1940, as *Amblyomma auriculare*) are misidentifications of *Amblyomma coelebs*, while Guzmán-Cornejo et al. (2011) found *Amblyomma auricularium* identified as *Amblyomma parvum*. Guglielmone et al. (2003a) erroneously stated that the nymph of *Amblyomma auricularium* was described in Robinson (1926) under the name of *Amblyomma concolor*. In fact, it is unclear whether Robinson (1926) described a female or a nymph of *Amblyomma concolor*.

Amblyomma curruca of Schulze (1936b) is generally treated as a synonym of *Amblyomma auricularium* (Guglielmone et al. 2009; Guglielmone and Nava 2014), but Aragão and Fonseca (1953a) and Guglielmone et al. (2003b) treated *Amblyomma curruca* as a synonym of *Amblyomma parvum*. The latter authors based their synonymy on the presence of chitinous plates (carenas) extending posteriorly from the festoons (Fig. 2a in Schulze 1936b) in *Amblyomma curruca*, a morphological characteristic of *Amblyomma parvum* that is not mentioned in descriptions and redescriptions of *Amblyomma auricularium*. However, the trochanteral spurring that characterizes adults of *Amblyomma auricularium* and *Amblyomma parvum* is not mentioned in the description of *Amblyomma curruca* and is also absent in the redescription of the female of this tick by Hoffmann (1950), who later stated that her Mexican *Amblyomma curruca* were, in fact, *Amblyomma parvum* (Hoffmann 1962). Guglielmone et al. (2017) stated that a reevaluation of the status of *Amblyomma curruca* is needed, and the same can be said of *Amblyomma concolor*, another alleged synonym of *Amblyomma auricularium,* because any proposed conspecificities cannot be confirmed since there is no type material of *Amblyomma auricularium* available for comparison. Clearly, it is of taxonomic importance to review all data concerning *Amblyomma auricularium* and related species throughout their range, and, more important, to examine any material deposited in museum collections that will assist in resolving this matter. Meanwhile, records of *Amblyomma auricularium* and its alleged synonyms are here considered provisionally valid. See also *Amblyomma pseudoconcolor* for further discussion of this topic.

Floch and Fauran (1958) allegedly found *Amblyomma auricularium* in French Guiana, but its presence there was not confirmed by Binetruy et al. (2019). Given the current inadequate definition of this tick, French Guiana and other territories are provisionally included within the range of this species. Since the taxonomic status of *Amblyomma auricularium* is unclear, its geographic distribution and hosts are also uncertain, including the information in Guglielmone et al. (2003b), which merely indicates where ticks named *Amblyomma auricularium* had been found as of 2003.

Muñoz-García et al. (2019) listed *Tamandua mexicana* as a host for the larva of *Amblyomma auricularium*, but this could not be confirmed using the references provided by these authors. Therefore, *Tamandua mexicana* is not included here as a host for this stage of *Amblyomma auricularium.* Estrada-Peña et al. (2020) include the larva of *Amblyomma auricularium* in a cluster of ticks feeding on avian hosts, which may be a significant host association but one that is difficult to confirm due to the taxonomic uncertainties outlined here.

6. ***Amblyomma beaurepairei*** Vogelsang and Santos Dias, 1953

Vogelsang, E.G. & Santos Dias, J.A.T. (1953a) Contribución al estudio de la fauna ixodológica en Venezuela. *Revista de Medicina Veterinaria y Parasitología*, 12, 3–62.

Descriptions and Redescriptions: The male and female of *Amblyomma beaurepairei* were described in Vogelsang and Santos Dias (1953a).

There are no redescriptions of the male and female of *Amblyomma beaurepairei.*

The nymph and larva of *Amblyomma beaurepairei* remain unknown.

Geographical Distribution: *Amblyomma beaurepairei* is a northern South American species that has only been found in **Venezuela**: Aragua (Vogelsang and Santos Dias 1953a).

Hosts: The only known host for the male and female of *Amblyomma beaurepairei* is *Dasypus novemcinctus* (Cingulata: Dasypodidae) (Vogelsang and Santos Dias 1953a).

Human Parasitism: No.

Comments: *Amblyomma beaurepairei* is treated as a synonym of *Amblyomma auricularium* by Kolonin (2009), while Camicas et al. (1998) consider this species to be a synonym of *Amblyomma curruca*, but most workers regard *Amblyomma curruca* as a synonym of *Amblyomma auricularium*. However, Guglielmone et al. (2003a) state that *Amblyomma beaurepairei* is tentatively a valid species, closer to *Amblyomma inornatum* than to *Amblyomma auricularium*, because of the absence of the trochanteral spurring that characterizes *Amblyomma auricularium* as well as *Amblyomma parvum*, *Amblyomma pseudoconcolor,* and *Amblyomma pseudoparvum* but not *Amblyomma inornatum*. See also *Amblyomma auricularium* for further discussion of the problems involved in identifying this species and related taxa.

7. *Amblyomma boeroi* Nava, Mangold, Mastropaolo, Venzal, Oscherov and Guglielmone, 2009

Nava, S., Mangold, A.J., Mastropaolo, M., Venzal, J.M., Oscherov, E.B. & Guglielmone, A.A. (2009b) *Amblyomma boeroi* n. sp. (Acari: Ixodidae), a parasite of the Chacoan peccary *Catagonus wagneri* (Rusconi) (Artiodactyla: Tayassuidae) in Argentina. *Systematic Parasitology*, 73, 161–174.

Descriptions and Redescriptions: The male, female, nymph, and larva of *Amblyomma boeroi* were described in Nava et al. (2009b).

The male, female, and nymph were redescribed in Nava et al. (2017), while the nymph alone was redescribed by Martins et al. (2014c, key and figures).

There are no redescriptions of the larva of *Amblyomma boeroi*.

Geographical Distribution: *Amblyomma boeroi* is a southern South American species that has been found in **Argentina**: Salta (Nava et al. 2009b); and **Paraguay**: Boquerón (Mastropaolo et al. 2012).

Hosts: All parasitic stages of *Amblyomma boeroi* have been collected only from *Catagonus wagneri* (Artiodactyla: Tayassuidae).

Human Parasitism: No.

Comments: Beati and Klompen (2019) claim that *Amblyomma boeroi* as well as *Amblyomma neumanni* and *Amblyomma parvitarsum* form the basal lineage of South American *Amblyomma* that evolved more than 60 million years ago. However, *Amblyomma boeroi* is a parasite of Tayassuidae, one of the many families of mammals of northern origin that invaded South America during the Great American Interchange. This event occurred about 3 million years ago, when a land bridge connected Central and South America, favoring the mammal invasion of South America and the radiation of *Amblyomma* on these new hosts. Consequently, *Amblyomma boeroi* is treated here as a rather recent species in the context of geological time, with little likelihood that it was part of a basal South American *Amblyomma* lineage.

8. *Amblyomma boulengeri* Hirst and Hirst, 1910

Hirst, S. & Hirst, L.F. (1910) Description of five new species of ticks (Ixodidae). *Annals and Magazine of Natural History, Series 8*, 6, 299–308.

Descriptions and Redescriptions: Hirst and Hirst (1910) described and figured the male and female of *Amblyomma boulengeri.*

Robinson (1926) redescribed the male and female of *Amblyomma boulengeri* but provided no figures.

The nymph and larva of *Amblyomma boulengeri* remain unknown.

Geographical Distribution: *Amblyomma boulengeri* has been found on two islands belonging to **Ecuador**: Galápagos (Hirst and Hirst 1910).

Hosts: *Bona fide* specimens of males and females of *Amblyomma boulengeri* have been found on *Microlophus delanonis* (Squamata: Tropiduridae). The host profile of *Amblyomma boulengeri* is shown in Table 2.6, which is based on data from Hirst and Hirst (1910).

Human Parasitism: No.

Comments: Santos Dias (1958a) incorrectly considered *Amblyomma boulengeri* a synonym of *Amblyomma pilosum*, but Keirans et al. (1973) examined the types of *Amblyomma boulengeri,* finding that this is a valid species.

Camicas et al. (1998) list Iguanidae as the only hosts of *Amblyomma boulengeri*.

9. *Amblyomma brasiliense* Aragão, 1908

Aragão, H. B. (1908a) Algunas novas especies de carrapatos brazileiros. *Brazil Medico*, 22, 111–115.

Descriptions and Redescriptions: Aragão, H. B. (1908a) described but did not figure the male, female, and nymph of *Amblyomma brasiliense*, while the larva was described by Sanches et al. (2009).

The male, female, and nymph were redescribed in Aragão (1911) and Nava et al. (2017). The male and female were redescribed in Robinson (1926), Boero (1957), Aragão and Fonseca (1961a), Voltzit (2007), Dantas-Torres et al. (2019b) and Rodríguez et al. (2019). The nymph of *Amblyomma brasiliense* was redescribed in Sanches et al. (2009) and Martins et al. (2010).

There are no redescriptions of the larva of *Amblyomma brasiliense.*

Table 2.6 Hosts for males (M) and females (F) of *Amblyomma boulengeri*

REPTILIA			
SQUAMATA: Iguanidae		**Tropiduridae**	
*Conolophus subcristatus**	F	*Microlophus barringtonensis**	F
		Microlophus delanonis	MF

* Ticks from these hosts are treated as close to *Amblyomma boulengeri* by Hirst & Hirst (1910)

Geographical Distribution: *Amblyomma brasiliense* is a South American species that has been found in **Argentina**: Misiones (Welschen et al. 2012); **Bolivia**: Beni (Rodríguez et al. 2019); **Brazil**: Amapá, Amazonas, Espírito Santo, Minas Gerais, Pará, Paraná, Rio de Janeiro, Santa Catarina, and São Paulo (Bequaert 1926; Aragão 1936; Keirans 1985b; Labruna et al. 2005a); and **Paraguay**: San Pedro (Keirans 1985b).

Hosts: Males, females, nymphs, and larvae of *Amblyomma brasiliense* have mostly been found on a variety of mammals, with no clear preference for any of them; there are also some records of immature stages collected from several birds. The host profile of *Amblyomma brasiliense* is shown in Table 2.7, which is based on data from Aragão (1908a, 1918), Szabó et al. (2013), Lamattina et al. (2014, 2018a), Martins et al. (2015a, 2017), Acosta et al. (2016), Luz et al. (2017b, 2018a), Nava et al. (2017) and Kmetiuk et al. (2019) and Teixeira et al. (2020).

Human Parasitism: Guglielmone and Robbins (2018) list *Amblyomma brasiliense* as a sporadic parasite of humans. All parasitic stages cause human parasitism, but the nymph is the most common stage found on people. This tick has been collected from humans in Argentina (Misiones), Brazil (Espírito Santo, Minas Gerais, Paraná, Rio de Janeiro, and São Paulo), and Paraguay (San Pedro). One female and 11 nymphs,

Table 2.7 Hosts for adults of undetermined sex (A), males (M), females (F), nymphs (N), and larvae (L) of *Amblyomma brasiliense*

MAMMALIA			
ARTIODACTYLA: Cervidae		**CARNIVORA: Procyonidae**	
Mazama americana	NL	*Nasua nasua*	ANL
Mazama gouazoubira	N	**DIDELPHIMORPHIA: Didelphidae**	
ARTIODACTYLA: Suidae		*Didelphis aurita*	NL
Feral pigs	MFN	**PERISSODACTYLA: Tapiridae**	
ARTIODACTYLA: Tayassuidae		*Tapirus terrestris*	MFN
Pecari tajacu	MF	**PILOSA: Myrmecophagidae**	
Tayassu pecari	MFNL	*Myrmecophaga tridactyla*	MFN
CARNIVORA: Canidae		*Tamandua tetradactyla*	N
Cerdocyon thous	N	**RODENTIA: Caviidae**	
Chrysocyon brachyurus	N	*Hydrochoerus hydrochaeris*	FN
Domestic dog	FNL	**RODENTIA: Cuniculidae**	
Urocyon cinereoargenteus	N	*Cuniculus paca*	N
CARNIVORA: Felidae		**RODENTIA: Dasyproctidae**	
*Panthera onca**	N	*Dasyprocta azarae*	NL
*Puma concolor**	N		
AVES			
ACCIPITRIFORMES: Accipitridae		**PASSERIFORMES: Passerellidae**	
Spizateus melanoleucus	N	*Zonotrichia capensis*	L
CORACIIFORMES: Momotidae		**PASSERIFORMES: Pipridae**	
Baryphthengus ruficapillus	N	*Chiroxiphia caudata*	N
GALLIFORMES: Cracidae		**PASSERIFORMES: Thamnophilidae**	
Penelope obscura	N	*Pyriglena leucoptera*	N
Penelope superciliaris	N	**PASSERIFORMES: Thraupidae**	
GRUIFORMES: Rallidae		*Tachyphonus coronatus*	N
Aramides saracura	N	*Trichothraupis melanops*	L

* Lamattina *et al.* (2014) state that these nymphs are morphologically indistinguishable from nymphs of *Amblyomma incisum*

and three larvae were recently collected in the Brazilian state of Paraná by Santos et al. (2020) and Valente et al. (2020).

Comments: Camicas et al. (1998) do not recognize the description and redescription of the nymph of *Amblyomma brasiliense* in Aragão (1908a) and Aragão (1911), respectively.

Amblyomma superbrasiliense is treated as a synonym of *Amblyomma brasiliense* in Camicas et al. (1998), but Schulze (1941b) described *Amblyomma superbrasiliense* with one spur on coxae II–IV while *Amblyomma brasiliense* is characterized by two spurs on these coxae, and Guglielmone et al. (2003a) and Guglielmone and Nava (2014) treat *Amblyomma superbrasiliense* as a synonym of *Amblyomma incisum*, a position also accepted here. Lamattina et al. (2014) describe the difficulties involved in morphologically separating the nymph of *Amblyomma brasiliense* from *Amblyomma incisum*, underscoring the importance of molecular taxonomy in resolving such problems.Unclear evidence of the occurrence of *Amblyomma brasiliense* in Bolivia was provided by Squire (1972), but its presence in that country was confirmed by Rodríguez et al. (2019). Early records of *Amblyomma brasiliense* in northwestern Argentina's Salta Province (Dios and Knopoff 1930, 1934), repeated in several studies thereafter, are based on doubtful specimens, as detailed in Welschen et al. (2012). The presence of *Amblyomma brasiliense* in the Brazilian state of Rio Grande do Sul, based on unknown tick stages collected on *Dasypus septemcinctus* by Correa (1955), requires confirmation, but Andreotti et al. (2018) listed Rio Grande do Sul as within the geographical distribution of this tick, although they provided no new records of *Amblyomma brasiliense* for this Brazilian state. Therefore, Rio Grande do Sul is provisionally excluded from the range of *Amblyomma brasiliense*.

Estrada-Peña et al. (2020) include adults of *Amblyomma brasiliense* in a cluster of species feeding mostly on Artiodactyla, while the nymph is placed in a cluster feeding on carnivorous hosts. However, the range of the principal hosts of this tick appears to be broader than these authors' proposed host profiles.

10. *Amblyomma cajennense* (Fabricius, 1787) under the name *Acarus cajennensis* and given its current status in Koch (1844)

Fabricius, J.C. (1787) *Mantisa insectorum sistens species nuper detectas adiectis synonymis, observationibus, descriptionibus, emendationibus*. Hafniae, 2, 382 pp.

Koch, C.L. (1844) Systematische Übersicht über die Ordnung der Zecken. *Archiv für Naturgeschichte*, 10, 217-239.

Descriptions and Redescriptions: Fabricius (1787) described but did not figure the male under the name *Acarus cajennensis*, while Koch (1844) described the female, also without illustrations, and the nymph of *Amblyomma cajennense* was described in Martins et al. (2016b).

The male and female of *Amblyomma cajennense* were redescribed by Tonelli Rondelli (1937), Floch and Abonnenc (1940, under the name *Amblyomma cayennense*, a *lapsus*), Floch and Fauran (1958, under the name *Amblyomma cayennense*, a *lapsus*), and Nava et al. (2014a).

There are no redescriptions of the nymph of *Amblyomma cajennense.*

The larva of *Amblyomma cajennense* remains undescribed, but Soares et al. (2015) have been able to recognize this stage from field-collected material.

Geographical Distribution: A Neotropical species established on the northern South American mainland. *Amblyomma cajennense* has been found in **Brazil**: Amazonas, Goiás, Maranhão, Mato Grosso, Pará, Rondônia, Roraima, and Tocantins (Nava et al. 2014a; Martins et al. 2016b; Rosas et al. 2016); **French Guiana**: Cayena, and Saint Laurent du Maroni (Binetruy et al. 2019); **Guyana**: Potaro-Supurani, Upper Demerara-Berbice, and Upper Takutu-Upper Essequibo (Nava et al. 2014a), **Suriname**: Brokopondo, and Paramaribo (Santos Dias 1989a; Nava et al. 2014a); and **Venezuela**: Bolívar (Nava et al. 2014a).

Hosts: Most males, females, and nymphs of *Amblyomma cajennense* have been found on several domestic and wild mammals, and adults have rarely been collected on Aves and Testudines. Nymphs have been collected from several orders of mammals and a few birds (Galliformes and Passeriformes). Larvae have been taken from one mammal and a few species of birds by Soares et al. (2015) and Binetruy et al. (2020b), but additional information is needed to determine the principal hosts for the larva of this tick. The host profile of *Amblyomma cajennense* is shown in Table 2.8, which is based on data from Nava et al. (2014a), Minervino et al. (2015), Soares et al. (2015), Martins et al. (2016b), Witter et al. (2016), Binetruy et al. (2019, 2020b), Colle et al. (2020), Costa et al. (2020) and Teixeira et al. (2020).

Human Parasitism: Guglielmone and Robbins (2018) treat *Amblyomma cajennense* as a frequent parasite of humans. All parasitic stages (records of nymphs and larvae are provisional) have been found feeding on people, with records from Brazil (Pará, and Maranhão), French Guiana (Cayena), Guyana (Potaro-Siparuni, Upper Demerara-Berbice, and Upper Takutu-Upper Essequibo), and Suriname (Brokopondo, and Paramaribo).

Comments: *Amblyomma cajennense* has historically been regarded as ranging from northern Argentina in the Neotropics to the southern USA in the Nearctic Zoogeographic Region. However, Tonelli Rondelli (1937), in her redescriptions of this species and redescriptions of *Amblyomma mixtum* and *Amblyomma sculptum*, clearly demonstrated that several species exist under the name *Amblyomma cajennense*, although her view was not accepted by Vogelsang and Santos Dias (1953a) and Aragão and Fonseca (1953a), among others. Guglielmone et al. (1992b) analyzed the life cycle of *Amblyomma cajennense* from different American countries and proposed that the name *cajennense* may represent a species complex, but analyses of cuticular hydrocarbons from several populations failed to support this view (Estrada-Peña et al. 1993a). Years later, it was demonstrated that the name

Table 2.8 Hosts for males (M), females (F), nymphs (N), and larvae* (L) of *Amblyomma cajennense*

MAMMALIA			
ARTIODACTYLA: Bovidae		**DIDELPHIMORPHIA: Didelphidae**	
Cattle	MF	*Caluromys philander*	N
Syncerus caffer	MF	*Didelphis albiventris*	N
ARTIODACTYLA: Cervidae		*Didelphis imperfecta*	N
Mazama gouazoubira	N	*Didelphis marsupialis*	NL
ARTIODACTYLA: Suidae		*Metachirus nudicaudatus*	N
Domestic pig	MF	**PERISSODACTYLA: Equidae**	
ARTIODACTYLA: Tayassuidae		Donkey	MF
Pecari tajacu	MFN	Horse	MFN
Tayassu pecari	MFN	**PERISSODACTYLA: Tapiridae**	
CARNIVORA: Canidae		*Tapirus terrestris*	MFN
Cerdocyon thous	N	**PILOSA: Bradypodidae**	
Domestic dog	MFN	*Bradypus tridactylus*	MFN
CARNIVORA: Felidae		**PILOSA: Myrmecophagidae**	
Panthera onca	FN	*Myrmecophaga tridactyla*	MF
CARNIVORA: Mustelidae		*Tamandua tetradactyla*	F
Pteronura brasiliensis	N	**RODENTIA: Caviidae**	
CARNIVORA: Procyonidae		*Hydrochoerus hydrochaeris*	MF
Nasua nasua	NL	**RODENTIA: Cuniculidae**	
CHIROPTERA: Phyllostomidae		*Cuniculus paca*	MF
Choeroniscus minor	F	**RODENTIA: Dasyproctidae**	
CINGULATA: Chlamyphoridae		*Dasyprocta* sp.	F
Cabassous unicinctus	M	*Dasyprocta azarae*	N
CINGULATA: Dasypodidae			
Dasypus novemcinctus	MFN		
AVES			
ACCIPITRIFORMES: Accipitridae		**PASSERIFORMES: Pipridae**	
Harpia harpyja	N	*Chiroxiphia pareola*	N
GALLIFORMES: Cracidae		**PASSERIFORMES: Turdidae**	
Crax fasciolata	N	*Turdus leucomelas*	L
Mitu tuberosum	NL	**STRIGIFORMES: Strigidae**	
Penelope superciliaris	NL	*Asio clamator*	A
REPTILIA			
TESTUDINES: Testudinidae			
Chelonoidis denticulata	MF		

* Provisional diagnosis because the larva of *Amblyomma cajennense* remains undescribed

Amblyomma cajennense comprises several species (Nava et al. 2014a), confirming the early view of Tonelli Rondelli (1937). *Amblyomma cajennense* is the type species of a complex consisting of *Amblyomma interandinum*, *Amblyomma mixtum*, *Amblyomma patinoi*, *Amblyomma sculptum,* and *Amblyomma tonelliae*, as discussed in Nava et al. (2014a).

According to Nava et al. (2014a) and Dantas-Torres et al. (2019a, b), the males of *Amblyomma cajennense* and *Amblyomma sculptum* are indistinguishable, although the females of these species can be separated by morphological differences in the external morphology of the genital aperture. The difficulties involved in identifying species belonging to the *Amblyomma cajennense* complex were recognized by Nava et al. (2014a), who stated that sometimes it might be necessary to invoke morphological, distributional, and molecular data when working with this group.

Andreotti et al. (2018) reduced the geographical distribution of *Amblyomma cajennense* in Brazil, removing the states of Amazonas, Goiás, and Roraima, but no reasons were provided for this decision.

Vogelsang and Cordero (1939) allegedly found *Amblyomma cajennense* on *Hydrochoerus hydrochaeris* in Uruguay, but Guglielmone et al. (2003a) suggested that their ticks were, in fact, *Amblyomma dubitatum*. Recent records of *Amblyomma cajennense* from *Rhinella marina* (Anura: Bufonidae) in Trinidad and Tobago by Kamilar et al. (2020) are treated as diagnostic errors. A record of *Amblyomma cajennense* on the tortoise *Chelonoidis denticulata* is based on supplementary information in Martins et al. (2016b).

Estrada-Peña et al. (2020) include adults of *Amblyomma cajennense* in a cluster of species feeding mostly on Artiodactyla, while the nymph and larva are placed in a cluster of Neotropical ticks chiefly parasitizing carnivorous mammals. But while there are several records of adults of this tick collected from Artiodactyla, males and females of *Amblyomma cajennense* appear to have a much wider range of principal hosts than artiodactyls. The role of Carnivora as hosts of the immature stages of *Amblyomma cajennense* is regarded here as unconfirmed.

Because *Amblyomma cajennense sensu lato* now includes six defined species, the literature on these ticks will need to be revisited, and collections around the world will have to be reevaluated both taxonomically and with regard to the hosts and distribution of *Amblyomma cajennense* and related species.

11. *Amblyomma calcaratum* Neumann, 1899

Neumann, L.G. (1899) Révision de la famille des ixodidés (3e mémoire). *Mémoires de la Société Zoologique de France*, 12, 107–294.

Descriptions and Redescriptions: The male and female of *Amblyomma calcaratum* were described but not figured by Neumann (1899); the nymph was described by Martins et al. (2010), and the larva by Barbieri et al. (2013).

The male, female, and nymph were redescribed in Nava et al. (2017); the male and female of *Amblyomma calcaratum* were redescribed in Robinson (1926), Floch and Abonnenc (1941), Floch and Fauran (1958), Aragão and Fonseca (1961a), Boero and Delpietro (1971), Onofrio et al. (2006b, keys and figures), Voltzit (2007), Guzmán-Cornejo et al. (2011, keys and figures), Bermúdez et al. (2018a) and Dantas-Torres et al. (2019b, keys and figures); the nymph alone was redescribed by Martins et al. (2014c, key and figures).

There are no redescriptions of the larva of *Amblyomma calcaratum*.

Geographic Distribution: *Amblyomma calcaratum* has a wide Neotropical range encompassing lands from southern Mexico to Argentina as follows: **Argentina**: Chaco, Formosa, and Misiones (Guglielmone and Nava 2006), **Belize**: Orange Walk (Fairchild et al. 1966); **Bolivia**: Santa Cruz (Fonseca 1959); **Brazil**: Acre, Amazonas, Bahia, Maranhão, Mato Grosso, Mato Grosso do Sul, Minas Gerais, Pará, Paraná, Rio de Janeiro, Rio Grande do Sul, Rondônia, Santa Catarina, and São

Paulo (Guimarães et al. 2001; Labruna et al. 2005d; Ogrzewalska et al. 2011, 2013; Costa et al. 2020); **Colombia**: Boyacá, Caldas, Chocó, and Meta (Reyes 1938; Fairchild et al. 1966; Acevedo-Gutiérrez et al. 2020; Martínez-Sánchez et al. 2020); **Costa Rica**: Cartago, Guanacaste, Limón, and San José (Álvarez et al. 2005; Ogrzewalska et al. 2015); **Ecuador**: Morona Santiago (Tonelli Rondelli 1931); **French Guiana**: Cayena, and Saint Laurent du Maroni (Floch and Fauran 1958; Binetruy et al. 2020a); **southern Mexico**: Veracruz (Guzmán-Cornejo et al. 2011); **Panama**: Bocas del Toro, Cocle, Colón, Darién, Panamá, and Panamá Oeste (Fairchild et al. 1966; Miller et al. 2016; Gittens et al. 2020); **Paraguay**: Itapuá (Ogrzewalska et al. 2014); **Peru**: Amazonas, and Iquitos (Cáceres et al. 2002; Ogrzewalska et al. 2012b); **Suriname** (Keirans 1985b); **Trinidad and Tobago**: Couva-Tabaquite-Tabaro, and Río Claro-Mayaro (Aitken et al. 1968; Fairchild et al. 1966); and **Venezuela**: Amazonas, Apure, Distrito Federal, Guárico, and Zulia (Díaz Ungría 1957; Jones et al. 1972).

Hosts: Males and females of *Amblyomma calcaratum* are usually found on Pilosa: Myrmecophagidae, but the majority of specimens have been collected from *Tamandua tetradactyla*, a South American mammal, although the range of this tick is wider than the geographic distribution of the preferred host. This is a consequence of the fact that most records of *Amblyomma calcaratum* are from South America. It is possible that *Tamandua mexicana* plays an important role as a host of *Amblyomma calcaratum* outside South America. Larvae and nymphs have been collected from six orders of Aves, but several families of Passeriformes are common hosts for nymphs and larvae, although the species parasitized by nymphs are more numerous than those hosting larvae. The host list of *Amblyomma calcaratum* includes numerous species of vertebrates, especially birds, as shown in Table 2.9, which is based on data from Neumann (1899), Fairchild et al. (1966), Jones et al. (1972), Keirans (1982, 1985b), Arzua et al. (2005), Bermúdez et al. (2010a, 2018a), Luz et al. (2016b, 2017a, b), Nava et al. (2017), Zeringóta et al. (2017), Domínguez et al. (2019), Binetruy et al. (2020a), Martínez-Sánchez et al. (2020) and Teixeira et al. (2020).

Human Parasitism: Guglielmone and Robbins (2018) list *Amblyomma calcaratum* as a very rare parasite of people, with one case of a presumably adult tick causing human parasitism in Trinidad and Tobago (Smith 1974). A recent case was added by Valente et al. (2020), who found a nymph of *Amblyomma calcaratum* on a human in the Brazilian state of Paraná.

Comments: The scutum of the female of *Amblyomma calcaratum* is inornate in the redescription of Voltzit (2007), but other authors state that it is ornate, with variations in ornamentation patterns in both male and female ticks. Jones et al. (1972) and Voltzit (2007) state that the scutum of the female of *Amblyomma calcaratum* is wider than long, while authors such as Robinson (1926), Boero and Delpietro (1971), Nava et al. (2017), and Bermúdez et al. (2018a) found the female scutum longer than wide, raising uncertainties about the identification of *Amblyomma calcaratum*.

Table 2.9 Hosts for adults of undetermined sex (A), males (M), females (F), nymphs (N), and larvae (L) of *Amblyomma calcaratum**

MAMMALIA			
ARTIODACTYLA: Cervidae		**PERISSODACTYLA: Tapiridae**	
Mazama temama	M	*Tapirus terrestris*	FNL
CARNIVORA: Canidae		**PILOSA: Choloepodidae**	
Canidae		*Choloepus hoffmanni*	F
Domestic dog	F	**PILOSA: Cyclopedidae**	
CARNIVORA: Procyonidae		*Cyclopes didactylus*	MF
Procyon cancrivorus	A	**PILOSA: Myrmecophagidae**	
CHIROPTERA: Phyllostomidae		*Myrmecophaga tridactyla*	MFN
Macrophyllum macrophyllum	M	*Tamandua mexicana*	MFN
Platyrrhinus umbratus	M	*Tamandua tetradactyla*	MFN
AVES			
ACCIPITRIFORMES: Accipitridae		*Thamnophilus caerulescens*	NL
Rupornis magnirostris	N	*Thamnophilus pelzelni*	N
CAPRIMULGIFORMES: Trochilidae		*Willisornis poecilinotus*	N
Phaethornis eurynome	N	**PASSERIFORMES: Thraupidae**	
Phaethornis superciliosus	NL	*Coryphospingus cucullatus*	N
CORACIIFORMES: Momotidae		*Poospiza nigrorufa*	N
Baryphthengus martii	N	*Ramphocelus carbo*	N
Baryphthengus ruficapillus	N	*Saltator similis*	N
GALBULIFORMES: Bucconidae		*Sporophila frontalis*	N
Malacoptila panamensis	N	*Sporophila funerea*	N
PASSERIFORMES: Cardinalidae		*Tachyphonus coronatus*	N
Cyanoloxia cyanoides	N	*Tangara cayana*	N
PASSERIFORMES: Conopophagidae		*Thlypopsis pyrrhocoma*	N
Conopophaga lineata	NL	*Thraupis palmarum*	N
Conopophaga melanops	NL	*Thraupis sayaca*	N
PASSERIFORMES: Furnariidae		*Tiaris fuliginosus*	N
Anabacerthia lichtensteini	N	*Trichothraupis melanops*	NL
Anabazenops fuscus	NL	**PASSERIFORMES: Tityridae**	
Automolus leucophthalmus	N	*Schiffornis virescens*	N
Dendrocolaptes platyrostris	N	**PASSERIFORMES: Troglodytidae**	
Furnarius rufus	N	*Cantorchilus modestus*	L
Glyphorynchus spirurus	N	*Microcerculus marginatus*	N
Hylexetastes perrotii	NL	*Troglodytes aedon*	L
Syndactyla rufosuperciliata	N	**PASSERIFORMES: Turdidae**	
Xiphocolaptes albicollis	N	*Catharus ustulatus*	L
Xiphorhynchus ocellatus	N	*Hylocichla mustelina*	N
PASSERIFORMES: Mimidae		*Turdus albicollis*	N
Mimus saturninus	N	*Turdus amaurochalinus*	N
PASSERIFORMES: Parulidae		*Turdus grayi*	N
Basileuterus culicivorus	N	*Turdus leucomelas*	N
Geothlypis aequinoctialis	N	*Turdus rufiventris*	N
PASSERIFORMES: Passerellidae		*Turdus subalaris*	N
Arremon flavirostris	N	**PASSERIFORMES: Tyrannidae**	
Arremon taciturnus	N	*Attila rufus*	L
PASSERIFORMES: Pipridae		*Cnemotriccus fuscatus*	N
Ceratopipra mentalis	NL	*Hemitriccus margaritaceiventer*	N
Chiroxiphia caudata	N	*Lathrotriccus euleri*	N
Dixiphia pipra	N	*Leptopogon amaurocephalus*	N
Lepidothrix coronata	NL	*Mionectes oleaginous*	N
Manacus manacus	NL	*Mionectes rufiventris*	N
Pipra fasciicauda	N	*Myiarchus ferox*	N
PASSERIFORMES: Thamnophilidae		*Myiarchus tyrannulus*	N
Cymbilaimus lineatus	N	*Platyrinchus mystaceus*	NL
Drymophila malura	N	*Poecilotriccus latirostris*	N
Dysithamnus mentalis	N	*Poecilotriccus plumbeiceps*	NL
Gymnopithys leucaspis	N	*Ramphotrigon megacephalum*	N
Hylophylax naevius	N	**Vireonidae**	
Mackenziaena severa	N	*Cyclarhis gujanensis*	N

(continued)

Table 2.9 (continued)

Pyriglena leucoptera	N	*Hylophilus poicilotis*	N
Terenura maculata	N	**STRIGIFORMES**	
Thamnophilus aethiops	N	**Strigidae**	
Thamnophilus atrinucha	N	*Pulsatrix perspicillata*	N

* The identification of nymphs and larvae, and, to a lesser extent, adults of *Amblyomma calcaratum*, should be considered provisional because of the risk of confusing these stages with those of *Amblyomma nodosum*

The diagnosis of *Amblyomma calcaratum* should be considered provisional. Fairchild et al. (1966) warn about confusing adults of this species with those of *Amblyomma nodosum*, another tick usually collected from pilosan hosts; these authors also infer that separating the immature stages of the two species is even more problematic. See *Amblyomma nodosum* for further discussion concerning the identification of that tick and *Amblyomma calcaratum.*

The Brazilian states of Maranhâo, Minas Gerais, and Rio de Janeiro were excluded from the geographical distribution of *Amblyomma calcaratum* in Andreotti et al. (2018), but no reasons were given for this exclusion.

Recent records of *Amblyomma calcaratum* from *Rhinella marina* (Anura: Bufonidae) in Trinidad and Tobago by Kamilar et al. (2020) are here treated as diagnostic errors.

12. *Amblyomma coelebs* Neumann, 1899

Neumann, L.G. (1899) Révision de la famille des ixodidés (3e mémoire). *Mémoires de la Société Zoologique de France*, 12, 107–294.

Descriptions and Redescriptions: Neumann (1899) described the male of *Amblyomma coelebs* but provided no figures; the female was described in Neumann (1906), again without figures. The nymph was described in Martins et al. (2010).

The male, female and nymph were redescribed in Nava et al. (2017); the male and female of *Amblyomma coelebs* were redescribed by Robinson (1926), Floch and Abonnenc (1941), Boero and Prosen (1955), Floch and Fauran (1958), Onofrio et al. (2006b, keys and figures), Voltzit (2007), Guzmán-Cornejo et al. (2011, keys and figures), Bermúdez et al. (2018a) and Dantas-Torres et al. (2019b, key and figures). The nymph alone was redescribed in Martins et al. (2014c, key and figures).

The larva of *Amblyomma coelebs* remains undescribed, but several authors have been able to identify this stage from field-collected material.

Geographical Distribution: *Amblyomma coelebs* has a wide Neotropical range from southern North America to southern South America, with a few records from the Nearctic portion of Mexico (Guanuajato) (Guzmán-Cornejo et al. 2011). It has been found in **Argentina**: Misiones (Lamattina et al. 2014); **Belize**: Cayo, and Stann Creek (Lopes et al. 2016); **Bolivia**: Cochabamba, and Pando (Mastropaolo et al. 2014); **Brazil**: Acre, Amazonas, Espírito Santo, Goiás, Mato Grosso, Mato Grosso do Sul, Minas Gerais, Pará, Paraná, Rondônia, Roraima, and São Paulo (Arzua et al. 2005; Labruna et al. 2005a; Guglielmone et al. 2006a; Martins et al. 2011; Luz et al. 2015; Acosta et al. 2016; Gianizella et al. 2018b); **Colombia**: Antioquía, and Meta (Acevedo-Gutiérrez et al. 2020); **Costa Rica**: Guanacaste, Limón, and Puntarenas (Hernández-Divers et al. 2005; Ogrzewalska et al. 2015); **French Guiana**: Cayena, and Saint Laurent du Maroni (Binetruy et al. 2019); **Guatemala** (Monroy Lefebre and Cajas González 1988); **Guyana**: Essequibo Islands-West Demerara, and Upper Demerara-Berbice (Tonelli Rondelli 1939); **Honduras**: Colón (Gorham 1995); **southern Mexico**: Chiapas, and Jalisco (Guzmán-Cornejo et al. 2011); **Nicaragua** (Fairchild et al. 1966); **Panama**: Chiriquí, Bocas del Toro, Darién, Guna Yala, and Panamá (Dunn 1934; Fairchild et al. 1966); **Paraguay**: Guairá and San Pedro (Nava et al. 2007); **Peru**: Cusco, Loreto, and Madre de Dios (Need et al. 1991; Mendoza-Uribe and Chávez-Chorocco 2004; Labruna et al. 2010b); **Suriname** (Keirans 1985b); and **Venezuela**: Amazona, and Bolívar (Jones et al. 1972).

Hosts: Males and females of *Amblyomma coelebs* have been found on several mammals, but most records are from Perissodactyla: Tapiridae, with the South American *Tapirus terrestris* as the principal host for adult ticks. It should be noted that the geographic distribution of *Amblyomma coelebs* is wider than the range of this tapir, a consequence of the fact that most records of *Amblyomma coelebs* are from South America. It is possible that *Tapirus bairdii* plays an important role as a host for adults of *Amblyomma coelebs* outside South America. Larvae and nymphs have been collected from several orders of mammals but also from various birds belonging to five orders; however, *Nasua nasua* (Carnivora: Procyonidae) and *Didelphis albiventris*, *Didelphis aurita,* and *Didelphis marsupialis* (Didelphimorphia: Didelphidae) appear to be more important hosts for immature stages of this tick than other vertebrates (Lamattina et al. 2018a, b; Colle et al. 2020) and are treated here as provisional principal hosts of this tick. The host profile of *Amblyomma coelebs* is shown in Table 2.10, which is based on data from Floch and Fauran (1958), Fairchild et al. (1966), Keirans (1985b); Labruna et al. (2005d), Ogrzewalska et al. (2009), Martins et al. (2014a), Lopes et al. (2016), Witter et al. (2016), Cançado et al. (2017), Nava et al. (2017), Lamattina et al. (2018a, c), Colle et al. (2020) and Teixeira et al. (2020).

Human Parasitism: Guglielmone and Robbins (2018) list *Amblyomma coelebs* as a frequent parasite of humans in Argentina (Misiones), Belize (Cayo), Brazil (Amazonas, Mato Grosso do Sul, Rondônia, Roraima, and São Paulo), Costa Rica, Mexico (Chiapas), and Paraguay (Guairá). Additionally, Binetruy et al. (2019) found larvae, nymphs, and adults on people in French Guiana (Cayena), and Santos et al. (2020) found 18 nymphs feeding on humans in the State of Paraná (Brazil). Molaei et al. (2019) reported a nymph of *Amblyomma coelebs* collected from a USA resident returning from a trip to Costa Rica and Panama. All parasitic stages of *Amblyomma coelebs* have been found on humans (larval diagnoses are provisional), but nymphs appear to prevail as human parasites (Guglielmone and Robbins 2018).

Table 2.10 Hosts for adult ticks of undetermined sex (A), males (M), females (F), nymphs (N), and larvae* (L) of *Amblyomma coelebs*

	MAMMALIA		
ARTIODACTYLA: Cervidae		*Metachirus nudicaudatus*	L
Mazama americana	L	**LAGOMORPHA: Leporidae**	
ARTIODACTYLA: Suidae		*Sylvilagus brasiliensis*	N
Domestic pig	A	**PERISSODACTYLA: Equidae**	
ARTIODACTYLA: Tayassuidae		Horse	F
Pecari tajacu	A	**PERISSODACTYLA: Tapiridae**	
Tayassu pecari	MFN	*Tapirus bairdii*	MF
CARNIVORA: Canidae		*Tapirus terrestris*	MFN
Cerdocyon thous	N	**PILOSA: Myrmecophagidae**	
CARNIVORA: Felidae		*Tamandua tetradactyla*	N
Herpailurus yagouaroundi	N	**PRIMATES: Cebidae**	
Panthera onca	MNL	*Sapajus microcephalus*	F
Puma concolor	ANL	**RODENTIA: Caviidae**	
CARNIVORA: Mustelidae		*Hydrochoerus hydrochaeris*	MFN
Eira barbara	NL	**RODENTIA: Cricetidae**	
CARNIVORA: Procyonidae		*Neacomys spinosus*	N
Nasua narica	N	*Rhipidomys macrurus*	N
Nasua nasua	NL	**RODENTIA: Cuniculidae**	
CINGULATA: Dasypodidae		*Cuniculus paca*	FN
Dasypus novemcinctus	N	**RODENTIA: Dasyproctidae**	
DIDELPHIMORPHIA: Didelphidae		*Dasyprocta* sp.	A
Didelphis albiventris	NL	*Dasyprocta azarae*	NL
Didelphis aurita	NL	**RODENTIA: Echimyidae**	
Didelphis marsupialis	NL	*Thrichomys pachyurus*	N
Lutreolina crassicaudata	N		
	AVES		
ACCIPITRIFORMES: Accipitridae		**PASSERIFORMES: Mitrospingidae**	
Buteo nitidus	N	*Mitrospingus cassinii*	N
Buteogallus lacernulatus	N	**PASSERIFORMES: Pipridae**	
COLUMBIFORMES: Columbidae		*Ceratopipra mentalis*	N
Leptotila verreauxi	L	*Chiroxiphia linearis*	N
CORACIIFORMES: Momotidae		**PASSERIFORMES: Thamnophilidae**	
Baryphthengus ruficapillus	N	*Phlegopsis nigromaculata*	L
Momotus momota	N	*Pyriglena leuconota*	L
GALBULIFORMES: Bucconidae		*Thamnophilus pelzelni*	N
Bucconidae		**PASSERIFORMES: Turdidae**	
Malacoptila panamensis	N	*Turdus albicollis*	N
Malacoptila striata	N	*Turdus amaurochalinus*	N
PASSERIFORMES: Conopophagidae		*Turdus leucomelas*	N
Conopophaga lineata	N	*Turdus rufiventris*	N

* Provisional diagnosis because the larva of *Amblyomma coelebs* remains undescribed

Comments: Camicas et al. (1998) consider *Amblyomma coelebs* an exclusively Neotropical tick, but there are a few records of this species from the Nearctic Region, including the specimens used in the original description by Neumann (1899).

Enríquez et al. (2020) allegedly found a male of *Amblyomma coelebs* on cattle in Ecuador, but their accompanying figures do not depict a *bona fide* specimen, and Ecuador is not included within the range of this tick. Floch and Fauran (1958) state that adults of *Amblyomma auricularium* in Floch and Abonnenc (1940, under the name *Amblyomma auriculare*), are misidentifications of *Amblyomma coelebs*. Records of *Amblyomma coelebs* from northwestern Argentina (Salta Province), including alleged records from humans, are now considered to represent

Amblyomma hadanii, as explained in Nava et al. (2014b). *Amblyomma bispinosum*, a synonym of *Amblyomma coelebs* in Guglielmone and Nava (2014), is treated as a valid species by Keirans and Hillyard (2001).

The Brazilian states of Amazonas, Espírito Santo, Goiás, Minas Gerais, and Pará are not included in the geographical distribution of *Amblyomma coelebs* published by Andreotti et al. (2018), but no reasons were provided for this exclusion.

Nava and Guglielmone (2013) found that adults of *Amblyomma coelebs* are not highly specific in host selection, and Nava et al. (2017) do not support a previous hypothesis that Tapiridae are the usual hosts for adults of *Amblyomma coelebs*. Also, no host specificity has been demonstrated for nymphs and larvae of this tick. Nevertheless, it is provisionally argued here that Tapiridae are the usual hosts for males and females *Amblyomma coelebs* because the great majority of records have been from *Tapirus*. Nymphs and larvae of *Amblyomma coelebs* have a broad host range, but individual vertebrates are generally infested with very low numbers of ticks. Data published by Lamattina et al. (2018a, b) and Colle (2020) appear to indicate that Didelphidae and Procyonidae are more important hosts of immature stages; consequently, mammals belonging to these families are hypothetically considered to be the principal hosts of *Amblyomma coelebs* larvae and nymphs, at least within its South American range. Fairchild et al. (1966) and Guzmán-Cornejo et al. (2011) listed cattle as a host of *Amblyomma coelebs*, but tick stages found were not provided, and this domestic host is not included in Table 2.10. Estrada-Peña et al. (2020) found that larvae of *Amblyomma coelebs* are prone to feed on Aves, but the low numbers of this stage collected from birds do not appear to support that hypothesis.

13. *Amblyomma crassum* Robinson, 1926

Robinson, L.E. (1926) *Ticks. A monograph of the Ixodoidea. Part IV. The genus* Amblyomma. Cambridge University Press, London, 302 pp.

Descriptions and Redescriptions: Robinson (1926) described and figured the female of *Amblyomma crassum.*

The female of *Amblyomma crassum* was redescribed by Méndez Arocha and Ortiz (1957) and Voltzit (2007).

The male, nymph, and larva of *Amblyomma crassum* remain unknown.

Geographic Distribution: *Amblyomma crassum* is a Neotropical species that has been found in southern Central America and northern South America as follows: **Colombia**: Chocó (Fairchild et al. 1966); **Panama**: tentatively in Darién (Robinson 1926); **Peru**: Loreto, and Ucayali (Fairchild et al. 1966; Need et al. 1991); and **Venezuela**: Apure, Bolívar, and Delta Amacuro (Fiasson 1949; Méndez Arocha and Ortiz 1957).

Hosts: Females of *Amblyomma crassum* have only been collected from Testudines, but most of the limited records of this species have been reported from members of

Table 2.11 Hosts for females (F) of *Amblyomma crassum**

REPTILIA			
TESTUDINIDAE: Geoemydidae		**TESTUDINIDAE: Testudinidae**	
Rhinoclemmys annulata	F	*Chelonoidis carbonarius*	F
TESTUDINIDAE: Kinosternidae		*Chelonoidis denticulata*	F
Kinosternon leucostomum	F		

* Identification of *Amblyomma crassum* is difficult; accordingly, the above records are tentative

the family Testudinidae, and the two species of this family (*Chelonoidis carbonarius* and *Chelonoidis denticulata*) are tentatively treated as principal hosts for this tick. The host profile of *Amblyomma crassum* is shown in Table 2.11, which is based on data from Fiasson (1949), Medem (1956), and Garcés-Restrepo et al. (2013).

Human Parasitism: No.

Comments: *Amblyomma crassum*, *Amblyomma humerale,* and *Amblyomma sabanerae* are morphologically similar taxa. *Amblyomma crassum* is difficult to identify, and Santos Dias (1958a) lists this species and *Amblyomma sabanerae* as synonyms of *Amblyomma humerale*, although later Santos Dias (1989b) accepted *Amblyomma sabanerae* as a valid species, while Fairchild et al. (1966) suspect that *Amblyomma crassum* is a synonym of *Amblyomma sabanerae.* Méndez Arocha and Ortiz (1957) depict coxa IV of the female as having two spurs, while Robinson (1926) and Voltzit (2007) figure the same coxa with one spur. Jones et al. (1972) state that one critical morphological characteristic separating the female of *Amblyomma crassum* from females of *Amblyomma humerale* and *Amblyomma humerale* is the large size of the scutum of *Amblyomma crassum* (4 mm wide) compared with related species. Osorno Mesa (1942) included the male of *Amblyomma crassum* in his key for *Amblyomma*, and Méndez Arocha and Ortiz (1957) described the male, but Fairchild et al. (1966) suggested that these authors had been working with the male of *Amblyomma sabanerae.* The male of *Amblyomma crassum* is treated here as unknown, and the Colombian records in Osorno Mesa (1942) based on diagnoses of male ticks are considered doubtful, although those records are treated as valid in Acevedo-Gutiérrez et al. (2020), a view not shared here.

Robinson (1926) states that the holotype of *Amblyomma crassum* was collected in "Darien Country, Colombia," but Fairchild et al. (1966) note that this locality may actually be within Panamanian territory; nevertheless, Bermúdez et al. (2018a) treat as doubtful the presence of this tick in Panama, and this country is tentatively included in the range of *Amblyomma crassum.*

Land tortoises and *Eunectes murinus* (Squamata: Boidae) are listed as hosts of *Amblyomma crassum* in Brazil by Sauter et al. (1999) and Amorim et al. (2013), but morphological differences between this and related species are not discussed, rendering these records doubtful. Additionally, Amorim et al. (2013) erroneously interpret the host and distribution data in Fairchild et al. (1966).

According to Guerrero (1996), *Amblyomma humerale* is established in central and northern South America, *Amblyomma sabanerae* is a Central American tick, and

Amblyomma crassum is found within the ecotone between both species. Nevertheless, Fairchild (1943) believes that extensive studies of *Amblyomma crassum*, *Amblyomma humerale,* and *Amblyomma sabanerae* are needed to determine whether these are different species. In summary, *Amblyomma crassum* and all the above data are treated here as provisionally valid, pending additional studies of related taxa.

14. *Amblyomma cruciferum* Neumann, 1901

Neumann, L.G. (1901) Révision de la famille des ixodidés (4e mémoire). *Mémoires de la Société Zoologique de France*, 14, 249–372.

Descriptions and Redescriptions: Neumann (1901) described the male and nymph of *Amblyomma cruciferum* but provided no figures of these stages; the female was described and illustrated by Bequaert (1932b).

Voltzit (2007) redescribed the male and female of *Amblyomma cruciferum*, and Robinson (1926) redescribed the male.

There are no redescriptions of the nymph of *Amblyomma cruciferum.*

The larva of *Amblyomma cruciferum* remains unknown.

Geographic Distribution: *Amblyomma cruciferum* is a Caribbean species that has been found in **Haiti** (Bequaert 1932b) and **Puerto Rico**: Mayagüez (Bequaert 1932b).

Hosts: Males, females, and nymphs of *Amblyomma cruciferum* have only been found on *Cyclura cornuta* and *Cyclura stejnegeri* (Squamata: Iguanidae) (Bequaert 1932b; Morel 1967), which are considered the principal hosts of this tick.

Human Parasitism: No.

Comments: The scutal ornamentation, shape of the prosoma, porose areas, and spurs of coxae II–IV of the female of *Amblyomma cruciferum* in Voltzit (2007) differ from those shown in the original description by Bequaert (1932b). Neumann (1901) and Robinson (1926) described and redescribed, respectively, the male of *Amblyomma cruciferum* as having one spur on coxa IV, but Voltzit (2007) redescribed a male having two spurs on coxa IV. Morel (1967) lists *Amblyomma torrei* as a synonym of *Amblyomma cruciferum*, while Guglielmone et al. (2003a) suggest that comparison of the types of *Amblyomma cruciferum*, *Amblyomma torrei,* and related species would be useful to better define these species. Barros-Battesti et al. (2009) state that *Amblyomma cruciferum* and *Amblyomma torrei* are morphologically very close but differ markedly in scutal ornamentation. Most records of *Amblyomma cruciferum* should be considered provisionally valid.

The presence of this tick on Puerto Rico's Mona Island, located by Voltzit (2007) in Brazil, is a *lapsus.* Continuing uncertainties surrounding the diagnosis of *Amblyomma cruciferum* indicate that a revision of this species is needed.

15. *Amblyomma darwini* Hirst and Hirst, 1910

Hirst, S. & Hirst, L.F. (1910) Description of five new species of ticks (Ixodidae). *Annals and Magazine of Natural History, Series 8*, 6, 299–308.

Descriptions and Redescriptions: Hirst and Hirst (1910) described and illustrated the male and female of *Amblyomma darwini*; Bequaert (1932b) described the nymph.

Robinson (1926) and Voltzit (2007) redescribed the male and female.

There are no redescriptions of the nymph of *Amblyomma darwini.*

The larva of *Amblyomma darwini* remains undescribed, but some authors have been able to identify this stage from field-collected material.

Geographic Distribution: *Amblyomma darwini* is a Neotropical species that has only been found on several islands belonging to **Ecuador**: Galápagos (Gadsen and Guerra 1991).

Hosts: Males, females, nymphs, and the undescribed larva of *Amblyomma darwini* are usually collected from *Amblyrhynchus cristatus* (Squamata: Iguanidae) (Gadsen and Guerra 1991), while adults have occasionally been found on *Microlophus* sp. (Squamata: Tropiduridae) (Banks 1924) and nymphs on *Microlophus albemarlensis* (Bequaert 1932b).

Human Parasitism: No.

Comments: Camicas et al. (1998) list the nymph of *Amblyomma darwini* as undescribed, but the description of this stage in Bequaert (1932b) is treated as valid in Guglielmone et al. (2003a) and tentatively valid in Guglielmone et al. (2014) and here.

Schulze (1936a) identified ticks collected from birds (Charadriiformes: Scolopacidae) as a subspecies of *Amblyomma darwini*, but this record is treated as doubtful in Guglielmone et al. (2014) and here.

16. *Amblyomma dissimile* Koch, 1844

Koch, C.L. (1844) Systematische Übersicht über die Ordnung der Zecken. *Archiv für Naturgeschichte*, 10, 217–239.

Descriptions and Redescriptions: The male and female of *Amblyomma dissimile* were described but not figured by Koch (1844), the nymph was described accompanied by one figure in Neumann (1899), and the larva was described and illustrated by Hooker et al. (1912).

There are many redescriptions of *Amblyomma dissimile*. The male, female, and nymph were redescribed in Hooker et al. (1912), Cooley and Kohls (1944), and Nava et al. (2017); the male and female were redescribed in Robinson (1926), Boero (1957), Barros-Battesti et al. (2009), Bermúdez et al. (2018a) and Dantas-Torres et al. (2019b, keys and figures), among others. The nymph alone was described in Keirans and Durden (1998) and Martins et al. (2010), while the larva of *Amblyomma dissimile* alone was described in Clifford et al. (1961) and Amorim and Serra-Freire (1994a).

Geographic Distribution: *Amblyomma dissimile* is a Nearctic and Neotropical species, although most of its wide range encompasses Neotropical territories as follows: **Antigua and Barbuda** (Hooker 1909); **Argentina**: Chaco, Corrientes, Formosa, and Santiago del Estero (Guglielmone and Nava 2006); **Bahamas**: Mangrove Cay (Durden and Knapp 2005); **Barbados**: Saint Philip (Kohls 1969c); **Belize**: Cayo, Orange Walk, and Stann Creek (Varma 1973; Rainwater et al. 2000; Ernst and Ernst 1977); **Brazil**: Acre, Amapá, Amazonas, Bahia, Mato Grosso, Mato Grosso do Sul, Pará, Rondônia, Roraima, and São Paulo (Guimarães et al. 2001; Alcántara et al. 2018; Zimmermann et al. 2018; Mendoza-Roldán et al. 2020); **Colombia**: Antioquía, Atlántico, Caldas, Casanare, César, Córdoba, Huila, La Guájira, Magdalena, Meta, Quindío, Sucre, Tolima, and Valle del Cauca (Dunn 1929; Stebbins and Hendrickson 1959; Carrascal et al. 2009; Verbel-Vergara et al. 2015; Benavides-Montaño et al. 2018; Rivera-Páez et al. 2018; Santodomingo et al. 2018; Martínez-Sánchez et al. 2020); **Costa Rica**: Alajuela, Guanacaste, Heredia, Limón, Puntarenas, and San José (Tonn et al. 1963; Álvarez et al. 2005); **Cuba**: Artemisa, Cienfuegos, Isla de la Juventud, Mayabeque, Pinar del Río, and Santiago de Cuba (Černý 1969a, b; Barros-Battesti et al. 2009; Álvarez et al. 2020); **Curaçao** (Santos Dias 1989a); **Dominican Republic** (Santos Dias 1958c); **Ecuador** (Guglielmone et al. 2003a); **El Salvador** (Navarrete Abarca et al. 2014); **French Guiana**: Cayena, and Saint Laurent du Maroni (Binetruy et al. 2019); **Grenada**: Carriacou, and Saint George (Kohls 1969c; Drake et al. 2014); **Guadeloupe**: Basse-Terre, and Grande-Terre (Audebad and Courmes 1962); **Guatemala**: Retahuleu (Stoll 1894); **Guyana** (Robinson 1926); **Haiti** (Jakowska 1972); **Honduras**: Atlántida, and Islas de la Bahía (Bequaert 1932a); **Jamaica**: Clarendon, Hanover, Saint Andrews, and Saint James (Newstead 1909); **southern Mexico**: Campeche, Chiapas, Guerrero; Jalisco, Michoacán, Nayarit, Oaxaca, Quintana Roo, Tabasco, Veracruz, and Yucatán (Keirans 1985b; Guzmán-Cornejo et al. 2011; Charruau et al. 2016); **Nicaragua**: Granada, León, Managua, and Masaya (Zepeda et al. 2006); **Panama**: Chiriquí, Colón, Los Santos, and Panamá (Bermúdez et al. 2010a; Domínguez et al. 2020a; Gittens et al. 2020); **Paraguay**: Central, and Guairá (Nava et al. 2007): **Peru** (Keirans 1985b); **Puerto Rico** (Tate 1941); **Saint Lucia** (Keirans 1985b); **Suriname**: Brokopondo, and Paramaribo (Santos Dias 1989a); **Trinidad and Tobago**: Río Claro-Mayaro, Saint George, Saint Patrick, and Tobago (Turk 1948; Morel 1967; Voltzit 2007); and **Venezuela**: Amazonas, Anzoátegui, Apure, Aragua, Barinas, Carabobo, Cojedes, Distrito Federal, Falcón, Guárico, Lara, Miranda, Monagas, Nueva Esparta, Sucre, Trujillo, Yaracuy, and Zulia (Lampo et al. 1998; Jones et al. 1972; Clavijo et al. 2009).

Hosts: The principal hosts of males, females, nymphs, and larvae of *Amblyomma dissimile* are *Rhinella marina* (Anura: Bufonidae), *Boa constrictor* (Squamata: Boidae), and *Iguana iguana* (Iguanidae), but these stages have also been found on a large number of Neotropical vertebrates, including occasional records from mammals and birds, and the host list for *Amblyomma dissimile* increases substantially when records from the Nearctic Region are included. Moreover, there are many records from captive vertebrates. Some Neotropical hosts of *Amblyomma dissimile* may be more important for the maintenance of this tick than currently realized. The diverse range of hosts is shown in Table 2.12, where captive Neotropical hosts of *Amblyomma dissimile* are excluded. This table is based on data from Aragão (1912b), Dunn (1918), Robinson (1926), Joan (1930), Bequaert (1932a), Pearse (1936), Santos Dias (1958c), Fairchild et al. (1966), Kohls (1969c), Jones et al. (1972), Ernst and Ernst (1977), Keirans (1985b), Guerrero (1996), Rainwater et al. (2000), Brum and Rickes (2003), Burridge and Simmons (2003), Durden and Knapp (2005), Dantas-Torres et al. (2008), Rivas (2008), Bermúdez et al. (2010a, 2018a), Guglielmone and Nava (2010), Lopes et al. (2010), Guerrero and Rodríguez-Acosta (2014), Novakova et al. (2015), Charruau et al. (2016), Esser et al. (2016a, b), Witter et al. (2016), Cançado et al. (2017), Kelehear et al. (2017a), Nascimento et al. (2017), Nava et al. (2017), Benavides-Montaño et al. (2018), Cotes-Perdomo et al. (2018), Gianizella et al. (2018b), Luz et al. (2018b), Rivera-Páez et al. (2018), Santodomingo et al. (2018), Torres et al. (2018), Binetruy et al. (2019), Ogrzewalska et al. (2019), Martínez-Sánchez et al. (2020), Pedroso-Santos et al. (2020), Mendoza-Roldán et al. (2020) and Miranda et al. (2020).

Human Parasitism: Guglielmone and Robbins (2018) state that nymphs and females of *Amblyomma dissimile* have sporadically been found on people in Brazil (Pará), Colombia (Antioquía), Mexico, Panama (Panamá), and Venezuela (Yaracuy).

Comments: Diagnosis of *Amblyomma dissimile* is sometimes difficult because various synonyms of this species are still considered valid by some authors, and this tick has been confused with other species. Santos Dias (1958a, c, 1989a) and Camicas et al. (1998), among others, treat *Amblyomma dissimile* as a synonym of *Amblyomma bibroni*, but this synonymy is rejected in Guglielmone et al. (2003a, 2014), Guglielmone and Nava (2010, 2014) and here. Camicas et al. (1998) also accept as valid *Amblyomma trinitatis* collected in Trinidad and Tobago (Turk 1948), while Schulze (1941d) described *Amblyomma cubanum* and *Aponomma thumbi* from Cuba, and Méndez Arocha (1956) described *Amblyomma margaritae* from Venezuelan specimens. However, all these names are synonyms of *Amblyomma dissimile*, as discussed in Guglielmone and Nava (2010, 2014), who also included several currently unused names described before 1900 as synonyms of this tick. Thus, the record of *Amblyomma dissimile* in Guatemala by Stoll (1894) appears under the name *Ixodes boarum*, a synonym of *Amblyomma dissimile*.

Bodkin (1918) confused *Amblyomma dissimile* with *Amblyomma rotundatum*, stating that the former species is parthenogenetic (Aragão 1935), whereas the opposite is true. Lampo et al. (1997) stress the difficulties involved in differentiating these species and described genetic markers for doing so while also pointing out differences in the spurring of coxae I to IV and in the pattern of scutal punctations that are useful for morphologically separating both species, points later confirmed by Luz et al. (2018b), who emphasized the importance of differences in coxa I and the scutum when attempting to separate females of *Amblyomma dissimile* and

Table 2.12 Hosts for adult ticks of undetermined sex (A), males (M), females (F), nymphs (N), and larvae (L) of *Amblyomma dissimile*

	AMPHIBIA		
ANURA: Bufonidae		*Rhinella jimi*	MF
Peltophryne fustiger	A	*Rhinella major*	FN
Peltophryne peltocephala	MFNL	*Rhinella margaritifera*	AN
Rhinella bergi	MF	*Rhinella marina*	MFNL
Rhinella granulosa	ANL	*Rhinella schneideri*	MFN
	REPTILIA		
CROCODILIA: Alligatoridae		**SQUAMATA: Elapidae**	
Caiman crocodilus	MFN	*Micrurus averyi*	F
Paleosuchus trigonatus	MF	*Micrurus lemniscatus*	MFN
CROCODILIA: Crocodylidae		**SQUAMATA: Gekkonidae**	
Crocodylus acutus	F	*Hemidactylus mabouia*	N
Crocodylus moreletii	F	**SQUAMATA: Iguanidae**	
SQUAMATA: Boidae		*Ctenosaura bakeri*	MFN
Boa constrictor	MFNL	*Ctenosaura pectinata*	M
Chilabothrus strigilatus	MNL	*Ctenosaura similis*	F
Corallus hortulanus	MN	*Cyclura cornuta*	F
Corallus ruschenbergerii	N	*Cyclura cychlura*	MFNL
Epicrates cenchria	F	*Iguana iguana*	MFNL
Epicrates maurus	MFN	*Iguana rhinolopha*	MF
Eunectes murinus	MFN	*Iguana tuberculata*	M
Eunectes notaeus	N	**SQUAMATA: Teiidae**	
SQUAMATA: Colubridae		*Ameiva ameiva*	FN
Chironius carinatus	A	*Ameiva bifrontata*	N
Chironius laevicollis	MFN	*Ameiva praesignis*	N
Chironius laurenti	MF	*Cnemidophorus gaigei*	N
Chironius scurrulus	MF	*Tupinambis teguixin*	MFNL
Drymarchon caudomaculatus	N	**SQUAMATA: Tropiduridae**	
Mastigodryas bifossatus	M	*Leiocephalus carinatus*	N
Mastigodryas boddaerti	FN	*Tropidurus hispidus*	N
Oxybelis aeneus	MNL	**SQUAMATA: Viperidae**	
Phrynonax poecilonotus	MF	*Bothrops asper*	MFNL
Spilotes pullatus	MFN	*Bothrops atrox*	MFNL
SQUAMATA: Corytophanidae		*Bothrops matogrossensis*	AL
Basiliscus basiliscus	N	*Crotalus durissus*	MFN
SQUAMATA: Dactyloidae		*Lachesis muta*	MN
Anolis auratus	N	*Porthidium lansbergii*	M
SQUAMATA: Dipsadidae		**TESTUDINES: Chelidae**	
Dipsas indica	FN	*Acanthochelys macrocephala*	MFN
Erythrolamprus melanotus	FL	*Phrynops geoffroanus*	F
Erythrolamprus reginae	N	**TESTUDINES: Emydidae**	
Helicops polylepis	N	*Terrapene mexicana*	F
Hydrodynastes gigas	MFN	*Trachemys callirostris*	MFN
Imantodes cenchoa	N	*Trachemys scripta*	MFNL
Leptodeira rhombifera	N	**TESTUDINES: Geoemydidae**	
Mussurana bicolor	MN	*Rhinoclemmys areolata*	M
Phalotris matogrossensis	F	*Rhinoclemmys melanosterna*	MFN
Phimophis guerini	F	*Rhinoclemmys pulcherrima*	M
Phimophis guianensis	N	*Rhinoclemmys rubida*	M
Pseudoboa sp.	N	**TESTUDINES: Kinosternidae**	
Pseudoboa neuwiedii	F	*Kinosternon acutum*	F
Pesudoboa nigra	MFN	*Kinosternon leucostomum*	F
Thamnodynastes gambotensis	L	*Kinosternon scorpioides*	MFN
Xenodon merremii	MF	**TESTUDINES: Testudinidae**	
Xenodon severus	A	*Chelonoidis carbonarius*	F
	MAMMALIA		
ARTIODACTYLA: Bovidae		**PILOSA: Bradypodidae**	
Cattle	AN	*Bradypus tridactylus*	MFN

(continued)

Table 2.12 (continued)

Sheep	A	**PILOSA: Choloepodidae**	
CARNIVORA: Procyonidae		*Choloepus didactylus*	MFN
Nasua narica	F	**RODENTIA: Caviidae**	
DIDELPHIMORPHIA: Didelphidae		*Hydrochoerus hydrochaeris*	MN
Didelphis marsupialis	NL	**RODENTIA: Dasyproctidae**	
Marmosa demerarae	NL	*Dasyprocta leporina*	M
Marmosa murina	L	*Dasyprocta punctata*	M
Philander opossum	N	**RODENTIA: Echimyidae**	
PERISSODACTYLA: Equidae		*Proechimys semispinosus*	MNL
Horse	L	**RODENTIA: Erethizontidae**	
		Coendou sp.	F
	AVES		
PASSERIFORMES: Parulidae		*Thamnophilus atrinucha*	NL
Oporornis agilis	L	**PELECANIFORMES: Ardeidae**	
PASSERIFORMES: Thamnophilidae		*Cochlearius cochlearius*	M
Myrmotherula axillaris	N		

Amblyomma rotundatum. Until recently, *Ixodes humanus* had been treated as a synonym of *Amblyomma dissimile* (Guglielmone and Nava 2010), but this name was later found to apply to another as yet undetermined species of *Amblyomma* (Guglielmone and Nava 2014).

The geographical distribution of *Amblyomma dissimile* in the Neotropics is vast, but Capriles and Gaud (1977) claim that this tick appears to be extinct in Puerto Rico. Álvarez et al. (2020) exclude the Province of Cienfuegos from the range of *Amblyomma dissimile* in Cuba, but Černý (1969b) listed this tick from a locality that is still within the limits of this province. Records outside the Western Hemisphere, as in the Philippines (Neumann 1899) or India (Keirans 1985b), are most probably due to incorrect sample labeling. Pietzsch et al. (2006) recorded the introduction of *Amblyomma dissimile* in the Palearctic Region, but there is no evidence that this tick is established there; nevertheless, Burridge (2011) stresses the ability of this tick to infest exotic hosts introduced to the New World, meaning that *Amblyomma dissimile* may possess the potential to develop permanent populations outside the Americas.

The geographical distribution of *Amblyomma dissimile* in Brazil merits further analysis. Polo et al. (2021) proposed to eliminate all records of *Amblyomma dissimile* from eastern Brazil, encompassing the states of Bahia, Maranhão, Minas Gerais, Pernambuco, Rio de Janeiro, and São Paulo, after examination of specimens classified as *Amblyomma dissimile* that were reidentified as *Amblyomma rotundatum.* A record from the State of Piauí, probably the two females, three males and one nymph found on *Iguana iguana* in this state by Alcántara et al. (2018), was confirmed as *Amblyomma dissimile,* but the origin of the host was questioned, and Piauí State has also been removed from the range of this tick. The decision was supported by 14 references, but eight of them were not related or contradict the statement concerning the absence of *Amblyomma dissimile* from the Brazilian states named above due to typographical errors in the final version of the paper (personal communication of Professor Hermes R. Luz to Alberto

A. Guglielmone). The reidentifications of *Amblyomma dissimile* as *Amblyomma rotundatum* in Polo et al. (2021) are treated as valid here, but so are the diagnoses of eight males and 20 females of *Amblyomma dissimile* collected from *Rhinella jimi* in the State of Bahia and those of four males and four females collected from *Pseudoboa nigra* in São Paulo State by Mendoza-Roldán et al. (2020). Therefore, Bahía and São Paulo states are still regarded as being within the range of *Amblyomma dissimile*, at least until new studies find that these specimens belong to another species.

The list of hosts in Table 2.12 differs from the host profile in Guglielmone and Nava (2010) and Guglielmone et al. (2014) because vertebrates found infested with *Amblyomma dissimile* in the Nearctic Region are excluded. Moreover, the extensive list of hosts of *Amblyomma dissimile* in Guglielmone and Nava (2010) includes vertebrates used in experimental infestations as well as hosts infested under various conditions of captivity, but no such vertebrates are included in the current study.

Neumann (1911a) stated that *Amblyomma cooperi* (a synonym of *Amblyomma dubitatum* to most tick workers), a parasite of *Hydrochoerus hydrochaeris* (Rodentia: Caviidae), is a synonym of *Amblyomma dissimile*, and this statement has been repeated in several studies, giving the impression that *Amblyomma dissimile* is a common parasite of capibaras. However, this is an error—there is only one record of this species from capibaras (Joan 1930). The many records of *Amblyomma dissimile* from mammals in Botelho et al. (2002) were treated as provisionally valid in Guglielmone and Nava (2010) but not in Guglielmone et al. (2014) or here; the same applies to records from donkeys in Onofre-Barragán (2015) because these host-parasite relationships require confirmation.

Doss et al. (1974a) cite Boero (1944b) for supposedly listing five mammal species as hosts of *Amblyomma dissimile*, but the sole host family named in Boero (1944b) is Boidae. Rawlins et al. (1993) list mammals as the only hosts of *Amblyomma dissimile*, and Camacho Escobar and Pérez-Lara (2009) depict a non-tick acarine as the nymph of *A. dissimile.* Scott and Durden (2015b) claim that a nymph of *Amblyomma dissimile* was collected from *Catharus fuscescens* in Canada, but there are differences between their figure and *bona fide* descriptions of the nymph of this species in Martins et al. (2010, 2014c). This controversial record is repeated in Scott et al. (2018).

Neotropical records of *Amblyomma dissimile* on birds have been reported from Panama (Fairchild et al. 1966; Bermúdez et al. 2018a), and there is one record from Colombia (Martínez-Sánchez et al. 2020). Dunn (1918) found that some engorged larvae of *Amblyomma dissimile* molt to nymphs and engorge on their hosts before detaching, resulting in a two-host life cycle, a phenomenon also occasionally seen in *Amblyomma rotundatum.* However, no study has been found that confirms Dunn's (1918) observations.

17. *Amblyomma dubitatum* Neumann, 1899

Neumann, L.G. (1899) Révision de la famille des ixodidés (3e mémoire). *Mémoires de la Société Zoologique de France*, 12, 107–294.

Descriptions and Redescriptions: Neumann (1899) described but did not figure the female of *Amblyomma dubitatum*; Nuttall and Warburton (1908) described and illustrated the male under the name *Amblyomma cooperi*, a synonym of *Amblyomma dubitatum*, as discussed in Guglielmone and Nava (2014). The nymph was described in Estrada-Peña et al. (2002) and the larva in Joan (1930, under the name *Amblyomma cooperi*).

The male, female, and nymph of *Amblyomma dubitatum* were redescribed in Nava et al. (2017); the male and female were redescribed by Estrada-Peña et al. (2002), Onofrio et al. (2006b, keys and figures), Voltzit (2007), and Dantas-Torres et al. (2019b, keys and figures); the nymph was redescribed in Martins et al. (2010), and the larva was redescribed by Amorim and Serra-Freire (1999a, as *Amblyomma cooperi*).

Geographic Distribution: *Amblyomma dubitatum* is a South American species that has been found in **Argentina**: Chaco, Corrientes, Formosa, Entre Ríos, Misiones, and Santa Fe (Nava et al. 2010a); **Bolivia** (Keirans and Brewster 1981); **Brazil**: Acre, Amazonas, Bahia, Espírito Santo, Goiás, Mato Grosso, Mato Grosso do Sul, Minas Gerais, Pará, Paraíba, Pernambuco, Paraná, Rio de Janeiro, Rio Grande do Sul, Rondônia, Santa Catarina, São Paulo, and Tocantins (Nava et al. 2010a; Dantas-Torres et al. 2010b; Martins et al. 2014a; Gianizella et al. 2018b; Santos et al. 2018; Gruhn et al. 2019; Martins et al. 2020b); **Paraguay**: Concepción, Misiones, and Presidente Hayes (Massi Pallarés and Benítez Usher 1982; Nava et al. 2010a); and **Uruguay**: Canelones, Durazno, Flores, Lavalleja, Rocha, Tacuarembó, and Treinta y Tres (Nava et al. 2010a; Martins et al. 2014d).

Hosts: Males, females, nymphs and larvae of *Amblyomma dubitatum* are usually found on *Hydrochoerus hydrochaeris* (Rodentia: Caviidae), but the list of hosts of this tick is rather long, especially for the immature stages, involving several mammals and a few birds, including a record of an adult tick on a *Megascops choliba* (Strigiformes) in Teixeira et al. (2020). The host profile of *Amblyomma dubitatum* is shown in Table 2.13, which is based on data from Labruna et al. (2005a), Dantas-Torres et al. (2010a, b, 2012), Nava et al. (2010a, 2017), Medeiros et al. (2011), Debárbora et al. (2012), Flores et al. (2014), Sponchiado et al. (2015), Acosta et al. (2016), Coelho et al. (2016), Eberhardt et al. (2016), Witter et al. (2016), Blanco et al. (2017), Fontalvo et al. (2017), Martins et al. (2017), Silva et al. (2018a, b), Barbieri et al. (2019), Lourenço et al. (2020), Ramos et al. (2020) and Teixeira et al. (2020).

Human Parasitism: According to Guglielmone and Robbins (2018), *Amblyomma dubitatum* is a sporadic parasite of humans, and all parasitic stages have been found on people in Argentina (Corrientes, and Misiones), Brazil (São Paulo), and Uruguay (Tacuarembó). This list was enlarged by Reck et al. (2018), who reported cases of human parasitism by nymphs of *Amblyomma dubitatum* at two localities in Rio Grande do Sul (Brazil).

Comments: *Amblyomma yucumense* and *Amblyomma dubitatum* are morphologically close and related to one another at the molecular level, with the result that

Table 2.13 Hosts for adult ticks of undetermined sex (A), males (M), females (F), nymphs (N), and larvae (L) of *Amblyomma dubitatum*

MAMMALIA			
ARTIODACTYLA: Bovidae		**PERISSODACTYLA: Tapiridae**	
Domestic buffalo	N	*Tapirus terrestris*	F
Cattle	N	**PILOSA: Myrmecophagidae**	
ARTIODACTYLA: Cervidae		*Myrmecophaga tridactyla*	MFNL
Axis axis	MNL	*Tamandua tetradactyla*	NL
Mazama gouazoubira	FL	**PRIMATES: Atelidae**	
ARTIODACTYLA: Suidae		*Alouatta caraya*	N
Domestic pig	MFNL	*Alouatta guariba*	N
ARTIODACTYLA: Tayassuidae		**RODENTIA: Caviidae**	
Tayassu pecari	F	*Cavia aperea*	NL
CARNIVORA: Canidae		*Hydrochoerus hydrochaeris*	MFNL
Cerdocyon thous	ANL	**RODENTIA: Chinchillidae**	
Chrysocyon brachyurus	N	*Lagostomus maximus*	N
Domestic dog	MN	**RODENTIA: Cricetidae**	
CARNIVORA: Felidae		*Akodon azarae*	NL
Puma concolor	N	*Cerradomys maracajuensis*	N
CARNIVORA: Procyonidae		*Hylaeamys megacephalus*	N
Nasua nasua	N	*Lundomys molitor*	NL
Procyon cancrivorus	N	*Necromys lasiurus*	N
CHIROPTERA: Phyllostomidae		*Nectomys rattus*	NL
Glossophaga soricina	F	*Oligoryzomys* sp.	N
DIDELPHIMORPHIA: Didelphidae		*Oligoryzomys flavescens*	L
Didelphis albiventris	NL	*Oxymycterus delator*	N
Didelphis aurita	N	*Rhipidomys* sp.	N
Didelphis marsupialis	N	*Scapteromys aquaticus*	L
Lutreolina crassicaudata	N	*Scapteromys tumidus*	NL
Monodelphis dimidiata	NL	**RODENTIA: Echimyidae**	
LAGOMORPHA: Leporidae		*Thrichomys pachyurus*	N
Lepus europaeus	NL	**RODENTIA: Erethizontidae**	
PERISSODACTYLA: Equidae		*Coendou spinosus*	N
Horse	MF	**RODENTIA: Myocastoridae**	
		Myocastor coypus	N
AVES			
PASSERIFORMES: Furnariidae		**RHEIFORMES: Rheidae**	
Furnarius rufus	L	*Rhea americana*	N
PASSERIFORMES: Passerellidae		**STRIGIFORMES: Strigidae**	
Zonotrichia capensis	L	*Megascops choliba*	A
PASSERIFORMES: Turdidae			
Turdus rufiventris	L		

adults are easily confused, and separation of nymphs is even more difficult (Krawczak et al. 2015). Boero (1957) depicted two patterns of scutal ornamentation in Argentinean *Amblyomma dubitatum* males under the name *Amblyomma cooperi.*

Amblyomma dubitatum was described by Neumann (1899) from a specimen allegedly collected from a bird in Spain, but later Neumann (1911a) doubted the locality record, and this species had been considered an African or Oriental species by different authors until Camicas et al. (1998) listed it as a Neotropical species, as discussed in Guglielmone et al. (2003a). Camicas et al. (1998) state that the Neotropical *Amblyomma cooperi*, *Amblyomma lutzi,* and *Amblyomma ypsilophorum* are synonyms of *Amblyomma dubitatum,* a fact later confirmed by Estrada-Peña et al. (2002). Nevertheless, Neumann's earlier (1911a) synonymization of

Amblyomma dubitatum (under the name *Amblyomma cooperi*) with *Amblyomma dissimile* distorted the host profile of the latter species, as discussed in Guglielmone and Nava (2010a) and in the entry for *Amblyomma dissimile* above.

Nava et al. (2017) doubt the presence of *Amblyomma dubitatum* in Bolivia, but Guglielmone et al. (2003a) treat as valid the Bolivian record in Keirans and Brewster (1981, under the name *Amblyomma cooperi* and without further data), although the first mention of the presence of this tick in Bolivia was by Squire (1972 as *Amblyomma cooperi*). Mastropaolo et al. (2014) also listed *Amblyomma dubitatum* as being present in Bolivia, and this view is provisionally accepted here. Vogelsang and Cordero (1940, also as *Amblyomma cooperi*) claim that this tick has been found in Venezuela, a fact unconfirmed by Jones et al. (1972), and Venezuela is provisionally excluded from the range of *Amblyomma dubitatum*. The geographic distribution of this species in Brazil, as documented by Andreotti et al. (2018), excludes the states of Goiás, Pará, Paraíba, Pernambuco, and Rondônia, but no reasons were given for doing so.

18. *Amblyomma fulvum* Neumann, 1899

Neumann, L.G. (1899) Révision de la famille des ixodidés (3e mémoire). *Mémoires de la Société Zoologique de France*, 12, 107–294.

Descriptions and Redescriptions: Neumann (1899) described the male of *Amblyomma fulvum*, but no illustrations were provided.

Robinson (1926) redescribed the male of *Amblyomma fulvum* by translating the original description, while Keirans (1972) redescribed and illustrated the holotype male, and Voltzit (2007) reproduced the figures in Keirans (1972) in her redescription of this tick.

The female, nymph, and larva of *Amblyomma fulvum* remain unknown.

Geographic Distribution: Northern South America (Keirans 1972).

Hosts: *Amblyomma fulvum* is known from the male holotype collected on *Eunectes murinus* (Squamata: Boidae) (Keirans 1972).

Human Parasitism: No.

Comments: Guglielmone et al. (2003a) and Guglielmone et al. (2014) did not question the redescription of *Amblyomma fulvum* in Keirans (1972). However, Keirans (1972) states that a distinctive morphological feature of the male of *Amblyomma fulvum* is the presence of ventral posterointernal tubercles on each festoon except the central, although this character is not shown in the corresponding figure. Voltzit (2007) repeats the text and figure of Keirans (1972). It is important to note that neither Neumann (1899) nor Robinson (1926) described such distinctive tubercles, raising doubts about the diagnosis of *Amblyomma fulvum*. A redescription of the type of *Amblyomma fulvum* is needed.

The type locality for the only known specimen of *Amblyomma fulvum* is uncertain because Neumann (1899) stated that the host was collected in "Brésil?," but Keirans (1972) considers its host, *Eunectes murinus*, to be absent from Brazil, which is an error because *Eunectes murinus* is established there. The presence of *Amblyomma fulvum* in Brazil remains unconfirmed, and the alleged record of a male of *Amblyomma fulvum* collected on the ground in Brazil by Figueiredo et al. (1999) is treated here as a diagnostic error. Kelehear et al. (2017a) also named *Amblyomma fulvum* as present in French Guiana, but no information has been found to support that statement.

19. ***Amblyomma fuscum*** Neumann, 1907

Neumann, L.G., (1907) Quatre espèces nouvelles d'ixodidés. *Notes from the Leyden Museum*, 29, 88–100.

Descriptions and Redescriptions: The male of *Amblyomma fuscum* was described and illustrated by Neumann (1907), while the female was described but not figured by Aragão (1936); Martins et al. (2010) described the nymph.

The male and female of *Amblyomma fuscum* were redescribed in Barros-Battesti et al. (2005b), Onofrio et al. (2006b, keys and figures), Voltzit (2007), and Dantas-Torres et al. (2019b, keys and figures), while the male alone was redescribed in Robinson (1926) and Schulze (1936b).

There are no redescriptions of the nymph of *Amblyomma fuscum.*

The larva of *Amblyomma fuscum* remains undescribed, but some authors have been able to identify this stage from field-collected material.

Geographic Distribution: *Amblyomma fuscum* is a Neotropical species that has only been found in **Brazil**: Pernambuco, Rio Grande do Sul, Santa Catarina, and São Paulo (Aragão 1936; Barros-Battesti et al. 2005b; Dantas-Torres et al. 2008).

Hosts: *Amblyomma fuscum* has a peculiar host profile because most males and females have been collected from *Dasypus novemcinctus* and *Dasypus septemcinctus* (Cingulata: Dasypodidae), and *Boa constrictor* (Squamata: Boidae), although there are also a few records from other squamatan, anuran and mammalian hosts. Nymphs and the undescribed larva of *Amblyomma fuscum* have been found on several species of Didelphimorphia and Rodentia, although some larvae have also been collected from dogs. Nevertheless, the results of the studies of Martins et al. (2009b) and Szabó et al. (2013) appear to indicate that *Didelphis aurita* (Didelphidae) is the preferred host for the nymphs and larvae of *Amblyomma fuscum*, and this marsupial is hypothetically treated here as the principal host for immature stages of this tick. The host profile of *Amblyomma fuscum* is shown in Table 2.14, which is based on data from Sinkoc and Brum (1997), Barros-Battesti et al. (2005b), Dantas-Torres et al. (2008, 2012, 2019a), Martins et al. (2009b), Aléssio et al. (2012), Szabó et al. (2013), Gomes et al. (2015), and Blanco et al. (2017).

Table 2.14 Hosts for males (M), females (F), nymphs (N), and larvae* (L) of *Amblyomma fuscum*

	AMPHIBIA		
ANURA: Bufonidae			
Rhinella arenarum	MF		
	MAMMALIA		
CARNIVORA: Canidae		**RODENTIA: Cricetidae**	
Cerdocyon thous	F	*Akodon* sp.	L
Domestic dog	L	*Akodon montensis*	N
CINGULATA: Dasypodidae		*Cerradomys subflavus*	N
Dasypus novemcinctus	MF	*Euryoryzomys russatus*	L
Dasypus septemcinctus	MF	*Nectomys rattus*	N
DIDELPHIMORPHIA: Didelphidae		*Oxymycterus* sp.	L
Didelphis albiventris	N	**RODENTIA: Echimyidae**	
Didelphis aurita	NL	*Thrichomys laurentius*	N
Marmosa demerarae	NL	**RODENTIA: Sciuridae**	
Metachirus nudicaudatus	NL	*Sciurus aestuans*	N
Monodelphis domestica	N		
	REPTILIA		
CROCODILIA: Alligatoridae		**Dipsadidae**	
Caiman latirostris	M	*Clelia clelia*	F
SQUAMATA: Boidae		**Teiidae**	
Boa constrictor	MF	*Salvator merianae*	M
		Tupinambis teguixin	M

* Tentative diagnoses because the larva of *Amblyomma fuscum* remains undescribed

Human Parasitism: Guglielmone and Robbins (2018) list *Amblyomma fuscum* as a very rare parasite of humans, with two Brazilian records (São Paulo, and probably Santa Catarina) caused by female ticks. Reck et al. (2018) added two new cases of human parasitism by a male and a female of *Amblyomma fuscum* in Rio Grande do Sul.

Comments: Keirans (1992) and Horak et al. (2002) exclude *Amblyomma fuscum* from their lists of valid species, but this omission is not supported by current evidence.

Cingulata are not included as relevant hosts for adults of *Amblyomma fuscum* in the study of Estrada-Peña et al. (2020).

20. *Amblyomma geayi* Neumann, 1899

Neumann, L.G. (1899) Révision de la famille des ixodidés (3e mémoire). *Mémoires de la Société Zoologique de France*, 12, 107–294.

Descriptions and Redescriptions: Neumann (1899, 1901) described but did not illustrate the male and female of *Amblyomma geayi*, respectively. The nymph was described by Martins et al. (2013). Amorim and Serra-Freire (1999b) described the larva, although this description is considered provisionally valid here.

The male and female of *Amblyomma geayi* were redescribed in Robinson (1926), Floch and Fauran (1958), Aragão and Fonseca (1961a), Onofrio et al. (2006b, keys and figures), Voltzit (2007), Labruna et al. (2009), Bermúdez et al. (2018a) and Dantas-Torres et al. (2019b, keys and figures). The male alone was described by Urushiyama et al. (2020).

There are no redescriptions of the nymph and larva of *Amblyomma geayi*.

Geographic Distribution: *Amblyomma geayi* is a Neotropical species that has been found in Central and South America as follows: **Brazil**: Acre, Amazonas, Maranhão, Pará, Rondônia, and Tocantins (Labruna et al. 2009; Gianizella et al. 2018b; Lima et al. 2018; Zimmermann et al. 2018; Costa et al. 2020; Martins et al. 2020b); **Colombia**: Chocó (Acevedo-Gutiérrez et al. 2020); **Costa Rica**: Alajuela, Heredia, and Limón (Hun et al. 1991; Troyo et al. 2016; Dolz et al. 2019); **French Guiana**: Cayena, and Saint Laurent du Maroni (Floch and Fauran 1958); **Guyana** (Keirans 1985b; Urushiyama et al. 2020); **Panama**: Bocas del Toro, Colón, Darién (tentative), Panamá, and Panamá Oeste (Osorno Mesa 1942; Fairchild et al. 1966; Miller et al. 2016); **Peru**: Loreto (Fairchild et al. 1966); **Suriname** (Fairchild et al. 1966); and **Venezuela** (Clavijo et al. 2009).

Hosts: Males and females of *Amblyomma geayi* are usually found on *Bradypus tridactylus* and *Bradypus variegatus* (Pilosa: Bradypodidae), with occasional records from other mammals and rarely on species of Reptilia. Nymphs of this species were found prone to feed on Bradypodidae by Gianizella et al. (2018b) in Brazil, but there are also several records from other mammalian and avian hosts, with Passeriformes (several families) being the most common order of birds infested by this tick. The larva of *Amblyomma geayi* has been recorded from a few mammals, but this stage has often been collected from Aves; of 46 records of larvae of this tick, 36 have been reported from several families of Passeriformes. Therefore, while the preferred hosts for immature stages of *Amblyomma geayi* are uncertain, it is speculated here that nymphs are common on Bradypodidae, and larvae usually parasitize Passeriformes (several families). The host profile of *Amblyomma geayi* is shown in Table 2.15, which is based on data from Fairchild et al. (1966), Ogrzewalska et al. (2010), Martins et al. (2014b), Soares et al. (2015), Esser et al. (2016a, b), Bermúdez et al. (2018a, 2020), Gianizella et al. (2018b), Lima et al. (2018), Dolz et al. (2019) and Binetruy et al. (2020b).

Human Parasitism: No.

Comments: Neumann (1911a) stated that *Amblyomma perpunctatum*, originally named *Ixodes perpunctatus* in Packard (1869), is probably a senior synonym of *Amblyomma geayi*, and Santos Dias (1961) reinstated *Amblyomma perpunctatum* with *Amblyomma geayi* as its synonym, an opinion supported by Camicas et al. (1998), although Fairchild et al. (1966), Guglielmone and Nava (2014) and Guglielmone et al. (2014) disagree with Santos Dias (1961). While Guglielmone et al. (2003a) treated *Amblyomma geayi* as valid, these authors also proposed type comparison as a means to resolve the ongoing taxonomic controversy, a view endorsed here.

Amblyomma geayi, *Amblyomma longirostre*, *Amblyomma parkeri,* and *Amblyomma romarioi* are morphologically and phylogenetically related species, as shown in studies by Labruna et al. (2009) and Martins et al. (2019b). Alleged specimens of *Amblyomma geayi* from southern Brazil, mostly collected from Rodentia: Erethizontidae, were reidentified as *Amblyomma parkeri* (Labruna et al.

Table 2.15 Hosts for males (M), females (F), nymphs (N), and larvae (L) of *Amblyomma geayi*

MAMMALIA			
CHIROPTERA: Phyllostomidae		**PILOSA: Myrmecophagidae**	
Artibeus lituratus	L	*Tamandua mexicana*	F
DIDELPHIMORPHIA: Didelphidae		*Tamandua tetradactyla*	MF
Caluromys lanatus	N	**PRIMATES: Atelidae**	
Didelphis marsupialis	MF*NL	*Alouatta nigerrima*	L
Marmosa robinsoni	L	*Alouatta palliata*	N
Philander opossum	N	**PRIMATES: Callitrichidae**	
PILOSA: Bradypodidae		*Saguinus bicolor*	N
Bradypus tridactylus	MFN	**RODENTIA: Echimyidae**	
Bradypus variegatus	MFN	*Proechimys* sp.	N
PILOSA: Choloepodidae		*Proechimys semispinosus*	L
Choloepus didactylus	MFN	**RODENTIA: Erethizontidae**	
Choloepus hoffmanni	MF	*Coendou prehensilis*	M
PILOSA: Cyclopodidae			
Cyclopes didactylus	N		
AVES			
COLUMBIFORMES: Columbidae		**PASSERIFORMES: Thamnophilidae**	
Leptotila cassinii	NL	*Gymnopithys leucaspis*	L
CORACIIFORMES: Momotidae		*Percnostola rufifrons*	L
Baryphthengus martii	NL	*Thamnomanes caesius*	NL
PASSERIFORMES: Cardinalidae		*Thamnophilus atrinucha*	NL
Cyanoloxia cyanoides	L	*Thamnophilus doliatus*	L
PASSERIFORMES: Fringillidae		*Willisornis poecilonotus*	L
Euphonia laniirostris	L	**PASSERIFORMES: Thraupidae**	
PASSERIFORMES: Furnariidae		*Eucometis penicillata*	L
Dendrocolaptes hoffmannsi	L	**PASSERIFORMES: Turdidae**	
Glyphorynchus spirurus	L	*Turdus albicollis*	L
Hylexetastes perrotii	L	*Turdus ignobilis*	N
Xiphorhynchus pardalotus	L	*Turdus leucomelas*	L
PASSERIFORMES: Passerellidae		**PASSERIFORMES: Tyrannidae**	
Arremonops conirostris	N	*Attila spadiceus*	L
PASSERIFORMES: Pipridae		*Leptopogon amaurocephalus*	L
Ceratopipra mentalis	NL	*Mionectes oleaginous*	L
Chiroxiphia pareola	L	*Rhynchocyclus olivaceus*	L
Dixiphia pipra	NL	**PICIFORMES: Ramphastidae**	
Manacus manacus	L	*Ramphastos vitellinus*	N
Manacus vitellinus	NL	**STRIGIFORMES: Strigidae**	
Pipra aureola	L	*Megascops choliba*	NL
Pipra filicauda	N		
REPTILIA			
CROCODILIA: Alligatoridae		**SQUAMATA: Iguanidae**	
Caiman crocodilus	M	*Iguana iguana*	F
SQUAMATA: Colubridae		**TESTUDINES: Testudinidae**	
Spilotes pullatus	MF	*Chelonoidis denticulata*	MF

* tentative diagnosis of adults of *Amblyomma geayi* in Fairchild *et al.* (1966)

2009). Therefore, the southern Brazilian states are not included within the range of *Amblyomma geayi*, and Rodentia: Erethizontidae are no longer treated as preferred hosts, as stated previously in Guglielmone et al. (2014). Larvae of *Amblyomma geayi* collected from birds in Costa Rica by Dolz et al. (2019) show a 4% difference in 16S rDNA sequences in comparison with Brazilian *Amblyomma geayi*, and these Costa Rican specimens are treated as provisionally belonging to *Amblyomma geayi*.

The description of the larva of *Amblyomma geayi* by Amorim and Serra-Freire (1999b) is tentative because specimens for their study were obtained at the Rio de

Janeiro zoo (southeasterrn Brazil), an area that, according to Labruna et al. (2009), lies outside the range of *Amblyomma geayi*.

Fairchild et al. (1966) believe it probable that *Amblyomma geayi* is present in Darién (Panamá); as well, these authors provide a tentative diagnosis of adults of this tick, based on specimens supposedly collected from *Didelphis marsupialis*. Andreotti et al. (2018) reduced the Brazilian distribution of *Amblyomma geayi* to the states of Amazonas and Pará, but the Brazilian range of this tick is thought to be broader than proposed by this author.The provisional principal hosts for the nymph of *Amblyomma geayi* are based on Gianizella et al. (2018b), who studied several Brazilian tick collections, finding that Bradypodidae are usually infested with nymphs in numbers not matched by other hosts infested with this tick stage. The provisional hosts for larvae are several families of Passeriformes, as proposed by Martins et al. (2019b), but few larvae of *Amblyomma geayi* have been obtained from any hosts. Estrada-Peña et al. (2020) include the nymph and larva of *Amblyomma geayi* in a cluster of Neotropical species feeding mostly on avian hosts.

A human infestation with a nymph of *Amblyomma geayi* was reported in Esser et al. (2016a, p. 300), but in Table 2 of the same paper, the specimen is assigned to another primate (*Allouatta palliata*). Ellen Esser (the senior author of that paper), in a personal communication to Alberto A. Guglielmone, confirmed that the nymph was collected from *Allouatta palliata*.

Urushiyama et al. (2020) provided morphological and molecular evidence for the introduction into Japan of a male of *Amblyomma geayi* found on a *Choloepus didactylus* imported from Guyana.

21. *Amblyomma goeldii* Neumann, 1899

Neumann, L.G. (1899) Révision de la famille des ixodidés (3e mémoire). *Mémoires de la Société Zoologique de France*, 12, 107–294.

Descriptions and Redescriptions: Neumann (1899) described but did not figure the male of *Amblyomma goeldii*, and the female was described and figured by Floch and Abonnenc (1941) under the name *Amblyomma ininii*, partly a synonym of *Amblyomma goeldii*, as discussed in Guglielmone and Nava (2014). The nymph was described by Martins et al. (2015c).

The male and female of *Amblyomma goeldii* were redescribed in Floch and Fauran (1958), Barros-Battesti et al. (2005b), Onofrio et al. (2006b, keys and figures), Voltzit (2007), and Dantas-Torres et al. (2019b, keys and figures); the male alone was redescribed by Robinson (1926).

There are no redescriptions of the nymph of *Amblyomma goeldii*.

The larva of *Amblyomma goeldii* remains unknown.

Geographical Distribution: *Amblyomma goeldii* is a Neotropical species that has been found in northern South America as follows: **Brazil**: Amazonas, and Pará (Martins et al. 2015c); **French Guiana**: Cayena, and Saint Laurent du Maroni (Floch and Fauran 1958); and **Suriname**: Marowijne (Voltzit 2007).

Table 2.16 Hosts for males (M) and females (F) of *Amblyomma goeldii*

MAMMALIA			
CARNIVORA: Canidae		**PILOSA: Myrmecophagidae**	
Domestic dog	F	*Tamandua tetradactyla*	MF
PILOSA: Bradypodidae			
Bradypus tridactylus	MF		
REPTILIA			
SQUAMATA: Boidae		**Viperidae**	
Boa constrictor	M	*Lachesis muta*	MF
AMPHIBIA			
ANURA: Bufonidae			
Rhinella sp.	M		

Hosts: Males and females of *Amblyomma goeldii* are usually found on *Tamandua tetradactyla* (Pilosa: Myrmecophagidae), but they have occasionally been collected from reptilian hosts, and one male was recorded as infesting *Rhinella* sp. (Anura: Bufonidae) in the description of *Amblyomma goeldii* by Neumann (1899). The host profile of *Amblyomma goeldii* is shown in Table 2.16, which is based on data from Neumann (1899), Floch and Fauran (1958), and Gianizella et al. (2018b). The nymph of *Amblyomma goeldii* is known, but its hosts are not because this stage was described by Martins et al. (2015c) from laboratory-reared material.

Human Parasitism: No.

Comments: There are taxonomic and nomenclatural problems associated with *Amblyomma goeldii*, which is thought to be partly a synonym of *Amblyomma rotundatum* because of confusion between the female of *Amblyomma rotundatum* and that of *Amblyomma goeldii* in the original description of Neumann (1899) (Floch and Fauran, 1958), as well as in the redescription of Robinson (1926). Authors such as Bequaert (1926) and Aragão (1936) recognized the similarities between both species, and Aragão (1936) changed his previous diagnosis of *Amblyomma goeldii* to *Amblyomma rotundatum*. The confusion in Neumann's original (1899) description was also noted by Keirans (1985b), who examined a female tick from the syntype series of *Amblyomma goeldii* and confirmed that it was, in fact, a specimen of *Amblyomma rotundatum*. According to Floch and Abonnenc (1942), the first description of the female of *Amblyomma goeldii* is in Floch and Abonnenc (1941) under the name *Amblyomma ininii*.

Santos Dias (1958a) compared the syntypes (all females) of *Ixodes fuscomaculatus* described by Lucas (1873) with the syntype females of *Amblyomma goeldii*, concluding that they were identical and that Lucas' (1873) species, under the name *Amblyomma fuscomaculatum*, therefore has priority over *Amblyomma goeldii*, a view repeated in Santos Dias (1989b). However, Santos Dias overlooked the opinions of Floch and Fauran (1958) and Keirans (1985b), who consider the female of *Amblyomma goeldii* in Neumann (1899) to be a synonym of *Amblyomma rotundatum* described by Koch (1844), with the result that *Amblyomma fuscomaculatum* becomes a junior synonym of *Amblyomma rotundatum*.

The geographic distribution of *Amblyomma goeldii* presented here is based on specimens thought to be *bona fide* representatives of this species; therefore, alleged records of *Amblyomma goeldii* females from Colombia and Guyana in Robinson (1926) are considered to actually represent *Amblyomma rotundatum*. Jamaican records of *Amblyomma goeldii* in Thompson (1950) were found to be *Amblyomma rotundatum* by Kohls (1969c). *Amblyomma goeldii* or a closely related species was found in southern Brazil by Evans et al. (2000), a record also excluded from the above list.

The dog is listed as host for *Amblyomma goeldii* in Floch and Fauran (1958) but not in Floch and Fauran (1959a); therefore, this host-parasite relationship is provisionally included in Table 2.16.

22. *Amblyomma hadanii* Nava, Mastropaolo, Mangold, Martins, Venzal and Guglielmone, 2014

Nava, S., Mastropaolo, M., Mangold, A.J., Martins, T.F., Venzal, J.M. & Guglielmone, A.A. (2014b) *Amblyomma hadanii* n. sp. (Acari: Ixodidae), a tick from northwestern Argentina previously confused with *Amblyomma coelebs* Neumann, 1899. *Systematic Parasitology*, 88, 261–272.

Descriptions and Redescriptions: The male, female, larva, and nymph of *Amblyomma hadanii* were described and illustrated in Nava et al. (2014b).

The male, female, and nymph were redescribed in Nava et al. (2017).

There are no redescriptions of the larva of *Amblyomma hadanii.*

Geographic Distribution: *Amblyomma hadanii* has only been found in southern South America, in **Argentina**: Jujuy and Salta (Nava et al. 2014b).

Hosts: There are few records of *Amblyomma hadanii,* but they encompass three orders and four families of mammals. Males, females, and nymphs have only been collected from cattle (Bovidae), which are treated here as provisional principal hosts of *Amblyomma hadanii*. The hosts of the larvae of *Amblyomma hadanii* remain unknown because all specimens have been collected from vegetation. The host profile of *Amblyomma hadanii* is shown in Table 2.17, which is based on data from Nava et al. (2014b) and Saracho-Bottero et al. (2018).

Table 2.17 Hosts for males (M), females (F), and nymphs (N) of *Amblyomma hadanii*

MAMMALIA			
ARTIODACTYLA: Bovidae		**PERISSODACTYLA: Equidae**	
Cattle	MFN	Horse	FN
CARNIVORA: Canidae		**PERISSODACTYLA: Tapiridae**	
Domestic dog	N	*Tapirus terrestris*	M

Human Parasitism: Guglielmone and Robbins (2018) list *Amblyomma hadanii* as a sporadic parasite of people in Argentina (Salta and Jujuy), with records of females (tentative), nymphs, and larvae feeding on humans.

Comments: *Amblyomma hadanii* has only been found in northwestern Argentina, where it was previously identified as *Amblyomma coelebs*, as shown in Nava et al. (2014b); thus, the record of *Amblyomma coelebs* collected from a human in Salta Province by Beldoménico et al. (2003b) is provisionally treated here as *Amblyomma hadanii*. Krawczak et al. (2015) stated that *Amblyomma hadanii* is morphologically close to *Amblyomma yucumense*.

23. ***Amblyomma hirtum*** Neumann, 1906

Neumann, L.G. (1906) Notes sur les Ixodidés. IV. *Archives de Parasitologie*, 10, 195–219.

Descriptions and Redescriptions: Neumann (1906) described and illustrated the male and female of *Amblyomma hirtum*.

Robinson (1926) redescribed the male and female of *Amblyomma hirtum* using the figures in Neumann (1906).

The larva and nymph of *Amblyomma hirtum* remain unknown.

Geographic Distribution: *Amblyomma hirtum* is known from specimens collected on an undetermined Galápagos island of **Ecuador**: Galápagos (Neumann 1906).

Hosts: Unknown.

Human parasitism: No.

Comments: Keirans (1992) excluded *Amblyomma hirtum* in his list of legitimate tick species, but Camicas et al. (1998) treat it as a valid taxon. Perhaps Keirans (1992) overlooked the description of *Amblyomma hirtum* in Neumann (1906), whose figures show a tick very similar to *Amblyomma macfarlandi*. Surprisingly, J.E. Keirans in Guglielmone et al. (2003a) seems to support the validity of *Amblyomma hirtum*, noting that type specimens of female *Amblyomma hirtum* possess very large porose areas and long setae on the dorsum, with whitish-ivory ornamentation confined to the scapular region, morphological characters not found in the corresponding types of *Amblyomma macfarlandi*. Moreover, the male of *Amblyomma hirtum* is notably smaller than that of *Amblyomma macfarlandi*, and its dentition is 2/2 versus 3/3 in *Amblyomma macfarlandi*. Santos Dias (1958a) considers *Amblyomma hirtum* a synonym of *Amblyomma pilosum*, but both species are valid, as is evident from the description of *Amblyomma hirtum* in Neumann (1906) and the redescription of *Amblyomma pilosum* in Keirans et al. (1973).

Neumann (1906) described *Amblyomma hirtum* from specimens collected in the Galápagos Islands and "l'île St. Paul." The latter locality is thought by Guglielmone et al. (2003a, b) to be Saint Paul's Rocks, close to Brazil's northeast coast, but this has not been confirmed, and Dantas-Torres et al. (2009a) treat the presence of

Amblyomma hirtum in Brazil as doubtful. Hooker (1909) listed this tick as found on the Caribbean island of Guadeloupe, a record repeated in Thompson (1950), but Guadeloupe is another doubtful locality for *Amblyomma hirtum*, which is considered an exclusively Ecuadorian species here.

Schulze (1936a) stated that *Amblyomma hirtum* had been found on Aves, while Camicas et al. (1998) claim that this species has been collected from Testudines, but these hosts were not included in the original description of Neumann (1906). The actual hosts of *Amblyomma hirtum* are treated here as unknown.

24. *Amblyomma humerale* Koch, 1844

Koch, C.L. (1844) Systematische Übersicht über die Ordnung der Zecken. *Archiv für Naturgeschichte*, 10, 217–239.

Descriptions and Redescriptions: The male of *Amblyomma humerale* was described but not figured by Koch (1844); the female was described and illustrated by Cooper and Robinson (1908) under the name *Amblyomma longirostrum*, a synonym of *Amblyomma humerale*, as discussed in Guglielmone and Nava (2014). The nymph of *Amblyomma humerale* was described by Martins et al. (2010).

The male and female of *Amblyomma humerale* were redescribed in Robinson (1926), Floch and Abonnenc (1940), Floch and Fauran (1958), Onofrio et al. (2006b, keys and figures), Voltzit (2007), and Dantas-Torres et al. (2019b, keys and figures), while the female alone was redescribed in Santos Dias (1989b).

There are no redescriptions of the nymph of *Amblyomma humerale*.

The larva of *Amblyomma humerale* remains undescribed, but some authors have been able to identify this stage from field-collected material.

Geographical Distribution: *Amblyomma humerale* is found in northern South America and a Caribbean island as follows: **Bolivia**: Santa Cruz (Robbins et al. 2003a); **Brazil**: Acre, Amapá, Amazonas, Bahia, Espírito Santo, Mato Grosso, Pará, Roraima, Rondônia, and Tocantins (Labruna et al. 2005d); **Colombia**: Meta, and Valle del Cauca (Robbins et al. 2003a; Acevedo-Gutiérrez et al. 2020); **Ecuador**: Napo (Robbins et al. 2003a); **French Guiana**: Cayena, Saint Laurent du Maroni (Floch and Fauran 1958; Binetruy et al. 2020a); **Guyana**: East Berbice-Corentyne, and Mazaruni-Potaro (Keirans 1985b; Voltzit 2007); **Peru**: Huánuco, Loreto, Madre de Dios, and Puno (Need et al. 1991; Robbins et al. 2003a); **Suriname** (Keirans 1985b); **Trinidad and Tobago**: Río Claro-Mayaro (Robbins et al. 2003a); and **Venezuela**: Amazonas, and Bolívar (Guerrero 1996).

Hosts: The usual hosts for males and females of *Amblyomma humerale* are *Chelonoidis denticulata* (Testudines: Testudinidae), with occasional records from Mammalia and Crocodilia. The host profile of the nymph is complex because this stage has been collected from several families of Passeriformes, Mammalia (several orders), and Squamata (several families) without any definite indication that one type of host prevails over the others, and there are additional records from Coraciiformes:

Table 2.18 Hosts for males (M), females (F), nymphs (N), and larvae* (L) of *Amblyomma humerale*

MAMMALIA			
CARNIVORA: Procyonidae		**PILOSA: Cyclopedidae**	
Nasua nasua	N	*Cyclopes didactylus*	N
CHIROPTERA: Phyllostomidae		**PILOSA: Myrmecophagidae**	
Chrotopterus auritus	N	*Tamandua tetradactyla*	MFN
CINGULATA: Dasypodidae		**RODENTIA: Caviidae**	
Dasypus novemcinctus	N	*Hydrochoerus hydrochaeris*	N
DIDELPHIMORPHIA: Didelphidae		**RODENTIA: Cricetidae**	
Didelphis marsupialis	NL	*Euryoryzomys* sp.	N
Marmosa sp.	N	*Neacomys spinosus*	N
Marmosa constantiae	N	**RODENTIA: Dasyproctidae**	
Marmosops sp.	L	*Myoprocta acouchy*	N
Metachirus nudicaudatus	NL	**RODENTIA: Echimyidae**	
Philander opossum	NL	*Proechimys* sp.	N
PILOSA: Bradypodidae		*Proechimys roberti*	L
Bradypus tridactylus	MF		
AVES			
CORACIIFORMES: Momotidae		*Myrmelastes leucostigma*	L
Momotus momota	N	*Myrmotherula axillaris*	N
PASSERIFORMES: Furnariidae		*Pyriglena leuconota*	L
Dendrocolaptes hoffmannsi	NL	*Thamnomanes caesius*	N
Sclerurus sp.	N	*Thamnomanes schistogynus*	N
Sclerurus caudacutus	L	*Willisornis poecilonotus*	NL
Sittasomus griseicapillus	N	**PASSERIFORMES: Thraupidae**	
Xiphorhynchus guttatus	N	*Tachyphonus surinamus*	NL
PASSERIFORMES: Thamnophilidae		**PASSERIFORMES: Tyrannidae**	
Hylophylax naevius	L	*Rhynchocyclus olivaceus*	N
REPTILIA			
CROCODILIA: Alligatoridae		**SQUAMATA: Teiidae**	
Caiman crocodilus	M	*Kentropyx calcarata*	N
Paleosuchus trigonatus	N	*Kentropyx pelviceps*	N
SQUAMATA: Colubridae		**SQUAMATA: Tropiduridae**	
Chironius carinatus	N	*Plica plica*	N
SQUAMATA: Dipsadidae		*Plica umbra*	N
Erythrolamprus typhlus	N	*Uranoscodon superciliosus*	N
Xenodon werneri	N	**TESTUDINES: Testudinidae**	
		Chelonoidis carbonarius	MF
		Chelonoidis denticulata	MFN

* Tentative diagnosis because the larva of *Amblyomma humerale* remains undescribed

Momotidae and Crocodilia and Testudines. The hosts of the larva of *Amblyomma humerale* are known from a few specimens listed in the studies of Ogrzewalska et al. (2010), Binetruy et al. (2019, 2020a), and Colle et al. (2020), who found this stage on Passeriformes, Didelphimorphia, and Rodentia. It is hypothesized that the nymph of this tick has the ability to sustain itself on a variety of mammalian, passeriform, and squamatan hosts, while the principal hosts of the larva are regarded as basically undetermined. The host profile of *Amblyomma humerale* is shown in Table 2.18, which is based on data from Guerrero (1996), Labruna et al. (2002c, 2005d), Ogrzewalska et al. (2010), Witter et al. (2016), Gianizella et al. (2018b), Lima et al. (2018), Zimmermann et al. (2018), Binetruy et al. (2019, 2020a), Gruhn et al. (2019), Colle (2020) and Souza et al. (2020).

Human Parasitism: Guglielmone and Robbins (2018) do not list *Amblyomma humerale* as a human parasite. Thereafter, Binetruy et al. (2019) recognized larvae of this species collected from people in French Guiana based on molecular methods, a record treated as provisionally valid here. Later, Binetruy et al. (2020a) added nymphal and adult infestations of humans in French Guiana (Cayena and Saint Laurent du Maroni).

Comments: *Amblyomma crassum*, *Amblyomma humerale*, and *Amblyomma sabanerae* are morphologically close, and Santos Dias (1958a) treats *Amblyomma crassum* and *Amblyomma sabanerae* as synonyms of *Amblyomma humerale*, although later Santos Dias (1989b) accepted *Amblyomma sabanerae* as a valid species. *Amblyomma humerale* appears to be restricted to central and northern South America and Trinidad and Tobago. Panamanian records of this tick cited as *Amblyomma humerli* (*lapsus*) in Evans (1947) are doubtful, and Fairchild et al. (1966) state that *Amblyomma humerale* has not been found in Panama. Tonelli Rondelli (1939) found that the description of *Amblyomma humerale* from Paraguay by Berlese (1888) corresponds to *Amblyomma dissimile*. Neumann (1899) described *Amblyomma gypsatum*, a synonym of *Amblyomma humerale*, from specimens collected in "Uruguay, nord de Goyaz," where "Goyaz" may be a misspelling of Goiás (a Brazilian State) or Goya (an Argentinean city in Corrientes Province), but most probably the specimens used to describe *Amblyomma gypsatum* are of Brazilian origin. Vargas (1955) cites this species from Mexico, while Freire (1972), Brum et al. (1987), and Sinkoc and Brum (1997) allegedly found *Amblyomma humerale* on dogs and Bufonidae in southern Brazil. None of these records are included here in the geographic distribution and host profile of *Amblyomma humerale*. The Nearctic record of *Amblyomma humerale* on a migratory bird in Morshed et al. (2005) appears to correspond, in fact, to *Amblyomma sabanerae* (Ogrzewalska et al. 2010).

The information above indicates that mammalian, passeriform, and squamatan hosts are to be expected for the nymph of *Amblyomma humerale*, while field data for the larvae remain insufficient for identifying principal hosts. Laboratory studies by Martins et al. (2020c) demonstrated the capacity of larvae and nymphs of *Amblyomma humerale* to feed on different types of hosts; this was especially evident for the larvae of this tick. New information concerning hosts of the larva of *Amblyomma humerale* will probably demonstrate that the laboratory study of Martins et al. (2020c) correlates with data from the field.

Estrada-Peña et al. (2020) include the nymph and larva of *Amblyomma humerale* in a cluster of Neotropical species that feed mostly on birds, but we consider the host profile of the nymph to extend well beyond Aves, and more records of larval *Amblyomma humerale* are needed to confirm the principal hosts for this stage.

25. *Amblyomma incisum* Neumann, 1906

Neumann, L.G. (1906) Notes sur les Ixodidés. IV. *Archives de Parasitologie*, 10, 195–219.

Descriptions and Redescriptions: Neumann (1906) described and illustrated the male of *Amblyomma incisum*, while the female was described and figured by Aragão (1911), and the nymph was described by Martins et al. (2010).

The male, female, and nymph of *Amblyomm incisum* were redescribed in Nava et al. (2017), and the male and female were redescribed by Robinson (1926), Floch and Abonnenc (1941), Floch and Fauran (1958), Labruna et al. (2005c), Onofrio et al. (2006b, keys and figures), Voltzit (2007) and Dantas-Torres et al. (2019b, keys and figures). The female and the nymph were redescribed in Boero and Prosen (1955) and Martins et al. (2014c, key and figures), respectively.

The larva of *Amblyomma incisum* remains undescribed, but some authors have been able to identify this stage from field-collected material.

Geographical Distribution: *Amblyomma incisum* is a South American species that has been found in **Argentina**: Misiones (Lamattina et al. 2014); **Bolivia**: Cochabamba (Boero and Prosen 1955); **Brazil**: Acre, Amazonas, Espírito Santo, Mato Grosso, Mato Grosso do Sul, Minas Gerais, Paraná, Rio de Janeiro, Rio Grande do Sul, Rondônia, and São Paulo (Aragão 1911, 1936; Labruna et al. 2005c; Andreotti et al. 2018); **Paraguay**: Villarica (Nava et al. 2007); and **Peru**: Cusco, Madre de Dios, and Puno (Labruna et al. 2005c).

Hosts: Males and females of *Amblyomma incisum* are usually found on *Tapirus terrestris* (Perissodactyla: Tapiridae). It can be difficult to identify the nymph of *Amblyomma incisum*, and the few records of this stage involve specimens taken from mammals, while the undescribed larva was collected on one occasion from Didelphidae by Lamattina et al. (2018a). Accordingly, the host range of larvae and nymphs of this tick is treated here as essentially undetermined. The host profile of this species is shown in Table 2.19, which is based on data from Labruna et al. (2005c), Acosta et al. (2016), Nava et al. (2017), and Lamattina et al. (2014, 2018a).

Human Parasitism: Guglielmone and Robbins (2018) list *Amblyomma incisum* as a sporadic parasite of humans in Argentina (Misiones), Brazil (São Paulo), and Paraguay (Guairá). Subsequent records of two females and one male of *Amblyomma*

Table 2.19 Hosts for males (M), females (F), nymphs (N), and larvae* (L) of *Amblyomma incisum*

MAMMALIA			
ARTIODACTYLA: Cervidae		**DIDELPHIMORPHIA: Didelphidae**	
Mazama bororo	N	*Didelphis aurita*	L
Mazama gouazoubira	N	**PERISSODACTYLA: Equidae**	
CARNIVORA: Canidae		Horse	M
Cerdocyon thous	N*	**PERISSODACTYLA: Tapiridae**	
CARNIVORA: Felidae		*Tapirus terrestris*	MFN
Panthera onca	N**	**RODENTIA: Caviidae**	
Puma concolor	N**	*Hydrochoerus hydrochaeris*	M
CARNIVORA: Procyonidae			
Nasua nasua	N**		

* Provisional diagnosis because the larva of *Amblyomma incisum* remains undescribed

** Lamattina *et al.* (2014) state that these nymphs of *Amblyomma incisum* are indistinguishable from the nymphs of *Amblyomma brasiliense* *incisum* found on humans in Rio Grande do Sul (Brazil) by Reck et al. (2018), and four females, and four nymphs collected from people in the Brazilian state of Paraná by Santos et al. (2020) and Valente et al. (2020), can now be added to that list. All parasitic stages of *Amblyomma incisum* have been collected from people, but the nymph is the prevalent stage causing human parasitism.

Comments: Labruna et al. (2005c) state that *Amblyomma incisum* has been extensively confused with *Amblyomma latepunctatum.* Therefore, records of *Amblyomma incisum* from Ecuador, French Guiana, and Peru (Huánuco and Loreto) require confirmation and have been excluded from this species' geographic range, while all specimens of alleged *Amblyomma incisum* from Guyana and Venezuela, as revised by Labruna et al. (2005c), have been reclassified as *Amblyomma latepunctatum.* As a result, human records of *Amblyomma incisum* from French Guiana compiled by Guglielmone et al. (2006a) are now doubtful. Lamattina et al. (2014) state that there are difficulties involved in morphologically separating the nymphs of *Amblyomma incisum* and *Amblyomma brasiliense*, and molecular taxonomy may serve as a decisive tool in their diagnosis.

Northern populations of *Amblyomma incisum* from the Brazilian and Peruvian Amazon show molecular variation and morphological differences in scutal punctations, coxal spurring, and the development of the cornua, in contrast to southern populations from Argentina, southern Brazil, and Paraguay (Labruna et al. 2005c). Perhaps these populations represent different species. Labruna et al. (2005c) examined the two Bolivian male syntypes of *Amblyomma incisum*, finding that one male corresponded to the northern population of this tick, but the other male was, in fact, a specimen of *Amblyomma scalpturatum*, probably its holotype, which had been thought lost, a situation earlier anticipated by Fonseca and Aragão (1952b).

Colombian records of *Amblyomma incisum* in Acevedo-Gutiérrez et al. (2020) require confirmation, and Colombia is not included within the range of this tick. The Brazilian states of Mato Grosso, Rio de Janeiro, and Rio Grande do Sul are not included in the range of *Amblyomma incisum* by Andreotti et al. (2018), but no reasons were given for this exclusion.

Estrada-Peña et al. (2020) include the nymph of *Amblyomma incisum* in a cluster of Neotropical *Amblyomma* prone to feed mostly on artiodactyls, but existing information on the principal hosts of the immature stages of this tick is considered here to be insufficient to support this view.

26. *Amblyomma inornatum* (Banks, 1909) under the name *Aponomma inornata*, and given its current status by Fairchild (1943)

Banks, N. (1909) Three new ticks from the United States. *Proceedings of the Entomological Society of Washington*, 10, 170–173.

Fairchild, G.B. (1943) An annotated list of the bloodsucking insects, ticks and mites known from Panama. *American Journal of Tropical Medicine and Hygiene*, 23, 569–591.

Descriptions and Redescriptions: Banks (1909) described and figured the female of *Amblyomma inornatum* under the name *Aponomma inornata*; the male was described by Cooley and Kohls (1939) under the name *Amblyomma philipi*, a synonym of *Amblyomma inornatum*, as proposed by Cooley and Kohls (1944), while the nymph and larva were described by Eads and Borom (1975).

The male and female of *Amblyomma inornatum* were redescribed in Cooley and Kohls (1944), Voltzit (2007), and Guzmán-Cornejo et al. (2011, keys and figures). The female alone was redescribed in Cooley and Kohls (1939) under the name *Amblyomma philipi*. Keirans and Durden (1998) redescribed the nymph.

There are no redescriptions of the larva of *Amblyomma inornatum*.

Geographic Distribution: *Amblyomma inornatum* is largely a Nearctic species with controversial Neotropical records from **Costa Rica**: Guanacaste, and Puntarenas (Hermans et al. 1994; Álvarez et al. 2005); **Guatemala**: El Progreso, Huehuetenango, and Jutiapa (Eads and Borom 1975; Instituto Interamericano de Cooperación para la Agricultura 1988); and **southern Mexico**: Chiapas, Guerrero, Jalisco, Michoacán, Morelos, Nayarit, Oaxaca, Tabasco, Nayarit, and Yucatán (Guzmán-Cornejo et al. 2011; Light et al. 2020).

Hosts: *Amblyomma inornatum* has been found on a variety of hosts in the Nearctic Region, where males, females, nymphs, and larvae have been collected from several orders of mammals, Cuculiformes: Cuculidae, and Passeriformes: Icteridae and Mimidae; males, females, and nymphs have been recovered from Artiodactyla: Tayassuidae, Carnivora: Canidae and Felidae, and Rodentia: Sciuridae; adults alone have been collected from Carnivora: Mephitidae, Mustelidae and Procyonidae; while nymphs have been taken from Perissodactyla: Equidae, Galliformes: Phasianidae, and Passeriformes: Turdidae (Gladney et al. 1977; Guglielmone et al. 2014; Medlin et al. 2015; Scott et al. 2019). This host profile is partly an emendation of the profile provided in Guglielmone et al. (2014) and Guglielmone and Robbins (2018). Data concerning hosts of *Amblyomma inornatum* in the Neotropical Region are usually not accompanied by information on the tick stages found and thus differ from Nearctic records. Only three Neotropical reports of this species unambiguously address the tick stages found on hosts: Tovar (1944) found male and female ticks on "hare"; Eads and Borom (1975) listed an adult tick ex-*Sigmodon hispidus*, but the presence of this host in Mexico is doubtful and is here treated as *Sigmodon* sp.; and Guzmán-Cornejo et al. (2011) found males and females of *Amblyomma inornatum* on domestic dogs. It is uncertain whether any of these hosts are more prone to infestation by adults of this tick than other hosts. The Neotropical host profile of adults of *Amblyomma inornatum* is treated here as basically undetermined, while the Neotropical hosts of the nymphs and larvae remain unknown.

Human Parasitism: Guglielmone and Robbins (2018) list *Amblyomma inornatum* as a rare parasite of humans, based on one Neotropical record from southern Mexico (Veracruz) of undetermined tick stages reported in Hoffmann and López-Campos (2000) and treated as provisionally valid here.

Comments: Guglielmone et al. (2003a) state that Neotropical specimens of *Amblyomma inornatum* from Central America should be compared with *Amblyomma parvum*, *Amblyomma auricularium*, and even *Amblyomma beaurepairei* because of close morphological similarities among these species. Additional problems have since arisen because the identification of *Amblyomma auricularium* is currently uncertain, as discussed above, and alleged *Amblyomma parvum* from Central America most probably represents a species different from *bona fide* South American *Amblyomma parvum* (Lado et al. 2016). Consequently, all published information concerning Neotropical records of *Amblyomma inornatum* should be considered provisionally valid.

Delabra-Vaca et al. (1996) claim that *Amblyomma inornatum* has been found in French Guiana and Guyana, but no data were provided to support their statement. Muñoz and Casanueva (2001) wrongly state that *Amblyomma inornatum* has been found in Venezuela. Fairchild (1943) included this tick as part of the ixodid fauna of Panama, but later Fairchild et al. (1966) and Bermúdez et al. (2018a) excluded *Amblyomma inornatum* in their studies of Panamanian ticks. Arana-Guardia et al. (2015) allegedly collected *Amblyomma inornatum* in Mexico (Yucatán), but the morphological support for their records is questionable, and their data are excluded from this analysis.

Romero-Castañon et al. (2008) claim that they found the first specimens of *Amblyomma inornatum* from Mexico, ignoring many previous records for this country, and in the same paper stated that *Hyalomma* ticks had been collected from cattle in Chiapas (southern Mexico), without providing further information. The specimens identified by Romero-Castañon et al. (2008) need to be reexamined.

27. *Amblyomma interandinum* Beati, Nava and Cáceres, 2014 in Nava et al. (2014a)

Nava, S., Beati, L., Labruna, M.B., Cáceres, A.G., Mangold, A.J. & Guglielmone, A.A. (2014a) Reassessment of the taxonomic status of *Amblyomma cajennense* (Fabricius, 1787) with the description of three new species, *Amblyomma tonelliae* n. sp., *Amblyomma interandinum* n. sp., and *Amblyomma patinoi* n. sp., and reinstatement of *Amblyomma mixtum* Koch, 1844 and *Amblyomma sculptum* Berlese, 1888 (Ixodida: Ixodidae). *Ticks and Tick-borne Diseases*, 5, 252–276.

Descriptions and Redescriptions: The male and female of *Amblyomma interandinum* were described in Nava et al. (2014a).

There are no redescriptions of the male and female of *Amblyomma interandinum*.

The nymph and larva of *Amblyomma interandinum* remain unknown.

Geographical Distribution: *Amblyomma interandinum* has been found in **Peru**: Cajamarca (Nava et al. 2014a).

Hosts: The only known host for males and females of *Amblyomma interandinum* is "deer" (Artiodactyla: Cervidae).

Human Parasitism: No.

Comments: *Amblyomma interandinum* is a member of the *Amblyomma cajennense* species complex, which also comprises *Amblyomma mixtum*, *Amblyomma patinoi*, *Amblyomma sculptum,* and *Amblyomma tonelliae*. Further discussion of this complex is provided under *Amblyomma cajennense* above.

28. *Amblyomma latepunctatum* Tonelli Rondelli, 1939

Tonelli Rondelli, M. (1939) Ixodoidea. Parte II. Contributo alla conoscenza della fauna ixodologica Sud-americana. *Rivista di Parassitologia*, 3, 39–55.

Descriptions and Redescriptions: The female of *Amblyomma latepunctatum* was described and illustrated by Tonelli Rondelli (1939), while the male was described by Labruna et al. (2005c); the nymph was described by Martins et al. (2010).

The male and female of *Amblyomma latepunctatum* were redescribed in Onofrio et al. (2006b, keys and figures), Voltzit (2007), and Dantas-Torres et al. (2019b, keys and figures), while the female alone was redescribed by Labruna et al. (2005c).

There are no redescriptions of the nymph of *Amblyomma latepunctatum.*

The larva of *Amblyomma latepunctatum* remains undescribed, but Binetruy et al. (2019), using molecular methods, have been able to identify larvae collected from a human.

Geographical Distribution: *Amblyomma latepunctatum* is a Neotropical species that has only been found in northern South America in **Brazil**: Amazonas, Pará, and Rondônia (Labruna et al. 2005c); **Ecuador**: Pastaza, and Sucumbíos (Labruna et al. 2005c); **French Guiana**: Cayena (Binetruy et al. 2019); **Guyana**: East Berbice-Corentyne, and Upper Demerara-Berbice (Tonelli Rondelli 1939; Labruna et al. 2005c); **Peru**: Cusco, Huánuco, Loreto, Madre de Dios, and Puno (Labruna et al. 2005c); and **Venezuela**: Amazonas, and Bolívar (Labruna et al. 2005c).

Hosts: The only known host for males and females of *Amblyomma latepunctatum* is *Tapirus terrestris* (Perissodactyla: Tapiridae). A few nymphs have been recorded from three orders of mammals, but the host range of this stage is treated here as basically undetermined, while the hosts for larvae of *Amblyomma latepunctatum* remain unknown. The host profile of *Amblyomma latepunctatum* is shown in Table 2.20, which is based on data from Labruna et al. (2010a, b), Martins et al. (2014a), and Gianizella et al. (2018b).

Human Parasitism: Guglielmone and Robbins (2018) list *Amblyomma latepunctatum* as a very rare parasite of humans, with two records of adult ticks (a male and a female) found feeding on people in Brazil (Amazonas) (Gianizella et al. 2018b). Later, Binetruy et al. (2019) found two nymphs and, allegedly, one larva of *Amblyomma latepunctatum* on humans in French Guiana, and this larval diagnosis is treated as provisionally valid here.

Table 2.20 Hosts for males (M), females (F), and nymphs (N) *Amblyomma latepunctatum*

MAMMALIA			
ARTIODACTYLA: Tayassuidae		**RODENTIA: Caviidae**	
Tayassu pecari	N	*Hydrochoerus hydrochaeris*	N
DIDELPHIMORPHIA: Didelphidae		**RODENTIA: Dasyproctidae**	
Didelphis maruspialis	N	*Dasyprocta fuliginosa*	N
PERISSODACTYLA: Tapiridae			
Tapirus terrestris	MF		

Comments: Labruna et al. (2005c) reinstated *Amblyomma latepunctatum*, a species that had been extensively confused with *Amblyomma incisum* and *Amblyomma scalpturatum*. Prior to the study of Labruna et al. (2005c), *Amblyomma latepunctatum* was treated as a synonym of *Amblyomma scalpturatum* by Camicas et al. (1998) and others.

29. *Amblyomma longirostre* (Koch, 1844) under the name *Haemalastor longirostris* and given its current status in Neumann (1905)

Koch, C.L. (1844) Systematische Übersicht über die Ordnung der Zecken. *Archiv für Naturgeschichte*, 10, 217–239.

Neumann, L.G. (1905) Notes sur les ixodidés. III. *Archives de Parasitologie*, 9, 225–241.

Descriptions and Redescriptions: Koch (1844) described the female of *Amblyomma longirostre* under the name *Haemalastor longirostris*, while Karsch (1880) described the male under the name *Haemalastor crassitarus*, a synonym of *Amblyomma longirostre*, as stated in Neumann (1911a). The nymph was described by Marx and Neumann in Neumann (1899) under the name *Amblyomma avicola*, another synonym of *Amblyomma longirostre* (Neumann 1911a). None of these authors included illustrations with their descriptions. The larva of *Amblyomma longirostre* was described by Barros-Battesti et al. (2005a).

The male, female and nymph of *Amblyomma longirostre* were redescribed by Nava et al. (2017). There are numerous redescriptions of the male and female, as in Robinson (1926), Aragão and Fonseca (1961a), Boero and Delpietro (1970), Onofrio et al. (2006b, keys and figures), Labruna et al. (2009), Guzmán-Cornejo et al. (2011, keys and figures), Bermúdez et al. (2018a) and Dantas-Torres et al. (2019b, keys and figures), among others. Enríquez et al. (2020) redescribed the female alone, while the nymph of *Amblyomma longirostre* was redescribed in Floch and Abonnenc (1940), Keirans and Durden (1998), Venzal et al. (2003a), and Martins et al. (2010).

There are no redescriptions of the larva of *Amblyomma longirostre.*

Geographic Distribution: *Amblyomma longirostre* is a widespread Neotropical tick that has been found in southern North America, Central and South America and one Caribbean island as follows: **Argentina**: Misiones (Boero and Delpietro 1970); **Belize**: Cayo District (Cooley and Kohls 1944); **Bolivia**: Cochabamba (Keirans 1982); **Brazil**: Acre, Amapá, Amazonas, Bahía, Ceará, Espírito Santo, Goiás,

Maranhão, Mato Grosso, Mato Grosso do Sul, Minas Gerais, Pará, Paraíba, Paraná, Pernambuco, Rio de Janeiro, Rio Grande do Sul, Rondônia, Santa Catarina, and São Paulo (Guimarães et al. 2001; Labruna et al. 2002d; Arzua et al. 2005; Ogrzewalska et al. 2010; Martins et al. 2014a; Lugarini et al. 2015; Moerbeck et al. 2016; Witter et al. 2016; Oniki-Willis and Willis 2018; Costa et al. 2020); **Colombia**: Arauca, Meta, and Boyacá (Acevedo-Gutiérrez et al. 2020; Cardona-Romero et al. 2020); **Costa Rica**: Alajuela, Cartago, Guanacaste, Limón, Puntarenas, and San José (Tonn et al. 1963; Álvarez et al. 2005); **Ecuador**: Zamora Chinchipe (Enríquez et al. 2020); **French Guiana**: Cayena, and Saint Laurent du Maroni (Binetruy et al. 2019); **Honduras**: Atlántida (Novakova et al. 2015); **southern Mexico**: Chiapas, and Veracruz (Guzmán-Cornejo et al. 2011); **Panama**: Bocas del Toro, Cocle, Colón, Darién, Panamá, Panamá Oeste, and Veraguas (Fairchild et al. 1966; Miller et al. 2016); **Paraguay**: Caa-Guazú, Canindeyú, and Itapuá (Massi Pallarés and Benítez Usher 1982; Oniki-Willis and Willis 2018); **Peru**: San Martín (Nava et al. 2010b); **Trinidad and Tobago**: Rio Claro-Mayaro, and Saint Andrew (Aitken et al. 1968; Voltzit 2007); **Uruguay**: Rivera, and Tacuarembó (Venzal et al. 2003a, 2005a); and **Venezuela**: Aragua, Apure or Carabobo, Barinas, Distrito Federal, Falcón, Lara, Mérida, Monagas, Yaracuy, and Zulia (Robinson 1926; Jones et al. 1972; Verea and Aponte 2004; Rodríguez-Peraza et al. 2014).

Hosts: Males and females of *Amblyomma longirostre* are usually found on Rodentia: Erethizontidae, with most records from *Coendou prehensilis*, *Coendou quichua,* and *Coendou spinosus*, but these stages have been collected from seven other species of this family, and some of them may be more important as hosts of this tick than currently realized, because the three species of *Coendou* cited above as principal hosts are not found over the entire range of this tick. Adults of *Amblyomma longirostre* have occasionally been collected on other mammals and rarely from birds. Larvae and nymphs are frequently collected from several families of Passeriformes, and more than 230 species of Passeriformes have been found carrying this tick. However, birds from 12 other orders have been found infected with the immatures of *Amblyomma longirostre.* Overall, it appears that nymphs and larvae of *Amblyomma longirostre* do not show a clear preference for particular avian families other than Passeriformes, which seem to be the most important hosts for the immature stages of this tick. There are also records of larvae or nymphs collected from a few mammals. The host profile of *Amblyomma longirostre* is shown in Table 2.21, which is based on data from Tonn et al. (1963), Fairchild et al. (1966), Boero and Delpietro (1970), Jones et al. (1972), Everard and Tikansingh (1973), Keirans (1982), Barros and Baggio (1992), Guerrero (1996), Silva et al. (1999), Verea and Aponte (2004), Arzua et al. (2005), Labruna et al. (2007a), Voltzit (2007), Silveira et al. (2008), Moraes et al. (2009), Tolesano-Pascoli et al. (2010), Ogrzewalska et al. (2012a, 2014, 2015), Pacheco et al. (2012), Sanches et al. (2013), Martins et al. (2014b), Maturano et al. (2015), Esser et al. (2016b), Luz et al. (2016b, 2017a, b, 2018a), Witter et al. (2016), Martins et al. (2017), Nascimento et al. (2017), Nava et al. (2017), Zeringóta et al. (2017), Bermúdez et al. (2018a), Lamattina et al. (2018a), Lima et al. (2018), Oniki-Willis and Willis

Table 2.21 Hosts for adult ticks of undetermined sex (A), males (M), females (F), nymphs (N), and larvae (L) of *Amblyomma longirostre**

MAMMALIA			
ARTIODACTYLA: Cervidae		**RODENTIA: Cricetidae**	
Blastocerus dichotomus	F	*Hylaeamys megacephalus*	N
CARNIVORA: Canidae		**RODENTIA: Echimyidae**	
Cerdocyon thous	A	*Proechimys guyannensis*	N
Domestic dog	MF	*Trinomys dimidiatus*	F
CARNIVORA: Felidae		**RODENTIA: Erethizontidae**	
Leopardus tigrinus	N	*Chaetomys subspinosus*	MF
CARNIVORA: Mustelidae		*Coendou bicolor*	MF
Eira barbara	M	*Coendou insidiosus*	MF
CHIROPTERA: Phyllostomidae		*Coendou mexicanus*	MF
Artibeus lituratus	N	*Coendou nycthemera*	A
DIDELPHIMORPHIA: Didelphidae		*Coendou prehensilis*	MFN
Didelphis marsupialis	L	*Coendou quichua*	MFN
PERISSODACTYLA: Equidae		*Coendou spinosus*	MFN
Horse	A	*Coendou vestitus*	M
PILOSA: Bradypodidae		**RODENTIA: Sciuridae**	
Bradypus torquatus	F	*Sciurus granatensis*	N
Bradypus tridactylus	N		
AVES			
ACCIPITRIFORMES: Accipitridae		*Phaenostictus macleannani*	NL
Accipiter bicolor	N	*Phlegopsis nigromaculata*	L
Buteo platypterus	L	*Pyriglena leuconota*	NL
Buteogallus lacernulatus	N	*Pyriglena leucoptera*	NL
Rupornis magnirostris	NL	*Taraba major*	NL
Spizaetus melanoleucus	N	*Thamnomanes caesius*	L
CAPRIMULGIFORMES: Caprimulgidae		*Thamnophilus aethiops*	N
Nyctidromus albicollis	MN	*Thamnophilus atrinucha*	NL
CAPRIMULGIFORMES: Trochilidae		*Thamnophilus bridgesi*	N
Amazilia tzacatl	N	*Thamnophilus caerulescens*	MNL
Aphantochroa cirrochloris	N	*Thamnophilus doliatus*	NL
Campylopterus longipennis	L	*Thamnophilus multistriatus*	N
Chlorostilbon lucidus	N	*Thamnophilus nigriceps*	N
Clytolaema rubricauda	N	*Thamnophilus pelzelni*	NL
Eupetomena macroura	N	*Thamnophilus punctatus*	NL
Florisuga fusca	N	*Willisornis poecilonotus*	NL
Leucochloris albicollis	N	**PASSERIFORMES: Thraupidae**	
Phaethornis guy	L	*Coereba flaveola*	NL
Phaethornis superciliosus	N	*Dacnis cayana*	NL
Thalurania glaucopis	NL	*Eucometis penicillata*	NL
COLUMBIFORMES: Columbidae		*Haplospiza unicolor*	N
Columbina talpacoti	N	*Nemosia pileata*	N
Geotrygon montana	N	*Pipraeidea melanonota*	N
Leptotila rufaxilla	N	*Ramphocelus bresilius*	NL
Leptotila verreauxi	N	*Ramphocelus carbo*	NL
CORACIIFORMES: Alcedinidae		*Ramphocelus sanguinolentus*	N
Chloroceryle aenea	N	*Saltator atriceps*	L
CORACIIFORMES: Momotidae		*Saltator maximus*	NL
Baryphthengus martii	N	*Saltator similis*	NL
Baryphthengus ruficapillus	NL	*Sicalis flaveola*	N
Momotus momota	L	*Sporophila angolensis*	N
CUCULIFORMES: Cuculidae		*Sporophila caerulescens*	N
Crotophaga ani	N	*Sporophila corvina*	N
Crotophaga major	N	*Sporophila leucoptera*	N
Piaya cayana	N	*Sporophila lineola*	N
FALCONIFORMES: Falconidae		*Sporophila nigricollis*	NL
Herpetotheres cachinnans	N	*Tachyphonus coronatus*	NL
Micrastur ruficollis	N	*Tachyphonus cristatus*	NL
GALBULIFORMES: Bucconidae		*Tachyphonus luctuosus*	N
Malacoptila panamensis	N	*Tachyphonus rufus*	NL
Malacoptila striata	NL	*Tachyphonus surinamus*	L
GALBULIFORMES: Galbulidae		*Tangara cayana*	N
Galbula ruficauda	L	*Tangara peruviana*	N

(continued)

Table 2.21 (continued)

GALLIFORMES: Cracidae	
Penelope obscura	F
GRUIFORMES: Rallidae	
Laterallus albigularis	L
PASSERIFORMES: Cardinalidae	
Cyanoloxia brissonii	N
Cyanoloxia cyanoides	NL
Habia fuscicauda	N
Habia gutturalis	L
Habia rubica	NL
Piranga flava	NL
PASSERIFORMES: Conopophagidae	
Conopophaga lineata	NL
Conopophaga melanops	L
PASSERIFORMES: Corvidae	
Cyanocorax cristatellus	N
PASSERIFORMES: Cotingidae	
Pyroderus scutatus	N
Querula purpurata	N
PASSERIFORMES: Fringillidae	
Euphonia hirundinacea	L
Euphonia laniirostris	NL
Euphonia pectoralis	NL
Euphonia violacea	N
Euphonia xanthogaster	N
PASSERIFORMES: Furnariidae	
Anabacerthia amaurotis	N
Anabazenops fuscus	NL
Automolus leucophthalmus	NL
Automolus ochrolaemus	L
Berlepschia rikeri	N
Campylorhamphus falcularius	NL
Campylorhamphus trochilirostris	N
Certhiasomus stictolaemus	L
Certhiaxis cinnamomeus	N
Deconychura longicauda	NL
Dendrocincla anabatina	N
Dendrocincla fuliginosa	NL
Dendrocinca merula	NL
Dendrocincla turdina	L
Dendrocolaptes certhia	N
Dendrocolaptes hoffmannsi	NL
Dendrocolaptes picumnus	N
Dendroplex picus	NL
Glyphorynchus spirurus	NL
Hylexetastes perrotii	NL
Lepidocolaptes angustirostris	N
Lepidocolaptes souleyetii	NL
Lepidocolaptes squamatus	L
Philydor atricapillus	N
Philydor rufum	NL
Sittasomus griseicapillus	NL
Synallaxis ruficapilla	NL
Synallaxis spixi	N
Syndactyla rufosuperciliata	N
Xenops minutus	NL
Xiphocolaptes albicollis	N
Xiphorhynchus elegans	NL
Xiphorhynchus fuscus	NL
Xiphorhynchus guttatus	NL
Xiphorhynchus lachrymosus	N
Xiphorhynchus pardalotus	L
Xiphorhynchus susurrans	NL
PASSERIFORMES: Icteridae	
Tangara seledon	N
Tersina viridis	NL
Thlypopsis pyrrhocoma	N
Thlypopsis sordida	N
Thraupis ornata	N
Thraupis palmarum	NL
Thraupis sayaca	NL
Tiaris olivaceus	N
Trichothraupis melanops	NL
PASSERIFORMES: Tityridae	
Pachyramphus castaneus	L
Pachyramphus polychopterus	NL
Schiffornis turdina	NL
Schiffornis virescens	NL
PASSERIFORMES: Troglodytidae	
Cantorchilus longirostris	N
Cantorchilus modestus	L
Cantorchilus nigricapillus	NL
Cyphorhinus phaeocephalus	N
Henicorhina leucosticta	L
Microcerculus marginatus	N
Pheugopedius atrogularis	L
Thryophilus rufalbus	L
Troglodytes aedon	NL
PASSERIFORMES: Turdidae	
Catharus ustulatus	NL
Turdus albicollis	NL
Turdus amaurochalinus	NL
Turdus flavipes	N
Turdus fumigatus	NL
Turdus grayi	NL
Turdus hauxwelli	N
Turdus ignobilis	N
Turdus leucomelas	NL
Turdus nigriceps	N
Turdus nudigensis	N
Turdus obsoletus	L
Turdus rufiventris	NL
Turdus sanchezorum	N
Turdus subalaris	NL
PASSERIFORMES: Tyrannidae	
Attila rufus	N
Attila spadiceus	N
Camptostoma obsoletum	N
Casiornis rufus	N
Cnemotriccus fuscatus	N
Elaenia cristata	MN
Elaenia flavogaster	NL
Elaenia mesoleuca	N
Elaenia parvirostris	N
Empidonax alnorum	L
Empidonax virescens	L
Hemitriccus margaritaceiventer	N
Hemitriccus minor	L
Hemitriccus nidipendulus	N
Lathrotriccus euleri	N
Leptopogon amaurocephalus	NL
Leptopogon superciliaris	L
Machetornis rixosa	N
Megarynchus pitangua	N
Mionectes macconnelli	L
Mionectes oleaginous	NL
Mionectes olivaceus	L
Mionectes rufiventris	NL

(continued)

Table 2.21 (continued)

Cacicus cela	N	*Myiarchus ferox*	NL
Cacicus haemorrhous	N	*Myiarchus tuberculifer*	N
Cacicus uropygialis	NL	*Myiobius barbatus*	L
Icterus chrysater	N	*Myiodynastes maculatus*	N
Icterus icterus	N	*Myiopagis viridicata*	L
PASSERIFORMES: Parulidae		*Myiophobus fasciatus*	L
Basileuterus culicivorus	NL	*Myiozetetes similis*	N
Basileuterus rufifrons	FNL	*Oncostoma olivaceum*	NL
Cardellina canadensis	L	*Onychorhynchus coronatus*	L
Geothlypis aequinoctialis	N	*Phylloscartes kronei*	N
Myiothlypis flaveola	NL	*Phylloscartes ventralis*	N
Myiothlypis leucoblephara	N	*Pitangus sulphuratus*	N
Setophaga castanea	L	*Platyrinchus cancrominus*	L
PASSERIFORMES: Passerellidae		*Platyrinchus coronatus*	N
Arremonops conirostris	N	*Platyrinchus leucoryphus*	L
PASSERIFORMES: Pipridae		*Platyrinchus mystaceus*	NL
Antilophia galeata	NL	*Poecilotriccus latirostris*	L
Ceratopipra erythrocephala	NL	*Poecilotriccus sylvia*	L
Ceratopipra mentalis	NL	*Rhynchocyclus brevirostris*	N
Chiroxiphia caudata	NL	*Rhynchocyclus olivaceus*	NL
Chiroxiphia linearis	L	*Todirostrum cinereum*	L
Chiroxiphia pareola	NL	*Tolmomyias poliocephalus*	L
Dixiphia pipra	NL	*Tolmomyias sulphurescens*	FNL
Ilicura militaris	NL	*Tyrannus melancholicus*	N
Lepidothrix coronata	NL	*Tyrannus savanna*	N
Machaeropterus pyrocephalus	N	*Xolmis cinereus*	N
Machaeropterus regulus	N	*Xolmis velatus*	N
Manacus candei	NL	**PASSERIFORMES: Vireonidae**	
Manacus manacus	NL	*Cyclarhis gujanensis*	N
Neopelma pallescens	NL	*Vireo chivi*	L
Pipra aureola	NL	**PICIFORMES: Capitonidae**	
Pipra fasciicauda	NL	*Capito auratus*	N
PASSERIFORMES: Thamnophilidae		**PICIFORMES: Picidae**	
Biatas nigropectus	N	*Celeus flavescens*	N
Cercomacra tyrannina	L	*Dryobates maculifrons*	N
Cymbilaimus lineatus	N	*Picumnus cirratus*	N
Drymophila ferruginea	N	*Picumnus exilis*	NL
Dysithamnus mentalis	NL	*Picumnus temminckii*	N
Dysithamnus stictothorax	N	**PICIFORMES: Ramphastidae**	
Epinecrophylla leucophthalma	N	*Pteroglossus castanotis*	N
Formicivora melanogaster	N	*Ramphastos dicolorus*	N
Gymnopithys leucaspis	NL	*Ramphastos vitellinus*	N
Hylophylax naevioides	N	**STRIGIFORMES: Strigidae**	
Hylophylax naevius	L	*Asio clamator*	NL
Hypocnemoides maculicauda	NL	*Asio stygius*	N
Mackenziaena severa	N	*Megascops choliba*	N
Myrmoderus ferrugineus	L	*Pulsatrix koeniswaldiana*	N
Myrmotherula axillaris	NL	**TROGONIFORMES: Trogonidae**	
Myrmotherula longipennis	L	*Trogon viridis*	N
Percnostola rufifrons	NL		

* Some records of males and females of *Amblyomma longirostre* may have been confused with those for *Amblyomma geayi* or *Amblyomma parkeri*, and the likehood of confusion among these species increases for the larvae and nymphs

(2018), Binetruy et al. (2019, 2020a, b), Dolz et al. (2019), Domínguez et al. (2019), Pacheco et al. (2019), Cardona-Romero et al. (2020), Enríquez et al. (2020), Costa et al. (2020), Fecchio et al. (2020) and Teixeira et al. (2020).

Human Parasitism: *Amblyomma longirostre* is listed as a rare parasite of humans in Guglielmone and Robbins (2018), because only a nymph, a male, and few female ticks have been found on people in Brazil (Paraná, and São Paulo), French Guiana (Cayena) and Venezuela (Yaracuy). Reck et al. (2018) added a female tick collected from a person in Rio Grande do Sul (Brazil), while Valente et al. (2020) found two nymphs feeding on humans in the Brazilian state of Paraná.

Comments: There can be problems in identifying this species. Labruna et al. (2009) described several cases where adults of *Amblyomma geayi* and *Amblyomma parkeri* were confused with *Amblyomma longirostre*; therefore, it is quite possible that some of the records of males and females listed above are misidentifications. The nymphs of these three species are also morphologically close (Martins et al. 2010, 2013), so it is almost certain that some nymphal records of *Amblyomma longirostre* actually represent *Amblyomma geayi* or *Amblyomma parkeri*. Larvae are probably also extensively misidentified. *Amblyomma romarioi*, another species belonging to this group (Martins et al. 2019b), may add to the confusion.

The Belize record of *Amblyomma longirostre* in Cooley and Kohls (1944) was published under the name *Amblyomma avecolens*, a synonym of *Amblyomma longirostre*, as stated in Jones et al. (1972).

Rawlins et al. (1993) list Guyana and Jamaica as being within the range of *Amblyomma longirostre* but provided no additional information and these countries are not included within the geographic distribution of this tick. There are several records of nymphs and larvae of *Amblyomma longirostre* collected from migratory birds in the Nearctic Region, but there is no evidence that this tick is established there. Andreotti et al. (2018) do not include the Brazilian states of Bahia, Mato Grosso, Mato Grosso do Sul, and Paraíba within the geographical distribution of *Amblyomma longirostre*, but no reasons were given for their exclusion.

Marx and Neumann in Neumann (1899) described *Amblyomma longirostre* (as *Amblyomma avicola*) from a nymph found on a "gobe-mouche" (= *Muscicapa* spp.) from Trinidad, but this genus of birds is not found in the Neotropics (Nava et al. 2010b). Records of *Amblyomma longirostre* from Echimyidae and Cricetidae in Everard and Tikansingh (1973) were not included in the host profile of this species in Guglielmone et al. (2014), but Luz et al. (2018a) also found *Amblyomma longirostre* on Echimyidae, prompting a reevaluation of the records of Everard and Tikansingh (1973), which are now treated as provisionally valid, including their record from Cricetidae.

Hundreds of specimens of *Amblyomma longirostre* allegedly collected from people by Serra-Freire (2010) require confirmation.

30. *Amblyomma macfarlandi* Keirans, Hoogstraal and Clifford, 1973

Keirans, J.E., Hoogstraal, H. & Clifford, C.M. (1973) The *Amblyomma* (Acarina: Ixodidae) parasitic on giant tortoises (Reptilia: Testudinidae) of the Galapagos Islands. *Annals of the Entomological Society of America*, 66, 673–688.

Descriptions and Redescriptions: The male of *Amblyomma macfarlandi* was described by Keirans et al. (1973), while the female was described by Robinson (1926) under the name *Amblyomma pilosum.* The larva and nymph of *Amblyomma macfarlandi* were described by Keirans et al. (1973).

Voltzit (2007) redescribed the male and female of *Amblyomma macfarlandi.*

There are no redescriptions of the nymph and larva of *Amblyomma macfarlandi.*

Geographical Distribution: *Amblyomma macfarlandi* has only been found on several islands belonging to **Ecuador**: Galápagos (Keirans et al. 1973).

Hosts: *Amblyomma macfarlandi* is a parasite of Testudines: Testudinidae. Males, females, nymphs, and larvae have been collected from *Chelonoidis vicina* (under the name *Geochelone elephantopus vicina*), while males, females, and nymphs have been found on *Chelonoidis porteri* (under the name *Geochelone elephantopus porteri*); these two tortoises are considered principal hosts of *Amblyomma macfarlandi.* Males and females of this tick have also been recovered from *Chelonodis* sp., named as "*Geochelone elephantopus* subspecies." Host information for *Amblyomma macfarlandi* is from Keirans et al. (1973).

Human Parasitism: Guglielmone et al. (2006a) listed a nymph of *Amblyomma macfarlandi* found on a human, but it is uncertain whether the tick was biting.

Comments: Keirans et al. (1973) examined the specimen used by Robinson (1926) in his redescription of *Amblyomma pilosum* and the type specimen utilized in Neumann's (1899) description of this tick, finding that they represent different species. In fact, the female tick described by Robinson (1926) as *Amblyomma pilosum* corresponds to the female of *Amblyomma macfarlandi.* The figures of the male and female of *Amblyomma macfarlandi* in Keirans et al. (1973) are quite similar to the corresponding figures of Neumann (1906) in his description of *Amblyomma hirtum.* However, Keirans, J.E., in Guglielmone et al. (2003a), described the differences between these species, as discussed under *Amblyomma hirtum* above.

31. *Amblyomma maculatum* Koch, 1844

Koch, C.L. (1844) Systematische Übersicht über die Ordnung der Zecken. *Archiv für Naturgeschichte*, 10, 217–239.

Descriptions and Redescriptions: The male of *Amblyomma maculatum* was described but not figured by Koch (1844), while the female was described and illustrated in Hunter and Hooker (1907); the larva and nymph were described and figured in Hooker et al. (1912).

The male, female, nymph, and larva were redescribed by Cooley and Kohls (1944), Estrada-Peña et al. (2005), and Lindquist et al. (2016); Sonenshine (1979) redescribed the female, nymph, and larva. Banks (1908), Hooker et al. (1912), Kohls (1956b), Guzmán-Cornejo et al. (2011, keys and figures), and Lado et al. (2018) redescribed the male and female, while Hunter and Hooker (1907) redescribed the male. Dubie et al. (2017) redescribed the nymph and larva of *Amblyomma maculatum*, Keirans and Durden (1998) redescribed the nymph, and Clifford et al. (1961) and Coley (2015) redescribed the larva.

Geographical Distribution: *Amblyomma maculatum* is a Nearctic and Neotropical species but its range in the Neotropics is only provisionally understood, with its southern limit in Peru as follows: **Belize**: Cayo (Polsomboon et al. 2017); **Colombia**: Bogotá, Caldas, Casanare, Cundinamarca, Meta, Nariño, Tolima, and Valle del Cauca (Osorno Mesa 1942; Wells et al. 1981; Keirans 1985b; Acero et al. 2011; Lado et al. 2018; Rivera-Páez et al. 2018; Acevedo-Gutiérrez et al. 2020); **Costa Rica**: Alajuela, Limón, Puntarenas, and San José (Álvarez et al. 2005); **Ecuador**: Santo Domingos de los Tsáchilas (Maya-Delgado et al. 2020); **Guatemala**: Alta Verapaz, Chimaltenango, Escuintla, and Izabal (Instituto Interamericano de Cooperación para la Agricultura 1988; Lado et al. 2018); **Honduras**: Atlántida (Lado et al. 2018); **southern Mexico**: Campeche, Chiapas, Jalisco, Michoacán, Nayarit, Tabasco, Veracruz, and Yucatán (Espinoza-Gómez et al. 2011; Guzmán-Cornejo et al. 2011; Vargas-Sandoval et al. 2014; Rodríguez-Vivas et al. 2016); **Nicaragua**: Carazo, Costa Caribe Norte, Costa Caribe Sur, Nueva Segovia, Río San Juan, and Rivas (Düttmann et al. 2016; Lado et al. 2018); **Peru**: Cusco, Ica, and Piura (Mendoza-Uribe and Chávez-Chorocco 2004); and **Venezuela**: Aragua, Distrito Federal, Guárico, Monagas, Lara, and Zulia (Vogelsang and Cordero 1940; Kohls 1956b; Jones et al. 1972; Manilla 1984; Manzanilla et al. 2002; Clavijo et al. 2009).

Hosts: The host range of *Amblyomma maculatum* is broad in the Nearctic Region, where males, females, nymphs, and larvae have been recovered from several orders of mammals, but adult ticks are usually found on large-sized mammals, while nymphs and larvae are commonly recovered from smaller mammals and Aves (several orders), but mostly Galliformes: Odontophoridae and Passeriformes (several families), with some odd records of adult ticks from Squamata: Teiidae. The situation is different in the Neotropical Region, where records of larvae and nymphs are few, and the host range of *Amblyomma maculatum* immatures is basically undetermined. Our knowledge of hosts for adult ticks in the Neotropics is based on more than 50 records, but the great majority of these are from Artiodactyla: Bovidae, Carnivora: Canidae, and Perissodactyla: Equidae. However, of about 587 specimens collected from such hosts, 437 (74% of the total) were from the domestic dog (Canidae), which is provisionally considered the principal host of

Table 2.22 Hosts for males (M), females (F), nymphs (N), and larvae (L) *Amblyomma maculatum**

MAMMALIA			
ARTIODACTYLA: Bovidae		**CARNIVORA: Canidae**	
Cattle	MF	*Cerdocyon thous*	F
Sheep	MFNL	Domestic dog	MFN
ARTIODACTYLA: Cervidae		**PERISSODACTYLA: Equidae**	
"Deer"	F	Donkey	M
ARTIODACTYLA: Suidae		Horse	MF
Domestic pig	MF	**RODENTIA: Cricetidae**	
		Sigmodon sp.	FN
AVES			
PASSERIFORMES: Pipridae			
Manacus candei	N		

* Provisional data because of uncertainties concerning the identification of *Amblyomma maculatum*

Amblyomma maculatum in the Neotropics. The Neotropical host profile of *Amblyomma maculatum* is shown in Table 2.22, which is based on data from Fiasson (1949), Jones et al. (1972), Keirans (1985b), Mendoza-Uribe and Chávez-Chorocco (2004), Vargas-Sandoval et al. (2014), Lira-Amaya et al. (2015), Ogrzewalska et al. (2015), and Polsomboon et al. (2017).

Human Parasitism: *Amblyomma maculatum* is listed as a frequent parasite of humans in Guglielmone and Robbins (2018), but most records are from the Nearctic Region. Merten and Durden (2000) stressed that this species is one of the eight principal human-biting ticks in the USA, but records of human parasitism by *Amblyomma maculatum* in the Neotropical Region consist of just two cases, one in Belize (Cayo) caused by adult ticks, and the other in Mexico (Jalisco) where two females were found on a person.

Comments: *Amblyomma maculatum*, *Amblyomma tigrinum,* and *Amblyomma triste* are morphologically and phylogenetically related species, and for decades the latter two species were treated as synonyms of *Amblyomma maculatum* until Kohls (1956b) reinstated them as valid species. Several redescriptions of *Amblyomma maculatum* are uncertain, starting with the redescriptions of the male and female and the description of the nymph in Neumann (1899) because these cannot be differentiated from descriptions of the related species. Floch and Abonnenc (1940) and Boero (1957) redescribed the male and female of *Amblyomma maculatum*, but in fact their accounts refer to *Amblyomma tigrinum*. Walker and Olwage (1987) did not formally redescribe the male and female of *Amblyomma maculatum* but include a figure of the female praised by Guglielmone et al. (2003a) because it clearly shows the chitinous tubercles along the posterior body margin. Nevertheless, Mendoza-Uribe and Chávez-Chorocco (2004) refer to the difficulties involved in morphologically separating *Amblyomma maculatum* from *Amblyomma triste*, while Estrada-Peña et al. (2005) state that it is extremely difficult to correctly identify nymphs and especially larvae of *Amblyomma maculatum*, *Amblyomma tigrinum* and *Amblyomma triste* using morphological characters. According to

Nava et al. (2017), there is a possibility that *Amblyomma maculatum* and *Amblyomma triste* are conspecific, and Lado et al. (2018) provided molecular evidence in this regard but divided the *Amblyomma maculatum* group into four morphotypes, with *Amblyomma maculatum sensu stricto* described as morphotype II. More conclusive evidence is needed to confirm the synonymy of *Amblyomma maculatum* and *Amblyomma triste*. Polsomboon et al. (2017) found specimens of *Amblyomma* near *maculatum* but presented no further information. Allerdice et al. (2020) found a new morphotype of *Amblyomma maculatum* in Arizona (USA) that is reproductively incompatible with other populations of *Amblyomma maculatum* from Georgia (USA). See also *Amblyomma tigrinum* and *Amblyomma triste* for further discussion of the confusion of these species with *Amblyomma maculatum*.The geographic distribution of *Amblyomma maculatum* in the Neotropics is as controversial as its morphological definition. Need et al. (1991), Hoffmann and López-Campos (2000), and Muñoz and Casanueva (2001) state that the southern limit is in Argentina, but alleged records of *Amblyomma maculatum* from Argentina, Bolivia, Chile, Brazil, and Uruguay are provisionally treated as cases of confusion with *Amblyomma tigrinum* or *Amblyomma triste* (Evans et al. 2000; Guglielmone et al. 2003a; Dantas-Torres et al. 2009a; Guglielmone and Robbins 2018). The four Argentinean records of *Amblyomma maculatum* in Keirans (1982) correspond, in fact, to *Amblyomma tigrinum*, as discussed in Guglielmone et al. (2003a). *Amblyomma maculatum* has been reported from Jamaica (Newstead 1909; Thompson 1950), but Kohls (1969c) did not mention its presence there, while Morel (1966) considered the presence of *Amblyomma maculatum* in Caribbean islands a consequence of imports of cattle, since this species is not established there. Morel's (1966) theory is accepted here, but Walker and Olwage (1987) provide a map of the range of *Amblyomma maculatum* that includes the Jamaican record. Berlese (1888) described *Amblyomma complanatum* as a Brazilian species that Jones et al. (1972) treated as a synonym of *Amblyomma maculatum*, while Camicas et al. (1998) listed it as a synonym of *Amblyomma tigrinum*, but Guglielmone and Nava (2014) consider *Amblyomma complanatum* to be a name *incertae sedis*. There are also some records of *Amblyomma maculatum* in countries outside the New World, such as Pakistan (Khattak et al. 2012) and Nigeria (Eyo et al. 2014), among others, that are considered misidentifications. Romero-Castañon et al. (2008), who allegedly found *Hyalomma* ticks in southern Mexico, believed their Mexican record of *Amblyomma maculatum* to be the first for that country, ignoring many previous records from Mexico, starting with Neumann (1901). Arana-Guardia et al. (2015) allegedly collected *Amblyomma maculatum* in Mexico (Quintana Roo), but the morphological support for their record is questionable. Finally, although the ixodid fauna of Panama is well known, neither *Amblyomma maculatum* nor related species have been found there, a situation that appears unusual, considering the geographic distribution of this tick discussed above.

The only report of larval infestation by *Amblyomma maculatum* in the Neotropical Region is by Lira-Amaya et al. (2015), who collected this stage from sheep in southern Mexico. This record is treated as provisionally valid here, but the host profile for larvae of *Amblyomma maculatum* in the Neotropics remains basically undetermined.

It is clear that all Neotropical information concerning *Amblyomma maculatum* is provisional, and only additional definitive data on the morphology and ecology of this species will clarify the picture. This problem extends to the host profile, because while many Neotropical collections include clearly identified hosts, the tick stages found were not recorded. Humans have supposedly been parasitized by *Amblyomma maculatum* in Argentina (Boero, 1955) and Chile (Donoso, 1953), cases that presumably involved adult ticks, but these may, in fact, have been *Amblyomma tigrinum* or *Amblyomma triste*.

32. *Amblyomma mixtum* Koch, 1844

Koch, C.L. (1844) Systematische Übersicht über die Ordnung der Zecken. *Archiv für Naturgeschichte*, 10, 217–239.

Descriptions and Redescriptions: The male and female of *Amblyomma mixtum* were described but not figured by Koch (1844), while the larva and nymph were described and illustrated by Hooker et al. (1912) under the name *Amblyomma cajennense*.

Cooley and Kohls (1944) redescribed the male, female, nymph, and larva under the name *Amblyomma cajennense*; Hooker et al. (1912) and Barros-Battesti et al. (2009) redescribed the male and female, also under the name *Amblyomma cajennense*, while Tonelli Rondelli (1937), Nava et al. (2014a), Rivera-Páez et al. (2016), Bermúdez et al. (2018a) and Encinosa et al. (2021) redescribed the male and female under the correct name *Amblyomma mixtum*. The nymph was redescribed by Keirans and Durden (1998, as *Amblyomma cajennense*), and Coley (2015) redescribed the larva as *Amblyomma mixtum*.

Geographic Distribution: *Amblyomma mixtum* is a Nearctic and Neotropical species whose Neotropical range encompasses lands from southern North America to northern South America and some Caribbean islands. *Amblyomma mixtum* is the only species of the *Amblyomma cajennense* group found in Central America, the Caribbean Islands, and southern Mexico. Therefore, all prior records of *Amblyomma cajennense* from those territories are treated here as *Amblyomma mixtum*, and the range of this tick is considered to include: **Belize**: Belize, Cayo, Corozal, Orange Walk, and Stan Creek (Varma 1973; Polsomboon et al. 2017); **Bermuda** (Bishopp and Trembley 1945); **Colombia**: Arauca, Caldas, Casanare, Magdalena and Meta (Rivera-Páez et al. 2018; Cotes-Perdomo et al. 2020); **Costa Rica**: Alajuela, Cartago, Guanacaste, Limón, Puntarenas, and San José (Álvarez et al. 2005; Nava et al. 2014a; Troyo et al. 2016); **Cuba**: Artemisa, Camagüey, Guantánamo, Holguín, Isla de la Juventud, Matanzas, Mayabeque, Sancti Spíritus, Santiago de Cuba, Pinar del Río, and Villa Clara (Prieto 1974; Danielová et al. 1982; Nava et al. 2014a; Noda et al. 2015; Encinosa et al. 2021); **Ecuador**: El Oro, Guayas, Los Ríos, Manabí, and Pichincha (Nava et al. 2014a); **El Salvador** (Payne and Scott 1982), **Guatemala**:

Alta Verapaz, Baja Verapaz, Chimaltenango, Chiquimula, El Progreso, Escuintla, Guatemala, Huehuetenango, Izabal, Jalapa, Jutiapa, Quiché, Petén, Retalhuleu, Sacatepéquez, San Marcos, Suchitepéquez, and Zacapa (Stoll 1894; Instituto Interamericano de Cooperación para la Agricultura 1988; Nava et al. 2014a); **Honduras**: Atlántida, Colón, Lempira, Olancho, and Yoro (Bequaert 1925; Linthicum et al. 1991; Nava et al. 2014a; Novakova et al. 2015); **Jamaica**: Clarendon, Portland, Saint Andrew, Saint Ann, Saint Catherine, Saint Mary, Saint Thomas, and Trelawny (Thompson 1950; Morel 1966; Rawlins and Mansingh 1977; Wilson 1980; Keirans 1982; Nava et al. 2014a); **southern Mexico**: Campeche, Chiapas, Colima, Guerrero, Jalisco, Michoacán, Nayarit, Oaxaca, Quintana Roo, Tabasco, Veracruz, and Yucatán (Guzmán-Cornejo et al. 2011; Nava et al. 2014a); **Nicaragua**: Boaco, Chinandega, Costa Caribe Norte, Costa Caribe Sur, Estelí, Granada, Jinotega, León, Madriz, Managua, Matagalpa, Nueva Segovia, and Rivas (Woke 1947; Maes et al. 1989; Düttmann et al. 2016); **Panama**: Chiriquí, Coclé, Colón, Darién, Herrera, Panamá, Panamá Oeste, Los Santos, and Veraguas (Bermúdez and Miranda 2011b; Nava et al. 2014a); and **Trinidad and Tobago**: Río Claro-Mayaro, Sangre Grande, and Siparia (Basu et al. 2012; Nava et al. 2014a).

Hosts: Males, females, nymphs and larvae of *Amblyomma mixtum* are usually found on several mammalian hosts, including many records from domestic mammals, sometimes in great numbers, but this tick is also common on wild mammals, with odd records from Anura (nymphs), Aves (all tick stages) and Reptilia (adults). The host profile of *Amblyomma mixtum* for the Neotropical Region is shown in Table 2.23, which is based on data from Neumann (1899), Pearse (1936), Hoffmann (1962), Černý (1969a), Varma (1973), Ernst and Ernst (1977), Hermans et al. (1994), Cruz-Aldán et al. (2006), Guzmán-Cornejo et al. (2011), Rodríguez-Vivas et al. (2013, 2016), García et al. (2014), Nava et al. (2014a), Bermúdez et al. (2015a, 2018a), Domínguez et al. (2016), Esser et al. (2016a, b), Polsomboon et al. (2017), Aguilar-Domínguez et al. (2018), Muñoz-García et al. (2019), Sánchez-Montes et al. (2019), Canto-Osorio et al. (2020), Cardona-Romero et al. (2020), Cotes-Perdomo et al. (2020), Domínguez et al. (2020c) and Ulloa-García et al. (2020).

Human Parasitism: *Amblyomma mixtum* is listed as a frequent parasite of people in Guglielmone and Robbins (2018), with human collections in the Neotropics from Belize (Cayo), Colombia (Antioquía, based on specimens provisionally treated as *Amblyomma mixtum*), Costa Rica (Guanacaste), Cuba (Artemisa), Guatemala (Escuintla), Honduras (Atlántida), Jamaica (Clarendon, Portland, Saint Ann), southern Mexico (Chiapas, Colima, Guerrero, Jalisco, Nayarit, Oaxaca, Tabasco, Veracruz, Yucatán), Nicaragua (Chinandega, Costa Caribe Sur), Panama (Colón, Panamá), and Trinidad and Tobago (Río Claro-Mayaro). All parasitic stages of *Amblyomma mixtum* have been collected from humans. Additionally, Chitimia-Dobler et al. (2020) detected a case of human parasitism by a nymph of *Amblyomma mixtum* that occurred in Cuba but was detected in Germany.

Comments: *Amblyomma mixtum* is a member of the *Amblyomma cajennense* species complex, which also comprises *Amblyomma interandinum*, *Amblyomma*

Table 2.23 Hosts for adults of undetermined sex (A), males (M), females (F), nymphs (N), and larvae (L) of *Amblyomma mixtum**

AMPHIBIA			
ANURA: Bufonidae			
Rhinella sp.	N		
MAMMALIA			
ARTIODACTYLA: Bovidae		**CINGULATA: Dasypodidae**	
Domestic buffalo	MF	*Dasypus novemcinctus*	FNL
Cattle	MFNL	**DIDELPHIMORPHIA: Didelphidae**	
Goat	M	*Didelphis marsupialis*	NL
Sheep	MFN	*Didelphis virginiana*	MFNL
ARTIODACTYLA: Cervidae		*Metachirus nudicaudatus*	L
Cervus elaphus	MFNL	**PERISSODACTYLA: Equidae**	
Mazama americana	F	Donkey	MFN
Odocoileus virginianus	M	Horse	MFNL
ARTIODACTYLA: Suidae		Mule	MFN
Domestic pig	AL	**PERISSODACTYLA: Tapiridae**	
ARTIODACTYLA: Tayassuidae		*Tapirus bairdii*	MF
Pecari tajacu	MFL	**PILOSA: Myrmecophagidae**	
Tayassu pecari	F	*Tamandua mexicana*	MFNL
CARNIVORA: Canidae		**RODENTIA: Caviidae**	
Canis latrans	MFN	*Hydrochoerus isthmius*	MFNL
Domestic dog	MFNL	**RODENTIA: Cricetidae**	
Urocyon cinereoargenteus	N	*Transandinomys bolivaris*	NL
CARNIVORA: Felidae		**RODENTIA: Cuniculidae**	
Domestic cat	MF	*Cuniculus paca*	A
Panthera onca	M	**RODENTIA: Echimyidae**	
CARNIVORA: Procyonidae		*Hoplomys gymnurus*	N
Nasua narica	MF	*Proechimys semispinosus*	N
Procyon sp.	N	**RODENTIA: Muridae**	
CHIROPTERA: Phyllostomidae		*Rattus* sp.	L
Artibeus jamaicensis	L	**RODENTIA: Sciuridae**	
		Sciurus aureogaster	M
AVES			
ACCIPITRIFORMES: Accipitridae		**GALLIFORMES: Phasianidae**	
Rupornis magnirostris	A	Chicken	NL
CUCULIFORMES: Cuculidae		Turkey	N
Coccyzus merlini	NL	**PELECANIFORMES: Ardeidae**	
GALLIFORMES: Cracidae		"Heron"	FN
Ortalis ruficauda	N		
REPTILIA			
CROCODILIA: Crocodylidae		**TESTUDINES: Emydidae**	
Crocodylus acutus	F	*Terrapene carolina*	F
SQUAMATA: Iguanidae		*Trachemys scripta*	MF
Ctenosaura similis	F	**TESTUDINES: Geoemydidae**	
Iguana iguana	F	*Rhinoclemmys areolata*	F

* Records of *Amblyomma cajennense* from southern Mexico, Central America, and the Caribbean Islands are treated as *Amblyomma mixtum*

patinoi, *Amblyomma sculptum,* and *Amblyomma tonelliae*, and morphological separation of these species is difficult. Further discussion of this complex is provided under *Amblyomma cajennense* above. Nava et al. (2014a) redescribed the male and female of *Amblyomma mixtum*, but not the nymph and larva; regardless, the nymph and larva of this tick are known, having been described and redescribed under the name *Amblyomma cajennense* on several occasions.

Amblyomma mixtum is the only species of the *cajennense* complex found in Mexico, Central America, and some Caribbean islands. The record of *Amblyomma sculptum* from Nicaragua in Vogel et al. (2018), later corrected in the online version, is a diagnostic error. The situation is different in northern South America, where other species of this complex are established, and many records of *Amblyomma cajennense sensu lato* are awaiting specific definitions. Most probably, the range of *Amblyomma mixtum* and related species will be modified in the near future.

The host range of *Amblyomma mixtum* is almost certainly wider than shown above, especially in northern South America, where the geographic limits of this tick remain undefined. Estrada-Peña et al. (2020) include adults of *Amblyomma mixtum* in a cluster of Neotropical ticks that feed mostly on Artiodactyla, while the larva of this tick is listed in a cluster of species feeding mostly on Didelphimorphia or Rodentia. However, the variety of principal hosts presented here exceeds that of Estrada-Peña et al. (2020). Arana-Guardia et al. (2015) allegedly collected *Amblyomma cajennense* on *Mazama temama* in Mexico (Quintana Roo), but the morphological support for their record is questionable, and their host data have therefore been excluded from this analysis.

33. *Amblyomma multipunctum* Neumann, 1899

Neumann, L.G. (1899) Révision de la famille des ixodidés (3e mémoire). *Mémoires de la Société Zoologique de France*, 12, 107–294.

Descriptions and Redescriptions: The male of *Amblyomma multipunctum* was described but not figured by Neumann (1899), and the female was described and illustrated by Fonseca and Aragão (1953). The nymph of *Amblyomma multipunctum* was described by Labruna et al. (2013).

Onofrio et al. (2006b, key and figures) and Voltzit (2007) redescribed the male and female of *Amblyomma multipunctum*, while Robinson (1926) and Fonseca and Aragão (1953) redescribed the male.

There are no redescriptions of the nymph of *Amblyomma multipunctum.*

The larva of *Amblyomma multipunctum* remains unknown.

Geographic Distribution: *Amblyomma multipunctum* has been found in three countries in northwestern South America: **Colombia**: Huila (Kohls 1956a), Caquetá, Huila or Meta (Voltzit 2007); **Ecuador**: Napo, and Pichincha (Voltzit 2007; Labruna et al. 2013); and **Venezuela** (Neumann 1911a).

Hosts: Females, males, and nymphs of *Amblyomma multipunctum* have been collected from *Tapirus pinchaque* (Perissodactyla: Tapiridae) (Kohls 1956a; Labruna et al. 2013; Pesquera et al. 2015), but these records should be considered provisional because there are difficulties involved in identifying this tick. Therefore, *Tapirus pinchaque* is provisionally considered the principal host for all known parasitic stages of *Amblyomma multipunctum.*

Human Parasitism: No.

Comments: It can be difficult to correctly identify *Amblyomma multipunctum.* Labruna et al. (2013) state that Boero and Prosen (1959) confused the male of *Amblyomma multipunctum* with the male of *Amblyomma scalpturatum*, and the latter authors' redescription of *Amblyomma multipunctum* is not included above. There is consensus among the authors named above for describing and redescribing the male of *Amblyomma multipunctum* that coxae II and III have two spurs, but Jones et al. (1972) listed the male as having one spur on these coxae, rendering their key useless for the identification of male *Amblyomma multipunctum.* There are also other differences among authors concerning adult coxal spurring and the hypostomal dentition that cause difficulties in precisely defining *Amblyomma multipunctum* morphologically, and Robinson (1926) suggested that this species may be a variety of *Amblyomma calcaratum.* Perhaps the most reliable redescription of the male of *Amblyomma multipunctum* is in Fonseca and Aragão (1953) because it is based on the type series (two males with small differences between them), while Robinson (1926) used one male of the type series for his redescription of this species. It would be worthwhile to compare the few specimens of *Amblyomma multipunctum* deposited in collections to confirm or reject the view of Guglielmone et al. (2003a), who suggested that more than one species may be included under the name *Amblyomma multipunctum.* The alleged synonymy of this species with *Amblyomma guianense* is treated in the section "Comments on *Amblyomma.*"

The range of *Amblyomma multipunctum* is also not well defined. Neumann (1899) described this species from two males collected in North America (Nearctic Region); one specimen was from *Dicranocerus furcatus* (= *Antilocapra americana*, Artiodactyla: Antilocapridae), a monotypic genus exclusive to the Nearctic Region, and the other was collected from *Tapirus* (Perissodactyla: Tapiridae), a genus long extinct in North America. Clearly, these data conflict with our current understanding of this species' hosts and range. Thereafter, Neumann (1911a) included Venezuela within the geographic distribution of *Amblyomma multipunctum*, which would then support *Tapirus* as a host for this tick, as further confirmed by additional authors' records from this type of host. The presence of *Amblyomma multipunctum* in the Nearctic Region has not been confirmed, and Guglielmone et al. (2014) doubt its presence there; therefore, this tick is treated here as exclusively Neotropical. Vargas (1955) listed *Amblyomma multipunctum* as a Mexican species without providing a host or locality, and this record is not considered valid here. The record of *Amblyomma multipunctum* in Bolivia by Boero and Prosen (1959) is not treated as valid in Labruna et al. (2013), and this country is provisionally excluded from its range. Fonseca and Aragão (1953) stated that the specimens found by them in a museum collection are most probably from an uncertain South American country. Dantas-Torres et al. (2019b) deleted *Amblyomma multipunctum* from the list of ticks of Brazil; records of *Amblyomma multipunctum* from this country from unusual hosts ("Edentata") reported by Teixeira et al. (2013) are not considered valid, and Brazil is excluded from the range of this species. In brief, *Amblyomma multipunctum* is treated here as a species found in Colombia, Ecuador, and Venezuela, but this distribution is provisional.

All specimens of *Amblyomma multipunctum* treated as valid here have been collected from *Tapirus*. Therefore, controversial records from tortoises and cingulatan and/or pilosan hosts in Keirans and Durden (2001) and Teixeira et al. (2013), respectively, require confirmation and have been excluded from the host list for *Amblyomma multipunctum*.

34. *Amblyomma naponense* (Packard, 1869) as *Ixodes naponensis*, and given its current status in Neumann (1911a)

Packard, A.S. (1869) Report of the curator of Articulata. *First Annual Report of the Trustees of the Peabody Academy of Sciences*, pp. 56–69.
Neumann, L.G. (1911a) *Ixodidae*. Das Tierreich, (26), 169 pp.

Descriptions and Redescriptions: The male of *Amblyomma naponense* was described but not figured by Packard (1869) under the name *Ixodes naponensis*, and the female was described by Aragão (1908b) under the name *Amblyomma mantiquirense*, a synonym of *Amblyomma naponense*, as discussed in Guglielmone and Nava (2014). Martins et al. (2010) described the nymph of *Amblyomma naponense.*

Robinson (1926) and Floch and Fauran (1958) redescribed the male and female of *Amblyomma naponense* under the name *Amblyomma mantiquirense*. Onofrio et al. (2006b, keys and figures), Voltzit (2007), Bermúdez et al. (2018a), Dantas-Torres et al. (2019b, keys and figures) and Rodríguez et al. (2019) also redescribed the male and female; the male was redescribed by Floch and Abonnenc (1940, as *Amblyomma mantiquirense*), and the female was redescribed in Aragão (1911) and Floch and Abonnenc (1942) as *Amblyomma mantiquirense.*

There are no redescriptions of the nymph of *Amblyomma naponense.*

The larva of *Amblyomma naponense* remains undescribed, but some authors have been able to recognize this stage from field-collected material.

Geographic Distribution: *Amblyomma naponense* is a Neotropical tick that has been found in Central and South America as follows: **Bolivia**: Beni, and Santa Cruz (Robbins et al. 1998; Rodríguez et al. 2019); **Brazil**: Acre, Amazonas, Bahia, Espírito Santo, Goiás, Maranhão, Mato Grosso, Mato Grosso do Sul, Minas Gerais, Pará, Paraná, Rio de Janeiro, Rondônia, Sâo Paulo, and Tocantins (Barros and Baggio 1992; Labruna et al. 2005d; Onofrio 2007; Acosta et al. 2016; Witter et al. 2016; Gruhn et al. 2019; Oliveira et al. 2020; Costa et al. 2020); **Colombia**: Bolívar (Fairchild et al. 1966); **Costa Rica**: San José (Keirans 1985b); **Ecuador**: Manabí, and Pastaza (Packard 1869; Labruna et al. 2013); **French Guiana**: Cayena, and Saint Laurent du Maroni (Binetruy et al. 2019); **Guatemala** (Monroy Lefebre and Cajas González 1988); **Guyana**: Essequibo Islands-West Demerara (Keirans 1985b); **Honduras**: Atlántida (Keirans 1985b); **Panama**: Bocas del Toro, Chiriquí, Colón, Guna Yala, Panamá, and Panamá Oeste (Dunn 1934; Fairchild et al. 1966; Miller et al. 2016); **Peru**: Cusco, Loreto, Madre de Dios, and Ucayali (Need et al.

1991; Mendoza-Uribe and Chávez-Chorocco 2004; Labruna et al. 2010b); **Suriname** (Keirans 1985b); and **Venezuela**: Amazonas, Apure, Bolívar, and Falcón (Jones et al. 1972).

Hosts: Males and females of *Amblyomma naponense* are usually found on Artiodactyla: Tayassuidae, with a few records on other types of hosts. A total of 120 nymphs of this tick have been recorded from several orders of mammals and birds, but the great majority of specimens were collected from mammals. A few specimens of larvae of *Amblyomma naponense* have also been found on several orders of mammals and birds, but it is uncertain whether there is a preferred host for the larval stage, and more studies are needed to determine the host range of *Amblyomma naponense* larvae, which is treated here as basically undefined. The host profile of this tick is shown in Table 2.24, which is based on data from Floch and Fauran (1958), Jones et al. (1972), Keirans (1982, 1985b), Labruna et al. (2005d), Ogrzewalska et al. (2009), Martins et al. (2011, 2014a), Soares et al.

Table 2.24 Hosts for adults of undetermined sex (A), males (M), females (F), nymphs (N), and larvae (L)* of *Amblyomma naponense***

MAMMALIA			
ARTIODACTYLA: Cervidae		**PERISSODACTYLA: Tapiridae**	
Mazama gouazoubira	N	*Tapirus terrestris*	MFN
Odocoileus virginianus	A	**PILOSA: Myrmecophagidae**	
ARTIODACTYLA: Suidae		*Myrmecophaga tridactyla*	MFN
Domestic pig	A	*Tamandua mexicana*	FNL
ARTIODACTYLA: Tayassuidae		*Tamandua tetradactyla*	N
Pecari tajacu	MFN	**RODENTIA: Caviidae**	
Tayassu pecari	MFN	*Hydrochoerus hydrochaeris*	MN
CARNIVORA: Canidae		**RODENTIA: Cricetidae**	
Cerdocyon thous	N	*Oryzomys* sp.	NL
Domestic dog	FN	**RODENTIA: Cuniculidae**	
CARNIVORA: Mustelidae		*Cuniculus paca*	MN
Galictis sp.	N	**RODENTIA: Dasyproctidae**	
CARNIVORA: Procyonidae		*Dasyprocta* sp.	A
Nasua narica	F	*Dasyprocta azarae*	N
Nasua nasua	N	*Dasyprocta fuliginosa*	N
Procyon sp.	N	*Dasyprocta leporina*	N
Procyon cancrivorus	F	*Dasyprocta punctata*	N
DIDELPHIMORPHIA: Didelphidae		*Myoprocta acouchy*	MFN
Didelphis marsupialis	NL	**RODENTIA: Echimyidae**	
Marmosa robinsoni	N	*Hoplomys gymnurus*	L
		Proechimys semispinosus	NL
AVES			
CORACIIFORMES: Momotidae		**PASSERIFORMES: Thamnophilidae**	
Baryphthengus martii	NL	*Pyriglena leucoptera*	N
GALLIFORMES: Cracidae		*Thamnophilus pelzelni*	N
Mitu tuberosum	N	**PASSERIFORMES: Thraupidae**	
PASSERIFORMES: Furnariidae		*Tachyphonus coronatus*	L
Dendrocincla fuliginosa	N	**PASSERIFORMES: Troglodytidae**	
Sclerurus guatemalensis	N	*Cyphorhinus phaeocephalus*	NL
PASSERIFORMES: Pipridae		**PASSERIFORMES: Tyrannidae**	
Ceratopipra mentalis	N	*Attila rufus*	N
Lepidothrix coronata	NL		

* Tentative diagnosis because the larva of *Amblyomma naponense* remains undescribed

** Several diagnoses of *Amblyomma naponense* have been published under the name *Amblyomma mantiquirense*

(2015), Acosta et al. (2016), Esser et al. (2016a, b), Witter et al. (2016), Luz et al. (2017b), Nascimento et al. (2017), Bermúdez et al. (2018a), Gianizella et al. (2018b), Barbieri et al. (2019), Gruhn et al. (2019), Domínguez et al. (2019), Bermúdez et al. (2020) and Costa et al. (2020).

Human Parasitism: *Amblyomma naponense* is listed as a sporadic parasite of humans in Guglielmone and Robbins (2018), with records from Brazil (Amazonas, Espírito Santo, Pará, and Rondônia), French Guiana (Cayena), and Panama (Panamá). Binetruy et al. (2019, 2020a, b) found additional cases of human parasitism by males and nymphs of *Amblyomma naponense* in Cayena (French Guiana). All parasitic stages of this tick have been collected from humans.

Comments: Acevedo-Gutiérrez et al. (2020) listed *Amblyomma naponense* as found in four Colombian departments on several hosts based on Fairchild et al. (1966), but these authors recorded this tick from only one host in Bolívar Department; consequently, the distributional data for *Amblyomma naponense* in Acevedo-Gutiérrez et al. (2020) have been excluded from this analysis. Barros and Baggio (1992) listed the State of Paraná in southern Brazil as within the range of *Amblyomma naponense* (under the name *Amblyomma mantiquirense*), but Guimarães et al. (2001) and Labruna et al. (2005d) excluded Paraná from this species' range. In arguing to exclude Paraná, Guimarães et al. (2001) stated that specimens of this tick have been collected from *Tapirus*, which Fairchild et al. (1966) reported to be rarely parasitized by *Amblyomma naponense* in Panama—a weak argument for exclusion. Therefore, Paraná State is provisionally maintained as being within the range of *Amblyomma naponense*. The State of Mato Grosso is excluded from the geographic distribution of *Amblyomma naponense* in Brazil by Andreotti et al. (2018), but no reasons were given for this exclusion.

Estrada-Peña et al. (2020) include the nymph of *Amblyomma naponense* in a cluster of Neotropical species that feed mostly on artiodactyls, while the larva is listed among species prone to parasitize Didelphimorphia and Rodentia. However, the nymph of *Amblyomma naponense* is treated here as having many more principal hosts than Artiodactyla, while the records of larval *Amblyomma naponense* are here considered insufficient to support these authors' hypothesis. Several records of *Amblyomma naponense* have been published under the name *Amblyomma mantiquirense*, including the study of Serra-Freire et al. (1996), who allegedly found all parasitic stages of this tick on Cervidae. Guglielmone et al. (2014) regard records of *Amblyomma naponense* in Serra-Freire et al. (1996) as doubtful; consequently, those records have not been included in the above host list. Camicas et al. (1998) include the Australasian Tachyglossidae as hosts for adults of this tick, probably due to a typographical error.

35. *Amblyomma neumanni* Ribaga, 1902

Ribaga, C. (1902) Acari sudamericani. *Zoologischer Anzeiger*, 25, 502–508.

Descriptions and Redescriptions: The male of *Amblyomma neumanni* was described but not figured by Ribaga (1902), and the female was described, again without illustrations, by Dönitz (1909) under the name *Amblyomma furcula*, a synonym of *Amblyomma neumanni*, as stated in Joan (1928). The nymph and larva were described by Estrada-Peña et al. (1993b).

The male, female, nymph, and larva were redescribed by Estrada-Peña et al. (2005); the male, female, and nymph were redescribed in Nava et al. (2017); while Joan (1920), Boero (1957), and López and Parra (1985) redescribed the male and female. The male was also redescribed in Dönitz (1909) and the female in Robinson (1926), both under the name *Amblyomma furcula*. The nymph of *Amblyomma neumanni* was redescribed in Martins et al. (2014c, key and figures).

Geographic Distribution: *Amblyomma neumanni* is a Neotropical species found only in two South American countries, **Argentina**: Catamarca, Chaco, Córdoba, Formosa, Jujuy, La Rioja, Salta, San Juan, Santiago del Estero, and Tucumán (Guglielmone and Nava 2006); and **Colombia**: Antioquía (López and Parra 1985).

Hosts: Males, females, nymphs, and larvae of *Amblyomma neumanni* are usually found on cattle (Artiodactyla: Bovidae), but they have also been collected from other types of mammals and are common on several domestic species. The host profile of *Amblyomma neumanni* is shown in Table 2.25, which is based on data from Aragão (1936, 1938), Guglielmone and Hadani (1981), Zerpa et al. (2003b), Autino et al. (2006), Guglielmone and Nava (2006), Nava et al. (2006a), Guglielmone et al. (2007a), Mastropaolo et al. (2010), and Ortiz et al. (2011).

Human Parasitism: Guglielmone and Robbins (2018) listed *Amblyomma neumanni* as a very frequent parasite of humans. In Argentina, this tick has been found on people in Córdoba, Jujuy, Salta, and Tucumán. All parasitic stages have been collected from humans, but the nymph of *Amblyomma neumanni* is the stage most often recovered from people.

Comments: *Amblyomma neumanni* was recognized in Neumann (1911a), but this author ignored the name *Amblyomma furcula*, a synonym of *Amblyomma neumanni*, while Robinson (1926) treated *Amblyomma furcula* as a valid species and *Amblyomma neumanii* as doubtful. When considering the validity of both names,

Table 2.25 Hosts for males (M), females (F), nymphs (N), and larvae (L) of *Amblyomma neumanni*

MAMMALIA			
ARTIODACTYLA: Bovidae		**CARNIVORA: Canidae**	
Cattle	MFNL	*Cerdocyon thous*	MF
Goat	MFN	Domestic dog	MFNL
Sheep	MN	*Lycalopex gymnocercus*	MFNL
ARTIODACTYLA: Cervidae		**PERISSODACTYLA: Equidae**	
Mazama americana	MFNL	Donkey	MFN
Mazama gouazoubira	MFNL	Horse	MFNL
ARTIODACTYLA: Suidae		Mule	MFN
Domestic pig	MFN	**PILOSA: Myrmecophagidae**	
ARTIODACTYLA: Tayassuidae		*Myrmecophaga tridactyla*	MN
Catagonus wagneri	F		
Pecari tajacu	MFN		

Joan (1928) was the first to recognize the error in Robinson (1926), concluding that *Amblyomma furcula* is a synonym of *Amblyomma neumanni*. Many years later, the synonymy proposed by Joan (1928) was confirmed by Santos Dias (1961), albeit while ignoring the precedence of Joan's (1928) study. Camicas et al. (1998) erroneously stated that the female of *Amblyomma neumanni* is undescribed.The above situation, plus the lack of illustrations in the descriptions of *Amblyomma neumanni* by Ribaga (1902) and *Amblyomma furcula* by Dönitz (1909), has probably contributed to the misidentification of *Amblyomma neumanni* in several studies. The male of *Amblyomma neumanni* is characterized by a marginal groove that is incomplete (absent along the paracentral festoons) and a spur on metatarsi II–IV; this last morphological character is also seen in females of *Amblyomma neumanni*, but these characters have been overlooked in several redescriptions of *Amblyomma neumanni* although they are clearly depicted in Robinson (1926) under the name *Amblyomma furcula*, as discussed in Guglielmone and Hadani (1985). Thus, Pinto and Lins de Almeida (1939) described as *Amblyomma neumanni* a female tick belonging to the *Amblyomma cajennense* species complex. The key to the *Amblyomma* of the Western Hemisphere by Jones et al. (1972) fails to recognize these morphological characters and is consequently of no use in identifying *Amblyomma neumanni.* López and Parra (1985) correctly describe the male and female of *Amblyomma neumanni* but, in the same study, the key by Jones et al. (1972) is translated without corrections, leading users astray. The adults of *Amblyomma neumanni* described in Voltzit (2007) are in fact *Amblyomma parvitarsum* in the opinion of Nava et al. (2009a, 2017), but the orbited eyes of the latter species are not depicted.

Vogelsang (1928) reported the presence of *Amblyomma neumanni* in Uruguay, and this country was included within the range of this tick in Guglielmone et al. (2003a). Nevertheless, Venzal et al. (2003c) consider it unlikely that *Amblyomma neumanni* occurs there, and we have accordingly limited the geographic distribution of this species to Argentina and Colombia. Doss et al. (1978) listed *Amblyomma neumanni* (under the name *Amblyomma furcula*) as a tick found in French Guiana. Their statement was based on Sénevet (1940), but Sénevet merely used morphological characters of *Amblyomma furcula* for comparative purposes when describing *Amblyomma bouthieri*, a synonym of *Amblyomma tigrinum*, while not stating that *Amblyomma furcula* occurs in French Guiana.A Colombian record of unknown stages of *Amblyomma neumanni* collected from *Hydrochoerus* in López (2017) and repeated in Acevedo-Gutiérrez et al. (2020) was not included in Table 2.25. Cattle are now considered to be the usual hosts for *Amblyomma neumanni*, but this is thought to be an adaptation following the introduction of cattle into the range of this tick about 400 years ago. Nava et al. (2006a) hypothesized that *Amblyomma neumanni* originally parasitized large primeval South American artiodactyls (Camelidae, Cervidae, and Tayassuidae), facilitating the survival of this tick when cattle, another artiodactyl, displaced these artiodactyls.

Kolonin (2009) states that *Amblyomma neumanni* is a rare parasite of people, but human parasitism by this tick frequently occurs (Guglielmone and Robbins 2018). Another characteristic of this peculiar species is the high number of gynandromorphs that have been found relative to other Neotropical species of Ixodidae (Guglielmone et al. 1999).

Beati and Durden (2019) claim that *Amblyomma neumanni* as well as *Amblyomma boeroi* and *Amblyomma parvitarsum* form the basal lineage of South American *Amblyomma* that evolved more than 60 million years ago. However, *Amblyomma neumanni* is currently a parasite of cattle, and its original hosts are thought to have been South American artiodactyls of northern origin that invaded South America during the Great American Interchange. This event occurred about three million years ago, when a land bridge connected Central and South America, favoring the mammal invasion and the radiation of *Amblyomma* upon these new hosts. Consequently, *Amblyomma neumanni* is treated here as an evolutionarily recent species, with little likelihood that it could be part of a basal South American *Amblyomma* lineage.

36. *Amblyomma nodosum* Neumann, 1899

Neumann, L.G. (1899) Révision de la famille des ixodidés (3e mémoire). *Mémoires de la Société Zoologique de France*, 12, 107–294.

Descriptions and Redescriptions: The male and female of *Amblyomma nodosum* were described but not figured by Neumann (1899); the nymph and larva were described by Martins et al. (2010) and Amorim and Serra-Freire (1994b), respectively.

Nava et al. (2017) redescribed the male, female, and nymph of *Amblyomma nodosum*. Robinson (1926), Aragão and Fonseca (1961a), Onofrio et al. (2006b, keys and figures), Voltzit (2007), Bermúdez et al. (2018a) and Dantas-Torres et al. (2019b, keys and figures) redescribed the male and female, while Vogelsang and Santos Dias (1953a) and Guzmán-Cornejo et al. (2011, key and figures) redescribed the male alone. Martins et al. (2014c, key and figures) redescribed the nymph of *Amblyomma nodosum*. Bitencourth et al. (2007) redescribed the larva focusing on the morphology of Haller's organ.

Geographic Distribution: *Amblyomma nodosum* is a species with a wide range, covering a great part of the Neotropical mainland and one Caribbean island as follows: **Argentina**: Chaco, Corrientes, and Formosa (Guglielmone and Nava 2006; Debárbora et al. 2012); **Belize**: Cayo (Garnham and Lewis 1959); **Bolivia**: Santa Cruz (Fonseca 1959); **Brazil**: Acre, Bahia, Ceará, Espírito Santo, Goiás, Maranhão, Mato Grosso, Mato Grosso do Sul, Minas Gerais, Pará, Paraíba, Paraná, Pernambuco, Rio de Janeiro, Rio Grande do Sul, Rondônia, Santa Catarina, São Paulo, and Tocantins (Keirans 1982; Evans et al. 2000; Labruna et al. 2005d; Dantas-Torres et al. 2010b; Ogrzewalska et al. 2011; Lugarini et al. 2015; Soares et al. 2015; Acosta et al. 2016; Lima et al. 2018; Costa et al. 2020; Martins et al. 2020b); **Colombia**: Antioquía, Arauca, Caldas, Cundinamarca, Meta, Tolima, and Valle del Cauca (Osorno Mesa 1942; Luque Forero 1948; Benavides-Montaño et al. 2018; Acevedo-Gutiérrez et al. 2020; Cardona-Romero et al. 2020; Martínez-Sánchez et al. 2020; Robayo-Sánchez et al. 2020); **Costa Rica**: Cartago, and

Guanacaste (Álvarez et al. 2005); **Guatemala** (Robinson 1926); **Honduras**: Atlántida (Novakova et al. 2015); **southern Mexico**: Veracruz, and Yucatán (Keirans 1982; Rodríguez-Vivas et al. 2016); **Nicaragua** (Keirans and Brewster 1981); **Panama**: Bocas del Toro, Colón, Darién, Emberá Wounaan, and Panamá (Fairchild et al. 1966; Miller et al. 2016); **Paraguay**: Cordillera, and Ñeembucú (Massi Pallarés and Benítez Usher 1982; Petters et al. 2020); **Trinidad and Tobago** (Jones et al. 1972); and **Venezuela**: Amazonas, Bolívar, Carabobo, Guárico, Lara, Monagas, and Portuguesa (Díaz Ungría 1957; Jones et al. 1972; Clavijo et al. 2009; Forlano et al. 2013).

Hosts: Usual hosts for males and females of *Amblyomma nodosum* are *Tamandua tetradactyla* and *Tamandua mexicana* (Pilosa: Myrmecophagidae), while larvae and nymphs are commonly collected from several families of Passeriformes, with some avian records from other orders, but the number of species hosting nymphs greatly surpasses the species of birds parasitized by larvae of *Amblyomma nodosum*. There is also an unexpected record of males and females of *Amblyomma nodosum* collected from *Boa constrictor* (Reptilia: Squamata) in Mendoza-Roldán et al. (2020). The host profile of *Amblyomma nodosum* is shown in Table 2.26, which is based on data from Bechara et al. (2002), Ogrzewalska et al. (2009, 2011, 2012a), Tolesano-Pascoli et al. (2010), Luz et al. (2012, 2016b, 2017b), Mazioli et al. (2012), Pascoal et al. (2013), Torga et al. (2013), Lugarini et al. (2015), Maturano et al. (2015), Novakova et al. (2015), Ramos et al. (2015), Witter et al. (2016), Bermúdez et al. (2018a, 2020), Zeringóta et al. (2017), Lima et al. (2018), Domínguez et al. (2019), Cardona-Romero et al. (2020), Fecchio et al. (2020), Martins et al. (2020b), Martí-nez-Sánchez et al. (2020), Mendoza-Roldán et al. (2020) and Souza et al. (2020).

Human Parasitism: No.

Comments: As noted above, *Amblyomma calcaratum* has been confused with *Amblyomma nodosum*, and this confusion likely continues in museum tick collections, warranting further studies of both species throughout their ranges. *Amblyomma nodosum* is by definition a Costa Rican tick because Neumann (1899) described the species from 27 males, 12 females, and 22 nymphs collected in Costa Rica, plus one male from Brazil, but the nymph was not described by Neumann (1899). Subsequent to Neumann (1899), few adults and no immature stages of *Amblyomma nodosum* were reported from Costa Rica, while larvae and nymphs of *Amblyomma calcaratum* were collected from six species of Costa Rican birds by Ogrzewalska et al. (2015). This puzzling observation—the former presence of adult *Amblyomma nodosum* and the current presence of immature *Amblyomma calcaratum*—is, perhaps, an artifact of collecting, but an effort should be made to determine whether both tick species are present in Costa Rica. At this moment, it would appear that many records of *Amblyomma nodosum* and *Amblyomma calcaratum* should be treated as provisional.

The presence of *Amblyomma nodosum* in Belize was confirmed by Garnham and Lewis (1959), who found presumed adults of this tick on an unusual host called "gibnut" (Rodentia: Cuniculidae?). Jones et al. (1972) and Keirans and Brewster

Table 2.26 Hosts for adults of undetermined sex (A), males (M), females (F), nymphs (N), and larvae (L) of *Amblyomma nodosum**

	MAMMALIA		
CARNIVORA: Canidae		**PILOSA: Myrmecophagidae**	
Domestic dog	A	*Myrmecophaga tridactyla*	MF
CINGULATA: Chlamyphoridae		*Tamandua mexicana*	MF
Euphractus sexcinctus	MF	*Tamandua tetradactyla*	MF
PILOSA: Bradypodidae			
Bradypus variegatus	F		
	AVES		
ACCIPITRIFORMES: Accipitridae		*Thamnophilus atrinucha*	N
Buteogallus lacernulatus	N	*Thamnophilus caerulescens*	N
CAPRIMULGIFORMES: Trochilidae		*Thamnophilus doliatus*	N
Eupetomena macroura	N	*Thamnophilus pelzelni*	NL
COLUMBIFORMES: Columbidae		*Thamnophilus schistaceus*	N
Columbina talapacoti	N	*Thamnophilus torquatus*	N
Geotrygon montana	N	**PASSERIFORMES: Thraupidae**	
Leptotila verreauxi	N	*Coryphospingus cucullatus*	N
CORACIIFORMES: Momotidae		*Cyanerpes cyaneus*	N
Baryphthengus ruficapillus	N	*Eucometis penicillata*	N
CUCULIFORMES: Cuculidae		*Herpsilochmus atricapillus*	N
Crotophaga major	N	*Ramphocelus bresilius*	N
Piaya cayana	N	*Ramphocelus carbo*	N
PASSERIFORMES: Cardinalidae		*Saltator coerulescens*	N
Cyanoloxia cyanoides	N	*Saltator maximus*	N
Habia fuscicauda	N	*Saltator similis*	N
Habia rubica	N	*Saltator striatipectus*	N
PASSERIFORMES: Conopophagidae		*Sporophila angolensis*	N
Conopophaga lineata	NL	*Sporophila caerulescens*	N
Conopophaga melanops	N	*Sporophila intermedia*	N
PASSERIFORMES: Fringillidae		*Sporophila lineola*	N
Chlorophonia cyanea	N	*Tachyphonus coronatus*	N
PASSERIFORMES: Furnariidae		*Tachyphonus cristatus*	N
Anabazenops fuscus	NL	*Tachyphonus luctuosus*	N
Automolus leucophthalmus	L	*Tachyphonus phoenicius*	N
Cranioleuca vulpina	N	*Tachyphonus rufus*	N
Dendrocolaptes platyrostris	N	*Tangara cayana*	N
Dendroplex picus	NL	*Thraupis ornata*	N
Glyphorynchus spirurus	N	*Thraupis palmarum*	N
Lepidocolaptes angustirostris	N	*Thraupis sayaca*	N
Lepidocolaptes squamatus	L	*Trichothraupis melanops*	NL
Xenops minutus	N	**PASSERIFORMES: Tityridae**	
Xiphorhynchus guttatus	N	*Pachyramphus polychopterus*	N
Xiphorhynchus susurrans	N	**PASSERIFORMES: Troglodytidae**	
PASSERIFORMES: Icteridae		*Cantorchilus nigricapillus*	N
Cacicus haemorrhous	N	*Pheugopedius genibarbis*	N
Icterus chrysater	N	*Troglodytes aedon*	L
Molothrus bonariensis	N	**PASSERIFORMES: Turdidae**	
PASSERIFORMES: Parulidae		*Catharus ustulatus*	N
Basileuterus culicivorus	N	*Turdus albicollis*	N
Myiothlypis flaveola	N	*Turdus amaurochalinus*	N
PASSERIFORMES: Passerellidae		*Turdus grayi*	N
Arremon flavirostris	N	*Turdus ignobilis*	N
Arremonops conirostris	N	*Turdus leucomelas*	NL
PASSERIFORMES: Pipridae		*Turdus rufiventris*	NL
Ceratopipra mentalis	N	**PASSERIFORMES: Tyrannidae**	
Chiroxipohia caudata	NL	*Casiornis rufus*	N
Dixiphia pipra	NL	*Cnemotriccus fuscatus*	N
Lepidothrix coronata	N	*Elaenia chiriquensis*	N
Manacus manacus	N	*Elaenia cristata*	N

(continued)

Table 2.26 (continued)

Manacus vitellinus	N	*Elaenia flavogaster*	N
Pipra fasciicauda	N	*Elaenia mesoleuca*	N
PASSERIFORMES: Polioptilidae		*Elaenia obscura*	N
Microbates cinereiventer	N	*Empidonax alnorum*	N
PASSERIFORMES: Thamnophilidae		*Hemitriccus margaritaceiventer*	NL
Cymbilaimus lineatus	N	*Lathrotriccus euleri*	N
Drymophila squamata	N	*Leptopogon amaurocephalus*	N
Dysithamnus mentalis	L	*Myiarchus tyrannulus*	N
Dysithamnus stictothorax	N	*Myiozetetes similis*	N
Formicivora grisea	N	*Platyrinchus mystaceus*	NL
Formicivora rufa	N	*Poecilotriccus latirostris*	N
Gymnopithys leucaspis	N	*Ramphotrigon megacephalum*	N
Hylophylax naevioides	N	*Rhynchocyclus olivaceus*	N
Hypocnemoides maculicauda	N	*Todirostrum maculatum*	N
Myrmelastes hyperythrus	N	*Tolmomyias sulphurescens*	N
Myrmotherula schisticolor	N	**PICIFORMES: Capitonidae**	
Phaenostictus macleannani	N	*Capito auratus*	N
Phlegopsis nigromaculata	N	**STRIGIFORMES: Strigidae**	
Poliocrania exsul	N	*Pulsatrix perspicillata*	L
Sciaphylax hemimelaena	N	**TROGONIFORMES: Trogonidae**	
Taraba major	N	*Trogon viridis*	N
Thamnomanes schistogynus	N		
	REPTILIA		
Squamata: Boidae			
Boa constrictor	MF		

* The diagnosis of larvae and nymphs, and to a lesser extent adults of *Amblyomma nodosum* should be considered provisional because of the risk of confusing these stages with those of *Amblyomma calcaratum*

(1981) do not include Belize within the range of this tick; therefore, the occurrence of *Amblyomma nodosum* in Belize is treated as provisionally valid here. We also consider tentatively valid the record of *Amblyomma nodosum* from the State of Veracruz (Mexico) listed in Keirans (1982) but see the comments below. Mazioli et al. (2012) included Guyana within the geographic distribution of this tick, but no *bona fide* records from this country have been found.

Burridge (2011) lists Bovidae and Tapiridae as hosts for *Amblyomma nodosum*, but we were unable to confirm those records, and they are not included in the list of hosts for this tick. Muñoz-García et al. (2019) listed *Tamandua mexicana* as a host for larvae of *Amblyomma nodosum*, but this was not confirmed in the references provided by these authors. Similarly, the Colombian record of *Amblyomma nodosum* from the bird *Ramphocelus carbo* in Acevedo-Gutiérrez et al. (2020) is based on two references that do not support the authors' statement. However, there are *bona fide* Brazilian records of *Ramphocelus carbo* infested with *Amblyomma nodosum*, as in Luz et al. (2012). Keirans (1982) identified one male, three nymphs, and three larvae of *Amblyomma nodosum* collected in Veracruz, Mexico, during January 1911 "ex-*Myrmecophaga tetradactla*" [sic] corrected to *Myrmecophaga tridactyla*, but Keirans (1982) states that this diagnosis of the immature stages is tentative. *Myrmecophaga tridactyla* is not found in Mexico, but the name *Myrmecophaga tetradactyla* has been used in the past, although it is now a synonym of *Tamandua tetradactyla*. It is possible that the host for the male tick was, in fact, *Tamandua mexicana*, a close relative of *Tamandua tetradactyla* that occurs in Mexico. The larvae and nymphs found on this host are not treated here as *bona fide Amblyomma nodosum*.

37. *Amblyomma oblongoguttatum* Koch, 1844

Koch, C.L. (1844) Systematische Übersicht über die Ordnung der Zecken. *Archiv für Naturgeschichte*, 10, 217–239.

Descriptions and Redescriptions: The female of *Amblyomma oblongoguttatum* was described but not figured by Koch (1844), and the male was described and figured by Aragão (1911). The nymph and larva were described by Nuttall (1912) under the name *Amblyomma darlingi*, a synonym of *Amblyomma oblongoguttatum*, as stated in Robinson (1926).

The male and female of *Amblyomma oblongoguttatum* were redescribed by Robinson (1926), Floch and Abonnenc (1940), Floch and Fauran (1958), Aragão and Fonseca (1961a), Onofrio et al. (2006b, keys and figures), Voltzit (2007), Bermúdez et al. (2018a), Dantas-Torres et al. (2019b, keys and figures) and Rodríguez et al. 2019); the male alone was redescribed by Luque Forero (1949) and Guzmán-Cornejo et al. (2011, key and figures). The nymph was redescribed by Martins et al. (2010), and the larva of *Amblyomma oblongoguttatum* was redescribed by Barbieri et al. (2012a, b).

Geopraphic Distribution: *Amblyomma oblongoguttatum* has a wide Neotropical range, but it is not found on Caribbean islands. There is at least one record of *Amblyomma oblongoguttatum* from the Nearctic Zoogeographic Region in Mexico (Guglielmone et al. 2003a). The geographic distribution of this tick encompasses **Belize**: Belize, Cayo, and Stann Creek (Varma 1973; Lopes et al. 2016; Polsomboon et al. 2017); **Bolivia**: Beni, and Santa Cruz (Robbins et al. 1998; Rodríguez et al. 2019); **Brazil**: Acre, Amazonas, Espírito Santo, Goiás, Maranhão, Mato Grosso, Minas Gerais, Pará, Rio de Janeiro, Rondônia, Roraima, and Tocantins (Labruna et al. 2005d; Guglielmone et al. 2006a; Martins et al. 2009a; Costa et al. 2017); **Colombia**: Chocó, Meta, and Santander (Luque Forero 1949; Acevedo-Gutiérrez et al. 2020); **Costa Rica**: Limóm, and Puntarenas (Álvarez et al. 2005); **French Guiana**: Cayena, and Saint Laurent du Maroni (Binetruy et al. 2019); **Guatemala** (Monroy Lefebre and Cajas González 1988); **Guyana**: Cuyuni-Mazaruni, Mazaruni-Potaro, Potaro-Siparuni, and Upper Demerara-Berbice (Tonelli Rondelli 1939; Keirans 1985b; Guglielmone et al. 2006a); **southern Mexico**: Chiapas, Jalisco, Oaxaca, Tabasco, and Yucatán (Guzmán-Cornejo et al. 2011, 2020; Ojeda-Chi et al. 2019a, b); **Nicaragua** (Guglielmone et al. 2003a); **Panama**: Bocas del Toro, Chiriquí, Coclé, Colón, Darién, Guna Yala, Panamá, and Panamá Oeste (Dunn 1934; Bermúdez and Miranda 2011b; Bermúdez et al. 2011a, b, 2012a; Miller et al. 2016); **Peru**: Cusco, and Madre de Dios (Mendoza-Uribe and Chávez-Chorocco 2004; Labruna et al. 2010b); **Suriname**: Brokopondo (Santos Dias

1989a); and **Venezuela**: Amazonas, Apure, Barinas, Bolívar, and Lara (Jones et al. 1972; Clavijo et al. 2009).

Hosts: Males, females, and nymphs of *Amblyomma oblongoguttatum* have largely been collected from several orders of mammals, but no preferences for particular hosts have been discerned among approximately 160 records where host and tick stages were identified. Adults and nymphs of *Amblyomma oblongoguttatum* have also been recovered from a few birds. Larval records for this tick are too few to draw conclusions concerning the principal hosts for this tick stage, which are basically undetermined. The host profile of this tick is shown in Table 2.27, which is based on data from Dunn (1934), Fairchild et al. (1966), Varma (1973), Keirans (1985b), Guerrero (1996), Candanedo-Guerra and Serra-Freire (1999), Labruna et al. (2002b, 2005d), García et al. (2014), Martins et al. (2014a), Bermúdez et al. (2015a, 2018a),

Table 2.27 Hosts for adults of undetermined sex (A), males (M), females (F), nymphs (N), and larvae (L) of *Amblyomma oblongoguttatum**

MAMMALIA			
ARTIODACTYLA: Bovidae		**CHIROPTERA: Phyllostomidae**	
Cattle	MF	*Carollia perspicillata*	M
Goat	A	**DIDELPHIMORPHIA: Didelphidae**	
ARTIODACTYLA: Cervidae		*Chironectes minimus*	ANL
Mazama americana	F	*Didelphis marsupialis*	N
Mazama temama	MFN	*Philander opossum*	N
Odocoileus virginianus	MFNL	**PERISSODACTYLA: Equidae**	
ARTIODACTYLA: Suidae		Horse	MFN
Domestic pig	MFN	**PERISSODACTYLA: Tapiridae**	
ARTIODACTYLA: Tayassuidae		*Tapirus bairdii*	FN
Pecari tajacu	MFNL	*Tapirus terrestris*	MF
Tayassu pecari	MFN	**PILOSA: Choloepodidae**	
CARNIVORA: Canidae		*Choloepus hoffmanni*	N
Canis latrans	MF	**PILOSA: Myrmecophagidae**	
Cerdocyon thous	MN	*Myrmecophaga tridactyla*	MF
Domestic dog	MFNL	*Tamandua mexicana*	MFL
Urocyon cinereoargenteus	MF	**PRIMATES: Cebidae**	
CARNIVORA: Felidae		*Cebus capucinus*	ANL
Domestic cat	MF	**RODENTIA: Caviidae**	
Herpailurus yagouaroundi	AL	*Hydrochoerus hydrochaeris*	MF
Panthera onca	F	*Hydrochoerus isthmius*	MF
Puma concolor	MFL	**RODENTIA: Cricetidae**	
CARNIVORA: Mustelidae		*Oryzomys* sp.	N
Eira barbara	A	**RODENTIA: Cuniculidae**	
CARNIVORA: Procyonidae		*Cuniculus paca*	AN
Nasua narica	MFN	**RODENTIA: Dasyproctidae**	
Nasua nasua	N	*Dasyprocta azarae*	N
Procyon sp.	N	*Dasyprocta fuliginosa*	N
Procyon cancrivorus	A	*Dasyprocta punctata*	N
Procyon lotor	F	**RODENTIA: Echimyidae**	
		Proechimys semispinosus	NL
AVES			
CATHARTIFORMES: Cathartidae		**PASSERIFORMES: Cotingidae**	
Coragyps atratus	N**	*Rupicola rupicola*	F
GALLIFORMES: Cracidae		**PASSERIFORMES: Thamnophilidae**	
Crax rubra	MN	*Thamnophilus atrinucha*	N

* Several specimens have been identified as *Amblyomma* sp. cf. *A. oblongoguttum*
** Identified as *Amblyomma darlingi*

Acosta et al. (2016), Esser et al. (2016a, b), Witter et al. (2016), Gianizella et al. (2018b), Binetruy et al. (2019), Domínguez et al. (2019) and Martins et al. (2020b).

Human Parasitism: *Amblyomma oblongoguttatum* is listed as a sporadic parasite of humans in Guglielmone and Robbins (2018). This species has been collected from humans in Brazil (Amazonas, Espírito Santo, Minas Gerais, Pará, and Rondônia,), Colombia (Santander), French Guiana (Cayena), Guyana (Potaro-Siparuni Region, and Upper Takutu-Upper Essequibo), Panama (Colón, and Panamá), Suriname (Brokopondo), and Venezuela (Amazonas, and Apure). Binetruy et al. (2019) reported several additional cases of human parasitism by adults and nymphs of *Amblyomma oblongoguttatum* in Cayena and one case from Saint Laurent du Maroni (French Guiana), while Costa et al. (2020) listed a female tick collected from a human in the Brazilian state of Maranhão. Guglielmone and Robbins (2018) state that it is unclear whether the specimens of *Amblyomma oblongoguttatum* found on humans in Panama and listed in Esser et al. (2016a) were adult ticks, immatures, or both. Ellen Esser (personal communication to Alberto A. Guglielmone) confirmed that all specimens found on people were adult ticks. Males, females, and nymphs of *Amblyomma oblongoguttatum* have all been collected from humans, but most cases have involved adult ticks. Molaei et al. (2020) reported a male of *Amblyomma oblonguttatum* collected from a USA resident after returning from Costa Rica.

Comments: Lopes et al. (2016) collected specimens of *Amblyomma oblongoguttatum* in Belize that differed slightly in morphology from *bona fide Amblyomma oblongoguttatum* from Brazil but showed substantial differences at the molecular level, implying that more than one species may be included under this name. Aware of the data in Lopes et al. (2016), Bermúdez et al. (2018a) named specimens of this tick collected in Panama as *Amblyomma* circa *oblongoguttatum*, while Ojeda-Chi et al. (2019a, b) and Guzmán-Cornejo et al. (2020, who found molecular differences between ticks from Belize and Brazil) classified their Mexican specimens as *Amblyomma* cf. *oblongoguttatum*. This new scenario implies that records from Mexico and Central America should be treated as provisional, it being possible that the Panamanian *Amblyomma darlingi* of Nuttall (1912), currently treated as a synonym of *Amblyomma oblongoguttatum* (Guglielmone and Nava 2014), could eventually be reinstated as a valid species. The alleged synonymy of *Amblyomma oblongoguttatum* with *Amblyomma guianense* is treated under "Comments on *Amblyomma*."

Camicas et al. (1998) list *Amblyomma oblongoguttatum* as a Neotropical species, but this tick has also been found in the Nearctic portion of Mexico, as discussed in Guglielmone et al. (2003a). Arzua et al. (2005) identified as *Amblyomma oblongoguttatum* a specimen collected from a human in Parana State (Brazil) that was subsequently determined as *Amblyomma scalpturatum* by Onofrio et al. (2010a). The presence of *Amblyomma oblongoguttatum* there was based solely on this record; consequently, Paraná State is not included in the range of this tick.

The Brazilian states of Maranhâo and Tocantins are excluded from the range of *Amblyomma oblongoguttatum* in Andreotti et al. (2018), but no reasons were given for this exclusion.

Estrada-Peña et al. (2020) include adults and nymphs of *Amblyomma oblongoguttatum* in a cluster of Neotropical species that feed mostly on artiodactyls, but the range of principal hosts of these stages is regarded here as broader than Artiodactyla.

38. *Amblyomma ovale* Koch, 1844

Koch, C.L. (1844) Systematische Übersicht über die Ordnung der Zecken. *Archiv für Naturgeschichte*, 10, 217–239.

Descriptions and Redescriptions: The male of *Amblyomma ovale* was described but not figured by Koch (1844), and the female was also described, without illustrations, by Koch (1844) but under the name *Amblyomma oblongum*, a synonym of *Amblyomma ovale*, as discussed in Guglielmone and Nava (2014). The nymph was described by Martins et al. (2010) and the larva by Barbieri et al. (2008a).

The male, female, and nymph of *Amblyomma ovale* were redescribed in Nava et al. (2017); the male and female were redescribed in Vogelsang and Cordero (1953b), Aragão and Fonseca (1961b), Voltzit (2007), Guzmán-Cornejo et al. (2011, keys and figures), Bermúdez et al. (2018a) and Dantas-Torres et al. (2019b, keys and figures); Neumann (1899) and Tonelli Rondelli (1937) redescribed the male, and Martins et al. (2014c, key and figures) redescribed the nymph.

There are no redescriptions of the larva of *Amblyomma ovale*.

Geographic Distribution: *Amblyomma ovale* is a Nearctic and Neotropical species. Nevertheless, Nearctic records (northern Mexico and the USA) are few in comparison with those from the Neotropics, where this tick has been found from southern Mexico to southern South America and in one Caribbean country as follows: **Argentina**: Chaco, Corrientes, Entre Ríos, Formosa, Jujuy, Mendoza, Misiones, La Rioja, Salta, and Santa Fe (Guglielmone and Nava 2006; Debárbora et al. 2011; Tarragona et al. 2012); **Belize**: Belize, Cayo, Orange Walk, and Stann Creek (Varma 1973; Lopes et al. 2016; Polsomboon et al. 2017); **Bolivia**: Cochabamba, and Santa Cruz (Mastropaolo et al. 2014); **Brazil**: Acre, Amapá, Amazonas, Bahia, Ceará, Distrito Federal, Espírito Santo, Goiás, Maranhão, Mato Grosso, Mato Grosso do Sul, Minas Gerais, Pará, Paraná, Pernambuco, Rio de Janeiro, Rio Grande do Norte, Rio Grande do Sul, Rondônia, Santa Catarina, São Paulo, and Tocantins (Aragão 1936; Labruna et al. 2005d; Martins et al. 2009a; Dantas-Torres et al. 2010b; Cansi et al. 2012; Ferreira et al. 2013; Acosta et al. 2016; Costa et al. 2017); **Colombia**: Antioquía, Caldas, Chocó, Córdoba, Guaviare, Magdalena, Meta, Nariño, Sucre, Tolima, and Valle del Cauca (Paternina et al. 2009; Londoño et al. 2014; Rivera-Páez et al. 2018; Acevedo-Gutiérrez et al. 2020; Uribe et al. 2020); **Costa Rica**: Cartago, Guanacaste, Heredia, Limón, Puntarenas, and San José (Álvarez et al. 2005; Springer et al. 2018); **Ecuador**: Napo, and Sucumbíos (Zerpa et al. 2003a); **El Salvador** (Navarrete Abarca et al. 2014); **French Guiana**: Cayena, and Saint Laurent du Maroni (Floch and Fauran 1958); **Guatemala**: Alta Verapaz, and Petén (Instituto Interamericano de Cooperación para la Agricultura 1988); **Guyana**: Upper Demerara-Berbice, and Potaro-Sipurani (Tonelli Rondelli 1937;

Keirans 1985b); **southern Mexico**: Chiapas, Guerrero, Jalisco, Oaxaca, Quintana Roo, Tabasco, Veracruz, and Yucatán (Guzmán-Cornejo et al. 2011; Rodríguez-Vivas et al. 2016; Light et al. 2019); **Nicaragua**: Boaco, Jinotega, and León (Maes et al. 1989; Springer et al. 2018); **Panama**: Bocas del Toro, Chiriquí, Cocle, Colón, Darién, Gnobe-Buglé, Guna Yala, Herrera, Panamá, Panamá Oeste, Los Santos, and Veraguas (Murgas et al. 2013; Miller et al. 2016); **Paraguay**: Caa-Guazú, Central, Concepción, Guairá, Ñeembucú, San Pedro, and Villa Hayes (Nava et al. 2007; Ogrzewalska et al. 2014); **Peru**: Amazonas, Cajamarca, Cusco, Loreto, and Madre de Dios (Need et al. 1991; Mendoza-Uribe and Chávez-Chorocco 2004; Labruna et al. 2010b); **Suriname** (Keirans 1985b); **Trinidad and Tobago**: Río Claro-Mayaro, Sangre Grande, Tobago, and Tunapuna-Piarco (Basu et al. 2012; Sameroff et al. 2019; Charles et al. 2020); and **Venezuela**: Amazonas, Apure, Araguá, Barinas, Bolívar, Carabobo, Falcón, Lara, Miranda, Sucre, Yaracuy, and Zulia (Jones et al. 1972; Forlano et al. 2008; Clavijo et al. 2009).

Hosts: Males and females of *Amblyomma ovale* are usually found on Carnivora (Canidae and Felidae), and to a lesser extent on other Carnivora, and Perissodactyla: Tapiridae, and rarely on birds. Larvae and especially nymphs are commonly collected from several species of Rodentia: Cricetidae and Echimyidae. However, the host range of this species is wide, including records from several birds, mainly Passeriformes. The host profile of *Amblyomma ovale* is shown in Table 2.28, which is based on data from Floch and Fauran (1958), Aragão and Fonseca (1961b), Fairchild et al. (1966), Jones et al. (1972), Keirans (1982), Ivancovich and Luciani (1992), Guerrero (1996), Sinkoc et al. (1998), Labruna et al. (2000, 2005d), Guglielmone et al. (2003c), Arzua et al. (2005), Ogrzewalska et al. (2009, 2015), Guzmán-Cornejo et al. (2011), Lavina et al. (2011, 2015), Martins et al. (2011, 2015b, 2016a), Cansi et al. (2012), Luz et al. (2012, 2017a, b, 2018a), Pacheco et al. (2012), Saraiva et al. (2012), Di Benedetto et al. (2013), Murgas et al. (2013), Flores et al. (2014), Lamattina et al. (2014, 2018a), Londoño et al. (2014), Ramos et al. (2014, 2020), Bermúdez et al. (2015a, 2018a), Gomes et al. (2015), Maturano et al. (2015), Ramos et al. (2015), Soares et al. (2015), Sponchiado et al. (2015), Esser et al. (2016a, b), Blanco et al. (2017), Gianizella et al. (2018b), Lareschi et al. (2019), Szabó et al. (2019), Benítez-Ibalo et al. (2020), Bermúdez et al. (2020), Martínez-Sánchez et al. (2020), Nogueira et al. (2020) and Teixeira et al. (2020).

Human Parasitism: Guglielmone and Robbins (2018) list *Amblyomma ovale* as a frequent parasite of humans, and this tick has been found infesting people in Argentina (Chaco, and Misiones); Brazil (Ceará, Minas Gerais, Paraná, Rio de Janeiro, Rio Grande do Sul, Rondônia, Santa Catarina, and São Paulo); Colombia (Antioquía); Costa Rica (northeastern); Ecuador (Napo, and Sucumbíos); French Guiana (Cayena); Mexico (Veracruz); Panama (Colón, and Panamá); Paraguay (Villarica); Suriname; and Venezuela (Amazonas, Apure, and Delta Amacuro). Moreover, 18 specimens of *Amblyomma ovale* have been collected from humans in undetermined American countries. Additionally, Reck et al. (2018) collected six males and three females from people in Rio Grande do Sul (Brazil), Oniki-Willis and Willis (2018) listed one female tick causing human parasitism in São Paulo (Brazil),

Table 2.28 Hosts for adults of undetermined sex (A), males (M), females (F), nymphs (N), and larvae (L) of *Amblyomma ovale**

MAMMALIA			
ARTIODACTYLA: Cervidae		**PERISSODACTYLA: Tapiridae**	
Blastocerus dichotomus	M	*Tapirus bairdii*	MF
Mazama americana	MF	*Tapirus terrestris*	MF
Mazama gouazoubira	F	**PILOSA: Myrmecophagidae**	
ARTIODACTYLA: Suidae		*Myrmecophaga tridactyla*	F
Domestic pig	FN	**PRIMATES: Atelidae**	
Feral pig	MF	*Alouatta guariba*	A
ARTIODACTYLA: Tayassuidae		**PRIMATES: Cebidae**	
Pecari tajacu	A	*Sapajus apella*	F
Tayassu pecari	A	**RODENTIA: Caviidae**	
CARNIVORA: Canidae		*Hydrochoerus hydrochaeris*	MF
Canis latrans	MFN	**RODENTIA: Cricetidae**	
Cerdocyon thous	MFN	*Akodon* sp.	NL
Chrysocyon brachyurus	MF	*Akodon montensis*	N
Domestic dog	MFL	*Euryoryzomys russatus*	NL
Lycalopex gymnocercus	MF	*Holochilus sciureus*	N
Speothos venaticus	MF	*Hylaeamys megacephalus*	NL
Urocyon cinereoargenteus	MN	*Melanomys caliginosus*	N
CARNIVORA: Felidae		*Necromys lasiurus*	N
Domestic cat	MFN	*Nectomys squamipes*	L
Herpailurus yagouaroundi	MF	*Oligoryzomys flavescens*	NL
Leopardus geoffroyi	MF	*Oligoryzomys nigripes*	NL
Leopardus pardalis	MFN	*Oxymycterus rufus*	N
Leopardus tigrinus	F	*Sigmodon* sp.	MN
Leopardus wiedii	F	*Sooretamys angouya*	N
Panthera onca	MFN	*Thaptomys nigrita*	N
Puma concolor	MFN	*Transandinomys bolivaris*	NL
CARNIVORA: Mephitidae		*Transandinomys talamancae*	MN
Conepatus semistriatus	F	*Zygodontomys brevicauda*	MNL
CARNIVORA: Mustelidae		**RODENTIA: Cuniculidae**	
Eira barbara	MFN	*Cuniculus paca*	A
Galictis cuja	MF	**RODENTIA: Dasyproctidae**	
Galictis vittata	MF	*Dasyprocta* sp.	A
Lontra longicaudis	MF	*Dasyprocta azarae*	L
CARNIVORA: Procyonidae		*Dasyprocta punctata*	N
Nasua narica	MFN	**RODENTIA: Echimyidae**	
Nasua nasua	MFNL	*Hoplomys gymnurus*	NL
Procyon cancrivorus	MFN	*Proechimys canicollis*	NL
Procyon lotor	MFN	*Proechimys guyannensis*	N
CINGULATA: Dasypodidae		*Proechimys quadruplicatus*	N
Dasypus novemcinctus	MF	*Proechimys semispinosus*	FNL
DIDELPHIMORPHIA: Didelphidae		*Thrichomys apereoides*	N
Didelphis albiventris	FNL	*Thrichomys pachyurus*	N
Didelphis aurita	NL	*Trinomys setosus*	N
Didelphis marsupialis	NL	**RODENTIA: Erethizontidae**	
Marmosa robinsoni	NL	*Coendou spinosus*	M
Philander opossum	N	**RODENTIA: Heteromyidae**	
PERISSODACTYLA: Equidae		*Heteromys anomalus*	N
Donkey	M	**RODENTIA: Muridae**	
Horse	MF	*Rattus rattus*	ANL
Mule	M		
AVES			
CATHARTIFORMES: Cathartidae		**PASSERIFORMES: Thamnophilidae**	
Coragyps atratus	A	*Drymophila rubricollis*	N
CORACIIFORMES: Momotidae		*Formicivora grisea*	L
Baryphthengus martii	NL	*Mackenziaena severa*	N
GALLIFORMES: Cracidae		*Poliocrania exsul*	NL
Mitu tuberosum	L	*Thamnophilus pelzelni*	N
Penelope superciliaris	A	**PASSERIFORMES: Thraupidae**	
GRUIFORMES: Rallidae		*Sporophila nigricollis*	MN

(continued)

Table 2.28 (continued)

"Saracura"	A	*Tachyphonus coronatus*	L
PASSERIFORMES: Conopophagidae		*Trichothraupis melanops*	L
Conopophaga lineata	L	**PASSERIFORMES: Tityridae**	
PASSERIFORMES: Formicariidae		*Schiffornis turdina*	NL
Formicarius analis	N	**PASSERIFORMES: Troglodytidae**	
PASSERIFORMES: Furnariidae		*Cyphorhinus phaeocephalus*	L
Sittasomus griseicapillus	L	*Troglodytes aedon*	N
PASSERIFORMES: Icteridae		**PASSERIFORMES: Turdidae**	
Molothrus bonariensis	L	*Turdus albicollis*	NL
PASSERIFORMES: Parulidae		*Turdus amaurochalinus*	N
Parkesia noveboracensis	N	*Turdus leucomelas*	N
PASSERIFORMES: Passerellidae		*Turdus rufiventris*	NL
Arremon flavirostris	NL	**PASSERIFORMES: Tyrannidae**	
Arremonops conirostris	F	*Myiarchus ferox*	N
		Tyrannus melancholicus	N

* Several records of *Amblyomma ovale* published before 1961 and a few after that year have been reported under the name *Amblyomma fossum*

Bitencourth et al. (2019) found a female of *Amblyomma ovale* on a person in Ceará (Brazil), and Sevá et al. (2019) reported two cases of human parasitism by females of *Amblyomma ovale* in the Brazilian state of Bahia. Valente et al. (2020) collected three males and two females of *Amblyomma ovale* from people at two localities in the State of Paraná (Brazil). Guglielmone and Robbins (2018) state that human parasitism by *Amblyomma ovale* is caused by male and female ticks, with no obvious preponderance of one sex over the other as a parasite of people.

Comments: *Amblyomma aureolatum* and *Amblyomma ovale* are morphologically similar species that have been confused with each other or considered to be synonyms, a problem discussed above under *Amblyomma aureolatum*, where the research of Aragão and Fonseca (1961b) was instrumental in resolving this issue. However, the correct identification of *Amblyomma ovale* is not an easy task. Robinson (1926) listed *Amblyomma aureolatum* (under the name *Amblyomma striatum*) as a synonym of *Amblyomma ovale,* but his figures correspond solely to the latter name. Neumann (1899, 1911a) and Tonelli Rondelli (1937) recognized *Amblyomma ovale* and *Amblyomma fossum*, a name usually treated as a synonym of *Amblyomma ovale*, as distinct species. Aragão and Fonseca (1961b) listed *Amblyomma fossum* as a synonym of *Amblyomma ovale* but clearly stated that specimens from northern and west-central Brazil are conspicuously larger than specimens from southern Brazil, and Fournier et al. (2019) found important molecular differences in specimens of *Amblyomma ovale* from northern and southern Brazil, implying that such populations may represent different species, while Miller et al. (2016) presented genomic evidence concerning the presence of a species close to *Amblyomma ovale* in Panama. Most probably, the name *Amblyomma ovale* represents more than one species. Reexamination of supposed *Amblyomma ovale* populations throughout this taxon's range and redefinition of the several names considered to be synonyms of *Amblyomma ovale* would constitute important scientific contributions to tick ecology, nomenclature, and taxonomy. According to Guglielmone and Nava (2014), the following names are synonyms of *Amblyomma* *ovale*: *Amblyomma auronitens*, *Amblyomma confine*, *Amblyomma fossum*, *Amblyomma oblongum*, *Amblyomma ovale kriegi*, *Amblyomma quasistriatum*, and *Amblyomma striatum fossum*.

Labruna et al. (2005c) stated that the female of *Amblyomma beccarii* closely resembles the female of *Amblyomma ovale*. This view was not disputed by Guglielmone and Nava (2014). Nevertheless, the female described by Tonelli Rondelli (1939) bears noticeable cornua on the basis capituli, possesses a scutum that is clearly wider than long, and peculiar scutal punctations that are not typical of *Amblyomma ovale*. Examination of the female types of *Amblyomma beccarii* is needed to determine whether *Amblyomma beccarii* is a synonym of *Amblyomma ovale*.

The marginal groove of male *Amblyomma ovale* is usually treated as complete, a morphological character that is important in separating *Amblyomma ovale* males from those of *Amblyomma aureolatum*, where the marginal groove is absent (Aragão and Fonseca 1961b). However, Voltzit (2007) states that the marginal groove of *Amblyomma ovale* is incomplete.

Camicas et al. (1998) listed *Amblyomma ovale* as a Neotropical species, but the small number of records of this tick from the Nearctic Zoogeographic Region are treated as valid here. According to Andreotti et al. (2018), the Brazilian states of Bahia, Ceará, Espírito Santo, Maranhão, Pará, and Tocantins lie outside the geographic range of *Amblyomma ovale*, but no reasons were provided for excluding these states.

Records of *Amblyomma ovale* from *Bufo* sp. (= *Rhinella* sp.?) (Anura: Bufonidae) from Santa Catarina State, Brazil, and from an unknown Costa Rican locality in Lavina et al. (2015) and Álvarez et al. (2005), respectively, as well as a record from an undetermined tortoise in French Guiana reported by Floch and Fauran (1958), are not included in Table 2.28. Recent records of *Amblyomma ovale* from *Rhinella marina* (Anura: Bufonidae) in Trinidad and Tobago by Kamilar et al. (2020) are treated as diagnostic errors.

Guglielmone et al. (2006a) cited Keirans (1985b) in support of a record of human parasitism by *Amblyomma ovale* in Guyana, but there is no such record in Keirans (1985b), while Robinson (1926) listed *Amblyomma ovale* as a human parasite but made no distinction between *Amblyomma ovale* and *Amblyomma aureolatum* or its synonym *Amblyomma striatum*.

39. *Amblyomma pacae* Aragão, 1911

Aragão, H.B. (1911) Notas sobre ixódidas brazileiros. *Memórias do Instituto Oswaldo Cruz*, 3, 145–195.

Descriptions and Redescriptions: The male and female of *Amblyomma pacae* were described and illustrated by Aragão (1911). The nymph was described by Martins et al. (2010), and the larva of *Amblyomma pacae* was described by Barbieri et al. (2008b).

The male, female and nymph of *Amblyomma pacae* were redescribed in Nava et al. (2017), and the male and female were redescribed in Robinson (1926), Aragão and Fonseca (1961a), Onofrio et al. (2006b, keys and figures), Voltzit (2007), Bermúdez et al. (2018a) and Dantas-Torres et al. (2019b, keys and figures), while Guzmán-Cornejo et al. (2011, key and figures) redescribed the female alone.

There are no redescriptions of the larva of *Amblyomma pacae*.

Geographical Distribution: *Amblyomma pacae* has a wide Neotropical range, encompassing mainland territories from southern Mexico to Paraguay as follows: **Belize**: Cayo (Lopes et al. 2016); **Brazil**: Acre, Mato Grosso, Minas Gerais, Pará, Rio de Janeiro, Rondônia, and São Paulo (Labruna et al. 2005d; Soares et al. 2015; Gruhn et al. 2019); **Colombia**: Boyacá, and Meta (Osorno Mesa 1942; Acevedo-Gutiérrez et al. 2020); **Costa Rica** (Álvarez et al. 2005); **French Guiana**: Cayena (Binetruy et al. 2019); **Guyana** (Tonelli Rondelli 1939); **southern Mexico**: Chiapas (Guzmán-Cornejo et al. 2011); **Panama**: Bocas del Toro, Chiriquí, and Panamá (Fairchild et al. 1966); **Paraguay**: Cordillera (Keirans and Brewster 1981); **Peru**: Loreto (Hayman 2006); **Suriname**: Brokopondo, and Sipalawani (Santos Dias 1989a; Jones et al. 1972); and **Venezuela**: Amazonas, Barinas, Falcón, Miranda, and Zulia (Jones et al. 1972).

Hosts: Males and females of *Amblyomma pacae* have been found on several hosts, but these stages are usually collected from *Cuniculus paca* (Rodentia: Cuniculidae); the preferred hosts of larvae and nymphs are unclear because records of these stages are from a few specimens collected on three species of Didelphimorphia: Didelphidae and four species representing three families of Rodentia, including the principal host of adult ticks. Consequently, the principal hosts for the immature stages of *Amblyomma pacae* are regarded as basically undetermined. The host profile of *Amblyomma pacae* is shown in Table 2.29, which is based on data from Fairchild et al. (1966), Jones et al. (1972), Santos Dias (1989a), Labruna et al. (2005d), Hayman (2006), Guzmán-Cornejo et al. (2011), Esser et al. (2016a, b) and Gruhn et al. (2019).

Human Parasitism: Guglielmone and Robbins (2018) list *Amblyomma pacae* as a very rare parasite of humans, with one record of a male tick in Suriname (Sipalawini). Additionally, Binetruy et al. (2019) collected 13 larvae of *Amblyomma pacae* from humans in Cayena (French Guiana) that were identified by molecular

Table 2.29 Hosts for adults of undetermined sex (A), males (M), females (F), nymphs (N), and larvae (L) of *Amblyomma pacae*

MAMMALIA			
ARTIODACTYLA: Suidae		**PILOSA: Myrmecophagidae**	
Feral pig	M	*Tamandua tetradactyla*	F
ARTIODACTYLA: Tayassuidae		**RODENTIA: Caviidae**	
“Peccary”	MF	*Hydrochoerus hydrochaeris*	FN
CARNIVORA: Canidae		**RODENTIA: Cuniculidae**	
Domestic dog	A	*Cuniculus paca*	MFN
DIDELPHIMORPHIA: Didelphidae		**RODENTIA: Dasyproctidae**	
Didelphis marsupialis	NL	*Dasyprocta fuliginosa*	A*
Metachirus nudicaudatus	L	*Dasyprocta punctata*	MF
Philander opossum	NL	**RODENTIA: Echimyidae**	
PERISSODACTYLA: Tapiridae		*Hoplomys gymnurus*	L
Tapirus bairdii	F	*Proechimys semispinosus*	NL

* Tentative diagnosis of *Amblyomma pacae* cited in Hayman (2006)

analysis, a record treated as provisionally valid here. Another nymph found attached to a human in French Guiana (Cayena) is listed in Binetruy et al. (2020a).

Comments: The only record of *Amblyomma pacae* from Guyana was reported by Tonelli Rondelli (1939) under the name *Amblyomma nigrum*, a name that is listed as a synonym of *Amblyomma pacae* in Guglielmone and Nava (2014). González-Martín del Campo et al. (2018) included the Mexican state of Campeche within the range of *Amblyomma pacae,* but this record is treated as a diagnostic error, as discussed below. The only known record of *Amblyomma pacae* from Peru is in Hayman (2006), who reported adult ticks collected from *Dasyprocta fuliginosa* as probable *Amblyomma pacae*, a record treated as provisionally valid here.

Estrada-Peña et al. (2020) include the nymph of *Amblyomma pacae* in a cluster of Neotropical species prone to feed on Didelphimorphia and Rodentia, but this host-parasite association requires confirmation, while the larva of this tick is included in a cluster of species that chiefly parasitize Pilosa, but this statement appears to be a typographical error because there are no records of larvae of *Amblyomma pacae* feeding on Pilosa. There are two records of *Amblyomma pacae* from Aves, one in Teixeira et al. (2008) from a captive host, the other in González-Martín del Campo et al. (2018) from an adult tick collected on a Mexican bird. In the study of González-Martín del Campo et al. (2018), it is stated that an adult *Hyalomma* sp., a representative of a genus that is of paramount medical and veterinary importance but that is not known to be established in the Americas, was found in southern Mexico, but insufficient taxonomic information was presented to support this statement, and neither of these avian records is included in Table 2.29. Nava et al. (2017) listed *Rhipidomys macrurus* (Rodentia: Cricetidae) as a host of the nymph of *Amblyomma pacae*, but this record was not confirmed and, consequently, this rodent is not included in the table above.

40. *Amblyomma parkeri* Fonseca and Aragão, 1952

Fonseca, F. & Aragão, H.B. (1952a) Notas de ixodología. II. Uma nova espécie do gênero *Amblyomma* e uma nova espécie do gênero *Ixodes* (Acari, Ixodidae). *Memórias do Instituto Oswaldo Cruz*, 50, 713–726.

Descriptions and Redescriptions: The female, nymph, and larva of *Amblyomma parkeri* were described by Fonseca and Aragão (1952a), and the male was described by Onofrio et al. (2006b, key and figures).

The male and female of *Amblyomma parkeri* were redescribed by Labruna et al. (2009) and Dantas-Torres et al. (2019b, keys and figures); the female was redescribed by Onofrio et al. (2006b, key and figures), and the nymph by Martins et al. (2010, 2013).

There are no redescriptions of the larva of *Amblyomma parkeri.*

Geographic Distribution: *Amblyomma parkeri* is a Neotropical species that has only been found in **Brazil**: Bahia, Mato Grosso, Mato Grosso do Sul, Minas Gerais, Paraná, Rio Grande do Sul, Rio de Janeiro, Santa Catarina, and São Paulo (Witter et al. 2016; Reck et al. 2018).

Hosts: Males and females of *Amblyomma parkeri* are usually found on *Coendou prehensilis* and *Coendou spinosus* (Rodentia: Erethizontidae), with a few records from other mammals. Nymphs have been found on Didelphimorphia, Primates, Rodentia, and Aves (Accipitriformes, Galliformes, Passeriformes, and Strigiformes), but of the few specimens that have been collected, most have been taken from *Alouatta guariba* (Primates: Atelidae) and *Coendou spinosus* (Rodentia: Erethizontidae). Two larvae of *Amblyomma parkeri* were found on Erethizontidae, another specimen was collected on *Thalurania glaucopis* (Caprimulgiformes: Trochilidae), but the remainder of the more than 60 larvae that have been reported were taken from several families of Passeriformes. Therefore, *Coendou spinosus* and *Alouatta guariba* are provisionally considered principal hosts for the nymphs, while Passeriformes (several families) are regarded as principal hosts for the larvae of *Amblyomma parkeri*. The host profile of *Amblyomma parkeri* is shown in Table 2.30, which is based on data from Fonseca and Aragão (1952a), Labruna et al. (2009), Ogrzewalska et al. (2008, 2011, 2012a), Pacheco et al. (2012), Martins et al. (2013, 2017), Maturano et al. (2015), Sponchiado et al. (2015), Witter et al. (2016), Zeringóta et al. (2017), Luz et al. (2017a, b), and Teixeira et al. (2017, 2020).

Human Parasitism: Guglielmone and Robbins (2018) listed *Amblyomma parkeri* as a very rare parasite of people, with just one record of a nymph collected from a person in São Paulo (Brazil) by Martins et al. (2013). Thereafter, Reck et al. (2018) collected 17 nymphs from humans at several localities in the Brazilian state of Rio

Table 2.30 Hosts for males (M), females (F), nymphs (N), and larvae (L) of *Amblyomma parkeri*

MAMMALIA			
CARNIVORA: Procyonidae		*Alouatta guariba*	N
Nasua nasua	F	**PRIMATES: Pitheciidae**	
DIDELPHIMORPHIA: Didelphidae		*Callicebus nigrifrons*	N
Didelphis albiventris	N	**RODENTIA: Erethizontidae**	
Didelphis marsupialis	N	*Coendou* sp.	L
PRIMATES: Atelidae		*Coendou prehensilis*	MF
Alouatta sp.	F	*Coendou spinosus*	MFN
AVES			
ACCIPITRIFORMES: Accipitridae		**PASSERIFORMES: Thamnophilidae**	
Leptodon cayanensis	N	*Mackenziaena severa*	N
Rupornis magnirostris	N	*Thamnophilus caerulescens*	L
Spizaetus tyrannus	N	**PASSERIFORMES: Thraupidae**	
CAPRIMULGIFORMES: Trochilidae		*Saltator maximus*	L
Thalurania glaucopis	L	*Tachyphonus coronatus*	N
GALLIFORMES: Cracidae		*Tiaris fuliginosus*	L
Penelope obscura	N	*Trichothraupis melanops*	NL
PASSERIFORMES: Conopophagidae		**PASSERIFORMES: Troglodytidae**	
Conopophaga lineata	L	*Troglodytes aedon*	L
Conopophaga melanops	L	**PASSERIFORMES: Turdidae**	
PASSERIFORMES: Furnariidae		*Turdus albicollis*	L
Anabazenops fuscus	L	**PASSERIFORMES: Tyrannidae**	
Dendrocincla turdina	L	*Leptopogon amaurocephalus*	L
Lepidocolaptes squamatus	L	*Myiarchus ferox*	N
Philydor atricapillus	L	*Platyrinchus mystaceus*	L
Sittasomus griseicapillus	L	*Tolmomyas poliocephalus*	L
PASSERIFORMES: Pipridae		**STRIGIFORMES: Strigidae**	
Chiroxiphia caudata	L	*Megascops choliba*	N
Manacus manacus	L		

Grande do Sul, and Borsoi et al. (2019) found 11 nymphs on people in Rio Grande do Sul, Paraná, and Santa Catarina (Brazil), while Valente et al. (2020) collected five nymphs of *Amblyomma parkeri* from humans at four localities in the State of Paraná (Brazil).

Comments: *Amblyomma parkeri* was not recognized as valid in Jones et al. (1972), Keirans (1992), Horak et al. (2002), and Voltzit (2007), while Kolonin (2009) doubted its validity. However, Camicas et al. (1998) consider *Amblyomma parkeri* to be a good species, and Guglielmone et al. (2003a) treat this tick as provisionally valid. Labruna et al. (2009) presented sound evidence to further confirm the validity of *Amblyomma parkeri.*

Amblyomma geayi, *Amblyomma longirostre*, *Amblyomma parkeri,* and *Amblyomma romarioi* are morphologically similar species. Many records of *Amblyomma parkeri* adults have been confused with *Amblyomma longirostre* or *Amblyomma geayi*, as discussed in detail in Labruna et al. (2009). It is also quite probable that some records of larvae and nymphs identified as *Amblyomma longirostre* may in fact be specimens of *Amblyomma parkeri* (or *Amblyomma geayi*) or *Amblyomma romarioi*, another member of this species group (Martins et al. 2019b).

Estrada-Peña et al. (2020) include the nymph of *Amblyomma parkeri* in a cluster of Neotropical species that feed mostly on Carnivora, but this may be a typographical error because no species of carnivorous mammal has been recorded as a host for the nymph of this tick.

41. *Amblyomma parvitarsum* Neumann, 1901

Neumann, L.G. (1901) Révision de la famille des ixodidés (4e mémoire). *Mémoires de la Société Zoologique de France*, 14, 249–372.

Descriptions and Redescriptions: Neumann (1901) described and figured the female of *Amblyomma parvitarsum*; the male was briefly described without figures in Neumann (1911a). The nymph was described by Martins et al. (2014c), and the larva of *Amblyomma parvitarsum* was described by Estrada-Peña et al. (2005).

The male, female, and nymph of *Amblyomma parvitarsum* were redescribed in Nava et al. (2017); the male and female were redescribed in Boero (1957) and Estrada-Peña et al. (2005), while Robinson (1926) redescribed the female alone.

There are no redescriptions of the larva of *Amblyomma parvitarsum.*

Geographic Distribution: *Amblyomma parvitarsum* is a South American species that has been found in: **Argentina**: Catamarca, Chubut, Jujuy, Mendoza, Neuquén, Río Negro, Salta, San Juan, and Santa Cruz (Guglielmone and Nava 2006; Muñoz-Leal et al. 2014; Fantozzi et al. 2018); **Bolivia**: La Paz, and Oruro (Muñoz-Leal et al. 2014); **Chile**: Antofagasta, Arica y Parinacota, and Atacama (Muñoz-Leal et al. 2014); and **Peru**: Arequipa, and Ayacucho (Muñoz-Leal et al. 2014).

Hosts: Usual hosts for males and females of *Amblyomma parvitarsum* are domestic (alpaca, llama) and wild South American camelids (Artiodactyla: Camelidae), with occasional records from other vertebrates, including other domestic mammals. All records of larvae and nymphs are from lizards (Squamata: Liolaemidae), and the species listed below are treated as provisional principal hosts for the immature stages of this tick. The host profile of *Amblyomma parvitarsum* is shown in Table 2.31, which is based on data from Boero (1955), Becker et al. (1997), González-Acuña et al. (2004b), Muñoz-Leal et al. (2014, 2016), Fantozzi et al. (2018), Castillo et al. (2019) and Cerdeña et al. (2019).

Human Parasitism: No.

Comments: Robinson (1926) reproduced the brief description of the male of *Amblyomma parvitarsum* in Neumann (1911a) but considered it insufficient and treated the male of this species as unknown. Boero (1957) redescribed both sexes of *Amblyomma parvitarsum* while ignoring the two spines on tibiae II–IV that are useful when identifying this species. Nava et al. (2017) stated that Voltzit (2007) redescribed *Amblyomma parvitarsum* under the name *Amblyomma neumanni*, although some morphological characters of *Amblyomma parvitarsum,* such as orbited eyes, are not depicted in the description of Voltzit (2007).

A female of *Amblyomma parvitarsum* was collected from a penguin (*Spheniscus magellanicus*) caught along Brazil's southern coast (Becker et al. 1997), probably a migrant from southern Argentina. This record is considered exceptional, and Brazil is not included within the range of this tick. Ivancovich and Luciani (1992) recorded *Amblyomma parvitarsum* from a horse in Córdoba Province (Argentina) at a locality that is unusual for this species. This record is treated as doubtful in Guglielmone and Nava (2006) and provisionally valid in Guglielmone et al. (2014), Muñoz-Leal et al.

Table 2.31 Hosts for adults of undetermined sex (A), males (M), females (F), nymphs (N), and larvae (L) of *Amblyomma parvitarsum*

	MAMMALIA		
ARTIODACTYLA: Bovidae		**ARTIODACTYLA: Camelidae**	
Cattle	A	Alpaca	MF
Goat	MF	*Lama guanicoe*	A
Sheep	A	Llama	MF
		Vicugna vicugna	MF
	AVES		
RHEIFORMES: Rheidae		**SPHENISCIFORMES: Spheniscidae**	
Rhea pennata	F	*Spheniscus magellanicus*	F
	REPTILIA		
SQUAMATA: Liolaemidae*		*Liolaemus jamesi*	L
Liolaemus alticolor	L	*Liolaemus nigriceps*	L
Liolaemus andinus	L	*Liolaemus patriciaiturrae*	L
Liolaemus annectens	NL	*Liolaemus pleopholis*	NL
Liolaemus copiapensis	L	*Liolaemus puritamensis*	N
Liolaemus eleodori	NL		

* There is controversy concerning the validity of several species of *Liolaemus*; the species names above should be considered tentative

(2014), and Nava et al. (2017). Córdoba Province and horses are not included within the range and host profile of *Amblyomma parvitarsum*. The species of lizards constituting the genus *Liolaemus* are disputed (San-Martín-Órdenes et al. 2019); therefore, the species of *Liolaemus* infested with *Amblyomma parvitarsum* in Table 2.31 should be considered tentative.

A male of *Amblyomma parvitarsum* about 8000 years old (Middle Holocene Period) was recovered along with numerous skeletal remains of *Vicugna vicugna* in a rock shelter used by hunter-gatherer human groups in northern Argentina (Nava et al. 2013).

Beati and Durden (2019) claim that *Amblyomma parvitarsum* as well as *Amblyomma boeroi* and *Amblyomma neumanni* form the basal lineage of South American *Amblyomma* that evolved more than 60 million years ago. However, adults of *Amblyomma parvitarsum* are parasites of Camelidae, one of the many families of mammals of northern origin that invaded South America during the Great American Interchange. This event occurred about 3 million years ago, when a land bridge connected Central and South America, favoring the mammal invasion and the radiation of *Amblyomma* upon these newly available hosts. Consequently, *Amblyomma parvitarsum* is treated here as a historically rather recent species, with little evidence that it was part of a basal South American *Amblyomma* lineage.

42. *Amblyomma parvum* Aragão, 1908

Aragão, H. B. (1908a) Algunas novas especies de carrapatos brazileiros. *Brazil Medico*, 22, 111–115.

Descriptions and Redescriptions: The male of *Amblyomma parvum* was described but not figured in Aragão (1908a), and Aragão (1911) described and figured the female. The nymph and larva of *Amblyomma parvum* were described by Guglielmone et al. (1990).

The male, female, and nymph of *Amblyomma parvum* were redescribed by Nava et al. (2017), while Robinson (1926), Boero (1957), Guglielmone et al. (1990), Onofrio et al. (2006b, keys and figures), Voltzit (2007), Guzmán-Cornejo et al. (2011, keys and figures), Bermúdez et al. (2018a) and Dantas-Torres et al. (2019b, keys and figures) redescribed the male and female; Aragão (1911) redescribed the male. Martins et al. (2010) redescribed the nymph alone. The larva of *Amblyomma parvum* was redescribed in Amorim and Serra-Freire (2000a).

Geographic Distribution: *Amblyomma parvum* has an allegedly wide distribution in the Neotropical Region (southern North America, Central and South America), although currently many records of this species appear to be uncertain. Its supposed range includes **Argentina**: Catamarca, Chaco, Córdoba, Formosa, Jujuy, La Rioja, Salta, San Luis, and Santiago del Estero (Nava et al. 2008; Ortiz et al. 2011); **Bolivia**:.Santa Cruz (Nava et al. 2008); **Brazil**: Amazonas, Bahia, Ceará, Distritio Federal, Goiás, Maranhão, Mato Grosso, Mato Grosso do Sul, Minas Gerais, Pernambuco, Piauí, Rio de Janeiro, Rio Grande do Norte, São Paulo, and Tocantins (Ferreira et al. 2008; Nava et al. 2008; Martins et al. 2009a; Pinheiro et al. 2014;

Moerbeck et al. 2016; Costa et al. 2017); **Colombia**: Magdalena (Santodomingo et al. 2019); **Costa Rica**: Guanacaste (Nava et al. 2008); **El Salvador** (Payne and Scott 1982); **Guatemala**: Alta Verapaz, Escuintla, Huehuetenango, and Petén (Instituto Interamericano de Cooperación para la Agricultura 1988); **southern Mexico**: Campeche, Chiapas, Guerrero, Jalisco, Michoacán, Nayarit, Oaxaca, Tabasco, and Yucatán (Guzmán-Cornejo et al. 2011); **Nicaragua**: León, Managua, Nuevo Segovia, and Rivas (Nava et al. 2008; Düttmann et al. 2016); **Panama**: Coclé, Herrera, Los Santos, Panamá, and Panamá Oeste (Nava et al. 2008; Bermúdez et al. 2013); **Paraguay**: Alto Paraguay, and Boquerón (Nava et al. 2008); and **Venezuela**: Anzoátegui, Aragua, Apure, Distrito Federal, Falcón, Guárico, and Nueva Esparta (Rodríguez and Flores 2003; Nava et al. 2008).

Hosts: Males and females of *Amblyomma parvum* have been found on a wide variety of mammals without any obviously preferred hosts and occasionally on birds. Larvae and nymphs have also been found on several mammals and birds, but the preferred host for South American populations of *Amblyomma parvum* is *Galea musteloides* (Rodentia: Caviidae) (Nava et al. 2006a, 2017), while the principal hosts for immature stages of northern populations of this tick are undetermined. Table 2.32 depicts the host profile of *Amblyomma parvum* and is based on data from Yunker et al. (1975), Guglielmone et al. (1990), Szabó et al. (2007), Nava et al. (2008, 2017), Medri et al. (2010), Guzmán-Cornejo et al. (2011), Bermúdez et al. (2013), Lugarini et al. (2015), Esser et al. (2016a, b), Kluyber et al. (2016), Rodríguez-Vivas et al. (2016), Witter et al. (2016), Cançado et al. (2017), Sousa et al. (2017), Maia et al. (2018), Barbieri et al. (2019), Canto-Osorio et al. (2020), Costa et al. (2020) and Oliveira et al. (2020).

Human Parasitism: Guglielmone and Robbins (2018) listed *Amblyomma parvum* as a frequent parasite of humans that has been found on people in Argentina (Catamarca, Chaco, Córdoba, Salta, and Santiago del Estero); Bolivia (Santa Cruz); Brazil (Bahia, Goiás, Maranhão, Mato Grosso do Sul, Piauí, and Rio Grande do Norte); Guatemala (Escuintla); southern Mexico (Jalisco, and Yucatán); and Panama (Panamá Oeste). Aside from one instance of infestation caused by nymphs of *Amblyomma parvum*, cases of human parasitism have been due to male and female ticks. Szabó et al. (2020) collected several male and female ticks but also three nymphs from people in the State of Minas Gerais, Brazil, while Oliveira et al. (2020) found three nymphs of *Amblyomma parvum* on people in the Brazilian state of Pernambuco.

Comments: Guglielmone et al. (2003a) and Guglielmone et al. (2014) suggest that *Amblyomma parvum* may have occasionally been confused with *Amblyomma inornatum*. Hoffmann (1950) redescribed *Amblyomma curruca*, an alleged synonym of *Amblyomma auricularium*, from specimens collected in Mexico, but later Hoffmann (1962) reversed her prior diagnosis *Amblyomma parvum*. Nava et al. (2008) found molecular differences between *Amblyomma parvum* from Argentina and Brazil and hypothetized that those populations represent different species, but the studies of Gerardi et al. (2013) and Nava et al. (2016) failed to confirm this

Table 2.32 Hosts for adults of undetermined sex (A), males (M), females (F), nymphs (N), and larvae (L) of *Amblyomma parvum**

MAMMALIA			
ARTIODACTYLA: Bovidae		**CINGULATA: Dasypodidae**	
Domestic buffalo	MF	*Dasypus kappleri*	M
Cattle	MFN	*Dasypus novemcinctus*	MF
Goat	MF	**DIDELPHIMORPHIA: Didelphidae**	
Sheep	MF	*Didelphis* sp.	MF
ARTIODACTYLA: Cervidae		*Didelphis albiventris*	NL
Mazama gouazoubira	MFN	*Didelphis virginiana*	MFNL
Odocoileus virginianus	MF	*Gracilinanus agilis*	N
ARTIODACTYLA: Suidae		*Monodelphis domestica*	NL
Domestic pig	AN	*Thylamys karimii*	L
Feral pig	MFN	*Thylamys macrurus*	N
ARTIODACTYLA: Tayassuidae		**LAGOMORPHA: Leporidae**	
Catagonus wagneri	MF	*Sylvilagus floridanus*	MF
Pecari tajacu	A	**PERISSODACTYLA: Equidae**	
CARNIVORA: Canidae		Donkey	F
Canis latrans	MF	Horse	MF
Cerdocyon thous	MFN	**PERISSODACTYLA: Tapiridae**	
Chrysocyon brachyurus	A	*Tapirus terrestris*	MF
Domestic dog	MFNL	**PILOSA: Myrmecophagidae**	
Lycalopex gymnocercus	MF	*Myrmecophaga tridactyla*	MFN
Urocyon cinereoargenteus	M	*Tamandua mexicana*	F
CARNIVORA: Felidae		*Tamandua tetradactyla*	MFN
Domestic cat	MF	**PRIMATES: Cebidae**	
Herpailurus yagouaroundi	M	*Sapajus apella*	F
Leopardus geoffroyi	MF	**RODENTIA: Caviidae**	
Leopardus pardalis	MF	*Galea musteloides*	NL
Panthera onca	MF	*Galea spixii*	NL
Puma concolor	MF	*Hydrochoerus hydrochaeris*	F
CARNIVORA: Mustelidae		*Kerodon rupestris*	MF
Galictis vittata	MF	**RODENTIA: Chinchillidae**	
CARNIVORA: Procyonidae		*Lagostomus maximus*	MF
Nasua narica	M	**RODENTIA: Cricetidae**	
Nasua nasua	MFN	*Cerradomys marinhus*	L
Procyon sp.	MF	*Graomys chacoensis*	L
Procyon cancrivorus	A	*Oecomys mamorae*	N
CHIROPTERA: Phyllostomidae		*Sigmodon* sp.	MF
Carollia subrufa	F	**RODENTIA: Dasyproctidae**	
CINGULATA: Chlamyphoridae		*Dasyprocta azarae*	N
Chaetophractus vellerosus	A	**RODENTIA: Echimyidae**	
Euphractus sexcinctus	MFN	*Clyomys laticeps*	N
Priodontes maximus	MF	*Thrichomys apereoides*	NL
Tolypeutes matacus	MF	*Thrichomys inermis*	N
		RODENTIA: Muridae	
		Mus musculus	N
AVES			
CATHARTIFORMES: Cathartidae		*Saltator similis*	L
Coragyps atratus	MF	*Saltatricula multicolor*	NL
COLUMBIFORMES: Columbidae		*Sicalis flaveola*	L
Leptotila verreauxi	N	*Sporophila caerulescens*	N
PASSERIFORMES: Cardinalidae		*Sporophila nigricollis*	N
Cyanoloxia brissonii	NL	*Tachyphonus rufus*	N
PASSERIFORMES: Parulidae		**PASSERIFORMES: Troglodytidae**	
Myiothlypis flaveola	NL	*Cantorchilus longirostris*	N
PASSERIFORMES: Passerellidae		*Troglodytes aedon*	N
Zonotrichia capensis	NL	**PASSERIFORMES: Turdidae**	
PASSERIFORMES: Thamnophilidae		*Turdus amaurochalinus*	NL
Myrmorchilus strigilatus	N	*Turdus rufiventris*	N
Sakesphorus cristatus	N	**PASSERIFORMES: Tyrannidae**	
Taraba major	N	*Casiornis rufus*	N
Thamnophilus doliatus	N	*Cnemotriccus fuscatus*	N
Thamnophilus pelzelni	N	*Myiarchus tyrannulus*	NL
PASSERIFORMES: Thraupidae		*Pitangus sulphuratus*	N

(continued)

Table 2.32 (continued)

Coryphospingus pileatus	L	*Stigmatura napensis*	N
Dacnis cayana	N	*Tyrannus savanna*	N
Ramphocelus carbo	N	*Xolmis velatus*	N
Saltator aurantiirostris	N		

* Many records of *Amblyomma parvum* should be considered tentative because southern and northern populations most probably represent different species

hypothesis. Later, Lado et al. (2016) obtained strong molecular evidence indicating that the northern and southern lineages of *Amblyomma parvum* most probably represent different species. Clearly, there is some degree of uncertainty concerning the identity of *bona fide Amblyomma parvum*, and many records should probably be considered provisionally valid. Camicas et al. (1998) treat *Amblyomma minutum* as a synonym of *Amblyomma parvum*, but the former name is a *nomen nudum*, as discussed in Guglielmone et al. (2003a) and Guglielmone and Nava (2014).

The range of *Amblyomma parvum* will change dramatically if, as expected, southern North American and Central American populations are redefined as separate species. Authors such as Bermúdez et al. (2013, 2018a) treat Panamanian tick populations as *Amblyomma* near *parvum*. Lado et al. (2016) suggest that *Amblyomma parvum* from Costa Rica (Cartago, and Guanacaste), El Salvador (Cabañas), southern Mexico (Yucatán), and Panama (Chiriquí, Colón, Herrera, Panamá, and Los Santos) differ molecularly from *bona fide Amblyomma parvum*. Therefore, the range of *Amblyomma parvum sensu stricto* may be restricted to South America, with valid records for Argentina, Brazil, and Paraguay. However, this species is also found in other South American countries, namely, Bolivia, Colombia, and Venezuela (not French Guiana, as discussed below), yet there are no data confirming that all these populations represent *bona fide Amblyomma parvum*. Further studies are needed to properly define the taxonomic status and range of *Amblyomma parvum*, whose populations are provisionally considered to encompass lands from southern Mexico to northern Argentina at this time.

Records of *Amblyomma parvum* from the states of Paraná and Rio Grande do Sul (Brazil) in Serra-Freire and Mello (1993) and Freire (1972) require confirmation and were not included in the above range of this tick. Andreotti et al. (2018) exclude the states of Amazonas, Pernambuco, Rio de Janeiro, Rio Grande do Norte, Sâo Paulo, and Tocantins from the geographic distribution of *Amblyomma parvum*, but no reasons were given for this decision. Acevedo-Gutiérrez et al. (2020) listed this tick as found in the Colombian department of Meta, but based on a publication lacking locality records; consequently, Meta is not included in the range of *Amblyomma parvum*.

Estrada-Peña et al. (2020) include adults of *Amblyomma parvum* in a cluster of Neotropical species that feed mostly on Carnivora, while the larvae and nymphs are listed in a cluster of ticks that chiefly parasitize Aves, in clear disagreement with the host profile for *Amblyomma parvum* presented above. Nevertheless, many hosts of *Amblyomma parvum* should be treated as provisionally valid, in line with current uncertainties about this species' range. Table 2.32 provides a host profile that differs slightly from similar compilations in Guglielmone et al. (2014), Nava et al. (2017), and Guglielmone and Robbins (2018) because, apart from specimens collected on novel hosts, several records of nymphs and larvae published prior to the descriptions of these stages in Guglielmone et al. (1990) were excluded, given the high risk of misidentification. Thus, Floch and Fauran (1958) claimed to have found a nymph of *Amblyomma parvum* on a rodent in French Guiana, but Binetruy et al. (2019) were unable to confirm this species' presence there; consequently, this host and country are here excluded from the host profile and range of *Amblyomma parvum*. Mullins et al. (2004) cited Boero (1944c) to argue that Squamata and Testudines are hosts for *Amblyomma parvum*, but this is a mistake because Mullins et al. (2004) confused the hosts of *Amblyomma parvum* in Boero (1944c) with the hosts of *Amblyomma rotundatum*. Teixeira et al. (2008) found adults of *Amblyomma parvum* on a captive *Rhea americana*, but this record is not included in Table 2.32. Caviidae are listed above as principal hosts for larvae and nymphs of southern populations of *Amblyomma parvum*, but Caviidae are restricted to the South American mainland; therefore, the principal hosts for northern populations of larvae and nymphs remain unknown because there are few records from Central America and southern Mexico.

Nava et al. (2006a, b) found that in Argentina the parasitic cycle of *Amblyomma parvum* in part involves surrogate host utilization, with adult ticks feeding mostly on introduced domestic artiodactyls that displaced primeval hosts, while the larvae and nymphs remain dependent on endemic rodents.

43. *Amblyomma patinoi* Labruna, Nava and Beati, 2014 in Nava et al. (2014a)

Nava, S., Beati, L., Labruna, M.B., Cáceres, A.G., Mangold, A.J. & Guglielmone, A.A. (2014a) Reassessment of the taxonomic status of *Amblyomma cajennense* (Fabricius, 1787) with the description of three new species, *Amblyomma tonelliae* n. sp., *Amblyomma interandinum* n. sp., and *Amblyomma patinoi* n. sp., and reinstatement of *Amblyomma mixtum* Koch, 1844 and *Amblyomma sculptum* Berlese, 1888 (Ixodida: Ixodidae). *Ticks and Tick-borne Diseases*, 5, 252–276.

Descriptions and Redescriptions: The male and female of *Amblyomma patinoi* were described in Nava et al. (2014a).

There are no redescriptions of *Amblyomma patinoi*.

The nymph and larva of *Amblyomma patinoi* remain unknown.

Geographic Distribution: *Amblyomma patinoi* has only been found in **Colombia**: Antioquia, Cauca, and Cundinamarca (Nava et al. 2014a; Acevedo-Gutiérrez et al. 2020; Quintero et al. 2021).

Hosts: Males and females of *Amblyomma patinoi* are known only to feed on cattle (Artiodactyla: Bovidae) (Nava et al. 2014a).

Human Parasitism: Quintero et al. (2021) found 22 females and 23 males of *Amblyomma patinoi* feeding on people in the Colombian department of Antioquia.

Comments: *Amblyomma patinoi* is a member of the *Amblyomma cajennense* species complex, which also comprises *Amblyomma interandinum*, *Amblyomma mixtum*, *Amblyomma sculptum,* and *Amblyomma tonelliae*. Further discussion of this complex is provided under *Amblyomma cajennense* above.

44. *Amblyomma pecarium* Dunn, 1933

Dunn, L.H. (1933) Two new species of ticks from Panama (*Amblyomma tapirellum* and *A. pecarium*). *Parasitology*, 25, 353–358.

Descriptions and Redescriptions: The male and female of *Amblyomma pecarium* were described and illustrated by Dunn (1933).

The male and female of *Amblyomma pecarium* were redescribed by Voltzit (2007), Guzmán-Cornejo et al. (2011) and Bermúdez et al. (2018a).

The nymph of *Amblyomma pecarium* remains undescribed, but some authors have been able to identify this stage from field-collected material.

The larva of *Amblyomma pecarium* remains unknown.

Geographical Distribution: *Amblyomma pecarium* is a Neotropical species found from southern Mexico to central South America as follows: **Belize**: Cayo or Stann Creek (Fairchild et al. 1966); **Bolivia**: Santa Cruz (Robbins et al. 1998); **Colombia** (provisional) (Acevedo-Gutiérrez et al. 2020); **Costa Rica**: San José (Álvarez et al. 2005); **southern Mexico**: Chiapas, Oaxaca, and Tabasco (Guzmán-Cornejo et al. 2011); **Panama**: Bocas del Toro, Colón, Darién, Guna Yala, Panamá, and Panamá Oeste (Fairchild et al. 1966; Bermúdez et al. 2018b); and **Peru**: Loreto (Hayman 2006).

Hosts: Males and females of *Amblyomma pecarium* are usually found on peccaries (Artiodactyla: Tayassuidae) with occasional records from other mammals. The preferred host of the undescribed nymph is basically unknown because this stage has only been recorded twice from different orders of mammals, Artiodactyla and Carnivora. The host profile of *Amblyomma pecarium* is shown in Table 2.33, which is based on data from Robbins et al. (1998), Hayman (2006), Esser et al. (2016a, b), and Bermúdez et al. (2018b).

Human Parasitism: Guglielmone and Robbins (2018) listed *Amblyomma pecarium* as a rare parasite of people, with records of human infestation in Costa Rica and Panama. The nymph of *Amblyomma pecarium* is the only stage reportedly recovered from people, a diagnosis that should be considered tentative because the nymph of this tick remains undescribed.

Comments: Sauter et al. (1999) allegedly found *Amblyomma pecarium* in the field at a location in Brazil, a record that needs confirmation. Guglielmone et al. (2003a) ignored Belize as falling within the range of *Amblyomma pecarium*, but there is no reason to question the Belizean record in Fairchild et al. (1966).

Table 2.33 Hosts for adults of undetermined sex (A), males (M), females (F), and nymphs (N)* of *Amblyomma pecarium*

MAMMALIA			
ARTIODACTYLA: Cervidae		**CARNIVORA: Procyonidae**	
Mazama temama	A	*Nasua narica*	N
ARTIODACTYLA: Tayassuidae		**PERISSODACTYLA: Tapiridae**	
Pecari tajacu	MFN	*Tapirus terrestris*	M
Tayassu pecari	MF		

* Tentative diagnosis because the nymph of *Amblyomma pecarium* remains undescribed

45. *Amblyomma pictum* Neumann, 1906

Neumann, L.G. (1906) Notes sur les Ixodidés. IV. *Archives de Parasitologie*, 10, 195–219.

Descriptions and Redescriptions: Neumann (1906) described but did not figure the male and female of *Amblyomma pictum.*

Robinson (1926) and Floch and Fauran (1958) redescribed the male and female of *Amblyomma pictum*, while Aragão (1912a, 1913, under the name *Amblyomma conspicuum*, a synonym of *Amblyomma pictum*, as stated in Robinson (1926)), Floch and Abonnenc (1942) and Voltzit (2007) redescribed the male alone. The first figure of the male of *Amblyomma pictum* is in Aragão (1913, as *Amblyomma conspicuum*), but there are no illustrations of the female of *Amblyomma pictum* in any of the redescriptions named above.

The nymph and larva of *Amblyomma pictum* remain unknown.

Geographic Distribution: *Amblyomma pictum* is a Neotropical species that has been found in South America and is known from one Central American record as follows: **Brazil**: Amazonas, Bahía, Goiás, Mato Grosso, and Pará (Guimarães et al. 2001); **French Guiana**: Cayena (Floch and Fauran 1958); **Guyana**: Essequibo Islands-West Demerara (Keirans 1985b); and **Panama**: Guna Yala (Fairchild et al. 1966).

Hosts: Males and females of *Amblyomma pictum* are considered to be prone to feed on Pilosa: Myrmecophagidae, with both sexes recorded only from *Myrmecophaga tridactyla*, which is treated here as this species' principal host. There is also a record of a male of *Amblyomma pictum* from a dog, under the name *Amblyomma conspicuum* in Aragão (1912a, 1913). The host profile of *Amblyomma pictum* is shown in Table 2.34, which is based on data from Neumann (1906), Aragão (1912a), Floch and Fauran (1958), and Fairchild et al. (1966).

Human Parasitism: No.

Comments: The male of *Amblyomma pictum* in Neumann (1906), Floch and Abonnenc (1942), Floch and Fauran (1958), and Aragão (1912a, 1913, as *Amblyomma conspicuum*) is described as having a pseudoscutum and numerous scutal punctations, but these morphological characters are lacking in the redescription of Voltzit (2007). Additionally, Neumann (1906) describes the male dental

Table 2.34 Hosts for males (M) and females (F) of *Amblyomma pictum*

MAMMALIA			
CARNIVORA: Canidae		**PILOSA: Myrmecophagidae**	
Domestic dog	M*	*Myrmecophaga tridactyla*	MF
		Tamandua sp.	M

* Identified as *Amblyomma conspicuum*

formula as 4/4, but other authors give the dentition as 3/3. A redescription of the types of *Amblyomma pictum* (and *Amblyomma conspicuum*) should help to better understand the morphology of this tick.

Guglielmone et al. (2003a) treat the Argentinian record of *Amblyomma pictum* in Dios and Knopoff (1930, 1934) as a misidentification, a view maintained here. The last field collection of *Amblyomma pictum* was made decades ago, and authors such as Onofrio et al. (2006b) and Dantas-Torres et al. (2009a, 2019b) in Brazil and Bermúdez et al. (2018a) in Panama doubt the presence of this tick in these countries. *Amblyomma pictum* was not collected in the last tick survey carried out in French Guiana by Binetruy et al. (2019). Reexamination of specimens deposited in tick collections may shed light on this species' geographic distribution, which currently should be considered provisional.

A record of *Amblyomma pictum* on a host described as "probably *Cavia* sp." in Keirans (1985b) is excluded from the host list for *Amblyomma pictum.*

46. *Amblyomma pilosum* Neumann, 1899

Neumann, L.G. (1899) Révision de la famille des ixodidés (3e mémoire). *Mémoires de la Société Zoologique de France*, 12, 107–294.

Descriptions and Redescriptions: Neumann (1899) described but did not illustrate the female of *Amblyomma pilosum.* The male, nymph, and larva were described by Keirans et al. (1973).

Voltzit (2007) redescribed the male and female of *Amblyomma pilosum*, and Keirans et al. (1973) redescribed the male.

There are no redescriptions of the nymph and larva of *Amblyomma pilosum.*

Geographic Distribution: *Amblyomma pilosum* is a Neotropical species that has only been found on four islands belonging to **Ecuador**: Galápagos (Keirans et al. 1973).

Hosts: Males, females, nymphs, and larvae of *Amblyomma pilosum* have been found on *Chelonoidis duncanensis*, under the name *Geochelone elephantopus ephippium*, and *Chelonoidis darwini*, under the name *Geochelone elephantopus darwini* (Testudines: Testudinidae), and these two tortoises are treated as principal hosts of *Amblyomma pilosum*. Information on hosts of *Amblyomma pilosum* is from Keirans et al. (1973), who provided subspecies names for tortoises listed as *Geochelone elephantopus*.

Human Parasitism: No.

Comments: According to Keirans et al. (1973), both Santos Dias (1958a) and Robinson (1926) confused the type of *Amblyomma pilosum* with a specimen of a female tick that was later described as *Amblyomma macfarlandi.* Keirans and Hillyard (2001) also state that the description of the female of *Amblyomma pilosum* in Robinson (1926) corresponds, in fact, to *Amblyomma macfarlandi.* Moreover, the redescriptions of the male and female of *Amblyomma pilosum* by Voltzit (2007) include illustrations of the coxal spurs and hypostomal dentition that conflict with the figures in Keirans et al. (1973).

47. *Amblyomma pseudoconcolor* Aragão, 1908

Aragão, H. B. (1908c) Mais um novo carrapato brazileiro. *Brazil Medico*, 22, 431–432.

Descriptions and Redescriptions: Aragão (1908c) described but did not figure the male and female of *Amblyomma pseudoconcolor*; the nymph was described by Martins et al. (2010).

The male, female, and nymph of *Amblyomma pseudoconcolor* were redescribed in Nava et al. (2017); the male and female were also redescribed in Aragão (1911), Robinson (1926), Boero (1957), Onofrio et al. (2006b, keys and figures), Voltzit (2007) and Dantas-Torres et al. (2019b, keys and figures), while Boero (1947) redescribed the female alone. The nymph of *Amblyomma pseudoconcolor* was redescribed in Martins (2014c, key and figures).

The larva of *Amblyomma pseudoconcolor* remains undescribed, but some authors have been able to identify this stage from field-collected material.

Geographic Distribution: *Amblyomma pseudoconcolor* is a South American species that has been found in **Argentina**: Buenos Aires, Catamarca, Chaco, Chubut, Córdoba, Formosa, La Rioja, Mendoza, Salta, and Santiago del Estero (Ivancovich and Luciani 1992; Superina et al. 2004; Guglielmone and Nava 2006); **Bolivia**: Santa Cruz (Mastropaolo et al. 2014); **Brazil**: Bahia, Goiás, Maranhão, Mato Grosso, Mato Grosso do Sul, Minas Gerais, Paraná, Pernambuco, Rio de Janeiro, Rio Grande do Sul, São Paulo, and Tocantins (Fonseca 1958b; Bothelo et al. 1989; Arzua et al. 2005; Silva et al. (2018a, b); Martins et al. 2020b); **French Guiana**: Cayena (Floch and Fauran 1958); **Paraguay**: Alto Paraguay, Boquerón, and Cordillera (Nava et al. 2007); **Suriname** (Buitendijk 1945); and **Uruguay**: Artigas, and Cerro Largo (Martins et al. 2014d).

Hosts: Males and females of *Amblyomma pseudoconcolor* are usually found on Cingulata: Chlamyphoridae and Dasypodidae, but there are also records from other mammals and rarely from birds. Those cingulatan hosts carrying males and females of *Amblyomma pseudoconcolor* are treated here as principal hosts for this tick. Information concerning immature stages is based on very few records from

Table 2.35 Hosts for adults of undetermined sex (A), males (M), females (F), nymphs (N), and larvae (L)* of *Amblyomma pseudoconcolor*

MAMMALIA			
ARTIODACTYLA: Bovidae		*Euphractus sexcinctus*	MF
Domestic buffalo	F	*Priodontes maximus*	MF
Cattle	F	*Tolypeutes* sp.	MFL
ARTIODACTYLA: Cervidae		*Tolypeutes matacus*	MF
Mazama gouazoubira	A	*Zaedyus pichiy*	MNL
CARNIVORA: Canidae		**CINGULATA: Dasypodidae**	
Canidae		*Dasypus hybridus*	F
Domestic dog	MF	*Dasypus novemcinctus*	MF
CARNIVORA: Mustelidae		**DIDELPHIMORPHIA: Didelphidae**	
Galictis sp.	MN	*Didelphis albiventris*	N
CINGULATA: Chlamyphoridae		**PERISSODACTYLA: Tapiridae**	
Cabassous tatouay	MF	*Tapirus terrestris*	MF
Cabassous unicinctus	F	**PILOSA Myrmecophagidae**	
Calyptophractus retusus	MN	*Myrmecophaga tridactyla*	MF
Chaetophractus vellerosus	MFL	*Tamandua tetradactyla*	F
Chaetophractus villosus	MF		
AVES			
TINAMIFORMES: Tinamidae		*Nothura maculosa*	MNL
Nothura boraquira	MNL	*Rhynchotus rufescens*	MNL

* Tentative diagnosis because the larva of *Amblyomma pseudoconcolor* remains undescribed

mammals and birds; this is especially true of the larva of *Amblyomma pseudoconcolor* because avian records of this stage are based solely on Aragão (1936). Therefore, host preferences of the immature stages of this tick are treated as basically undetermined. Hosts of *Amblyomma pseudoconcolor* are shown in Table 2.35, which is based on data from Aragão (1936, 1938), Floch and Fauran (1958), Bothelo et al. (1989), Ivancovich and Luciani (1992), Guglielmone et al. (2003b), Robbins et al. (2003b), Superina et al. (2004), Arzua et al. (2005), Guglielmone and Nava (2006), Martins et al. (2011) and Nava et al. (2017).

Human Parasitism: Guglielmone and Robbins (2008) list *Amblyomma pseudoconcolor* as a very rare parasite of humans, with just one case of human parasitism by an adult tick in Argentina (Santiago del Estero) (Tomassone et al. 2010).

Comments: The chief morphological character separating *Amblyomma auricularium* from *Amblyomma pseudoconcolor* is the scutal ornamentation of the latter species, which is not seen in *Amblyomma auricularium*, according to Jones et al. (1972), Guzmán-Cornejo et al. (2011), Nava et al. (2017) and Bermúdez et al. (2018a, b). Sometimes the ornamentation in *Amblyomma pseudoconcolor* is indistinct, especially in male ticks, leading to confusion between the species. By carefully recording the presence or absence of ornamentation, Guglielmone and Nava (2006) were able to reclassify several collections of Argentinean *Amblyomma auricularium* as *Amblyomma pseudoconcolor*.

Confusing the issue is a key constructed by Onofrio et al. (2006b), where the male and female of *Amblyomma auricularium* are described as inornate, unlike *Amblyomma pseudoconcolor* whose males and females have ornamented scuta, but the corresponding figures of *Amblyomma auricularium* show that both sexes of this tick possess scutal ornamentation. This ambiguity is exacerbated in the dichotomous morphological diagnostic key of Dantas-Torres et al. (2019b), where males and females of *Amblyomma auricularium* are defined as inornate or ornate, the difference between this species and the ornate *Amblyomma pseudoconcolor* being that the scuta are "faintly ornate" in *Amblyomma auricularium*, as opposed to the "pale spots" seen in *Amblyomma pseudoconcolor*. Further confusing this matter, Dantas-Torres et al. (2019b) then repeat the figures of Onofrio et al. (2006b). The seemingly ambiguous nature of ornamentation in *Amblyomma auricularium* and *Amblyomma pseudoconcolor* in Onofrio et al. (2006b) and Dantas-Torres et al. (2019b) favors the position of Camicas et al. (1998), who treat *Amblyomma pseudoconcolor* as a synonym of *Amblyomma auricularium*, an opinion that, in any event, cannot be confirmed because the types of *Amblyomma auricularium* are not available and no *bona fide* specimens have been collected for a neotype designation. Nonetheless, Costa et al. (2020) claim that Brazilian specimens from Maranhão identified as *Amblyomma pseudoconcolor* in Arzua et al. (2005) are in fact *Amblyomma auricularium*, a statement that requires confirmation, and we provisionally consider the diagnosis of Arzua et al. (2005) to be correct. See *Amblyomma auricularium* for further discussion of this matter.

Arzua et al. (2005) state that the last known record of *Amblyomma pseudoconcolor* from southern Brazil is from 1953, and this appears unchanged. Floch and Fauran (1958) collected several specimens of this tick in French Guiana, but its presence there was not confirmed in the study of Binetruy et al. (2019), and French Guiana is provisionally included in the range of *Amblyomma pseudoconcolor*. The southern Mexican record of *Amblyomma pseudoconcolor* in Romero-Castañon et al. (2008) is treated as a diagnostic error, and this country is excluded from the range of this tick. This decision is based on the lack of proper taxonomic support for tick diagnoses in a paper where alleged *Hyalomma* specimens were collected from cattle in southern Mexico without any further elaboration.

Estrada-Peña et al. (2020) place the nymph and larva of *Amblyomma pseudoconcolor* in a cluster of Neotropical species that feed mainly on Cingulata, Lagomorpha, and Tinamiformes, but information on the immature stages of this tick is considered insufficient to determine principal hosts. Nava et al. (2017) list the nymph as the only stage found on Tinamiformes, but Aragão (1936) recorded males, nymphs, and, allegedly, larvae of *Amblyomma pseudoconcolor* from these birds, a statement treated as valid here considering that Aragão was the describer of this species.

48. *Amblyomma pseudoparvum* Guglielmone, Mangold and Keirans, 1990

Guglielmone, A.A., Mangold, A.J. & Keirans, J.E. (1990) Redescription of the male and female of *Amblyomma parvum* Aragão, 1908, and description of the nymph and larva, and description of all stages of *Amblyomma pseudoparvum* sp. n. (Acari: Ixodida: Ixodidae). *Acarologia*, 31, 143–159.

Descriptions and Redescriptions: The male, female, nymph, and larva of *Amblyomma pseudoparvum* were described by Guglielmone et al. (1990).

The male, female, and nymph of *Amblyomma pseudoparvum* were redescribed in Nava et al. (2017), and the male and female were redescribed by Voltzit (2007). Martins et al. (2014c, key and figures) redescribed the nymph of *Amblyomma pseudoparvum* alone.

There are no redescriptions of the larva of *Amblyomma pseudoparvum.*

Geographic Distribution: *Amblyomma pseudoparvum* is a Neotropical species found only in southern South America in **Argentina**: Chaco, Formosa, Salta, and Santiago del Estero (Guglielmone and Nava 2006).

Hosts: Males and females of *Amblyomma pseudoparvum* are usually found on *Dolichotis salinicola* (Rodentia: Caviidae), but there are also records from other mammals. There are only two *bona fide* records of nymphs of this species; therefore, the host range for this stage is basically undetermined. The host profile of *Amblyomma pseudoparvum* is shown in Table 2.36, which is based on data from Guglielmone et al. (1990) and Ivancovich and Luciani (1992). The larva of *Amblyomma pseudoparvum* is known (Guglielmone et al. 1990) but not the natural hosts, because this stage was described from laboratory-reared material.

Human Parasitism: Guglielmone and Robbins (2018) list *Amblyomma pseudoparvum* as a rare parasite of humans in Argentina (Chaco and Salta), where males and females have been collected from people but it is unclear whether one sex prevails over the other as a parasite of humans.

Comments: *Amblyomma pseudoparvum* was initially confused with *Amblyomma parvum* in Guglielmone and Hadani (1980), who separated males of the latter species into two groups, based on the presence or absence of carenas. Those males without carenas would later become specimens used in the description of *Amblyomma pseudoparvum* in Guglielmone et al. (1990). Ivancovich and Luciani (1992) named this species *Amblyomma parvum parvum.*

Guglielmone et al. (2003a) included Brazil within the range of *Amblyomma pseudoparvum*, an opinion that has not been confirmed, and Dantas-Torres et al. (2009a) treat its presence there as doubtful. Accordingly, Brazil is excluded from the geographic distribution of this tick.

Table 2.36 Hosts for males (M), females (F), and nymphs (N) of *Amblyomma pseudoparvum*

MAMMALIA			
ARTIODACTYLA: Bovidae		**CINGULATA: Chlamyphoridae**	
Cattle	MFN	*Chaetophractus villosus*	MF
Goat	F	*Tolypeutes matacus*	MF
ARTIODACTYLA: Cervidae		**RODENTIA: Caviidae**	
Mazama gouazoubira	MF	*Dolichotis salinicola*	MFN
CARNIVORA: Canidae		**RODENTIA: Chinchillidae**	
Domestic dog	F	*Lagostomus maximus*	MF

Guglielmone et al. (2014) state that all records of *Amblyomma pseudoparvum* nymphs from several mammals, including a number of records from domestic Bovidae and humans in Ivancovich and Luciani (1992, as *Amblyomma parvum parvum*), are doubtful because these authors included no discussion of how they separated their specimens from the morphologically similar *Amblyomma parvum*, and their data are therefore not included in this analysis. Voltzit (2007) confused the principal host of *Amblyomma pseudoparvum*, *Dolichotis salinicola* ("conejo del palo"), with *Dolichotis patagonum* ("Patagonian mara"). Estrada-Peña et al. (2020) include the nymph of *Amblyomma pseudoparvum* in a cluster of Neotropical species that feed mostly on Didelphimorphia and Rodentia, while the larva is listed in a cluster of ticks that feed mostly on artiodactyls, but these statements require confirmation due to the uncertainties involved in identifying the immature stages of this tick. In fact, the natural hosts of the larva of *Amblyomma pseudoparvum* remain unknown.

49. *Amblyomma quadricavum* (Schulze, 1941) as *Aponomma quadricavum* and given its current status in Keirans and Klompen (1996)

Schulze, P. (1941a) Ein neues *Amblyomma* und ein neues *Aponomma* mit Augenrudimenten aus Haiti. *Zoologischer Anzeiger*, 133, 225–229.

Keirans, J.E. & Klompen, J.S.H. (1996) *Amblyomma quadricavum* (Schulze) (new combination), and *Amblyomma arianae* Keirans and Garris, a new junior synonym of *Amblyomma quadricavum* (Acari: Ixodidae). *Proceedings of the Entomological Society of Washington*, 98, 164–165.

Descriptions and Redescriptions: Schulze (1941a) described and figured the female, and the male was described and illustrated in Černý (1966b), while Anderson et al. (1981) described the larva. All these authors used the name *Aponomma quadricavum.*

The male and female of *Amblyomma quadricavum* were redescribed in Santos Dias (1993) under the name *Aponomma quadricavum*, Voltzit (2007), Barros-Battesti et al. (2009), and Nowak (2010) as *Amblyomma quadricavum*. The male alone was described by Keirans and Garris (1986) under the name *Amblyomma arianae*, a synonym of *Amblyomma quadricavum*, as discussed in Keirans and Klompen (1996).

There are no redescriptions of the larva of *Amblyomma quadricavum*.

The nymph of *Amblyomma quadricavum* remains unknown.

Geographic Distribution: *Amblyomma quadricavum* is a Neotropical species that has only been found on Caribbean Islands as follows: **Cuba**: Artemisa, Cienfuegos, Matanzas, Mayabaque, and Pinar del Río (Černý 1966b, 1969b; Álvarez et al. 2020); **Haiti** (Schulze 1941a); **Jamaica** (Keirans and Klompen 1996); and **Puerto Rico**: Manatí, and Vega Baja (Keirans and Garris 1980).

Hosts: All male and female specimens of *Amblyomma quadricavum* have been collected from Squamata: Boidae and Dipsadidae, but in Cuba most records are from *Chilabothrus angulifer* (Boidae), a principal host of this tick, although the range of

Table 2.37 Hosts for males (M) and females (F) of *Amblyomma quadricavum*

REPTILIA			
SQUAMATA: Boidae		**SQUAMATA: Dipsadidae**	
Chilabothrus angulifer	MF	*Borikenophis portoricensis*	M*
Chilabothrus striatus	M	*Cubophis cantherigerus*	MF
Chilabothrus subflavus	F		

* Identified as *Amblyomma arianae*

Amblyomma quadricavum is wider than the range of this snake, and preferred hosts of this tick probably vary among islands. The larva of *Amblyomma quadricavum* is known from laboratory-reared material. The host profile of *Amblyomma quadricavum* is shown in Table 2.37, which is based on data from Schulze (1941a), Černý (1966b, 1969b), Keirans and Garris (1986), and Voltzit (2007).

Human Parasitism: No.

Comments: Morel (1967) believes that *Amblyomma quadricavum* is a synonym of *Amblyomma albopictum*.

The Cuban province of Matanzas is excluded from the range of *Amblyomma quadricavum* by Álvarez et al. (2020), but the records of this tick in Ciénaga de Zapata by Černý (1966b) support its presence in that province.

There is a record of *Amblyomma quadricavum* from *Iguana iguana* (Squamata: Iguanidae) based on specimens collected in Poland from captive hosts imported from El Salvador (Siuda et al. 2005; Nowak 2010). Because uncertainties attend the origin of this infestation, both El Salvador and *Iguana iguana* are excluded from this analysis.

Guglielmone et al. (2014) list Squamata: Colubridae as hosts for larvae of *Amblyomma quadricavum*, but such a host-parasite relationship was not confirmed during the current review. Information concerning the families of hosts of *Amblyomma quadricavum* in Barros-Battesti et al. (2009) is confusing (Guglielmone et al. 2014), and data from that paper have not been included in this analysis.

50. *Amblyomma romarioi* Martins, Luz and Labruna, 2019 in Martins et al. (2019b)

Martins, T.F., Luz, H.R., Muñoz-Leal, S., Ramírez, D.G., Milanelo, L., Marques, S., Sanches, T.C., Onofrio, V.C., Acosta, I.C.L., Benatti, H.R., Maturano, R., Oliveira, P.B., Albuquerque, G.R., Marcili, A., Flausino, W., Silveira, L.F., McIntosh, D., Faccini, J.L.H. & Labruna, M.B. (2019b) A new species of *Amblyomma* (Acari: Ixodidae) associated with monkeys and passerines of the Atlantic rainforest biome, southeastern Brazil. *Ticks and Tick-borne Diseases*, 10 (6) (article 101259) 10 pp.

Descriptions and Redescriptions: Martins et al. (2019b) described the male and female of *Amblyomma romarioi*.

There are no redescriptions of the male and female of *Amblyomma romarioi*.

Table 2.38 Hosts for males (M), females (F), nymphs (N), and larvae (L) of *Amblyomma romarioi*

MAMMALIA			
PRIMATES: Pitheciidae			
Callicebus nigrifrons	MFN		
AVES			
PASSERIFORMES: Cardinalidae		**Thamnophilidae**	
Cyanoloxia brissonii	L	*Pyriglena leucoptera*	L
Habia rubica	L	**Thraupidae**	
Conopophagidae		*Ramphocelus bresilius*	L
Conopophaga lineata	L	*Tachyphonus coronatus*	L
Furnariidae		*Tricothraupis melanops*	L
Automolus leucophthalmus	L	**Tityridae**	
Dendrocincla turdina	L	*Schiffornis virescens*	L
Lepidocolaptes falcinellus	L	**Turdidae**	
Sclerurus scansor	L	*Turdus albicollis*	L
Sittasomus griseicapillus	L	*Turdus amaurochalinus*	L
Xenops minutus	L	*Turdus flavipes*	L
Xiphorynchus fuscus	L	*Turdus leucomelas*	L
Parulidae		*Turdus rufiventris*	L
Basileuterus culicivorus	L	**Tyrannidae**	
Pipridae		*Attila rufus*	L
Chiroxiphia caudata	L	*Leptopogon amaurocephalus*	L
		Platyrinchus mystaceus	L

The nymph and larva of *Amblyomma romarioi* remain undescribed, but Martins et al. (2019b) recognized these stages from field-collected material.

Geographic Distribution: *Amblyomma romarioi* is a Neotropical species that has been found in **Brazil**: Minas Gerais, Rio de Janeiro, and São Paulo (Martins et al. 2019b).

Hosts: Males, females, and nymphs of *Amblyomma romarioi* have been collected from *Callicebus nigrifrons* (Primates: Pitheciidae), while larvae have been found on several families of Passeriformes, and these vertebrates are treated as principal hosts for adults and immature stages, respectively. The host profile of *Amblyomma romarioi* is shown in Table 2.38, which is based on data from Martins et al. (2019b) and Ramírez et al. (2020).

Human Parasitism: No.

Comments: *Amblyomma romarioi* is morphologically close to *Amblyomma geayi*, *Amblyomma longirostre,* and *Amblyomma parkeri.* See *Amblyomma longirostre* for a discussion of probable confusion among these taxa, especially in the case of immatures.

Larvae of *Amblyomma romarioi* were first collected from Passeriformes and identified as *Amblyomma* sp. haplotype Nazaré by Ogrzewalska et al. (2012a) and others, as explained in Martins et al. (2019b).

51. *Amblyomma romitii* Tonelli Rondelli, 1939

Tonelli Rondelli, M. (1939) Ixodoidea. Parte II. Contributo alla conoscenza della fauna ixodologica Sud-americana. *Rivista di Parassitologia*, 3, 39–55.

Descriptions and Redescriptions: The male, female, and nymph of *Amblyomma romitii* were described and illustrated by Tonelli Rondelli (1939), and the larva was described by Barros-Battesti et al. (2012).

The male and female of *Amblyomma romitii* were redescribed in Floch and Abonnenc (1940) under the name *Amblyomma tasquei*, a synonym of *Amblyomma romitii*, as stated in Barros-Battesti et al. (2007a), Floch and Fauran (1958), Onofrio et al. (2006b, keys and figures), Barros-Battesti et al. (2007a) and Dantas-Torres et al. (2019b, keys and figures). The nymph of *Amblyomma romitii* was redescribed in Martins et al. (2010).

There are no redescriptions of the larva of *Amblyomma romitii.*

Geographic Distribution: *Amblyomma romitii* is a north-central South American species found in **Brazil**: Mato Grosso, Pará, and Rondônia (Witter et al. 2016); **French Guiana**: Cayena (Binetruy et al. 2019); **Guyana**: Upper Demerara-Berbice (Tonelli Rondelli et al. 1939); **Suriname** (Jones et al. 1972); and **Venezuela**: Bolívar, and Monagás (Jones et al. 1972).

Hosts: All records of males, females, and nymphs of *Amblyomma romitii* are from *Hydrochoerus hydrochaeris* (Rodentia: Caviidae), with the exception of a male tick recovered from *Didelphis marsupialis* (Didelphimorphia: Didelphidae) and listed in Jones et al. (1972) under the name *Amblyomma extraoculatum*, as are several other records of *Amblyomma romitii.* The larva of this species was described by Barros-Battesti et al. (2012), but its natural hosts are undetermined because this stage is only known from laboratory-reared material and some larvae collected from humans that are not included in this host-tick analysis.

Human Parasitism: Guglielmone and Robbins (2018) listed *Amblyomma romitii* as a rare parasite of humans, with adult ticks (sex undetermined, with one exception below) and larvae collected from people in Brazil (Pará). One record of a male tick collected from a person was reported under the name *Amblyomma tasquei* in Aragão and Fonseca (1961a), as stated in Barros-Battesti et al. (2007a).

Comments: Santos Dias (1955) considered the type locality of *Amblyomma extraoculatum* in Singapore (Oriental Zoogeographic Region) a labeling error and synonymized *Amblyomma romitii* (Neotropical) with *Amblyomma extraoculatum*, but Barros-Battesti et al. (2007a) demonstrated the validity of both species, and *Amblyomma extraoculatum* is now considered an Oriental species. However, Kolonin (2009) incorrectly treats *Amblyomma extraoculatum* as a synonym of *Amblyomma romitii.*

A record of *Amblyomma romitii* (under the name *Amblyomma tasquei*) from Testudines, published by Amorim et al. (1999), is treated as doubtful in Guglielmone et al. (2014) and here; therefore, Testudines are not included as hosts of this tick.

52. *Amblyomma rotundatum* Koch, 1844

Koch, C.L. (1844) Systematische Übersicht über die Ordnung der Zecken. *Archiv für Naturgeschichte*, 10, 217–239.

Descriptions and Redescriptions: Koch (1844) described but did not figure the female of *Amblyomma rotundatum*. Keirans and Oliver (1993) described a male with gynandromorphic characters, and the first normal male of *Amblyomma rotundatum* was described by Labruna et al. (2005b). The nymph and larva of *Amblyomma rotundatum* were described by Aragâo (1912b) under the name *Amblyomma agamum*, a synonym of *Amblyomma rotundatum*, as stated in Robinson (1926).

Nava et al. (2017) redescribed the male, female and nymph of *Amblyomma rotundatum*, while Voltzit (2007) redescribed the male and female; Robinson (1926), Floch and Abonnenc (1940), Boero (1957), Floch and Fauran (1958), Guzmán-Cornejo et al. (2011, key and figures), Bermúdez et al. (2018a) and Dantas-Torres et al. (2019b, key and figures) redescribed the female alone. Keirans and Oliver (1993) redescribed the nymph and larva; the nymph alone was redescribed by Keirans and Durden (1998) and Martins et al. (2010). The larva of *Amblyomma rotundatum* was redescribed by Barbará and Dios (1918) under the name *Amblyomma agamum*, and by Amorim and Serra-Freire (1995).

Geographic Distribution: *Amblyomma rotundatum* is found on most of the Neotropical mainland and several Caribbean islands, but also in the Nearctic Region (southern USA), and it is present on an island in the Pacific Ocean (Kelehear et al. 2017b). The geographic distribution of *Amblyomma rotundatum* in the Neotropical Region encompasses the following countries: **Argentina**: Corrientes and Formosa (Guglielmone and Nava 2006; Smith et al. 2008); **Bolivia** (Jones et al. 1972); **Brazil**: Acre, Amapá, Amazonas, Bahia, Ceará, Distrito Federal, Espírito Santo, Goiás, Maranhâo, Mato Grosso, Mato Grosso do Sul, Minas Gerais, Pará, Paraíba, Paraná, Pernambuco, Rio de Janeiro, Río Grande do Norte, Rio Grande do Sul, Rondônia, Santa Catarina, Sâo Paulo, Sergipe, and Tocantins (Labruna et al. 2005d, 2007b; Onofrio 2007; Dantas-Torres et al. 2010a; Antonucci et al. 2012; Luz et al. 2013a, 2018b; Acosta et al. 2016; Lima et al. 2018); **Colombia**: Caldas, Chocó, Meta, and Valle del Cauca (Osorno Mesa 1942; Keirans 1985b; Benavides-Montaño et al. 2018; Acevedo-Gutiérrez et al. 2020); **Costa Rica**: Cartago (Troyo et al. 2016); **Dominica** (Durden et al. 2015); **French Guiana**: Cayena, and Saint Laurent du Maroni (Floch and Fauran 1958); **Grenada** (Kohls 1969c); **Guadeloupe**: Basse-Terre, and Point-à-Pitre (Morel 1967); **Guatemala**: Quiché, Retalhuleu, and Suchitepeque (Bequaert 1938); **Jamaica**: Saint Andrew, and Saint Thomas (Kohls 1969c); **Martinique** (Jones et al. 1972); **southern Mexico**: Chiapas, Jalisco, Michoacán, and Yucatán (Guzmán-Cornejo et al. 2011); **Montserrat** (Rawlins et al. 1993); **Panama**: Panamá (Ogrzewalska and Bermúdez 2019); **Paraguay**: Central (Nava et al. 2007); **Peru**: Loreto (Need et al. 1991); **Suriname** (Keirans 1985b); **Trinidad and Tobago** (Jones et al. 1972); and **Venezuela**: Amazonas, Araguá, Bolívar, and Guárico (Díaz Ungría 1957; Jones et al. 1972).

Hosts: Usual hosts for the female, nymph, and larva of *Amblyomma rotundatum* are *Rhinella marina* (Anura: Bufonidae) and *Boa constrictor* (Squamata: Boidae), but this tick has also been collected from several Neotropical reptiles of different orders, and some of them may be more important hosts of this tick than currently realized. *Amblyomma rotundatum* has occasionally been collected from a few species of mammals, but the host list for this tick is significantly longer, including a record of a nymph from Aves (Passeriformes: Turdidae) (Scott and Durden 2015a, b), implying that *Amblyomma rotundatum* may be carried by avian hosts outside the Neotropical Region, and many records are from captive vertebrates and laboratory hosts (Guglielmone and Nava 2010). This great diversity of hosts is reflected in Table 2.39, where captive and laboratory hosts carrying *Amblyomma rotundatum* in Neotropical areas are excluded. Our host profile is based on data from Boffy and Almeida-Santos (2010), Guglielmone and Nava (2010), Horta et al. (2015), Martins et al. (2015a), Acosta et al. (2016, 2019), Quirino et al. (2016), Silva et al. (2016a, b), Witter et al. (2016), Nava et al. (2017), Bermúdez et al. (2018a), Gianizella et al. (2018a), Oda et al. (2018), Alcántara et al. (2018), Zimmermann et al. (2018), Gruhn et al. (2019), Nossa et al. (2019), Ogrzewalska et al. (2019), Fonseca et al. (2020), Costa et al. (2020), Martins et al. (2020b) and Mendoza-Roldán et al. (2020).

Human Parasitism: Guglielmone and Robbins (2018) list *Amblyomma rotundatum* as a very rare parasite of humans, with one Brazilian record (Pará State) of female ticks feeding on people (Serra-Freire et al. 1995).

Comments: Identification of *Amblyomma rotundatum* is sometimes difficult, as evidenced by its confusion with *Amblyomma goeldii* and, especially, *Amblyomma dissimile*; see also the comments under these species.

Vogelsang (1928) recorded the presence of *Amblyomma rotundatum* (under the name *Amblyomma agamum*) in Uruguay, but its presence in that country has not been confirmed, as discussed in Venzal et al. (2003c), and Uruguay is not included within the range of this tick. Several old Argentinean records in the provinces of Córdoba, Salta, and Santiago del Estero in Boero (1955) and others are considered valid in Guglielmone and Nava (2006); however, after several studies, the presence of this tick there was not confirmed, and they are now regarded as outside the range of *Amblyomma rotundatum*. Honduras and the Peruvian department of Ucayali are also excluded from the geographic distribution of *Amblyomma rotundatum*, as discussed below. Andreotti et al. (2018) exclude the states of Amapá, Bahia, Ceará, Paraíba, Rio Grande do Norte, and Rondônia from the Brazilian range of *Amblyomma rotundatum*, but no reasons were given for this exclusion.

Amblyomma rotundatum is a parthenogenetic species whose males are seldom encountered in laboratory and field-collected material. Nevertheless, Pietzsch et al. (2006) allegedly found 25 males and nine females on *Rhinella marina* from Honduras, while Rojas et al. (2010) claim to have found 14 *Amblyomma rotundatum* males on eight *Rhinella marina* in Ucayali Department, Peru. None of these authors discuss their records, which are treated as doubtful here; consequently, Honduras and the Peruvian department of Ucayali are excluded from the range of *Amblyomma rotundatum*. Similarly, Rodríguez-Vivas et al. (2016) found three females and two

Table 2.39 Hosts for adults of undetermined sex (A), males (M)*, females (F), nymphs (N), and larvae (L) of *Amblyomma rotundatum***

	AMPHIBIA		
ANURA: Bufonidae		*Rhinella margaritifera*	FN
Anaxyrus terrestris	L	*Rhinella marina*	FNL
Rhaebo guttatus	FN	*Rhinella ornata*	FN
Rhinella bergi	FN	*Rhinella pygmaea*	F
Rhinella crucifer	FN	*Rhinella schneideri*	FNL
Rhinella diptycha	F	**ANURA: Leptodactylidae**	
Rhinella gildae	N	*Leptodactylus pentadactylus*	F
Rhinella granulosa	FN	*Physalaemus nattereri*	F
Rhinella icterica	FNL	**ANURA: Pipidae**	
Rhinella jimi	FNL	*Pipa pipa*	F
Rhinella major	F		
	REPTILIA		
CROCODILIA: Alligatoridae		**SQUAMATA: Scincidae**	
Caiman latirostris	F	*Mabuya mabouya*	NL
Paleosuchus palpebrosus	F	*Trachylepis atlantica*	N
Paleosuchus trigonatus	FN	**SQUAMATA: Teiidae**	
SQUAMATA: Boidae		*Ameiva ameiva*	F
Boa constrictor	MFNL	*Salvator* sp.	L
Corallus hortulanus	FN	*Salvator merianae*	N
Epicrates cenchria	FN	**SQUAMATA: Tropiduridae**	
SQUAMATA: Colubridae		*Plica plica*	F
Chironius flavolineatus	N	*Tropidurus* sp.	MFNL
Chironius multiventris	FNL	*Tropidurus hispidus*	F
Chironius scurrulus	FN	*Tropidurus torquatus*	N
Drymarchon corais	N	*Uranoscodon superciliosus*	FN
Leptophis ahaetulla	N	**SQUAMATA: Viperidae**	
Mastigodryas boddaerti	N	*Bothrops alternatus*	FN
Spilotes pullatus	FN	*Bothrops asper*	A
SQUAMATA: Dipsadidae		*Bothrops atrox*	FN
Atractus guentheri	N	*Bothrops erythromelas*	N
Dipsas neuwiedi	FN	*Bothrops insularis*	FN
Dipsas turgida	FN	*Bothrops jararaca*	FN
Helicops carinicaudus	F	*Bothrops jararacussu*	FN
Hydrodynastes gigas	F	*Bothrops lanceolatus*	F
Leptodeira annulata	FN	*Bothrops leucurus*	FN
Lygophis dilepis	N	*Bothrops moojeni*	F
Oxyrhopus guibei	F	*Bothrops neuwiedi*	F
Oxyrhopus melanogenys	N	*Crotalus durissus*	FN
Oxyrhopus trigeminus	N	*Lachesis muta*	FNL
Philodryas nattereri	N	**TESTUDINES: Chelidae**	
Philodryas olfersii	FN	*Hydromedusa tectifera*	F
Philodryas viridissima	N	*Mesoclemmys tuberculata*	F
Pseudoboa nigra	N	*Mesoclemmys vanderhaegei*	F
Xenodon merremii	FN	*Platemys platycephala*	F
Xenopholis scalaris	N	**TESTUDINES: Emydidae**	
SQUAMATA: Elapidae		*Trachemys dorbigni*	F
Micrurus sp.	FNL	**TESTUDINES: Geoemydidae**	
Micrurus corallinus	N	*Heosemys annandalii*	F
Micrurus ibiboboca	FN	*Rhinoclemmys areolata*	F
Micrurus lemniscatus	N	**TESTUDINES: Kinosternidae**	
SQUAMATA: Iguanidae		*Kinosternon scorpioides*	FN
Iguana iguana	MFN	**TESTUDINES: Testudinidae**	
SQUAMATA: Phrynosomatidae		*Chelonoidis carbonarius*	FN
Phrynosoma sp.	FN	*Chelonoidis denticulata*	MFNL
	MAMMALIA		
ARTIODACTYLA: Suidae		**PILOSA: Bradypodidae**	
Domestic pig	F	*Bradypus variegatus*	F
CARNIVORA: Procyonidae		**PILOSA: Myrmecophagidae**	
Nasua nasua	F	*Tamandua tetradactyla*	F

(continued)

Table 2.39 (continued)

CHIROPTERA: Phyllostomidae		**RODENTIA: Caviidae**	
Choeroniscus minor	F	*Hydrochoerus hydrochaeris*	N
CINGULATA: Dasypodidae			
Dasypus novemcinctus	F		

* *Amblyomma rotundatum* is a parthenogenetic species with a few natural records of males in field-collected material
** Captive and laboratory hosts are not included in this table

males of this tick in México, but this unusual finding was not addressed by these authors; therefore, hosts listed in that study, including such alleged new hosts of *Amblyomma rotundatum* as *Crocodylus acutus* and *Terrapene carolina*, are excluded from Table 2.39. The few male specimens of *Amblyomma rotundatum* treated as *bona fide* here are those recorded in Brazil by Labruna et al. (2005b) on Squamata: Tropiduridae, Martins et al. (2014a) on Squamata: Boidae, Silva et al. (2016a, b) on Squamata: Iguanidae, and Gianizella et al. (2018a) on Testudines: Testudinidae.

The list of hosts of *Amblyomma rotundatum* in Table 2.39 differs from those shown in Guglielmone and Nava (2010) and Nava et al. (2017). This is due to the discovery of new host-parasite relationships, but also because laboratory and captive hosts as well as hosts found outside the Neotropical Region are excluded in the present study. Barros and Baggio (1992) listed a female of *Amblyomma rotundatum* collected from a *Tamandua tetradactyla*, but Arzua et al. (2005) found this identification erroneous. The record from *Tamandua tetradactyla* in Table 2.38 is based on Neumann's (1899) description of the female of *Amblyomma goeldii*, which later was found to be, in fact, a redescription of the female of *Amblyomma rotundatum.* This record is treated as provisionally valid. Burridge (2011) states that *Dasypus novemcinctus* is an important host for *Amblyomma rotundatum*; however, there is one record from this type of host in Floch and Fauran (1958), which has been cited several times in the absence of new collections from this host. Costa et al. (2017) found a specimen of *Amblyomma rotundatum* on a dog, but the tick was not attached; this record is apparently repeated in Costa et al. (2020), who stated that a female of this tick had been collected from a dog, but no indication was provided as to whether or not the tick was attached, and dogs are provisionally excluded as hosts of *Amblyomma rotundatum.*Oliver (1989) states that *Amblyomma rotundatum* can revert to a two-host life cycle from the typical three-host cycle of this and other *Amblyomma* species, a phenomenon confirmed by Rodrigues et al. (2010).

53. *Amblyomma sabanerae* Stoll, 1894

Stoll, O. (1894) *Arachnida Acaridae*. Biologia Centrali-Americana, Serie Zoologia, pp. 1–55.

Descriptions and Redescriptions: Stoll (1894) described and illustrated the female of *Amblyomma sabanerae*, and Schulze (1937a) described and figured the male.

The male and female of *Amblyomma sabanerae* were redescribed by Voltzit (2007), Guzmán-Cornejo et al. (2011, keys and figures) and Bermúdez et al. (2018a), while the female alone was redescribed in Neumann (1899), Santos Dias (1989b) and Robinson (1926).

The nymph and larva of *Amblyomma sabanerae* remain undescribed, but several workers have been able to identify these stages from field-collected material.

Geographic Distribution: *Amblyomma sabanerae* is a Neotropical species, with most records from Central America, but this tick has also been recorded from southern North America and northern South America as follows: **Belize**: Belize, Corozal, and Orange Walk (Ernst and Ernst 1977; Robbins et al. 2001a); **Colombia**: Antioquía, Chocó, Meta, and Valle del Cauca (Ernst and Ernst 1977; Garcés-Restrepo et al. 2013; Acevedo-Gutiérrez et al. 2020); **Costa Rica**: Guanacaste, and Limón (Álvarez et al. 2005); **El Salvador**: La Libertad, and San Miguel (Fairchild et al. 1966; Barbieri et al. 2012a, b); **Guatemala**: Jutiapa, and Retalhuleu (Stoll 1894; Ernst and Ernst 1977); **Honduras**: Olancho (Ernst and Ernst 1977); **southern Mexico**: Campeche, Chiapas, Oaxaca, Quintana Roo, and Yucatán (Paredes-León et al. 2008; Guzmán-Cornejo et al. 2011); **Nicaragua** (Ernst and Ernst 1977); **Panama**: Bocas del Toro, Coclé, Colón, Darién, Guna Yala, Panamá, and Panamá Oeste (Ernst and Ernst 1977; Voltzit 2007; Miller et al. 2016); and **Suriname** (Ernst and Ernst 1977).

Hosts: Males and females of *Amblyomma sabanerae* are usually found on tortoises (Testudines: Geoemydidae). Most specimens of the undescribed larva and nymph have been collected from Geoemydidae and mammals (Didelphimorphia and Rodentia), with a few records from anuran, avian, and squamatan hosts. Nevertheless, the numbers of larvae and nymphs collected from Geoemydidae by Ernst and Ernst (1977), Robbins et al. (2001a), and Garcés-Restrepo et al. (2013) greatly surpass those found on other hosts; consequently, Geoemydidae are provisionally regarded as the preferred hosts for the larva and nymph of *Amblyomma sabanerae,* especially the five species of *Rhinoclemmys*, which have been found infested with all parasitic stages, as shown below. The host profile of *Amblyomma sabanerae* is detailed in Table 2.40, which is based on data from Hoffmann (1962), Fairchild et al. (1966), Ernst and Ernst (1977), Robbins et al. (2001a), Bermúdez et al. (2010a, 2012a, 2018a), Garcés-Restrepo et al. (2013), Ogrzewalska et al. (2015), Esser et al. (2016a, b) and Rodríguez-Vivas et al. (2016).

Human Parasitism: Guglielmone and Robbins (2018) list *Amblyomma sabanerae* as a very rare parasite of humans, citing two records from Panama (Bermúdez et al. 2012b). Quintero et al. (2021) collected a nymph from a human in the Colombian department of Antioquia, a record treated as provisionally valid here.

Comments: *Amblyomma crassum*, *Amblyomma humerale,* and *Amblyomma sabanerae* are morphologically similar taxa, and their identification is difficult. Thus, Santos Dias (1958a) treats *Amblyomma sabanerae* (and *Amblyomma crassum*) as synonyms of *Amblyomma humerale*, although, later, Santos Dias (1989b) accepted *Amblyomma sabanerae* as valid. Fairchild et al. (1966) also suspect that

Table 2.40 Hosts for males (M), females (F), nymphs (N)*, and larvae (L)* of *Amblyomma sabanerae*

	AMPHIBIA		
ANURA: Bufonidae			
Rhinella alata	N		
	REPTILIA		
SQUAMATA: Colubridae		*Rhinoclemmys areolata*	MFNL
Leptophis depressirostris	N	*Rhinoclemmys funerea*	MFNL
SQUAMATA: Iguanidae		*Rhinoclemmys melanosterna*	MFN
Iguana iguana	MF	*Rhinoclemmys nasuta*	MFNL
TESTUDINES: Emydidae		*Rhinoclemmys pulcherrima*	MFNL
Terrapene carolina	MF	*Rhinoclemmys punctularia*	N
Terrapene mexicana	MF	**TESTUDINES: Kinosternidae**	
Trachemys scripta	M	*Kinosternon* sp.	MFN
TESTUDINES: Geoemydidae		*Kinosternon scorpioides*	F
Rhinoclemmys annulata	MFNL		
	AVES		
COLUMBIFORMES: Columbidae		**PASSERIFORMES: Thamnophilidae**	
Geotrygon montana	L	*Gymnopithys leucaspis*	N
PASSERIFORMES: Pipridae		*Thamnophilus atrinucha*	N
Manacus candei	N		
	MAMMALIA		
CARNIVORA: Canidae		*Metachirus nudicaudatus*	NL
Dog	M	*Philander opossum*	NL
CARNIVORA: Procyonidae		**RODENTIA: Cricetidae**	
Procyon cancrivorus	N	*Transandinomys talamancae*	NL
DIDELPHIMORPHIA: Didelphidae		**Cricetidae: Echimyidae**	
Didelphis marsupialis	NL	*Hoplomys gymnurus*	NL
Marmosa robinsoni	L	*Proechimys semispinosus*	NL

* records of the nymph and larva of *Amblyomma sabanerae* should be treated as provisionally valid because both stages remain undescribed

Amblyomma crassum is a synonym of *Amblyomma sabanerae*, and these authors treat the male of *Amblyomma crassum* described in Méndez Arocha and Ortiz (1957) as probably the male of *Amblyomma sabanerae.*

Kelehear et al. (2017a) listed *Amblyomma sabanerae* as present in French Guiana, but no information has been found to support that statement.

Ogrzewalska et al. (2010) state that the alleged record of a nymph of *Amblyomma humerale* collected from a migratory bird in Canada (Morshed et al. 2005) likely represents *Amblyomma sabanerae*, while Guglielmone et al. (2014) doubt the identity of a supposed *Amblyomma sabanerae* specimen collected from a bird in Canada (Scott et al. 2001, 2010). Although the nymph of *Amblyomma sabanerae* remains undescribed, Garcés-Restrepo et al. (2013) studied its hypostomal dentition, finding that its dental formula varies from 2/2 to 3/3, further complicating the identification of this species.

González-Martín del Campo et al. (2018) used a questionable technique to identify nymphs of *Amblyomma sabanerae* allegedly collected from *Hylocichla mustelina* (Passeriformes: Turdidae) in Mexico; this record is treated as doubtful, and that bird is excluded as a host of *Amblyomma sabanerae*. Further, González-Martín del Campo et al. (2018) stated that an adult *Hyalomma* sp., a genus of paramount medical and veterinary importance not known to be established in the Americas, was found on a bird in southern Mexico, but no further data were provided to support this diagnosis. All tick specimens reported by González-Martín del Campo et al. (2018) should be reexamined.

54. *Amblyomma scalpturatum* Neumann, 1906

Neumann, L.G. (1906) Notes sur les Ixodidés. IV. *Archives de Parasitologie*, 10, 195–219.

Descriptions and Redescriptions: Neumann (1906) described but did not figure the female of *Amblyomma scalpturatum*, and the male was described by Schulze (1933b) under the name *Amblyomma myrmecophagae*, a synonym of *Amblyomma scalpturatum*, as discussed in Guglielmone and Nava (2014). The nymph of *Amblyomma scalpturatum* was described by Martins et al. (2010).

The male and female of *Amblyomma scalpturatum* were redescribed by Floch and Fauran (1958), Boero and Prosen (1960), Labruna et al. (2005c), Voltzit (2007), and Dantas-Torres et al. (2019b, keys and figures), while Robinson (1926) redescribed the female alone.

There are no redescriptions of the nymph of *Amblyomma scalpturatum*.

The larva of *Amblyomma scalpturatum* remains undescribed, but Colle et al. (2020) have been able to molecularly identify this stage.

Geographic Distribution: *Amblyomma scalpturatum* is a Neotropical species found in central and northern South America as follows: **Bolivia**: Cochabamba (Boero and Prosen 1960); **Brazil**: Acre, Amazonas, Mato Grosso, Pará, Paraná, Rondônia, and Roraima (Labruna et al. 2005d; Onofrio et al. 2010a; Lima et al. 2018); **Colombia**: Casanare, and Meta (Labruna et al. 2005c); **Ecuador**: Napo, Pastaza, and Sucumbíos (Zerpa et al. 2003a; Labruna et al. 2005c); **French Guiana**: Cayena (Binetruy et al. 2019); **Guyana** (Labruna et al. 2005c); **Peru**: Cusco, and Madre de Dios (Labruna et al. 2005c, 2010b); **Suriname** (Schulze 1933b); and **Venezuela**: Amazonas, and Bolívar (Labruna et al. 2005c).

Hosts: The usual host for males and females of *Amblyomma scalpturatum* is *Tapirus terrestris* (Perissodactyla: Tapiridae), with a few records from several mammals belonging to other orders. The hosts of the nymph of *Amblyomma scalpturatum* are known from ten specimens collected from different mammals, while the hosts of the larva are known from three larvae collected from *Didelphis marsupialis* (Didelphimorphia: Didelphidae) by Colle et al. (2020). The principal hosts of the nymph and larva of *Amblyomma scalpturatum* are treated here as basically undetermined. The host profile of *Amblyomma scalpturatum* is shown in Table 2.41, which is based on data from Labruna et al. (2005c, d), Martins et al. (2014a), Witter et al. (2016), Gianizella et al. (2018b), Binetruy et al. (2019) and Colle et al. (2020).

Human Parasitism: Guglielmone and Robbins (2018) listed *Amblyomma scalpturatum* as a sporadic parasite of humans. All cases have been recorded in Brazil (Amazonas, Mato Grosso, Pará, and Rondônia), where males, females, and nymphs have been collected from people, but Guglielmone and Robbins (2018)

Table 2.41 Hosts for adults of undetermined sex (A), males (M), females (F), nymphs (N), and larvae (L)* of *Amblyomma scalpturatum*

MAMMALIA			
ARTIODACTYLA: Suidae		**PERISSODACTYLA: Tapiridae**	
Domestic pig	FN	*Tapirus terrestris*	MF
ARTIODACTYLA: Tayassuidae		**PILOSA: Myrmecophagidae**	
Pecari tajacu	AN	*Myrmecophaga tridactyla*	MF
CARNIVORA: Canidae		**RODENTIA: Caviidae**	
Domestic dog	AN	*Hydrochoerus hydrochaeris*	MN
CARNIVORA: Felidae		**RODENTIA: Cuniculidae**	
Panthera onca	N	*Cuniculus paca*	AN
DIDELPHIMORPHIA: Didelphidae		**RODENTIA: Dasyproctidae**	
Didelphis marsupialis	NL	*Dasyprocta azarae*	N
		Dasyprocta fuliginosa	N

* Provisional diagnosis because the larva of *Amblyomma scalpturatum* remains undescribed

overlooked a case of human parasitism by a male of this tick in Paraná State (Brazil), described in Onofrio et al. (2010a). Aguirre et al. (2019) added a nymph collected from a person in the Brazilian state of Rondônia.

Comments: Labruna et al. (2005c) found that some records of *Amblyomma scalpturatum* from Brazil, Ecuador, and Venezuela corresponded in fact to *Amblyomma latepunctatum*, while Labruna et al. (2013) believe that the male of *Amblyomma multipunctum* from Bolivia redescribed in Boero and Prosen (1959) is actually *Amblyomma scalpturatum*.

Campos Pereira et al. (2000) list the Brazilian state of Rio Grande do Sul as within the range of *Amblyomma scalpturatum*, but this record has not been cited by subsequent workers studying this species, and Rio Grande do Sul is excluded from the geographic distribution of this tick. Rodríguez and Flores (2003) state that *Amblyomma* cf. *A. scalpturatum* was collected from a goat in the Venezuelan state of Falcón, but this identification requires confirmation and is not included in our analysis.

Estrada-Peña et al. (2020) place the nymph of *Amblyomma scalpturatum* in a cluster of Neotropical species that feed mostly on Didelphimorphia and Rodentia, but the principal host for this stage is here regarded as undetermined.

55. *Amblyomma sculptum* Berlese, 1888

Berlese, A. (1888) Acari austro-americani. *Bolletino della Societá di Entomologia Italiana*, 20, 171–242.

Descriptions and Redescriptions: Berlese (1888) described and figured the male and female of *Amblyomma sculptum*. The nymph was described by Martins et al. (2010) under the name *Amblyomma cajennense*, as explained in Martins et al. (2014c), while the larva was described by Famadas et al. (1997), also under the name *Amblyomma cajennense*, as stated in Dantas-Torres et al. (2019b).

The male, female, and nymph of *Amblyomma sculptum* were redescribed by Nava et al. (2017), while, Tonelli Rondelli (1937), Nava et al. (2014a), and Dantas-Torres et al. (2019b, keys and figures) redescribed the male and female. Martins et al. (2014c) redescribed the nymph.

There are no redescriptions of the larva of *Amblyomma sculptum*.

Geographic Distribution: *Amblyomma sculptum* is a Neotropical species that has been found in central and southern South America as follows: **Argentina**: Chaco, Corrientes, Formosa, Jujuy, and Salta (Tarragona et al. 2018); **Bolivia**: Beni (Nava et al. 2014a); **Brazil**: Bahia, Ceará, Distrito Federal, Espírito Santo, Goiás, Maranhão, Mato Grosso, Mato Grosso do Sul, Minas Gerais, Pará, Paraná, Pernambuco, Piauí, Rio de Janeiro, Rondônia, São Paulo, and Tocantins (Martins et al. 2016b; Moerbeck et al. 2016); and **Paraguay**: Alto Paraguay, and Concepción (Nava et al. 2014a).

Hosts: Males, females, nymphs, and larvae of *Amblyomma sculptum* have been found on a great variety of mammals and birds and rarely on Reptilia; nevertheless, it is hypothesized that the life cycle of this species is chiefly dependent on several orders of mammals, including a variety of domestic mammals. The host profile of *Amblyomma sculptum*, excluding captive hosts, is shown in Table 2.42, which is based on data from Silva et al. (2015), Sponchiado et al. (2015), Martins et al. (2016b, 2017), Kluyber et al. (2016), Luz et al. (2016a, 2017b, 2018a), Osava et al. (2016), Witter et al. (2016), Rocha et al. (2017), Sousa et al. (2017), Teixeira et al. (2017, 2020), Saracho-Bottero et al. (2018), Szabó et al. (2018, 2019), Tarragona et al. (2018), Barbieri et al. (2019), Fecchio et al. (2020), Mendoza-Roldán et al. (2020), Martins et al. (2020a) and Ramos et al. (2020).

Human Parasitism: Guglielmone and Robbins (2018) listed *Amblyomma sculptum* as a frequent human parasite that has been found on people in Argentina (Corrientes, Formosa, Jujuy, and Salta) and Brazil (Distrito Federal, Espírito Santo, Goiás, Maranhão, Mato Grosso, Mato Grosso do Sul, Minas Gerais, Paraná, Piauí, Rio de Janeiro, Rondônia, São Paulo, and Tocantins). Additionally, Bitencourth et al. (2017) found a case of human parasitism by a female tick in the Brazilian state of Tocantins, while Szabó et al. (2018, 2020) collected males, females, and many nymphs from persons in Minas Gerais, also in Brazil, and Kmetiuk et al. (2019) recovered males and females from humans in the Brazilian state of Goiás. Valente et al. (2020) collected eight males, two females and 11 nymphs of *Amblyomma sculptum* from people at four localities in the State of Paraná (Brazil); additionally, these authors stated that human records of *Amblyomma cajennense* in Paraná State published by Arzua et al. (2005) correspond, in fact, to *Amblyomma sculptum*. All parasitic stages have been found feeding on people, but the nymph appears to prevail as a parasite of humans (Guglielmone and Robbins 2018).

Comments: *Amblyomma sculptum* is a member of the *Amblyomma cajennense* species complex, which also comprises *Amblyomma interandinum*, *Amblyomma mixtum*, *Amblyomma patinoi*, and *Amblyomma tonelliae*. Further discussion of this complex is provided under *Amblyomma cajennense* above. With the exception of Tonelli Rondelli (1937), *Amblyomma sculptum* was treated as a synonym of *Amblyomma cajennense* until it was reinstated by Nava et al. (2014a).

Table 2.42 Hosts for adults of undetermined sex (A), males (M), females (F), nymphs (N), and larvae (L) of *Amblyomma sculptum**

MAMMALIA			
ARTIODACTYLA: Bovidae		**DIDELPHIMORPHIA: Didelphidae**	
Domestic buffalo	MFNL	*Didelphis albiventris*	FNL
Cattle	MFNL	*Didelphis aurita*	FN
Sheep	MF	*Gracilinanus agilis*	NL
ARTIODACTYLA: Cervidae		*Philander frenatus*	N
Blastocerus dichotomus	MF	**LAGOMORPHA: Leporidae**	
Mazama americana	MF	*Sylvilagus brasiliensis*	N
Mazama gouazoubira	MFN	**PERISSODACTYLA: Equidae**	
ARTIODACTYLA: Suidae		Donkey	MF
Domestic pig	MFNL	Horse	MFNL
Feral pig	MFN	Mule	MF
ARTIODACTYLA: Tayassuidae		**PERISSODACTYLA: Tapiridae**	
Pecari tajacu	MFN	*Tapirus terrestris*	MFN
Tayassu pecari	MFN	**PILOSA: Bradypodidae**	
CARNIVORA: Canidae		*Bradypus torquatus*	F
Cerdocyon thous	MFNL	*Bradypus variegatus*	F
Chrysocyon brachyurus	MFNL	**Myrmecophagidae**	
Domestic dog	MFNL	*Myrmecophaga tridactyla*	MFNL
Lycalopex vetulus	N	*Tamandua tetradactyla*	MFN
Speothos venaticus	MF	**PRIMATES: Atelidae**	
CARNIVORA: Felidae		*Alouatta guariba*	MFN
Domestic cat	MF	**RODENTIA: Caviidae**	
Leopardus pardalis	MFNL	*Hydrochoerus hydrochaeris*	MFNL
Panthera onca	MFN	**RODENTIA: Chinchillidae**	
Puma concolor	MFN	*Lagostomus maximus*	L
CARNIVORA: Mephitidae		**RODENTIA: Cricetidae**	
Conepatus semistriatus	N	*Calomys callosus*	N
CARNIVORA: Procyonidae		*Holochilus* sp.	F
Nasua nasua	MFN	*Holochilus sciureus*	N
Procyon cancrivorus	MFN	*Rhipidomys macrurus*	N
CINGULATA: Chlamyphoridae		*Sooretamys angouya*	N
Cabassous unicinctus	FN	**RODENTIA: Dasyproctidae**	
Euphractus sexcinctus	F	*Dasyprocta azarae*	MFN
Priodontes maximus	MFN	**RODENTIA: Echimyidae**	
Tolypeutes tricinctus	MF	*Thrichomys pachyurus*	N
CINGULATA: Dasypodidae		**RODENTIA: Erethizontidae**	
Dasypus novemcinctus	MFN	*Coendou prehensilis*	F
		Coendou spinosus	FN
AVES			
ACCIPTRIFORMES: Accipitridae		*Crax fasciolata*	NL
Busarellus nigricollis	A	*Penelope obscura*	N
Buteo brachyurus	N	**GRUIFORMES: Aramidae**	
Buteogallus coronatus	N	*Aramus guarauna*	N
Buteogaluus meridionalis	AN	**PASSERIFORMES: Furnariidae**	
Parabuteo unicinctus	N	*Dendrocincla turdina*	L
Rostrhamus sociabilis	N	**PASSERIFORMES: Thraupidae**	
Rupornis magnirostris	N	*Tachyphonus coronatus*	L
Spizaetus tyrannus	N	**PASSERIFORMES: Turdidae**	
CAPRIMULGIFORMES: Caprimulgidae		*Turdus leucomelas*	N
Setopagis parvula	N	**PELECANIFORMES: Threskiornithidae**	
CARIAMIFORMES: Cariamidae		*Theristicus caudatus*	N
Cariama cristata	MFNL	**PSITTACIFORMES: Psittacidae**	
CATHARTIFORMES: Cathartidae		*Amazona aestiva*	N
Coaragyps atratus	N	**RHEIFORMES: Rheidae**	
FALCONIFORMES: Falconidae		*Rhea americana*	F
Caracara plancus	AN	**STRIGIFORMES: Strigidae**	
Falco sparverius	N	*Asio clamator*	N
Milvago chimachima	N	*Megascops choliba*	AN
GALLIFORMES: Cracidae		**STRIGIFORMES:Tytonidae**	
Crax blumenbachii	F	*Tyto alba*	N
REPTILIA			
SQUAMATA: Iguanidae		**SQUAMATA: Teiidae**	
Iguana iguana	F	*Salvator merianae*	M

* Captive hosts are not included in this table

Rodríguez et al. (2019) allegedly collected additional specimens of *Amblyomma sculptum* in Bolivia but published a figure of a male that belongs to another species.

The known hosts and localities for *Amblyomma sculptum*, although numerous, have likely been undercounted because there are many records from Argentina, Bolivia, Brazil, and Paraguay of ticks classified as *Amblyomma cajennense sensu lato*, and surely a significant number of these in fact represent *Amblyomma sculptum*. Vogel et al. (2018) allegedly found *Amblyomma sculptum* in Nicaragua, a record considered here to be erroneous, and that was subsequently amended in the online version of that paper.

Estrada-Peña et al. (2020) include adults and nymphs of *Amblyomma sculptum* in a cluster of Neotropical species that feed mostly on pilosan hosts, while the larva of this tick is placed in a cluster of species that feed mostly on artiodactyls. However, the range of principal hosts for all stages of *Amblyomma sculptum* exceeds the hypothesis of Estrada-Peña et al. (2020). Records of this tick from captive hosts in Acosta et al. (2016), who found a female tick on *Spheniscus magellanicus* (Spheniscidae), Teixeira et al. (2017), who collected a male tick from *Phrynops geoffroanus* (Testudines: Chelidae), and Araújo et al. (2019), who collected nymphs of *Amblyomma sculptum* from *Dromaius novaehollandiae* (Casuariiformes: Casuariidae), are not included in the list above.

56. *Amblyomma scutatum* Neumann, 1899

Neumann, L.G. (1899) Révision de la famille des ixodidés (3e mémoire). *Mémoires de la Société Zoologique de France*, 12, 107–294.

Descriptions and Redescriptions: Neumann (1899) described but did not figure the male and female of *Amblyomma scutatum*. The nymph and larva were described by Hoffmann (1946) under the name *Amblyomma boneti*, a synonym of *Amblyomma scutatum*, as stated in Keirans and Clifford (1984), a conclusion treated as provisionally valid here.

The male and female of *Amblyomma scutatum* were redescribed by Hoffmann (1946) under the name *Amblyomma boneti*, and by Robinson (1926), Voltzit (2007), and Guzmán-Cornejo et al. (2011), while Whittick (1939) redescribed the female alone.

There are no redescriptions of the larva and nymph of *Amblyomma scutatum*.

Geographic Distribution: *Amblyomma scutatum* is a Neotropical species with a few records from the Nearctic in northern Mexico. In the Neotropics this tick has been found in southern North America and Central America as follows: **Costa Rica**: Guanacaste, and Puntarenas (Álvarez et al. 2005); **El Salvador** (Navarrete Abarca et al. 2014); **Guatemala**: Retalhuleu (Neumann 1899); **Honduras** (Voltzit 2007); **southern Mexico**: Guerrero, Jalisco, Michoacán, Morelos, Oaxaca, Veracruz, and Yucatán (Paredes-León et al. 2008; Guzmán-Cornejo et al. 2011; Rodríguez-Vivas et al. 2016); and **Nicaragua**: Managua (Maes et al. 1989).

Hosts: Males, females, nymphs, and larvae of *Amblyomma scutatum* have been collected from *Ctenosaura pectinata* (Squamata: Iguanidae), which is treated as the

Table 2.43 Hosts for adults of undetermined sex (A), males (M), females (F), nymphs (N), and larvae (L) of *Amblyomma scutatum*

REPTILIA			
SQUAMATA: Iguanidae		*Ctenosaura similis*	MF
Ctenosaura acanthura	A*	*Iguana iguana*	A
Ctenosaura pectinata	MFNL	*Iguana tuberculata*	MF

* A gynander specimen cited in Schulze (1933a)

principal host for this tick, but the range of *Amblyomma scutatum* is broader than the range of this reptile, and other iguanids are probably also principal hosts of this tick. The host profile of *Amblyomma scutatum* is shown in Table 2.43, which is based on data from Neumann (1899), Schulze (1933a), Paredes-León et al. (2008) and Rodríguez-Vivas et al. (2016).

Human Parasitism: No.

Comments: There are some differences in the morphological descriptions of *Amblyomma scutatum* because the above authors described the female tick as having an ornamented scutum, with the exception of Voltzit (2007), who depicts the scutum as inornate. Additionally, Robinson (1926) illustrates coxa II of the male as having two spurs, but other describers cite only one spur. The hypostomal dentition of the male tick in Hoffmann (1946, under the name *Amblyomma boneti*) does not totally agree with the dentition described by other workers. And Camacho Escobar and Pérez-Lara (2009) claim to provide an illustration of a male of *Amblyomma scutatum* but, in fact, their figure appears to show a nymph or even a female tick.

The range of *Amblyomma scutatum* is also controversial. Kelehear et al. (2017a) listed *Amblyomma scutatum* as present in French Guiana, but no information has been found to support that statement. Camicas et al. (1998) and Guglielmone et al. (2003a, 2014) consider this species to be established only in the Neotropical Region, but Guzmán-Cornejo et al. (2011) clearly state that *Amblyomma scutatum* has been found a few times in the Nearctic portion of Mexico. Moreover, Guglielmone et al. (2003a) state that the geographic distribution of this tick includes Brazil, Costa Rica, El Salvador, Guatemala, Honduras, southern Mexico, Paraguay, and Venezuela, but Guglielmone et al. (2015) excluded South American countries from this distribution for the reasons discussed below.

Neumann (1899) described *Amblyomma scutatum* from four males and seven females collected from a "lézard" and 17 males and one female collected from *Iguana tuberculata* in Guatemala. In the same paper, Neumann listed nymphs of this tick collected from *Noctilio albiventris* in Paraguay and from *Cathartes uruba* (= *Cathartes aura*), *Dasyprocta croconota*, and *Didelphis pusilla* (= *Thylamys pusillus*) in Brazil, but the nymph of *Amblyomma scutatum* was not described by Neumann (1899), and his nymphal records were regarded as doubtful by Robinson (1926), Dantas-Torres et al. (2009a), and Nava et al. (2017), leading to the conclusion that the presence of this species in Brazil and Paraguay is doubtful. Fiasson (1949) and Díaz Ungría (1957) list *Amblyomma scutatum* as established in

Venezuela, but its presence there was not confirmed by Jones et al. (1972), and Guglielmone et al. (2015) state that Venezuelan records of *Amblyomma scutatum* require confirmation. Navarrete Abarca et al. (2014) listed Anura as hosts of *Amblyomma scutatum* in El Salvador, and Oliveira et al. (2014) allegedly found this tick on Didelphidae in Brazil; however, all these records are treated here as doubtful, and they are not included in our list of hosts of *Amblyomma scutatum*.

In summary, Guglielmone et al. (2015) treated *Amblyomma scutatum* as a parasite of Squamata: Iguanidae and as restricted to Mexico and Central America, an opinion also accepted here.

57. *Amblyomma tapirellum* Dunn, 1933

Dunn, L.H. (1933) Two new species of ticks from Panama (*Amblyomma tapirellum* and *A. pecarium*). *Parasitology*, 25, 353–358.

Descriptions and Redescriptions: Dunn (1933) described and illustrated the male and female of *Amblyomma tapirellum*.

Voltzit (2007) and Bermúdez et al. (2018a) redescribed the male and female of *Amblyomma tapirellum*.

The nymph of *Amblyomma tapirellum* remains undescribed, but some workers have been able to identify this stage from field-collected material.

The larva of *Amblyomma tapirellum* remains unknown.

Geographic Distribution: *Amblyomma tapirellum* is a Neotropical species that has been found in Central America, with a record from northern South America as follows: **Belize** (Varma 1973); **Colombia**: Chocó (Acevedo-Gutiérrez et al. 2020), **Costa Rica**: Puntarenas (Jiménez et al. 2015); **Nicaragua** (Fairchild et al. 1966); and **Panama**: Chiriquí, Colón, Panamá, and Panamá Oeste (Fairchild et al. 1966; Bermúdez et al. 2010a, 2018b).

Hosts: Males and females of *Amblyomma tapirellum* have been collected from several orders of mammals; however, from a total of 440 adult specimens listed in the literature, 390 (87%) were found on *Tapirus bairdii*. A few nymphs of this tick have been collected from several orders of mammals, but the principal hosts for this stage are regarded as basically undetermined. The host profile of *Amblyomma tapirellum* is shown in Table 2.44, which is based on data from Fairchild et al. (1966), Bermúdez et al. (2010a, 2015a, b), García et al. (2014), and Esser et al. (2016a, b).

Human Parasitism: Guglielmone and Robbins (2018) list *Amblyomma tapirellum* as a sporadic parasite of people that has been found on humans in Panama (Panamá, and Panamá Oeste in Barro Colorado Island), and in an undetermined American country. Adults and provisionally nymphs of *Amblyomma tapirellum* have been collected from humans.

Comments: Voltzit (2007) referred to chitinous plates under each festoon of the male of *Amblyomma tapirellum* but this morphological character is not depicted in the corresponding figure. This author also redescribed the male and female of

Table 2.44 Hosts for males (M), females (F), and nymphs (N)* of *Amblyomma tapirellum*

MAMMALIA			
ARTIODACTYLA: Bovidae		*Procyon lotor*	FN
Cattle	MF	**CHIROPTERA: Phyllostomidae**	
ARTIODACTYLA: Cervidae		*Carollia perspicillata*	M
Odocoileus virginianus	M	**DIDELPHIMORPHIA: Didelphidae**	
ARTIODACTYLA: Suidae		*Didelphis marsupialis*	N
Domestic pig	A	**PERISSODACTYLA: Equidae**	
ARTIODACTYLA: Tayassuidae		Horse	MF
Pecari tajacu	MFN	**PERISSODACTYLA: Tapiridae**	
Tayassu pecari	MF	*Tapirus bairdii*	MFN
CARNIVORA: Canidae		**PILOSA: Myrmecophagidae**	
Domestic dog	A	*Myrmecophaga tridactyla*	MF
CARNIVORA: Felidae		**RODENTIA: Caviidae**	
Domestic cat	MF	*Hydrochoerus isthmius*	M
CARNIVORA: Procyonidae		**RODENTIA: Dasyproctidae**	
Nasua narica	MN	*Dasyprocta punctata*	N

* Provisional diagnosis because the nymph of *Amblyomma tapirellum* remains undescribed

Amblyomma tapirellum as having two spurs on coxae II and III, but Dunn (1933) described this species with one spur on these coxae, although the corresponding figure of the female in Dunn (1933) shows no obvious spurs on coxae II and III. Bermúdez et al. (2018a) describe coxae II and III of the male and female of *Amblyomma tapirellum* as having one spur each, but this feature is not clearly shown in the accompanying figures. A redescription of the type material of *Amblyomma tapirellum* is needed to clarify the morphology of this tick.

Fairchild et al. (1966) state that *Amblyomma tapirellum* has been found in Venezuela, but as noted in Guglielmone et al. (2003a), Jones et al. (1972) state that this tick has not been detected there. Therefore, Guglielmone and Robbins (2018) list its presence in Venezuela as unconfirmed, and Venezuela is provisionally excluded from the range of *Amblyomma tapirellum.* Acevedo-Gutiérrez et al. (2020) listed this tick as found in the Colombian department of Chocó based on "Rocky Mountain Laboratory" information, and that department is tentatively included within the range of *Amblyomma tapirellum*, pending confirmation.

58. ***Amblyomma tenellum*** Koch, 1844

Koch, C.L. (1844) Systematische Übersicht über die Ordnung der Zecken. *Archiv für Naturgeschichte*, 10, 217–239.

Descriptions and Redescriptions: Koch (1844) described but did not figure the male of *Amblyomma tenellum*, and Kohls (1958) described the female and nymph under the name *Amblyomma imitator*, a synonym of *Amblyomma tenellum*, as discussed in Nava et al. (2014c).

The male and female of *Amblyomma tenellum* were redescribed by Voltzit (2007) and Guzmán-Cornejo et al. (2011, keys and figures) as *Amblyomma imitator*. Tonelli Rondelli (1937), Kohls (1958, under the name *Amblyomma imitator*), and Nava et al. (2014c) redescribed the male. Keirans and Durden (1998) redescribed the nymph under the name *Amblyomma imitator*.

The larva of *Amblyomma tenellum* remains unknown.

Geographic Distribution: *Amblyomma tenellum* is a Nearctic and Neotropical species with a Neotropical geographic distribution encompassing southern Mexico and Central America as follows: **Costa Rica**: Heredia (Álvarez et al. 2005); **Guatemala**: Alta Verapaz, Chiquimula, El Progreso, Escuintla, Jalapa, Jutiapa, Petén, and Zacapa (Instituto Interamericano de Cooperación para la Agricultura 1988); **Honduras** (Becklund 1959); **southern Mexico**: Campeche, Colima, Guerrero, Jalisco, Michoacán, Morelos, Nayarit, Oaxaca, Quintana Roo, Tabasco, Veracruz, and Yucatán (Guzmán-Cornejo et al. 2011; Nava et al. 2014c; Vargas-Sandoval et al. 2014; Coronel-Benedett et al. 2018); and **Nicaragua**: Costa Caribe Norte, Jinotega, Managua, Matagalpa, and Río San Juan (Düttmann et al. 2016).

Hosts: Males and females of *Amblyomma tenellum* have been collected from domestic mammals in the Neotropical Region, but the great majority of specimens have been found on cattle and horses (data not shown). These mammals are provisionally treated as the principal hosts of this tick in the Neotropics. There are no records of *Amblyomma tenellum* nymphs collected from Neotropical vertebrates. The host profile of *Amblyomma tenellum* appears in Table 2.45, which is based on data from Guzmán-Cornejo et al. (2011), Nava et al. (2014c), Düttmann et al. (2016), and Coronel-Benedett et al. (2018). The host profile of *Amblyomma tenellum* in the Nearctic Region differs from that of the Neotropical Region because it includes adults and nymphs collected from Artiodactyla: Tayassuidae, and Perissodactyla: Equidae, a male tick found on Didelphimorphia: Didelphidae, and nymphs recovered from Carnivora: Canidae, Rodentia: Sciuridae, Galliformes: Phasianidae, and Passeriformes: Mimidae and Turdidae (Guglielmone et al. 2014, under the name *Amblyomma imitator*).

Human Parasitism: Guglielmone and Robbins (2018) list *Amblyomma tenellum* as a sporadic parasite of people, but all cases of human parasitism have been reported from the Nearctic.

Comments: Neumann (1899) treated *Amblyomma tenellum* as a synonym of *Amblyomma maculatum* but later changed his view and listed it as a synonym of *Amblyomma cajennense*. Tonelli Rondelli (1937) solidly reinstated *Amblyomma tenellum*, but her opinion was largely ignored, and Aragão and Fonseca (1953a) maintained this tick as a synonym of *Amblyomma cajennense*, an opinion that was supported by tick workers for decades. After examination of the type specimens of

Table 2.45 Hosts for males (M) and females (F) of *Amblyomma tenellum*

MAMMALIA			
ARTIODACTYLA: Bovidae		**PERISSODACTYLA: Equidae**	
Cattle	MF	Donkey	M
Sheep	MF*	Horse	MF
CARNIVORA: Canidae			
Domestic dog	MF		

* Identified as *Amblyomma imitator*

Amblyommma tenellum (two males), Kohls (1958) described *Amblyomma imitator*, although stating that his new species had morphological characters of *Amblyomma tenellum* ("pale and small") which "are suggestive of *Amblyomma imitator*" but, surprisingly, Kohls (1958) maintained *Amblyomma tenellum* as a synonym of *Amblyomma cajennense*. Thereafter, Nava et al. (2014c) examined and redescribed the types of *Amblyomma tenellum* and produced a detailed comparison of morphological characters with *Amblyomma imitator*, concluding that *Amblyomma tenellum* is valid and *Amblyommma imitator* is its synonym.

Serra-Freire and Mello (2006) state that *Amblyomma tenellum* (under the name *Amblyomma imitator*) occurs in Brazil, but *bona fide* specimens of this tick have never been found in that country (Dantas-Torres et al. 2009a); consequently, Brazil is excluded from the geographic distribution of *Amblyomma tenellum*.

Arana-Guardia et al. (2015) allegedly collected adults of *Amblyomma tenellum* (under the name *Amblyomma imitator*) from *Odocoileus virginianus* and *Panthera onca* in Quintana Roo (Neotropical Mexico), but the morphological support for their records is questionable, and the data concerning this tick in Arana-Guardia et al. (2015) are excluded from this analysis.

59. *Amblyomma tigrinum* Koch, 1844

Koch, C.L. (1844) Systematische Übersicht über die Ordnung der Zecken. *Archiv für Naturgeschichte*, 10, 217–239.

Descriptions and Redescriptions: Koch (1844) described but did not figure the male and female of *Amblyomma tigrinum*; the female was described under the name *Amblyomma rubripes*, a synonym of *Amblyomma tigrinum*, as discussed in Guglielmone and Nava (2014). The larva and nymph of *Amblyomma tigrinum* were described by Estrada-Peña et al. (1993b).

The male, female, nymph, and larva of *Amblyomma tigrinum* were redescribed in Estrada-Peña et al. (2005), while Nava et al. (2017) redescribed the male, female, and nymph. Kohls (1956b), Floch and Fauran (1958), Onofrio et al. (2006b, keys and figures), Voltzit (2007), and Dantas-Torres et al. (2019b, keys and figures) redescribed the male and female. Martins et al. (2010, 2014c) redescribed the nymph of *Amblyomma tigrinum*.

Geographic Distribution: *Amblyomma tigrinum* is a Neotropical tick that has been found only in South America as follows: **Argentina**: Buenos Aires, Catamarca, Chaco, Chubut, Córdoba, Corrientes, Entre Ríos, Formosa, Jujuy, La Pampa, Mendoza, Misiones, Neuquén, Salta, San Juan, San Luis, Santa Fe, Santiago del Estero, and Tucumán (Guglielmone and Nava 2006); **Bolivia**: Cochabamba, Chuquisaca, and Santa Cruz (Mastropaolo et al. 2014); **Brazil**: Amazonas, Ceará, Distrito Federal, Goiás, Mato Grosso, Mato Grosso do Sul, Minas Gerais, Pará, Paraná, Pernambuco, Rio de Janeiro, Rio Grande do Sul, Rondônia, Santa Catarina, São Paulo, and Tocantins (Kohls 1956a; Labruna et al. 2005a, d; Martins et al. 2009a; Cansi et al. 2012; Luz et al. 2016a, b; Moerbeck et al. 2016); **Chile**: Araucanía, Aysén, Bío Bío, Coquimbo, Metropolitana, Ñuble, and Valparaíso (González-Acuña and Guglielmone 2005; Mastropaolo et al. 2008; Abarca et al. 2016; **French Guiana**: Cayena (Floch and Fauran 1958); **Paraguay**: Boquerón, Central, Concepción, Cordillera, Itapuá, Misiones, Ñeembucú, Paraguarí, and Villa Hayes (Nava et al. 2007); **Peru**: Cusco, and Puno (Mendoza-Uribe and Chávez-Chorocco 2004); **Uruguay**: Artigas, Canelones, Cerro Largo, Colonia, Durazno, Flores, Florida, Lavalleja, Maldonado, Montevideo, Paysandú, Rocha, Salto, San José, and Soriano (Martins et al. 2014d); and **Venezuela**: Amazonas, Apure, Bolívar, Carabobo, and Falcón (Jones et al. 1972).

Hosts: Usual hosts for males and females of *Amblyomma tigrinum* are Carnivora: Canidae. Nymphs of this tick are frequently recovered from Rodentia: Caviidae (*Galea musteloides* in Argentina), while larvae are often found on Rodentia: Cricetidae; however, ground-feeding birds (mainly several families of Passeriformes) are also frequently infested with nymphs and larvae (Nava et al. 2006b). Therefore, these families of rodents and Passeriformes are treated as principal hosts for the immature stages of *Amblyomma tigrinum*. Nevertheless, the range of hosts of this tick is broad and includes several orders of mammals and birds that may be more important as hosts for this tick than currently recognized. The host profile of *Amblyomma tigrinum* is shown in Table 2.46, which is based on data from Floch and Fauran (1958), Jones et al. (1972), Guglielmone et al. (1982, 2007a), Ivancovich and Luciani (1992), Serra-Freire et al. (1996), Robbins and Deem (2002), Venzal et al. (2003c), Nava et al. (2007, 2017), Martins et al. (2011), Luz et al. (2016b), Millán et al. (2016) and Barbieri et al. (2019).

Human Parasitism: Guglielmone and Robbins (2018) list *Amblyomma tigrinum* as a sporadic parasite of people that has been found on humans in Argentina (Buenos Aires, Chaco, Córdoba, and Santiago del Estero), Brazil (Ceará, and Rio Grande do Sul), French Guiana (Cayena), Paraguay and Uruguay (Soriano). Males, females, and nymphs have been found causing human parasitism, but female ticks appear to prevail as parasites of humans.

Comments: *Amblyomma maculatum*, *Amblyomma tigrinum,* and *Amblyomma triste* are morphologically similar species, and for decades the latter two species were treated as synonyms of *Amblyomma maculatum* until Kohls (1956b) reinstated them as valid taxa. Floch and Abonnenc (1940) and Boero (1957) redescribed the male and female of *Amblyomma maculatum*, but in fact their accounts refer to *Amblyomma tigrinum*. Moreover, the illustrations of *Amblyomma tigrinum* in Aragão and Fonseca (1961a) actually represent *Amblyomma triste*. The four Argentinean records of supposed *Amblyomma maculatum* in Keirans (1982) correspond to *Amblyomma tigrinum*, as discussed in Guglielmone et al. (2003a). According to Estrada-Peña et al. (2005), the morphological identification of nymphs and especially larvae of *Amblyomma maculatum*, *Amblyomma tigrinum,* and *Amblyomma triste* is extremely difficult. See also *Amblyomma maculatum* and *Amblyomma triste* for further discussion of the confusion of these species with *Amblyomma tigrinum.*

Graham et al. (1975) list *Amblyomma tigrinum* as present in Mexico, but this has not been confirmed (Guzmán-Cornejo et al. 2011), and Mexico is excluded from the

Table 2.46 Hosts for adults of undetermined sex (A), males (M), females (F), nymphs (N), and larvae (L) of *Amblyomma tigrinum*

MAMMALIA			
ARTIODACTYLA: Bovidae		*Puma concolor*	FL
Cattle	MF	**CARNIVORA: Procyonidae**	
Goat	F	*Nasua nasua*	A
Sheep	MF	**PERISSODACTYLA: Equidae**	
ARTIODACTYLA: Cervidae		Horse	MF
Blastocerus dichotomus	MF	**RODENTIA: Caviidae**	
Mazama gouazoubira	M	*Galea musteloides*	NL
ARTIODACTYLA: Suidae		*Galea spixii*	N
Feral pig	MF	*Hydrochoerus hydrochaeris*	MF
ARTIODACTYLA: Tayassuidae		**RODENTIA: Cricetidae**	
"Peccary"	F	*Akodon dolores*	NL
CARNIVORA: Canidae		*Akodon molinae*	NL
Cerdocyon thous	MFN	*Akodon spegazzini*	NL
Chrysocyon brachyurus	MF	*Calomys laucha*	N
Domestic dog	MFN	*Calomys musculinus*	L
Lycalopex culpaeus	MF	*Calomys venustus*	NL
Lycalopex griseus	MF	*Graomys chacoensis*	NL
Lycalopex gymnocercus	MF	*Necromys benefactus*	NL
Lycalopex vetulus	MF	*Oligoryzomys fulvescens*	AN
CARNIVORA: Felidae		*Oligoryzomys longicaudatus*	L
Domestic cat	MF	*Phyllotis xanthopygus*	L
Leopardus pardalis	F	**RODENTIA: Muridae**	
Panthera onca	MFL	*Rattus rattus*	L
AVES			
COLUMBIFORMES: Columbidae		**PASSERIFORMES: Thraupidae**	
Columbina picui	L	*Coryphospingus cucullatus*	NL
Zenaida auriculata	NL	*Lophospingus pusillus*	NL
CUCULIFORMES: Cuculidae		*Microspingus torquatus*	L
Guira guira	NL	*Paroaria coronata*	N
GALLIFORMES: Odontophoridae		*Ramphocelus bresilius*	N
Callipepla californica	NL	*Saltator aurantiirostris*	NL
PASSERIFORMES: Furnariidae		*Saltatricula multicolor*	NL
Asthenes baeri	L	*Sicalis flaveola*	L
Coryphistera alaudina	NL	*Sicalis luteola*	L
Drymornis bridgesii	L	*Volatinia jacarina*	L
Furnarius cristatus	NL	**PASSERIFORMES: Troglodytidae**	
Furnarius rufus	NL	*Troglodytes aedon*	NL
Pseudoseisura lophotes	NL	**PASSERIFORMES: Turdidae**	
Synallaxis albescens	NL	*Turdus amaurochalinus*	L
Tarphonomus certhioides	NL	**PASSERIFORMES: Tyrannidae**	
PASSERIFORMES: Icteridae		*Pitangus sulphuratus*	N
Agelaioides badius	NL	**STRIGIFORMES: Strigidae**	
Molothrus bonariensis	NL	*Asio clamator*	N
PASSERIFORMES: Passerellidae		**TINAMIFORMES: Tinamidae**	
Rhynchospiza strigiceps	NL	*Eudromia elegans*	L
Zonotrichia capensis	NL	*Nothoprocta perdicaria*	NL
PASSERIFORMES: Polioptilidae		*Nothura maculosa*	N
Polioptila dumicola	NL	*Rhynchotus rufescens*	N
PASSERIFORMES: Rhinocryptidae			
Rhinocrypta lanceolata	N		

range of this species. Kolonin (2009) included Guyana within the range of *Amblyomma tigrinum,* but its presence there needs confirmation. Acevedo-Gutiérrez et al. (2020) listed the presence of this tick in one Colombian department based on a publication lacking both locality records and discussion of morphological or molecular diagnoses, and Colombia is therefore provisionally excluded from the range of *Amblyomma tigrinum.* The Brazilian states of Ceará, Pará, Rio de Janeiro, Santa Catarina, and Tocantins were excluded from the range of *Amblyomma tigrinum* by Andreotti et al. (2018), but no reasons were given for this exclusion. *Dermacentor triangulatus* from Argentina is treated as a synonym of *Amblyomma tigrinum* by Camicas et al. (1998), but Aragão (1935) doubted the validity of *Dermacentor triangulatus* (named as *Dermacentor triangulus*, a *lapsus*), while Guglielmone and Nava (2014) list it as a *nomen dubium*, an opinion accepted here. Berlese (1888) described *Amblyomma complanatum* as a Brazilian species, and Jones et al. (1972) treated it as a synonym of *Amblyomma maculatum*, while Camicas et al. (1998) listed it as a synonym of *Amblyomma tigrinum*, but Guglielmone and Nava (2014) consider *Amblyomma complanatum* to be a name *incertae sedis*, and that opinion is accepted here.Massi Pallarés and Benítez Usher (1982) allegedly found nymphs and larvae of *Amblyomma tigrinum* on several mammals and a bird in Paraguay, but these records are treated as doubtful because they were published prior to the descriptions of the nymph and larva by Estrada-Peña et al. (1993b). Estrada-Peña et al. (2020) include the larva and nymph of *Amblyomma tigrinum* in a cluster of Neotropical species feeding mostly on avian hosts. Indeed, birds are important hosts for the immature stages of *Amblyomma tigrinum*, but rodents are also significant hosts for larvae and nymphs of this tick.

60. ***Amblyomma tonelliae*** Nava, Beati and Labruna, 2014 in Nava et al. (2014a)

Nava, S., Beati, L., Labruna, M.B., Cáceres, A.G., Mangold, A.J. & Guglielmone, A.A. (2014a) Reassessment of the taxonomic status of *Amblyomma cajennense* (Fabricius, 1787) with the description of three new species, *Amblyomma tonelliae* n. sp., *Amblyomma interandinum* n. sp. and *Amblyomma patinoi* n. sp., and reinstatement of *Amblyomma mixtum* Koch, 1844 and *Amblyomma sculptum* Berlese, 1888 (Ixodida: Ixodidae). *Ticks and Tick-borne Diseases*, 5, 252–276.

Descriptions and Redescriptions: The male and female of *Amblyomma tonelliae* were described by Nava et al. (2014a), while the nymph was described by Martins et al. (2014c).

The male, female, and nymph of *Amblyomma tonelliae* were redescribed in Nava et al. (2017).

The larva of *Amblyomma tonelliae* remains undescribed, but several workers have been able to identify this stage from field-collected material.

Geographic Distribution: *Amblyomma tonelliae* is a southern South American species found in **Argentina**: Chaco, Formosa, Jujuy, Salta, and Santiago del Estero (Tarragona et al. 2015); **Bolivia**: Santa Cruz (Nava et al. 2014a); and **Paraguay**: Boquerón (Nava et al. 2014a).

Table 2.47 Hosts for males (M), females (F), nymphs (N), and larvae (L)* of *Amblyomma tonelliae*

MAMMALIA			
ARTIODACTYLA: Bovidae		**CARNIVORA: Felidae**	
Cattle	MFNL	*Puma concolor*	F
ARTIODACTYLA: Cervidae		**PERISSODACTYLA: Equidae**	
Mazama sp.	M	Horse	MFNL
ARTIODACTYLA: Tayassuidae		**PILOSA: Myrmecophagidae**	
Catagonus wagneri	MF	*Myrmecophaga tridactyla*	MF
Pecari tajacu	MF	*Tamandua tetradactyla*	F
Tayassu pecari	MF	**RODENTIA: Caviidae**	
CARNIVORA: Canidae		*Hydrochoerus hydrochaeris*	MF
Domestic dog	M		

* Provisional diagnosis because the larva of *Amblyomma tonelliae* remains undescribed

Hosts: Males and females of *Amblyomma tonelliae* have been collected from a variety of mammals, but most specimens have been taken from cattle; larvae and nymphs of this tick have been found on domestic hosts (cattle and horses). Cattle (Artiodactyla: Bovidae) are provisionally treated as the principal hosts for all parasitic stages of this species. The host profile of *Amblyomma tonelliae* is shown in Table 2.47, which is based on data from Nava et al. (2014a, 2017).

Human Parasitism: Guglielmone and Robbins (2008) list *Amblyomma tonelliae* as a sporadic parasite of humans that has been found on people in Argentina (Jujuy and Salta) and Paraguay (Boquerón). Larvae (provisional diagnosis), nymphs, males, and females have been found on humans, but it is uncertain which stage prevails as a parasite of people.

Comments: *Amblyomma tonelliae* is a member of the *Amblyomma cajennense* species complex, which also comprises *Amblyomma interandinum*, *Amblyomma mixtum*, *Amblyomma patinoi,* and *Amblyomma sculptum.* Further discussion of this complex is provided under *Amblyomma cajennense* above.

The list of hosts and localities for *Amblyomma tonelliae* is longer than currently recognized because many records from Argentina, Bolivia, and Paraguay are classified as *Amblyomma cajennense sensu lato*, and a significant number of these may represent *Amblyomma tonelliae.*

Cattle are treated as principal hosts for all parasitic stages of *Amblyomma tonelliae*, but this host was introduced into South America about 400 years ago, displacing the original hosts of this tick and thus establishing a surrogate parasitic life cycle.

61. *Amblyomma torrei* Pérez Vigueras, 1934

Pérez Vigueras, I. (1934) On the ticks of Cuba, with description of a new species, *Amblyomma torrei* from *Cyclura macleayi* Gray. *Psyche*, 41, 13–18.

Descriptions and Redescriptions: The male of *Amblyomma torrei* was described but not illustrated by Pérez Vigueras (1934), while the female and nymph were described and figured by Whittick (1939); the larva was described by Černý (1969b).

Table 2.48 Hosts for males (M), females (F), nymphs (N), and larvae (L) of *Amblyomma torrei**

REPTILIA			
SQUAMATA: Iguanidae		*Cyclura stejnegeri*	M
Cyclura cychlura	MFNL	**SQUAMATA: Tropiduridae**	
Cyclura nubila	MFN	*Leiocephalus macropus*	M

* Records of *Amblyomma torrei* should be considered provisionally valid due to uncertainties in the morphological definition of all parasitic stages

The male and female of *Amblyomma torrei* were redescribed by Voltzit (2007) and Barros-Battesti et al. (2009); Whittick (1939) redescribed the male, and Černý (1966b) redescribed the female.

There are no redescriptions of the larva and nymph of *Amblyomma torrei.*

Geographic Distribution: *Amblyomma torrei* is a Neotropical species that has been found only on Caribbean islands as follows: **Bahamas**: Black Point (Durden and Knapp 2005); **Cayman Islands**: Little Cayman (Whittick 1939), **Cuba**: Camagüey, Guantánamo, La Habana, Isla de la Juventud, and Pinar del Río (Barros-Battesti et al. 2009; Černý 1969b) and **Puerto Rico**: Mayagüez (Capriles and Gaud 1977).

Hosts: Males, females, nymphs, and larvae of *Amblyomma torrei* have been found on *Cyclura cychlura* (Squamata: Iguanidae), which is treated as a principal host of this tick, but the range of *Amblyomma torrei* exceeds that of *Cyclura cychlura* and preferred hosts probably vary from island to island. The host profile of *Amblyomma torrei* is shown in Table 2.48, which is based on data from Černý (1969b), Durden and Knapp (2005), and Voltzit (2007).

Human Parasitism: No.

Comments: Considerable problems may be encountered when attempting to identify *Amblyomma torrei* with certainty. Camicas et al. (1998) treat *Amblyomma torrei* as valid but recognize the male as the only described stage, therefore tacitly treating the description of the female and nymph in Whittick (1939) and the larva in Černý (1969b) as erroneous, although these descriptions are here considered provisionally valid. The male of *Amblyomma torrei* was described by Pérez Vigueras (1934), whose description was followed by three redescriptions, but among the authors involved, there is no agreement concerning the spurring of coxae I–IV. In the case of the female, the situation is even more problematic because Whittick's (1939) sketch of the dorsum of the basis capituli on page 438 is rectangular in outline with diverging porose areas, while the figure on page 439 depicts a triangular basis capituli and porose areas that are almost parallel. Later, Černý (1966b) redescribed the female as possessing a triangular basis capituli with large porose areas in parallel, while Voltzit (2007) and Barros-Battesti et al. (2009) both depicted a basis capituli that is rectangular, although with porose areas in parallel in Voltzit (2007) but diverging in Barros-Battesti et al. (2009). As in the male, there is no agreement concerning the spurring of coxae I–IV among the authors who redescribed the female of *Amblyomma torrei.* Additionally, Morel (1967) listed *Amblyomma torrei* as a synonym of *Amblyomma cruciferum*, while Guglielmone et al. (2003a) suggested that comparison of the types of *Amblyomma cruciferum*, *Amblyomma torrei*, and related ticks would help in morphologically defining these species, and Barros-Battesti et al. (2009) stated that *Amblyomma cruciferum* and *Amblyomma torrei*, though morphologically very close, differ markedly in scutal ornamentation. Until these taxonomic problems are resolved, all records of *Amblyomma torrei* should be considered provisionally valid.

Camicas et al. (1998) and Guglielmone et al. (2003a) listed *Amblyomma torrei* as a Neotropical species, but Guglielmone et al. (2014) considered this tick to be present in the Nearctic Region as well. Further examination of data indicates that *Amblyomma torrei* is a Neotropical species. Černý (1969b) found *Amblyomma torrei* at localities in the provinces of Guantánamo and Isla de la Juventud, but these provinces are excluded from the range of this tick in Álvarez et al. (2020), who gave no reasons for this exclusion.

62. *Amblyomma triste* Koch, 1844

Koch, C.L. (1844) Systematische Übersicht über die Ordnung der Zecken. *Archiv für Naturgeschichte*, 10, 217–239.

Descriptions and Redescriptions: Koch (1844) described but did not figure the female, and Kohls (1956b) described the male of *Amblyomma triste*. Estrada-Peña et al. (2002) described the nymph and larva.

The male, female, nymph, and larva of *Amblyomma triste* were redescribed by Estrada-Peña et al. (2005). The male, female, and nymph were also redescribed by Nava et al. (2017); while Onofrio et al. (2006b, keys and figures), Voltzit (2007), Guzmán-Cornejo et al. (2011, keys and figures), and Dantas-Torres et al. (2019b, keys and figures) redescribed the male and female; Martins et al. (2010) redescribed the nymph of *Amblyomma triste*.

Geographic Distribution: *Amblyomma triste* is chiefly a South American species, with records from the southern Nearctic. Its range in the Neotropics includes **Argentina**: Buenos Aires, Ciudad Autónoma de Buenos Aires, Corrientes, Entre Ríos, and Formosa (Cicuttin et al. 2013; Guglielmone et al. 2013; Colombo et al. 2016a); **Bolivia**: Beni, and Santa Cruz (Mastropaolo et al. 2014; Lado et al. 2018); **Brazil**: Goiás, Maranhão, Mato Grosso, Mato Grosso do Sul, Minas Gerais, Pará, Paraná, Rio Grande do Sul, and São Paulo (Kohls 1956b; Evans et al. 2000; Arzua et al. 2005; Labruna et al. 2005a; Barbieri et al. 2019; Costa et al. 2020); **Chile**: Arica y Parinacota (Abarca et al. 2012); **Colombia**: "eastern plains region" (Corrier et al. 1979); **Ecuador**: El Oro, and Guayaquil (Keirans 1985b; Voltzit 2007); **Paraguay**: Boquerón (Nava et al. 2007); **Peru** (Guglielmone et al. 2003a); **Uruguay**: Canelones, Colonia, Maldonado, Montevideo, Rocha, and San José (Martins et al. 2014d); and **Venezuela**: Amazonas, and Yaracuy (Guerrero 1996; Forlano et al. 2008).

Hosts: *Blastocerus dichotomus* is considered the original host for adults of *Amblyomma triste*, but other large wild and domestic mammals are also currently common hosts for this tick, as stated by Nava et al. (2017), who also note that larvae and nymphs are usually collected from Rodentia: Caviidae and Cricetidae. However, most records of nymphs and larvae are from several species of Cricetidae, which are together regarded here as principal hosts because there is no one species of Cricetidae that clearly prevails as the preferred host for this tick. The range of vertebrates parasitized by *Amblyomma triste* is rather broad, including birds as occasional hosts for the immature stages. The host profile of *Amblyomma triste* is shown in Table 2.49, which is based on data from Venzal et al. (2003b, c), Arzua et al. (2005), Nava et al. (2007, 2011, 2017), Silva et al. (2007), Martins et al. (2011), Debárbora et al. (2012), Mastropaolo et al. (2014), Silveira et al. (2015), Witter et al. (2016), Barbieri et al. (2019), Szabó et al. (2019).

Table 2.49 Hosts for adults of undetermined sex (A), males (M), females (F), nymphs (N), and larvae (L) of *Amblyomma triste**

MAMMALIA			
ARTIODACTYLA: Bovidae		**PERISSODACTYLA: Equidae**	
Cattle	MF	Horse	MF
Goat	MF	**PERISSODACTYLA: Tapiridae**	
ARTIODACTYLA: Cervidae		*Tapirus terrestris*	MF
Blastocerus dichotomus	MF	**PILOSA: Myrmecophagidae**	
Mazama sp.	M	*Myrmecophaga tridactyla*	MF
Ozotoceros bezoarticus	MF	**RODENTIA: Caviidae**	
ARTIODACTYLA: Suidae		*Cavia aperea*	NL
Domestic pig	MF	*Hydrochoerus hydrochaeris*	MFN
CARNIVORA: Canidae		**Cricetidae**	
Cerdocyon thous	FN	*Akodon azarae*	NL
Chrysocyon brachyurus	MF	*Calomys callosus*	NL
Domestic dog	MFN	*Calomys tener*	L
Lycalopex vetulus	MF	*Cerradomys marinhus*	L
CARNIVORA: Felidae		*Holochilus brasiliensis*	NL
Domestic cat	M	*Holochilus sciureus*	NL
Herpailurus yagouaroundi	A	*Necromys lasiurus*	NL
Leopardus colocolo	N	*Necromys obscurus*	N
Panthera onca	MFL	*Oligoryzomys flavescens*	NL
Puma concolor	ML	*Oligoryzomys nigripes*	NL
CHIROPTERA: Vespertilionidae		*Oxymycterus delator*	N
Myotis albescens	N	*Oxymycterus nasutus*	NL
DIDELPHIMORPHIA: Didelphidae		*Oxymycterus rufus*	NL
Didelphis marsupialis	A	*Rhipidomys* sp.	NL
Gracililanus agilis	L	*Scapteromys aquaticus*	NL
Monodelphis dimidiata	NL	*Scapteromys tumidus*	NL
Monodelphis domestica	L	**Echimyidae**	
		Thrichomys apereoides	NL
AVES			
GRUIFORMES		**Passerellidae**	
Rallidae		*Zonotrichia capensis*	NL
Anurolimnas viridis	L	**Thraupidae**	
PASSERIFORMES		*Poospiza nigrorufa*	NL
Furnariidae		*Sporophila caerulescens*	L
Furnarius leucopus	L	**Turdidae**	
Furnarius rufus	NL	*Turdus rufiventris*	L
Phacellodomus ruber	NL		

* The validity of *Amblyomma triste* is under investigation; therefore, all identifications should be considered provisionally valid

Human Parasitism: Guglielmone and Robbins (2018) list *Amblyomma triste* as a sporadic parasite of humans, with records from Argentina (Buenos Aires), Uruguay (Canelones, Maldonado, Montevideo, and San José) and Venezuela (Amazonas). There are also cases of human parasitism by *Amblyomma triste* in the Nearctic Region. Males and females of *Amblyomma triste* have been found on people, with no preponderance of one sex over the other as a parasite of humans.

Comments: As stated under *Amblyomma maculatum*, *Amblyomma tigrinum* and *Amblyomma triste* are morphologically similar species. It has been hypothesized that *Amblyomma triste* and *Amblyomma tigrinum* are conspecific with *Amblyomma maculatum*, so it is unsurprising that there had been a long history of misidentifying these species prior to Kohls (1956b), who redescribed all of them. Mendoza-Uribe and Chávez-Chorocco (2004) discussed the difficulties involved in morphologically separating *Amblyomma maculatum* from *Amblyomma triste*, while Estrada-Peña et al. (2005) state that to correctly identify nymphs and especially larvae of *Amblyomma maculatum*, *Amblyomma tigrinum* and *Amblyomma triste* using morphological characters is extremely difficult. According to Nava et al. (2017), there is a possibility that *Amblyomma maculatum* and *Amblyomma triste* are indeed conspecific, and Lado et al. (2018) provided molecular evidence for this, although they divided *Amblyomma maculatum* into four morphotypes, with *Amblyomma triste* described as morphotype I. More conclusive evidence is needed to confirm the synonymy of *Amblyomma maculatum* and *Amblyomma triste*. The illustrations of *Amblyomma tigrinum* in Aragão and Fonseca (1961a) correspond in fact to *Amblyomma triste*.

The geographic distribution of *Amblyomma triste* is disjunct, with South American and southern Nearctic populations but no *bona fide* records from Central America because the Nicaraguan specimens of this tick in Vogel et al. (2018) require confirmation. Graham et al. (1975) and Woodham et al. (1983) treat *Amblyomma triste* as established in the Neotropics of Mexico, but records of this tick in Guzmán-Cornejo et al. (2011) include only Mexican localities in the Nearctic portion of that country. Nava et al. (2017) explain that the type locality of *Amblyomma triste* is Montevideo (Uruguay), and the tick populations in Argentina, Brazil, and Uruguay are morphologically and molecularly similar, representing *bona fide* specimens of *Amblyomma triste*. It has been speculated that *bona fide Amblyomma triste*, the morphotype I of Lado et al. (2018), reflect the original distribution of the marsh deer, *Blastocerus dichotomous*, in parts of Argentina, Bolivia, Brazil, Paraguay, Uruguay, and a small area of southeastern Peru, as depicted by the International Union for Conservation of Nature (2018). However, it should be noted that Fig. 1 in Lado et al. (2018) includes as *Amblyomma triste*-morphotype I Argentinean ticks collected in Santiago del Estero (sample 24), Salta (sample 25), and Córdoba (sample 26), but this species has never been found in those Argentinean provinces. The Brazilian state of Pará is excluded from the range of *Amblyomma triste* by Andreotti et al. (2018), but no reasons are given for this exclusion.

Estrada-Peña et al. (2020) include adults of *Amblyomma triste* in a cluster of Neotropical species prone to feeding on Carnivora, but it is maintained here that the range of adults of this tick exceeds that of carnivorous mammals.

Serra-Freire and Leal (2009) in Brazil allegedly found concurrent human infestations with adults, nymphs, and larvae of *Amblyomma triste*, but this requires confirmation given the differences in the temporal prevalence of the parasitic stages of this species, as discussed in Nava et al. (2011). This record is therefore excluded from our analysis.

63. *Amblyomma usingeri* Keirans, Hoogstraal and Clifford, 1973

Keirans, J.E., Hoogstraal, H. & Clifford, C.M. (1973) The *Amblyomma* (Acarina: Ixodidae) parasitic on giant tortoises (Reptilia: Testudinidae) of the Galapagos Islands. *Annals of the Entomological Society of America*, 66, 673–688.

Descriptions and Redescriptions: The male, female, nymph, and larva of *Amblyomma usingeri* were described by Keirans et al. (1973).

Voltzit (2007) redescribed the male and female of *Amblyomma usingeri.*

There are no redescriptions of the nymph and larva of *Amblyomma usingeri.*

Geographic Distribution: *Amblyomma usingeri* has only been found on two islands belonging to **Ecuador**: Galápagos (Keirans et al. 1973).

Hosts: Most records of males, females, nymphs, and larvae of *Amblyomma usingeri* are from Testudines: Testudinidae, but these tick stages have also been collected from Squamata: Iguanidae. The host profile of *Amblyomma usingeri* is shown in Table 2.50, which is based on data from Keirans et al. (1973). Information concerning testudine hosts of *Amblyomma usingeri* was obtained from Keirans et al. (1973), who named *Chelonoidis becki* as *Geochelone elephantopus becki*, *Chelonoidis microphyes* as *Geochelone elephantopus microphyes*, and *Chelonoidis vanderburghi* as *Geochelone elephantopus vanderburghi.* These tortoises are regarded as the principal hosts of *Amblyomma usingeri.*

Human Parasitism: There is a record of males and females of *Amblyomma usingeri* on humans in Guglielmone et al. (2006a, b), but it is uncertain whether the specimens were feeding or simply crawling, and Guglielmone and Robbins (2018) provisionally exclude this species as a parasite of humans.

Comments: No controversial issues related to *Amblyomma usingeri* have been found.

Table 2.50 Hosts for males (M), females (F), nymphs (N), and larvae (L) of *Amblyomma usingeri*

REPTILIA			
SQUAMATA: Iguanidae		**TESTUDINES: Testudinidae**	
Conolophus subcristatus	MF	*Chelonoidis becki*	MFNL
“Land iguana”	MFNL	*Chelonoidis microphyes*	MFNL
		Chelonoidis vanderburghi	MFNL

64. *Amblyomma variegatum* (Fabricius, 1798), under the name *Acarus variegatus* and given its current status in Gerstäcker (1873)

Fabricius, J.C. (1798) *Supplementum Entomologiae Systematicae*. Hafniae, 572 pp.
Gerstäcker, A. (1873) Gliderthiere (Insekten, Arachniden, Myriapoden und Isopoden) *In*: O. Kersten (Ed.), *Baron Carl Claus von der Decken's Reisen in Ost Africa in den Jahre 1859 bis 1865*. 3, Wissenschaftlicher Theil, C.J. Winter'sche Verlaghandlung, Leipzig und Heidelberg, 542 pp.

Descriptions and Redescriptions: The male of *Amblyomma variegatum* was described but not figured by Fabricius (1798) under the name *Acarus variegatus*; the female was also described without illustrations by Neumann (1899), who briefly described the nymph and larva.

The male, female, nymph, and larva were redescribed by Uilenberg et al. (1979) and Voltzit and Keirans (2003). The male, female, and larva were redescribed by Theiler (1943), while the male and female were redescribed by Dönitz (1910), Robinson (1926), Tendeiro (1952), Santos Dias (1953), Hoogstraal (1956), Elbl and Anastos (1966), Matthysse and Colbo (1987), Walker et al. (2003) and Horak et al. (2018). The nymph and larva were redescribed by Borght-Elbl (1977), while the female alone was redescribed by Neumann (1899). The larva of *Amblyomma variegatum* alone was redescribed by Camicas (1970).

Geographic Distribution: *Amblyomma variegatum* is an Afrotropical species that is also established in the Neotropical Region where it has been found on several Caribbean Islands as follows (although inclusion here does not necessarily mean that these islands are currently infested): **Anguilla**: Blowing Point (Barré et al. 1995); **Antigua and Barbuda**: Saint George, Saint John, Saint Paul, Saint Peter, and Saint Philip (Corn et al. 1994); **Barbados**: Saint Lucy (Barré et al. 1995); **Dominica**: Saint Andrew, and Saint Patrick (Barré et al. 1995); **Guadeloupe**: Base Terre, and Pointe-à-Pitre (Morel 1966; Barré et al. 1995); **Martinique**: Fort-de-France, Le-Marin, and Saint Pierre (Morel 1966; Barré et al. 1995); **Montserrat**: Saint Anthony, Saint Georges, and Saint Peter (Barré et al. 1995); **Puerto Rico**: Aguas Buenas, Aibonito, Barranquitas, Cabo Rojo, Camuy, Cayey, Cidra, Comerio, Culebra, Humacao, Lares, Naranjito, Ponce, Rincón, and Vieques (Garris 1987; Garris et al. 1989; Bokma and Shaw 1993; Barré et al. 1995); **Saba** (Barré et al. 1995); **Saint Eustatius** (Barré et al. 1995); **Saint Kitts and Nevis** (Barré et al. 1995); **Saint Lucia**: Castries, Dauphin, Gros Islet, Micoud, and Vieux Fort (Garris and Scotland 1985; Charles et al. 2020); **Saint Martin**: Dutch area, and French area (Barré et al. 1995); **Saint Vincent and the Grenadines** (Barré et al. 1995); and **USA Virgin Islands**: Saint Croix (Barré et al. 1995).

Hosts: Males, females, nymphs, and larvae of *Amblyomma variegatum* are usually found on Artiodactyla: Bovidae (mainly cattle and goats), but the host range of this tick in the Neotropics includes other mammals and a few birds. The Neotropical host profile of *Amblyomma variegatum* is shown in Table 2.51, which is based on data from Morel (1966), Garris (1987), Barré et al. (1988), Corn et al. (1994), and

Table 2.51 Hosts for males (M), females (F), nymphs (N), and larvae (L) of *Amblyomma variegatum*

	MAMMALIA		
ARTIODACTYLA: Bovidae		**CARNIVORA: Felidae**	
Cattle	MFNL	Feral cat	N
Goat	MFNL	**CARNIVORA: Herpestidae**	
Sheep	MFNL	*Herpestes auropunctatus*	NL
ARTIODACTYLA: Suidae		**PERISSODACTYLA: Equidae**	
Domestic pig	MFNL	Horse	MFNL
CARNIVORA: Canidae		**RODENTIA: Muridae**	
Feral dog	MFNL	*Mus musculus*	L
		Rattus rattus	L
	AVES		
COLUMBIFORMES: Columbidae		**PELECANIFORMES**	
Columbina passerina	L	**Ardeidae**	
PASSERIFORMES: Icteridae		*Bubulcus ibis*	NL
Quiscalus lugubris	NL	**Pelecanidae**	
PASSERIFORMES: Thraupidae		*Pelecanus occidentalis*	M
Tiaris bicolor	L		

L'Hostis et al. (1998). However, the Neotropical host profile of *Amblyomma variegatum* contains fewer hosts than are found in the Afrotropical Region, where all parasitic stages occur on a wide variety of mammals and, to a lesser extent, birds, with occasional records from Squamata (Guglielmone and Robbins 2018).

Human Parasitism: Guglielmone and Robbins (2018) list *Amblyomma variegatum* as a frequent parasite of people in Africa, but there is only a single human record from the Neotropics in Floch and Fauran (1959a), who found this tick on people in the Caribbean Islands but provided no further information.

Comments: *Amblyomma variegatum* is widely cited as having been described in 1794, e.g., Camicas et al. (1998), Horak et al. (2002, 2018), Guglielmone et al. (2003a, 2014), and Kolonin (2009), among many others, but Fabricius described this species in 1798, as explained in Guglielmone and Nava (2014) and Guglielmone et al. (2015).

Amblyomma variegatum was introduced with Senegalese zebu cattle imported to the Caribbean Islands and allegedly to French Guiana (South American mainland), but there are no *bona fide* records of the presence of this tick in French Guiana (Uilenberg et al. 1984); therefore, this territory is not included within the range of *Amblyomma variegatum*. Neumann (1899) listed a female tick from Guatemala, but the presence of this tick there has not been confirmed, and Guatemala is also excluded from its geographic range. Supposed records of *Amblyomma variegatum* from Venezuela and, again, French Guiana, in Doss and Anastos (1974a), repeated in Capriles and Gaud (1977) and Doss et al. (1978), are due to misreadings of Fiasson (1949) and Floch and Fauran (1959a), respectively, as subsequently pointed out by Uilenberg et al. (1984). Additionally, Capriles and Gaud (1977) state that *Amblyomma variegatum* was reported from Suriname, but no prior record has been found to support this claim. Maes et al. (1989) list this tick from Nicaragua, a record also cited by De La Cruz (2001), but Düttmann et al. (2016) found the specimens identified as *Amblyomma variegatum* were, in fact, *Amblyomma mixtum*. Rawlins et al. (1993) state that *Amblyomma variegatum* has been found on Trinidad and Tobago, an unconfirmed record, as discussed in Basu et al. (2012). Rawlins et al. (1993) also listed *Amblyomma variegatum* as present in the United Kingdom Virgin Islands, a record repeated in Basu and Charles (2017) but treated here as unverified. Machado-Ferreira et al. (2016) listed an alleged female of *Amblyomma variegatum* collected from an undetermined host in Barranquilla, Colombia. Ortiz et al. (2020) treat this record as valid and situate it in Barranquilla City, the capital of Atlántico Department, but there are several localities named "Barranquilla" in Colombia. This enigmatic Colombian record of *Amblyomma variegatum* requires confirmation, given the epidemiological and economic importance of this tick. In the interim, Colombia is provisionally excluded from the range of *Amblyomma variegatum*.The Caribbean geographic distribution of *Amblyomma variegatum* detailed here is historical and does not necessarily match the actual range of this species because attempts to eradicate this tick have probably led to its elimination from some of the Caribbean territories listed above as infested.

65. *Amblyomma varium* Koch, 1844

Koch, C.L. (1844) Systematische Übersicht über die Ordnung der Zecken. *Archiv für Naturgeschichte*, 10, 217–239.

Descriptions and Redescriptions: The male of *Amblyomma varium* was described but not figures by Koch (1844), while the female was described but not figured by Neumann (1901). Martins et al. (2010) described the nymph, and the larva was described by Amorim and Serra-Freire (1996) under the name *Amblyomma varius* (*lapsus*).

The male and female of *Amblyomma varium* were redescribed by Robinson (1926), Boero (1957), Floch and Fauran (1958), Aragão and Fonseca (1961a), Voltzit (2007), Onofrio et al. (2008), Bermúdez et al. (2018a) and Dantas-Torres et al. (2019b, keys and figures). The male alone was redescribed by Enríquez et al. (2020), while the nymph of *Amblyomma varium* was redescribed by Martins et al. (2014c, key and figures).

There are no redescriptions of the larva of *Amblyomma varium*.

Geographic Distribution: *Amblyomma varium* is a Neotropical species found in Central and South America (central and northern) as follows: **Brazil**: Acre, Alagoas, Amazonas, Bahia, Espírito Santo, Mato Grosso, Minas Gerais, Pará, Paraíba, Pernambuco, Rio de Janeiro, Rondônia, and São Paulo (Acosta et al. 2016; Witter et al. 2016; Gruhn et al. 2019); **Colombia**: Atlántico, Caldas, Chocó, Quindío, and Tolima (Robinson 1926; Torres-Mejía and de la Fuente 2006; Rivera-Páez et al. 2018; Acevedo-Gutiérrez et al. 2020; Martínez-Sánchez et al. 2020); **Costa Rica**: Alajuela, Limón, and San José (Álvarez et al. 2005); **Ecuador**: Pichincha, and Zamora Chinchipe (Enríquez et al. 2020); **French Guiana**: Cayena, and Saint Laurent du Maroni (Floch and Fauran 1958); **Guatemala** (Guglielmone et al. 2003a); **Guyana**: Essequibo Islands-West Demerara (Robinson 1926); **Nicaragua**: Chontales (Stoll

1894); **Panama**: Bocas del Toro, Chiriquí, Colón, Darién, Panamá, Panamá Oeste, and Veraguas (Cooley and Kohls 1942a, Miller et al. 2016); **Peru**: Loreto (Fairchild et al. 1966; Need et al. 1991); **Suriname**: Wanica (Robinson 1926); and **Venezuela**: Aragua, and Carabobo (Vogelsang and Cordero 1940; Jones et al. 1972).

Hosts: Males and females of *Amblyomma varium* are usually found on *Bradypus tridactylus* and *Bradypus variegatus* (Pilosa: Bradypodidae), with some records from Pilosa: Megalonychidae and less often from other mammals; adult ticks have rarely been found on Anura and Reptilia (Santos Dias 1958c; Gianizella et al. 2018b). There are only a few records of nymphs of *Amblyomma varium* and even fewer for the larvae. Nymphs have been reported from Passeriformes and Bradypodidae but also from other mammals and birds, while the larva of this tick has been found a few times on Aves and Mammalia. The principal hosts for nymphs and larvae of *Amblyomma varium* are regarded here as basically undetermined. The host profile of *Amblyomma varium* is shown in Table 2.52, which is based on data from Santos Dias (1958c), Fairchild et al. (1966), Keirans (1985b), Need et al.

Table 2.52 Hosts for adults of undetermined sex (A), males (M), females (F), nymphs (N), and larvae (L) of *Amblyomma varium*

	AMPHIBIA		
ANURA: Bufonidae			
Rhinella marina	M		
	MAMMALIA		
ARTIODACTYLA: Tayassuidae		*Bradypus variegatus*	MF
Tayassu sp.	M	**PILOSA: Choloepodidae**	
CARNIVORA: Canidae		*Choloepus didactylus*	MF
Domestic dog	A	*Choloepus hoffmanni*	MF
CARNIVORA: Procyonidae		**PILOSA: Myrmecophagidae**	
Nasua narica	F	*Tamandua tetradactyla*	M
DIDELPHIMORPHIA: Didelphidae		**RODENTIA: Dasyproctidae**	
Didelphis marsupialis	FNL	*Dasyprocta leporina*	M
PILOSA: Bradypodidae		**RODENTIA: Echimyidae**	
Bradypus torquatus	MFN	*Proechimys semispinosus*	NL
Bradypus tridactylus	MFN		
	AVES		
COLUMBIFORMES: Columbidae		**PASSERIFORMES: Pipridae**	
Leptotila rufaxilla	L	*Ceratopipra mentalis*	N
GALBULIFORMES: Bucconidae		*Manacus vitellinus*	NL
Malacoptila panamensis	N	*Neopelma pallescens*	N
PASSERIFORMES: Cardenalidae		**PASSERIFORMES: Thamnophilidae**	
Habia fuscicauda	N	*Gymnopithys leucaspis*	NL
PASSERIFORMES: Furnariidae		**PASSERIFORMES: Troglodytidae**	
Automolus ochrolaemus	N	*Cyphorhinus phaeocephalus*	NL
Dendrocincla fuliginosa	NL	**PASSERIFORMES: Turdidae**	
Glyphorynchus spirurus	NL	*Catharus aurantiirostris*	N
Xiphorhynchus guttatus	N	**PASSERIFORMES: Tyrannidae**	
PASSERIFORMES: Icteridae		*Zimmerius chrysops*	L
Molothrus bonariensis	L		
	REPTILIA		
SQUAMATA: Boidae		**SQUAMATA: Iguanidae**	
Boa constrictor	MF	*Ctenosaura acanthura*	F
Eunectes murinus	M		

(1991), Marques et al. (2002), Ogrzewalska et al. (2012b), Bermúdez et al. (2015a, 2018a), Lugarini et al. (2015), Esser et al. (2016a, b), Luz et al. (2017b), Gianizella et al. (2018b), Rivera-Páez et al. (2018), Binetruy et al. (2019, 2020b), Domínguez et al. (2019), Bermúdez et al. (2020) and Martínez-Sánchez et al. (2020).

Human Parasitism: Serra-Freire et al. (2011) allegedly found hundreds of *Amblyomma varium* parasitizing humans in Brazil, but according to Guglielmone and Robbins (2018) these records require confirmation, and this tick was listed as a very rare parasite of humans, with one case of human parasitism in Costa Rica and another in Panama, each caused by nymphal infestation. Later, Binetruy et al. (2019) recorded human parasitism by one nymph and five larvae in Cayena (French Guiana).

Comments: Among the authors above, there are differences in descriptions of the female dentition of *Amblyomma varium*, but discrepancies in male morphology are even greater, involving scutal coloration, size of the cornua, and the spurring of coxa IV. Neumann (1899) described a mutilated male of controversial origin, as discussed below, naming it *Amblyomma varium albida* to emphasize the pale color of the scutum (Guglielmone et al. 2003a; Onofrio et al. 2008). The latter authors examined the types of *Amblyomma varium* and *Amblyomma varium albida*, while conducting molecular and morphological studies of Brazilian populations, including measurements of the spur on coxa IV, but concluded that any differences were due to intraspecific variation.

Onofrio et al. (2008) described as "new" the synonymization of *Amblyomma gertschi* of Cooley and Kohls (1942a) with *Amblyomma varium*, but this had been anticipated by Fairchild et al. (1966).

There is some controversy concerning the geographic distribution of *Amblyomma varium*. The female of this tick was described by Neumann (1901) from a specimen collected from an unknown host in the province of Corrientes, Argentina. However, the principal hosts of *Amblyomma varium* have never been established in that Argentinean province, and records from that location in Lahille (1920), repeated in Roveda and Ringuelet (1947), as well as other Argentinean records, are considered doubtful in Nava et al. (2017), who believe that *Amblyomma varium* is not part of the hard tick fauna of Argentina, an opinion shared here. Therefore, while we accept the description of the female of *Amblyomma varium* in Neumann (1901), the geographic location of the specimen used by that author is most probably incorrect. A similar situation is encountered in Neumann's (1899) description of *Amblyomma varium albida*, allegedly collected in Chile, because the principal host of this species has never been found there, and Chile is therefore not included within the range of *Amblyomma varium*. Mexico is said to fall within the range of this tick in Vargas (1955), but its presence in that country has not been confirmed, as discussed in Guzmán-Cornejo et al. (2011); consequently, Mexico is also excluded from the geographic distribution of *Amblyomma varium*. The presence of *Amblyomma varium* in Nicaragua was supported by Stoll (1894), who described *Amblyomma crassipunctatum*, a synonym of *Amblyomma varium* in Onofrio et al. (2008) and Guglielmone and Nava (2014). The Brazilian states of Espírito Santo, Mato Grosso,

Minas Gerais, Paraíba, and Pernambuco are not included within the range of *Amblyomma varium* published by Andreotti et al. (2018), but no reasons were given for this exclusion.

The recent dramatic increase in commerce and communication between the Neotropical Region and Asia is evidenced by the collection of two male specimens of *Amblyomma varium* found on a male and female of *Choloepus didactylus* (country of origin unknown) that had been shipped to Wildlife Reserves Singapore during 2012. Both tick specimens, identified by RGR, have been deposited in the permanent research collections of the Peabody Museum of Natural History at Yale University, accession numbers YPM-ENT 300786–300787.

Fairchild et al. (1966) reported a male of *Amblyomma varium* from a "wild pig" in Peru, but Need et al. (1991) classified this host as *Tayassu* sp., and the latter host name is used here.

66. *Amblyomma williamsi* Banks, 1924

Banks, N. (1924) Arachnida of the Williams Galapagos Expedition. *Zoologica (New York)*, 5, 93–99.

Descriptions and Redescriptions: Banks (1924) described and illustrated the male and female of *Amblyomma williamsi*.

The male and female of *Amblyomma williamsi* were redescribed by Voltzit (2007).

The nymph and larva of *Amblyomma williamsi* remain unknown.

Geographic Distribution: *Amblyomma williamsi* is a Neotropical species that has only been found on five islands belonging to **Ecuador**: Galápagos (Banks 1924; Wikelski 1999; Voltzit 2007).

Hosts: Males and females of *Amblyomma williamsi* have been collected from *Conolophus subcristatus* and *Amblyrhynchus* sp. (Squamata: Iguanidae) (Banks 1924; Voltzit 2007), which are treated here as potential principal hosts for this tick.

Human Parasitism: No.

Comments: No controversial issues related to *Amblyomma williamsi* have been found.

67. *Amblyomma yucumense* Krawczak, Martins and Labruna, 2015 in Krawczak et al. (2015)

Krawczak, F.S., Martins, T.F., Oliveira, C.S., Binder, L.C., Costa, F.B., Nunes, P. H., Gregori, F. & Labruna, M.B. (2015) *Amblyomma yucumense* n. sp. (Acari: Ixodidae), a parasite of wild mammals in southern Brazil. *Journal of Medical Entomology*, 52, 28–37.

Table 2.53 Hosts for nymphs (N) and larvae (L) of *Amblyomma yucumense*

MAMMALIA			
DIDELPHIMORPHIA: Didelphidae		*Euryoryzomys russatus*	N
Didelphis aurita	NL	*Oxymycterus rufus*	N
RODENTIA: Cricetidae		*Sooretamys angouya*	N
Akodon montensis	N		

Descriptions and Redescriptions: Krawczak et al. (2015) described the male, female, and nymph of *Amblyomma yucumense*.

Lamattina et al. (2018b) redescribed the male and nymph of *Amblyomma yucumense*.

There are no redescriptions of the female of *Amblyomma yucumense*.

There are no descriptions of the larva of *Amblyomma yucumense*, but Krawczak et al. (2015) have been able to recognize this stage from field-collected material and used engorged larvae to obtain unfed nymphs of this tick.

Geographic Distribution: *Amblyomma yucumense* is a Neotropical species that has been found only in **Argentina**: Misiones (Lamattina et al. 2018b); and **Brazil**: Rio Grande do Sul (Krawczak et al. 2015).

Hosts: The hosts of males and females of *Amblyomma yucumense* are unknown. The apparent hosts of the larva and nymph of this tick are surmised from 33 larvae collected from Didelphimorphia: Didelphidae, and 64 nymphs, 53 of them from Didelphidae and 11 from Rodentia: Cricetidae. *Didelphis aurita* is provisionally considered the principal host for larvae and nymphs of *Amblyomma yucumense*. The host profile of *Amblyomma yucumense* is shown in Table 2.53, which is based on data from Krawczak et al. (2015).

Human Parasitism: No.

Comments: *Amblyomma yucumense* and *Amblyomma dubitatum* are similar morphologically and related at the molecular level, with the result that adults are easily confused with each other, and separation of nymphs can be very difficult. Lamattina et al. (2018b) found that the shape of the spiracular plates is an important morphological character for the separation *Amblyomma dubitatum* and *Amblyomma yucumense*. Adults of *Amblyomma hadanii* are also morphologically close to *Amblyomma yucumense* (Krawczak et al. 2015).

Males and females of *Amblyomma yucumense* are known from specimens collected on vegetation and from ticks that molted from engorged nymphs.

Comments on *Amblyomma*

Table 2.54 summarizes the known stages, zoogeographic distribution, and principal hosts for the 67 species of *Amblyomma* found in the Neotropics, 54 of which are endemic to the Neotropics, while 13 species are also found in other zoogeographic regions. One of these 13 ticks, *Amblyomma rotundatum*, is a Nearctic and Neotropical species that has also been found on one Pacific island; another, *Amblyomma variegatum*, is an Afrotropical and Neotropical species, while the remaining 11 taxa

Table 2.54 Known stages, zoogeographic distribution, and orders of principal hosts (or potential principal hosts) of Neotropical species of *Amblyomma* and their parasitic stages. Hosts listed are for parasitic associations determined in Neotropical localities. M = male, F = female, N = nymph, L = larva. "Unknown" refers to tick stages with unidentified natural hosts. "Undetermined" refers to tick stages with at least one host identified but principal hosts are uncertain

Species	Known stages	Zoogeographic Region	Principal hosts for adults	Principal hosts for nymphs	Principal hosts for larvae
1. *A. albopictum*	MFL	Neotropical	Squamata	Squamata*	Squamata
2. *A. antillorum*	MFNL	Neotropical	Squamata	Squamata	Squamata
3. *A. argentinae*	MFNL	Neotropical	Testudines	Testudines	Testudines
4. *A. aureolatum*	MFN	Neotropical	Carnivora	Passeriformes	Passeriformes*
5. *A. auricularium*	MFNL	Nearctic, Neotropical	Cingulata	Cingulata	Cingulata
6. *A. beaurepairei*	MF	Neotropical	Cingulata	Unknown	Unknown
7. *A. boeroi*	MFNL	Neotropical	Artiodactyla	Artiodactyla	Artiodactyla
8. *A. boulengeri*	MF	Neotropical	Squamata	Unknown	Unknown
9. *A. brasiliense*	MFNL	Neotropical	Mammalia several orders	Mammalia several orders	Mammalia several orders
10. *A. cajennense*	MFN	Neotropical	Mammalia several orders	Mammalia several orders	Undetermined
11. *A. calcaratum*	MFNL	Neotropical	Pilosa	Passeriformes	Passeriformes
12. *A. coelebs*	MFN	Nearctic, Neotropical	Perissodactyla	Carnivora, Didelphimorphia	Carnivora, Didelphimorphia*
13. *A. crassum*	F	Neotropical	Testudines	Unknown	Unknown
14. *A. cruciferum*	MFN	Neotropical	Squamata	Squamata	Unknown
15. *A. darwini*	MFN	Neotropical	Squamata	Squamata	Squamata*
16. *A. dissimile*	MFNL	Nearctic, Neotropical	Anura, Squamata	Anura, Squamata	Anura, Squamata
17. *A. dubitatum*	MFNL	Neotropical	Rodentia	Rodentia	Rodentia
18. *A. fulvum*	M	Neotropical	Squamata	Unknown	Unknown
19. *A. fuscum*	MFN	Neotropical	Cingulata, Squamata	Didelphimorphia	Didelphimorphia*
20. *A. geayi*	MFNL	Neotropical	Pilosa	Pilosa	Passeriformes
21. *A. goeldii*	MFN	Neotropical	Pilosa	Unknown	Unknown
22. *A. hadanii*	MFNL	Neotropical	Artiodactyla	Artiodactyla	Unknown
23. *A. hirtum*	MF	Neotropical	Unknown	Unknown	Unknown
24. *A. humerale*	MFN	Neotropical	Testudines	Mammalia several orders, Passeriformes, Squamata	Undetermined
25. *A. incisum*	MFN	Neotropical	Perissodactyla	Undetermined	Undetermined
26. *A. inornatum*	MFNL	Nearctic, Neotropical	Undetermined	Unknown	Unknown
27. *A. interandinum*	MF	Neotropical	Artiodactyla	Unknown	Unknown
28. *A. latepunctatum*	MFN	Neotropical	Perissodactyla	Undetermined	Unknown
29. *A. longirostre*	MFNL	Neotropical	Rodentia	Passeriformes	Passeriformes
30. *A. macfarlandi*	MFNL	Neotropical	Testudines	Testudines	Testudines
31. *A. maculatum*	MFNL	Nearctic, Neotropical	Canidae	Undetermined	Undetermined
32. *A. mixtum*	MFNL	Nearctic, Neotropical	Mammalia several orders	Mammalia several orders	Mammalia several orders
33. *A. multipunctum*	MFN	Neotropical	Perissodactyla	Perissodactyla	Unknown
34. *A. naponense*	MFN	Neotropical	Artiodactyla	Mammalia several orders	Undetermined

(continued)

Table 2.54 (continued)

35. *A. neumanni*	MFNL	Neotropical	Artiodactyla	Artiodactyla	Artiodactyla
36. *A. nodosum*	MFNL	Neotropical	Pilosa	Passeriformes	Passeriformes
37. *A. oblongoguttatum*	MFNL	Nearctic, Neotropical	Mammalia several orders	Mammalia several orders	Undetermined
38. *A. ovale*	MFNL	Nearctic, Neotropical	Carnivora	Rodentia	Rodentia
39. *A. pacae*	MFNL	Neotropical	Rodentia	Undetermined	Undetermined
40. *A. parkeri*	MFNL	Neotropical	Rodentia	Primates, Rodentia	Passeriformes
41. *A. parvitarsum*	MFNL	Neotropical	Artiodactyla	Squamata	Squamata
42. *A. parvum*	MFNL	Neotropical	Mammalia several orders	Rodentia	Rodentia
43. *A. patinoi*	MF	Neotropical	Artiodactyla	Unknown	Unknown
44. *A. pecarium*	MF	Neotropical	Artiodactyla	Undetermined	Unknown
45. *A. pictum*	MF	Neotropical	Pilosa	Unknown	Unknown
46. *A. pilosum*	MFNL	Neotropical	Testudines	Testudines	Testudines
47. *A. pseudoconcolor*	MFN	Neotropical	Cingulata	Undetermined	Undetermined
48. *A. pseudoparvum*	MFNL	Neotropical	Rodentia	Undetermined	Unknown
49. *A. quadricavum*	MFL	Neotropical	Squamata	Unknown	Unknown
50. *A. romarioi*	MF	Neotropical	Primates	Primates*	Passeriformes*
51. *A. romitii*	MFNL	Neotropical	Rodentia	Rodentia	Unknown
52. *A. rotundatum*	MFNL	Nearctic, Neotropical, Pacific Island	Anura, Squamata	Anura, Squamata	Anura, Squamata
53. *A. sabanerae*	MF	Neotropical	Testudines	Testudines	Testudines
54. *A. scalpturatum*	MFN	Neotropical	Perissodactyla	Undetermined	Undetermined
55. *A. sculptum*	MFN	Neotropical	Mammalia several orders	Mammalia several orders	Mammalia several orders
56. *A. scutatum*	MFNL	Nearctic, Neotropical	Squamata	Squamata	Squamata
57. *A. tapirellum*	MF	Neotropical	Perissodactyla	Undetermined	Unknown
58. *A. tenellum*	MFN	Nearctic, Neotropical	Artiodactyla, Perissodactyla	Unknown	Unknown
59. *A. tigrinum*	MFNL	Neotropical	Carnivora	Passeriformes, Rodentia	Passeriformes, Rodentia
60. *A. tonelliae*	MFN	Neotropical	Artiodactyla	Artiodactyla	Artiodactyla *
61. *A. torrei*	MFNL	Neotropical	Squamata	Squamata	Squamata
62. *A. triste*	MFNL	Nearctic, Neotropical	Mammalia several orders	Rodentia	Rodentia
63. *A. usingeri*	MFNL	Neotropical	Testudines	Testudines	Testudines
64. *A. variegatum*	MFNL	Afrotropical, Neotropical	Artiodactyla	Artiodactyla	Artiodactyla
65. *A. varium*	MFNL	Neotropical	Pilosa	Undetermined	Undetermined
66. *A. williamsi*	MF	Neotropical	Squamata	Unknown	Unknown
67. *A. yucumense*	MFN	Neotropical	Unknown	Didelphimorphia	Didelphimorphia

* Provisional result because the tick stage remains undescribed

have been found in both the Nearctic and Neotropical Zoogeographic Regions. The total number of *Amblyomma* species established in the Neotropics represents an increase of 17% relative to 2003 (Guglielmone et al. 2003a), when 57 species were found to be established in this region, 47 of them found only in the Neotropics, and the remaining ten taxa also present elsewhere. Two new species are now included as occurring in both the Nearctic and Neotropical Regions, *Amblyomma scutatum* and *Amblyomma triste.* Five species have been reinstated: the Neotropical and Nearctic *Amblyomma mixtum* and *Amblyomma tenellum*, and the Neotropical *Amblyomma latepunctatum*, *Amblyomma romitii,* and *Amblyomma sculptum*, while seven new endemic species have been added to the 2003 list of Neotropical *Amblyomma*: *Amblyomma boeroi*, *Amblyomma hadanii*, *Amblyomma interandinum*, *Amblyomma patinoi*, *Amblyomma romarioi*, *Amblyomma tonelliae,* and *Amblyomma yucumense*.

There are 31 *Amblyomma* species (or 46% of the total) with at least one undescribed stage, which has the effect of clouding our understanding of tick ecology and life histories, with implications for our ability to assess the importance of Neotropical *Amblyomma* species as vectors of tick-borne pathogens, a point also made in the case of Neotropical *Ixodes*.

The importance of Mammalia as hosts for *Amblyomma* is evident because 36 of the 67 Neotropical species are known only to parasitize mammals, although no tendency to feed on a particular order of mammals is seen. Thus, the most common mammalian order parasitized by Neotropical *Amblyomma* is Artiodactyla, with eight species feeding chiefly on these hosts, whereas five *Amblyomma* (*Amblyomma brasiliense*, *Amblyomma cajennense*, *Amblyomma mixtum*, *Amblyomma oblongoguttatum,* and *Amblyomma sculptum*) are catholic feeders, with all known parasitic stages found on members of several orders of mammals, and the remaining 23 species are parasites of different orders of mammals in various combinations (Tables 2.54, 5.2 and 5.6). Aves are also important as hosts of Neotropical *Amblyomma*, but unlike the genus *Ixodes* discussed above, no species feed exclusively on birds. However, there are eight species of *Amblyomma* whose immature stages parasitize Aves while the adults feed on mammals, and in the case of *Amblyomma humerale*, adult ticks feed chiefly on Testudines, but the principal hosts of the nymphs are mammals, squamatans, and passeriforms (Table 5.6). From an evolutionary viewpoint, it is hypothetized that *Amblyomma* were able to colonize mammalian and passeriform habitats throughout the Neotropics following the radiation of mammals and birds during the Cenozoic, a topic that is further treated in Chap. 5. Reptilia also play an important role as hosts for *Amblyomma*, with 16 species known to parasitize these vertebrates, the most frequent host association being with Squamata (10 species, or 15% of the total) (Tables 2.54, 5.2 and 5.6). Additionally, *Amblyomma dissimile* and *Amblyomma rotundatum* are peculiar because they are the only species of Ixodidae worldwide known to complete their life cycles by feeding principally on Amphibia and Reptilia. These two species are considered to be of Neotropical origin but their current ranges extend beyond this zoogeographic region.

Contrary to other genera of Ixodidae found in the Neotropics, species of *Amblyomma* are known to parasitize all four classes of principal hosts evaluated in this study (Amphibia, Reptilia, Aves, and Mammalia), and the number of *Amblyomma* host associations is higher than in other genera of Neotropical Ixodidae (Tables 2.54 and 5.2).

Human parasitism by *Amblyomma* in the Neotropics is a common event—39 species have been found feeding on people, by far the largest number among Neotropical ixodids. This represents a sharp increase from Guglielmone et al. (2003a), when 26 species of *Amblyomma* were found infesting humans in the Neotropics.

Additional information on tick geographical distribution and host associations is provided in Chaps. 4 and 5, and in the appendix.

Chapter 3
Metastriata: Genera *Dermacentor*, *Haemaphysalis*, and *Rhipicephalus*

Genus *Dermacentor*

Worldwide there are 42 species in the genus *Dermacentor*. Fifteen species (36% of the total) are known from the Palearctic Region, 11 of them occurring only there, while an additional 15 species (36%) are found in the Oriental Region, ten of them occurring only in that region. Nine species of *Dermacentor* or 21% of all taxa in the genus are found in the Neotropics, but only four of these are exclusively Neotropical species.

1. *Dermacentor albipictus* (Packard, 1869) as *Ixodes albipictus* and given its current status by Banks (1907)

Packard, A.S. (1869) Report of the curator of Articulata. *First Annual Report of the Trustees of the Peabody Academy of Sciences*, pp. 56–69.

Banks, N. (1907) A catalogue of the Acarina, or mites, of the United States. *Proceedings of the United States National Museum*, 32, 595–625.

Descriptions and Redescriptions: Packard (1869) described but did not figure the female, under the name *Ixodes albipictus*; Marx and Neumann (1897) in Neumann (1897) described and illustrated the male, under the name *Dermacentor variegatus*, a synonym of *Dermacentor albipictus*, as stated in Neumann (1911a). Stiles (1910) provided a very brief description without figures of the nymph, under the name *Dermacentor salmoni*, another synonym of *Dermacentor albipictus*, as stated in Cooley (1938). Cameron and Fulton (1927) described and figured the larva of *Dermacentor albipictus*.

The male, female, nymph, and larva of *Dermacentor albipictus* were redescribed by Cooley (1938), Arthur (1960a), Brinton et al. (1965), Furman and Loomis (1984), and Lindquist et al. (2016). The male, female, and nymph were redescribed by Cameron and Fulton (1927) and Gregson (1956). The female, nymph, and larva were redescribed by Sonenshine (1979). The male and female were redescribed in Yunker

A. A. Guglielmone et al., *Neotropical Hard Ticks (Acari: Ixodida: Ixodidae)*,
https://doi.org/10.1007/978-3-030-72353-8_3

et al. (1986), and the larva alone was redescribed by Clifford et al. (1961), Coley (2015), and Dubie et al. (2017).

Geographic Distribution: *Dermacentor albipictus* is mainly a Nearctic species, with Neotropical records from southern North America and northern Central America as follows: **Guatemala**: Alta Verapaz, Chiquimula, Escuintla, Huehuetenango, Petén, Quiché, and Zacapa (Instituto Interamericano de Cooperación para la Agricultura 1988); **southern Mexico**: Campeche, Chiapas, Guerrero, Jalisco, Michoacán, Nayarit, Quintana Roo, Tabasco, Veracruz, and Yucatán (Guzmán-Cornejo et al. 2016).

Hosts: *Dermacentor albipictus* is typically a parasite of Artiodactyla: Bovidae and Cervidae, and Perissodactyla: Equidae, but it has also been collected from mammals of several orders, with a few records from Aves, in the Nearctic Region (Guglielmone and Robbins 2018). Most hosts of Neotropical *Dermacentor albipictus* are the same for this tick in the Nearctic; however, with the exception of a female tick collected on cattle in the State of Veracruz (Mexico) (Guzmán-Cornejo et al. 2016), tick stages found on Neotropical hosts have not been recorded, with the result that there is insufficient information to construct a Neotropical host profile for *Dermacentor albipictus*.

Human Parasitism: Guglielmone and Robbins (2018) list *Dermacentor albipictus* as a frequent parasite of people, but no cases of human parasitism have been recorded in the Neotropics.

Comments: *Dermacentor albipictus* is a one-host tick. Several Nearctic and Neotropical records have been published under the name *Dermacentor nigrolineatus*, which is generally recognized as a synonym of *Dermacentor albipictus* (Cooley 1938; Arthur 1960b; Yunker et al. 1986), a position further reinforced by Ernst and Gladney (1975), who demonstrated that reciprocal crosses of *Dermacentor albipictus* and *Dermacentor nigrolineatus* produce fertile progeny. Nevertheless, Camicas et al. (1998) list *Dermacentor nigrolineatus* as a valid species, and in the same year, Crosbie et al. (1998) found molecular differences in 16S rDNA gene sequences indicating the presence of more than one species under the name *Dermacentor albipictus*. Leo et al. (2010) confirmed the results of Crosbie et al. (1998) but added analysis of molecular genes, endosymbiont markers, and morphological variation to conclude that there is insufficient evidence for the presence of sibling species; although no information on reciprocal crosses was provided. Nevertheless, the results of Ernst and Gladney (1975) provisionally support the synonymy of *Dermacentor nigrolineatus* with *Dermacentor albipictus*.

2. *Dermacentor dispar* Cooley, 1937

Cooley, R.A. (1937) Two new dermacentors [sic] from Central America. *Journal of Parasitology*, 23, 259–264.

Descriptions and Redescriptions: Cooley (1937) described and illustrated the male and female of *Dermacentor dispar*.

Table 3.1 Hosts for males (M) and females (F) of *Dermacentor dispar*

MAMMALIA			
ARTIODACTYLA: Cervidae		**CINGULATA: Dasypodidae**	
"Deer"	MF*	*Dasypus novemcinctus*	MF*
ARTIODACTYLA: Tayassuidae			
Tayassu pecari	MF		

* Provisional hosts

Arthur (1960a) and Yunker et al. (1986) redescribed the male and female of *Dermacentor dispar*.

The nymph and larva of *Dermacentor dispar* remain unknown.

Geographic Distribution: *Dermacentor dispar* is a Neotropical species found in northern Central America as follows: **Belize** (Yunker et al. 1986); and **Guatemala**: Petén (Cooley 1937).

Hosts: The only bona fide records of males and females of *Dermacentor dispar* are from *Tayassu pecari* (Artiodactyla: Tayassuidae), as detailed in the description of the species, and that tayassuid is treated here as the principal host of this tick. The host profile of *Dermacentor dispar* is shown in Table 3.1, constructed using information provided by Cooley (1937) and Yunker et al. (1986), but the host records from "deer" and *Dasypus novemcinctus* in Yunker et al. (1986) are treated here as provisional.

Human Parasitism: No.

Comments: Yunker et al. (1986) stated that "deer" and "peccary" are the hosts for adults *Dermacentor dispar*, but these authors claim to have redescribed the male and female of this tick using specimens collected from *Dasypus novemcinctus*, an entirely unexpected host. It is uncertain whether the specimens used to redescribe the adults of *Dermacentor dispar* were actually collected from *Dasypus novemcinctus*; the possibility exists that this is a *lapsus*. Consequently, as noted above, the alleged new hosts for adults of *Dermacentor dispar* ("deer" and *Dasypus novemcinctus*) in Yunker et al. (1986) are treated here as provisionally valid. These unusual records in Yunker et al. (1986) were ignored in Guglielmone et al. (2003a), but they were recognized as valid in Guglielmone et al. (2014, 2020).

3. *Dermacentor dissimilis* Cooley, 1947

Cooley, R.A. (1947) *Dermacentor dissimilis*, a new species of tick from southern Mexico. *American Museum Novitates*, (1332), 1–3.

Descriptions and Redescriptions: The female and nymph of *Dermacentor dissimilis* were described and illustrated by Cooley (1947), and the male was described and figured in Kohls and Dalmat (1952).

Table 3.2 Hosts for adult ticks of undetermined sex (A), males (M), females (F), nymphs (N), and larvae (L)* of *Dermacentor dissimilis*

MAMMALIA			
ARTIODACTYLA: Bovidae		**PERISSODACTYLA: Equidae**	
Cattle	ANL	Horse	MFNL
CARNIVORA: Canidae			
Domestic dog	F		

* Provisional diagnosis because the larva of *Dermacentor dissimilis* remains undescribed

The male, female, and nymph were redescribed by Arthur (1960a), and both sexes of *Dermacentor dissimilis* were also redescribed by Yunker et al. (1986).

The larva of *Dermacentor dissimilis* remains undescribed, but Kohls and Dalmat (1952) were able to identify this stage from field-collected material.

Geographic Distribution: *Dermacentor dissimilis* is basically a Neotropical species found in Central America and southern North America, there being a few records from the Nearctic portion of Mexico. The Neotropical records are as follows: **El Salvador** (Navarrete Abarca et al. 2014); **Guatemala**: Chimaltenango, Jalapa, and Santa Rosa (Kohls and Dalmat 1952; Instituto Interamericano de Cooperación para la Agricultura 1988); **southern Mexico**: Chiapas, Oaxaca, Guerrero, Michoacán, and Veracruz (Guzmán-Cornejo et al. 2016); and **Nicaragua**: Estelí, Jinotega, Madriz, Matagalpa, and Nueva Segovia (Düttmann et al. 2016).

Hosts: Males, females, nymphs, and larvae of *Dermacentor dissimilis* are usually found on horses, although Kohls and Dalmat (1952) state that all stages of this tick are also common on cattle, a host that requires confirmation. All records of *Dermacentor dissimilis* are from domestic mammals, but the principal host for this tick is here thought to be the horse (Perissodactyla: Equidae). The host profile of *Dermacentor dissimilis* is shown in Table 3.2, which is based on data from Kohls and Dalmat (1952) and Guzmán-Cornejo et al. (2016).

Human Parasitism: No.

Comments: Males and females of *Dermacentor dissimilis* may or may not possess eyes, and the dentition of the hypostome may vary among specimens, as detailed in Cooley (1947) and Kohls and Dalmat (1952).

Cooley (1947) believes that *Dermacentor dissimilis* is a one-host tick, and Kohls and Dalmat (1952) support this conclusion, while Domínguez et al. (2020b) present evidence that nymphs molt to adult ticks while on their hosts. This developmental characteristic is most probably the reason that Camicas et al. (1998) classified this species as *Anocentor dissimilis*, a genus not recognized here.

The record of *Dermacentor nigrolineatus*, a provisional synonym of *Dermacentor albipictus*, from Guatemala in fact corresponds to *Dermacentor dissimilis*, as explained in Kohls and Dalmat (1952).

Kohls and Dalmat (1952) stated that *Dermacentor dissimilis* is a common parasite of cattle and horses in Guatemala; however, Düttmann et al. (2016) studied the ticks of cattle and horses in Nicaragua and found *Dermacentor dissimilis* on horses but not on cattle. Therefore, it is provisionally accepted that this tick is prone to feed on horses rather than cattle.

4. *Dermacentor halli* McIntosh, 1931

McIntosh, A. (1931) A new species of tick from the Texas peccary. *Journal of Parasitology*, 18, 124.

Descriptions and Redescriptions: McIntosh (1931) described but did not figure the male and female of *Dermacentor halli*, while the larva and nymph were described and illustrated in Cooley (1938).

The male, female, nymph, and larva of *Dermacentor halli* were redescribed in Arthur (1960a) and Brinton et al. (1965). The male and female were also redescribed by McIntosh (1932) and Cooley (1938).

Geographic Distribution: *Dermacentor halli* is chiefly a Nearctic species, with a few Neotropical records from southern North America as follows: **southern Mexico**: Chiapas, Jalisco, Veracruz, and Yucatán (Guzmán-Cornejo et al. 2016).

Hosts: There are seven Neotropical records of *Dermacentor halli*, but hosts and tick stages collected are known for just one of them in Cruz-Aldán et al. (2006), who found males and females on *Tapirus bairdii* in Chiapas (southern Mexico). Adults of this tick have been collected on Artiodactyla: Bovidae and Tayassuidae, Carnivora: Canidae and Mephitidae, and Perissodactyla: Equidae in the Nearctic Region (Guglielmone and Robbins 2018). When compared with host records from the Nearctic, it would appear that the one Neotropical record of *Dermacentor halli* from *Tapirus bairdii* is insufficient to conclude that this mammal is a principal host for *Dermacentor halli* in the Neotropics.

Human Parasitism: No (Guglielmone and Robbins 2018), but listed as a human parasite in Guglielmone et al. (2014).

Comments: Fairchild et al. (1966) stated that specimens of *Dermacentor halli* from the Neotropical Region differ morphologically from Nearctic specimens, indicating that more than one species may be included under the name *Dermacentor halli*, a hypothesis confirmed by Apanaskevich and Bermúdez (2013) in their description of *Dermacentor panamensis* from specimens previously identified as *Dermacentor halli*. The latter authors state that figures of *Dermacentor halli* in Yunker et al. (1986) in fact, represent *Dermacentor panamensis*.

All records of *Dermacentor halli* from Costa Rica, Guatemala, and Panama published prior to the study of Apanaskevich and Bermúdez (2013) were confirmed by these authors to be *Dermacentor panamensis*.

Guglielmone et al. (2014) listed *Dermacentor halli* as a parasite of humans in Mexico, based on Hoffmann and López-Campos (2000), who, in turn, relied on information provided by Fairchild et al. (1966). Guglielmone and Robbins (2018) analyzed the supposed record of *Dermacentor halli* from a human in Fairchild et al. (1966) but found no definitive evidence supporting human parasitism by this tick.

5. *Dermacentor imitans* Warburton, 1933

Warburton, C. (1933) On five new species of ticks (Arachnida Ixodoidea). *Ixodes petauristae*, *I. ampullaceus*, *Dermacentor imitans*, *Amblyomma laticaudae* and *Aponomma draconis*, with notes on three previously described species, *Ornithodorus* [sic] *franchinii* Tonelli Rondelli, *Haemaphysalis cooleyi* Bedford and *Rhipicephalus maculatus* Neumann. *Parasitology, 24*, 558–568.

Descriptions and Redescriptions: The male and female of *Dermacentor imitans* were described and illustrated in Warburton (1933).

The male and female of *Dermacentor imitans* were redescribed by Arthur (1960a), Yunker et al. (1986) and Bermúdez et al. (2018a).

The nymph and larva of *Dermacentor imitans* remain unknown.

Geographic Distribution: *Dermacentor imitans* is a Neotropical species that has been found in southern North America, Central America and northern South America as follows: **Colombia**: Chocó (Guglielmone et al. 2006a); **Costa Rica**: Cartago (Warburton 1933); **Guatemala**: Zacapa (Fairchild et al. 1966); **southern Mexico**: Chiapas, and Oaxaca (Guzmán-Cornejo et al. 2016); **Panama**: Bocas del Toro, Chiriquí, Darién, and Guna Yala (Fairchild et al. 1966; Bermúdez et al. 2018b); and **Venezuela**: Lara (Crosbie et al. 1998).

Hosts: Males and females of *Dermacentor imitans* have been collected from Artiodactyla: Tayassuidae and Cervidae, but 11 of the 12 bona fide records of this species were reported from *Pecari tajacu* and *Tayassu pecari* (Artiodactyla: Tayassuidae), which are regarded as the principal hosts for adults of this tick. The host profile of *Dermacentor imitans* is shown in Table 3.3, which is based on data from Fairchild et al. (1966) and Bermúdez et al. (2018a, b).

Human Parasitism: Guglielmone and Robbins (2018) list *Dermacentor imitans* as a very rare parasite of humans, citing a record of a female tick collected from a

Table 3.3 Hosts for males (M) and females (F) of *Dermacentor imitans*

MAMMALIA			
ARTIODACTYLA: Cervidae		**ARTIODACTYLA: Tayassuidae**	
Mazama sp.	M	*Pecari tajacu*	MF
		Tayassu pecari	MF

person in Chocó (Colombia) and a second female tick found on a human in Chiriquí (Panama).

Comments: Guglielmone et al. (2003a) noted that the figure used by Warburton (1933) to describe the female of *Dermacentor imitans* is unsatisfactory because the eyes were omitted and the coxae were not illustrated.

Bentley and Rodríguez (2001) allegedly found *Dermacentor imitans* in Honduras under the common names of "garrapata chata" ("thick (blood-filled) tick" [sic]), "garrapata menudita" ("small tick") or "coloradita" ("small red tick") without further data. Honduras is not included within the geographic distribution of *Dermacentor imitans*.

Camicas et al. (1998) stated that hosts for the larva and nymph of *Dermacentor imitans* are burrowing mammals, but this assertion was not confirmed in Guglielmone et al. (2014), nor has it been supported subsequently.

6. *Dermacentor latus* Cooley, 1937

Cooley, R.A. (1937) Two new dermacentors [sic] from Central America. *Journal of Parasitology*, 23, 259–264.

Descriptions and Redescriptions: The male of *Dermacentor latus* was described and illustrated in Cooley (1937), and the female was described and figured in Clifford and Kohls (1962).

The male and female of *Dermacentor latus* were redescribed by Arthur (1960a), Yunker et al. (1986) and Bermúdez et al. (2018a).

The nymph and larva of *Dermacentor latus* remain unknown.

Geographical Distribution: *Dermacentor latus* is a Neotropical tick found in Central America as follows: **Costa Rica**: Guanacaste (Cooley 1937); and **Panama**: Bocas del Toro, and Chiriquí (Fairchild et al. 1966).

Hosts: *Tapirus bairdii* (Perissodactyla: Tapiridae) is the usual host for males and females of *Dermacentor latus*, with two records of male ticks from dogs (Fairchild et al. 1966).

Human Parasitism: Guglielmone and Robbins (2018) listed *Dermacentor latus* as a very rare parasite of humans, with just two records of adult ticks (one male, one female) found on persons in Panama (Bocas del Toro, and Chiriqui) and reported in Fairchild et al. (1966).

Comments: Guzmán-Cornejo et al. (2016) state that the southern Mexican record of *Dermacentor latus* in Cruz-Aldán et al. (2006) needs confirmation, and this portion of the Neotropics is not included within the range of this tick.

7. *Dermacentor nitens* Neumann, 1897

Neumann, L.G. (1897) Révision de la famille des ixodidés (2e mémoire). *Mémoires de la Société Zoologique de France*, 10, 324–420.

Descriptions and Redescriptions: Neumann (1897) described the male and female of *Dermacentor nitens*, including a figure of the stigmal plate of the male, while Hooker et al. (1912) described and illustrated the nymph and larva.

The male, female, nymph, and larva of *Dermacentor nitens* were redescribed by Cooley (1938, under the name *Otocentor nitens*) and Arthur (1960a, under the name *Anocentor nitens*). The male, female, and nymph were redescribed in Nava et al. (2017); and the male and female were redescribed by Banks (1908), Hooker et al. (1912), Floch and Fauran (1958), Schulze (1937b, under the name *Anocentor columbianus*, a synonym of *Dermacentor nitens*, as discussed in Guglielmone and Nava (2014)), Yunker et al. (1986), Barros-Battesti et al. (2009) and Bermúdez et al. (2018a). Amorim et al. (1997) redescribed the larva of *Dermacentor nitens* under the name *Anocentor nitens*.

Geographical Distribution: *Dermacentor nitens* is a Nearctic and Neotropical tick that has been found from the southern USA to northern Argentina, with a broad distribution in the Neotropics including Caribbean Islands as follows: **Antigua and Barbuda** (Keirans 1985b); **Argentina**: Jujuy, and Salta (Guglielmone and Nava 2005); **Bahamas**: Cat Island, Harbor Island (Clarkson and Owen 1959); **Barbados**: Christ Church, Saint Andrew, Saint George, Saint James, Saint John, Saint Joseph, Saint Lucy, Saint Michel, Saint Peter, Saint Phillip, and Saint Thomas (Drummond et al. 1981); **Belize**: Cayo, Corozal, and Orange Walk (Polsomboon et al. 2017); **Bolivia**: Beni, Chuquisaca, and Santa Cruz (Mastropaolo et al. 2014); **Brazil**: Acre, Amazonas, Bahia, Ceará, Espírito Santo, Goiás, Maranhão, Mato Grosso, Mato Grosso do Sul, Minas Gerais, Pará, Paraná, Pernambuco, Rio de Janeiro, Rondônia, Roraima, Santa Catarina, São Paulo, and Tocantins (Aragão and Fonseca 1953b; Leite et al. 2001; Guerra and Brito 2004; Arzua et al. 2005; Labruna et al. 2005a, d; Dantas-Torres et al. 2009a; Martins et al. 2009a; Cunha et al. 2014; Lavina et al. 2014; Acosta et al. 2016; Moerbeck et al. 2016; Gruhn et al. 2019); **Colombia**: Amazonas, Antioquía, Arauca, Atlántico, Bolívar, Boyacá, Caldas, Casanare, Cauca, Córdoba, Cundinamarca, Distrito Capital, Guaviare, Huila, Magdalena, Meta, Nariño, Risaralda, Santander, Sucre, Tolima, and Valle del Cauca (Guerrero 1931; Osorno Mesa 1942; Arias et al. 1991; Betancourt et al. 1992; Guglielmone et al. 2006a; Valencia et al. 2017; Rivera-Páez et al. 2018; Santodomingo et al. 2019; Acevedo-Gutiérrez et al. 2020); **Costa Rica**: Cartago, Guanacaste, Heredia, Limón, Puntarenas, and San José (Álvarez et al. 2005; Troyo et al. 2016); **Cuba**: Artemisa, Camagüey, Guantánamo, Holguín, Isla de la Juventud, La Habana, Matanzas, Mayabeque, Pinar del Río, and Villa Clara (Pérez Vigueras 1934; Morel 1966; Becklund 1968; Černý 1969c; Danielová et al. 1982; Noda et al. 2015; Díaz-Sánchez et al. 2021); **Dominica** (Bishopp and Trembley 1945); **Dominican Republic**: Distrito Nacional, and Monte Cristi (Neumann 1897; Santos Dias 1958c); **Ecuador**: Galápagos, and Guayas (Keirans 1985b; Yunker et al. 1986); **El Salvador**

(Navarrete Abarca et al. 2014); **French Guiana**: Cayena (Binetruy et al. 2019); **Grenada** (Keirans 1985b); **Guadeloupe**: Basse-Terre, and Grand Terre (Morel 1966); **Guatemala**: Alta Verapaz, El Progreso, Escuintla, Huehuetenango, Izabal, Jalapa, Jutiapa, Petén, Quetzaltenango, Quiche, Santa Rosa, Suchitepequez, and Zacapa (Instituto Interamericano de Cooperación para la Agricultura 1988); **Guyana** (Myers 1935); **Haiti** (Floch and Fauran 1958); **Honduras**: Lempira (Linthicum et al. 1991); **Jamaica**: Hanover, Kingston, Manchester, Portland, Saint Andrew, Saint Catherine, Saint Elizabeth, Saint James, Saint Mary, and Westmoreland (Newstead 1909; Thompson 1950; Morel 1966; Wilson 1980; Keirans 1985b); **Martinique**: Fort de France, La Trinité, Le Marin, and Saint Pierre (Morel 1966); **southern Mexico**: Campeche, Chiapas, Guerrero, Michoacán, Nayarit, Oaxaca, Quintana Roo, Tabasco, Veracruz, and Yucatán (Guzmán-Cornejo et al. 2016); **Montserrat** (Keirans 1985b); **Nicaragua**: Chinandega, Costa Caribe Norte, Costa Caribe Sur, Estelí, Granada, Jinotega, León, Madriz, Managua, Matagalpa, Nueva Segovia, and Rivas (Woke 1947; Maes et al. 1989; Düttmann et al. 2016); **Panama**: Chiriquí, Colón, Darién, Panamá, and Panamá Oeste (Dunn 1934; Bermúdez et al. 2011, 2012a; Miller et al. 2016; Keirans 1985b); **Paraguay**: Caa-Guazú, Central, Concepción, Guairá, Paraguarí, and San Pedro (Nava et al. 2007); **Peru**: Amazonas, Lambayeque, Piura, and San Martín (Need et al. 1991; Blair et al. 2004; Billeter et al. 2011); **Puerto Rico**: Culebra, Guaynabo, Mayaguez, and Trujillo Alto (McChesney and Jones 1965; Fox and León 1983; Keirans 1985b; Crom 1992); **Saint Kitts and Nevis** (Morel 1966); **Saint Lucia**: Gros Islet, and Vieux Fort (Garris and Scotland 1985); **Saint Martin** (Morel 1966); **Saint Vincent and the Grenadines** (Morel 1966); **Trinidad and Tobago** (Floch and Fauran 1958); **United Kingdom Virgin Islands**: Tortola Island (Hall 1951); **USA Virgin Islands**: Saint Croix, Saint John, and Saint Thomas (Bishopp and Trembley 1945; Nelson et al. 2017); and **Venezuela**: Anzoátegui, Apure, Araguá, Barinas, Bolívar, Cojedes, Delta Amacuro, Distrito Federal, Falcón, Guárico, Lara, Miranda, Monagas, Portuguesa, Sucre, Táchira, Yaracuy, and Zulia (Vergani 1956; Guerrero 1996).

Hosts: Horses (Perissodactyla: Equidae) are the usual hosts for males, females, nymphs, and larvae of *Dermacentor nitens*, but this tick has also been collected from a variety of wild and domestic mammals, with an unusual record of a female tick found on an amphibian host (Morel 1966). The host profile of *Dermacentor nitens* is shown in Table 3.4, constructed using data provided by Hooker et al. (1912), Vogelsang and Santos Dias (1953b), Morel (1966), Jones et al. (1972), Varma (1973), Keirans (1985b), Serra-Freire et al. (1996), Labruna et al. (2005a, d), Cruz-Aldán et al. (2006), Corn et al. (2009), Martins et al. (2015a, b), Guzmán-Cornejo et al. (2016), Witter et al. (2016), Rivera-Páez et al. (2018), Martins et al. (2020b) and Ramos et al. (2020).

Human Parasitism: Guglielmone and Robbins (2018) listed *Dermacentor nitens* as a sporadic human parasite, with all cases restricted to the Neotropics in Bolivia (Beni), Colombia (Antioquía, Casanare, and Distrito Capital), and Venezuela, but Szabó et al. (2020) collected one male from a person in Minas Gerais, Brazil. All

Table 3.4 Hosts for adult ticks of undetermined sex (A), males (M), females (F), nymphs (N), and larvae (L) of *Dermacentor nitens*

AMPHIBIA			
ANURA: Bufonidae			
Peltophryne peltocephala	F		
MAMMALIA			
ARTIODACTYLA: Bovidae		*Leopardus tigrinus*	A
Cattle	MFNL	*Panthera onca*	AN
Goat	F	*Puma concolor*	MFN
ARTIODACTYLA: Cervidae		**PERISSODACTYLA: Equidae**	
Blastocerus dichotomus	MF	Donkey	MFNL
Mazama gouazoubira	MFNL	Horse	MFNL
Odocoileus virginianus	MFNL	Mule	MFNL
ARTIODACTYLA: Suidae		**PERISSODACTYLA: Tapiridae**	
Domestic pig	A	*Tapirus bairdii*	MF
CARNIVORA: Canidae		*Tapirus terrestris*	F
Cerdocyon thous	FN	**RODENTIA: Caviidae**	
Domestic dog	MFN	*Hydrochoerus hydrochaeris*	MF
Lycalopex vetulus	MFN	**RODENTIA: Cuniculidae**	
CARNIVORA: Felidae		*Cuniculus paca*	MF
Domestic cat	MF		

parasitic stages of *Dermacentor nitens* have been found on humans, but the larva is the chief parasite of people.

Comments: *Dermacentor nitens* is another one-host member of its genus. This species was described by Neumann in the genus *Dermacentor*, but Cooley (1938) created the genus *Otocentor* with *Dermacentor nitens* as the type species, and most earlier authors used *Anocentor*, created by Schulze (1937b), as the genus name for this tick. Currently, the original designation of *Dermacentor nitens* proposed by Neumann (1897) is preferred.

According to Morel (1966), the unusual record of this species from Anura: Bufonidae is based on a specimen collected from *Peltophryne peltocephala* (under the name *Bufo peltocephalus*) in Cuba, as recorded in Schulze (1941d) and classified as *Anocentor columbianus*; this record is treated here as provisionally valid. Černý (1966b) found a larva of *Dermacentor nitens* on *Glaucidium siju* (Strigiformes: Strigidae), but the specimen was not attached.

Serra-Freire (2010, 2014), under the name *Anocentor nitens*, allegedly collected hundreds of *Dermacentor nitens* from people in Brazil, but these records require confirmation, while Guglielmone et al. (2006a, b) mentioned *Dermacentor nitens* on persons in Rondônia (Brazil), but this record was not confirmed in the study of Rondonian ticks by Labruna et al. (2005d). However, Szabó et al. 2020 found a case of human parasitism by *Dermacentor nitens* in Brazil.

8. ***Dermacentor panamensis*** Apanaskevich and Bermúdez, 2013

Apanaskevich, D.A. & Bermúdez, S.E. (2013) Description of a new *Dermacentor* (Acari: Ixodidae) species, a parasite of wild mammals in Central America. *Journal of Medical Entomology*, 50, 1190–1201.

Descriptions and Redescriptions: The male, female, nymph and larva of *Dermacentor panamensis* were described in Apanaskevich and Bermúdez (2013).

The male and female of *Dermacentor panamensis* were redescribed in Bermúdez et al. (2018a).

There are no redescriptions of the larva and nymph of *Dermacentor panamensis.*

Geographic Distribution: *Dermacentor panamensis* is a Neotropical species that has been found in Central America as follows: **Costa Rica**: Puntarenas, and San José; **Guatemala**: Chimaltenango; **Honduras**: Olancho; **Nicaragua**: Matagalpa; and **Panama**: Chiriquí, and Los Santos (Apanaskevich and Bermúdez 2013).

Hosts: Males and females of *Dermacentor panamensis* are usually found on *Coendou mexicanus* (Rodentia: Erethizontidae), while nymphs and larvae are commonly collected from several species of Rodentia: Cricetidae, it being uncertain which species are preferred as hosts. Immature stages of this tick have also been recovered from other mammals. The host profile of *Dermacentor panamensis* is shown in Table 3.5, constructed using information provided by Apanaskevich and Bermúdez (2013) and Bermúdez et al. (2018a).

Human Parasitism: No.

Comments: *Dermacentor panamensis* had been confused with *Dermacentor halli* prior to its description by Apanaskevich and Bermúdez (2013), an outcome anticipated by Fairchild et al. (1966), who stated that more than one species might exist under the name *Dermacentor halli*. The figures of *Dermacentor halli* in Yunker et al.

Table 3.5 Hosts for males (M), females (F), nymphs (N), and larvae (L) of *Dermacentor panamensis*

MAMMALIA			
CHIROPTERA: Vespertilionidae		*Reithrodontomys mexicanus*	NL
Myotis sp.	N	*Scotinomys teguina*	N
RODENTIA: Cricetidae		*Tylomys nudicaudus*	N
Isthmomys flavidus	NL	**RODENTIA: Eretizhontidae**	
Nephelomys devius	N	*Coendou mexicanus*	MFN
Nyctomys sumichrasti	L	**RODENTIA: Heteromyidae**	
Oligoryzomys sp.	N	*Heteromys adspersus*	L
Oryzomys sp.	L	*Heteromys desmarestianus*	N
Peromyscus aztecus	NL	**RODENTIA: Sciuridae**	
Peromyscus mexicanus	NL	*Sciurus granatensis*	NL
Reithrodontomys creper	NL		

(1986) in fact represent *Dermacentor panamensis*, as discussed in Apanaskevich and Bermúdez (2013).

9. *Dermacentor variabilis* (Say, 1821), originally named *Ixodes variabilis* and given its current status by Banks (1907)

Say, T. (1821) An account of the arachnides of the United States. *Journal of the Academy of Natural Sciences of Philadelphia*, 2, 59–82.

Banks, N. (1907) A catalogue of the Acarina, or mites, of the United States. *Proceedings of the United States National Museum*, 32, 595–625.

Descriptions and Redescriptions: Say (1821) described but did not figure the male of *Dermacentor variabilis* under the name *Ixodes variabilis*, while the female was described but not illustrated by Koch (1844) under the name *Dermacentor electus*, a synonym of *Dermacentor variabilis*, as stated in Neumann (1911a). Hooker et al. (1912) described and illustrated the nymph and larva.

The male, female, nymph, and larva were redescribed by Cooley (1938), Brinton et al. (1965), Arthur (1960a), Furman and Loomis (1984), and Lindquist et al. (2016). The male and female were redescribed in Banks (1908), Hooker et al. (1912), and Yunker et al. (1986). The larva and nymph were redescribed by Dubie et al. (2017), while the larva of *Dermacentor variabilis* was redescribed by Clifford et al. (1961) and Coley (2015).

Geographic Distribution: *Dermacentor variabilis* is found in the Nearctic Region with a few Neotropical records from **southern Mexico**: Chiapas, Oaxaca, and Yucatán (Guzmán-Cornejo et al. 2016).

Hosts: Adults of *Dermacentor variabilis* are usually found on several families of artiodactyls (including domestic animals), Perissodactyla: Equidae, and Carnivora (several families, but especially dogs), while immature stages are commonly collected from Rodentia (several families), but *Dermacentor variabilis* has also been recovered from mammals of various orders and, sometimes, from birds (Apodiformes, Galliformes, Passeriformes) and squamatans (Agamidae) in the Nearctic Region (Guglielmone and Robbins 2018). The limited records of *Dermacentor variabilis* from the Neotropical Region are of specimens collected from cattle, rabbits, and deer, but parasitic stages found on those hosts were not provided; therefore, the Neotropical host profile for *Dermacentor variabilis* is considered basically undetermined.

Human Parasitism: Guglielmone and Robbins (2018) listed *Dermacentor variabilis* as a frequent parasite of humans, but no bona fide cases of local human parasitism have been reported from the Neotropical Region.

Comments: *Dermacentor variabilis* is a tick of medical and veterinary importance; therefore, additional research is needed to determine the range and hosts of this tick in the Neotropical Region.

Alvarez-Robles et al. (2018) allegedly found *Dermacentor variabilis* in Guatemala, but the diagnosis was based on Fairchild et al. (1966) and Voltzit (2007), studies that do not provide information concerning this tick. Consequently, Guatemala is not included within the range of *Dermacentor variabilis*.

There is a case of human parasitism by *Dermacentor variabilis* involving a child from the USA who visited Panama (Bermúdez et al. 2010b), but the tick, a partially engorged female, was detected several days after the child's arrival in Panama. The tick probably infested the child in the USA, but the presence of a Panamian population of *Dermacentor variabilis* should also be considered. See Chap. 4, "Invasive hard tick species found in the Neotropics."

Comments on *Dermacentor*

The zoogeographic distribution and principal hosts for the nine species of *Dermacentor* found in the Neotropics are summarized in Table 3.6, with four species endemic to the Neotropics (*Dermacentor dispar*, *Dermacentor imitans*, *Dermacentor latus,* and *Dermacentor panamensis*) and five species also found in the Nearctic Region (*Dermacentor albipictus*, *Dermacentor dissimilis*, *Dermacentor halli*, *Dermacentor nitens,* and *Dermacentor variabilis*). The total number of species of Neotropical *Dermacentor* represents an increase of 12% since 2003 (Guglielmone et al. 2003a) when eight species were found to be established in this region, four found only in the Neotropics and four found in the Nearctic. A new species, *Dermacentor panamensis*, endemic to the Neotropics, is now confirmed as part of the Neotropical fauna. *Dermacentor nitens* has a wide Neotropical distribution but other species are restricted to southern Mexico and Central America, with the exception of *Dermacentor imitans*, which has also been collected in northwestern South America.

There are four species of Neotropical *Dermacentor* (44% of the total) in which at least one life-history stage is undescribed, a situation that merits additional study. Amphibia, Reptilia, and Aves are essentially irrelevant as hosts of Neotropical *Dermacentor* because all species preferentially parasitize Mammalia. One species, *Dermacentor panamensis*, is restricted to Rodentia, but other species, such as *Dermacentor albipictus*, *Dermacentor dissimilis*, *Dermacentor nitens,* and *Dermacentor variabilis*, are usually parasites of large and medium-sized domestic mammals, being chiefly found on Artiodactyla and Perissodactyla. Nevertheless, there are few Neotropical records for most species, and the host profile of Neotropical *Dermacentor* is in need of clarification to better understand the ecology of this genus.

Three of the nine species in this genus (*Dermacentor imitans*, *Dermacentor latus*, and *Dermacentor nitens*) are recognized as feeding on people in the Neotropics, but cases of human parasitism in the region are few.

Additional information on tick geographical distribution and host associations is provided in Chaps. 4 and 5, and in the appendix.

Table 3.6 Known stages, zoogeographic distribution, and orders of principal hosts (or potential principal hosts) for Neotropical species of *Dermacentor*, *Haemaphysalis*, *Rhipicephalus*, and their parasitic stages. Hosts listed are for parasitic associations determined in Neotropical localities. M = male, F = female, N = nymph, L = larva. "Unknown" refers to tick stages with unidentified natural hosts. "Undetermined" refers to tick stages with at least one host identified but principal hosts are uncertain

Species	Known stages	Zoogeographic Region	Principal hosts for adults	Principal hosts for nymphs	Principal hosts for larvae
			Dermacentor		
1. *D. albipictus*	MFNL	Nearctic, Neotropical	Undetermined	Unknown	Unknown
2. *D. dispar*	MF	Neotropical	Artiodactyla	Unknown	Unknown
3. *D. dissimilis*	MFN	Nearctic, Neotropical	Perissodactyla	Perissodactyla	Perissodactyla*
4. *D. halli*	MFNL	Nearctic, Neotropical	Undetermined	Unknown	Unknown
5. *D. imitans*	MF	Neotropical	Artiodactyla	Unknown	Unknown
6. *D. latus*	MF	Neotropical	Perissodactyla	Unknown	Unknown
7. *D. nitens*	MFNL	Nearctic, Neotropical	Perissodactyla	Perissodactyla	Perissodactyla
8. *D. panamensis*	MFNL	Neotropical	Rodentia	Rodentia	Rodentia
9. *D. variabilis*	MFNL	Nearctic, Neotropical	Unknown	Unknown	Unknown
			Haemaphysalis		
1. *H. cinnabarina*	F	Neotropical	Unknown	Unknown	Unknown
2. *H. juxtakochi*	MFNL	Nearctic, Neotropical	Artiodactyla	Mammalia several orders	Mammalia several orders, Passeriformes
3. *H. leporispalustris*	MFNL	Nearctic, Neotropical	Lagomorpha	Lagomorpha	Lagomorpha
			Rhipicephalus		
1. *R. microplus*	MFNL	Pantropical	Artiodactyla	Artiodactyla	Artiodactyla
2. *R. sanguineus s.s.*	MFNL	Worldwide?	Carnivora	Carnivora	Carnivora
3. *R. sanguineus s.l.*	????	Worldwide	Carnivora	Carnivora	Carnivora

* Provisional diagnosis because the larva of *Dermacentor dissimilis* remains undescribed

Genus *Haemaphysalis*

There are 176 species in the genus *Haemaphysalis*, most found in the Oriental Zoogeographic Region, where 97 taxa (55% of the total) are established, 65 of them occurring nowhere else. The Afrotropical Region contains the second largest number of *Haemaphysalis* with 47 species (27%), 44 of them established exclusively in that region. Only three species of *Haemaphysalis*, or 2% of all taxa in this genus, have been found in the Neotropics, and only one of these is exclusively Neotropical.

1. *Haemaphysalis cinnabarina* Koch, 1844

Koch, C.L. (1844) Systematische Übersicht über die Ordnung der Zecken. *Archiv für Naturgeschichte*, 10, 217–239.

Descriptions and Redescriptions: The female of *Haemaphysalis cinnabarina* was described but not figured by Koch (1844).

Hoogstraal (1973) redescribed the female of *Haemaphysalis cinnabarina*.

The male, nymph, and larva of *Haemaphysalis cinnabarina* remain unknown.

Geographic Distribution: *Haemaphysalis cinnabarina* is a Neotropical species recorded from **Brazil**: Pará (Moritz and Fischer 1981).

Hosts: Unknown.

Human Parasitism: No.

Comments: *Haemaphysalis cinnabarina* is known from two female ticks collected in Brazil about 175 years ago. One specimen was used to describe *Haemaphysalis cinnabarina*, and the second specimen was described by Koch (1844) as *Haemaphysalis sanguinolenta*, a synonym of *Haemaphysalis cinnabarina*, as discussed by Hoogstraal (1973). Gregson (1956) states that *Haemaphysalis cinnabarina* in Nuttall and Warburton (1915) and a few other publications are in fact *Haemaphysalis chordeilis.* Keirans and Restifo (1993), Guglielmone et al. (2003a), and Kolonin (2009), among others, regard *Haemaphysalis cinnabarina* as a synonym of *Haemaphysalis punctata.* However, Hoogstraal (1973) examined the holotype of *Haemaphysalis cinnabarina*, concluding that it is a valid species, and this opinion was supported by Barros-Battesti et al. (2008), who also studied the holotype of *Haemaphysalis cinnabarina.* Thereafter, Guglielmone and Nava (2014) and Guglielmone et al. (2014, 2015) treated this species as a valid Neotropical taxon, a conclusion provisionally supported here.

Reyne (1923) allegedly found *Haemaphysalis cinnabarina* on a rodent in Suriname, while Camicas et al. (1998) list this tick as a parasite of Neotropical ungulates, and Guimarães et al. (2001) claim that *Haemaphysalis cinnabarina* is a parasite of birds, but no data have been found to support these claims.

2. *Haemaphysalis juxtakochi* Cooley, 1946

Cooley, R.A. (1946) The genera *Boophilus*, *Rhipicephalus*, and *Haemaphysalis* (Ixodoidea) of the New World. *National Institute of Health Bulletin*, (187) 54 pp.

Descriptions and Redescriptions: The male and female were described but not figured by Aragão (1908a) under the name *Haemaphysalis kochi*, a name preoccupied by *Haemaphysalis concinna kochi*, as discussed in Aragão and Fonseca (1951b). The nymph and larva of *Haemaphysalis juxtakochi* were described by Kohls (1960).

The male, female, nymph, and larva of *Haemaphysalis juxtakochi* were redescribed by Egizi et al. (2019), while Nava et al. (2017) redescribed the male, female, and nymph. The male and female were redescribed in Cooley (1946), who provided the first descriptions under the name *Haemaphysalis juxtakochi*. Boero (1957) also redescribed the male and female, but under the name *Haemaphysalis kohlsi*, a synonym of *Haemaphysalis juxtakochi*, as proposed by Kohls (1960). The latter author and Bermúdez et al. (2018a) redescribed the male and female under the correct name, *Haemaphysalis juxtakochi*.

Geographic Distribution: *Haemaphysalis juxtakochi* is a Nearctic and Neotropical species found over a vast area from the USA to Argentina and Uruguay, as well as one Caribbean island. Its geographic distribution in the Neotropical Region encompasses the following countries: **Argentina**: Buenos Aires, Catamarca, Chaco, Corrientes, Entre Ríos, Formosa, Jujuy, Misiones, Salta, Santa Fe, and Tucumán (Guglielmone and Nava 2005; Lamattina et al. 2018a); **Belize** (Varma 1973); **Bolivia** (Mastropaolo et al. 2014); **Brazil**: Amazonas, Espírito Santo, Maranhão, Mato Grosso, Mato Grosso do Sul, Pará, Paraná, Pernambuco, Rio Grande do Sul, Rondônia, São Paulo, and Tocantins (Labruna et al. 2005d; Ogrzewalska et al. 2010; Acosta et al. 2016; Gianizella et al. 2018b; Souza et al. 2018; Martins et al. 2020b); **Colombia**: Meta (Kohls 1960); **Costa Rica**: Guanacaste (Álvarez et al. 2005); **Ecuador** (Keirans and Restifo 1993); **French Guiana**: Cayena, and Saint Laurent du Maroni (Floch and Fauran 1958); **Guyana**: Essequibo Islands-West Demerara, and Upper Demarara-Berbice (Tonelli Rondelli 1939; Keirans 1985b); **southern Mexico**: Chiapas, Guerrero, and Yucatán (Kohls 1960; Onofre-Barragán 2015; Rodríguez-Vivas et al. 2016); **Panama**: Chiriquí, Colón, Panamá, and Panamá Oeste (Kohls 1960; Miller et al. 2016); **Paraguay**: Itapuá, Ñeembucú, and Presidente Hayes (Nava et al. 2007); **Peru**: Arequipa (Guglielmone et al. 2005a); **Suriname** (Jones et al. 1972); **Trinidad and Tobago**: Couva-Tabaquite-Talparo, and Sangre Grande (Kohls 1960; Keirans 1982); **Uruguay**: Artigas, Lavalleja, Maldonado, Rocha, and Tacuarembó (Venzal et al. 2003c, 2005a); and **Venezuela**: Amazonas, Aragua, Bolívar, Falcón, Guárico, Nueva Esparta, and Táchira (Vogelsang and Cordero 1940; Jones et al. 1972).

Hosts: Usual hosts for males and females of *Haemaphysalis juxtakochi* are *Mazama* species and *Odocoileus virginianus* (Artiodactyla: Cervidae), but adult ticks have also been found on several orders of mammals and one bird. Larvae and nymphs have been collected from a variety of mammals, including Cervidae, while three orders of birds have been found infested with immature stages. But while *Turdus albicollis* and *Turdus rufiventris* (Passeriformes: Turdidae) appear to be important hosts for larvae, this is not the case with nymphs of *Haemaphysalis juxtakochi*; moreover, the range of this tick is broader than the range of these birds. It is provisionally believed that several orders of mammals serve as principal hosts for nymphs and larvae, and the species of *Turdus* named here are considered provisional principal hosts of *Haemaphysalis juxtakochi* larvae. The host profile of *Haemaphysalis juxtakochi* is shown in Table 3.7, constructed using data provided by Aragão (1935), Cooley (1946), Kohls (1960), Fairchild et al. (1966), Jones et al. (1972), Varma (1973), Yunker et al. (1975), Massi Pallarés and Benítez Usher (1982), Keirans (1985b), Guglielmone et al. (1992a, 2005a), Ivancovich and Luciani (1992), Serra-Freire et al. (1996), Beldoménico et al. (2003b), Venzal et al. (2003c, 2005a), Arzua et al. (2005), Szabó et al. (2006), Martins et al. (2007), Nava et al. (2007, 2011), Labruna et al. (2010a), Mastropaolo et al. (2010), Ogrzewalska et al. (2010), Ortiz et al. (2011), Debárbora et al. (2012), Lamattina et al. (2014, 2018a), Soares et al. (2015), Bermúdez et al. (2015a, 2018a), Martins et al. (2015b), Esser et al. (2016a, b), Domínguez et al. (2019) and Teixeira et al. (2020).

Human Parasitism: Guglielmone and Robbins (2018) listed *Haemaphysalis juxtakochi* as a sporadic parasite of humans in Argentina (Misiones, Salta), Brazil (Amazonas, Paraná, Rondônia, São Paulo), French Guiana (Cayena), Panama (Colón, Panamá), Uruguay (Tacuarembó) and Venezuela (Bolívar), the nymph being the stage most commonly associated with human parasitism, but males and females have also been collected from people. Recently, Reck et al. (2018) collected a nymph from a person in Rio Grande do Sul (Brazil), and Binetruy et al. (2019, 2020a) found four larvae and one nymph of *Haemaphysalis juxtakochi* on people in Cayena (French Guiana).

Comments: Sánchez-Montes et al. (2020) and Thompson et al. (2020) presented molecular evidence indicating that populations of *Haemaphysalis juxtakochi* from North and South America may represent different species, but these authors did not further discuss their results.

Camicas et al. (1998) treat *Haemaphysalis juxtakochi* as a Neotropical species, but there are also a few records of this tick from the Nearctic Region (Guglielmone et al. 2014).

Estrada-Peña et al. (2020) include the nymph and larva of *Haemaphysalis juxtakochi* in a cluster of Neotropical species that feed mostly on artiodactyls, but the host profile of the immature stages of this tick is regarded here as broader, including several orders of mammals and birds. Arana-Guardia et al. (2015) allegedly collected *Haemaphysalis juxtakochi* in Mexico (Yucatán), but the morphological support for their records is questionable, and their data are excluded from this analysis. A record of adults of *Haemaphysalis juxtakochi* from Chiroptera in Venezuela is listed in Jones et al. (1972), but these authors doubted the identity of the hosts, and this parasite association also is not included in the above list. Records of

Table 3.7 Hosts for adult ticks of undetermined sex (A), males (M), females (F), nymphs (N), and larvae (L) of *Haemaphysalis juxtakochi*

MAMMALIA			
ARTIODACTYLA: Bovidae		**DIDELPHIMORPHIA: Didelphidae**	
Cattle	MFNL	*Didelphis aurita*	NL
Goat	MF	*Didelphis marsupialis*	NL
Sheep	N	*Marmosa* sp.	L
ARTIODACTYLA: Camelidae		*Metachirus nudicaudatus*	N
Alpaca	F	**LAGOMORPHA: Leporidae**	
ARTIODACTYLA: Cervidae		*Sylvilagus brasiliensis*	NL
Axis axis	FN	*Sylvilagus floridanus*	F
Mazama americana	MFNL	**PERISSODACTYLA: Equidae**	
Mazama bororo	MFNL	Horse	MFN
Mazama gouazoubira	MFNL	Mule	F
Mazama nana	MF	**PERISSODACTYLA: Tapiridae**	
Mazama temama	M	*Tapirus bairdii*	MFN
Odocoileus virginianus	MFN	*Tapirus terrestris*	MF
Ozotoceros bezoarticus	F	**PILOSA: Myrmecophagidae**	
ARTIODACTYLA: Suidae		*Tamandua mexicana*	N
Domestic pig	F	**PRIMATES: Cebidae**	
Feral pig	MFN	*Cebus olivaceus*	M
ARTIODACTYLA: Tayassuidae		**RODENTIA: Cricetidae**	
Pecari tajacu	MFNL	*Akodon azarae*	L
Tayassu pecari	F	*Oxymycterus rufus*	L
CARNIVORA: Canidae		*Tylomys nudicaudus*	N
Canis latrans	N	**RODENTIA: Dasyproctidae**	
Cerdocyon thous	N	*Dasyprocta azarae*	NL
Chrysocyon brachyurus	N	*Dasyprocta fuliginosa*	NL
Domestic dog	MN	*Dasyprocta punctata*	N
CARNIVORA: Felidae		**RODENTIA: Echimyidae**	
Puma concolor	F	*Hoplomys gymnurus*	NL
CARNIVORA: Procyonidae		*Proechimys semispinosus*	NL
Nasua narica	MNL	**RODENTIA: Eretizhontidae**	
Nasua nasua	NL	*Coendou quichua*	AN
CINGULATA: Chlamyphoridae		**RODENTIA: Sciuridae**	
Tolypeutes matacus	M	*Sciurus aestuans*	N
		Sciurus igniventris	L
AVES			
ACCIPITRIFORMES: Accipitridae		*Willisornis poecilonotus*	L
Harpia harpyja	N	**PASSERIFORMES: Thraupidae**	
Rupornis magnirostris	N	*Thlypopsis pyrrhocoma*	F
GALLIFORMES: Cracidae		**PASSERIFORMES: Turdidae**	
Penelope obscura	L	*Turdus albicollis*	NL
PASSERIFORMES: Corvidae		*Turdus leucomelas*	L
Cyanocorax chrysops	NL	*Turdus rufiventris*	NL
PASSERIFORMES: Passerellidae		**PASSERIFORMES: Tyrannidae**	
Arremon flavirostris	L	*Cnipodectes subbrunneus*	N
PASSERIFORMES: Thamnophilidae			
Hylophylax naevius	NL		

Haemaphysalis juxtakochi from *Rhinella marina* (Anura: Bufonidae) in Trinidad and Tobago by Kamilar et al. (2020) are treated here as diagnostic errors.

3. *Haemaphysalis leporispalustris* Packard, 1869 as *Ixodes leporispalustris* and given its current name in Nuttall and Warburton (1915)

Packard, A.S. (1869) Report of the curator of Articulata. *First Annual Report of the Trustees of the Peabody Academy of Sciences*, pp. 56–69.

Nuttall, G.H.F. & Warburton, C. (1915) *Ticks. A monograph of the Ixodoidea. Part III. The genus* Haemaphysalis. Cambridge University Press, London, pp. 349–550.

Descriptions and Redescriptions: Packard (1869) described but did not figure the female of *Haemaphysalis leporispalustris*, under the name *Ixodes leporispalustris*, and Neumann (1897) described but also did not figure the male, nymph, and larva under the name *Haemaphysalis leporis*, a synonym of *Haemaphysalis leporispalustris*, as discussed in Guglielmone and Nava (2014).

The male, female, nymph, and larva of *Haemaphysalis leporispalustris* were redescribed by Hooker et al. (1912), Nuttall and Warburton (1915), Cooley (1946), Furman and Loomis (1984), Lindquist et al. (2016), and Egizi et al. (2019). The male, female, and nymph were redescribed by Nava et al. (2017), while Sonenshine (1979) redescribed the female, nymph, and larva. Banks (1908), Vogelsang and Santos Dias (1953b) and Bermúdez et al. (2018a) redescribed the male and female. The larva alone of *Haemaphysalis leporispalustris* was redescribed by Clifford et al. (1961) and Coley (2015).

Geographic Distribution: *Haemaphysalis leporispalustris* is a Nearctic and Neotropical species with a broad range in the Neotropics, including Caribbean islands, as follows: **Argentina**: Misiones, Salta, Santa Fe, and Tucumán (Guglielmone and Nava 2005; Lamattina et al. 2018a); **Aruba**: Noord, and Oranjestad (Kohls 1969c); **Bolivia**: Santa Cruz (Fonseca 1959); **Brazil**: Goiás, Minas Gerais, Pernambuco, Rio de Janeiro, and São Paulo (Aragão 1936; Dantas-Torres et al. 2012; Zeringóta et al. 2016); **Colombia**: Bolívar, and Caldas (Kohls 1960; Martínez-Sánchez et al. 2020); **Costa Rica**: Alajuela, Guanacaste, Heredia, and Limón (Álvarez et al. 2005); **Curação**: Ronde Klip, and Willemstad (Kohls 1969c); **Guatemala**: Sacatepequez (Kohls 1960); **southern Mexico**: Guerrero, Jalisco, Michoacán, Morelos, Nayarit, Quintana Roo, Veracruz, and Yucatán (Hoffmann 1962; Woodham et al. 1983; Rodríguez-Vivas et al. 2016; Sánchez-Montes et al. 2020); **Panama**: Chiriquí, Guna Yala, Los Santos, and Panamá (Fairchild et al. 1966); **Paraguay**: Caaguazú, and Cordillera (Nava et al. 2007); **Peru**: Lima, Piura, and Tumbes (Fonseca 1960; Kohls 1960); **Saint Martin** (Kohls 1969c); and **Venezuela**: Aragua, Bolívar, Falcón, and Nueva Esparta (Vogelsang and Santos Dias 1953b; Kohls 1960; Jones et al. 1972).

Table 3.8 Hosts for males (M), females (F), nymphs (N), and larvae (L) of *Haemaphysalis leporispalustris*

MAMMALIA			
DIDELPHIMORPHIA: Didelphidae		**RODENTIA: Cricetidae**	
Metachirus nudicaudatus	N	*Peromyscus* sp.	L
LAGOMORPHA: Leporidae		**RODENTIA: Dasyproctidae**	
Domestic rabbit	MFN	*Dasyprocta azarae*	M
Lepus europaeus	MF	*Dasyprocta punctata*	FL
Sylvilagus brasiliensis	MFNL	**RODENTIA: Echimyidae**	
Sylvilagus floridanus	MFNL	*Proechimys guyannensis*	N
Sylvilagus gabbi	FNL		
AVES			
PASSERIFORMES: Corvidae		**PASSERIFORMES: Troglodytidae**	
Cyanocorax chrysops	L	*Cantorchilus nigricapillus*	L
PASSERIFORMES: Parulidae		*Pheugopedius genibarbis*	L
Myiothlypis bivittata	L	*Troglodytes aedon*	L
PASSERIFORMES: Passerellidae		**PASSERIFORMES: Turdidae**	
Arremon flavirostris	L	*Catharus dryas*	L
Arremon semitorquatus	NL	*Turdus albicollis*	NL
Arremonops conirostris	L	*Turdus rufiventris*	NL
Zonotrichia capensis	NL	**PASSERIFORMES: Tyrannidae**	
PASSERIFORMES: Thamnophilidae		*Corythopis delalandi*	L
Poliocrania exsul	NL	*Empidonax virescens*	L
Thamnophilus caerulescens	L	*Fluvicola nengeta*	L
PASSERIFORMES: Thraupidae		*Myiarchus tyrannulus*	L
Saltator similis	N	**TINAMIFORMES: Tinamidae**	
Trichothraupis melanops	L	*Crypturellus tataupa*	N
Volatinia jacarina	L		

Hosts: Males, females, nymphs, and larvae of *Haemaphysalis leporispalustris* are usually found on Lagomorpha: Leporidae, and *Sylvilagus brasiliensis* (not treated here as restricted to coastal northeastern Brazil) and *Sylvilagus floridanus* are provisionally regarded as principal hosts for this tick. All parasitic stages of *Haemaphysalis leporispalustris* have also been recovered from other mammals, while larvae and nymphs have been found on several species of Passeriformes, with one record of a nymph from Tinamiformes in Flores et al. (2014). The host profile of *Haemaphysalis leporispalustris* is shown in Table 3.8, constructed using data provided by Kohls (1960, 1969c), Tonn et al. (1963), Fairchild et al. (1966), Arnold (1970), Keirans (1985b), Oniki (1999), Beldoménico et al. (2003b), Barros-Battesti and Labruna (2005), Nava et al. (2007), Flores et al. (2014), Zeringóta et al. (2016), Bermúdez et al. (2018a), Martínez-Sánchez et al. (2020) and Ramírez et al. (2020). However, the host profile of this tick is broader than shown in Table 3.8, because other host taxa have been found infested with *Haemaphysalis leporispalustris* in the Nearctic Region, where all parasitic stages have been collected on several orders of mammals, as well as on Galliformes, Passeriformes, and Strigiformes, while immature stages have been found on several orders of Aves (Guglielmone et al. 2014).

Human Parasitism: Guglielmone and Robbins (2018) listed *Haemaphysalis leporispalustris* as a sporadic parasite of humans, but human parasitism in the Neotropical Region has only been detected in the Argentine provinces of Misiones and Santa Fe.

Comments: Neumann (1897) erroneously treated *Haemaphysalis leporispalustris* (under the name *Haemaphysalis leporis*) as a species found beyond the Nearctic and Neotropical Regions and considered *Haemaphysalis chordeilis* (under the name *Ixodes chordeilis*) its synonym. However, Neumann's descriptions are based on specimens collected on the typical host of *Haemaphysalis leporispalustris* in the USA; therefore, the descriptions in Neumann (1897) are treated here as bona fide, and the first descriptions of the male, nymph, and larva of *Haemaphysalis leporispalustris* are considered to be in that publication.

Molecular data provided in Miller et al. (2016) suggest the existence of a species close to *Haemaphysalis leporispalustris* in Panama, while records of this tick in Nigeria (Afrotropical Zoogeographic Region) in Isaac et al. (2019) are treated here as misidentifications.

Camicas et al. (1998) and others regard *Haemaphysalis proxima* Aragão as a synonym of *Haemaphysalis leporispalustris*, but the name *Haemaphysalis proxima* of Aragão in Rohr (1909) is a *nomen dubium*, as explained in Guglielmone and Nava (2014); therefore, the Colombian records of *Haemaphysalis proxima* from Boyacá and Cundinamarca by Osorno Mesa (1942) are not treated as bona fide records of *Haemaphysalis leporispalustris* here, although Kohls (1960) and Acevedo-Gutiérrez et al. (2020) consider these records valid. The latter authors add the presence of this tick in the departments of Antioquia, Córdoba, and Valle del Cauca based on references that do not support their statement. Thus, the Colombian departments where bona fide specimens of *Haemaphysalis leporispalustris* have been collected are considered here to be Bolívar (Kohls, 1960) and Caldas (Martínez-Sánchez et al. 2020).

Sylvilagus brasiliensis is regarded here as one of the principal hosts of *Haemaphysalis leporispalustris*, although some specialists believe that this mammal is restricted to a small area in Brazil, while others regard *Sylvilagus brasiliensis* as having a broad geographical distribution in the Neotropics. The latter view has been adopted here, but workers such as Bermúdez et al. (2018a, b) list only *Sylvilagus gabbi* as a host of *Haemaphysalis leporispalustris* in Panama. This problem is unresolved, and the argument that *Sylvilagus brasiliensis* is the principal Neotropical host of *Haemaphysalis leporispalustris* should be considered provisional. Estrada-Peña et al. (2020) include the larva of *Haemaphysalis leporispalustris* in a cluster of Neotropical species that feed mostly on birds, but the principal hosts for the larvae of this tick are here considered to be species of Lagomorpha: Leporidae. A Brazilian record of *Haemaphysalis leporispalustris* in Freire (1972) from *Mazama americana* in Rio Grande do Sul requires confirmation and is not included in the above list.

Comments on *Haemaphysalis*

The zoogeographic distribution and principal hosts for the limited Neotropical *Haemaphysalis* fauna are summarized in Table 3.6, with one species (*Haemaphysalis cinnabarina*) present only in the Neotropics, while the remaining two species are also found in the Nearctic Region. The total number of species represents an increase of 33% since 2003 (Guglielmone et al. 2003a) because these authors treated *Haemaphysalis cinnabarina* as a synonym of *Haemaphysalis punctata* (Palearctic) rather than as a provisionally valid Neotropical taxon, as recognized here.

There are no records of Amphibia and Reptilia being parasitized by Neotropical *Haemaphysalis*. The two species of this genus whose hosts are known principally parasitize Mammalia, with several records of immature stages parasitizing Aves in the case of *Haemaphysalis juxtakochi* and, to a lesser extent, *Haemaphysalis leporispalustris*.

Haemaphysalis juxtakochi and *Haemaphysalis leporispalustris* have occasionally been found feeding on people in the Neotropical Region.

Additional information on tick geographical distribution and host associations is provided in Chaps. 4 and 5 and in the appendix.

Genus *Rhipicephalus*

There are 87 species in the genus *Rhipicephalus* worldwide; the great majority of them, 72 species (83% of the total), are found in the Afrotropical Region, with 65 species known only from that region. The Palearctic Region accounts for 16 species (18%), seven of which are found nowhere else. Only three species of *Rhipicephalus* are known from the Neotropics, but none are exclusive to this region. However, all three Neotropical *Rhipicephalus* are of economic and medical importance.

1. *Rhipicephalus microplus* (Canestrini, 1888) was originally named as *Haemaphysalis micropla* but transferred to the genus *Rhipicephalus* by Canestrini (1890, p. 493); a few years later, Lahille (1905) placed this tick in the genus *Boophilus*, but it was returned to the genus *Rhipicephalus* by Murrell and Barker (2003)

Canestrini, G. (1888) Intorno da alcuni Acari ed Opilonidi dell'America. *Atti della Società Veneto-Trentina di Scienze Naturali Residente in Padova*, 11, 100–109.

Canestrini, G. (1890) *Prospetto dell'acarofauna italiana. IV. Famiglie: Tetranychini, Ixodini, Argasini.* Edizione Prosperini, Padova, pp. 427–540.

Lahille, F. (1905) Contribution à l'étude des ixodidés de la République Argentine. *Anales del Ministerio de Agricultura, República Argentina*, 2, 166 pp.

Murrell, A. & Barker, S.C. (2003) Synonymy of *Boophilus* Curtice, 1891 with *Rhipicephalus* Koch, 1844 (Acari: Ixodidae). *Systematic Parasitology*, 56,169–172.

Descriptions and Redescriptions: The male and female were described and figured by Canestrini (1888), under the name *Haemaphysalis micropla*; the nymph and larva were described by Lahille (1905) under the name *Boophilus microplus.*

There are many redescriptions of *Rhipicephalus microplus,* and some of the most relevant of them are cited here. The male, female, nymph, and larva of *Rhipicephalus microplus* were redescribed in Cooley (1946), Kohls (1957a), Arthur (1960a), Gothe (1967), Yamaguti et al. (1971), Londt and Arthur (1975), Nuñez et al. (1982) and Teng and Jiang (1991), all under the name *Boophilus microplus*. The male, female, and nymph were redescribed by Nava et al. (2017). The male and female were redescribed by Lahille (1905), Hoogstraal (1956), Boero (1957), Keegan and Toshioka (1957), Floch and Fauran (1958), Yamaguti and Kitaoka (1980), and Matthysse and Colbo (1987), again under the name *Boophilus microplus,* followed by Walker et al. (2003), Bermúdez et al. (2018a), and Horak et al. (2018), who used the name *Rhipicephalus microplus*. Heyne (1986) redescribed the male, while Bedford (1934), Kadarsan (1971), Bohmfalk and Price (1979), and Fujita and Takada (2007) redescribed the larva, all under the name *Boophilus microplus.*

Geographic Distribution: *Rhipicephalus microplus* is a pantropical species found in the Afrotropical, Australasian, Nearctic, Neotropical, Oriental, and Palearctic Zoogeographic Regions; additionally, this tick is present on islands around the world. The range of this species in the Neotropical Region is vast, from southern Mexico to Argentina and Uruguay, including Caribbean islands as follows: **Antigua and Barbuda**: Antigua (Corn et al. 1994); **Argentina**: Buenos Aires, Chaco, Córdoba, Corrientes, Entre Ríos, Formosa, Jujuy, Misiones, Salta, Santa Fe, Santiago del Estero, and Tucumán (Ringuelet 1948; Guglielmone and Mangold 1987), but this is a historical distribution because this tick has been eradicated in the southern part of its range (Guglielmone and Nava 2005); **Bahamas**: New Providence (Clarkson and Owen 1959); **Barbados**: Christ Church, Saint Andrew, Saint George, Saint James, Saint Joseph, Saint Lucy, Saint Michel, Saint Peter, Saint Phillip, and Saint Thomas (Drummond et al. 1981); **Belize**: Cayo, and Orange Walk (Varma 1973; Polsomboon et al. 2017); **Bolivia**: Beni, Cochabamba, Chuquisaca, La Paz, Pando, Santa Cruz, and Tarija (Mastropaolo et al. 2014); **Brazil**: Acre, Alagoas, Amazonas, Bahia, Ceará, Distrito Federal, Espírito Santo, Goiás, Maranhão, Mato Grosso, Mato Grosso do Sul, Minas Gerais, Pará, Paraíba, Paraná, Pernambuco, Piauí, Rio de Janeiro, Rio Grande do Norte, Rio Grande do Sul, Rondônia, Santa Catarina São Paulo, Sergipe, and Tocantins (Aragão 1936; Famadas and Faccini 1989; Labruna et al. 2005a, d; Martins et al. 2009a); **Colombia**: Amazonas, Antioquía, Arauca, Atlántico, Bolívar, Boyacá, Caldas, Caquetá, Casanare, Cauca, César, Córdoba, Cundinamarca, Distrito Capital, Guaviare, Huila, La Guajira, Magdalena, Meta, Nariño, Putumayo, Risaralda, Santander, Sucre, Tolima, and Valle del Cauca (Osorno Mesa 1942; Arias et al. 1991; Betancourt et al. 1992; Guglielmone et al. 2006a; Vecino et al. 2010; Acero et al. 2011; Rivera-Páez et al. 2018; Acevedo-Gutiérrez et al. 2020); **Costa Rica**: Aljuela, Cartago, Guanacaste, Heredia, Limón, Puntarenas, and San José (Álvarez et al. 2005); **Cuba**: Artemisa, Camagüey, Granma, Isla de la Juventud, La Habana, Matanzas,

Mayabeque, Pinar del Río, Santiago de Cuba, and Villa Clara (Minning 1934; Pérez Vigueras 1934; Danielová et al. 1982; Barros-Battesti et al. 2009; Obregón et al. 2020); **Dominica**: Saint George (Keirans 1985b); **Dominican Republic**: Distrito Nacional (Vásquez et al. 2011); **Ecuador**: Galápagos, Manabí, Napo, Pastaza, Pichincha, Santo Domingo de los Tsáchilas, and Sucumbíos (Schulze 1936a; Suárez-González 2004; Pesquera et al. 2015; Rodríguez-Hidalgo et al. 2017; Valle et al. 2020); **El Salvador** (Payne and Scott 1982); **French Guiana**: Cayena, and Saint Laurent du Maroni (Floch and Fauran 1958); **Grenada** (Camus and Barre 1995); **Guadeloupe**: Basse-Terre, Grande-Terre, and Marie Galante (Floch and Fauran 1958; Morel 1966); **Guatemala**: Alta Verapaz, Baja Verapaz, Chiquimula, El Progreso, Escuintla, Huehuetenango, Izabal, Jalapa, Jutiapa, Petén, Quetzaltenango, Quiché, Retalhuleu, San Marcos, Santa Rosa, Sololá, Suchitepéquez, Totonicapán, and Zacapa (Instituto Interamericano de Cooperación para la Agricultura 1988); **Guyana**: Demarara-Mahiaca (Keirans 1985b); **Haiti** (Camus and Barre 1995); **Honduras**: regions Centro Occidental, Centro Oriental, Litoral Atlántico, Nor-Oriental, Norte, Occidental, and Sur (Instituto Interamericano de Cooperación para la Agricultura 1985); **Jamaica**: Saint Andrew, Saint Catherine, Saint Thomas, and Westmoreland (Rawlins and Mansingh 1977, 1978; Rawlins 1979); **Martinique**: Fort de France, La Trinité, Le Marin, and Saint Pierre (Morel 1966); **southern Mexico**: Campeche, Chiapas, Colima, Guerrero, Jalisco, Michoacán, Morelos, Nayarit, Oaxaca, Quintana Roo, Tabasco, Veracruz, and Yucatán (Hoffmann 1962); **Montserrat** (Morel 1966); **Nicaragua**: Boaco, Chinandega, Chontales, Costa Caribe Norte, Costa Caribe Sur, Estelí, Granada, Jinotega, León, Madriz, Managua, Matagalpa, Nueva Segovia, Río San Juan, and Rivas (Woke 1947; Maes et al. 1989; Vallejos-Rayos 2011; Düttmann et al. 2016); **Panama**: Chiriquí, Colón, Darién, Guna Yala, Panamá, and Panamá Oeste (Keirans 1985b; Bermúdez et al. 2012a; Miller et al. 2016; Ferrell et al. 2017; Gittens et al. 2020); **Paraguay**: Boquerón, Central, Concepción, Guairá, Misiones, Paraguarí, Presidente Hayes, and San Pedro (Nava et al. 2007); **Peru**: Amazonas, Cajamarca, Cusco, Huánuco, Junín, La Libertad, Lima, Madre de Dios, Pasco, Piura, and Ucayali (Need et al. 1991; Glenny et al. 2004; Valle-Mendoza et al. 2018); **Puerto Rico**: Aibonito, Arecibo, Barceloneta, Cabo Rojo, Caguas, Camuy, Carolina, Cayey, Ceiba, Cidra, Coamo, Corozak, Culebra, Dorado, Fajardo, Florida, Gurabo, Hatillo, Humacao, Juncos, Lajas, Las Piedras, Luquillo, Manatí, Mayaguez, Moca, Morovis, Naguabo, Quebradillas, Rincón, San Lorenzo, San Sebastián, Toa Alta, Toa Baja, Utuado, Vega Alta, and Vega Baja (Van Volkenberg 1939; Fox and León 1983; Garris and George 1983, 1985; Zimmerman and Garris 1985; León 1986; León et al. 1987; Crom 1992; Cortés et al. 2005); **Saba** (Morel 1966); **Saint Eustatius** (Morel 1966); **Saint Kitts and Nevis**: Nevis, and Saint Kitts (Rawlins and Mansingh 1978; Hadrill and Walker 1994); **Saint Lucia**: Anse La Raye, Castries, Choiseul, Dauphin, Dennery, Gros Islet, Laborie, Micoud, Praslin, Soufirière, and Vieux Fort (Garris and Scotland 1995); **Saint Martin** (Morel 1966); **Saint Vincent and the Grenadines**: Saint Vincent (Saunders 1914); **Suriname**: Paramaribo (Minning 1934); **Trinidad and Tobago**: Arima, Chaguanas, Couva-Tabaquite-Talparo, Penal-Debe, Princes Town, Río Claro-Mayaro, San Juan-Laventille, Sangre Grande,

Siparia, Tobago, and Tunapuna-Piarco (Smith 1974; Rawlins and Mansingh 1978); **Uruguay**: Artigas, Canelones, Cerro Largo, Durazno, Florida, Lavalleja, Minas, Montevideo, Paysandú, Río Negro, Rivera, Rocha, Salto, San José, Soriano, and Tacuarembó (Wolffhügel 1916; Minning 1934; Petraccia et al. 1983; Cardozo 1996), but this is a historical distribution because this tick has been eradicated in the southern and southwestern parts of its range (Venzal et al. 2003c); **USA Virgin Islands**: Saint Croix, and Saint John (Corn et al. 2009; Nelson et al. 2017); and **Venezuela**: Anzoátegui, Apure, Aragua, Barinas, Bolívar, Carabobo, Cojedes, Distrito Federal, Falcón, Guárico, Lara, Mérida, Miranda, Monagas, Nueva Esparta, Portuguesa, Sucre, Táchira, Trujillo, Yaracuy, and Zulia (Rivas 1919; Vergani 1956; Díaz Ungría 1957; Power et al. 1988; Suárez et al. 2016).

Hosts: Males, females, nymphs, and larvae of *Rhipicephalus microplus* are typically parasites of cattle (Artiodactyla: Bovidae), but this species has also been collected from several other mammals and rarely from Aves. The host profile of *Rhipicephalus microplus* is shown in Table 3.9, constructed using information provided by Thompson (1950), Jones et al. (1972), Machado et al. (1985), Barré et al. (1988), Ivancovich and Luciani (1992), Corn et al. (1994, 2009), Ito et al. (1998), L'Hostis et al. (1998), Labruna et al. (2005a, d), Szabó et al. (2007, 2019), Mastropaolo et al. (2010), Martins et al. (2011, 2015b), Debárbora et al. (2012), Ramos et al. (2014, 2020), Lavina et al. (2015), Bastos et al. (2016), Rodríguez-Vivas et al. (2016), Witter et al. (2016), Sousa et al. (2017), Teixeira et al. (2017, 2020), Valle-Mendoza et al. (2018), Zimmermann et al. (2018) and Domínguez et al. (2020c).

Human Parasitism: Guglielmone and Robbins (2018) listed *Rhipicephalus microplus* as a sporadic parasite of humans and recorded several Neotropical cases of human parasitism in Argentina (Corrientes, Entre Ríos, Formosa, Misiones), Bolivia (Beni), Brazil (Paraíba, Rio de Janeiro, Rondônia), Colombia (Antioquía, Casanare), Costa Rica, El Salvador, southern Mexico (Yucatán), Peru (Junín) and Venezuela (Cojedes). Additionally, Szabó et al. (2020) collected two male ticks from people in the Brazilian state of Minas Gerais. All parasitic stages of *Rhipicephalus microplus* have been found feeding on people.

Comments: *Rhipicephalus australis*, previously recognized as *Rhipicephalus microplus*, was reinstated by Estrada-Peña et al. (2012), but several authors (e.g., Roy et al. 2018) found that other species exist under the name *Rhipicephalus microplus* in Asia, although there is no evidence that a similar situation exists in the Neotropical Region.

Clifford et al. (1961) allegedly described the larva of *Rhipicephalus microplus* (under the genus *Boophilus*) using material from Australia; therefore, this description refers to *Rhipicephalus australis* and not to *Rhipicephalus microplus*, which is not present in Australia (Estrada-Peña et al. 2012). Several early records of *Rhipicephalus microplus* were published under the name *Boophilus annulatus*, a synonym of *Rhipicephalus annulatus*, as noted by Fox and León (1983) in Puerto Rico, Guglielmone et al. (2003a) in Brazil, and Düttmann et al. (2016) in Nicaragua,

Table 3.9 Hosts for adult ticks of undetermined sex (A), males (M), females (F), nymphs (N), and larvae (L) of *Rhipicephalus microplus*

MAMMALIA			
ARTIODACTYLA: Bovidae		**CARNIVORA: Felidae**	
Domestic buffalo	MFN	Domestic cat	A
Cattle	MFNL	*Leopardus colocolo*	AN
Goat	MFNL	*Leopardus pardalis*	A
Sheep	MFNL	*Leopardus tigrinus*	A
ARTIODACTYLA: Cervidae		*Panthera onca*	FN
Axis axis	FN	*Puma concolor*	MFNL
Blastocerus dichotomus	MFN	**CARNIVORA: Herpestidae**	
Cervus elaphus	ANL	*Herpestes auropunctatus*	L
Mazama americana	MF	**CARNIVORA: Procyonidae**	
Mazama gouazoubira	MFNL	*Nasua nasua*	F
Odocoileus virginianus	MFNL	**CHIROPTERA: Phyllostomidae**	
Ozotoceros bezoarticus	MFNL	*Sturnira lilium*	L
ARTIODACTYLA: Suidae		**LAGOMORPHA: Leporidae**	
Domestic pig	F	*Lepus europaeus*	NL
Feral pig	F	**PERISSODACTYLA: Equidae**	
ARTIODACTYLA: Tayassuidae		Donkey	F
Catagonus wagneri	MF	Horse	MFNL
Pecari tajacu	MFN	**PERISSODACTYLA: Tapiridae**	
Tayassu pecari	F	*Tapirus terrestris*	MFN
CARNIVORA: Canidae		**PILOSA: Myrmecophagidae**	
Cerdocyon thous	MFNL	*Myrmecophaga tridactyla*	MFNL
Chrysocyon brachyurus	M	**PRIMATES: Atelidae**	
Domestic dog	MFNL	*Alouatta seniculus*	F
Lycalopex gymnocercus	N	**RODENTIA: Caviidae**	
Lycalopex vetulus	MFNL	*Hydrochoerus hydrochaeris*	MFN
AVES			
ACCIPITRIFORMES: Accipitridae		**PELECANIFORMES: Ardeidae**	
Rupornis magnirostris	L	*Bubulcus ibis*	L
FALCONIFORMES: Falconidae			
Milvago chimachima	A		

among others. Aragão (1935) observed that Dios and Knopoff (1930, 1934) named as *Rhipicephalus sanguineus* specimens of *Rhipicephalus microplus* (under the genus *Boophilus*) collected from horses in northern Argentina, while Santos Dias (1958c) found that supposed females of *Rhipicephalus bursa* from Saint Vincent and the Grenadines were in fact *Rhipicephalus microplus* (as *Boophilus microplus*). Historically, most authors have referred to *Rhipicephalus microplus* as *Boophilus microplus*, but some identifications appear in the literature as *Boophilus australis*, *Boophilus annulatus australis*, *Boophilus annulatus microplus*, *Boophilus cyclops*, *Margaropus microplus*, *Margaropus annulatus australis,* or *Margaropus annulatus microplus*.

Rhipicephalus microplus has an unusually broad distribution in the Neotropics, as indicated above; nevertheless, collection data are lacking for several countries. Some workers have defined the range of this tick simply as "widespread" or in similar vague terms, and such descriptors are not included here.

Keirans (1985b) listed eight unusual records of *Rhipicephalus microplus* (under the genus *Boophilus*) collected in Guyana and deposited in the Nuttall Tick Collection. Seven of these records are of specimens collected by "Dr. K.S. Wise" in 1912, and the remaining record is of specimens gathered by "G.E. Bodkin" in 1913, for a total of 37 male and 22 female ticks found on "chameleon" (a host not found naturally in the Americas, but the term probably refers to lizards of the genus *Anolis*), "marmoset" (Primates: Callitrichidae?), "hackia" (a tree?), "water hass" (?), "grison" (Mustelidae), "raccoon" (Procyonidae), "kibihee" (Procyonidae?) and *Bufo marinus* (= *Rhinella marina*, Anura: Bufonidae). These records likely reflect errors in host identification or misunderstandings in use of the terms applied to them, and such alleged hosts of *Rhipicephalus microplus* have been excluded from our list. Jones et al. (1972) identify as *Rhipicephalus microplus* (under the name *Boophilus microplus*) males, females and nymphs supposedly collected from *Carollia perspicillata* (Chiroptera: Phyllostomidae) but express doubt about the identity of the host, which has also been excluded from the list in Table 3.9. Records of *Rhipicephalus microplus* from *Rhinella marina* (Anura: Bufonidae) in Trinidad and Tobago by Kamilar et al. (2020) are treated here as diagnostic errors.

2. *Rhipicephalus sanguineus* (Latreille, 1806) *sensu stricto* under the name *Ixodes sanguineus* and given its current status by Koch (1844)

Latreille, P.A. (1806) *Genera crustaceorum et insectorum secundum ordinem naturalem in familia disposita, iconibus exemplisque plurimis explicata*, volume 1, Parisiis et Argentorati, 302 pp.

Koch, C.L. (1844) Systematische Übersicht über die Ordnung der Zecken. *Archiv für Naturgeschichte*, 10, 217–239.

Descriptions and Redescriptions: Latreille (1806, under the name *Ixodes sanguineus*) published a brief, uninformative description of a tick whose sex cannot be determined using the data provided by the author and its type specimen is unavailable. Nevertheless, Nava et al. (2015) maintain the validity of this species under the name *Rhipicephalus sanguineus*, because of its wide usage in scientific literature referring to a tick of economic and public health importance. The same authors listed numerous morphological descriptions of *Rhipicephalus sanguineus* from 1847 to 2013 and genetic studies of populations all over the world, finding evidence that several species were included under this name. Nava et al. (2015) therefore proposed that a neotype be selected together with descriptions of all parasitic stages of the neotype population and supplemented with molecular characterization of this population. Thereafter, Nava et al. (2018) selected a male neotype from a population of *Rhipicephalus sanguineus* at a kennel in Montpellier, France, described the male, female, nymph, and larva, and provided the molecular profile using sequences of nuclear and mitochondrial genes. The descriptions presented in

Nava et al. (2018) are considered the only bona fide accounts of *Rhipicephalus sanguineus sensu stricto*.

Geographic Distribution: *Rhipicephalus sanguineus sensu stricto* has been found in the Nearctic, Neotropical, and Palearctic Regions. The Neotropical geographic distribution encompasses **Argentina**, **Brazil**, **Chile,** and **Uruguay**.

Hosts: Although Nava et al. (2018) did not analyze the host profile of *Rhipicephalus sanguineus sensu stricto*, it is generally recognized that males, females, nymphs, and larvae of this species preferentially parasitize dogs.

Human Parasitism: *Rhipicephalus sanguineus sensu stricto* is not listed as a human parasite in Guglielmone and Robbins (2018), who provide a long list of cases of human parasitism worldwide due to *Rhipicephalus sanguineus sensu lato*. Most probably, some of those cases can be attributed to *Rhipicephalus sanguineus sensu stricto*, but we currently lack sufficient information to make such a judgment.

Comments: The presence of *Rhipicephalus sanguineus sensu stricto* in the Neotropical Region has been confirmed. Nevertheless, additional efforts are needed to precisely define its range and to characterize related cryptic species present in the region but currently hidden as *Rhipicephalus sanguineus sensu lato* or *Rhipicephalus sanguineus* "tropical" and "temperate" lineages, despite morphological, biological, and molecular differences (Oliveira et al. 2005; Sanches et al. 2016).

Rhipicephalus sanguineus sensu stricto has been found in the Nearctic, Neotropical and Palearctic Zoogeographic Regions (Nava et al. 2018), but it is expected that this species will soon also be found in the Afrotropical, Australasian and Oriental Regions. Coimbra-Dores et al. (2020) stated that *Rhipicephalus sanguineus sensu stricto* had been collected from dogs in Angola (Afrotropical Zoogeographic Region) and on a variety of domestic and wild mammals and birds in Portugal, as well as from dogs and cats in Spain (Palearctic Region). However, a molecular analysis of COI and 16S gene sequences of *Rhipicephalus sanguineus sensu stricto* in Coimbra-Dores et al. (2020) does not include a comparison with the corresponding sequences of the French population used to describe *Rhipicephalus sanguineus sensu stricto*, from which a neotype of this tick was selected by Nava et al. (2018).

3. ***Rhipicephalus sanguineus*** *sensu lato*. A name applied to ticks that are morphologically close to *Rhipicephalus sanguineus sensu stricto* but cannot be included under that name as defined in Nava et al. (2018)

Nava, S., Beati, L., Venzal, J.M., Labruna, M.B., Szabó, M.P.J., Petney, T., Saracho-Bottero, M.N., Tarragona, E.L., Santos Silva, M.M., Mangold, A.J., Guglielmone, A.A. & Estrada-Peña, A. (2018) *Rhipicephalus sanguineus* (Latreille, 1806): neotype designation, morphological re-description of all parasitic stages and molecular characterization. *Ticks and Tick-borne Diseases*, 9, 1573–1585.

Descriptions and Redescriptions: Latreille (1806, under the name *Ixodes sanguineus*) published a brief, uninformative description of a tick whose sex cannot be determined by using the data provided by the author, and its type specimen is unavailable. Nevertheless, Nava et al. (2015) maintain the validity of this species under the name *Rhipicephalus sanguineus* because of its wide usage in scientific literature referring to a tick of economic and public health importance. The same authors listed numerous morphological descriptions of *Rhipicephalus sanguineus* from 1847 to 2013 and genetic studies of populations all over the world, finding discrepancies rather than congruence among studies. These populations of *Rhipicephalus sanguineus* are here recognized as *Rhipicephalus sanguineus sensu lato*, which is more or less close (data not shown) to *Rhipicephalus sanguineus sensu stricto* as defined by Nava et al. (2018) and discussed under the preceding species.

Geographic Distribution: *Rhipicephalus sanguineus sensu lato* has a worldwide distribution, including a vast range in the Neotropics as follows: **Antigua and Barbuda**: Antigua, and Barbuda Island (Keirans 1985b); **Argentina**: Buenos Aires, Ciudad Autónoma de Buenos Aires, Chaco, Chubut, Córdoba, Corrientes, Formosa, Jujuy, Mendoza, Misiones, Salta, San Juan, San Luis, Santa Fe, and Tucumán (Mallo 1990; Guglielmone and Nava 2005; Nava et al. 2012); **Aruba** (Bool and Sutmöller 1957); **Bahamas** (Clarkson and Owen 1959); **Barbados** (Keirans 1985b); **Belize**: Cayo, Corozal, and Orange Walk (Polsomboon et al. 2017); **Bolivia**: Santa Cruz (Mastropaolo et al. 2014); **Bonaire** (Kohls 1969c); **Brazil**: Acre, Amapá, Amazonas, Bahia, Ceará, Distrito Federal, Espírito Santo, Goiás, Maranhão, Mato Grosso, Mato Grosso do Sul, Minas Gerais, Pará, Paraíba, Paraná, Pernambuco, Piauí, Rio de Janeiro, Rio Grande do Norte, Rio Grande do Sul, Rondônia, Roraima, Santa Catarina, São Paulo, Sergipe, and Tocantins (Aragão 1936; Ribeiro et al. 1996; Yoshizawa et al. 1996; Martins et al. 2009a; Moraes-Filho et al. 2011; Fernandes et al. 2018; Gianizella et al. 2018b); **Chile**: Antofagasta, Araucanía, Arica y Parinacota, Atacama, Concepción, Coquimbo, Maule, Metropolitana, Ñuble, Libertador Bernardo O'Higgins, Tarapacá, and Valparaíso (Salas and González-Acuña 2015); **Colombia**: Amazonas, Antioquía, Arauca, Atlántico, Bolívar, Caldas, Casanare, Córdoba, Cundinamarca, Distrito Capital, Magdalena, Meta, Nariño, Quindío, Risaralda, Sucre, Tolima, and Valle del Cauca (Osorno Mesa 1942; Wells et al. 1981; Torres-Mejía and de la Fuente 2006; Paternina et al. 2009; Miranda and Mattar 2015; Rivera-Páez et al. 2018; Acevedo-Gutiérrez et al. 2019; Cotes-Perdomo et al. 2020); **Costa Rica**: Alajuela, Cartago, Guanacaste, Heredia, Limón, Puntarenas, and San José (Barrantes-González et al. 2016; Troyo et al. 2016; Springer et al. 2018); **Cuba**: Artemisa, Camagüey, Isla de la Juventud, La Habana, Matanzas, Mayabeque, Pinar del Río, and Villa Clara (Pérez Vigueras 1934; Barros-Battesti et al. 2009; Noda et al. 2015; Silva et al. 2016a, b); **Curaçâo** (Oudemans 1906); **Dominica**: Saint George (Keirans 1985b); **Dominican Republic**: Distrito Nacional (Morel 1966); **Ecuador**: Galápagos, and Guayas (Landívar et al. 1985; Díaz et al. 2016); **El Salvador** (Navarrete Abarca et al. 2014); **French Guiana**: Cayena (Floch and Fauran 1958); **Grenada** (Yabsley et al. 2008); **Guadeloupe**: Basse-Terre, Grande-Terre, and Marie

Galante (Morel 1966); **Guatemala**: Alta Verapaz, Escuintla, Huehuetenango, Izabal, Jutiapa, Petén, Suchitepéquez, and Zacapa (Instituto Interamericano de Cooperación para la Agricultura 1988); **Guyana**: Demerara-Mahaica (Keirans 1985b); **Haiti** (Morel 1966); **Honduras** (Instituto Interamericano para la Cooperación en Agricultura 1984); **Jamaica**: Clarendon, Kingston, Portland, and Westmoreland (Thompson 1950; Wilson 1980); **Martinique**: Fort de France, La Trinité, Le Marin, and Saint Pierre (Morel 1966; L'Hostis et al. 1998); **southern Mexico**: Campeche, Chiapas, Colima, Guerrero, Jalisco, Michoacán, Morelos, Nayarit, Oaxaca, Quintana Roo, Tabasco, Veracruz, and Yucatán (Silva-Goytia and Elizondo 1952; Hoffmann 1962; Woodham et al. 1983; Morales-Soto and Cruz-Vázquez 1998; Light et al. 2019); **Montserrat** (Keirans 1985b); **Nicaragua**: Chinandega, Chontales, Jinotega, León, Managua, Masaya, Matagalpa, and Rivas (Woke 1947; De La Cruz 1995; Vallejos-Rayos 2011; Springer et al. 2018); **Panama**: Bocas del Toro, Chiriquí, Coclé, Colón, Darién, Guna Yala, Herrera, Panamá, Panamá Oeste, Los Santos, and Veraguas (Bermúdez and Miranda 2011b); **Paraguay**: Central, Ñeembucú, and Paraguarí (Nava et al. 2007); **Peru**: Amazonas, Cajamarca. Cusco, Huánuco, Lamabayeque, Lima, Loreto, Puna, and Puno (Escalante 1981; Need et al. 1991; Glenny et al. 2004; Billeter et al. 2011; Huerto-Medina and Dámaso-Mata 2015; Kocher et al. 2016; Cervantes et al. 2020); **Puerto Rico**: Mayaguez, and San Juan (Van Volkenberg 1939; Morel 1966); **Saint Kitts and Nevis**: Saint Kitts (Saunders 1914); **Saint Lucia** (Keirans 1985b); **Saint Martin** (Kohls 1969c); **Saint Vincent and the Grenadines** (Saunders 1914); **Suriname**: Brokopondo (Santos Dias 1989a); **Trinidad and Tobago**: Arima, Chaguanas, Couva-Tabaquite-Talparo, Diego Martin, Port of Spain, Río Claro-Mayaro, San Fernando, San Juan-Laventille, Sangre Grande, Siparia, Tobago, and Tunapura-Piarco (Sameroff et al. 2019); **Uruguay**: Artigas, Canelones, Maldonado, Montevideo, and Salto (Venzal et al. 2007a; Nava et al. 2012; Carvalho et al. 2017); **USA Virgin Islands**: Saint Croix (Beatty 1944); and **Venezuela**: Anzoátegui, Apure, Aragua, Bolívar, Cojedes, Distrto Federal, Falcón, Guárico, Lara, Miranda, Mérida, Portuguesa, Sucre, Táchira, Trujillo, and Zulia (Vogelsang and Cordero 1940; Vergani 1956; Guerrero 1996; Román et al. 2002; Ramírez-Barrios et al. 2008; Clavijo et al. 2009; Moraes-Filho et al. 2011).

Hosts: Males, females, nymphs, and larvae of *Rhipicephalus sanguineus sensu lato* are usually found on dogs, but collections have also been made from other mammals, as well as a few birds, and there are two records from Reptilia. The host profile of *Rhipicephalus sanguineus sensu lato* is shown in Table 3.10, which is based on data from Tate (1941), Fox (1950), Fairchild et al. (1966), Keirans (1985b), Guglielmone et al. (1989), Ivancovich and Luciani (1992), Diogo et al. (2003), González-Acuña et al. (2003a), Zerpa et al. (2003b), Labruna et al. (2005a, d), Szabó et al. (2008, 2019), Clavijo et al. (2009), Bermúdez et al. (2010a, 2015a, 2018a), Dantas-Torres et al. (2010a, b), Figueiredo et al. (2010), Luz et al. (2012), Witter et al. (2016), González et al. (2017), Martins et al. (2017), Rocha et al. (2017), Gianizella et al. (2018b) Rivera-Páez et al. (2018), Silva et al. (2018b), Martins et al. (2020b), Fonseca et al. (2020), Mora-Rivera et al. (2020) and Teixeira et al. (2020). The

Table 3.10 Hosts for adult ticks of undetermined sex (A), males (M), females (F), nymphs (N), and larvae (L) of *Rhipicephalus sanguineus sensu lato*

MAMMALIA			
ARTIODACTYLA: Bovidae		**LAGOMORPHA: Leporidae**	
Cattle	MFN	Domestic rabbit	MFNL
Goat	MFNL	**PERISSODACTYLA: Equidae**	
ARTIODACTYLA: Cervidae		Donkey	A
"Deer"	MF	Horse	MF
ARTIODACTYLA: Suidae		**PERISSODACTYLA: Tapiridae**	
Domestic pig	M	*Tapirus terrestris*	F
CARNIVORA: Canidae		**PILOSA: Bradypodidae**	
Canis latrans	A	*Bradypus tridactylus*	F
Cerdocyon thous	MFN	**Myrmecophagidae**	
Chrysocyon brachyurus	A	*Tamandua mexicana*	F
Domestic dog	MFNL	*Tamandua tetradactyla*	MF
Lycalopex vetulus	FL	**PRIMATES: Callitrichidae**	
Urocyon cinereoargenteus	M	*Mico melanurus*	M
CARNIVORA: Felidae		*Saguinus bicolor*	F
Domestic cat	MF	*Saguinus geoffroyi*	L
Herpailurus yagouaroundi	F	**RODENTIA: Caviidae**	
Leopardus wiedii	A	*Cavia aperea*	F
Panthera onca	F	*Dolichotis patagonum*	M
Puma concolor	M	*Hydrochoerus hydrochaeris*	FN
CARNIVORA: Mustelidae		**RODENTIA: Muridae**	
Galictis cuja	A	*Rattus norvegicus*	MNL
CARNIVORA: Procyonidae		*Rattus rattus*	ANL
Nasua nasua	F		
AVES			
ACCIPITRIFORMES: Accipitridae		**PASSERIFORMES: Thraupidae**	
Geranoaetus albicaudatus	N	*Coereba flaveola*	M
COLUMBIFORMES: Columbidae		**STRIGIFORMES: Strigidae**	
Columba livia	F	*Asio clamator*	A
PASSERIFORMES: Icteridae			
Gnorimopsar chopi	N		
REPTILIA			
CROCODILIA: Crocodylidae		**SQUAMATA: Boidae**	
Caiman crocodilus	F	*Boa constrictor*	F

host range of *Rhipicephalus sanguineus sensu lato* is broader than that in the Neotropical Region because this tick has been collected on a variety of hosts worldwide (data not shown).

Human Parasitism: Guglielmone and Robbins (2018) listed *Rhipicephalus sanguineus sensu lato* as a very frequent parasite of humans, with adult ticks more often biting people than nymphs and larvae. Cases of human parasitism in the Neotropics have been recorded in Argentina (Corrientes, Misiones, and Salta), Brazil (Ceará, Goiás, Mato Grosso do Sul, Pernambuco, Recife, Rio de Janeiro, Rio Grande do Sul, and São Paulo), Chile (Metropolitana, and Valparaíso),

Colombia (Antioquía), Curação, French Guiana, Guyana, southern Mexico (Chiapas, Guerrero, and Yucatán), Panama (Colón, and Panamá), Peru, Uruguay (Canelones, Pando, and Montevideo), and Venezuela (Guglielmone and Robbins 2018). Costa et al. (2020) added a nymph of *Rhipicephalus sanguineus sensu lato* collected from a human in the Brazilian State of Maranhão, and Valente et al. (2020) collected a male of this tick from a person in the state of Paraná (Brazil).

Comments: All Neotropical records of *Rhipicephalus sanguineus*—with the exception of those records classified as *Rhipicephalus sanguineus sensu stricto*—are here designated as *Rhipicephalus sanguineus sensu lato*, although most authors cited above did not make this distinction and simply used the name *Rhipicephalus sanguineus*.

The Neotropical distribution of *Rhipicephalus sanguineus sensu lato* is broad, and additional species are surely included under this vague name. Some authors define populations of *Rhipicephalus sanguineus* from warmer regions as "tropical lineage," in contrast to those constituting a "temperate lineage." Populations of *Rhipicephalus sanguineus* from the "tropical lineage" are considered to differ from *Rhipicephalus sanguineus sensu stricto*, while *Rhipicephalus sanguineus* "temperate lineage" is thought to comprise populations of *Rhipicephalus sanguineus sensu stricto*, and perhaps species close to it. Morphological, biological, and molecular differences between both lineages have been found by Oliveira et al. (2005) and Sanches et al. (2016), among others. Nevertheless, most Neotropical populations remain undefined, and a major effort should be made to define the species included under *Rhipicephalus sanguineus sensu lato* and their ranges. Additionally, the name *Rhipicephalus sanguineus* was wrongly applied by Dios and Knopoff (1930, 1934) to *Rhipicephalus microplus* (under the name *Boophilus microplus*), as noted by Aragão (1935).

As discussed above under *Rhipicephalus microplus*, Keirans (1985b), working with the Nuttall Tick Collection, listed ten unusual records of *Rhipicephalus sanguineus sensu lato* from Guyana. All these records are from specimens collected by "Dr. K.S. Wise" in 1911 and 1912, for a total of 48 male and 37 female specimens of *Rhipicephalus sanguineus sensu lato*. There are two records from "turtles" (10 males and 11 females), one record of two males, one female and six nymphs collected from "salepenter" (probably meaning "salipenter," a vulgar name for lizards of the genus *Tupinambis*), a collection from a lizard (six males and three females), another from "alligator" (nine male and two female ticks), one record from "mongoose" (Herpestidae?, seven males and two females), another record from "labba" (*Cuniculus paca*?, two males and seven females), plus one record each from "crab dog," "hog" and "ocelot." These records likely reflect errors in host identification or misunderstandings in the use of the terms applied to them, and such alleged hosts have been excluded from the above list. Recent records of *Rhipicephalus sanguineus* from *Rhinella marina* (Anura: Bufonidae) in Trinidad and Tobago by Kamilar et al. (2020) are treated here as diagnostic errors.

Comments on *Rhipicephalus*

The zoogeographic distribution and principal hosts of Neotropical *Rhipicephalus* are summarized in Table 3.6, where three species are listed; however, the number of taxa actually present in this region remains unknown. Two taxa with worldwide ranges are recognized in the Neotropics. *Rhipicephalus microplus*, a specific parasite of cattle, and *Rhipicephalus sanguineus sensu stricto*, a parasite of dogs, were both introduced with imported animals. By contrast, the number of species that may exist under the name *Rhipicephalus sanguineus sensu lato* is unknown, although this tick is also recognized as being chiefly a domestic dog parasite with origins outside the Neotropics. Our current understanding of Neotropical *Rhipicephalus* is quite different from that described in Guglielmone et al. (2003a), when *Rhipicephalus microplus* was treated as *Boophilus microplus*, and the taxonomic problems related to the *Rhipicephalus sanguineus* complex were suspected but largely unaddressed. This situation changed dramatically after the redescription of *Rhipicephalus sanguineus sensu stricto* by Nava et al. (2018).

No Amphibia have been recorded as having been parasitized by *Rhipicephalus* in the Neotropics, while records from Aves (*Rhipicephalus microplus* and *Rhipicephalus sanguineus sensu lato*) and Reptilia (*Rhipicephalus sanguineus sensu lato*) should be considered exceptional, because worldwide the great majority of *Rhipicephalus* hosts are mammals (Guglielmone et al. 2014).

On several occasions, *Rhipicephalus microplus* and *Rhipicephalus sanguineus sensu lato* have been reported feeding on humans in the Neotropics.

Additional information on tick geographical distribution and host associations is provided in Chaps. 4 and 5 and in the appendix.

Chapter 4
Geographic Distribution, Invasive Species, and Incorrect Names Applied to Neotropical Ixodids

4.1 Geographic Distribution

Here we examine the numbers of ixodid ticks found in the Neotropical Region, which comprises both a vast continental area and numerous islands with their own unique faunas. Where relevant, the evolutionary origins of these faunas are discussed, a topic addressed in depth in Chap. 5.

4.1.1 Species of Ixodidae Found in Neotropical Countries and Territories

The species of ixodid ticks known from each Neotropical country are listed in Table 4.1, where it is apparent that Caribbean countries possess consistently depauperate faunas when compared with countries on the Neotropical mainland. The exception is Trinidad and Tobago, a Caribbean nation off the coast of northeastern South America, where 15 species of Ixodidae (nine *Amblyomma*, two *Ixodes*, two *Rhipicephalus*, and one species each of *Dermacentor* and *Haemaphysalis*) are known to occur. All these species are also found in South America, and it is speculated that most, if not all, taxa recorded in Trinidad and Tobago reached those islands as a consequence of invasion from the South American mainland. Cuba possesses the second largest number of ixodid tick taxa among Caribbean countries with nine species (five *Amblyomma*, two *Rhipicephalus*, one *Dermacentor*, and one endemic *Ixodes*), while eight species of ixodids have been recorded in Jamaica (four *Amblyomma*, two *Rhipicephalus*, one *Dermacentor*, and one endemic *Ixodes*), and Puerto Rico (five *Amblyomma*, two *Rhipicephalus*, and one *Dermacentor*, none endemic). The smaller Caribbean islands of Anguilla, Bermuda, the Cayman Islands, and Turks and Caicos Islands each have reported a single

A. A. Guglielmone et al., *Neotropical Hard Ticks (Acari: Ixodida: Ixodidae)*,
https://doi.org/10.1007/978-3-030-72353-8_4

Table 4.1 Geographic distributions of Neotropical Ixodidae, arranged by country/territory

Country/territory	Tick species
Anguilla (1)	*Amblyomma variegatum*
Antigua and Barbuda (5)	*Amblyomma dissimile*, *A. variegatum* (2)
	Dermacentor nitens
	Rhipicephalus microplus, *R. sanguineus s.l.* (2)
Argentina (43)	*Ixodes affinis*, *I. auritulus*, *I. longiscutatus*, *I. loricatus*, *I. luciae*, *I. neuquenensis*, *I. nuttalli*, *I. pararicinus*, *I. schulzei*, *I. sigelos*, *I. silvanus*, *I. uriae* (12)
	Amblyomma argentinae, *A. aureolatum*, *A. auricularium*, *A. boeroi*, *A. brasiliense*, *A. calcaratum*, *A. coelebs*, *A. dissimile*, *A. dubitatum*, *A. hadanii*, *A. incisum*, *A. longirostre*, *A. neumanni*, *A. nodosum*, *A. ovale*, *A. parvitarsum*, *A. parvum*, *A. pseudoconcolor*, *A. pseudoparvum*, *A. rotundatum*, *A. sculptum*, *A. tigrinum*, *A. tonelliae*, *A. triste*, *A. yucumense* (25)
	Dermacentor nitens
	Haemaphysalis juxtakochi, *H. leporispalustris* (2)
	Rhipicephalus microplus, *R. sanguineus s.s.*, *R. sanguineus s.l.* (3)
Aruba (2)	*Haemaphysalis leporispalustris*
	Rhipicephalus sanguineus s.l.
Bahamas (6)	*Amblyomma albopictum*, *A. dissimile*, *A. torrei* (3)
	Dermacentor nitens
	Rhipicephalus microplus, *R. sanguineus s.l.* (2)
Barbados (5)	*Amblyomma dissimile*, *A. variegatum* (2)
	Dermacentor nitens
	Rhipicephalus microplus, *R. sanguineus s.l.* (2)
Belize (22)	*Ixodes affinis*, *I. boliviensis*, *I. luciae* (3)
	Amblyomma auricularium, *A. calcaratum*, *A. coelebs*, *A. dissimile*, *A. longirostre*, *A. maculatum*, *A. mixtum*, *A. nodosum*, *A. oblongoguttatum*, *A. ovale*, *A. pacae*, *A. pecarium*, *A. sabanerae*, *A. tapirellum* (14)
	Dermacentor dispar, *D. nitens* (2)
	Haemaphysalis juxtakochi
	Rhipicephalus microplus, *R. sanguineus s.l.* (2)
Bermuda (1)	*Amblyomma mixtum*
Bolivia (30)	*Ixodes boliviensis*, *I. cooleyi*, *I. luciae* (3)

(continued)

Table 4.1 (continued)

	Amblyomma auricularium, A. brasiliense, A. calcaratum, A. coelebs, A. dubitatum, A. humerale, A. incisum, A. longirostre, A. naponense, A. nodosum, A. oblongoguttatum, A. ovale, A. parvitarsum, A. parvum, A. pecarium, A. pseudoconcolor, A. rotundatum, A. scalpturatum, A. sculptum, A. tigrinum, A. tonelliae, A. triste (22)
	Dermacentor nitens
	Haemaphysalis juxtakochi, H. leporispalustris (2)
	Rhipicephalus microplus, R. sanguineus s.l. (2)
Bonaire (1)	*Rhipicephalus sanguineus s.l.*
Brazil (52)	*Ixodes amarali, I. auritulus, I. bocatorensis, I. catarinensis, I. fuscipes, I. lasallei, I. longiscutatus, I. loricatus, I. luciae, I. paranaensis, I. schulzei, I. spinosus* (12)
	Amblyomma aureolatum, A. auricularium, A. brasiliense, A. cajennense, A. calcaratum, A. coelebs, A. dissimile, A. dubitatum, A. fuscum, A. geayi, A. goeldii, A. humerale, A. incisum, A. latepunctatum, A. longirostre, A. naponense, A. nodosum, A. oblongoguttatum, A. ovale, A. pacae, A.parkeri, A. parvum, A. pictum, A. pseudoconcolor, A. romarioi, A. romitii, A. rotundatum, A. scalpturatum, A. sculptum, A. tigrinum, A. triste, A. varium, A. yucumense (33)
	Dermacentor nitens
	Haemaphysalis cinnabarina, H. juxtakochi, H. leporispalustris (3)
	Rhipicephalus microplus, R. sanguineus s.s., R. sanguineus s.l. (3)
Cayman Islands (1)	*Amblyomma torrei*
Chile (14)	*Ixodes abrocomae, I. auritulus, I. chilensis, I. cornuae, I. neuquenensis, I. sigelos, I. stilesi, I. taglei, I. uriae* (9)
	Amblyomma parvitarsum, A. tigrinum, A. triste (3)
	Rhipicephalus sanguineus s.s., R. sanguineus s.l. (2)
Colombia (42)	*Ixodes affinis, I. auritulus, I. bocatorensis, I. boliviensis, I. lasallei, I. luciae, I. montoyanus, I. tapirus, I. tropicalis, I. venezuelensis* (10)
	Amblyomma auricularium, A. calcaratum, A. coelebs, A. crassum, A. dissimile, A. geayi, A. humerale, A. longirostre, A. maculatum, A. mixtum, A. multipunctum, A. naponense, A. neumanni, A. nodosum, A. oblongoguttatum, A. ovale, A. pacae, A. parvum, A. patinoi, A. pecarium, A. rotundatum, A. sabanerae, A. scalpturatum, A. tapirellum, A. triste, A. varium (26)
	Dermacentor imitans, D. nitens (2)
	Haemaphysalis juxtakochi, H. leporispalustris (2)

(continued)

Table 4.1 (continued)

	Rhipicephalus microplus, *R. sanguineus s.l.* (2)
Costa Rica (40)	*Ixodes affinis*, *I. auritulus*, *I. boliviensis*, *I. dicei*, *I. luciae*, *I. minor*, *I. sinaloa*, *I. tapirus*, *I. venezuelensis* (9)
	Amblyomma albopictum, *A. auricularium*, *A. calcaratum*, *A. coelebs*, *A. dissimile*, *A. geayi*, *A. inornatum*, *A. longirostre*, *A. maculatum*, *A. mixtum*, *A. naponense*, *A. nodosum*, *A. oblongoguttatum*, *A. ovale*, *A. pacae*, *A. parvum*, *A. pecarium*, *A. rotundatum*, *A. sabanerae*, *A. scutatum*, *A. tapirellum*, *A. tenellum*, *A. varium* (23)
	Dermacentor imitans, *D. latus*, *D. nitens*, *D. panamensis* (4)
	Haemaphysalis juxtakochi, *H. leporispalustris* (2)
	Rhipicephalus microplus, *R. sanguineus s.l.* (2)
Cuba (9)	*Ixodes capromydis*
	Amblyomma albopictum, *A. dissimile*, *A. mixtum*, *A. quadricavum*, *A. torrei* (5)
	Dermacentor nitens
	Rhipicephalus microplus, *R. sanguineus s.l.* (2)
Curaçao (3)	*Amblyomma dissimile*
	Haemaphysalis leporispalustris
	R. sanguineus s.l.
Dominica (7)	*Amblyomma albopictum*, *A. antillorum*, *A. rotundatum*, *A. variegatum* (4)
	Dermacentor nitens
	Rhipicephalus microplus, *R. sanguineus s.l.* (2)
Dominican Republic (5)	*Amblyomma albopictum*, *A. dissimile* (2)
	Dermacentor nitens
	Rhipicephalus microplus, *R. sanguineus s.l.* (2)
Ecuador (32)	*Ixodes affinis*, *I. auritulus*, *I. boliviensis*, *I. cornuae*, *I. fossulatus*, *I. galapagoensis*, *I. luciae*, *I. pomerantzi* (8)
	Amblyomma boulengeri, *A. calcaratum*, *A. darwini*, *A. dissimile*, *A. hirtum*, *A. humerale*, *A. latepunctatum*, *A. longirostre*, *A. macfarlandi*, *A. maculatum*, *A. mixtum*, *A. multipunctum*, *A. naponense*, *A. ovale*, *A. pilosum*, *A. scalpturatum*, *A. triste*, *A. usingeri*, *A. varium*, *A. williamsi* (20)

(continued)

Table 4.1 (continued)

	Dermacentor nitens
	Haemaphysalis juxtakochi
	Rhipicephalus microplus, *R. sanguineus s.l.* (2)
El Salvador (11)	*Amblyomma auricularium*, *A. dissimile*, *A. mixtum*, *A. ovale*, *A. parvum*, *A. sabanerae*, *A. scutatum* (7)
	Dermacentor dissimilis, *D. nitens* (2)
	Rhipicephalus microplus, *R. sanguineus s.l.* (2)
French Guiana (27)	*Ixodes luciae*
	Amblyomma aureolatum, *A. auricularium*, *A. cajennense*, *A. calcaratum*, *A. coelebs*, *A. dissimile*, *A. geayi*, *A. goeldii*, *A. humerale*, *A. latepunctatum*, *A. longirostre*, *A. naponense*, *A. oblongoguttatum*, *A. ovale*, *A. pacae*, *A. pictum*, *A. pseudoconcolor*, *A. romitii*, *A. rotundatum*, *A. scalpturatum*, *A. tigrinum*, *A. varium* (22)
	Dermacentor nitens
	Haemaphysalis juxtakochi
	Rhipicephalus microplus, *R. sanguineus s.l.* (2)
Grenada (5)	*Amblyomma dissimile*, *A. rotundatum* (2)
	Dermacentor nitens
	Rhipicephalus microplus. *Rhipicephalus sanguineus s.l.* (2)
Guadeloupe (6)	*Amblyomma dissimile*, *A. rotundatum*, *A. variegatum* (3)
	Dermacentor nitens
	Rhipicephalus microplus, *R. sanguineus s.l.* (2)
Guatemala (33)	*Ixodes affinis*, *I. auritulus*, *I. boliviensis*, *I. guatemalensis*, *I. luciae*, *I. minor*, *I. pomerantzi*, *I. rubidus*, *I. tecpanensis* (9)
	Amblyomma coelebs, *A. dissimile*, *A. inornatum*, *A. maculatum*, *A. mixtum*, *A. naponense*, *A. nodosum*, *A. oblongoguttatum*, *A. ovale*, *A. parvum*, *A. rotundatum*, *A. sabanerae*, *A. scutatum*, *A. tenellum*, *A. varium* (15)
	Dermacentor albipictus, *D. dispar*, *D. dissimilis*, *D. imitans*, *D. nitens*, *D. panamensis* (6)
	Haemaphysalis leporispalustris
	Rhipicephalus microplus, *R. sanguineus s.l.* (2)

(continued)

Table 4.1 (continued)

Guyana (19)	*Amblyomma auricularium, A. cajennense, A. coelebs, A. dissimile, A. geayi, A. humerale, A. latepunctatum, A. naponense, A. oblongoguttatum, A. ovale, A. pacae, A. pictum, A. romitii, A. scalpturatum, A. varium* (15)
	Dermacentor nitens
	Haemaphysalis juxtakochi
	Rhipicephalus microplus, R. sanguineus s.l. (2)
Haiti (6)	*Amblyomma cruciferum, A. dissimile, A. quadricavum* (3)
	Dermacentor nitens
	Rhipicephalus microplus, R. sanguineus s.l. (2)
Honduras (19)	*Ixodes bequaerti, I. boliviensis, I. luciae* (3)
	Amblyomma albopictum, A. auricularium, A. coelebs, A. dissimile, A. longirostre, A. maculatum, A. mixtum, A. naponense, A. nodosum, A. sabanerae, A. scutatum, A. tenellum (12)
	Dermacentor nitens, D. panamensis (2)
	Rhipicephalus microplus, R. sanguineus s.l. (2)
Jamaica (8)	*Ixodes copei*
	Amblyomma dissimile, A. mixtum, A. quadricavum, A. rotundatum (4)
	Dermacentor nitens
	Rhipicephalus microplus, R. sanguineus s.l. (2)
Martinique (5)	*Amblyomma rotundatum, A. variegatum* (2)
	Dermacentor nitens
	Rhipicephalus microplus, R. sanguineus s.l. (2)
Mexico (southern) (43)	*Ixodes affinis, I. bequaerti, I. boliviensis, I. cookei, I. cuernavacensis, I. guatemalensis, I. luciae, I. mexicanus, I. rubidus, I. scapularis, I. sinaloa, I. spinipalpis, I. tancitarius, I. texanus, I. woodi* (15)
	Amblyomma auricularium, A. calcaratum, A. coelebs, A. dissimile, A. inornatum, A. longirostre, A. maculatum, A. mixtum, A. nodosum, A. oblongoguttatum, A. ovale, A. pacae, A. parvum, A. pecarium, A. rotundatum, A. sabanerae, A. scutatum, A. tenellum (18)
	Dermacentor albipictus, D. dissimilis, D. halli, D. imitans, D. nitens, D. variabilis (6)

(continued)

Table 4.1 (continued)

	Haemaphysalis juxtakochi, H. leporispalustris (2)
	Rhipicephalus microplus, R. sanguineus s.l. (2)
Montserrat (5)	*Amblyomma rotundatum, A. variegatum* (2)
	Dermacentor nitens
	Rhipicephalus microplus, R. sanguineus s.l. (2)
Nicaragua (24)	*Ixodes bequaerti, I. boliviensis, I. luciae, I. sinaloa* (4)
	Amblyomma auricularium, A. coelebs, A. dissimile, A. maculatum, A. mixtum, A. nodosum, A. oblongoguttatum, A. ovale, A. parvum, A. sabanerae, A. scutatum, A. tapirellum, A. tenellum, A. varium (14)
	Dermacentor dissimilis, D. nitens, D. panamensis (3)
	Haemaphysalis leporispalustris
	Rhipicephalus microplus, R. sanguineus s.l. (2)
Panama (39)	*Ixodes affinis, I. auritulus, I. bequaerti, I. bocatorensis, I. boliviensis, I. luciae, I. pomerantzi, I. rubidus, I. tapirus, I. tiptoni, I. venezuelensis* (11)
	Amblyomma auricularium, A. calcaratum, A. coelebs, A. crassum, A. dissimile, A. geayi, A. longirostre, A. mixtum, A. naponense, A. nodosum, A. oblongoguttatum, A. ovale, A. pacae, A. parvum, A. pecarium, A. pictum, A. rotundatum, A. sabanerae, A. tapirellum, A. varium (20)
	Dermacentor imitans, D. latus, D. nitens, D. panamensis (4)
	Haemaphysalis juxtakochi, H. leporispalustris (2)
	Rhipicephalus microplus, R. sanguineus s.l. (2)
Paraguay (26)	*Ixodes loricatus*
	Amblyomma aureolatum, A. auricularium, A. boeroi, A. brasiliense, A. calcaratum, A. coelebs, A. dissimile, A. dubitatum, A. incisum, A. longirostre, A. nodosum, A. ovale, A. pacae, A. parvum, A. pseudoconcolor, A. rotundatum, A. sculptum, A. tigrinum, A. tonelliae, A. triste (20)
	Dermacentor nitens
	Haemaphysalis juxtakochi, H. leporispalustris (2)
	Rhipicephalus microplus, R. sanguineus s.l. (2)
Peru (39)	*Ixodes affinis, I. andinus, I. auritulus, I. boliviensis, I. downsi, I. lasallei, I. luciae, I. nectomys, I. nuttalli, I. pomerantzi, I. spinosus, I. tropicalis* (12)

(continued)

Table 4.1 (continued)

	Amblyomma calcaratum, A. coelebs, A. crassum, A. dissimile, A. geayi, A. humerale, A. incisum, A. interandinum, A. latepunctatum, A. longirostre, A. maculatum, A. naponense, A. oblongoguttatum, A. ovale, A. pacae, A. parvitarsum, A. pecarium, A. rotundatum, A. scalpturatum, A. tigrinum, A. triste, A. varium (22)
	Dermacentor nitens
	Haemaphysalis juxtakochi, H. leporispalustris (2)
	Rhipicephalus microplus, R. sanguineus s.l. (2)
Puerto Rico (8)	*Amblyomma cruciferum, A. dissimile, A. quadricavum, A. torrei, A. variegatum* (5)
	Dermacentor nitens
	Rhipicephalus microplus, R. sanguineus s.l. (2)
Saba (2)	*Amblyomma variegatum*
	Rhipicephalus microplus
Saint Eustatius (2)	*Amblyomma variegatum*
	Rhipicephalus microplus
Saint Kitts and Nevis (4)	*Amblyomma variegatum*
	Dermacentor nitens
	Rhipicephalus microplus, R. sanguineus s.l. (2)
Saint Lucia (5)	*Amblyomma dissimile, A. variegatum* (2)
	Dermacentor nitens
	Rhipicephalus microplus, R. sanguineus s.l.
Saint Martin (5)	*Amblyomma variegatum*
	Dermacentor nitens
	Haemaphysalis leporispalustris
	Rhipicephalus microplus, R. sanguineus s.l. (2)
Saint Vincent and the Grenadines (4)	*Amblyomma variegatum*
	Dermacentor nitens
	Rhipicephalus microplus, R. sanguineus s.l. (2)
Suriname (22)	*Ixodes luciae*
	Amblyomma auricularium, A. cajennense, A. calcaratum, A. coelebs, A. dissimile, A. geayi, A. goeldii, A. humerale, A. naponense, A.

(continued)

Table 4.1 (continued)

	oblongoguttatum, *A. ovale*, *A. pacae*, *A. pseudoconcolor*, *A. romitii*, *A. rotundatum*, *A. sabanerae*, *A. scalpturatum*, *A. varium* (18)
	Haemaphysalis juxtakochi
	Rhipicephalus microplus, *R. sanguineus s.l.* (2)
Trinidad and Tobago (15)	*Ixodes downsi*, *I. luciae* (2)
	Amblyomma auricularium, *A. calcaratum*, *A. dissimile*, *A. humerale*, *A. longirostre*, *A. mixtum*, *A. nodosum*, *A. ovale*, *A. rotundatum* (9)
	Dermacentor nitens
	Haemaphysalis juxtakochi
	Rhipicephalus microplus, *R. sanguineus s.l.* (2)
Turks and Caicos Islands (1)	*Amblyomma antillorum*
United Kingdom Virgin Islands (2)	*Amblyomma antillorum*
	Dermacentor nitens
Uruguay (15)	*Ixodes auritulus*, *I. fuscipes*, *I. longiscutatus*, *I. loricatus* (4)
	Amblyomma aureolatum, *A. auricularium*, *A. dubitatum*, *A. longirostre*, *A. pseudoconcolor*, *A. tigrinum*, *A. triste* (7)
	Haemaphysalis juxtakochi
	Rhipicephalus microplus, *R. sanguineus s.s.*, *R. sanguineus s.l.* (3)
USA Virgin Islands (4)	*Amblyomma variegatum*
	Dermacentor nitens
	Rhipicephalus microplus, *R. sanguineus s.l.* (2)
Venezuela (39)	*Ixodes auritulus*, *I. bocatorensis*, *I. downsi*, *I. jonesae*, *I. lasallei*, *I. luciae*, *I. montoyanus*, *I. venezuelensis* (8)
	Amblyomma auricularium, *A. beaurepairei*, *A. cajennense*, *A. calcaratum*, *A. coelebs*, *A. crassum*, *A. dissimile*, *A. geayi*, *A. humerale*, *A. latepunctatum*, *A. longirostre*, *A. maculatum*, *A. multipunctum*, *A. naponense*, *A. nodosum*, *A. oblongoguttatum*, *A. ovale*, *A. pacae*, *A. parvum*, *A. romitii*, *A. rotundatum*, *A. scalpturatum*, *A. tigrinum*, *A. triste*, *A. varium* (25)
	Dermacentor imitans, *D. nitens* (2)
	Haemaphysalis juxtakochi, *H. leporispalustris* (2)
	Rhipicephalus microplus, *R. sanguineus s.l.* (2)
Unknown South American Country (1)	*Amblyomma fulvum*

species of *Amblyomma*, while the only documented tick species on Bonaire is *Rhipicephalus sanguineus sensu lato*.

As expected, the vast nation of Brazil accounts for the highest number of Neotropical tick species. The current total of 52 species consists of 33 *Amblyomma* (three endemic), 12 *Ixodes* (four endemic), three *Haemaphysalis* (one endemic), three *Rhipicephalus*, and one species of *Dermacentor*. Southern Mexico follows with 43 species: 18 *Amblyomma*, 15 *Ixodes* (two endemic), six *Dermacentor*, two *Haemaphysalis*, and two *Rhipicephalus*; and Argentina, also with 43 species: 12 *Ixodes* (two endemic), 25 *Amblyomma* (three endemic), three *Rhipicephalus*, two *Haemaphysalis*, and one *Dermacentor*. However, some Central American countries with comparatively small territories in relation to Brazil, Mexico, and Argentina also possess a rich ixodid fauna. Thus, Costa Rica has recorded 40 species: nine *Ixodes* (one endemic), 23 *Amblyomma*, four *Dermacentor*, two *Haemaphysalis* and two *Rhipicephalus*; 39 species of hard ticks are known from Panama: 11 *Ixodes* (one endemic), 20 *Amblyomma*, four *Dermacentor*, two *Haemaphysalis*, and two *Rhipicephalus*; and 33 species occur in Guatemala: nine *Ixodes* (one endemic), 15 *Amblyomma*, six *Dermacentor*, one *Haemaphysalis*, and two *Rhipicephalus*. The Central American country with the least reported number of ixodid species is El Salvador, with 11 species: seven *Amblyomma*, two *Dermacentor*, and two *Rhipicephalus*. Moreover, El Salvador and the South American nation of Guyana, with 19 non-endemic species (15 *Amblyomma*, two *Rhipicephalus*, one *Dermacentor*, and one *Haemaphysalis*), are the only countries on the Neotropical mainland where no *Ixodes* species have been detected, but this is probably due to sampling error rather than the actual absence of *Ixodes* in these countries.

Although Brazil, Argentina and Mexico possess the three most diverse faunas among Neotropical countries, Colombia also has a rich ixodid fauna with 42 taxa (10 *Ixodes*, 26 *Amblyomma* (one endemic), two *Dermacentor*, two *Haemaphysalis*, and two *Rhipicephalus*). And two other countries, Venezuela and Peru, are similarly diverse, each having recorded 39. Venezuela's fauna comprises eight *Ixodes* (one endemic), 25 *Amblyomma* (one endemic), two *Dermacentor*, two *Haemaphysalis*, and two *Rhipicephalus*, while in Peru there are 12 *Ixodes* (two endemic), 22 *Amblyomma* (one endemic), two *Haemaphysalis*, two *Rhipicephalus*, and one *Dermacentor*. The lowest numbers of ixodid species occur in two southern South American countries, Uruguay with 15 taxa (four *Ixodes*, seven *Amblyomma*, three *Rhipicephalus*, and one *Haemaphysalis*), followed by Chile with 14 species (nine *Ixodes* (four endemic), three *Amblyomma*, and two *Rhipicephalus*). These small numbers of species are thought to reflect ecological conditions in the southern part of this subregion, which also includes Argentina. However, northern Argentina possesses numerous *Amblyomma* species that parasitize a diverse vertebrate fauna not found in Chile and Uruguay. The number of ixodid species in Argentina decreases with increasing latitude, eventually approximating the numbers seen in Chile and Uruguay. Although Chile's northern limit extends beyond its border with Argentina, its peculiar environment and limited vertebrate fauna, both determined by the Andes Mountains, conspire to limit the presence of *Amblyomma* species even in northern Chile. Thus, although the genus *Amblyomma* is the most speciose tick genus in the Neotropics, Chile is represented by just three non-endemic species (*Amblyomma parvitarsum*, *Amblyomma tigrinum*, and *Amblyomma triste*), whereas nine species of *Ixodes*, four of them endemic (*Ixodes abrocomae*, *Ixodes chilensis*, *Ixodes stilesi*, and *Ixodes taglei*) occur there.

The low numbers of *Amblyomma* species found in most parts of the Caribbean contrast with the situation on the Neotropical mainland. The genus *Amblyomma* is dominant on some large Caribbean islands (Bahamas, Cuba, Haiti, Jamaica, Puerto Rico, and Trinidad and Tobago) but not in the Dominican Republic, where four of the five species present consist of two *Amblyomma* and two *Rhipicephalus*. Amblyommine ticks are particularly poorly represented on the smaller Caribbean islands. Thus, two species each of *Rhipicephalus* and *Amblyomma* have been collected on Antigua and Barbuda, Barbados, Grenada, Martinique, and Montserrat. One species each of *Amblyomma* and *Rhipicephalus* have been found on Curação, Montserrat, Saba, and Saint Eustatius, while one *Amblyomma* and one *Dermacentor* have been recorded in the United Kingdom Virgin Islands. Two species of *Rhipicephalus* and one of *Amblyomma* have been found in Saint Kitts and Nevis, Saint Lucia, Saint Martin, and Saint Vincent and the Grenadines, as well as the United States Virgin Islands. No species of *Amblyomma* have been found in Aruba and Bonaire. It is likely that the small numbers or absence of *Amblyomma* on some of these islands is due to sampling error, especially in the case of reptilian hosts, rather than actual scarcity.

The genus *Ixodes* is well represented in the Neotropical Region, with southern Mexico possessing the highest number of species at 15 (two endemic and eight also found in the Nearctic Region), followed by Brazil, Argentina, and Peru with 12 species each (four endemic to Brazil, two species endemic to Argentina, and another two endemic to Peru), Panama (11 species, one endemic), and Colombia (also with ten species), while two Central American and one South American country each have nine species of *Ixodes*: Costa Rica (one endemic), Guatemala (one endemic), and Chile (four endemic). Venezuela possesses eight species of *Ixodes* (one endemic). No *Ixodes* have been found in El Salvador and Guyana, a condition that may change with additional tick sampling efforts. Only four species of *Ixodes* have been collected in the Caribbean, two in Trinidad and Tobago, and one endemic species each in Cuba and Jamaica.

The genus *Dermacentor* is represented in the Neotropical Region by nine species. The highest number of species has been detected in two contiguous territories, Guatemala and southern Mexico, with six species established in each territory, although three species of *Dermacentor* found in Guatemala and five species in southern Mexico are also present in the Nearctic Region. Two Central American countries, Costa Rica and Panama, possess four species of *Dermacentor*. The genus *Dermacentor* is poorly represented in South American countries, where only two species are known from Colombia and Venezuela (*Dermacentor imitans* and *Dermacentor nitens*), and this genus is almost nonexistent in the Caribbean, where only one species has been collected. Throughout South America and the Caribbean, if only one *Dermacentor* species is present, it invariably is *Dermacentor nitens*, which has a wide Neotropical distribution and is also established in the Nearctic.

This species has not been collected in the southern South American countries of Uruguay and Chile, probably as a consequence of adverse abiotic conditions, especially temperatures below those required for the off-host portion of the life cycle, but its absence from Suriname (northern South America) is likely due to insufficient sampling.

The genus *Haemaphysalis* is poorly represented in the Neotropics, with just three species found in the region. Brazil is the only Neotropical country where all three *Haemaphysalis* species, including the endemic *Haemaphysalis cinnabarina*, have been found, while Chile is the only South American country where no species of this genus has yet been collected, again underscoring the uniqueness of the Chilean ixodid fauna. *Haemaphysalis juxtakochi* and *Haemaphysalis leporispalustris* have been found in the South American countries of Argentina, Bolivia, Colombia, Paraguay, Peru, and Venezuela, and *Haemaphysalis juxtakochi* alone has been collected in Ecuador, French Guiana, Guyana, Suriname, and Uruguay. Both *Haemaphysalis juxtakochi* and *Haemaphysalis leporispalustris* have been found in southern Mexico and the Central American countries of Costa Rica and Panama, but only *Haemaphysalis leporispalustris* has been collected in Guatemala and Nicaragua, and only *Haemaphysalis juxtakochi* has been found in Belize. No representatives of this genus are known from El Salvador and Honduras, but again, the absence of haemaphysalids from these two Central American countries is probably due not to environmental restrictions but to poor tick sampling. *Haemaphysalis leporispalustris* alone has been found on the Caribbean islands of Aruba, Curação, and Saint Martin, and *Haemaphysalis juxtakochi* alone has been collected in Trinidad and Tobago. New tick studies in this region may yield *Haemaphysalis* species on other Caribbean islands.

The genus *Rhipicephalus* includes three species of great veterinary importance in the Neotropical Region. All these species have been found in Argentina, Brazil, and Uruguay, although this scenario may be a consequence of the recent definition of *Rhipicephalus sanguineus sensu stricto* by Nava et al. (2018), who used bona fide specimens from these countries and Chile in a study that included the redescription of this species. Therefore, the present geographic distribution of *Rhipicephalus sanguineus sensu stricto* in the Neotropics may be expected to expand as results are reported from ongoing regional tick studies. The remaining two taxa in this genus, *Rhipicephalus microplus* and *Rhipicephalus sanguineus sensu lato*, are widespread on the Neotropical mainland, again with the exception of Chile, where the cattle tick *Rhipicephalus microplus* is definitely known to be absent. *Rhipicephalus microplus* and *Rhipicephalus sanguineus sensu lato* have been found on the Caribbean islands of Antigua and Barbuda, Barbados, Grenada, Martinique, and Montserrat, Saint Kitts and Nevis, Saint Lucia, Saint Martin and Saint Vincent and the Grenadines, as well as the United States Virgin Islands, while one species of *Rhipicephalus* has been found on Aruba, Bonaire, Curação, Montserrat, Saba, and Saint Eustatius. No species of *Rhipicephalus* have been reported from Anguilla, Bermuda, the Cayman Islands, the Turks and Caicos Islands, and the United Kingdom Virgin Islands, but this surely reflects inadequate tick sampling.

4.1.2 *Species of Ixodidae Found in Neotropical Subregions*

The geographic distribution of Neotropical Ixodidae can be analyzed on the basis of those genera occurring within four subregions: (1) the Caribbean, (2) Central America and southern North America (referred to here as "southern Mexico"), (3) South America, and (4) the Galápagos Islands, as well as combinations of these subregions, as summarized in Table 4.2. Obviously, the Galápagos Islands constitute a limited area in comparison with the other subregions, but it was decided to treat these islands as a distinct subregion because their ixodid fauna is largely endemic, meriting special consideration.

It should be noted that only three tick species occur in all four Neotropical subregions—*Dermacentor nitens*, *Rhipicephalus microplus*, and *Rhipicephalus sanguineus sensu lato*—and all three invaded the Neotropics from elsewhere. Thus, *Dermacentor nitens* probably entered the Neotropics naturally from the Nearctic, and now is an important parasite of Neotropical domestic equids, while the presence of both rhipicephalids can be traced to human activity: *Rhipicephalus microplus* parasitizes cattle, and *Rhipicephalus sanguineus sensu lato* is chiefly a parasite of dogs.

4.1.2.1 Prostriata: Genus *Ixodes*

Within the Neotropical Zoogeographic Region, the genus *Ixodes* is strongly represented in the South American subregion. A total of 26 species of *Ixodes* are found only there; 25 are endemic to South America, while the remaining species, *Ixodes uriae*, is also present in the Northern Hemisphere.

A total of 16 *Ixodes* species have been collected only in the Central American-southern Mexico subregion; seven of these are endemic to this subregion, but nine species—*Ixodes cookei* (provisional), *Ixodes minor*, *Ixodes rubidus*, *Ixodes scapularis*, *Ixodes sinaloa*, *Ixodes spinipalpis*, *Ixodes tancitarius*, *Ixodes texanus*, and *Ixodes woodi*—also occur in the Nearctic Zoogeographic Region, and some of these species have probably spread into the Neotropics from the Nearctic.

Two species of *Ixodes* are found exclusively in the Caribbean, *Ixodes capromydis* in Cuba, and *Ixodes copei* in Jamaica, and both are endemic to this subregion.

Only the endemic *Ixodes galapagoensis* has been found in the Galápagos Islands. This tick is known from three specimens collected 45 years ago from a critically endangered host on Santa Fe Island (also known as Barrington Island).

Eight species of *Ixodes* have been reported from Central America-southern Mexico and South America, but two of these, *Ixodes auritulus* (in fact, a member of the *Ixodes auritulus* species group) and *Ixodes affinis* (also likely representing a species group), have ranges that exceed the Neotropical Region. Two species, *Ixodes boliviensis* and *Ixodes pomerantzi*, have disjunct geographic distributions that may be a consequence of diagnostic errors in the case of *Ixodes boliviensis* (see Chap. 1), and poor sampling for *Ixodes pomerantzi*. The remaining four species, *Ixodes*

Table 4.2 Geographic distributions of Neotropical Ixodidae arranged by main territorial divisions and their combinations

Territories	Tick species
Caribbean (7)	*Ixodes capromydis, I. copei* (2)
	Amblyomma antillorum, A. cruciferum, A. quadricavum, A. torrei, A. variegatum (5)
Central America-southern Mexico (26)	*Ixodes bequaerti, I. cookei, I. cuernavacensis, I. dicei, I. guatemalensis, I. mexicanus, I. minor, I. rubidus, I. scapularis, I. sinaloa, I. spinipalpis, I. tancitarius, I. tecpanensis, I. texanus, I. tiptoni, I. woodi* (16)
	Amblyomma inornatum, A. scutatum, A. tenellum (3)
	Dermacentor albipictus, D. dispar, D. dissimilis, D. halli, D. latus, D. panamensis, D. variabilis (7)
Galápagos Islands (8)	*Ixodes galapagoensis*
	Amblyomma boulengeri, A. darwini, A. hirtum, A. macfarlandi, A. pilosum, A. usingeri, A. williamsi (7)
South America (55)	*Ixodes abrocomae, I. amarali, I. andinus, I. catarinensis, I. chilensis, I. cooleyi, I. cornuae, I. fossulatus, I. fuscipes, I. jonesae, I. longiscutatus, I. loricatus, I. montoyanus, I. nectomys, I. neuquenensis, I. nuttalli, I. paranaensis, I. pararicinus, I. schulzei, I. sigelos, I. silvanus, I. spinosus, I. stilesi, I. taglei, I. tropicalis, I. uriae* (26)
	Amblyomma argentinae, A. aureolatum, A. beaurepairei, A. boeroi, A. brasiliense, A. cajennense, A. dubitatum, A. fulvum, A. fuscum, A. goeldii, A. hadanii, A. incisum, A. internadinum, A. latepunctatum, A. multipunctum, A. neumanni, A. parkeri, A. parvitarsum, A. patinoi, A. pseudoconcolor, A. pseudoparvum, A. romarioi, A. romitii, A. scalpturatum, A. sculptum, A. tigrinum, A. tonelliae, A. triste, A. yucumense (29)
Caribbean + Central America-southern Mexico (1)	*Amblyomma albopictum*
Caribbean + South America (2)	*Ixodes downsi*
	Amblyomma humerale
Caribbean + Central America-southern Mexico + South America (10)	*Ixodes luciae*
	Amblyomma auricularium, A. dissimile, A. longirostre, A. mixtum, A. nodosum, A. ovale, A. rotundatum (7)
	Haemaphysalis juxtakochi, H. leporispalustris (2)

(continued)

Table 4.2 (continued)

Central America-southern Mexico + South America (23)	*Ixodes affinis, I. auritulus, I. bocatorensis, I. boliviensis, I. lasallei, I. pomerantzi, I. tapirus, I. venezuelensis* (8)
	Amblyomma calcaratum, A. coelebs, A. crassum, A. geayi, A. maculatum, A. naponense, A. oblongoguttatum A. pacae, A. parvum, A. pecarium, A. pictum, A. sabanerae, A. tapirellum, A. varium (14)
	Dermacentor imitans
All territories (3)	*Dermacentor nitens*
	Rhipicephalus microplus, R. sanguineus s.l.

bocatorensis, *Ixodes lasallei*, *Ixodes tapirus* and *Ixodes venezuelensis*, are all established in southern Central America and northwestern South America, an area of evolutionary relevance.

Ixodes luciae is the only tick found in the Caribbean, Central America-southern Mexico and South America, while *Ixodes downsi* has been collected in the Caribbean and South American localities. Both species have been recorded on the same Caribbean island (Trinidad), near the mainland of northeastern South America, and their presence there is probably due to natural invasions from South America. *Ixodes luciae* has the widest distribution of any Neotropical *Ixodes*, occurring from northern Argentina to southern Mexico.

Overall, *Ixodes* is predominant on the South American mainland, where 36 of the 55 Neotropical species have been found, 25 of them endemic to this subregion. This contrasts with the total of 25 species of *Ixodes* established in Central America-southern Mexico, seven of them endemic, the four Caribbean species (two endemic), and the one endemic *Ixodes* present in the Galápagos Islands. The abundance of South American *Ixodes* is considered to be a consequence of invasion by rodents of the family Cricetidae that colonized almost all South American territories (including the Galápagos Islands), radiating into new species with their *Ixodes* parasites (see also Chap. 5), a process that extended to south-central Chile and Argentina, where the diversity of *Ixodes* species exceeds even that of *Amblyomma*.

Ixodes luciae is of special interest because its broad geographic range and host profile may indicate an invasion from South America into Central America extending to southern Mexico (see also next chapter), a reversal of the usual pattern in which *Ixodes* from Central America enter South America.

Finally, there are several species of *Ixodes* that have been recorded by various authors as occurring in the Neotropical Region, but are here considered absent from this region; they are: *Ixodes angustus*, *Ixodes conepati*, *Ixodes dentatus*, *Ixodes diversifossus*, *Ixodes frontalis*, *Ixodes granulatus*, *Ixodes hexagonus*, *Ixodes pacificus*, *Ixodes ricinus*, *Ixodes tamaulipas,* and *Ixodes tovari.* Each of these names is addressed in the fourth section of this chapter, which discusses tick species incorrectly or doubtfully collected in the Neotropics. Additionally, there is the special case of *Ixodes percavatus sensu lato*, which is found on pelagic marine birds in the Neotropics, although this tick reproduces on islands in the Atlantic and Indian Oceans far from the Neotropical Region. *Ixodes percavatus sensu lato* is not treated here as a genuine Neotropical species, but see also Chap. 1.

4.1.2.2 Metastriata

Genus *Amblyomma*

In the Neotropics, the genus *Amblyomma* is characterized by 28 species found only in the South American subregion, and one taxon, *Amblyomma triste* (a tick whose morphological diagnosis is uncertain), that has also been collected in the Nearctic Region. The South American mainland contains by far the greatest number of endemic *Amblyomma* species, followed by the Galápagos Islands, which, although small in area, account for seven species, all of them endemic and instructive with respect to tick life histories and evolution: *Amblyomma boulengeri*, *Amblyomma darwini*, *Amblyomma hirtum*, *Amblyomma macfarlandi*, *Amblyomma pilosum*, *Amblyomma usingeri*, and *Amblyomma williamsi.* With the exception of *Amblyomma hirtum*, whose hosts are unknown, all these species are parasites of reptiles.

Five *Amblyomma* species are confined to Caribbean Islands, and four of them—*Amblyomma antillorum*, *Amblyomma cruciferum*, *Amblyomma quadricavum*, *Amblyomma torrrei*—are endemic to this subregion, while the fifth species, *Amblyomma variegatum*, is an Afrotropical tick of veterinary importance that was introduced into the Caribbean by human activities. This last species is a menace not only to the Neotropical mainland, but also to the Nearctic Zoogeographic Region and, probably, elsewhere. As in the case of the Galápagos Islands, the *Amblyomma* species endemic to the Caribbean Islands preferentially parasitize reptiles.

Three Neotropical species—*Amblyomma inornatum*, *Amblyomma scutatum* and *Amblyomma tenellum*—have only been found in the Central America-southern Mexico subregion, but none of these species are endemic to this subregion because all of them are also established in the Nearctic Region. Therefore, within the Neotropical Zoogeographic Region, Central America-southern Mexico is the only subregion without endemic species of *Amblyomma*.

A total of 14 species of Neotropical *Amblyomma* have been recorded from Central America-southern Mexico and South America, three of which (*Amblyomma coelebs*, *Amblyomma maculatum*, and *Amblyomma oblongoguttatum*) are also found in the Nearctic Region.

Seven species of Neotropical *Amblyomma* have been found in the Caribbean, Central America-southern Mexico, and South America subregions, but four of these species, *Amblyomma auricularium*, *Amblyomma dissimile*, *Amblyomma mixtum*, and *Amblyomma ovale*, have also been found in the Nearctic Region, while

Amblyomma rotundatum has a geographic distribution that encompasses Nearctic territory and one remote Pacific island.

One species, *Amblyomma albopictum*, has been recorded in the Caribbean and Central America-southern Mexico subregions, while another, *Amblyomma humerale*, has been detected in the Caribbean and South American countries.

The abundance of *Amblyomma* species in South America would appear to indicate that the presence of this genus in Central America-southern Mexico is a consequence of invasion from South America, as hypothesized for species of the *Amblyomma cajennense* complex in Central America-southern Mexico after their presumed evolution in northern South America (Beati et al. 2013). However, many mammals currently established in the South American subregion arrived as a result of northern invasions, especially during the Great American Interchange. Therefore, it cannot be ruled out that species of Neotropical *Amblyomma* or their ancestors may have arrived in South America with northern invaders (see also Chap. 5). Nevertheless, the high number of *Amblyomma* in the Central America-southern Mexico subregion (25 species) contrasts with the fact that none of them are endemic to this large territory, suggesting that this subregion may be a corridor for *Amblyomma* of South American origin expanding northward to the Nearctic, in contradistinction to the probable role of Central America in facilitating the introduction of *Ixodes* to the South American subregion during the Great American Interchange.

The diversity of *Amblyomma* species on the South American mainland might be interpreted as indicating that this genus evolved there and dispersed to other Neotropical subregions and elsewhere thereafter. However, there is no evidence to support the argument that *Amblyomma* arose in Gondwanian South America, or that the presence of this genus there is a consequence of invasion from other Gondwanian territories, before the long period of South American isolation that lasted some 50 million years during the Cenozoic Era. But such theories should not be disregarded. Several workers support the theory that South American tortoises and hystricognathan rodents are of African origin, derived from ancestors that arrived by rafting between 50 and 30 million years ago. Moreover, an African origin for South American *Amblyomma* can readily be hypothesized because the genus *Amblyomma* is well represented in the Afrotropical Zoogeographic Region, with 26 species found only there. In Chap. 5, the hypothesis that mammals of Afrotropical origin carried *Amblyomma* ticks to South America is not supported, but the arrival of Afrotropical tortoises infested with *Amblyomma* is not discounted.

The several hypotheses concerning the origin of South American *Amblyomma* reflect a lack of solid information supporting the evolution of this genus. Recent studies of the *Amblyomma cajennense* species complex by Beati et al. (2013) and analyses of *Amblyomma maculatum* and related species by Lado et al. (2018) have shed much-needed light on this subject.

Finally, there are several species of *Amblyomma* that have been reported by various authors as occurring in the Neotropical Region but are here considered absent from this region; they are: *Amblyomma americanum*, *Amblyomma extraoculatum*, *Amblyomma gervaisi*, *Amblyomma hebraeum*, *Amblyomma nitidum*, *Amblyomma nuttalli*, and *Amblyomma tuberculatum*. Each of these names is treated in the fourth section of this chapter, which addresses incorrect or doubtful names applied to Neotropical records.

Genus *Dermacentor*

In the Neotropics, the genus *Dermacentor* is characterized by seven species found only in Central America-southern Mexico; three of these species, *Dermacentor dispar*, *Dermacentor latus*, and *Dermacentor panamensis* are endemic to this subregion, but the remaining four species, *Dermacentor albipictus*, *Dermacentor dissimilis*, *Dermacentor halli*, and *Dermacentor variabilis*, are also found in the Nearctic Region. Central America-southern Mexico is by far the subregion with the greatest diversity of Neotropical *Dermacentor* species.

Central America-southern Mexico and South America share one Neotropical species, *Dermacentor imitans*, but the range of this species in South America is restricted to the northwestern part of the subregion (Colombia and Venezuela).

There are no species of *Dermacentor* that are restricted to the Caribbean, Galápagos Islands, and South America subregions, but a species of veterinary importance, *Dermacentor nitens*, is found in those subregions as well as in Central America-southern Mexico. This tick has a wide range that includes the Nearctic Region, but its southern limit in South America is northern Argentina.

The presence of the genus *Dermacentor* in the Neotropical Region appears to be a consequence of invasion from the Nearctic, evidenced by the concentration of species in southern Mexico (southern North America) and Guatemala (northern Central America) with subsequent southern radiation, but not in South America, where *Dermacentor imitans* is confined to the northwest corner of the subregion, and *Dermacentor nitens* alone occurs from the subtropical Nearctic to the subtropical Neotropics. With the exception of *Dermacentor nitens*, the limited range of the genus *Dermacentor* in the Neotropics may indicate recency of arrival in this region.

Finally, there are four species of *Dermacentor* that have been reported by various authors as occurring in the Neotropical Region but are here considered absent from this region; they are: *Dermacentor andersoni*, *Dermacentor hunteri*, *Dermacentor parumapertus*, and *Dermacentor reticulatus*. Each of these names is treated in the fourth section of this chapter, which addresses incorrect or doubtful names applied to Neotropical records. However, it should be emphasized that the supposed presence of *Dermacentor andersoni*, a species of medical and veterinary importance, in southern Mexico has not been taxonomically confirmed outside the authors' statement. No bona fide description of specimens of *Dermacentor andersoni* from southern Mexico has been found; therefore, we considered the presence of this species in Neotropical localities doubtful pending further study.

Genus *Haemaphysalis*

South America is the only Neotropical subregion containing an endemic *Haemaphysalis* species, *Haemaphysalis cinnabarina*. This species was described in 1844, and no specimens of *Haemaphysalis cinnabarina* have since been found, leading some workers to claim that this tick is a synonym of the Palearctic *Haemaphysalis punctata*, and its Brazilian origin a labeling error.

The other two species of *Haemaphysalis* established in the Neotropical Region are *Haemaphysalis juxtakochi* and *Haemaphysalis leporispalustris*; both species are known to be established in the Caribbean, Central America-southern Mexico, and South America subregions, as well as in the Nearctic Region.

No *Haemaphysalis* ticks have yet been found in the Galápagos Islands.

Haemaphysalis is the largest genus in the Metastriata group, containing 176 species worldwide, but only five of these have been detected in the Americas (Nearctic and Neotropical Regions), including *Haemaphysalis longicornis*, a species currently known to be established in the USA (Nearctic). This genus most probably originated in the Oriental Zoogeographic Region because 97 species of *Haemaphysalis* are present there, 65 of them exclusively Oriental. It is unclear whether haemaphysalids arrived to the Neotropics from Africa and later reached the Nearctic Region, or if Palearctic haemaphysalids invaded the Nearctic and from there reached the Neotropics, but we consider the latter route to be more plausible. In any event, the poor representation of *Haemaphysalis* in the Americas indicates a rather recent invasion of the Western Hemisphere when measured in evolutionary terms.

Finally, one species, *Haemaphysalis chordeilis*, has been reported as having been found in the Neotropical Region but is here considered to be absent from this region. This name is discussed in the fourth section of this chapter, which addresses incorrect or doubtful names applied to Neotropical records.

Genus *Rhipicephalus*

There are at least three species of *Rhipicephalus* in the Neotropics: *Rhipicephalus microplus*, *Rhipicephalus sanguineus sensu stricto* and *Rhipicephalus sanguineus sensu lato* (this last probably comprising more than one species). All these species are of great veterinary importance and were introduced into the Neotropical Region as a result of human activities. Accordingly, their ranges extend well beyond the Neotropical Region.

South America is the only Neotropical subregion where *Rhipicephalus sanguineus sensu stricto* has been found, but the range of this tick also includes the Nearctic and Palearctic Regions. It is uncertain whether the Neotropical range of this species is restricted to South America because the morphological and molecular characters of this species have only recently been defined by Nava et al. (2018). Additional studies may find that *Rhipicephalus sanguineus sensu stricto* is present in other Neotropical subregions.

The Caribbean, Central America-southern Mexico, South America, and Galápagos Islands subregions all support populations of *Rhipicephalus microplus* and *Rhipicephalus sanguineus sensu lato*, which is expected because *Rhipicephalus microplus* is a pantropical species, while *Rhipicephalus sanguineus sensu lato* has a worldwide distribution.

Rhipicephalus microplus is thought to have been introduced with cattle imported from Africa or Asia by colonists and early traders, while the two remaining species of *Rhipicephalus* most probably arrived in the New World with domestic dogs from the Afrotropics, the Palearctic, or both regions, although this is conjecture. Almost certainly, there have been numerous introductions of all three *Rhipicephalus* species, especially in the case of canine ticks that accompany immigrants and tourists.

Finally, there are five species of *Rhipicephalus* that have been reported by various authors as having been found in the Neotropical Region but are here considered to be absent from this region; they are: *Rhipicephalus annulatus*, *Rhipicephalus australis*, *Rhipicephalus bursa*, *Rhipicephalus decoloratus*, and *Rhipicephalus turanicus*. Each of these names is treated in the fourth section of this chapter, which addresses incorrect or doubtful names applied to Neotropical records. It should be emphasized that *Rhipicephalus turanicus* is morphologically close to *Rhipicephalus sanguineus sensu stricto*, and it is therefore possible that specimens of *Rhipicephalus turanicus* may be found among Neotropical collections of *Rhipicephalus sensu lato*.

4.2 Invasive Hard Tick Species Found in the Neotropics

Invasion by exotic tick species is a problem worldwide, chiefly because of health implications but also out of concern for the ecological damage that so often accompanies such introductions. The international pet trade, particularly in reptiles, has long been linked to this threat (Burridge and Simmons 2003; González-Acuña et al. 2005a; Nowak 2010), but there are other ways in which exotic ticks may disperse far outside their natural ranges. Thus, Burridge (2011) analyzed how exotic ticks enter the USA, finding that they most often arrived on wild mammals, followed by reptiles, animal products (e.g., hides, trophies, hair), domestic animals, plant materials (e.g., flowers, plants, cork), birds, humans, amphibians, and baggage.

As a consequence of human activities, the Neotropical Region has suffered the introduction of such economically important invasive ticks as *Rhipicephalus microplus*, *Rhipicephalus sanguineus sensu stricto*, and *Rhipicephalus sanguineus sensu lato*, which are now common parasites of cattle and dogs in most parts of this region. It is considered of vital importance that governments vigilantly safeguard all ports of entry to prevent the establishment of potentially invasive Ixodidae, and lessons from the past may aid in this endeavor. Here we present a non-exhaustive list of species of invasive hard ticks that have been found in the Neotropical Region but that fortunately appear to have been unable to establish themselves.

Ixodes percavatus Neumann, 1906: A natural invader, this tick is a parasite of pelagic birds that breed on remote Atlantic Islands and perhaps also on islands in the Indian Ocean. Labruna et al. (2020b) found nymphs and larvae of *Ixodes percavatus sensu lato* on birds foraging in the Neotropics, but this tick is not considered a Neotropical species. See "The special Neotropical records of *Ixodes percavatus sensu lato*" in Chap. 1 for a discussion of this species, which does not appear to be a threat to the Neotropical avifauna.

Amblyomma flavomaculatum (Lucas, 1846), as *Amblyomma flavomacuatum* (*lapsus*) in Bermúdez and Miranda (2011a): an Afrotropical species found in Panama on a reptile that had been imported from the USA, as detailed in Bermúdez and Miranda (2011a). These authors suspect that the host of *Amblyomma* *flavomaculatum* became infested in the USA and not in Africa, because the reptile was captured in Togo toward the end of 2009, and arrived in the USA during January 2010, but was not transported to Panama until October of that year.

Amblyomma latum Koch, 1844: An Afrotropical species that has been introduced into Argentina with reptiles imported from Africa, and introduced into Chile from Africa via the USA (González-Acuña et al. 2005a). *Amblyomma latum* has also been found in Uruguay, but the origin of the specimens was not specified (Venzal et al. 2007b).

Amblyomma varanense (Supino, 1897): An Australasian and Oriental species that was recorded on reptiles imported to Argentina from Indonesia (Oriental Zoogeographic Region) (Cicuttin et al. 2019b).

Bothriocroton undatum (Fabricius, 1775): An Australasian species that was reported from a Brazilian zoo by Teixeira et al. (2003) under the name *Aponomma decorosum*, a synonym of *Bothriocroton undatum*, as discussed in Guglielmone and Nava (2014). The diagnosis of this tick in Teixeira et al. (2003) is treated as provisionally valid here.

Dermacentor andersoni Stiles, 1908: A Nearctic species found in Canada and the USA. Faccini-Martínez et al. (2021) found a female of *Dermacentor andersoni* on a Brazilian citizen after returning from a holiday in the USA. However, there are additional problems related to *Dermacentor andersoni* because Romero-Castañon et al. (2008) and Sosa-Gutiérrez et al. (2016) allegedly collected *Dermacentor andersoni* in the Neotropical part of Mexico without providing suitable taxonomic support for this important diagnosis. *Dermacentor andersoni* is not treated as a Neotropical species found in southern Mexico, and Guzmán-Cornejo et al. (2016) also doubt its presence in northern Mexico (Nearctic Region). Nevertheless, the economic and medical importance of this tick demands that additional efforts be made to confirm its presence in the Neotropics.

Dermacentor variabilis (Say, 1821): Chiefly a Nearctic species with a few records from mammals in the Neotropics of southern Mexico, but tick stages found on those mammals are unknown. *Dermacentor variabilis* is a tick of medical and veterinary importance, so the record of a partially engorged female *Dermacentor variabilis* found on a child that entered Panama from the USA (Bermúdez et al. 2010b) is significant. However, the specimen was detected 10 days after the child's arrival in Panama, highlighting the possibility that humans may serve as carriers of invasive ticks, or, alternatively, that *Dermacentor variabilis* may be present in Panama. See also *Dermacentor variabilis* in Chap. 3.

Hyalomma dromedarii Koch, 1844: An Afrotropical, Oriental, and Palearctic species that has been detected on camels imported to Argentina and Brazil from the Canary Islands (Lahille 1914; Massard et al. 2002).

Hyalomma marginatum Koch, 1844: A Palearctic species that was found in Brazil on horses imported from Portugal (Labruna et al. 2001).

Rhipicephalus annulatus (Say, 1821): Galli-Valerio (1909) reported this tick, under the name *Margaropus annulatus*, as having been found in Colombia, while Minning (1934) recorded it from Cuba, Ecuador, Guadeloupe, Guatemala, and Jamaica. Bequaert (1938) stated that this tick had been found in Guatemala, and

Hutson (1944) diagnosed this species using specimens collected in Antigua and Barbuda. Hall (1951) declared that this tick had been eradicated from the United Kingdom Virgin Islands; Clarkson and Owen (1959) diagnosed this species using specimens collected in the Bahamas; Seifert (1959), Keirans (1985b), and Need et al. (1991) claimed that this species is present in Peru, while Hoffmann (1962), among others, said that it is present in southern Mexico. Payno Balazans (1978) listed this tick from Bolivia, and Guglielmone (1995) vaguely stated that it had been found in "northern Central America." All the authors cited above (Galli Valerio excepted) referred to this species as *Boophilus annulatus*.

Reexamination of specimens identified as *Rhipicephalus annulatus* (under the genus *Boophilus*) resulted, in some cases, in a new diagnosis as *Rhipicephalus microplus* (also under the genus *Boophilus*), while on other occasions the presence of this species could be traced to importations, but the invasive ticks failed to become established. Thus, Fox and León (1983) stated that alleged *Rhipicephalus annulatus* collected in Puerto Rico were in fact *Rhipicephalus microplus*. Labruna, M.B. (personal communication to A.A. Guglielmone) changed the diagnosis of alleged Brazilian females of *Rhipicephalus annulatus* collected by Figueiredo et al. (1999) to *Rhipicephalus microplus*. Fairchild et al. (1966) and Černý (1969a) found *Rhipicephalus annulatus* on cattle imported to Panamá and Cuba, respectively, but, again, the ticks failed to become established in those countries. Camicas et al. (1998) listed this species as found in the Neotropical Region, but Guglielmone et al. (2003a) excluded it from this region. Smith (1974) regarded prior diagnoses of this tick in Trinidad and Tobago as misidentifications of *Rhipicephalus microplus* (under the genus *Boophilus*). No evidence of the presence of *Rhipicephalus annulatus* in the Neotropics has been presented since 2003, with the exception of one male tick collected in the Neotropical Mexican state of Tabasco by Coronel-Benedett et al. (2018) that needs confirmation. Therefore, the position of Guglielmone et al. (2003a) is followed here. Recent records of this tick on *Rhinella marina* (Anura: Bufonidae) in Trinidad and Tobago by Kamilar et al. (2020) are treated here as diagnostic errors.

In summary, the many alleged Neotropical records of *Rhipicephalus annulatus* are not supported by reliable collection data or clear redescriptions of specimens obtained from established Neotropical tick populations.

Rhipicephalus bursa Canestrini and Fanzago (1878): A Palearctic species with records from Cuba, Curaçāo, Haiti, and southern Mexico (Veracruz) (Neumann 1911a), French Guiana (Floch and Fauran 1958, 1959b), Uruguay (Rodríguez González and Lujambio 1954) and Venezuela (Rivas 1919), but presumably most if not all of these records are a consequence of diagnostic errors. Nevertheless, Labruna et al. (2001) found *Rhipicephalus bursa* in Brazil on horses imported from Portugal. Santos Dias (1958c) stated that female ticks from Saint Vincent and the Grenadines, classified as *Rhipicephalus bursa* by Neumann, were in fact *Rhipicephalus microplus* (under the name *Boophilus microplus*).

Rhipicephalus decoloratus Koch, 1844: Lahille (1914) allegedly collected two females of *Rhipicephalus decoloratus* (named as *Boophilus decoloratus*) from one camel imported to Argentina from the Canary Islands. These islands are outside the range of *Rhipicephalus decoloratus*, but *Rhipicephalus annulatus* is recognized as occurring in the Canary Islands. It is quite possible that the ticks collected by Lahille (1914) were in fact *Rhipicephalus annulatus*, or, alternatively, that they were from a locality other than the Canary Islands. Seifert (1959) stated that *Rhipicephalus decoloratus* (under the name *Boophilus decoloratus*) is present in Peru, but this is treated as a diagnostic error in Guglielmone et al. (2003a) and here. Doss et al. (1978) stated that Chavarría (1943) reported the presence of *Rhipicephalus decoloratus* (under the genus *Boophilus*) in Costa Rica; however, Chavarría (1943) made no such claim, instead merely listing some of the ticks considered to be probable vectors of anaplasmosis worldwide.

Rhipicephalus evertsi Neumann, 1897: An Afrotropical species (Guglielmone et al. 2014) recently found to be established in the Palearctic Region (Alanazi et al. 2019). Aragão (1936) recorded *Rhipicephalus evertsi* from goats under quarantine conditions in Brazil.

4.3 A Non-exhaustive List of Valid and Invalid Names of Ixodidae Incorrectly or Doubtfully Applied to Ixodids Found in Neotropical Territories

Historically, several valid and invalid species of Ixodidae have been incorrectly or doubtfully reported from the Neotropical Region, ascribed to Neotropical taxa under different names, or originally described under names that are no longer valid. A non-exhaustive list of such names is presented below, first for ticks belonging to the Prostriata group, followed by names in the Metastriata group, arranged by genus. Names of valid species appear in boldface.

4.3.1 Prostriata: Incorrect Names Associated with the Genus Ixodes

Acarus vibrans Fabricius, 1794: It is not possible to determine the tick stage described by Fabricius (1794) from material collected in French Guiana. *Acarus vibrans* was later placed in the genus *Ixodes* by Fabricius (1805); it is considered to be a *nomen dubium* in Camicas et al. (1998), Guglielmone and Nava (2014) and here.

Ixodes angustus Neumann, 1899: This Nearctic and Palearctic species was described from a mutilated female collected in the USA (Neumann 1899), and the morphological description was further elaborated in Neumann (1901) based on two female specimens allegedly collected in Argentina, which led authors such as Lahille (1905) to include *Ixodes angustus* as an Argentinean tick. Later, however, Neumann (1910) realized that the females used for his 1901 redescription of *Ixodes angustus* were, in fact, *Ixodes loricatus*. Despite this, authors such as Muñoz and Casanueva (2001) still treat *Ixodes angustus* as being present in the southern Neotropics. Estébanes-González and Cervantes (2005) allegedly found *Ixodes angustus* in the states of Chiapas and Oaxaca, located in the Neotropical part of Mexico (in the same paper, *Ixodes granulatus*, an Australasian, Oriental and Palearctic species, is also erroneously reported from the Neotropics). Guzmán-Cornejo et al. (2007) reexamined the supposed specimens of *Ixodes angustus* studied by Estébanes-González and Cervantes (2005) and concluded that they do not represent this species. Nevertheless, the erroneous records of *Ixodes angustus* in Estébanes-González and Cervantes (2005) were treated as valid in Rivas and Guzmán-Cornejo (2016) due to a typographical error (C. Guzmán-Cornejo, personal communication to A.A. Guglielmone).

Ixodes aragaoi Fonseca, 1935: See *Ixodes ricinus aragaoi*, below in this section.

Ixodes bicornis Neumann, 1906: This species was described from female ticks collected in southern Mexico. *Ixodes bicornis* is treated as valid in Neumann (1911a) and as a probable synonym of *Ixodes diversifossus* in Camicas et al. (1998), but *Ixodes bicornis* is generally regarded as a synonym of *Ixodes boliviensis* (Kohls 1956c; Guglielmone et al. 2003a; Guglielmone and Nava 2014). It is possible that *Ixodes bicornis* will be reinstated as a valid species, as discussed under *Ixodes bolivienis* above (Chap. 1).

Ixodes brunneus Koch, 1844 was described from a female tick collected on a North American passeriform bird (Moritz and Fischer 1981). This tick was confused with the Palearctic *Ixodes frontalis* by Nuttall and Warburton (1911), and their view was supported for many years by some European workers. Cooley and Kohls (1945) called Nuttall and Warburton's (1911) opinion into question and treated *Ixodes brunneus* as a Nearctic species with a single Neotropical record of a female specimen collected in Venezuela in 1936, while Anastos and Smith (1957) stated that Nuttall and Warburton (1911) had described a species other than *Ixodes brunneus*. Within the Neotropical Region, this tick has allegedly been found in Argentina, Colombia, Panama, and Venezuela, but only the record from Venezuela has been treated as valid in Guglielmone et al. (2014, 2020) and Guglielmone and Robbins (2018), based on Keirans and Webster (1981), who stated that the only bona fide Neotropical specimen of *Ixodes brunneus* had been collected in Venezuela. Earlier, however, Jones et al. (1972) had already doubted the presence of *Ixodes brunneus* in Venezuela, a view endorsed by Saracho Bottero et al. (2021), who examined another specimen of alleged *Ixodes brunneus* from Venezuela, concluding that the tick was not *Ixodes brunneus*. Saracho Bottero et al. (2021) discussed the alleged Argentinean, Colombian, and Panamanian records of *Ixodes brunneus*, and further reviewed the statements in Keirans and Brewster (1981) concerning the validity of the Venezuelan record of *Ixodes brunneus*, stressing that Keirans and Brewster (1981) provided no morphological support for their claim, and concluding that there is no solid evidence for the presence of *Ixodes brunneus* in the Neotropical Region. Consequently, *Ixodes brunneus* is treated here as a Nearctic species.

Ixodes conepati Cooley and Kohls, 1943: This tick is a Nearctic species (Guglielmone et al. 2014) described from two females collected in the USA. Doss et al. (1978) listed *Ixodes conepati* as found in Chile (Río El Ganso in Magallanes y de la Antártica Chilena), but these authors confused the locality, which in fact applies to a record of *Ixodes auritulus* in Kohls and Clifford (1966), as previously discussed in Guglielmone et al. (2003a).

Ixodes coxaefurcatus Neumann, 1899: This species was described from one male allegedly collected in Siwa (Egypt), but Nuttall and Warburton (1911), as well as Moritz and Fischer (1981), corrected the type locality to Serra Geral (Brazil). *Ixodes coxaefurcatus* was listed as a valid Brazilian species by Aragão and Fonseca (1961a) and Arzua et al. (1994) but in fact it is a synonym of *Ixodes loricatus*, as discussed in Guglielmone and Nava (2014).

Ixodes dentatus Marx, 1899 in Neumann (1899): The Costa Rican record of this tick in Santos Dias (1958c) is treated as a misidentification in Guglielmone et al. (2003a, 2014) and here.

Ixodes dentatus spinipalpis Nuttall and Hadwen, 1916 in Nuttall (1916): This is the original name for *Ixodes spinipalpis*, which was given its current status in Hubbard (1941). See also *Ixodes spinipalpis* in Chap. 1.

Ixodes didelphidis Fonseca and Aragão, 1952: This tick was described by Fonseca and Aragão (1952a) from males and females collected in west-central Brazil. *Ixodes didelphidis* was treated as provisionally valid in Barros-Battesti and Knysak (1999), but morphological and molecular data in Labruna et al. (2002a) indicate that *Ixodes didelphidis* is a synonym of *Ixodes loricatus*, an opinion that has since been supported by other tick workers.

Ixodes diversifossus Neumann, 1899: This species is known only from two mutilated females collected in New Mexico (USA), one of which is lost, according to Keirans and Clifford (1978). *Ixodes diversifossus* is a problematic species treated as Nearctic in Guglielmone et al. (2014) but as Nearctic and Neotropical in Camicas et al. (1998). The opinion of the latter authors is based on the probable synonymy of *Ixodes boliviensis* (Neotropical) with *Ixodes diversifossus*. Keirans and Clifford (1978) recognized that *Ixodes boliviensis* and *Ixodes diversifossus* are morphologically close but maintained the two names as valid, an opinion also supported by Guglielmone and Nava (2014) and Guglielmone et al. (2014). According to Keirans and Clifford (1978), *Ixodes dentatus* and *Ixodes spinipalpis* have often been confused with *Ixodes diversifossus*, and the first description of the male and female of *Ixodes spinipalpis* is in Bishopp (1912) under the name *Ixodes diversifossus*. Some uses of the name *Ixodes diversifossus* are difficult to interpret. Thus, Neumann (1899) described the female of *Ixodes diversifossus* using the two known specimens of this taxon, and Neumann (1911a) redescribed *Ixodes diversifossus* in a brief text that summarized the morphological characters detailed in Neumann (1899). However, authors such as Gregson (1956), Keirans and Clifford (1978), and Camicas et al. (1998) treat *Ixodes diversifossus* in Neumann (1899) as valid but consider the same name in Neumann (1911a) a synonym of *Ixodes spinipalpis*, which would appear to be inconsistent. See also *Ixodes boliviensis* and *Ixodes spinipalpis* in Chap. 1.

Ixodes elegans Neumann, 1910: This tick was described from female specimens collected in Chile. However, the name proposed by Neumann was preoccupied by

Ixodes elegans in Guérin-Méneville (1844), a synonym of *Amblyomma variegatum*, an Afrotropical and Neotropical species, as discussed in Guglielmone and Nava (2014). Therefore, the name *Ixodes elegans* Neumann, 1910 was replaced by *Ixodes stilesi* in Neumann, 1911b, meaning that the first description of the female of *Ixodes stilesi* appeared under the name *Ixodes elegans*. See also *Ixodes stilesi* in Chap. 1.

Ixodes frontalis (Panzer, 1798): Neumann (1911a) listed *Ixodes frontalis*, a Palearctic species, as found in southern Chile, probably because of confusion with an undetermined species of *Ixodes*.

Ixodes frontalis sulcatus Neumann, 1911: Neumann (1899) described an unnamed variety of *Ixodes frontalis* from a female tick of Colombian origin. This tick was later named *Ixodes frontalis sulcatus* in Neumann (1911a) and considered to be a synonym of *Ixodes brunneus* (currently regarded as a Nearctic species) in Camicas et al. (1998, under the name *Scaphixodes brunneus*), but treated as a name *incertae sedis* in Guglielmone and Nava (2014) and here.

Ixodes granulatus Supino, 1897: An Australasian, Oriental and Palearctic species diagnosed in the Neotropical area of Mexico by Estébanes-González and Cervantes (2005), but Guzmán-Cornejo et al. (2007) concluded that this diagnosis was erroneous. Consequently, *Ixodes granulatus* is not listed as a Neotropical tick.

Ixodes hexagonus Leach, 1815: Vogelsang and Cordero (1939) stated that *Ixodes hexagonus* was collected from cattle in Uruguay, a diagnosis treated as a misidentification by Guglielmone et al. (2003a) and here.

Ixodes hexagonus cookei Packard, 1869: the name *Ixodes hexagonus cookei* in Nuttall (1911) is a synonym of *Pholeoixodes cookei* in Camicas et al. (1998), but it is a synonym of *Ixodes cookei*, as stated in Keirans and Clifford (1978), an opinion accepted in Guglielmone and Nava (2014) and here. Nuttall (1916) published a record of a nymph of *Ixodes hexagonus cookei* (erroneously named *Ixodes angustus cookei* in Guglielmone et al. 2003a) from *Mustela frenata* in Costa Rica that is treated as doubtful in Guglielmone et al. (2003a) and here.

Ixodes imperfectus Neumann, 1899: This tick was described from two nymphs collected in Brazil, but it is treated as a *nomen dubium* in Camicas et al. (1998), Guglielmone and Nava (2014), and here.

Ixodes juvenis Neumann, 1899: This tick was described from six nymphs and one larva collected in New Grenada, a former territory that included parts of Central America and northern South America. *Ixodes juvenis* is considered to be a *nomen dubium* in Camicas et al. (1998), Guglielmone and Nava (2014), and here.

Ixodes loricatus spinosus Nuttall, 1910: This tick was described from females collected in southern Mexico and actually applies to the first description of the female of *Ixodes luciae*. However, the name *Ixodes loricatus spinosus* was preoccupied by *Ixodes spinosus* in Neumann (1899), a species considered to be a synonym of *Ixodes fuscipes* but reinstated as valid by Labruna et al. (2020a). The name *Ixodes loricatus spinosus* was changed to *Ixodes loricatus vogelsangi* by Santos Dias (1954). See also the following name and *Ixodes luciae* in Chap. 1.

Ixodes loricatus vogelsangi Santos Dias, 1954: This is a new name for *Ixodes loricatus spinosus* proposed by Santos Dias (1954) because the name was preoccupied by *Ixodes spinosus* Neumann, 1899, a species considered to be a synonym of

Ixodes fuscipes but reinstated as valid by Labruna et al. (2020a). The name *Ixodes loricatus vogelsangi* is a synonym of *Ixodes luciae*, as stated in Camicas et al. (1998), Keirans and Hillyard (2001), and Guglielmone and Nava (2014). See the preceding name and the discussion of *Ixodes luciae* in Chap. 1.

Ixodes neotomae Cooley, 1944b: This tick was described from males, females, and nymphs collected in California, but Norris et al. (1997) later found it to be a synonym of the Nearctic and Neotropical *Ixodes spinipalpis*. Hoffmann (1962) named as *Ixodes neotomae* Neotropical specimens of *Ixodes spinipalpis* found in southern Mexico. See also *Ixodes spinipalpis* in Chap. 1.

Ixodes obscurus Fabricius, 1805: It is not possible to determine the stage used by Fabricius (1805) to describe this tick, which was collected in French Guiana. *Ixodes obscurus* is considered to be a *nomen dubium* in Camicas et al. (1998), Guglielmone and Nava (2014), and here.

Ixodes pacificus Cooley and Kohls, 1943: Doss et al. (1974b) listed *Ixodes pacificus*, a Nearctic species, as present in French Guiana based on Floch et al. (1955), which is a misreading of the latter authors. Floch et al. (1955) cited morphological differences between *Ixodes pacificus* and *Ixodes luciae* without providing any record of the presence of *Ixodes pacificus* in French Guiana.

Ixodes ricinus (Linnaeus, 1758): There are several Neotropical references to *Ixodes ricinus*, a Palearctic species, due to incorrect identification of *Ixodes* specimens throughout the region, especially in papers published before 1960. Some cases are due to confusion with species of the *Ixodes ricinus* group, such as *Ixodes affinis*, *Ixodes fuscipes*, or *Ixodes pararicinus*, but most cases remain unsettled because the specimens involved are no longer available. In any event, there are no populations of *Ixodes ricinus sensu stricto* established in the Neotropics.

Ixodes ricinus aragaoi Fonseca, 1935: This is the original name for *Ixodes aragaoi*, spelled as *Ixodes ricinus aragãoi* in Fonseca (1935a) and as *Ixodes aragãoi* in Aragão and Fonseca (1951a). These names are currently synonyms of *Ixodes fuscipes*, as discussed in Labruna et al. (2020a). The description of the male of *Ixodes fuscipes* was published under the name *Ixodes ricinus aragaoi.* See also *Ixodes fuscipes* in Chap. 1.

Ixodes ricinus rochensis Calzada, 1936: This name was used by Calzada (1936) for Uruguayan specimens morphologically close to the Palearctic *Ixodes ricinus.* Camicas et al. (1998) list *Ixodes ricinus rochensis* as a synonym of *Ixodes aragaoi*, while Guglielmone et al. (2003a) consider it uncertain whether this name is a synonym of *Ixodes aragaoi* or *Ixodes pararicinus*, and Guglielmone and Nava (2014) define *Ixodes ricinus rochensis* as a name *incertae sedis.* Onofrio et al. (2014) demonstrated that *Ixodes ricinus rochensis* is a synonym of *Ixodes aragaoi*, but this name was subsequently relegated to the synonymy of *Ixodes fuscipes*, as discussed in Labruna et al. (2020a); consequently, *Ixodes ricinus rochensis* is a synonym of *Ixodes fuscipes.*

Ixodes scuticrenatus Vázquez, 1946: This tick was described from a male and a female collected in southern Mexico, but *Ixodes scuticrenatus* is currently considered to be a synonym of *Ixodes luciae* (Jones et al. 1972; Camicas et al. 1998; Guglielmone and Nava 2014).

Ixodes serrafreirei Amorim, Gazeta, Bossi and Linhares, 2003: This name was applied to specimens collected in Brazil, but it is a *nomen nudum* (Venzal et al. 2008; Guglielmone et al. 2009; Guglielmone and Nava 2014).

Ixodes tamaulipas Kohls and Clifford, 1966: Clifford et al. (1973) and Camicas et al. (1998) consider this species to be established in the Neotropical Region; however, *Ixodes tamaulipas* is known from only three females collected in southwestern Tamaulipas, Mexico, which is located in the Nearctic Region. This situation remains unchanged (Guzmán-Cornejo and Robbins 2010); therefore, *Ixodes tamaulipas* is treated as a Nearctic tick, following Guglielmone et al. (2003a, 2014).

Ixodes thoracicus Koch, 1844: This name was applied by Koch (1844) to describe a female tick collected in Brazil. Neumann (1899) used the name *Ixodes thoracicus* for four female ticks collected in southern Chile, but he later realized that the Chilean specimens were morphologically different from the tick described by Koch as *Ixodes thoracicus*. Consequently, Neumann (1904) changed the name of those ticks to *Ixodes auritulus*. Guglielmone and Nava (2014) treat *Ixodes thoracicus* as a *nomen dubium*, a designation also applied here. See also *Ixodes auritulus* in Chap. 1.

Ixodes tovari Cooley, 1945: Clifford et al. (1973) treated this species as occurring in the Neotropical Region, while Camicas et al. (1998) listed *Ixodes tovari* as a Nearctic and Neotropical species, but Guglielmone et al. (2003a, 2014) consider this species to be present only in the Nearctic. The latter authors are followed here because all records of *Ixodes tovari* are from the Mexican states of Guanuajuato and Nuevo León (Guzmán-Cornejo and Robbins 2010), which are in the Nearctic Region as defined in this study.

Ixodes uruguayensis Kohls and Clifford, 1967: This tick was described from nymphs and larvae collected in Uruguay. *Ixodes uruguayensis* has been treated as a valid species for many years and is still treated as valid in Barker and Murrell (2004, 2008), but it is a synonym of *Ixodes longiscutatus*, as demonstrated by Venzal et al. (2001). The first description of the nymph and larva of *Ixodes longiscutatus* was published by Kohls and Clifford (1967) under the name *Ixodes uruguayensis*. See also *Ixodes longiscutatus* in Chap. 1.

Ixodes vibrans (Fabricius, 1787): See *Acarus vibrans*, above in this section.

Ixodes vogelsangi Santos Dias, 1954: This name constitutes a new status for *Ixodes loricatus vogelsangi* in Morel and Pérez (1977), who applied the name *Ixodes vogelsangi* to a Neotropical species, but that name was found to be a synonym of *Ixodes luciae*, as stated in Camicas et al. (1998) and Guglielmone and Nava (2014).

Pholeoixodes chilensis (Kohls, 1956): This name is used in Camicas et al. (1998) for a species found in the Neotropical Region that was transferred to the genus *Pholeoixodes* by Morel and Pérez (1973), a move supported by Camicas et al. (1998). However, *Pholeoixodes chilensis* is a synonym of *Ixodes chilensis* in Guglielmone and Nava (2014) and here.

Pholeoixodes rubidus (Neumann, 1901): This name is used in Camicas et al. (1998) for a Neotropical species, but, as explained in Guglielmone and Nava (2014), *Pholeoixodes rubidus* is a synonym of *Ixodes rubidus*, a tick found in the Nearctic and Neotropical Regions. See also *Ixodes rubidus* in Chap. 1.

Scaphixodes auritulus (Neumann, 1904): This name is used in Camicas et al. (1998) for a species found in several zoogeographic regions, including the Neotropics, but *Scaphixodes auritulus* is a synonym of *Ixodes auritulus*, as explained in Guglielmone and Nava (2014). See also *Ixodes auritulus* in Chap. 1.

Scaphixodes brunneus (Koch, 1844): This name is used in Camicas et al. (1998) for a species that these authors consider to be present in the Nearctic and Neotropical Regions. However, the name *Ixodes brunneus* is treated as the correct name for this taxon in Guglielmone and Nava (2014) and here, and *Ixodes brunneus* is currently regarded as a Nearctic species. See also *Ixodes brunneus* above, this section.

Scaphixodes copei (Wilson, 1980): This name is used in Camicas et al. (1998) for a Neotropical species, but *Scaphixodes copei* is a synonym of *Ixodes copei*, as explained in Guglielmone and Nava (2014).

Scaphixodes cornuae (Arthur, 1960a): This name is used in Camicas et al. (1998) for a Neotropical species, but *Scaphixodes cornuae* is a synonym of *Ixodes cornuae* described in Arthur (1960b), as explained in Guglielmone and Nava (2014).

Scaphixodes downsi (Kohls, 1957): This name is used in Camicas et al. (1998) for a Neotropical species, but *Scaphixodes downsi* is a synonym of *Ixodes downsi* described in Kohls (1957b), as explained in Guglielmone and Nava (2014).

4.3.2 *Metastriata*

4.3.2.1 Incorrect Names Associated with the Genus *Amblyomma*

Acarus aureolatus Pallas, 1772: This is the original name for the Neotropical *Amblyomma aureolatum*, whose current status was established by Koch (1844). The original description of the female of *Amblyomma aureolatum* is in Pallas (1772), under the name *Acarus aureolatus*. See also *Amblyomma aureolatum* in Chap. 2.

Acarus cajennensis Fabricius, 1787: This is the original name for the Neotropical *Amblyomma cajennense*, whose current status was established by Koch (1844). The original description of the female of *Amblyomma cajennense* is in Fabricius (1787) under the name *Acarus cajennensis*. See also *Amblyomma cajennense* in Chap. 2.

Acarus grossus Pallas, 1772: Neumann (1899) stated that Pallas (1772) described the female of *Acarus grossus* from material collected in Suriname and placed this tick in the genus *Amblyomma* as *Amblyomma* (?) *grossum*. *Acarus grossus* is a *nomen dubium* in Camicas et al. (1998), Guglielmone and Nava (2014) and here, but is listed as valid in Rawlins et al. (1993) under the name *Amblyomma grossum*, a decision that is not supported here.

Acarus testudinis (Conil, 1877): This name is a *lapsus* in Santos Dias (1989a) for *Ixodes testudinis* Conil, 1877, a synonym of *Amblyomma argentinae*.

Amblyomma adspersum Koch, 1844: This tick was described from four females and a male collected in Colombia. Camicas et al. (1998) listed this name as a synonym of *Amblyomma bibroni* (Gervais, 1842, a *nomen nudum*), but Guglielmone and Nava (2014) follow Neumann (1911a), who treated *Amblyomma adspersum* as a synonym of *Amblyomma dissimile*.

Amblyomma agamum Aragão, 1912b: This tick was described from larvae, nymphs, and females collected in nature and reared in the laboratory in Brazil. *Amblyomma agamum* is regarded as a synonym of *Amblyomma rotundatum* by Robinson (1926), Camicas et al. (1998), Guglielmone et al. (2014), and here. See also *Amblyomma rotundatum* in Chap. 2.

Amblyomma altiplanum Dios, 1917: This tick was described from males and females collected in northwestern Argentina, but *Amblyomma altiplanum* is a synonym of the Neotropical *Amblyomma parvitarsum*, as stated in Aragão (1935), an opinion accepted by Camicas et al. (1998), Guglielmone and Nava (2014), and here.

Amblyomma americanum (Linnaeus, 1758): A Nearctic species originally named *Acarus americanus* and given its current status in Koch (1844). Guglielmone et al. (2003a) provided a long list of unconfirmed Neotropical records of *Amblyomma americanum* from Argentina, Bolivia, Brazil, Chile, Colombia, Guatemala, "Guiana," southern Mexico, Paraguay, Peru, Uruguay, and Venezuela, where, in some cases, *Amblyomma americanum* was confused with *Amblyomma ovale*, undetermined species of the *Amblyomma maculatum* group, or *Amblyomma parvum*, but usually it is uncertain which tick species have been incorrectly identified as *Amblyomma americanum*. Bequaert (1945) warned that published South American records of *Amblyomma americanum* are unreliable, and Fairchild et al. (1966) consider this tick restricted to the Nearctic Region, a consensus that remains unchanged. Guzmán-Cornejo et al. (2011) listed a record of *Amblyomma americanum* from the Mexican state of Guerrero, but the coordinates for the locality where this tick was found correspond to the State of Cohauila in northern Mexico.

Amblyomma arianae Keirans and Garris, 1986: This tick was described from specimens collected in Puerto Rico, but this name was found to be a synonym of *Amblyomma quadricavum* by Keirans and Klompen (1996). See also *Amblyomma quadricavum* in Chap. 2.

Amblyomma auronitens Berlese, 1888: A tick described from an undetermined number of males collected in Paraguay, but this name is considered to be a synonym of the Neotropical and, to a lesser extent, Nearctic *Amblyomma ovale* (Neumann 1899; Aragão and Fonseca 1961b; Camicas et al. 1998; Guglielmone and Nava 2014). See also *Amblyomma ovale* in Chap. 2.

Amblyomma avecolens Cooley and Kohls, 1944: This tick was described from nymphs collected in Belize, Panama, and the southern USA, but this name is another synonym of the Neotropical *Amblyomma longirostre* (Jones et al. 1972; Camicas et al. 1998; Guglielmone and Nava 2014). See also *Amblyomma longirostre* in Chap. 2.

Amblyomma avicola Marx and Neumann, 1899 in Neumann (1899): This tick was described from a nymph collected in Trinidad and Tobago, but this name represents another synonym of *Amblyomma longirostre* (Neumann 1911a; Camicas et al. 1998; Guglielmone and Nava 2014). The description of the nymph of *Amblyomma longirostre* is in Neumann (1899) under the name *Amblyomma avicola*. See also *Amblyomma longirostre* in Chap. 2.

Amblyomma beccarii Tonelli Rondelli, 1939: This tick was described from one male and two females collected in Guyana. Aragão and Fonseca (1953a) treated

Amblyomma beccarii as a synonym of *Amblyomma scalpturatum*, but Camicas et al. (1998) listed the male of *Amblyomma beccarii* as a synonym of both *Amblyomma scalpturatum* and *Amblyomma ovale*. Labruna et al. (2005c) specifically stated that the female of *Amblyomma beccarii* closely resembles the female of *Amblyomma ovale*, a view undisputed in Guglielmone and Nava (2014). Nevertheless, the female described by Tonelli Rondelli (1939) has some morphological features that are not shared by typical females of *Amblyomma ovale*, and a reexamination of the female types of *Amblyomma beccarii* would assist in confirming its synonymy with *Amblyoma ovale*. See also *Amblyomma ovale* in Chap. 2.

Amblyomma bibroni (Gervais, 1842): This tick was originally named as *Ixodes bibroni* by Gervais (1842) and later transferred to the genus *Amblyomma* by Blanchard (1909). *Amblyomma bibroni* (Gervais, 1842) is valid in Camicas et al. (1998), who treat *Amblyomma dissimile* as its synonym. However, the name *Ixodes bibroni* Gervais, 1842 is a name that lacks both a description and a figure, i.e., it is a *nomen nudum*. Accordingly, *Amblyomma bibroni* (Gervais, 1842) is also a *nomen nudum*, as explained by Guglielmone and Nava (2014). See also *Amblyomma dissimile* in Chap. 2.

Amblyomma bispinosum Neumann, 1906: This tick was described from two females collected in Venezuela and was regarded as representing a valid species in Neumann (1911a) and Keirans and Hillyard (2001). However, Santos Dias (1958a) examined the types of *Amblyomma bispinosum* and found them to be synonymous with the Neotropical and Nearctic *Amblyomma coelebs*, a view supported by Camicas et al. (1998), Guglielmone and Nava (2014), and here.

Amblyomma boneti Hoffmann, 1946: This tick was described from one larva, one female, two nymphs, and several males collected in southern Mexico, but this name is considered to be a synonym of the Neotropical and Nearctic *Amblyomma scutatum* (Keirans and Clifford 1984; Camicas et al. 1998; Guglielmone and Nava 2014). We provisionally believe that the first description of the nymph and larva of *Amblyomma scutatum* were published under the name *Amblyomma boneti.* See also *Amblyomma scutatum* in Chap. 2.

Amblyomma bouthieri Sénevet, 1940: This tick was described from one male and two females collected in French Guiana, but this name was later found to be a synonym of the Neotropical *Amblyomma tigrinum* (Floch and Fauran 1958; Camicas et al. 1998; Guglielmone and Nava 2014).

Amblyomma brasiliense guyanense Floch and Abonnenc, 1941: This tick was described from one male and one female collected in French Guiana, but this name was later found to be a synonym of the Neotropical *Amblyomma scalpturatum* (Floch and Fauran 1958; Camicas et al. 1998; Guglielmone and Nava 2014).

Amblyomma brimonti Neumann, 1913: This tick was described from three females collected in French Guiana, but this name was later found to be a synonym of the Neotropical *Amblyomma humerale* by Aragão and Fonseca (1953a), Camicas et al. (1998) and Guglielmone and Nava (2014).

Amblyomma cajennense chacoensis Ivancovich, 1973: This tick was described from a female collected in northern Argentina, but *Amblyomma cajennense chacoensis* was later found to be a synonym of the Neotropical *Amblyomma dubitatum*, as explained in Guglielmone and Nava (2014).

Amblyomma calcaratum leucozomum Schulze, 1936: This is a new name used in Santos Dias (1958b) for *Amblyomma leucozomum*, described by Schulze (1936b) from material collected in Brazil. Both names, *Amblyomma leucozomum* and *Amblyomma calcaratum leucozomum*, are synonyms of the Neotropical *Amblyomma calcaratum*, according to Camicas et al. (1998) and Guglielmone and Nava (2014). However, Guglielmone et al. (2017) have since suggested that new studies may be required to confirm this synonymy, which is treated as provisionally valid here.

Amblyomma calcaratum venezuelensis Vogelsang and Cordero, 1953: This tick was described by Vogelsang and Cordero (1953a) from several males and females collected in Venezuela, but *Amblyomma calcaratum venezuelensis* is treated as a synonym of *Amblyomma calcaratum* in Camicas et al. (1998) and Guglielmone and Nava (2014).

Amblyomma calcaris Nakatsudi, 1942: This tick was described from one female collected from a human in Manchuria (Palearctic Zoogeographic Region). Surprisingly, Camicas et al. (1998) listed *Amblyomma calcaris* as a synonym of the Neotropical *Amblyomma calcaratum*, a view apparently supported by Yamauchi (2010). Nevertheless, the tick described by Nakatsudi (1942) has no morphological relationship to *Amblyomma calcaratum*, and given the peculiar morphology of its palps and hypostome, it is unclear whether *Amblyomma calcaris* even belongs to the genus *Amblyomma* (Guglielmone et al. 2014). Therefore, Guglielmone and Nava (2014) and Guglielmone et al. (2014) regard *Amblyomma calcaris* as a name *incertae sedis*, a view also supported here.

Amblyomma carenatum Ivancovich, 1973: This Argentinean tick was originally named *Amblyomma parvum carenatus*, but Ivancovich and Luciani (1992) changed its name to *Amblyomma carenatum*, a synonym of the Neotropical *Amblyomma parvum* in Guglielmone and Nava (2014) and here.

Amblyomma castanedai Vargas and Hoffmann, 1952: This tick was described under the name *Amblyomma casteñedai* from several males collected in southern Mexico, but this name was found to be a synonym of *Amblyomma scutatum* by Camicas et al. (1998) and Guglielmone and Nava (2014).

Amblyomma complanatum Berlese, 1888: This tick was described from an unknown number of males collected in Brazil; it has been treated as a synonym of *Amblyomma maculatum* in Neumann (1911a) and Jones et al. (1972), and as a synonym of *Amblyomma tigrinum* in Camicas et al. (1998) and Dantas-Torres et al. (2009a). The name *Amblyomma complanatum* is considered *incertae sedis* by Guglielmone et al. (2014) and here. See also *Amblyomma maculatum* and *Amblyomma tigrinum* in Chap. 2.

Amblyomma concolor Neumann, 1899: This tick was described from one male and five females collected in northern Brazil. Neumann (1911a) and Camicas et al. (1998) listed *Amblyomma concolor* as a valid species, while Keirans and Hillyard (2001) and Guglielmone and Nava (2014) regarded this name as a synonym of the Neotropical and Nearctic *Amblyomma auricularium*. However, given current uncertainties concerning the morphological definition of *Amblyomma auricularium*, the status of *Amblyomma concolor* should be reexamined. See also the discussion of *Amblyomma auricularium* in Chap. 2.

Amblyomma confine Koch, 1844: This tick was described from one female collected in Brazil. *Amblyomma confine* is listed as a synonym of *Amblyomma aureolatum* in Neumann (1911a), Aragão and Fonseca (1961a, b), and Camicas et al. (1998), while it is a synonym of *Amblyomma ovale* in Dantas-Torres et al. (2009a). The type of *Amblyomma confine* has been examined by Santiago Nava, who confirmed that it is a synonym of *Amblyomma ovale*, as explained in Guglielmone and Nava (2014). See also *Amblyomma ovale* in Chap. 2.

Amblyomma conspicuum Aragão, 1912b: This tick, which was described from four males collected in northwestern Brazil, is regarded as a synonym of the Neotropical *Amblyomma pictum* (Robinson 1926; Camicas et al. 1998; Guglielmone and Nava 2014). See also *Amblyomma pictum* in Chap. 2.

Amblyomma cooperi Nuttall and Warburton, 1908: This tick was described from one male and five females collected in Paraguay. *Amblyomma cooperi* is a synonym of *Amblyomma dissimile* in Neumann (1911a), while Robinson (1926), Guglielmone and Viñabal (1994), Barker and Murrell (2004, 2008), and others treated this name as valid. The name *Amblyomma cooperi* is a synonym of *Amblyomma dubitatum*, as stated in Camicas et al. (1998) and confirmed by Estrada-Peña et al. (2002), and as explained in Guglielmone et al. (2003a). The first descriptions of the nymph and larva of *Amblyomma dubitatum* were made by Estrada-Peña et al. (2002) and Joan (1930), respectively, under the name *Amblyomma cooperi*. See also *Amblyomma dissimile* and *Amblyomma dubitatum* in Chap. 2.

Amblyomma crassipunctatum Stoll, 1894: This name was applied to a male tick collected in Nicaragua, but *Amblyomma crassipunctatum* has since been found to be a synonym of the Neotropical *Amblyomma varium* (Onofrio et al. 2008). See also *Amblyomma varium* in Chap. 2.

Amblyomma cubanum Schulze, 1941: This tick was described by Schulze (1941d) from a male collected in Cuba. *Amblyomma cubanum* is a synonym of *Amblyomma argentinae*, under the name *Amblyomma testudinis* in Santos Dias (1958b), and a synonym of *Amblyomma bibroni* (Gervais, 1842, a *nomen nudum*) in Camicas et al. (1998), but it is treated as a synonym of the Neotropical and Nearctic *Amblyomma dissimile* in Guglielmone et al. (2001), a view supported here. See also *Amblyomma argentinae* and *Amblyomma dissimile* in Chap. 2.

Amblyomma curruca Schulze, 1936: This tick was described by Schulze (1936b) from three males and one female collected in Venezuela. *Amblyomma curruca* is a valid name in Camicas et al. (1998), but it was treated as a synonym of *Amblyomma parvum* in Aragão and Fonseca (1953a) and Guglielmone et al. (2003b), while Hoffmann (1950) diagnosed all parasitic stages of *Amblyomma curruca* from specimens collected in the Neotropical portion of Mexico, but Hoffmann (1962) later changed her diagnosis to *Amblyomma parvum*. Nevertheless, *Amblyomma curruca* is regarded as a synonym of *Amblyomma auricularium* in Fairchild et al. (1966) as well as in Guglielmone et al. (2009) and Guglielmone and Nava (2014), which represents a change of opinion from that expressed in Guglielmone et al. (2003b). The conspecificity of *Amblyomma curruca* and *Amblyomma auricularium* cannot be confirmed because there are no types of *Amblyomma auricularium* for comparison. Therefore, such synonymy is treated as provisionally valid. *Amblyomma beaurepairei*, which is treated as a valid species here, is regarded as a synonym of *Amblyomma curruca* in Camicas et al. (1998). See also *Amblyomma auricularium*, *Amblyomma beaurepairei* and *Amblyomma parvum* in Chap. 2.

Amblyomma darlingi Nuttall, 1912: This tick was described from several males, females, nymphs, and larvae collected in Panama, but *Amblyomma darlingi* is regarded as a synonym of the Neotropical and Nearctic (one record) *Amblyomma oblongoguttatum* in Robinson (1926), Camicas et al. (1998) and Guglielmone and Nava (2014). The descriptions of the nymph and larva of *Amblyomma oblongoguttatum* were first reported under the name *Amblyomma darlingi.* See also *Amblyomma oblongoguttatum* for the possible reinstatement of *Amblyomma darlingi* (Chap. 2).

Amblyomma darwini wollebaeki Schulze, 1936: This subspecies was described by Schulze (1936a) from males and females collected in Ecuador (Galápagos Islands), but this name is treated as a provisional synonym of the Neotropical *Amblyomma darwini* by Guglielmone and Nava (2014). Avian hosts of this tick are treated as doubtful in Guglielmone et al. (2014) and here.

Amblyomma deminutivum Neumann, 1899: This tick was described from five females of Colombian origin. *Amblyomma deminutivum* is a valid name in Neumann (1911a), and a synonym of *Amblyomma bibroni* (Gervais, 1842, a *nomen nudum*) in Camicas et al. (1998), while *Amblyomma deminutivum* is regarded as a synonym of *Amblyomma dissimile* in Aragão and Fonseca (1953a), Guglielmone and Nava (2014), and here.

Amblyomma denticulatum Koch, 1844: The female of this tick was described from material collected in Brazil. *Amblyomma denticulatum* is treated as a *nomen dubium* in Camicas et al. (1998), Guglielmone and Nava (2014), and here.

Amblyomma diemeniae Schulze, 1950: This tick was described as an Australasian species from males and females collected on a squamatan host under captive conditions in Germany. Camicas et al. (1998) treated this tick as a probable synonym of the Neotropical *Amblyomma argentinae*, and Guglielmone and Nava (2014) confirmed this synonymy.

Amblyomma exophthalmum Méndez Arocha and Ortiz, 1958: This tick was described from a mutilated female collected in Venezuela. *Amblyomma exophthalmum* is considered valid in Guerrero (1996), listed as a synonym of *Amblyomma oblongoguttatum* in Camicas et al. (1998), and was treated as a name *incertae sedis* in Guglielmone et al. (2003a, b, c). The description of *Amblyomma exophthalmum* is vague and the type is lost. This name is a *nomen nudum*, as explained in Guglielmone and Nava (2014).

Amblyomma extraoculatum Neumann, 1899: This tick was described from a female collected in Singapore (Oriental Zoogeographic Region). Santos Dias (1955) stated that the locality of *Amblyomma extraoculatum* was incorrect because this species was identical to the Neotropical *Amblyomma romitii*, and proposed that the latter name be reduced to a synonym of the former. The opinion of Santos Dias (1955) prevailed for a long time until Barros-Battesti et al. (2007a) demonstrated that both species, *Amblyomma extraoculatum* and *Amblyomma romitii*, are valid. Serra-Freire et al. (2012) allegedly found *Amblyomma extraoculatum* on tortoises in Peru but depicted a species unlike *Amblyomma extraoculatum* and *Amblyomma romitii*, the alleged former synonym of *Amblyomma extraoculatum*. See also *Amblyomma romitii* in Chap. 2.

Amblyomma fiebrigi Robinson, 1912: This tick was described from a male and a female collected in Paraguay, but this name was found to be a synonym of the Neotropical *Amblyomma pacae* (Robinson 1926; Camicas et al. 1998; Guglielmone and Nava 2014).

Amblyomma finitimum Tonelli Rondelli, 1937: This tick was described from a male and several females collected in Guyana, but this name was found to be a synonym of the Neotropical *Amblyomma cajennense sensu stricto* in Nava et al. (2014a).

Amblyomma foreli Stoll, 1894: This tick was described from a female collected in Guatemala, and considered to be a synonym of *Amblyomma americanum* in Neumann (1911a), Robinson (1926), Camicas et al. (1998), and others. Guglielmone and Nava (2014) argued that the description of *Amblyomma foreli* is insufficient to warrant such a decision and regarded this name as *incertae sedis*, a view supported here.

Amblyomma fossum Neumann, 1899: This tick was described from several males and females collected in Costa Rica and one male from Argentina, but this name was found to be a synonym of *Amblyomma ovale* (Aragão and Fonseca 1961b; Camicas et al. 1998; Guglielmone and Nava 2014). See also *Amblyomma ovale* in Chap. 2.

Amblyomma fossum intemedium Ivancovich, 1973: This tick was described from one male and two females collected in northern Argentina, but this name was found to be a synonym of the Neotropical *Amblyomma calcaratum* (Guglielmone and Viñabal 1994; Guglielmone and Nava 2014).

Amblyomma furcula Dönitz, 1909: This tick was described from males and females collected in northwestern Argentina, but this name was found to be a synonym of the Neotropical *Amblyomma neumanni* (Joan 1928; Santos Dias 1961; Camicas et al. 1998; Guglielmone and Nava 2014). The first description of the female of *Amblyomma neumanni* was provided by Dönitz (1909) under the name *Amblyomma furcula*. Doss et al. (1978) listed *Amblyomma furcula* as present in French Guiana based on Sénevet (1940). However, Sénevet did not state that *Amblyomma furcula* had been found in French Guiana; rather, he used the morphological characters of *Amblyomma furcula* for comparative purposes in his description of *Amblyomma bouthieri*, a synonym of *Amblyomma tigrinum*. See also *Amblyomma neumanni* in Chap. 2.

Amblyomma fuscomaculatum (Lucas, 1873): See *Ixodes fuscomaculatus* below in this section.

Amblyomma gertschi Cooley and Kohls, 1942: This tick was described by Cooley and Kohls (1942a) from a male specimen collected in Panama, but this name was found to be a synonym of *Amblyomma varium* (Camicas et al. 1998; Onofrio et al. 2008; Guglielmone and Nava 2014). See also *Amblyomma varium* in Chap. 2.

Amblyomma gervaisi (Lucas, 1847): This is an Afrotropical and Oriental species that usually parasitizes squamatan hosts (Guglielmone et al. 2014). It was originally named *Ixodes gervaisi* and given its current status in Horak et al. (2002). Hooker (1909) listed a probable record of *Amblyomma gervaisi* (under the name *Aponomma gervaisi*) from Puerto Rico or the Galápagos Islands, most probably due to mislabeling or incorrect interpretation of the locality (Isabel Island), because there are no bona fide records of *Amblyomma gervaisi* from the Neotropical Region.

Amblyomma giganteum Neumann, 1899: This tick was described from a female collected in Trinidad and Tobago, but this name was found to be a synonym of *Amblyomma longirostre*, as stated in Neumann (1911a), Camicas et al. (1998), and Guglielmone and Nava (2014).

Amblyomma grossum (Pallas, 1772): This tick was originally named *Acarus grossus* and suspected of belonging to the genus *Amblyomma* by Neumann (1899). Rawlins et al. (1993) used the name *Amblyomma grossum* as a valid name for a species found in Suriname, but this name is a *nomen dubium*, as explained in Guglielmone and Nava (2014).

Amblyomma guianense Neumann, 1907: This tick was described by Neumann (1907) from two males and one female allegedly collected in Suriname. Jones et al. (1972) found morphological similarities between *Amblyomma guianense* and *Amblyomma multipunctum*, while Camicas et al. (1998) treated *Amblyomma guianense* as a synonym of *Amblyomma oblongoguttatum.* Guglielmone et al. (2003a) noted that the female of *Amblyomma guianense* is almost identical to the female of the Afrotropical species *Amblyomma trimaculatum*, described in Neumann (1908) from a female collected in Liberia, a fact recognized by Neumann (1908). Consequently, Guglielmone and Nava (2014) listed *Amblyomma guianense* as a name *incertae sedis*, an opinion supported here. See also *Amblyomma multipunctum* and *Amblyomma oblongoguttatum* in Chap. 2, and *Amblyomma trimaculatum* below in this section.

Amblyomma gypsatum Neumann, 1899: This tick was described from four males allegedly collected in Uruguay "nord de Goyaz," probably a reference to the Brazilian state of Goiás, but this name was found to be a synonym of *Amblyomma humerale* (Neumann 1911a; Camicas et al. 1998; Guglielmone and Nava 2014). See also *Amblyomma humerale* in Chap. 2.

Amblyomma haitianum Schulze, 1941: This tick was described by Schulze (1941a) from ten males allegedly collected in "Santiago de la Caballeros, Haití," most probably meaning Santiago de los Caballeros, a locality within the Dominican Republic. *Amblyomma haitianum* is a synonym of the Neotropical *Amblyomma albopictum* in Santos Dias (1958b), Camicas et al. (1998) and Guglielmone and Nava (2014). See also *Amblyomma albopictum* in Chap. 2.

Amblyomma hebraeum Koch, 1844: An Afrotropical species that has been found on a variety of hosts, but this tick usually parasitizes Artiodactyla: Bovidae. Guglielmone et al. (2003a) discussed the alleged Neotropical records of *Amblyomma hebraeum*, concluding that this tick is not established there, all such records having resulted from accidental importations, erroneous identifications, *lapsus calamorum*, or even incorrect locality labels.

Amblyomma humanum (Koch, 1844): See *Ixodes humanus* below in this section.

Amblyomma imitator Kohls, 1958: This tick was described from several males, females, and nymphs collected in the southern USA and in northern (Nearctic) and southern (Neotropical) Mexico, but this name was found to be a synonym of the Nearctic and Neotropical *Amblyomma tenellum* by Nava et al. (2014c). See also *Amblyomma tenellum* in Chap. 2.

Amblyomma inflatum Neumann, 1901: This tick was described from two nymphs collected in Chile. *Amblyomma inflatum* was regarded as a *nomen nudum* in González-Acuña and Guglielmone (2005) and a name *incertae sedis* in Guglielmone and Nava (2014), whose opinion is accepted here.

Amblyomma infumatum Koch, 1844: This tick was described from two females, one collected in Brazil and the other in Colombia or Venezuela (Moritz and Fischer 1981). *Amblyomma infumatum* is a name *incertae sedis* in Neumann (1911a), while Camicas et al. (1998) regarded this tick as a synonym of *Amblyomma bibroni* (Gervais, 1842, a *nomen nudum*). Robinson (1926) listed *Amblyomma infumatum* as a synonym of *Amblyomma dissimile*, and Guglielmone and Nava (2014) treat the synonymy proposed by Robinson (1926) as provisionally valid, an opinion also accepted here.

Amblyomma ininii Floch and Abonnenc, 1941: This tick was described from one male and one female collected in French Guiana. *Amblyomma ininii* is a synonym of *Amblyomma rotundatum* in Santos Dias (1958a), and a synonym of *Amblyomma goeldii* in Camicas et al. (1998). However, *Amblyomma ininii* is partly a synonym of both *Amblyomma goeldii* and *Amblyomma rotundatum*, as stated in Guglielmone et al. (2003a) and Guglielmone and Nava (2010), because Floch and Fauran (1958) found that the female of *Amblyomma ininii* is, in fact, the female of the Neotropical *Amblyomma goeldii*, and the male of *Amblyomma ininii* is a synonym of *Amblyomma rotundatum*, a tick found in the Nearctic and Neotropical Regions, as well as on a Pacific island. The description of the female of *Amblyomma goeldii* by Floch and Abonnenc (1941) was published under the name *Amblyomma ininii*. See also *Amblyomma goeldii* in Chap. 2.

Amblyomma irroratum Koch, 1844: This tick was described from a male collected in Brazil, and was regarded as a synonym of *Amblyomma bibroni* (Gervais, 1842, a *nomen nudum*) in Camicas et al. (1998), while Neumann (1911a) and Guglielmone and Nava (2014) listed *Amblyomma irroratum* as a synonym of *Amblyomma dissimile*, a view that is maintained here.

Amblyomma leucozomum Schulze, 1936: This tick was described by Schulze (1936b) from seven males and two females collected in Brazil. Tonelli Rondelli (1939) accepted *Amblyomma leucozomum* as valid, and Santos Dias (1958b) treated *leucozomum* as a subspecies of *Amblyomma calcaratum*. However, Camicas et al. (1998) and Guglielmone and Nava (2014) listed *Amblyomma leucozomum* as a synonym of *Amblyomma calcaratum*. Recently, Guglielmone et al. (2017) suggested that further studies are needed to definitely confirm the synonymy proposed by Camicas et al. (1998) and supported by Guglielmone and Nava (2014). Therefore, at this time *Amblyomma leucozomum* is provisionally considered to be a synonym of *Amblyomma calcaratum*.

Amblyomma longirostrum Cooper and Robinson, 1908: This tick was described from one female collected in Guyana. *Amblyomma longirostrum* was treated as a synonym of *Amblyomma longirostre* in Estrada-Peña and Castellá (1994), but this name is in fact a synonym of *Amblyomma humerale*, as stated in Robinson (1926), Camicas et al. (1998), and Guglielmone and Nava (2014), an outcome already anticipated by Cooper and Robinson (1908) when they said "it is possible that this tick is the hitherto unknown female of *Amblyomma humerale*." The description of the female of *Amblyomma humerale* appears in Cooper and Robinson (1908) under the name *Amblyomma humerale*. See also *Amblyomma humerale* in Chap. 2.

Amblyomma lutzi Aragão, 1908a: This tick was described from several males and females collected in Brazil. Santos Dias (1958a) stated that *Amblyomma lutzi* takes precedence over *Amblyomma cooperi*, but Rohr (1909) recognized *Amblyomma lutzi* as a synonym of *Amblyomma cooperi*. However, it was later found that *Amblyomma cooperi* is a synonym of *Amblyomma dubitatum* and, consequently, *Amblyomma lutzi* becomes a synonym of the Neotropical *Amblyomma dubitatum* (Camicas et al. 1998; Estrada-Peña et al. 2002; Guglielmone and Nava 2014). See also *Amblyomma dubitatum* in Chap. 2, and *Amblyomma cooperi* above in this section.

Amblyomma mantiquirense Aragão, 1908: This tick was described by Aragão (1908b) from several males and females collected in Brazil, but this name is treated as a synonym of the Neotropical *Amblyomma naponense* in Jones et al. (1972), Camicas et al. (1998), and Guglielmone and Nava (2014). The first description of the female of *Amblyomma naponense* was published by Aragão (1908a, b, c) under the name *Amblyomma mantiquirense*. See also *Amblyomma naponense* in Chap. 2.

Amblyomma margaritae Méndez Arocha, 1956: This tick was described from two females collected in Venezuela. *Amblyomma margaritae* is treated as a synonym of *Amblyomma bibroni* (Gervais, 1842, a *nomen nudum*) by Camicas et al. (1998), but this name is a synonym of *Amblyomma dissimile*, as proposed by Guerrero (1996) and Guglielmone and Nava (2014). See also *Amblyomma dissimile* in Chap. 2.

Amblyomma minutum Aragão, 1909 in Rohr (1909): This name is a synonym of *Amblyomma parvum* in Camicas et al. (1998) and Dantas-Torres et al. (2009a), but it is a name without description or figure, a *nomen nudum*, as stated in Guglielmone and Nava (2014).

Amblyomma myrmecophagae Schulze, 1933: This tick was described by Schulze (1933b) from three males collected in Suriname, but this name was found to be a synonym of the Neotropical *Amblyomma scalpturatum* by Fonseca and Aragão (1952b, under the name *Amblyomma myrmecophagium*), a view accepted by Camicas et al. (1998), Guglielmone and Nava (2014) and here. The first description of the male of *Amblyomma scalpturatum* is in Schulze (1933b) under the name *Amblyomma myrmecophagae*. See also *Amblyomma scalpturatum* in Chap. 2.

Amblyomma myrmecophagium Schulze, 1933: This tick was originally named *Amblyomma myrmecophagae*, a synonym of the Neotropical *Amblyomma scalpturatum*, but amended to *Amblyomma myrmecophagium* in Fonseca and Aragão (1952b).

Amblyomma nigrum Tonelli Rondelli, 1939: This tick was described from one female collected in Guyana, but this name was found to be a synonym of the Neotropical *Amblyomma pacae* (Aragão and Fonseca 1953a; Camicas et al. 1998; Guglielmone and Nava 2014). See also *Amblyomma pacae* in Chap. 2.

Amblyomma nitidum Hirst and Hirst, 2010: An Australasian and Oriental species that parasitizes squamatan hosts (Guglielmone et al. 2014). *Amblyomma nitidum* was listed as found in the Galápagos Islands (Ecuador) by Doss et al. (1974a), based on Vercammen-Grandjean (1966), who merely indicated that *Amblyomma nitidum* may have the ability to disperse with its hosts to the Galápagos.

Amblyomma nuttalli Dönitz, 1909: An Afrotropical species whose males and females usually parasitize tortoises (Guglielmone et al. 2014). Santos Dias (1958a) suggested that *Amblyomma nuttalli* is a synonym of the Neotropical *Amblyomma dubitatum*, but both species are valid (Guglielmone and Nava 2014; Guglielmone et al. 2014, 2015).

Amblyomma oblongum Koch, 1844: This tick was described from two females (Moritz and Fischer 1981) collected in Brazil. Guglielmone and Nava (2014) pointed out that Neumann (1911a), Aragão and Fonseca (1961b), Camicas et al. (1998), and Guglielmone and Viñabal (1994) all listed *Amblyomma oblongum* as a synonym of *Amblyomma aureolatum*, but Dantas-Torres et al. (2009a) regarded this tick as a synonym of *Amblyomma ovale*. One type of *Amblyomma oblongum* was examined by Santiago Nava and found to be a synonym of the Nearctic (few records) and Neotropical *Amblyomma ovale* (Guglielmone and Nava 2014). The description of the female of *Amblyomma ovale* is in Koch (1844) under the name *Amblyomma oblongum*. See also *Amblyomma ovale* in Chap. 2.

Amblyomma ovale kriegi Schulze, 1941: This tick was described by Schulze (1941c) from several males collected in Argentina, but this name was found to be a synonym of *Amblyomma ovale* in Aragão and Fonseca (1961b), Camicas et al. (1998) and Guglielmone and Nava (2014). See also *Amblyomma ovale* in Chap. 2.

Amblyomma ovatum Koch, 1844: This tick was described from a male specimen (Moritz and Fischer 1981) collected in Uruguay, but this name is treated as a synonym of the Neotropical *Amblyomma tigrinum* in Kohls (1956b), Camicas et al. (1998), Guglielmone and Nava (2014) and here.

Amblyomma parviscutatum Neumann, 1899: This tick was described from a female collected in Brazil. *Amblyomma parviscutatum* was treated as a synonym of *Amblyomma cajennense* in Robinson (1926), Camicas et al. (1998), and Dantas-Torres et al. (2009a), but this name is considered to be invalid in Guglielmone and Nava (2014), Nava et al. (2014a) and here.

Amblyomma parvum carenatus Ivancovich, 1973: This tick was described from several males and females collected in northern Argentina, but this name is treated as a synonym of the Neotropical *Amblyomma parvum* in Guglielmone and Viñabal (1994), Guglielmone and Nava (2014), and here.

Amblyomma perpunctatum (Packard, 1869): See *Ixodes perpunctatus* below in this section.

Amblyomma pygmaeum (Koch, 1844): See *Ixodes pygmaeus* below in this section.

Amblyomma quasicyprium Robinson, 1926: This tick was described from a female allegedly collected in Mexico. Santos Dias (1958a) and Camicas et al. (1998) treated *Amblyomma quasicyprium* as a synonym of *Amblyomma breviscutatum*. Later, Keirans (1985b) stated that the locality of the specimen used to describe *Amblyomma quasicyprium* was incorrect and treated this name as a synonym of *Amblyomma cyprium*, which in fact is a synonym of *Amblyomma breviscutatum*, a tick found in the Australasian and Oriental Regions, as discussed in Guglielmone and Nava (2014) and Guglielmone et al. (2014).

Amblyomma quasistriatum Tonelli Rondelli, 1937: This tick was described from one male collected in Paraguay, but it was found to be a synonym of *Amblyomma ovale* in Aragão and Fonseca (1961b), Camicas et al. (1998) and Guglielmone and Nava (2014). See also *Amblyomma ovale* in Chap. 2.

Amblyomma rubripes Koch, 1844: This tick was described from two females (Moritz and Fischer 1981) collected in Brazil. *Amblyomma rubripes* was treated as a synonym of *Amblyomma maculatum* in Neumann (1911a) and as a synonym of *Amblyomma tigrinum* in Camicas et al. (1998) and Dantas-Torres et al. (2009a). Santiago Nava examined the types of *Amblyomma rubripes* and confirmed that this name is a synonym of the Neotropical *Amblyomma tigrinum*, as explained in Guglielmone and Nava (2014). The description of the female of *Amblyomma tigrinum* is in Koch (1844) under the name *Amblyomma rubripes*. See also *Amblyomma tigrinum* in Chap. 2.

Amblyomma striatum Koch, 1844: This tick was described from a female (Moritz and Fischer 1981) collected in Brazil. *Amblyomma striatum* was long treated as a valid name. Schulze (1941b) carefully described the female of *Amblyomma aureolatum* but considered it a different species, although morphologically close to *Amblyomma striatum*. The prevalence of the name *Amblyomma aureolatum* over *Amblyomma striatum* was confirmed by Aragão and Fonseca (1961b) and supported by Camicas et al. (1998) and Guglielmone and Nava (2014), among others. Barker and Murrell (2004, 2008) continue to list *Amblyomma aureolatum* and *Amblyomma striatum* as separate and valid taxa. The first description of the male of *Amblyomma aureolatum* was published under the name *Amblyomma striatum* by Koch (1844). See also *Amblyomma aureolatum* in Chap. 2.

Amblyomma strobeli Berlese and Trouessart, 1889: This tick was described from one male collected in "Amérique tropicale" and was treated as a synonym of *Amblyomma oblongoguttatum* in Camicas et al. (1998). Nevertheless, Neumann (1911a) and Guglielmone and Nava (2014) regard *Amblyomma strobeli* as a name *incertae sedis*, and this opinion is maintained here.

Amblyomma superbrasiliense Schulze, 1941: This tick was described by Schulze (1941b) from a female collected in Brazil, the author suggesting that the specimen was a polyploid female of *Amblyomma brasiliense*. Fonseca and Aragão (1952b) treated *Amblyomma superbrasiliense* as a synonym of the Neotropical *Amblyomma incisum*, but Camicas et al. (1998) regarded it as a synonym of *Amblyomma brasiliense*. Guglielmone et al. (2003a) noted that the female of *Amblyomma superbrasiliense* was described as having one spur on coxae II–IV, whereas *Amblyomma brasiliense* has two spurs on these coxae. On this basis, Guglielmone et al. (2003a) supported the synonymy of *Amblyomma superbrasiliense* with *Amblyoma incisum*, as proposed by Fonseca and Aragão (1952b), a view accepted by Guglielmone and Nava (2014) and others. See also *Amblyomma brasiliense* in Chap. 2.

Amblyomma tapiri Tonelli Rondelli, 1937: This tick was described from one male and two females collected in Guyana, and this name was treated as valid in Rawlins et al. (1993). Nevertheless, *Amblyomma tapiri* was found to be a synonym of *Amblyomma cajennense sensu stricto* in the revision of the *Amblyomma cajennense* species complex by Nava et al. (2014a).

Amblyomma tasquei Floch and Abonnenc, 1940: This tick was described from five specimens collected in French Guiana, but this name was treated as a synonym of *Amblyomma extraoculatum* in Camicas et al. (1998) and Guglielmone et al. (2003a). However, *Amblyomma tasquei* is, in fact, a synonym of the Neotropical *Amblyomma romitii*, as discussed in Barros-Battesti et al. (2007a). See also *Amblyomma romitii* in Chap. 2.

Amblyomma testudinis (Conil, 1877): See *Ixodes testudinis* below in this section.

Amblyomma trimaculatum Neumann, 1908: This tick was described from a female collected in Liberia, and was later treated as a synonym of the Afrotropical *Amblyomma paulopunctatum* in Theiler and Salisbury (1959), Camicas et al. (1998) and others. However, *Amblyomma trimaculatum* of Neumann, 1908 is very similar to *Amblyomma guianense* of Neumann (1907) allegedly collected in Suriname, a fact recognized by Neumann (1908), as discussed in Guglielmone et al. (2003a). Both names, *Amblyomma trimaculatum* and *Amblyomma guianense*, are regarded as *incertae sedis* in Guglielmone and Nava (2014) and here. See also *Amblyomma guianense* above in this section.

Amblyomma trinitatis Turk, 1948: This tick was described from one female, four nymphs, and several larvae collected in Trinidad and Tobago. *Amblyomma trinitatis* is valid in Rawlins et al. (1993) and Camicas et al. (1998). Keirans and Hillyard (2001) listed *Amblyomma trinitatis* as a synonym of the Nearctic and Neotropical *Amblyomma dissimile*, a view accepted by Guglielmone and Nava (2014) and here. See also *Amblyomma dissimile* in Chap. 2.

Amblyomma tuberculatum Marx, 1894 in Hubbard (1894): This tick is treated as Nearctic and Neotropical in Camicas et al. (1998), but there are no bona fide populations of this species in the Neotropical Region (Guglielmone et al. 2003a, 2014, 2015). The specimens of *Amblyomma tuberculatum* listed from Cuba in Neumann (1899) may be the result of an introduction from the USA that failed to develop local populations on the island. Other Neotropical records, such as in Woodham et al. (1983) for Mexico, and Guatemala in Instituto Interamericano para la Cooperación en Agricultura (1988), are treated here as probable misidentifications.

Amblyomma uncatum Nuttall and Warburton, 1908: This tick was described from five males collected in South America, but this name was found to be a synonym of the Neotropical *Amblyomma nodosum* (Robinson 1926; Camicas et al. 1998; Guglielmone and Nava 2014).

Amblyomma v-notatum Nuttall, 1910: This tick was described from four females collected in Brazil. *Amblyomma v-notatum* is a synonym of *Amblyomma perpunctatum* in Camicas et al. (1998), but the name *Amblyomma perpunctatum* is invalid here. *Amblyomma v-notatum* is a synonym of the Neotropical *Amblyomma geayi* in Robinson (1926), a view supported by Guglielmone and Nava (2014) and here. See *Amblyomma perpunctatum* and *Ixodes perpunctatus* in this section.

Amblyomma varium albida Neumann, 1899: This tick was described from a male collected in Chile, but that locality is incorrect (González-Acuña and Guglielmone 2005). *Amblyomma varium albida* is a synonym of the Neotropical *Amblyomma varium* (Camicas et al. 1998; Onofrio et al. 2006a, b; Guglielmone and Nava 2014). See also *Amblyomma varium* in Chap. 2.

Amblyomma versicolor Nuttall and Warburton, 1908: This tick was described from one male and one female collected in Mexico. *Amblyomma versicolor* was found to be a synonym of the Nearctic and Neotropical *Amblyomma mixtum* in the study of the species constituting the *Amblyomma cajennense* group by Nava et al. (2014a).

Amblyomma vittatum Neumann, 1899: This tick was described from a female collected in Bolivia, but this name was found to be a synonym of the Nearctic (one record) and Neotropical *Amblyomma oblongoguttatum* (Neumann 1911a; Camicas et al. 1998; Guglielmone and Nava 2014).

Amblyomma ypsilophorum Schulze, 1941: This tick was described by Schulze (1941b) from one male and one female collected in Brazil. *Amblyomma ypsilophorum* is a synonym of *Amblyomma cooperi* in Fonseca and Aragão (1952b), Estrada-Peña and Castellá (1994), and others, but this name is actually a synonym of the Neotropical *Amblyomma dubitatum* (Camicas et al. 1998; Estrada-Peña et al. 2002; Guglielmone and Nava 2014). See also *Amblyomma dubitatum* in Chap. 2.

Aponomma decorosum (Koch, 1867): This name, treated as *Amblyomma decorosum* in Barker and Murrell (2004, 2008), is a synonym of *Bothriocroton undatum* in Klompen et al. (2002) and Guglielmone et al. (2009). See *Bothriocroton undatum* in this chapter (section on "Invasive ticks").

Aponomma elaphensis Price, 1959: This tick was described from several males, females, and nymphs collected in the southern USA, but the species was later transferred to the genus *Robertsicus* by Barker and Burger (2018) as *Robertsicus elaphensis.* Camicas et al. (1998) listed this species as *Aponomma elaphense* (a grammatically correct name in relation to *Aponomma elaphensis*), listing it as found in the Nearctic and Neotropical Regions, but no records of this tick from the Neotropics were confirmed in Guglielmone et al. (2003a) or by subsequent workers.

Aponomma inornata Banks, 1909: This tick was described from a female collected in the southern USA. The name *Aponomma inornata* is the original name for the Nearctic and Neotropical *Amblyomma inornatum*, whose status was recognized by Fairchild (1943) after collecting specimens of this tick in Panama. Nevertheless, considerable problems attend the correct identification of *Amblyomma inornatum*, as discussed under this species in Chap. 2.

Aponomma laevatum Schulze, 1936: This tick was described by Schulze (1936b) from one male and four females collected in Germany from a captive Neotropical host (*Boa constrictor*) and was treated as a South American representative of the former genus *Aponomma*, an opinion rejected by most tick workers, as discussed in Guglielmone et al. (2017). Estrada-Peña and Castellá (1994) listed *Aponomma laevatum* as a *lapsus* for *Aponomma latum*, a synonym of *Amblyomma latum*, while Santos Dias (1993) treated *Aponomma laevatum* as a synonym of *Aponomma latum* (=*Amblyomma latum*). Guglielmone and Nava (2014) regarded *Aponomma laevatum* as a *nomen nudum*, but this name is treated as *incertae sedis* in Kaufman (1972), Guglielmone et al. (2015, 2017) and here.

Aponomma laeve Neumann, 1899: This tick was described from six males and two females collected in Patagonia, "Santa Cruz?" probably meaning a locality in southern Argentina, but Aragão (1935) believes that Neumann (1899) received incorrect information about the type locality. *Aponomma laeve* is treated as a synonym of *Aponomma latum* (=*Amblyomma latum*) in Kaufman (1972) and Camicas et al. (1998), but is considered to be a name *incertae sedis* in Guglielmone and Nava (2014) and here.

Aponomma omissum Schulze, 1932: This tick was described from three males, three females, and two nymphs of alleged Brazilian origin but collected from a captive host in Germany, which renders the locality of this species doubtful. *Aponomma omissum* is a synonym of *Amblyomma varanense* in Kaufman (1972) and Camicas et al. (1998), under the genus *Aponomma* in both cases, and a synonym of *Amblyomma fuscolineatum* in Santos Dias (1993, under the currently invalid genus *Aponomma*). *Aponomma omissum* is treated as a name *incertae sedis* in Guglielmone and Nava (2014), Guglielmone et al. (2017), and here.

Aponomma patagonicum Schulze, 1936: This tick was described by Schulze (1936b) from a female found in the same vial that contained the specimens used by Neumann (1899) to describe *Aponomma laeve*, but in his description of *Aponomma patagonicum* Schulze (1936b) failed to question the incorrect Argentinean locality for *Aponomma laeve* (see above in this section). The name *Aponomma patagonicum* is a synonym of *Amblyomma flavomaculatum* in Santos Dias (1993), under the genus *Aponomma*, and a synonym of *Amblyomma gervaisi* in Kaufman (1972) and Camicas et al. (1998), also under the currently invalid genus *Aponomma*. *Amblyomma patagonicum* is treated as a name *incertae sedis* in Guglielmone and Nava (2014), Guglielmone et al. (2017), and here.

Aponomma thumbi Schulze, 1941: This tick was described in Schulze (1941d) from a female collected in Cuba. Santos Dias (1958b, c) treats *Aponomma thumbi* as a synonym of *Amblyomma argentinae*, under the name *Amblyomma testudinis*. *Amblyomma thumbi* is a synonym of *Amblyomma bibroni* (Gervais, 1842, a *nomen nudum*) in Camicas et al. (1998), but this name is treated as a synonym of *Amblyomma dissimile* in Estrada-Peña and Castellá (1994), Guglielmone et al. (2001), Guglielmone and Nava (2014), and here. See also *Amblyomma argentinae* and *Amblyomma dissimile* in Chap. 2.

Dermacentor triangulatus Neumann, 1901: This tick was described from one female collected in Argentina. *Dermacentor triangulatus* is treated as a synonym of

Amblyomma tigrinum in Camicas et al. (1998), but it is considered a *nomen dubium* in Guglielmone and Nava (2014) and here.

Haemalastor argentinae (Neumann, 1905): This name used in Oudemans (1939) is a synonym of *Amblyomma argentinae.*

Haemalastor aureolatus (Pallas, 1772): This name used in Oudemans (1939) is a synonym of *Amblyomma aureolatum.*

Haemalastor crassitarsus Karsch, 1880: This tick was described from six males (Moritz and Fischer 1981) collected in Venezuela, but this name was found to be a synonym of the Neotropical *Amblyomma longirostre* (Neumann 1911a; Camicas et al. 1998; Guglielmone and Nava 2014). The description of the male of *Amblyomma longirostre* is in Karsch (1880) under the name *Haemalastor crassitarsus.* See also *Amblyomma longirostre* in Chap. 2.

Haemalastor longirostris Koch, 1844: This tick was described from an unknown number of females collected in Brazil. The name *Haemalastor longirostris* is the original name for *Amblyomma longirostre*, whose current status was provided by Neumann (1905). See *Amblyomma longirostre* in Chap. 2.

Haemalastor maculatum (Koch, 1844): This name used in Oudemans (1936) is a synonym of *Amblyomma maculatum.*

Haemalastor rotundatum (Koch, 1844): This name used in Oudemans (1936) is a synonym of *Amblyomma rotundatum.*

Haemalastor varium (Koch, 1844): This name used in Oudemans (1936) is a synonym of *Amblyomma varium.*

Hyalomma cyclurae Pagenstecher, 1861: This tick was very briefly described from a specimen collected in Honduras that is probably a larval *Amblyomma*, according to Neumann (1911a). *Hyalomma cyclurae* is a *nomen dubium* in Camicas et al. (1998), Guglielmone and Nava (2014), and here.

Hyalomma longirostre (Koch, 1844): The name *Hyalomma longirostre* was first used in Neumann (1901), but this name is a synonym of the Neotropical *Amblyomma longirostre* (Neumann 1911a; Camicas et al. 1998; Guglielmone and Nava 2014). The name *Hyalomma longirostre* appears instead of *Amblyomma longirostre* in several older publications (Lahille 1905; Neumann 1905; Hooker 1909).

Hyalomma venustum (Koch, 1844): Barber (1895) stated that *Hyalomma venustum* was described by Koch, 1847 [sic] from a male collected in Senegal (Afrotropical Region), commenting that this species had become established in Antigua. The name *Hyalomma venustum* was proposed by Michael (1893) as new status for *Amblyomma venustum* described by Koch (1844) from a male collected in Senegal. Both names, *Amblyomma venustum* and *Hyalomma venustum*, are synonyms of the Afrotropical and Neotropical *Amblyomma variegatum*, as discussed in Guglielmone and Nava (2014).

Ixodes auricularius Conil, 1878: This tick was described from males and females collected in Argentina. *Ixodes auricularius* is the original name for the Nearctic and Neotropical *Amblyomma auricularium*, whose current status was provided by Lahille (1905). See also *Amblyomma auricularium* in Chap. 2.

Ixodes boarum Stoll, 1894: This tick was described from several alleged females collected in Guatemala, but Neumann (1899) found that the specimens of *Ixodes boarum* were in fact nymphs, an outcome somewhat anticipated by Stoll (1894), who noted "The rudimentary state of the genital orifice renders it possible that the specimens are not adult." *Ixodes boarum* is treated as a synonym of *Amblyomma bibroni* (Gervais, 1842, a *nomen nudum*) in Camicas et al. (1998), but this name is regarded as a synonym of the Nearctic and Neotropical *Amblyomma dissimile* in Neumann (1899, 1911a), Robinson (1926), Guglielmone and Nava (2014), and here. See also *Amblyomma dissimile* in Chap. 2.

Ixodes flavidus Koch, 1844: This tick was described from one female (Moritz and Fischer 1981) collected in Brazil, but that specimen was found to be a nymph of *Amblyomma dissimile* by Neumann (1899), although it was treated as a synonym of *Amblyomma bibroni* (Gervais 1842, a *nomen nudum*) in Camicas et al. (1998). However, *Ixodes flavidus* (under the name *Amblyomma flavidum*) is a name *incertae sedis* in Neumann (1911a), an opinion followed in Guglielmone and Nava (2014) and here.

Ixodes fuscomaculatus Lucas, 1873: This tick was described from five females collected in South America. Santos Dias (1958a, 1989b) found the females of *Ixodes fuscomaculatus* to be identical to females of *Amblyomma goeldii*; consequently, Santos Dias (1958a) renamed the species of Lucas (1873) as *Amblyomma fuscomaculatum* and treated *Amblyomma goeldii* of Neumann (1899) as its synonym. However, Floch and Fauran (1958) and Keirans (1985b) determined that the female of *Amblyomma goeldii* is a synonym of *Amblyomma rotundatum* of Koch (1844), so *Amblyomma fuscomaculatum* becomes a junior synonym of *Amblyomma rotundatum*, a tick found in the Nearctic and Neotropical Regions and on one Pacific Island. See also *Amblyomma goeldii* in Chap. 2.

Ixodes humanus Koch, 1844: This tick was described from one female (Moritz and Fischer 1981) collected in Brazil and considered to be a nymph of *Amblyomma dissimile* in Neumann (1899). *Ixodes humanus* was treated as a synonym of *Amblyomma bibroni* (Gervais, 1842, a *nomen nudum*) in Camicas et al. (1998). Neumann (1911a) named this tick as *Amblyomma humanum* but regarded it as *incertae sedis*. Santiago Nava examined the type nymph of *Ixodes humanus* and found it to be an *Amblyomma* different from *Amblyomma dissimile* while also treating this name as *incertae sedis* (Guglielmone and Nava, 2014). See also *Amblyomma dissimile* in Chap. 2.

Ixodes naponensis Packard, 1869: This tick was described from a male specimen collected in Ecuador. *Ixodes naponensis* is the original name for the Neotropical *Amblyomma naponense*, whose current status was provided by Neumann (1911a). See also *Amblyomma naponense* in Chap. 2.

Ixodes perpunctatus Packard, 1869: This tick was described from a male collected in South America, but it was placed in the genus *Amblyomma* as *Amblyomma perpunctatum* by Neumann (1911a), who stated that this tick was probably *Amblyomma geayi*. Santos Dias (1961) treated *Amblyomma geayi* as a synonym of *Amblyomma perpunctatum*, a view supported by Camicas et al. (1998). However, Robinson (1926) considered *Ixodes perpunctatus* a doubtful name, while Fairchild et al. (1966) rejected the opinion of Santos Dias (1961) because of the inadequate description of *Ixodes perpunctatus*. Guglielmone and Nava (2014) treated *Ixodes perpunctatus* as a *nomen dubium*. However, Guglielmone et al. (2003a, 2009) accepted the position of Fairchild et al. (1966) while considering it worthwhile to compare the types, a view that is supported here. See also *Amblyomma geayi* in Chap. 2.

Ixodes pulchellus Lucas, 1846: This tick was described from several males and females collected in French Guiana. *Ixodes pulchellus* is a synonym of *Amblyomma bibroni* (Gervais, 1842, a *nomen nudum*) in Camicas et al. (1998), but it is a synonym of the Nearctic and Neotropical *Amblyomma dissimile* in Neumann (1911a), Guglielmone and Nava (2010, 2014), and here.

Ixodes pygmaeus Koch, 1844: This tick was described from eight females (Moritz and Fischer 1981) collected in Brazil and Mexico, but this name was placed in the genus *Amblyomma* as *Amblyomma pygmaeum* by Neumann (1911a), who found that the female specimens used to describe *Ixodes pygmaeus* were, in fact, nymphs. *Ixodes pygmaeus* is a name *incertae sedis* in Neumann (1911a), Guglielmone and Nava (2014), and here.

Ixodes testudinis Conil, 1877: This tick was described from males and females collected in Argentina. The name *Ixodes testudinis* of Conil (1877) was preoccupied by *Ixodes testudinis* Leydig, 1855 (= *Hyalomma aegyptium*), and the correct name is *Amblyomma argentinae* Neumann, 1905, as explained in Guglielmone et al. (2001). Lahille (1905) transferred the species of Conil (1877) to *Amblyomma* as *Amblyomma testudinis*. Barbará and Dios (1918) found that the species described in Neumann (1905) was identical to *Ixodes testudinis* of Conil (1877), but treated the Neotropical *Amblyomma argentinae* as a synonym of *Amblyomma testudinis*. The view of Barbará and Dios (1918) prevailed for a long time but, in fact, the opposite is true (Camicas et al. 1998; Guglielmone et al. 2001). The larva of *Amblyomma argentinae* was described by Estrada-Peña et al. (2002) under the name *Amblyomma testudinis*. See also *Amblyomma argentinae* (Chap. 2) for further comments and records outside Argentina.

Ixodes variegatus Lucas, 1852: This tick was described from males collected in Cuba. *Ixodes variegatus* of Lucas (1852) is the original name for the Neotropical *Amblyomma albopictum*, because the name in Lucas (1852) was preoccupied by *Ixodes variegatus* (Fabricius, 1798), a synonym of the Afrotropical and Neotropical *Amblyomma variegatum*, as stated in Neumann (1911a). *Ixodes variegatus* was renamed *Amblyomma albopictum* in Neumann (1899). See also *Amblyomma albopictum* in Chap. 2.

Ixodes variolatus Gervais, 1844: This tick was briefly described from an unknown stage collected in Brazil. Neumann (1911a) stated that *Ixodes variolatus* is probably an *Amblyomma*, and this name is treated as a *nomen dubium* in Guglielmone and Nava (2014) and here.

4.3.2.2 Incorrect Names Associated with the Genus *Dermacentor*

Anocentor columbianus Schulze, 1937: This species and genus were described by Schulze (1937b) from males and females collected in Colombia, but *Anocentor columbianus* was found to be identical to the Nearctic and Neotropical *Anocentor nitens*, a synonym of *Dermacentor nitens*, by Aragão and Fonseca (1953b). See also *Dermacentor nitens* in Chap. 3.

Anocentor dissimilis (Cooley, 1947): This name was proposed by Camicas et al. (1998) as new status for the Nearctic and Neotropical *Dermacentor dissimilis*. However, *Anocentor dissimilis* was treated as a synonym of *Dermacentor dissimilis* in Yunker et al. (1986), Guglielmone et al. (2003a), Guglielmone and Nava (2014), and here. See also *Dermacentor dissimilis* in Chap. 3.

Anocentor nitens (Neumann, 1897): This tick was originally described as *Dermacentor nitens* by Neumann (1897), but Mohler (1939) placed *Dermacentor nitens* in the genus *Anocentor*, which was created by Schulze (1937b). The name *Anocentor nitens* has been treated as valid for a long time and some workers still consider this name valid. *Anocentor nitens* is considered a synonym of the Nearctic and Neotropical *Dermacentor nitens* here. See also *Dermacentor nitens* in Chap. 3.

Dermacentor andersoni Stiles, 1908: See this name in this chapter, section on "Invasive ticks."

Dermacentor hunteri Bishopp, 1912: This is a Nearctic species found in northern Mexico and the USA. Camicas et al. (1998) listed *Dermacentor hunteri* as a Nearctic and Neotropical species, but no Neotropical localities were found for this species by Guglielmone et al. (2003a) or subsequent workers.

Dermacentor nigrolineatus (Packard, 1869): This tick was described from a female collected in the USA. Camicas et al. (1998) treat *Dermacentor nigrolineatus* as valid, but it is a synonym of *Dermacentor albipictus* in Cooley (1938), Ernst and Gladney (1975), Guglielmone et al. (2003a), and Guglielmone and Nava (2014), a position provisionally supported here. Kohls and Dalmat (1952) found that Bequaert (1938) named as *Dermacentor nigrolineatus* specimens of the Nearctic and Neotropical *Dermacentor dissimilis* collected in Guatemala. See also *Dermacentor albipictus* and *Dermacentor dissimilis* in Chap. 3.

Dermacentor parumapertus Neumann, 1901: Camicas et al. (1998) listed *Dermacentor parumapertus* as a Nearctic and Neotropical species. However, no *bona fide* Neotropical records of *Dermacentor parumapertus* were listed in Guglielmone et al. (2003a, 2014), Guzmán-Cornejo et al. (2016), or subsequently.

Dermacentor reticulatus (Fabricius, 1794): This Palearctic species was allegedly found in Cuba, as stated in Černý (1969a), who regarded the Cuban record of *Dermacentor reticulatus* as a misidentification or the result of importation of animals from other continents.

Dermacentor venustus Banks, 1908: This name was suppressed to conserve the name *Dermacentor andersoni* Stiles, 1908, as stated in Keirans (1976). Barroso (1922) used the name *Dermacentor venustus* Banks instead of *Dermacentor andersoni* for a tick found in Brazil. The diagnosis of Barroso (1922) is erroneous, as stated in Guglielmone et al. (2003a), Dantas-Torres et al. (2009a), and Guglielmone and Nava (2014), because *Dermacentor andersoni* is not present in Brazil.

Otocentor nitens (Neumann, 1897): The genus *Otocentor* was created by Cooley (1938) for the Nearctic and Neotropical *Dermacentor nitens* described by Neumann (1897). However, the genus *Otocentor* was preceded by *Anocentor* of Schulze

(1937a, b), and *Otocentor nitens* has been regarded as a synonym of *Anocentor nitens* in several studies. Moreover, *Otocentor nitens* and *Anocentor nitens* are currently treated as synonyms of *Dermacentor nitens* by most workers. See also *Dermacentor nitens* in Chap. 3.

4.3.2.3 Incorrect Names Associated with the Genus *Haemaphysalis*

Gonixodes rostralis (Dugès, 1888): See *Ixodes rostralis* below.

Haemaphysalis chordeilis (Packard, 1869): Beatty (1944) allegedly found this tick, named as *Haemaphysalis chordeiles* (*lapsus*), in the USA Virgin Islands, and Camicas et al. (1998) list *Haemaphysalis chordeilis* as a Nearctic and Neotropical species. However, Hoogstraal (1973) and Keirans and Restifo (1993) stated that this species is not found outside the Nearctic Region, a situation that remains unchanged. Therefore, Guglielmone et al. (2003a, 2014, 2015) treat *Haemaphysalis chordeilis* as a Nearctic species, albeit one that has been confused with other species of *Haemaphysalis*. See also *Haemaphysalis cinnabarina* and *Haemaphysalis leporispalustris* in Chap. 3.

Haemaphysalis kochi Aragão, 1908: This tick was described by Aragão (1908a) from one male and two females collected in Brazil. *Haemaphysalis kochi* is identical to *Haemaphysalis juxtakochi*, and the first descriptions of the male and female of *Haemaphysalis juxtakochi* were published under the name *Haemaphysalis kochi* in Aragão (1908a). This name, however, was preoccupied by *Haemaphysalis concinna kochi* of Neumann (1905), and the species was later named *Haemaphysalis juxtakochi*, a Nearctic and Neotropical species. See also *Haemaphysalis juxtakochi* in Chap. 3.

Haemaphysalis kohlsi Aragão and Fonseca, 1951: This name was proposed by Aragão and Fonseca (1951b) to replace *Haemaphysalis kochi* Aragão (preoccupied). However, Kohls (1960) found that *Haemaphysalis kohlsi*, named in 1951, is a synonym of *Haemaphysalis juxtakochi*, described by Cooley in 1946. See also *Haemaphysalis juxtakochi* in Chap. 3.

Haemaphysalis lagotis (Gervais, 1849): See *Ixodes lagotis* below in this section.

Haemaphysalis leporis (Packard, 1869): This name was first used in Neumann (1897), and Neumann (1911a) simultaneously treated *Haemaphysalis leporis* as both a valid name and a synonym of *Haemaphysalis chordeilis*. However, *Haemaphysalis leporis* is treated as a synonym of the Nearctic and Neotropical *Haemaphysalis leporispalustris* in Cooley (1946), Camicas et al. (1998), and here. See also *Haemaphysalis leporispalustris* in Chap. 3 and *Haemaphysalis pinolillo* below in this section.

Haemaphysalis pinolillo Doss and Anastos, 1977, *Haemaphysalis*. This name appears in Doss and Anastos (1977) based on Dugès (1902) and is treated as a *nomen dubium* in Camicas et al. (1998). However, Dugès (1902) used the term "pinolillo" as a common name for the nymph of *Haemaphysalis leporis* (= *Haemaphysalis leporispalustris*), and no tick was described under the name *Haemaphysalis pinolillo* by Dugès (1902). Therefore, Guglielmone and Nava (2014) treat *Haemaphysalis pinolillo* as a *nomen nudum*.

Haemaphysalis proxima Aragão, 1909 in Rohr (1909): This name is treated as a synonym of *Haemaphysalis leporispalustris* in Camicas et al. (1998) and Dantas-Torres et al. (2009a). However, there is no formal text description for *Haemaphysalis proxima* Aragão, 1909, and the figures supporting validity are insufficient for its diagnosis. Therefore, Guglielmone and Nava (2014) regard *Haemaphysalis proxima* Aragão, 1909 as a *nomen dubium*. See also *Haemaphysalis leporispalustris* in Chap. 3.

Haemaphysalis sanguinolenta Koch, 1844: This tick was described from a female specimen collected in Brazil, but this name was found to be a synonym of *Haemaphysalis cinnabarina* by Hoogstraal (1973) and Camicas et al. (1998). See also *Haemaphysalis cinnabarina* in Chap. 3.

Ixodes lagotis Gervais, 1849: This tick was described from unknown stages collected in Chile. Neumann (1901, 1911a) stated that *Ixodes lagotis* is, in fact, a member of the genus *Haemaphysalis*. González-Acuña and Guglielmone (2005) followed Camicas et al. (1998) in declaring *Ixodes lagotis* and *Haemaphysalis lagotis nomina nuda*, but Guglielmone and Nava (2014) listed these names as *nomina dubia*. The specimens used to describe *Ixodes lagotis* are lost (González-Acuña and Guglielmone, 2005), rendering it impossible to determine the genus of this species.

Ixodes leporispalustris Packard, 1869: This tick was described from a female specimen collected in the USA, and assigned its current status as *Haemaphysalis leporispalustris* by Nuttall and Warburton (1915). The original description of the female of *Haemaphysalis leporispalustris* was published under the name *Ixodes leporispalustris*. See also *Haemaphysalis leporispalustris* in Chap. 3.

Ixodes rostralis Dugès, 1888: This tick was described from larvae, nymphs, and females (identified as males) collected in Mexico. Neumann (1897, 1911a), Camicas et al. (1998), and others referred to this tick as *Gonixodes rostralis*, but Dugès (1888) described it as *Ixodes rostralis*, the name *Gonixodes* being a subgenus. *Ixodes rostralis* is a synonym of *Haemaphysalis leporispalustris*, as explained in Guglielmone and Nava (2014).

4.3.2.4 Incorrect Names Associated with the Genus *Hyalomma*

Hyalomma sp.: González-Martín del Campo et al. (2018) allegedly collected an adult *Hyalomma* sp. from a bird in southern Mexico but without further data, a record that consequently has no taxonomic standing. The same comment applies to the records of *Hyalomma* sp. on cattle in Chiapas (southern Mexico) by Romero-Castañon et al. (2008) and the alleged parasitism of *Hyalomma* sp. on toads in Trinidad and Tobago by Kamilar et al. (2020). *Hyalomma* is a genus of paramount medical and veterinary importance but it is not known to be established in the Americas. The records above are treated here as diagnostic errors, but a reexamination of the alleged *Hyalomma* species is needed for a definitive identification. See also *Amblyomma pacae* and *Amblyomma sabanerae* in Chap. 2.

Hyalomma aegyptium (Linnaeus, 1758): There are several reports of the presence of this Palearctic species in the Neotropics. Galli-Valerio (1909) reported this tick from Brazil, where later Aragão (1936) allegedly collected *Hyalomma egyptium* (*lapsus*) from goats under quarantine conditions. Lahille (1914) diagnosed this species among ticks collected from camels imported from Sudan to Argentina, and Torregiani (1914) claimed that *Hyalomma aegyptium* had been found in Bolivia. Xiroudakis (1935) stated that *Hyalomma aegyptium* was present on the Caribbean island of Guadeloupe, probably confusing this species with the Afrotropical and Neotropical *Amblyomma variegatum*. The name *Hyalomma aegyptium* has been applied to different species of *Hyalomma* (Hoogstraal 1956), but it has also been confused with ticks from other genera (Keirans 1985b). None of the Neotropical records of *Hyalomma aegyptium* have been confirmed, and they are treated here as diagnostic errors or, at most, specimens that were introduced with imported domestic mammals. All such introductions appear to have failed.

4.3.2.5 Incorrect Names Associated with the Genus *Margaropus*

Margaropus winthemi Karsch, 1879: This tick was described from a male allegedly collected in Chile (Valparaíso), a locality label error, because *Margaropus winthemi*, as well as the other two species of *Margaropus*, is only found in the Afrotropical Region and has never been detected in the Neotropics. This error was perpetuated by authors such as Pinto (1945) and Bedford (1920), and the latter author stated that this tick was known in South Africa as the "Argentine tick" because it was believed to have been introduced with Argentine horses during the Boer War.

4.3.2.6 Incorrect Names Associated with the Genus *Rhipicephalus*

Boophilus annulatus australis (Fuller, 1899): This name was applied by some authors, such as Woke (1947), to Neotropical populations of *Rhipicephalus microplus*. See also *Rhipicephalus microplus* in Chap. 3.

Boophilus annulatus microplus (Canestrini, 1888): This name was applied by some authors, such as Tate (1941), to Neotropical populations of *Rhipicephalus microplus*. See also *Rhipicephalus microplus* in Chap. 3.

Boophilus australis (Fuller, 1899): This name is a synonym of *Rhipicephalus australis*, but it was treated as a synonym of *Rhipicephalus microplus* and equivalent names for a long time. This situation changed when *Rhipicephalus australis* was reinstated by Estrada-Peña et al. (2012). The name *Boophilus australis* was occasionally used instead of *Rhipicephalus microplus* by some authors, such as Ault (1946), when referring to Neotropical tick populations. See also *Rhipicephalus microplus* in Chap. 3.

Boophilus cyclops Minning, 1934: This tick was described from males and females collected in Costa Rica, Guatemala, and Panama, but it was found to be a synonym of *Rhipicephalus microplus*, as discussed in Guglielmone and Nava (2014). See also *Rhipicephalus microplus* in Chap. 3.

Boophilus decoloratus (Koch, 1844): See the name *Rhipicephalus decoloratus* in this chapter, section entitled "Invasive ticks."

Boophilus microplus (Canestrini, 1888): This name was used first by Lahille (1905) and was widely applied until 2003, when Murrell and Barker (2003) found that *Boophilus microplus* is a synonym of *Rhipicephalus microplus*, confirming an old proposal by Canestrini (1890). The nymph and larva of *Rhipicephalus microplus* were described by Lahille (1905) under the name *Boophilus microplus.* See also *Rhipicephalus microplus* in Chap. 3.

Haemaphysalis (spelled *Haemophysalis*) *micropla* Canestrini, 1888: This tick was described from males and females collected in Paraguay, but *Haemaphysalis micropla* was found to be a synonym of *Rhipicephalus microplus* and its original name, as discussed in Guglielmone and Nava (2014). See also *Rhipicephalus microplus* in Chap. 3.

Margaropus annulatus (Say, 1821): See the name *Rhipicephalus annulatus* in this chapter, section entitled "Invasive ticks."

Margaropus annulatus australis (Fuller, 1899): This name was used by authors such as Rivas (1919) for Neotropical populations of *Rhipicephalus microplus.* See also *Rhipicephalus microplus* in Chap. 3.

Margaropus annulatus microplus (Canestrini, 1888): This name was used by authors such as Rohr (1909) for Neotropical populations of *Rhipicephalus microplus.* See also *Rhipicephalus microplus* in Chap. 3.

Margaropus microplus (Canestrini, 1888): This name was used by authors such as Aragão (1911) for Neotropical populations of *Rhipicephalus microplus.* See also *Rhipicephalus microplus* in Chap. 3.

Rhipicephalus annulatus (Say, 1821): See this name in this chapter, section entitled "Invasive ticks."

Rhipicephalus australis Fuller, 1899: This tick was described from males and females collected in Australia. The name *Rhipicephalus australis* was long treated as a synonym of *Rhipicephalus microplus* and alternative names, but Estrada-Peña et al. (2012) reinstated *Rhipicephalus australis* as a valid species.

Rhipicephalus bursa Canestrini and Fanzago, 1878: See this name in this chapter, section entitled "Invasive ticks."

Rhipicephalus decoloratus Koch, 1844: See this name in this chapter, section entitled "Invasive ticks."

Rhipicephalus sanguineus punctatissimus Gerstäcker, 1873: This name is a synonym of *Rhipicephalus sulcatus* in Camicas et al. (1998) and a synonym of *Rhipicephalus sanguineus* in Walker et al. (2000), but a name *incertae sedis* in Guglielmone and Nava (2014) and here. *Rhipicephalus sanguineus punctatissimus* was identified in Venezuela by Rivas (1919), but this identification is treated here as erroneous.

Rhipicephalus sanguineus simus Koch, 1844: Díaz Ungría (1957), Doss and Anastos (1977), Camicas et al. (1998) and Guglielmone et al. (2003a) used the name *Rhipicephalus sanguineus simus* (a *lapsus*) for *Rhipicephalus simus simus*, a name used in Rivas (1919) for ticks collected in Venezuela. *Rhipicephalus simus simus* is a synonym of the Afrotropical *Rhipicephalus simus* (Guglielmone and Nava 2014) and the identification by Rivas (1919) is considered erroneous.

Rhipicephalus simus simus Koch, 1844: This name is a synonym of the Afrotropical species *Rhipicephalus simus*. Rivas (1919) erroneously identified as *Rhipicephalus simus simus* tick specimens collected in Venezuela.

Rhipicephalus turanicus Pomerantzev, 1940 in Pomerantzev et al. (1940): A controversial Palearctic species that is difficult to identify, with several incorrect and unconfirmed records worldwide, as discussed in Guglielmone et al. (2014). Thus, Bernasconi et al. (2002) listed specimens of this tick allegedly found in Costa Rica, and Dantas-Torres et al. (2009a) stated that its presence in Brazil requires confirmation, but *Rhipicephalus turanicus* is not mentioned in the latest revision of Brazilian ticks by Dantas-Torres et al. (2019b). Jarquín-Díaz et al. (2016) claimed that *Rhipicephalus turanicus* has been found on dogs in southern Mexico. All such Neotropical records of *Rhipicephalus turanicus*, a species morphologically close to *Rhipicephalus sanguineus sensu stricto*, have not been confirmed. Perhaps specimens of *Rhipicephalus turanicus* will be found in some of the many Neotropical collections of *Rhipicephalus sanguineus sensu lato*, but this will require examination of Neotropical specimens deposited in museum collections, together with sophisticated morphological and molecular analyses based on current knowledge of *Rhipicephalus turanicus*.

Any analysis of the current status of *Rhipicephalus turanicus* must include *Rhipicephalus afranicus*, described in Bakkes et al. (2020), who found that the Afrotropical records of *Rhipicephalus turanicus* in Pegram et al. (1987) in fact represent specimens of this new taxon. Walker et al. (2000) argued for the presence of Afrotropical populations of *Rhipicephalus turanicus* as described in Pegram et al. (1987), but the presence of this tick in the Afrotropics was considered doubtful in Guglielmone and Nava (2014), Guglielmone et al. (2014), Guglielmone and Robbins (2018), and Guglielmone et al. (2020), who treat *Rhipicephalus turanicus* as a Palearctic species. This view has been confirmed by the study of Bakkes et al. (2020), and *Rhipicephalus turanicus* remains a largely undefined species, although with bona fide Palearctic records. Curiously, bona fide specimens of *Rhipicephalus turanicus sensu stricto* in Bakkes et al. (2020) were not collected at the type locality, Tashkent, or its vicinity, nor were they found on the type host, *Ovis aries* (Filippova, 2008).

4.3.2.7 Incorrect Names Associated with the Genus *Robertsicus*

Robertsicus elaphensis (Price, 1959): See *Aponomma elaphensis* above in this section.

Chapter 5
Animal and Human Parasitism, and Lists of Hosts of Neotropical Ixodidae

5.1 Host–Tick Associations in Neotropical Ixodidae

Four classes of vertebrate hosts, Amphibia, Reptilia, Aves, and Mammalia, account for all principal or potentially principal hosts of Neotropical hard ticks. As expected, host utilization differs among tick genera, with the exception of the class Mammalia, whose members are parasitized by all genera. In this section, tick–host associations may initially appear to be definitive; however, even where numerous host records are available, the apparent trends in these associations are often hypothetical. Moreover, potential or provisional host records are often based on only one or a few records, and conclusions drawn from such scanty data may be radically altered by future research results. Readers are advised to consult the sections on "hosts" of particular species in Chaps. 1, 2, or 3 whenever doubts arise concerning the vertebrates that are key to a particular tick species' life.

The analyses below are based on Neotropical records where both the hosts and the tick stages found on them are known.

5.1.1 Amphibia as Hosts for Neotropical Ixodidae

Ixodid records from Amphibia in the Neotropical Region include specimens collected from one order, three families, and 23 species, as discussed alphabetically below.

A. A. Guglielmone et al., *Neotropical Hard Ticks (Acari: Ixodida: Ixodidae)*,
https://doi.org/10.1007/978-3-030-72353-8_5

5.1.1.1 Anura

This order contains about 30 families worldwide. Neotropical members of three of these families have all been found parasitized by ticks of the genus *Amblyomma*, with the exception of a single specimen of *Dermacentor*.

Bufonidae: This family contains more than 500 species with a worldwide range. At least 19 species have been found infested with ticks of the genus *Amblyomma*, plus one case of infestation by a species of *Dermacentor*. Bufonidae are of particular importance as hosts for two species of *Amblyomma*. Thus, *Rhinella marina* is one of the principal hosts of adults, nymphs and larvae of *Amblyomma dissimile* and *Amblyomma rotundatum*.

Additionally, a few males and females of *Amblyomma argentinae*, males of *Amblyomma goeldii* and nymphs of *Amblyomma mixtum* have been found on *Rhinella* sp. Males and females of *Amblyomma fuscum* have been taken from *Rhinella arenarum*, while males of *Amblyomma varium* have been recovered from *Rhinella marina*, and nymphs of *Amblyomma sabanerae* have been found on *Rhinella alata*. The remaining species of Bufonidae have been found parasitized by different stages of *Amblyomma dissimile* and *Amblyomma rotundatum*, and some may be more important as hosts of these ticks than currently recognized. A female of *Dermacentor nitens* was found on a *Peltophryne peltocephala*.

Leptodactylidae: This family comprises more than 180 species established in the Americas. Two of these species, *Leptodactylus pentadactylus* and *Physalaemus nattererei*, have occasionally been found infested with females of *Amblyomma rotundatum* in the Neotropical Region.

Pipidae: This family contains about 40 species found in the Afrotropical and Neotropical Regions, and there is one record of a female of *Amblyomma rotundatum* from *Pipa pipa* in the Neotropics.

5.1.1.2 Synopsis

Only one amphibian order (Anura) and three families have been found infested with ixodids in the Neotropics. With the exception of an odd record of parasitism by *Dermacentor nitens*, all cases of tick infestation have been caused by eight *Amblyomma* species, two of which show a strong preference for Bufonidae.

Amblyomma dissimile and *Amblyomma rotundatum* are prone to feed on Bufonidae in the Neotropics, their principal host for all parasitic stages being *Rhinella marina*, although these ticks also parasitize reptiles (Squamata). While relatively few anuran species are parasitized by ticks, this host-tick relationship demonstrates the plasticity of *Amblyomma* in colonizing niches ignored by other genera of Ixodidae. As well, this pattern of parasitism extends beyond the Neotropical Region because both *Amblyomma dissimile* and *Amblyomma rotundatum* are established in the Nearctic Zoogeographic Region, and *Amblyomma rotundatum* has even been found on a remote Pacific Island (Kelehear et al. 2017b). See also the appendix for a full list of amphibian host species and the ticks that parasitize them in the Neotropical Region.

5.1.2 Reptilia as Hosts for Neotropical Ixodidae

Reptilia are important Neotropical hosts for the genus *Amblyomma*, with several species exclusively parasitizing this class of vertebrates. Ixodid records from reptiles in the Neotropical Region include specimens collected from five orders, 21 families, and 179 species, as discussed alphabetically below.

5.1.2.1 Crocodilia

This order contains about 25 species in three families distributed in tropical and subtropical areas around the world. Members of two of these families have occasionally been found infested with ticks of the genus *Amblyomma* in the Neotropical Region.

Alligatoridae: This family contains eight species, most of them found in the Neotropical Region, but one species is found in the USA and another in China. Four Neotropical species have occasionally been found infested with ticks of the genus *Amblyomma.* Females of *Amblyomma rotundatum* have been collected from *Caiman latirostris*, *Paleosuchus palpebrosus*, and *Paleosuchus trigonatus* (this species has also been found infested with nymphs of *Amblyomma rotundatum*). Males and females of *Amblyomma dissimile* have been taken from *Caiman crocodilus* (also infested with nymphs of this tick) and *Paleosuchus trigonatus.* Males and nymphs of *Amblyomma humerale* have been found on *Caiman crocodilus* and *Paleosuchus trigonatus*, while males of *Amblyomma fuscum* have been recovered from *Caiman latirostris*, and males of *Amblyomma geayi* have been collected from *Caiman crocodilus.* A female of *Rhipicephalus sanguineus sensu lato* was collected from *Caiman crocodilus.*

Crocodylidae: This family comprises 14 species distributed in tropical and subtropical areas of the world. Two members of this family have occasionally been found infested with ticks of the genus *Amblyomma* in the Neotropics. Females of *Amblyomma dissimile* have been collected from *Crocodylus acutus* and *Crocodylus moreletii*, and an individual *Crocodylus acutus* was also found parasitized by a female of *Amblyomma mixtum.*

5.1.2.2 Squamata

This order is very large, containing about 7500–10,000 species distributed worldwide in circa 60 families. Members of 14 families in this order are important hosts of several species of *Amblyomma*. There is also a record of infestation with ticks of the genus *Ixodes*, and another case that involved a specimen in the genus *Rhipicephalus.*

Boidae: This family contains about 40 species with a worldwide distribution, but most are found in the Americas. A total of 12 species have been found infested with ticks of the genus *Amblyomma*, at least five species of which are highly dependent on boids as hosts. This family has also yielded one record of parasitism by a member of the genus *Rhipicephalus*. Thus, *Boa constrictor* is one of the principal hosts for males, females, nymphs, and larvae of *Amblyomma dissimile*; females, nymphs, and larvae of *Amblyomma rotundatum*; and males and females of *Amblyomma fuscum.* The Cuban boa, *Chilabothrus angulifer*, is the principal host for males and females of *Amblyomma quadricavum*, while *Eunectes murinus* is the only known host for the male of *Amblyomma fulvum* (the only stage known).

Additionally, a female of *Rhipicephalus sanguineus sensu lato* was found on a *Boa constrictor.* Adults of *Amblyomma argentinae* have been collected from *Boa constrictor* (also infested with nymphs of this tick), *Epicrates cenchria*, and *Eunectes notaeus*, while adults and nymphs of *Amblyomma albopictum* have been found on *Chilbothrus angulifer*, and adults of *Amblyomma varium* have been taken from *Boa constrictor* and *Eunectes murinus.* Males of *Amblyomma antillorum* and *Amblyomma goeldii* have been recovered from *Boa nebulosa* and *Boa constrictor*, respectively, and the latter hosts have also been found infested with adults of *Amblyomma nodosum.* Several other species of Boidae have also been found parasitized by *Amblyomma dissimile*, *Amblyomma quadricavum*, and *Amblyomma rotundatum.*

Colubridae: This family contains about 850 species with a worldwide distribution. Fifteen Neotropical species have been found carrying ticks of the genus *Amblyomma*, one species of which is dependent on colubrids as hosts, but the role of these snakes as hosts for *Amblyomma* in this region is probably more important than currently recognized. Thus, *Chironius carinatus* is considered one of the many important hosts for the nymph of *Amblyomma humerale.*

Additionally, males and females of *Amblyomma geayi* have been collected from *Spilotes pullatus*, and nymphs of *Amblyomma sabanerae* have been taken from *Leptophis depressirostris.* Both immature and adult stages of *Amblyomma dissimile* and *Amblyomma rotundatum* have been recovered from several species of Colubridae.

Corytophanidae: This family contains nine species found in the Americas. In the Neotropical Region one species, *Basiliscus basiliscus*, has occasionally been found infested with nymphs of *Amblyomma dissimile.*

Dactyloidae: This American family contains about 400 species. In the Neotropical Region, there is only one record of a species of Dactyloidae infested with ticks of the genus *Amblyomma*: a single *Anolis auratus* was found infested with nymphs of *Amblyomma dissimile*.

Dipsadidae: This family comprises about 700 species found only in the Americas but most of them are established in the Neotropical Region. At least 23 species have been found carrying ticks of the genus *Amblyomma*, one species of which is dependent on dipsadids, but the role of Dipsadidae as hosts for *Amblyomma* in this region is probably more important than currently recognized. Thus, *Erythrolamprus typhlus* and *Xenodon werneri* are considered principal hosts for the nymph of *Amblyomma humerale*.

Additionally, adults of *Amblyomma quadricavum* have been collected from *Borikenophis portoricensis* and *Cubophis cantherigerus*. Females of *Amblyomma fuscum* have been found on *Clelia clelia*, while larvae of *Amblyomma antillorum* have been taken from *Alsophis sibonius*. Adults and nymphs of *Amblyomma dissimile*, and females and nymphs of *Amblyomma rotundatum* have been recovered from several other species of Dipsadidae.

Elapidae: This family consists of about 350 species found in tropical and subtropical areas of the world. At least four species of Elapidae have occasionally been found infested with ticks of the genus *Amblyomma* in the Neotropical Region. Females, nymphs, and larvae of *Amblyomma rotundatum* have been collected from *Micrurus* sp., while females and nymphs, and nymphs alone have been taken from *Micrurus ibiboboca* and *Micrurus corallinus*, respectively, and nymphs alone have been collected from *Micrurus lemniscatus*. Males, females, and nymphs of *Amblyomma dissimile* have been found on *Micrurus lemniscatus*, while females of the same tick have been collected from *Micrurus averyi*.

Gekkonidae: This family contains about 900 species and is distributed worldwide. However, in the Neotropical Region, there is only one record of tick infestation: *Hemidactylus mabouia* was found infested with a nymph of *Amblyomma dissimile*.

Iguanidae: This family contains about 40 species, the great majority of them found only in the Americas. In the Neotropical Region, at least 16 species of iguanids have been found infested with ticks of the genus *Amblyomma*, eight species of which are dependent on Iguanidae as hosts. Thus, *Cyclura cychlura* and *Cyclura lewisi* are principal hosts for the male, female, nymph, and larva of *Amblyomma albopictum*, while *Cyclura pinguis* and *Iguana delicatissima* are preferred hosts for the male, female, nymph, and larva of *Amblyomma antillorum*. The marine iguana *Amblyrhynchus cristatus* is the principal host for the male, female, nymph, and larva of *Amblyomma darwini*, and *Iguana iguana* is one of the principal hosts for the male, female, nymph, and larva of *Amblyomma dissimile*, while *Ctenosaura pectinata* and *Cyclura cychlura* are principal hosts for the male, female, nymph, and larva of *Amblyomma scutatum* and *Amblyomma torrei*, respectively. *Cyclura cornuta* and *Cyclura stejnegeri* are important hosts for the male, female, and nymph of *Amblyomma cruciferum*, and *Conolophus subcristatus* and *Amblyrhynchus* sp. are principal hosts for the male and female of *Amblyomma williamsi.*

Additionally, different stages of nine species of *Amblyomma* have been collected from several species of Iguanidae.

Liolaemidae: This family contains about 250 species found only in South America. Ten species of Liolaemidae, some controversial, have been found infested with immatures of *Amblyomma parvitarsum*, and *Liolaemus* spp. are the principal hosts for the nymph and larva of this tick.

Phrynosomatidae: This family consists of about 140 species found in Central and North America. However, in the Neotropical Region there is only one record of tick infestation, a female and nymph of *Amblyomma rotundatum* having been found on *Phrynosoma* sp.

Scincidae: This family comprises about 1500 species and occurs worldwide. However, in the Neotropical Region there is just one record of scincid infestation by ixodid ticks: a specimen of *Mabuya mabouya* was found infested with nymphs and larvae of *Amblyomma rotundatum.*

Tropiduridae: This family contains about 140 species, all of which are found in the Neotropical Region. At least 12 species of Tropiduridae have been found infested with ticks of the genus *Amblyomma*, one of which is dependent on tropidurid hosts. Thus, *Plica plica*, *Plica umbra*, and *Uranoscodon superciliosus* are some of the many squamatans considered to be potential principal hosts for the nymph of *Amblyomma humerale.*

Additionally, different stages of six species of *Amblyomma* have been collected from several species of Tropiduridae.

Viperidae: This family contains about 250 species and is distributed worldwide. Twelve species of viperids have been found infested with four species of *Amblyomma* in the Neotropical Region, but Viperidae are not yet considered principal hosts for any of them. However, the role of Viperidae as hosts of Neotropical *Amblyomma* may be more important than currently recognized. Males, females, and nymphs of *Amblyomma argentinae* have been found on *Crotalus durissus*, and males of the same tick have been collected from *Bothrops neuwiedi*, while males and females of *Amblyomma goeldii* have been taken from *Lachesis muta.* All parasitic stages of *Amblyomma rotundatum* have been found on 11 species of Viperidae, and both adult and immature stages of *Amblyomma dissimile* have been collected from five species of Neotropical Viperidae.

5.1.2.3 Testudines

This order, which is found worldwide, contains about 350 species in 14 families. In the Neotropical Region, members of five families are parasitized by several species of *Amblyomma*, which often are dependent on testudines.

Chelidae: This family includes about 60 species found in Australasia, southern Asia, and South America. Seven species of Chelidae have been found infested with three species of *Amblyomma* in the Neotropical Region, but these tortoises are not considered important hosts for the species of *Amblyomma* involved. Females of *Amblyomma rotundatum* have been found on *Hydromedusa tectifera*, *Mesoclemmys tuberculata, Mesoclemmys vanderhaegei*, and *Platemys platycephala*, and females of *Amblyomma dissimile* have been collected from *Acantochelys macrocephala* and *Phrynops geoffroanus*, while females of *Amblyomma argentinae* have been taken from *Phrynops hilarii*.

Emydidae: This family contains about 50 species, the great majority of them in the Americas. In the Neotropical Region, four species of Emydidae have been found infested with males, females, nymphs, and larvae of *Amblyomma dissimile*, as well as males and females of *Amblyomma mixtum* and *Amblyomma sabanerae*, and females of *Amblyomma rotundatum* have been recovered from *Trachemys dorbigni*, but these tortoises are not considered important hosts for these species of *Amblyomma*.

Geoemydidae: This worldwide family contains about 70 species. Nine species of Geoemydidae have been found infested with five species of *Amblyomma* in the Neotropical Region, and these are important hosts for one *Amblyomma* species. Thus, males, females, nymphs, and larvae of *Amblyomma sabanerae* have been found on *Rhinoclemmys annulata*, *Rhinoclemmys areolata*, *Rhinoclemmys funerea*, *Rhinoclemmys nasuta*, and *Rhinoclemmys pulcherrima*, which are considered principal hosts for this tick.

Additionally, adults of *Amblyomma dissimile* have been found on *Rhinoclemmys areolata*, *Rhinoclemmys melanosterna* (also infested with nymphs of this tick), *Rhinoclemmys pulcherrima*, and *Rhinoclemmys rubida*. Females of *Amblyomma rotundatum* have been collected from *Heosemys annandalii* and *Rhinoclemmys areolata*, and females of *Amblyomma crassum* have been taken from *Rhinoclemmys annulata*, while females of *Amblyomma mixtum* have been recovered from *Rhinoclemmys areolata*.

Kinosternidae: About 25 species belong to this American family. Four of these have been found infested with ticks of the genus *Amblyomma*, but none of these tortoises are considered significant hosts for any species of *Amblyomma*. Females of *Amblyomma dissimile* have been taken from *Kinosternon acutum* and *Kinosternon leucostomum*, while adults of *Amblyomma sabanerae* have been collected from *Kinosternon* sp. (also infested with nymphs of this tick) and *Kinosternon scorpiodes*; the latter tortoise has also been found infested with females of *Amblyomma rotundatum*. Females of *Amblyomma crassum* have been recovered from *Kinosternon leucostomum*.

Testudinidae: This widely distributed family contains about 50 species, at least ten of which have been found infested with ticks of the genus *Amblyomma* in the Neotropical Region, testudines being significant hosts for six species. Thus, *Chelonoidis chilensis* is the principal host for males, females, nymphs, and larvae of *Amblyomma argentinae*, while *Chelonoidis porteri* and *Chelonoidis vicina* are the principal hosts for males, females, nymphs, and larvae of *Amblyomma macfarlandi*, and *Chelonoidis darwini* and *Chelonoidis duncanensis* are the preferred hosts for males, females, nymphs, and larvae of *Amblyomma pilosum*. Males, females, nymphs, and larvae of *Amblyomma usingeri* are usually found on *Chelonoidis becki*, *Chelonoidis microphyes*, and *Chelonoidis vanderburghi*; males and females of *Amblyomma humerale* are commonly found on *Chelonoidis denticulata*; and the principal hosts for females of *Amblyomma crassum* are *Chelonoidis carbonarius* and *Chelonoidis denticulata*.

Additionally, females of *Amblyomma rotundatum* have been found on *Chelonoidis carbonarius* and *Chelonoidis denticulata* (also infested with nymphs and larvae of this tick); females of *Amblyomma dissimile* have been collected from *Chelonoidis carbonarius*; and males and females of *Amblyomma cajennense* and *Amblyomma geayi* have been taken from *Chelonoidis denticulata*.

5.1.2.4 Synopsis

Reptilia have often been found infested with ixodids in the Neotropical Region, but with two exceptions, all infestations have been caused by ticks of the genus *Amblyomma*, which is the only Neotropical genus of Ixodidae for which reptiles are principal hosts. Three orders of Reptilia—Crocodilia, Squamata, and Testudines—are known to be parasitized by ixodids, but Crocodilia are never principal hosts (Tables 2.53 and 5.3).

Squamata are principal hosts for 15 species of *Amblyomma*. However, while this is strictly the case for ten species of *Amblyomma*, the remaining five species also parasitize hosts belonging to other classes. Thus, *Amblyomma dissimile* and *Amblyomma rotundatum* also feed on Amphibia as principal hosts for all parasitic stages, while Squamata and Artiodactyla are principal hosts of immatures and adults of *Amblyomma parvitarsum*, respectively, and adults of *Amblyomma fuscum* feed mainly on Squamata but also on Cingulata, while nymphs and larvae parasitize Didelphimorphia. And nymphs of *Amblyomma humerale* parasitize Squamata as well as mammals and birds, all of which are considered principal hosts for this stage, but adult ticks feed chiefly on Testudines.

Testudines are principal hosts for seven species of *Amblyomma*. Six of these (*Amblyomma argentinae*, *Amblyomma crassum*, *Amblyomma macfarlandi*, *Amblyomma pilosum*, *Amblyomma sabanerae*, and *Amblyomma usingeri*) utilize only Testudines as principal hosts. The remaining species, *Amblyomma humerale*, employs a complex of principal hosts, including Testudines, as noted above. See the appendix for a full list of reptile host species and the ticks that parasitize them in the Neotropical Region.

5.1.3 Aves as Hosts for Neotropical Ixodidae

Ixodid records from Aves in the Neotropical Region include specimens collected from 23 orders, 59 families, and 506 species, as discussed alphabetically below.

5.1.3.1 Accipitriformes

This order contains three families worldwide, two of them monotypic, and the cosmopolitan family Accipitridae with some members that have occasionally been found infested with ticks of the genera *Amblyomma*, *Ixodes*, and *Rhipicephalus* in the Neotropical Region.

Accipitridae: This family contains more than 240 species worldwide, and in the Neotropical Region 17 of these species have been reported carrying ticks from the genera *Amblyomma, Haemaphysalis*, and *Rhipicephalus*. Nymphs of *Haemaphysalis juxtakochi* have been found on *Harpya harpyja* and *Rupornis magnirostris*, while a larva of *Rhipicephalus microplus* was found on *Rupornis magnirostris*, and one nymph of *Rhipicephalus sanguineus sensu lato* was taken from *Geranoaetus albicaudatus*. Ten species of *Amblyomma* have been collected from members of this avian family, but none of the species of Accipitridae are principal hosts for the ticks named above.

5.1.3.2 Caprimulgiformes

This order contains more than 425 species in eight families worldwide; four families together account for 18 Neotropical species that have occasionally been found infested with ixodid ticks of the genus *Amblyomma*, but these birds are of major importance as hosts for three species of the genus *Ixodes*.

Apodidae: This family contains about 110 species worldwide, including five Neotropical taxa that are important hosts of two species of *Ixodes*. Thus, *Streptoprocne biscutata* is the principal host for all parasitic stages of *Ixodes paranaensis*, and *Streptoprocne semicollaris* is the only known host for the female of *Ixodes cuernavacensis*.

Additionally, *Cypseloides fumigatus* has been found parasitized by females of *Ixodes paranaensis*, while *Cypseloides senex* and *Streptoprocne zonaris* have occasionally been infested by females or nymphs and females, respectively, of the same tick species.

Caprimulgidae: This family consists of almost 100 taxa worldwide, and two Neotropical species, *Nyctidromus albicollis* and *Setopagis parvula*, have occasionally been found infested with males and nymphs of *Amblyomma longirostre*, and nymphs of *Amblyomma sculptum*, respectively.

Steatornithidae: This family contains one species, *Steatornis caripensis*, that is found only in the Neotropical Region, and is the only known host for the female of *Ixodes downsi.*

Trochilidae: This family contains about 350 members, all found in the Americas, and in the Neotropical Region 12 species are known to be occasionally infested with ticks of the genus *Amblyomma.* Nymphs and larvae of *Amblyomma longirostre* and larvae of *Amblyomma parkeri* have been collected from *Thalurania glaucopis*; nymphs and larvae of *Amblyomma calcaratum* have been taken from *Phaethornis superciliosus*, and nymphs of this tick have also been recovered from *Phaethornis eurynome.* Nymphs of *Amblyomma nodosum* have been found on *Eupetomena macroura*, while the remaining cases of parasitism in Neotropical species of Trochilidae have involved nymphs of *Amblyomma longirostre* and, to a lesser extent, larvae of this tick.

5.1.3.3 Cariamiformes

This order consists of one family that contains two Neotropical species, and one of these has occasionally been found infested with ticks of the genus *Amblyomma.*

Cariamidae: This South American family has one member, *Cariama cristata*, that has been found infested with adults, nymphs, and larvae of *Amblyomma sculptum*, but this bird is not considered crucial to the survival of this tick.

5.1.3.4 Cathartiformes

This order contains just one family with seven species found in the Americas; in the Neotropical Region, one of these species has occasionally been found infested with ticks of the genera *Ixodes* and *Amblyomma*.

Cathartidae: One member of this family, *Coragyps atratus*, has been found infested with females of *Ixodes affinis*, adults of *Amblyomma oblongoguttatum*, *Amblyomma ovale* and *Amblyomma parvum*, and nymphs of *Amblyomma sculptum*, but this bird is not a significant maintenance host of these ticks.

5.1.3.5 Charadriiformes

This order of mainly aquatic birds contains about 380 species constituting 19 families with a worldwide range. In the Neotropical Region, members of three of these families have occasionally been found parasitized by one species of *Ixodes.*

Haematopodidae: This family consists of 12 species with a worldwide distribution. One of these, *Haematopus ater*, has been found infested with nymphs of *Ixodes uriae.*

Laridae: This family contains almost 100 species distributed worldwide. One of these, *Leucophaeus scoresbii*, has been found parasitized by females of *Ixodes uriae*.

Stercorariidae: This family contains seven taxa worldwide. One species, *Stercorarius antarcticus*, has been found infested with females of *Ixodes uriae*.

5.1.3.6 Columbiformes

This order contains only one family comprising about 340 species distributed throughout the world, and in the Neotropics some members have occasionally been found infested with ticks of the genera *Ixodes*, *Amblyomma*, and *Rhipicephalus*.

Columbidae: Nine members of this family are known to be parasitized by nine species of *Amblyomma*, one *Ixodes*, and one species of *Rhipicephalus*, but none of these ticks depend on columbids for their survival. Thus, females of *Ixodes auritulus* have been found on *Columbina talpacoti*, while females of *Rhipicephalus sanguineus sensu lato* have been collected from *Columba livia*. Nymphs and larvae, and larvae of *Amblyomma tigrinun* have been taken from *Zenaida auriculata* and *Columbina picui*, respectively. Nymphs of *Amblyomma longirostre* and nymphs of *Amblyomma nodosum* have been recovered from *Columbina talpacoti*, *Geotrygon montana* and *Leptotila verreauxi*, and the last-named host has also been found infested with nymphs of *Amblyomma parvum* and larvae of *Amblyomma coelebs*. Nymphs of *Amblyomma longirostre* and larvae of *Amblyomma varium* have been found on *Leptotila rufaxilla*, while nymphs and larvae of *Amblyomma geayi* have been collected from *Leptotila cassinii*. Larvae of *Amblyomma variegatum* have been found on *Columbina passerina*, and larvae of *Amblyomma sabanerae* have been collected from *Geotrygon montana*.

5.1.3.7 Coraciiformes

This order comprises six families and about 150 species worldwide. In the Neotropics, members of two of these families have occasionally been found infested with ticks of the genus *Amblyomma*.

Alcedinidae: This family is cosmopolitan, with more than 110 species, but only one Neotropical taxon, *Chloroceryle aenea*, is known to be parasitized by ticks, in this case nymphs of *Amblyomma longirostre*.

Momotidae: This family is Neotropical and contains 14 species. Three momotids, *Baryphthengus martii*, *Baryphthengus ruficapillus*, and *Momotus momota*, have occasionally been found infested with immature stages of eight ticks in the genus *Amblyomma*.

5.1.3.8 Cuculiformes

This order contains one family with about 150 species worldwide. Some Neotropical members of this order have occasionally been found infested with ticks of the genus *Amblyomma*.

Cuculidae: This family includes five Neotropical species that have been found infested with immature stages of *Amblyomma*. Thus, nymphs and larvae of *Amblyomma mixtum*, and nymphs and larvae of *Amblyomma tigrinum* have been collected from *Coccyzus merlini* and *Guira guira*, respectively. Nymphs of *Amblyomma longirostre* and nymphs of *Amblyomma nodosum* have been found on *Crotophaga major* and *Piaya cayana*, while nymphs of *Amblyomma longirostre* have been taken from *Crotophaga ani*. None of these avian hosts are important in maintaining the *Amblyomma* species reported from them.

5.1.3.9 Falconiformes

This order contains one family of about 65 species with worldwide distribution; in the Neotropics, a few of these birds have occasionally been found infested with ticks of the genera *Ixodes* and *Amblyomma*.

Falconidae: This family includes six Neotropical taxa that have been found infested with one species of *Ixodes*, two species of *Amblyomma*, and one species of *Rhipicephalus*. Thus females, nymphs and larvae of *Ixodes auritulus* have been found on *Phalcoboenus australis*, while an adult of *Rhipicephalus microplus* was collected from a *Milvago chimachima*. Nymphs of *Amblyomma longirostre* have been collected from *Micrastur ruficollis* and *Herpetotheres cachinans*. Adults and nymphs of *Amblyomma sculptum* have been taken from *Caracara plancus* and nymphs of the same tick have been found on *Falco spaverius*. None of these birds are considered principal hosts of the tick species reported from them.

5.1.3.10 Galbuliformes

This Neotropical order comprises 54 species classified in two families, members of which have occasionally been found infested with ticks of the genus *Amblyomma*.

Bucconidae: This family contains 36 species; two of these, *Malacoptila panamensis* and *Malacoptila striata*, have been found parasitized by immature stages of *Amblyomma longirostre* and nymphs of *Amblyomma coelebs*. *Malacoptila panamensis* has also been found infested with nymphs of *Amblyomma calcaratum* and nymphs of *Amblyomma varium*.

Galbulidae: This family contains 18 species, one of which, *Galbula ruficauda*, has been found infested with larvae of *Amblyomma longirostre*.

5.1.3.11 Galliformes

This order contains about 290 species classified in five families with a worldwide range. Three of these families have been found infested with ticks of the genera *Ixodes*, *Amblyomma* and *Haemaphysalis*, and one family is apparently of importance in maintaining one Neotropical species of the genus *Ixodes*.

Cracidae: This family is basically Neotropical and consists of about 55 species, eight of which are parasitized by *Ixodes*, *Amblyomma*, and *Haemaphysalis* ticks, but cracids are not important in maintaining the life cycles of these ticks. Females of *Ixodes auritulus* have been found on *Oreophasis derbianus*, and immature stages of this tick have been found on *Penelope* sp. and *Penelope superciliaris*, while females and nymphs of *Ixodes boliviensis* have been collected from *Crax rubra*, and nymphs of *Haemaphysalis juxtakochi* have been taken from *Penelope obscura*. Additionally, different stages of eight species of *Amblyomma* have been recovered from Cracidae.

Odontophoridae: This family contains 33 species, most occurring in the Americas. In the Neotropical Region, two of these species, *Odontophorus guttatus* and *Callipepla californica*, have occasionally been found infested with females of *Ixodes auritulus*, and nymphs and larvae of *Amblyomma tigrinum*, respectively.

Phasianidae: This family contains about 155 species worldwide, but it is not well represented in the Neotropical Region, although some phasianids have been introduced into the Neotropics, where they have been found infested with two species of *Ixodes* and one species of *Amblyomma*. Phasianidae are an important host for one species of *Ixodes*: "caille" is the only known host for the female and nymph of *Ixodes cornuae*, but this bird is tentatively included in the Phasianidae, because it may actually belong to the Odontophoridae.

Additionally, "pheasant" have been found infested with females of *Ixodes boliviensis*, while nymphs and larvae of *Amblyomma mixtum* have been collected from "chicken."

5.1.3.12 Gruiformes

This order includes about 180 species in six families with a worldwide distribution; members of two of these families have occasionally been found parasitized by ticks of the genus *Amblyomma*.

Aramidae: This family contains one extant species, *Aramus guarauna*, an American bird that has been found infested with nymphs of *Amblyomma sculptum* in the Neotropics.

Rallidae: This family is cosmopolitan, containing around 150 species. At least three of these have been found parasitized by Neotropical *Amblyomma* ticks. Adults of *Amblyomma ovale* have been found on "saracura," and nymphs of *Amblyomma brasiliense* have been collected from *Aramides saracura*, while larvae of

Amblyomma longirostre and larvae of *Amblyomma triste* have been taken from *Laterallus albigularis* and *Anurolimnas viridis*, respectively.

5.1.3.13 Passeriformes

This speciose order of "perching birds" is found worldwide and comprises almost 150 families with more than 6500 species, representing more than half of all species of Aves. In the Neotropical Region, some passeriform families are important for the maintenance of tick species in the genera *Amblyomma* and *Ixodes*, as well as one tick in the genus *Haemaphysalis*.

Cardinalidae: This family contains about 50 species found only in the Americas, and in the Neotropics six of these have been found infested with immature stages of the genus *Amblyomma*. Thus, *Cyanoloxia cyanoides*, *Habia rubica*, and *Piranga flava* are all known to be parasitized by larvae and nymphs of *Amblyomma longirostre*, and *Habia rubica* and *Cyanoloxia cyanoides* are hosts for larvae of *Amblyomma romarioi*, the latter bird having also been found carrying larvae of *Amblyomma geayi*. All birds named above are considered important hosts for these ticks.

Additionally, immature stages of five other species of *Amblyomma* have been collected from Neotropical Cardinalidae.

Conopophagidae: This Neotropical family contains 11 species; two of them are known to be infested with immature stages of ticks in the genera *Ixodes* and *Amblyomma*, and they are important hosts for the latter genus. Thus, nymphs and larvae of *Amblyomma calcaratum*, *Amblyomma longirostre* and *Amblyomma nodosum*, and larvae of *Amblyomma parkeri* and *Amblyomma romarioi* have been found on *Conopophaga lineata*, while nymphs and larvae of *Amblyomma calcaratum*, and larvae of *Amblyomma parkeri* have been taken from *Conopophaga melanops*. Both species of *Conopophaga* are regarded as significant hosts of these five *Amblyomma* species.

Additionally, nymphs and larvae of *Ixodes auritulus*, nymphs and larvae of *Amblyomma aureolatum*, nymphs of *Amblyomma coelebs*, and larvae of *Amblyomma ovale* have been collected from *Conopophaga lineata*.

Corvidae: This family contains almost 130 species and has a worldwide geographic distribution. Two Neotropical species are known to be parasitized by immature stages of the genera *Amblyomma*, *Haemaphysalis*, and *Ixodes*. Thus, *Cyanacorax chrysops* has been found infested with larvae and nymphs of *Ixodes pararicinus* and *Haemaphysalis juxtakochi*, as well as larvae of *Haemaphysalis leporispalustris*, while nymphs of *Amblyomma longirostre* have been collected from *Cyanocorax cristatellus*. Neither bird is regarded as a principal host for these ticks.

Cotingidae: This Neotropical family comprises 65 species, three of which have occasionally been found infested with specimens of *Amblyomma*. Thus, *Rupicola*

rupicola has been found parasitized by females of *Amblyomma oblongoguttatum*, while nymphs of *Amblyomma longirostre* have been collected from *Pyroderus scutatus* and *Querula purpurata.*

Formicariidae: This Neotropical family contains 11 species, one of which, *Formicarius analis*, has been found parasitized by nymphs of *Amblyomma ovale*, although it is not a significant host for this tick.

Fringillidae: This family contains more than 220 species worldwide, and six Neotropical species have been found infested with immatures of the genus *Amblyomma*, including two species that are dependent on fringillids as hosts. Thus, *Euphonia laniirostris* and *Euphonia pectoralis* are two of the many species of passeriforms regarded as important hosts of *Amblyomma longirostre*, sustaining nymphs and larvae of this tick; the former bird has also been found infested with larvae of *Amblyomma geayi.*

Additionally, nymphs of *Amblyomma nodosum* have been taken from *Chlorophonia cyanea*, and nymphs of *Amblyomma longirostre* have been found on the remaining four species of Neotropical Fringillidae from which ticks have been collected, although these birds are not considered significant hosts.

Furnariidae: This family is Neotropical and large, containing about 300 species, 69 of which have been found infested with larvae and nymphs of the genus *Amblyomma*, as well as females, nymphs, and larvae of *Ixodes*. Several furnariids are considered important hosts for various species of *Amblyomma*. Thus, *Anabazenops fuscus* has been found parasitized by nymphs and larvae of *Amblyomma calcaratum*, *Amblyomma longirostre* and *Amblyomma nodosum*, and larvae of *Amblyomma parkeri*. The white-eyed foliage-gleaner *Automolus leucophthalmus*, together with *Xenops minutus* and *Xiphorynchus fuscus* have been found infested with nymphs and larvae of *Amblyomma longirostre* and larvae of *Amblyomma romarioi*. Nymphs and larvae of *Amblyomma longirostre* and larvae of *Amblyomma parkeri* have been collected from *Dendrocincla fuliginosa*, while nymphs and larvae of *Amblyomma humerale* and *Amblyomma longirostre* have been taken from *Dendrocolaptes hoffmannsi*. The rufous hornero, *Furnarius rufus*, is known to host nymphs and larvae of *Amblyomma tigrinum*, while *Hylexetastes perrotii* has been found infested with nymphs and larvae of *Amblyomma calcaratum* and *Amblyomma longirostre*. The furnariids *Campylorhamphus falcularius*, *Deconychura longicauda*, *Dendrocincla merula*, *Dendroplex picus*, *Glyphorynchus spirurus*, *Lepidocolaptes souleyetii*, *Philydor rufum*, *Sittasomus griseicapillus*, *Synallaxis ruficapilla*, *Xiphorynchus elegans*, *Xiphorynchus guttatus*, and *Xiphorynchus susurrans* all host nymphs and larvae of *Amblyomma longirostre*. Nymphs and larvae of *Amblyomma tigrinum* have been recovered from *Coryphistera alaudina*, *Furnarius cristatus*, *Pseudoseizura lophotes*, *Synallaxis albescens*, and *Tarphonomous certhioides*, while larvae of *Amblyomma romarioi* have been collected from *Dendrocincla turdina*, *Lepidocolaptes falcinellus*, *Sclerurus scansor*, and *Sittasomus griseicapillus*. Nymphs of *Amblyomma humerale* have been found on an undetermined species of *Sclerurus*. Larvae of *Amblyomma parkeri* have been found on *Lepidocolaptes squamatus*, *Philydor atricapillus*, and *Sitttasomus griseicapillus*, while larvae of *Amblyomma geayi* have been collected from *Dendrocolaptes hoffmannsi*, *Glyphorynchus spirurus*, *Hylexetastes perrotii*, and *Xiphorhynchus pardalotus*. All these host–tick associations are important in the life cycles of the named *Amblyomma* species.

Additionally, larvae and nymphs of *Ixodes fuscipes* have been found on both *Phacellodomus striaticollis* and *Syndactyla rufosuperciliata*, and nymphs and larvae of *Ixodes pararicinus* have been collected from the latter bird. All parasitic stages of *Ixodes auritulus* have been collected on 13 species of Furnariidae. Immature stages of 11 other species of *Amblyomma*, including nymphs of *Amblyomma geayi*, have been taken from several other furnariids.

Icteridae: This family contains more than 100 species found only in the Americas. In the Neotropical Region, 11 icterids have been found infested with the immature stages of *Amblyomma* and *Rhipicephalus* as well as adults and nymphs of *Ixodes*. The family Icteridae appears to include significant hosts for the immature stages of two species of *Amblyomma*, nymphs and larvae of *Amblyomma tigrinum* have been found on *Agelaioides badius* and *Molothrus bonariensis*, while nymphs and larvae of *Amblyomma longirostre* have been collected from *Cacicus uropygialis*.

Additionally, females of *Ixodes auritulus* have been obtained from *Curaeus curaeus* and *Leistes loyca*, and nymphs of *Rhipicephalus sanguineus sensu lato* have been recovered from *Gnorimopsar chopi*. Nymphs and larvae of *Amblyomma variegatum* have been found on *Quiscalis lugubris*, nymphs of *Amblyomma nodosum* have been collected from *Cacicus haemorrhous*, and larvae of *Amblyomma ovale* and *Amblyomma varium* have been taken from *Molothrus bonariensis*. However, none of these tick–host associations are regarded as important in maintaining populations of the ticks in these three genera.

Mimidae: This American family has about 30 species. In the Neotropical Region, two species have been found parasitized by ticks of the genera *Ixodes* and *Amblyomma*. *Mimus polyglottos* is an important host of *Ixodes copei*, whose female stage is known only from this bird.

Additionally, *Mimus saturninus* has been found infested with nymphs of *Amblyomma calcaratum*, although this bird is not considered a significant host.

Mitrospingidae: This Neotropical family consists of only four species, one of which, *Mitrospingus cassinii*, was found infested with nymphs of *Amblyomma coelebs* on a single occasion.

Parulidae: This American family contains about 110 species. Twelve species have been found infested with Neotropical ixodids belonging to the genera *Amblyomma*, *Haemaphysalis*, and *Ixodes*. Parulids are important hosts of two species of *Amblyomma:* nymphs and larvae of *Amblyomma longirostre* have been collected from *Basileuterus culicivorus*, *Basileuterus rufifrons*, and *Myiothlypis flaveola*, while larvae of *Amblyomma romarioi* have been taken from *Basileuterus culicivorus*.

Additionally, larvae of *Ixodes pararicinus* have been taken from *Myothlypis bivittata* and *Myioborus brunniceps*, while larvae of *Ixodes fuscipes* have been recovered from *Myiothlypis leucoblephara*, and immature stages of *Ixodes auritulus* have been found on four species of Parulidae. Nymphs and larvae of *Amblyomma aureolatum* have been found on *Myothlypis leucoblepharus*. A female of *Amblyomma longirostre* has been reported from *Basileuterus rufifrons*, and immature stages of four species of *Amblyomma* (*Amblyomma calcaratum*, *Amblyomma nodosum*, *Amblyomma parvum*, and *Amblyomma romarioi*), as well as a larva of *Haemaphysalis leporispalustris,* have been found on *Myothlypis bivittata*, although these bird–tick associations are not considered significant for the ixodid species involved.

Passerellidae: This American family contains more than 150 species. Twelve Neotropical species have been found infested with several species of *Amblyomma*, three *Ixodes*, and two species of *Haemaphysalis*, passerellids being important hosts for one species each of *Amblyomma* and *Ixodes*. Thus, nymphs and larvae of *Amblyomma tigrinum* have been found on *Rhynchospiza strigiceps* and *Zonotrichia capensis*, and *Junco phaeonotus* is one of two hosts of the female of *Ixodes mexicanus*. All these passerellid ixodid tick associations are considered significant for the ticks involved.

Additionally, nymphs and larvae of *Haemaphysalis leporispalustris* have been collected from *Zonotrichia capensis*, while larvae of *Haemaphysalis juxtakochi* have been taken from *Arremon flavirostris*. A female of *Amblyomma ovale* was found on *Arremonops conirostris*, and immature stages of 11 species of *Amblyomma*, *Haemaphysalis leporispalustris*, *Ixodes auritulus*, and *Ixodes pararicinus* have been found on several species of Passerellidae.

Pipridae: This Neotropical family contains almost 60 species, 18 of which have been found infested with ticks of the genus *Amblyomma*, piprids being important hosts for several species in this genus. Thus, *Chiroxiphia caudata* has been found parasitized by nymphs and larvae of *Amblyomma longirostre* and *Amblyomma nodosum*, as well as larvae of *Amblyomma parkeri* and *Amblyomma romarioi*. The red-capped manakin, *Ceratopipra mentalis*, as well as *Manacus manacus* have been found carrying larvae and nymphs of *Amblyomma calcaratum* and *Amblyomma longirostre*, and larvae of *Amblyomma geayi*, while *Dixiphia pipra* has been found infested with nymphs and larvae of *Amblyomma longirostre* and *Amblyomma nodosum*, and again, larvae of *Amblyomma geayi*. Nymphs and larvae of *Amblyomma longirostre* and larvae of *Amblyomma parkeri* have been collected from *Manacus manacus*, larvae of *Amblyomma geayi* have been found on *Chiroxiphia pareola* and *Manacus vitellinus*, while nymphs and larvae of *Amblyomma longirostre* have been taken from *Neopelma pallescens*. Nymphs and larvae of *Amblyomma longirostre* have been collected from *Antilophia galeata*, *Ceratopipra erythrocephala*, *Chiroxiphia pareola*, *Ilicura militaris*, *Manacus candei*, *Pipra aureola*, and *Pipra fasciicauda*. The piprid *Lepidothrix coronata* has been found infested with nymphs and larvae of *Amblyomma calcaratum* and

Amblyomma longirostre. These bird–tick associations are all regarded as important for the species of ixodids involved.

Additionally, immature stages of another nine species of *Amblyomma* have been found on Pipridae.

Polioptilidae: This American family consists of 15 species. In the Neotropical Region, two species are known to host immature stages of *Amblyomma*. Thus, *Polioptila dumicola* is regarded as an important host for nymphs and larvae of *Amblyomma tigrinum*.

Additionally, nymphs of *Amblyomma nodosum* have been found feeding on *Microbates cinereiventer*.

Rhinocryptidae: This family is Neotropical and contains about 60 species, but only two of these have occasionally been found parasitized with immature stages of *Amblyomma* and *Ixodes*. Thus, nymphs of *Amblyomma tigrinum* have been found on *Rhinocrypta lanceolata*, and larvae of *Ixodes auritulus* have been collected from *Scytalopus* sp.

Thamnophilidae: This is a Neotropical family of almost 240 species, 51 of which have been found infested with ticks of the genera *Amblyomma*, *Haemaphysalis,* and *Ixodes*, thamnophilids being important in the life cycle of several species of *Amblyomma*. Thus, *Pyriglena leucoptera* has been found infested with nymphs and larvae of *Amblyomma longirostre* and larvae of *Amblyomma romarioi*. Nymphs and larvae of *Amblyomma calcaratum* and *Amblyomma longirostre* (together with one male tick), and larvae of *Amblyomma parkeri* have been collected from *Thamnophilus caerulescens*, while *Pyriglena leuconota* has been found parasitized by nymphs and larvae of *Amblyomma longirostre*. The antbird *Willisornis poecilonotus* has been reported infested with nymphs and larvae of *Amblyomma longirostre*, larvae of *Amblyomma geayi* and nymphs of *Amblyomma humerale*. Nymphs and larvae of *Amblyomma longirostre* and larvae of *Amblyomma geayi* have been recovered from *Gymnopithys leucaspis*, *Percnostola rufifrons*, *Thamnophilus atrinucha*, and *Thamnophilus doliatus,* while nymphs and larvae of *Amblyomma longirostre* and *Amblyomma nodosum* have been taken from *Thamnophilus pelzelni*. Nymphs and larvae of *Amblyomma longirostre* have been found on *Dsyithamnus mentalis*, *Hypocnemoides maculicauda*, *Myrmotherula axillaris*, *Phaenostictus macleannani*, *Taraba major*, and *Thamnophilus punctatus*. Nymphs of *Amblyomma humerale* and larvae of *Amblyomma geayi* have been taken from *Thamnomanes caesius*, while nymphs of *Amblyomma humerale* have been collected from *Myrmotherula axillaris* and *Thamnomanes schistogynus*. All these bird–parasite associations are considered important for the *Amblyomma* species involved.

Additionally, females and immature stages of *Ixodes auritulus* have been found on *Thamnophilus caerulescens* and *Thamnophilus ruficapillus*, while immature stages of *Haemaphysalis juxtakochi* have been taken from *Hylophylax naevius* and *Willisornis poecilonotus*, nymphs and larvae of *Haemaphysalis leporispalustris* have been collected from *Poliocrania exsul*, and larvae of the latter tick have also been found on *Thamnophilus caerulescens*. Immature stages of another ten species of *Amblyomma* have been found infesting Thamnophilidae.

Thraupidae: This family is mostly Neotropical and contains about 375 species, 70 of which have been found infested with ticks of the genera *Amblyomma*, *Haemaphysalis*, *Ixodes*, and *Rhipicephalus*, thraupids being important in the life cycle of several *Amblyomma* species. Thus, nymphs and larvae of *Amblyomma longirostre* and *Amblyomma nodosum*, and larvae of *Amblyomma parkeri* and *Amblyomma romarioi* have been found on *Trichothraupis melanops*, while nymphs and larvae of *Amblyomma longirostre*, and larvae of *Amblyomma parkeri* and *Amblyomma romarioi* have been collected from *Tachyphonus coronatus*. The Brazilian tanager, *Ramphocelus bresilius*, is parasitized by nymphs and larvae of *Amblyomma longirostre*, and larvae of *Amblyomma romarioi*, while *Saltator maximus* has been found infested with nymphs and larvae of *Amblyomma longirostre*, and larvae of *Amblyomma parkeri*. Nymphs and larvae of *Amblyomma tigrinum* have been found on *Coryphospingus cucullatus*. Nymphs and larvae of *Amblyomma longirostre* have been collected from *Coereba flaveola*, *Dacnis cayana*, *Sporophila nigricollis*, *Ramphocelus carbo*, *Saltator similis*, *Tachyphonus cristatus*, *Tachyphonus rufus*, and *Thraupis palmarum*. Nymphs and larvae of *Amblyomma longirostre* and larvae of *Amblyomma geayi* have been recovered from *Eucometis penicillata*, while nymphs and larvae of *Amblyomma tigrinum* have been found on *Lophospingus pusillus*, *Saltator aurantiirostris*, and *Saltatricula multicolor*. Nymphs of *Amblyomma sabanerae* have been taken from *Tachyphonus surinamis*, and larvae of *Amblyomma parkeri* have been collected from *Tiaris fuliginosa*. All these bird–tick associations are important in maintaining the life cycles of the ticks involved.

Additionally, *Volatinia jacarina* was found infested with a female of *Ixodes bequaerti*, larvae of *Ixodes pararicinus* have been taken from *Coryphospingus cucullatus*, and females, nymphs and larvae of *Ixodes auritulus* have been collected from several species of Thraupidae. Females of *Haemaphysalis juxtakochi* have been found on *Thlypopsis pyrrhocoma*, while larvae of *Haemaphysalis leporispalustris* have been collected from *Trichothraupis melanops*, and a nymph of the latter tick was found on *Saltator similis*. A male of *Rhipicephalus sanguineus sensu lato* was recovered from *Coereba flaveola*, and a male of *Amblyomma ovale* was found on *Sporophila nigricollis*. Immature stages of another ten species of *Amblyomma* have been collected from Thraupidae.

Tityridae: This American family contains 34 species, most of them in the Neotropical Region. Five of these species have been found infested with immature stages of the genus *Amblyomma*, tityrids being important in the life cycle of two *Amblyomma* species. Thus, *Pachyramphus polychopterus* and *Schiffornis virescens* have been found parasitized by nymphs and larvae of *Amblyomma longirostre*, the latter bird also having been found infested with larvae of *Amblyomma romarioi*.

Additionally, nymphs and larvae of *Amblyomma ovale* have been found on *Schiffornis turdina*, while nymphs of *Amblyomma calcaratum* have been collected from *Schiffornis virescens*, and nymphs of *Amblyomma nodosum* have been taken from *Pachyramphus polychopterus*.

Troglodytidae: This family contains almost 90 species in the Americas and one species found in the Palearctic Region. In the Neotropics, 12 species in this family are known to have been parasitized by ticks of the genera *Amblyomma*, *Haemaphysalis*, and *Ixodes*, troglodytids being important in maintaining populations of three species of *Amblyomma* and two *Ixodes*. Thus, *Troglodytes aedon* is an important host of nymphs and larvae of *Amblyomma longirostre* and *Amblyomma tigrinum*, as well as larvae of *Amblyomma parkeri* and *Ixodes minor*. Nymphs and larvae of *Amblyomma longirostre* have been found on *Cantorchilus nigricapillus*. And *Campylorhynchus gularis* is one of two known hosts for the female of *Ixodes mexicanus*.

Additionally, larvae of *Ixodes pararicinus* have been found on *Troglodytes aedon*, and females, nymphs and larvae of *Ixodes auritulus* have been collected from several species of Troglodytidae. Larvae of *Haemaphysalis leporispalustris* have been taken from *Cantorchilus nigricapillus*, *Pheugopedius genibarbis*, and *Troglodytes aedon*. Immature stages of eight other species of *Amblyomma* have been recovered from Troglodytidae.

Turdidae: This large family contains about 175 species worldwide, and in the Neotropical Region 24 of these have been found parasitized by ticks of the genera *Amblyomma*, *Haemaphysalis*, and *Ixodes*, turdids being important as hosts of four species of *Ixodes*, five *Amblyomma*, and one species of *Haemaphysalis*. The family Turdidae is regarded here as the most important avian family in maintaining populations of Neotropical ixodid ticks. Thus, *Turdus albicollis*, *Turdus amaurochalinus*, *Turdus falcklandii*, and *Turdus rufiventris* are treated here as principal hosts for females, nymphs, and larvae of *Ixodes auritulus*, while *Turdus nigriceps* and *Turdus rufiventris* are the principal hosts for adults and the immature stages of *Ixodes silvanus*. The rufous-bellied thrush, *Turdus rufiventris*, is also a principal host for nymphs and larvae of *Ixodes pararicinus* and *Amblyomma aureolatum*. Nymphs and larvae of *Amblyomma longirostre* and larvae of *Haemaphysalis juxtakochi*, *Amblyomma parkeri*, and *Amblyomma romarioi* have been collected from *Turdus albicollis*, and this bird has also been found with larvae of *Amblyomma geayi*, while nymphs and larvae of *Amblyomma longirostre*, larvae of *Haemaphysalis juxtakochi*, and larvae of *Amblyomma romarioi* have been found on *Turdus rufiventris*. The creamy-bellied thrush, *Turdus amaurochalinus*, has been found infested with nymphs and larvae of *Amblyomma longirostre*, as well as larvae of *Amblyomma romarioi*, and *Turdus fumigatus* and *Turdus grayi* have been found parasitized by nymphs and larvae of *Amblyomma longirostre*. Nymphs and larvae of *Amblyomma longirostre*, and larvae of *Amblyomma romarioi* and *Amblyomma geayi* have been taken from *Turdus leucomelas*, while nymphs and larvae of *Ixodes longirostre* have been removed from *Catharus ustulatus* and *Turdus subalaris*. Larvae of *Amblyomma romarioi* have been found on *Turdus flavipes*. Larvae of *Ixodes minor* have been collected from *Catharus frantzii*, and *Turdus jamaicensis* is the only known host of *Ixodes copei*.

Additonally, a *Catharus* sp. has been found infested with females of *Ixodes bequaerti*, and nymphs and larvae of *Ixodes fuscipes* have been taken from *Turdus albicollis* and *Turdus rufiventris*. Immature stages of *Haemaphysalis leporispalustris* have been collected from *Catharus dryas*, *Turdus albicollis*, and *Turdus rufiventris*. Immature stages of 12 other species of *Amblyomma* have been recovered from Turdidae.

Tyrannidae: This American family is very large, with about 420 species. In the Neotropical Region, 63 species have been found infested with ixodids of the genera *Amblyomma*, *Haemaphysalis*, and *Ixodes*, tyrannids being of importance in maintaining the life cycles of seven species of *Amblyomma* and one species of *Haemaphysalis*. Thus, nymphs and larvae of *Amblyomma calcaratum*, *Amblyomma longirostre*, and *Amblyomma nodosum*, as well as larvae of *Amblyomma parkeri* and *Amblyomma romarioi*, have been collected from *Platyrinchus mistaceus*, while nymphs and larvae of *Amblyomma longirostre*, and larvae of *Amblyomma geayi*, *Amblyomma parkeri*, and *Amblyomma romarioi* have been taken from *Leptopogon amaurocephalus*. The olivaceus flatbill, *Rhynchocyclus olivaceus*, has been found parasitized by nymphs and larvae of *Amblyomma longirostre*, nymphs of *Amblyomma humerale* and larvae of *Amblyomma geayi*; and *Mionectes oleagineus* has been found infested with nymphs and larvae of *Amblyomma longirostre* and larvae of *Amblyomma geayi*. Nymphs and larvae of *Amblyomma longirostre* have been found on *Elaenia flavogaster*, *Mionectes rufiventris*, *Myiarchus ferox*, *Oncostoma olivaceum*, and *Tolmomyias sulphurescens*, along with a female specimen on the last-named bird. Larvae of *Amblyomma romarioi* and *Amblyomma geayi* have been found on *Attila rufus* and *Attila spidaceus*, respectively. Larvae of *Amblyomma parkeri* have been recovered from *Tolmomyias poliocephalus*.

Additionally, females, nymphs, and larvae of *Ixodes auritulus* have been found on *Knipolegus nigerrimus*, *Phylloscartes ventralis*, and *Elaenia albiceps*, respectively. Nymphs of *Haemaphysalis juxtakochi* have been collected from *Cnipodectes subbrunneus*, while larvae of *Haemaphysalis leporispalustris* have been found on *Corytophis delalandi*, *Fluvicola nengeta*, and *Myiarchus tyrannulus*. Immature stages of five other species of *Amblyomma* have been found on Tyrannidae.

Vireonidae: This family contains about 60 species, mostly in the Americas but a few in Southeast Asia. Three Neotropical species have occasionally been parasitized by immatures of the genus *Amblyomma*. Thus, nymphs and larvae of *Amblyomma longirostre* have been found on *Cyclarhis gujanensis* and *Vireo chivi*, while nymphs of *Amblyomma calcaratum* have been collected from *Cyclarhis gujanensis* and *Hylophilus poicilotis*, and larvae of *Amblyomma auricularium* have been taken from *Cyclarhis gujanensis*.

5.1.3.14 Pelecaniformes

This order contains five families with almost 110 species and has a worldwide distribution. In the Neotropical Region, members of three families have been found infested with ixodids, but none of these birds are important as hosts of the tick species that parasitize them.

Ardeidae: This family is the largest in the Pelecaniformes, with about 65 species and a cosmopolitan distribution. At least three species of Ardeidae have been found infested with Ixodidae in the Neotropics. *Bubulcus ibis* is known to be parasitized by larvae and nymphs of *Amblyomma variegatum* and larvae of *Rhipicephalus microplus*, while a male of *Amblyomma dissimile* was found on a *Cochlearius cochlearius*. The black-crowned night-heron, *Nycticorax nycticorax*, has been found infested with females and nymphs of *Ixodes auritulus*, while females and nymphs of *Amblyomma mixtum* have been collected from a "heron."

Pelecanidae: This family is cosmopolitan and contains eight species. In the Neotropical Region, one of these, *Pelecanus occidentalis*, was found parasitized by a male of *Amblyomma variegatum.*

Threskiornithidae: This family of 34 species is distributed in several regions of the world. In the Neotropics, one species, *Theristicus caudatus*, has been found infested with nymphs of *Amblyomma sculptum.*

5.1.3.15 Piciformes

This order contains more than 370 species in seven families with a worldwide geographical distribution. In the Neotropical Region, three families of Piciformes have been found infested with ixodids in the genera *Ixodes* and *Amblyomma*, but no piciformes are important hosts for these ticks.

Capitonidae: This family is Neotropical and contains 14 species; one of these, *Capito auratus*, has occasionally been found parasitized by nymphs of *Amblyomma longirostre* and *Amblyomma nodosum.*

Picidae: This family is large and cosmopolitan, with more than 200 species worldwide. In the Neotropical Region, six species of Picidae have been found infested with ticks of the genus *Amblyomma*, and one species has been found carrying an *Ixodes*. Thus, *Picumnus exilis* has been found infested with larvae and nymphs of *Amblyomma longirostre*, while nymphs of *Amblyomma longirostre* have been taken from *Celeus flavescens*, *Dryobates maculifrons*, *Picumnus cirratus*, and *Picumnus temmincki*, and a female of *Ixodes auritulus* was collected from *Colaptes rupicola.* None of these birds are considered principal hosts for either tick species.

Ramphastidae: This family contains 36 species, all of them restricted to the Neotropical Region. Three species of Ramphastidae are known hosts for ticks of the genus *Amblyomma*. Thus, *Pteroglossus castanotis*, *Ramphastos dicolorus*, and *Ramphastos vitellinus* have been found parasitized by nymphs of *Amblyomma longirostre* and the last-named bird has also been found infested with nymphs of *Amblymma geayi*, but none of these birds are considered principal hosts for either tick species.

5.1.3.16 Procellariiformes

This order of marine birds contains more than 130 species worldwide and consists of four families. Two species from one family have been found infested with *Ixodes* ticks in the Neotropics, but none of these birds are significant hosts for these ticks.

Procellariidae: This family contains about 90 species worldwide, two of which are known to be infested by ixodids in the Neotropics. Nymphs of *Ixodes uriae* have been found on *Daption capense*, while females, nymphs, and larvae of *Ixodes auritulus* have been collected from *Pelecanoides magellani.*

5.1.3.17 Psittaciformes

This order is characteristic of tropical and subtropical areas worldwide and contains about 370 species classified into four families. In the Neotropical Region, there is a single record of one psittaciform that was found infested with ticks of the genus *Amblyomma.*

Psittacidae: This family contains almost 170 species found in the Afrotropical and Neotropical Zoogeographic Regions. One species, *Amazona aestiva*, was found infested with nymphs of *Amblyomma sculptum.*

5.1.3.18 Rheiformes

A Neotropical order containing one family with two species found only in South America, and both species are known to be parasitized by ticks of the genus *Amblyomma.*

Rheidae: This family contains two members: *Rhea americana*, which has been found infested with nymphs of *Amblyomma dubitatum* and one female of *Amblyomma sculptum*, and *Rhea pennata*, which was found infested with a female of *Amblyomma parvitarsum.* Neither bird is considered an important host for these ticks.

5.1.3.19 Sphenisciformes

This order contains 18 species in one family, geographically distributed in South America, southern Africa, and southern Australasia. Five species of Spheniscidae have been found infested with ixodids, and these birds are principal hosts for one species of *Ixodes.*

Spheniscidae: This family contains five species, *Eudyptes chrysocome*, *Pygoscelis adeliae*, *Pygoscelis antarcticus*, *Pygoscelis papua*, and *Spheniscus magellanicus*, that are significant hosts for females, nymphs, and larvae of *Ixodes uriae.*

Additionally, a female of *Amblyomma parvitarsum* was collected from *Spheniscus magellanicus.*

5.1.3.20 Strigiformes

This order contains almost 230 species worldwide, classified in two extant families. In the Neotropical Region, six species have occasionally been found infested with immature stages of *Amblyomma.*

Strigidae: This family is cosmopolitan and comprises almost 210 species; six of these taxa have been found infested with species of *Rhipicephalus* and *Amblyomma.* A single adult of *Amblyomma cajennense* and *Rhipicephalus sanguineus sensu lato* was taken from *Asia clamator.* This species and *Asio stygius*, *Athene cunicularia*, *Megascops choliba*, *Pulsatrix koeniswaldiana*, and *Pulsatrix perspicillata* have been found infested with the immature stages of eight species of *Amblyomma*, and *Megascops choliba* was infested by an adult *Amblyomma dubitatum.* None of these birds are important hosts for these ticks.

Tytonidae: This family contains 18 species distributed in different parts of the world, and one of these, *Tyto alba*, has been found parasitized by nymphs of *Amblyomma aureolatum* and *Amblyomma sculptum*, but this bird is not important in sustaining these Neotropical ticks.

5.1.3.21 Suliformes

This order of marine birds contains about 60 species classified into four families with a worldwide distribution, but only one species of this order has been found infested with ixodids in the Neotropics.

Phalacrocoracidae: This family contains 40 species with a worldwide distribution, and one of these, *Phalacrocorax atriceps*, has been found infested with nymphs and larvae of *Ixodes uriae*, but this bird is not considered to be a principal host for this tick.

5.1.3.22 Tinamiformes

This order contains one family with 46 species, all restricted to the Neotropical Region, where several species have been found infested with ticks of the genera *Amblyomma*, *Haemaphysalis*, and *Ixodes*, but tinaform birds are not considered significant hosts for Neotropical ixodids.

Tinamidae: Seven species in this family have been found infested with ixodids. Thus, *Crypturellus tataupa* is known to be parasitized by nymphs of *Haemaphysalis leporispalustris*, while *Eudromia elegans* has been found infested with larvae of *Amblyomma tigrinum*, and females and one nymph of *Ixodes auritulus* have been collected from *Nothoprocta pentlandii.* Nymphs and larvae of *Amblyomma tigrinum* have been found on *Nothoprocta perdicaria*, while males, nymphs, and larvae of *Amblyomma pseudoconcolor* have been taken from *Nothura boraquira.* Males, nymphs, and larvae of *Amblyomma pseudoconcolor* and nymphs of *Amblyomma tigrinum* have been found on *Nothura maculosa* and *Rhynchotus rufescens.* Despite these records, as noted above, no species of Tinamidae appears to be an important host of ixodid ticks.

5.1.3.23 Trogonifomes

This order contains more than 40 species in one family; most of these are found in the Neotropics, but a few taxa occur in the Afrotropical and Oriental Zoogeographic Regions. Two Neotropical species are of importance for one species of *Ixodes* and another has been found parasitized by *Amblyomma.*

Trogonidae: This family has two species, *Pharomachrus mocinno* and *Trogon collaris*, that are principal hosts for the female of *Ixodes bequaerti.*

Additionally, nymphs of *Amblyomma longirostre* and *Amblyomma nodosum* have been collected from *Trogon viridis.*

All avian hosts and the ticks found on them are listed in the appendix to this book.

5.1.3.24 Synopsis

The number of avian orders infested with ixodid ticks greatly surpasses the number of orders in other classes of vertebrates that are known to be hosts of Neotropical Ixodidae. In total, 23 avian orders contain at least one species infested with ticks of the genera *Ixodes*, *Amblyomma*, *Haemaphysalis*, and *Rhipicephalus*, but no cases of avian parasitism by ticks of the genus *Dermacentor* have been reported. Further analysis shows that Neotropical birds are rarely infested with ticks of the genus *Rhipicephalus*. Thus, Aves are important hosts for ticks of the genera *Ixodes*, *Amblyomma* and *Haemaphysalis*, but the orders of Aves containing species considered to be significant hosts of Ixodidae are just five in number: Caprimulgiformes, Galliformes, Passeriformes, Sphenisciformes, and Trogoniformes. Even so, host utilization by these avian orders is not uniform, because birds are more important as hosts for prostriate (*Ixodes*) than for metastriate (*Amblyomma* and *Haemaphysalis*) ticks. The complex relationships between hosts and ixodids are summarized in Tables 1.29, 2.53, 3.6, 5.1, 5.2, and 5.3, but see also Tables 5.4 and 5.6, where differences in utilization of avian orders between the most abundant Neotropical genera, *Ixodes* (Prostriata) and *Amblyomma* (Metastriata), are shown.

The only ixodid genus containing species known to feed exclusively on Aves is *Ixodes*, which is also the only genus known to feed on Caprimulgiformes (*Ixodes cuernavacensis*, *Ixodes downsi* and *Ixodes paranaensis*), Galliformes (*Ixodes cornuae*), Sphenisciformes (*Ixodes uriae*), and Trogoniformes (*Ixodes bequaerti*). Additionally, four species—*Ixodes auritulus*, *Ixodes copei*, *Ixodes mexicanus*, and *Ixodes silvanus*—are known as parasites of Passeriformes (Tables 1.29 and 5.1). By contrast, the *Amblyomma* and *Haemaphysalis* specific to Aves (nine species of *Amblyomma* and *Haemaphysalis juxtakochi*) are restricted to Passeriformes, but in all these cases only immature stages utilize Aves as principal hosts, while adult ticks are parasites of mammals. Thus, Turdidae are among the principal hosts of the nymph and larva of *Haemaphysalis juxtakochi*, and Troglodytidae are principal hosts for the larva of *Amblyomma aureolatum*, while several families of Passeriformes are important hosts for the larvae of *Amblyomma geayi*, *Amblyomma romarioi*, and *Amblyomma parkeri*, and the nymph of *Amblyomma humerale*. Moreover, Turdidae are principal hosts for the larva and nymph of *Amblyomma aureolatum*, and several families of Passeriformes are regarded as important hosts for the larvae and nymphs of *Amblyomma calcaratum*, *Amblyomma nodosum*, *Amblyomma longirostre*, and *Amblyomma tigrinum* (Tables 2.53, 3.6, 5.2, and 5.3).

In summary, birds generally play an important role as hosts for *Ixodes*, while only Passseriformes are of importance for ticks of the genera *Amblyomma* and *Haemaphysalis*. To date birds are not known to be hosts of *Dermacentor* and are only rarely parasitized by *Rhipicephalus*. See also the appendix for a full list of avian host species and the ticks that parasitize them in the Neotropical Region.

5.1.4 *Mammalia as Hosts for Neotropical Ixodidae*

Mammalia are the most important hosts for all genera of Neotropical Ixodidae, and the life cycles of all Neotropical *Dermacentor* and *Rhipicephalus* species are exclusively tied to mammals. As well, 73% of the species of *Ixodes* and 54% of *Amblyomma* species principally parasitize mammals. Ixodid records from Mammalia in the Neotropical Region include specimens collected from 13 orders, 46 families, and 319 species, as discussed alphabetically below.

5.1.4.1 Artiodactyla (Cetacea Excluded)

This order contains about 230 species worldwide, classified in ten families. Artiodactyla belonging to five families, including domestic mammals, are significant hosts for several common or economically important tick species representing all genera of Ixodidae found in the Neotropical Region. However, artiodactyls have also been found infested by many lesser-known ixodids, as shown below, and some of these ticks may be more dependent on Artiodactyla than currently realized.

Table 5.1 Numbers of species of Neotropical *Ixodes* arranged according to classes, orders, and families of principal hosts, or potential principal hosts, and their combinations. Male = M, female = F, nymph = N, larva = L

AMBHIBIA 0	
AVES 9 (**16**%)	
CAPRIMULGIFORMES 2	
Apodidae	*Ixodes cuernavacensis* (F), *I. paranaensis* (FNL)
GALLIFORMES 1	
Phasianidae	*Ixodes cornuae* (FN)
PASSERIFORMES 4	
Mimidae + Turdidae	*Ixodes copei* (F) (L)*
Passerellidae + Troglodytidae	*Ixodes mexicanus* (F)
Turdidae	*Ixodes auritulus* (FNL), *I. silvanus* (FNL)
SPHENISCIFORMES 1	
Spheniscidae	*Ixodes uriae* (MFNL)
TROGONIFORMES 1	
Trogonidae	*Ixodes bequaerti* (F)
MAMMALIA 40 (**73**%)	
ARTIODACTYLA 4	
Cervidae	*Ixodes montoyanus* (FNL), *I. stilesi* (MFN), *I. taglei* (MF)
Bovidae, Cervidae	*Ixodes fuscipes* (MF) (MF)
CARNIVORA 4	
Canidae	*Ixodes boliviensis* (MF)
Felidae	*Ixodes affinis* (MF)
Procyonidae	*Ixodes texanus* (FN)

(continued)

Table 5.1 (continued)

Several families	*Ixodes rubidus* (FN)
CHIROPTERA 1 Vespertilionidae	*Ixodes chilensis* (FN)
LAGOMORPHA 2 Leporidae	*Ixodes dicei* (MF), *I. pomerantzi* (MF, N**)
MICROBIOTHERIA 1 Microbiotheriidae	*Ixodes neuquenensis* (FNL)
PERISSODACTYLA 1 Tapiridae	*Ixodes tapirus* (MF)
RODENTIA 20 Capromyidae	*Ixodes capromydis* (MFNL)
Chinchillidae	*Ixodes nuttalli* (MFNL)
Cricetidae	*Ixodes andinus* (FNL), *I. galapagoensis* (FN), *I. jonesae* (MFNL), *I. nectomys* (M), *I. schulzei* (F, NL inferred), *I. sigelos* (FNL), *I. tancitarius* (F), *I. woodi* (F)
Ctenomyidae	*Ixodes cooleyi* (F)
Cuniculidae	*Ixodes lasallei* (F)
Dasyproctidae	*Ixodes spinosus* (FNL)
Geomydae	*Ixodes tecpanensis* (F)
Sciuridae	*Ixodes guatemalensis* (F), *I. tiptoni* (MFNL)
Abrocomidae + Cricetidae	*Ixodes abrocomae* (M) (MF)
Caviidae + Cricetidae	*Ixodes longiscutatus* (F) (NL)

Cricetidae + Heteromyidae	*Ixodes sinaloa* (NL) (F)
Cuniculidae + Dasyproctidae	*Ixodes bocatorensis* (F) (F)
DIDELPHIMORPHIA + RODENTIA 6	
Didelphidae + Cricetidae	*Ixodes amarali* (FNL) (NL), *I. catarinensis* (F) (F), *I. loricatus* (MFNL) (NL), *I. luciae* (MFNL) (NL), *I. tropicalis* (F) (N**L**)
Didelphidae + Heteromyidae	*Ixodes venezuelensis* (NL) (F)
LAGOMORPHA + RODENTIA 1	
Leporidae + Cricetidae	*Ixodes spinipalpis* (F) (N)
REPTILIA 0	
AVES + MAMMALIA 3 (5%)	
CAPRIMULGIFORMES + CHIROPTERA 1	
Steatornithidae + Phyllostomidae	*Ixodes downsi* (F) (L)
PASSERIFORMES + RODENTIA 1	
Troglodytidae, Turdidae + Muridae	*Ixodes minor* (L) (L) (MF)
ARTIODACTYLA + PASSERIFORMES + RODENTIA 1	
Bovidae + Turdidae + Cricetidae	*Ixodes pararicinus* (MF) (N**, L**) (N**, L**)
UNKNOWN 3 (5%)	*Ixodes cookei*, *I. fossulatus*, *I. scapularis*
Total 55 species	

* Ticks stages in parentheses follow the alphabetical order of principal hosts
** Provisional result because the tick stage remains undescribed

Table 5.2 Numbers of species of Neotropical *Amblyomma* arranged by classes, orders, and families of principal hosts or potential principal hosts, and their combinations. M = male, F = female, N = nymph, L = larva

AMBHIBIA 0	
AVES 0	
MAMMALIA 36 (54%)	
ARTIODACTYLA 8	
Bovidae	*Amblyomma hadanii* (MF), *A. neumanni* (MFNL), *A. patinoi* (MF), *A. tonelliae* (MFN, L*), *A. variegatum* (MFNL)
Cervidae	*Amblyomma interandinum* (MF)
Tayassuidae	*Amblyomma boeroi* (MFNL), *A. pecarium* (MF)
CARNIVORA 1	
Canidae	*Amblyomma maculatum* (MF)
CINGULATA 3	
Dasypodidae	*Amblyomma beaurepairei* (MF)
Chlamyphoridae + Dasypodidae	*Amblyomma auricularium* (MFNL) (MFNL)**, *A. pseudoconcolor* (MF) (MF)
DIDELPHIMORPHIA 1	
Didelphidae	*Amblyomma yucumense* (NL)
PERISSODACTYLA 5	
Tapiridae	*Amblyomma incisum* (MF), *A. latepunctatum* (MF), *A. multipunctum* (MFN), *A. scalpturatum* (MF), *A. tapirellum* (MF)
PILOSA 3	
Bradypodidae	*Amblyomma varium* (MF)
Myrmecophagidae	*Amblyomma goeldii* (MF), *A. pictum* (MF)
RODENTIA 4	

Caviidae	*Amblyomma dubitatum* (MFNL), *A. pseudoparvum* (MF), *A. romitii* (MFN)
Cuniculidae	*Amblyomma pacae* (MF)
ARTIODACTYLA + PERISSODACTYLA 1	
Bovidae + Equidae	*Amblyomma tenellum* (MF) (MF)
ARTIODACTYLA + UNCERTAIN MAMMALS 1	
Tayassuidae + several orders	*Amblyomma naponense* (MF) (N)
CARNIVORA + RODENTIA 1	
Canidae, Felidae + Cricetidae, Echimyidae	*Amblyomma ovale* (MF) (MF) (NL) (NL)
CARNIVORA + DIDELPHIMORPHIA + PERISSODACTYLA 1	
Procyonidae + Didelphidae + Tapiridae	*Amblyomma coelebs* (N, L*) (N, L*) (MF)
UNCERTAIN MAMMALS 5	
Several orders	*Amblyomma brasiliense* (MFNL), *A. cajennense* (MFN), *A. mixtum* (MFNL), *A. oblongoguttatum* (MFN), *A. sculptum* (MFN, L*)
RODENTIA + UNCERTAIN MAMMALS 2	
Caviidae + several orders	*Amblyomma parvum* (NL) (MF)
Cricetidae + several orders	*Amblyomma triste* (NL) (MF)
REPTILIA 16 (24%)	
SQUAMATA 10	
Boidae	*Amblyomma fulvum* (M), *A. quadricavum* (MF)
Iguanidae	*Amblyomm albopictum* (MFL, N*), *A. antillorum* (MFNL), *A. cruciferum* (MFN), *A. darwini* (MFNL), *A. scutatum* (MFNL), *A. torrei* (MFNL), *A. williamsi* (MF)
Tropiduridae	*Amblyomma boulengeri* (MF)
TESTUDINES 6	
Testudinae	*Amblyomma argentinae* (MFNL), *A. crassum* (F), *A. macfarlandi* (MFNL), *A.*

(continued)

Table 5.2 (continued)

	pilosum (MFNL), *A. usingeri* (MFNL)
Geoemydidae	*Amblyomma sabanerae* (MFNL)
AMPHIBIA + REPTILIA 2 (3%)	
ANURA + SQUAMATA 2	
Bufonidae + Boidae	*A. rotundatum* (MFNL) (MFNL)**
Bufonidae + Boidae, Iguanidae	*Amblyomma dissimile* (MFNL) (MFNL) (MFNL)
AVES + MAMMALIA 8 (12%)	
CARNIVORA + PASSERIFORMES 1	
Canidae, Felidae, Procyonidae + Troglodytidae, Turdidae	*Amblyomma aureolatum* (MF) (MF) (MF) (L*) (N, L*)
PASSERIFORMES + PILOSA 3	
Several families + Bradypodidae	*Amblyomma geayi* (L) (MFN)
Several families + Myrmecophagidae	*Amblyomma calcaratum* (NL) (MF), *A. nodosum* (NL) (MF)
PASSERIFORMES + PRIMATES 1	
Several families + Pitheciidae	*Amblyomma romarioi* (L*) (MF, N*)
PASSERIFORMES + RODENTIA 1	
Several families + Erethizontidae	*Amblyomma longirostre* (NL) (MF)
CARNIVORA + PASSERIFORMES + RODENTIA 1	
Canidae + several families + Caviidae, Cricetidae	*Amblyomma tigrinum* (MF) (NL) (N) (L)
PASSERIFORMES + PRIMATES + RODENTIA 1	
Several families + Atelidae + Erethizontidae	*Amblyomma parkeri* (L) (N) (MFN)
MAMMALIA + REPTILIA 2 (3%)	
ARTIODACTYLA + SQUAMATA 1	
Camelidae + Liolaemidae	*Amblyomma parvitarsum* (MF) (NL)
CINGULATA + DIDELPHIMORPHIA + SQUAMATA 1	

Dasypodidae + Didelphidae + Boidae	*Amblyomma fuscum* (MF) (N, L*) (MF)
AVES + MAMMALIA + REPTILIA 1 (1%)	
UNCERTAIN MAMMALIA + PASSERIFORMES + SQUAMATA + TESTUDINES 1	
Several orders + several families + several families + Testudinidae	*Amblyomma humerale* (N) (N) (N) (MF)
UNKNOWN 2 (3%)	*Amblyomma hirtum*, *A. inornatum*
Total 67 species	

* Provisional result because the tick stage remains undescribed

** Tick stages in parentheses follow the alphabetical order of principal hosts

Table 5.3 Numbers of species of Neotropical *Dermacentor*, *Haemaphysalis*, and *Rhipicephalus* arranged according to classes, orders, and families of principal hosts or potential principal hosts, and their combinations. M = male, F = female, N = nymph, L = larva

	Dermacentor
AMBHIBIA 0	
AVES 0	
MAMMALIA 6 (66%)	
ARTIODACTYLA 2	
Tayassuidae	*Dermacentor dispar* (MF), *D. imitans* (MF)
PERISSODACTYLA 3	
Equidae	*Dermacentor dissimilis* (MFN, L*), *D. nitens* (MFNL)
Tapiridae	*Dermacentor latus* (MF)
RODENTIA 1	
Cricetidae + Erethizontidae	*Dermacentor panamensis* (NL) (MF)**
REPTILIA 0	
UNKNOWN 3 (33%)	*Dermacentor albipictus****, *D. halli****, *D. variabilis****
Total 9 species	
	Haemaphysalis
AMBHIBIA 0	
AVES 0	
MAMMALIA 1 (33%)	
LAGOMORPHA 1	
Leporidae	*Haemaphysalis leporispalustris* (MFNL)
MAMMALIA + AVES 1 (33%)	
ARTIODACTYLA + UNCERTAIN MAMMALIA + PASSERIFORMES	

Cervidae + several orders + Turdidae	*Haemaphysalis juxtakochi* (MF) (NL) (NL)
REPTILIA 0	
UNKNOWN 1 (33%)	*Haemaphysalis cinnabarina*
Total 3 species	
Rhipicephalus	
AMBHIBIA 0	
AVES 0	
MAMMALIA 3 (100%)	
ARTIODACTYLA 1	
Bovidae	*Rhipicephalus microplus* (MFNL)
CARNIVORA 2****	
Canidae	*Rhipicephalus sanguineus s.l* (MFNL), *R. sanguineus s.s.* (MFNL)
REPTILIA 0	
Total 3 species	

* Provisional diagnosis because the tick stage remains undescribed
** Tick stages in parentheses follow the alphabetical order of principal hosts
*** Neotropical records of *Dermacentor albipictus*, *D. halli* and *D. variabilis* are insufficient to assign them to a particular principal host in the Neotropical Region
**** This number includes *Rhipicephalus sanguineus sensu lato*, but more than one species may exist under this uncertain species definition

Table 5.4 Numbers of Neotropical species of *Ixodes* and *Amblyomma* utilizing Aves or the combinations Aves + Mammalia and Aves + Mammalia + Reptilia as principal hosts

Order of hosts	*Ixodes*	*Amblyomma*
1. Caprimulgiformes	2	0
2. Galliformes	1	0
3. Passeriformes	4	0
4. Sphensiciformes	1	0
5. Trogoniformes	1	0
6. Caprimulgiformes + Chiroptera	1	0
7. Passeriformes + Carnivora	0	1
8. Passeriformes + Pilosa	0	3
9. Passeriformes + Primates	0	1
10. Passeriformes + Rodentia	1	1
11. Passeriformes + Artiodactyla + Rodentia	1	0
12. Passeriformes + Carnivora + Rodentia	0	1
13. Passeriformes + Primates + Rodentia	0	1
14. Passeriformes + several Mammalia + Squamata + Testudines	0	1
Total	12	9
Number of species found in the Neotropics	55	67

Bovidae: This family contains about 140 species worldwide. Four long-established domestic bovids and one recently introduced species have been found infested with hard ticks from all Neotropical genera, but they are principal hosts for two species of *Ixodes*, five *Amblyomma*, and one species of *Rhipicephalus*. Thus, cattle are principal hosts for males and females of *Ixodes fuscipes* and *Ixodes pararicinus*, but cattle are also preferred hosts for all parasitic stages of *Amblyomma neumanni*, *Amblyomma tonelliae*, and *Amblyomma variegatum*, and for males, females, and nymphs of *Amblyomma hadanii* and males and females of *Amblyomma patinoi*. Cattle are the most important hosts for all parasitic stages of *Rhipicephalus microplus*, while goats are important hosts for all parasitic stages of *Amblyomma variegatum*.

Bovids help sustain populations of *Haemaphysalis juxtakochi* and also several *Amblyomma* species that do not otherwise exhibit a clear pattern of host preference because adults and/or immature stages feed on several orders of mammals. Thus, buffaloes have been found infested with males and females of *Amblyomma mixtum*, while males and females of *Amblyomma parvum* have been recovered from buffaloes, cattle, goats, and sheep; and all parasitic stages of *Amblyomma sculptum* have been collected from buffaloes and cattle. Males and females of *Amblyomma cajennense* have been taken from cattle and the exotic host *Syncerus caffer*; while adults of *Amblyomma oblongoguttatum* and *Amblyomma triste* have been found on cattle and goats, and all parasitic stages of *Amblyomma mixtum* have been collected from cattle. Goats have been parasitized by males of *Amblyomma mixtum*, but sheep have been infested by males, females and nymphs of the same tick. Adults of

Amblyomma sculptum have been recovered from goats. Immature stages of *Haemaphysalis juxtakochi* have been found on cattle and sheep.

Additionally, bovids have been found infested with adults of *Ixodes affinis*, *Ixodes boliviensis*, *Ixodes longiscutatus*, and *Haemaphysalis juxtakochi*. Females of *Dermacentor albipictus* and all parasitic stages of *Dermacentor dissimilis*, *Dermacentor nitens* and *Rhipicephalus sanguineus sensu lato* have been taken from Bovidae, but bovids are also known to be infested with different stages of 17 *Amblyomma* species.

Camelidae: This family is treated here as containing seven species naturally distributed in Africa, Asia, and South America. The Neotropical domestic alpaca and llama as well as *Lama guanicoe* and *Vicugna vicugna* are the principal hosts for males and females of *Amblyomma parvitarsum*.

Additionally, a female of *Haemaphysalis juxtakochi* was collected from an alpaca.

Cervidae: This family contains about 55 species worldwide. At least 13 species of Cervidae have been recorded as hosts of ixodid ticks representing all genera found in the Neotropical Region. Cervids are principal hosts for four species of *Ixodes*, and one species each of *Amblyomma* and *Haemaphysalis*. Thus, *Mazama gouazoubira* is a principal host for males and females of *Ixodes fuscipes*; the northern pudu, *Pudu mephistophiles*, is the principal host for females, nymphs, and larvae of *Ixodes montoyanus*; and *Pudu puda* is the preferred host for males, females, and nymphs of *Ixodes stilesi*. *Pudu puda* is also the only known host for males and females of *Ixodes taglei*. An unknown species of deer is the only recognized host for *Amblyomma interandinum*. Males and females of *Haemaphysalis juxtakochi* are usually collected from *Mazama americana*, *Mazama bororo*, *Mazama gouazoubira*, and *Odocoileus virginianus*.

Species of Cervidae also help sustain populations of *Haemaphysalis juxtakochi* and several *Amblyomma* species that do not otherwise exhibit a clear pattern of host preference because adults and/or immature stages feed on several orders of mammals. Thus, *Blastocerus dichotomus* and *Mazama americana* have been found infested with males and females of *Amblyomma sculptum*, while adults of *Amblyomma triste* have been recovered from *Blastocerus dichotomus*, *Mazama* sp., and *Ozotoceros bezoarticus*. All parasitic stages of *Amblyomma mixtum* have been found on *Cervus elaphus*, while immature stages of *Amblyomma brasiliense* have been collected from *Mazama americana* and *Mazama gouazoubira*. Adults of *Amblyomma mixtum* have been found on *Mazama americana* and *Odocoileus virginianus*. Females of *Amblyomma oblongoguttatum* have been collected from *Mazama americana*, but adults and immature stages of this tick have been taken from *Mazama temama* and *Odocoileus virginianus*. Males and females of *Amblyomma parvum* have been collected from *Mazama gouazoubira* and *Odocoileus virginianus*, while males, females, and nymphs of *Amblyomma sculptum* have been recovered from *Mazama gouazoubira*. Immature stages of *Haemaphysalis juxtakochi* have been found on *Axis axis*, *Mazama americana*, *Mazama bororo*, *Mazama gouazoubira*, and *Odocoileus virginianus*.

Additionally, *Mazama gouazoubira*, *Mazama temama*, and *Odocoileus virginianus* have been found infested with males and females of *Ixodes affinis*. Males of *Dermacentor imitans* have been recovered from *Mazama* sp., while all parasitic stages of *Dermacentor nitens* have been taken from *Odocoileus virginianus*, and males and females of this tick have been found on *Blastocerus dichotomus* and *Mazama gouazoubira*. Males and females of *Rhipicephalus sanguineus sensu lato* have been collected from "deer," and different parasitic stages of *Rhipicephalus microplus* have been found on seven species of Cervidae. Males, females, nymphs, and/or larvae of 21 *Amblyomma* species have been found on ten species of Cervidae in the Neotropical Region.

Suidae: This family is represented by domestic and feral pigs introduced by colonists in the Neotropical Region. These hosts have been found infested with ticks of the genera *Amblyomma*, *Dermacentor*, *Haemaphysalis*, and *Rhipicephalus*, but Suidae are not principal hosts for any species of Neotropical ixodid. Nevertheless, species of Suidae help sustain populations of *Haemaphysalis juxtakochi* and several *Amblyomma* species that do not otherwise exhibit a clear pattern of host preference because adults and/or immature stages feed on several orders of mammals. Thus, domestic and feral pigs have been found carrying adults and immature stages of *Amblyomma sculptum*, while adults and adults and nymphs of *Amblyomma parvum* have been taken from domestic and feral pigs, respectively. Males and females of *Amblyomma cajennense*, adults and larvae of *Amblyomma mixtum*, adults and nymphs of *Amblyomma oblongoguttatum*, and males and females of *Amblyomma triste* have been collected from domestic pigs. Males, females, and nymphs of *Amblyomma brasiliense* have been recovered from feral pigs, as well as nymphs of *Haemaphysalis juxtakochi*.

Additionally, adults of *Dermacentor nitens* and *Rhipicephalus sanguineus sensu lato* have been collected from domestic pigs, while adults of *Haemaphysalis juxtakochi* and *Rhipicephalus microplus* have been taken from both domestic and feral pigs, and nymphs of *Haemaphysalis juxtakochi* have also been collected from feral pigs. Different stages of 12 *Amblyomma* species have been found on both types of pigs present in the Neotropical Region.

Tayassuidae: This family contains three species, all found only in the Americas. With the exception of *Ixodes*, tayassuids have been found infested with all genera of hard ticks present in the Neotropics. Species of Tayassuidae are principal hosts for three species of *Amblyomma* and two species of *Dermacentor*. Thus, *Catagonus wagneri* is the only host for males, females, nymphs, and larvae of *Amblyomma boeroi*, while *Tayassu pecari* is the provisional principal host for males and females of *Dermacentor dispar*. Males and females of *Amblyomma naponense*, *Amblyomma pecarium* and *Dermacentor imitans* are usually collected from *Pecari tajacu* and *Tayassu pecari*.

Species of Tayassuidae also help sustain populations of *Haemaphysalis juxtakochi* and several *Amblyomma* species that do not otherwise exhibit a clear pattern of host preference because adults and/or immature stages feed on several orders of mammals. Thus, adults of *Amblyomma parvum* have been found on

Catagonus wagneri and *Pecari tajacu*, while all parasitic stages of *Amblyomma brasiliense* and males and females of this tick have been taken from *Pecari tajacu* and *Tayassu pecari*. These hosts have also been recorded carrying males, females, and nymphs of *Amblyomma cajennense*, *Amblyomma oblongoguttatum*, and *Amblyomma sculptum*. Males, females, and larvae of *Amblyomma mixtum* have been found on *Pecari tajacu*, and females of this tick have been collected from *Tayassu pecari*. Nymphs of *Amblyomma naponense* have been recovered from *Pecari tajacu* and *Tayassu pecari*, while nymphs and larvae of *Haemaphysalis juxtakochi* have been found on *Pecari tajacu*.

Additionally, different stages of *Rhipicephalus microplus* have been collected from the three species of Tayassuidac, while adults of *Haemaphysalis juxtakochi* have been found infesting *Pecari tajacu* and *Tayassu pecari*. Different stages of 12 *Amblyomma* species have been collected from Tayassuidae in the Neotropical Region.

5.1.4.2 Carnivora

This order contains about 300 species classified in 15 families with a worldwide geographic distribution. Carnivora from five families, including domestic carnivores, are important hosts for several Neotropical ixodid tick species in the genera *Ixodes*, *Amblyomma*, and *Rhipicephalus*. Carnivora have also been found infested with many other lesser-known species of ixodids, and these host–parasite relationships are detailed below.

Canidae: This family contains 38 species with a worldwide distribution. Ten Neotropical canids have been found infested with ixodids, and some of these carnivores are important hosts for four species of *Amblyomma*, two *Ixodes* and at least two species of *Rhipicephalus*. Thus, the domestic dog is the principal host for males and females of *Ixodes boliviensis*, and one of the principal hosts of males and females of *Amblyomma aureolatum*, *Amblyomma maculatum*, *Amblyomma ovale*, and *Amblyomma tigrinum*. Moreover, the dog is the only principal host of all parasitic stages of *Rhipicephalus sanguineus sensu stricto* and *sensu lato*. The coyote, *Canis latrans*, has been found infested with males and females of *Amblyomma ovale*, while *Cerdocyon thous* and *Chrysocyon brachyurus* are principal hosts for males and females of *Amblyomma aureolatum*, *Amblyomma ovale* and *Amblyomma tigrinum*. Males and females of *Amblyomma tigrinum* have been collected from four species of *Lycalopex*, and one of these, *Lycalopex gymnocercus*, has also been found infested with both sexes of *Amblyomma aureolatum* and *Amblyomma tigrinum*. Females of *Ixodes rubidus* have been collected from *Urocyon cinereoargenteus*.

Canids help sustain populations of *Haemaphysalis juxtakochi* and several *Amblyomma* species that do not otherwise exhibit a clear pattern of host preference because adults and/or immature stages feed on several orders of mammals. Thus, all parasitic stages of *Amblyomma brasiliense*, *Amblyomma mixtum*, and *Amblyomma sculptum*, adults and nymphs of *Amblyomma cajennense* and *Amblyomma oblongoguttatum*, males and females of *Amblyomma parvum* and *Amblyomma triste*, and nymphs of *Amblyomma naponense* and *Haemaphysalis juxtakochi* have been found on Canidae.

Additionally, the domestic dog has been found carrying adults of *Ixodes affinis*, *Ixodes fuscipes*, *Ixodes luciae*, and nymphs of *Ixodes pararicinus*, as well as adults and nymphs of *Dermacentor nitens*, and adults of *Dermacentor dissimilis* and *Dermacentor latus*. Adults and nymphs of *Dermacentor nitens* have been collected from *Lycalopex vetulus*. Tick stages of 23 *Amblyomma* species other than those named above have been found on members of the family Canidae.

Felidae: This family contains about 40 species worldwide, and in the Neotropics ten of these have been found infested with specimens of the five genera of Ixodidae occurring in this region, where felids are important hosts for one species of *Ixodes* and two species of *Amblyomma*. Thus, *Leopardus pardalis* and *Panthera onca* are the principal hosts for males and females of *Ixodes affinis*. The domestic cat, *Herpailurus yagouaroundi*, *Leopardus geoffroyi*, *Leopardus pardalis*, and *Puma concolor* have been found infested with males and females of *Amblyomma aureolatum* and *Amblyomma ovale*, and *Leopardus wiedii* and *Panthera onca* have also been found parasitized by males and females of *Amblyomma ovale*.

Species of Felidae also help sustain populations of several *Amblyomma* species that do not otherwise exhibit a clear pattern of host preference because adults and/or immature stages feed on several orders of mammals. Thus, adults of *Amblyomma parvum* have been found on *Herpailurus yagouaroundi*, *Leopardus geoffroyi*, *Leopardus pardalis*, *Panthera onca*, *Puma concolor*, and the domestic cat, while domestic cats along with *Herpailurus yagouaroundi*, *Panthera onca*, and *Puma concolor*, have been found parasitized by adults of *Amblyomma triste*, and adults of *Amblyomma oblongoguttatum* have been recovered from the domestic cat and *Panthera onca*. Males and females of *Amblyomma sculptum* have been recovered from the domestic cat, while adults and immature stages of this tick have been collected from *Leopardus pardalis*, *Panthera onca*, and *Puma concolor*. Adults and nymphs of *Amblyomma oblongoguttatum* have been found on *Herpailurus yagouaroundi* and *Puma concolor*, while nymphs of *Amblyomma brasiliense* have been recovered from *Panthera onca* and *Puma concolor*. Females and nymphs of *Amblyomma cajennense* and females of *Amblyomma mixtum* have been taken from *Panthera onca*.

Additionally, adults of *Ixodes boliviensis* have been collected from *Leopardus pardalis*, *Panthera onca*, and the domestic cat, and one female of *Ixodes lasallei* was found on *Leopardus pardalis*. Adults of *Dermacentor nitens* have been taken from the domestic cat, *Leopardus tigrinus* and *Puma concolor*, while adults and nymphs of the same tick have been collected from *Panthera onca*. A female of *Haemaphysalis juxtakochi* was found on *Puma concolor*, and several species of Felidae have been reported carrying *Rhipicephalus microplus* and *Rhipicephalus sanguineus sensu lato*. Tick stages of nine species of *Amblyomma* other than those named above have been found parasitizing Felidae.

Herpestidae: This family contains 35 species found in Africa, Asia, and Europe. In the Neotropical Region, the introduced species *Herpestes auropunctatus* has been found infested with nymphs and larvae of *Amblyomma variegatum* and larvae of *Rhipicephalus microplus*, but this host is not important in the life cycles of these two tick species.

Mephitidae: This family comprises 12 species found in Southeast Asia and the Americas. One Neotropical species, *Conepatus semistriatus*, is an important host for females and nymphs of *Ixodes rubidus*.

Additionally, *Conepatus semistriatus* has been found carrying males, females, nymphs, and larvae of *Amblyomma auricularium* and females of *Amblyomma ovale*, but this host is not considered vital to the survival of these species of *Amblyomma*.

Mustelidae: This family contains about 60 species with a worldwide distribution. At least six species of Neotropical Mustelidae have been found parasitized by several ixodid species, but mustelids are important hosts for only one species in the genus *Ixodes*. Thus, *Eira barbara* and, to a lesser extent, *Mustela frenata*, are important hosts of the female of *Ixodes rubidus*.

Additionally, females of *Ixodes guatemalensis* have been collected from *Mustela frenata*, and adults of *Rhipicephalus sanguineus sensu lato* have been taken from *Galictis cuja*. Different stages of nine *Amblyomma* species have been found on Neotropical Mustelidae.

Procyonidae: This family contains 14 species, all of them found in the Americas. At least seven procyonids are known to be infested with tick specimens representing all genera of Ixodidae found in the Neotropics, with the exception of the genus *Dermacentor*. Procyonidae are important hosts for two species of *Ixodes* and two species of *Amblyomma*. Thus, *Bassaricyon gabbii* and *Procyon cancrivorus* are hosts for females and nymphs of *Ixodes rubidus*, while *Bassariscus* sp. is the only known host for *Ixodes texanus* in the Neotropical Region. *Procyon cancrivorus* is an important host for males and females of *Amblyomma aureolatum*, while *Nasua nasua* is treated here as a significant host for nymphs and larvae of *Amblyomma coelebs*. Procyonidae may be more important hosts than currently recognized for *Amblyomma ovale*.

Procyonids help sustain populations of *Haemaphysalis juxtakochi* and several *Amblyomma* species that do not otherwise exhibit a clear pattern of host preference because adults and/or immature stages feed on several orders of mammals. Thus, nymphs and larvae of *Amblyomma brasiliense* have been taken from *Nasua nasua*, while males, females, and nymphs of *Amblyomma oblongoguttatum*, and males and nymphs of *Amblyomma mixtum* have been collected from *Nasua narica*. Males, females and nymphs of *Amblyomma sculptum* have been found on *Nasua nasua* and *Procyon cancrivorus*. Adults and nymphs of *Amblyomma parvum*, nymphs and larvae of *Amblyomma cajennense*, and nymphs of *Amblyomma humerale* and *Amblyomma naponense* have been recovered from *Nasua nasua*, while adults of *Amblyomma parvum* and *Amblyomma naponense* were found on *Nasua narica*. And nymphs and larvae of *Haemaphysalis juxtakochi* have been collected from *Nasua narica* and *Nasua nasua*.

Additionally, all parasitic stages of *Ixodes boliviensis* have been collected from *Procyon lotor*, and females of *Ixodes montoyanus* have been found on *Nasuella olivacea*. Females of *Rhipicephalus microplus* and *Rhipicephalus sanguineus sensu lato* have been taken from *Nasua nasua*. Different stages of 13 *Amblyomma* species have been found on several Neotropical members of Procyonidae.

Ursidae: This family contains eight species with a worldwide distribution, and one of these, *Tremarctos ornatus*, was found infested with a female of *Ixodes boliviensis* in the Neotropical Region.

5.1.4.3 Chiroptera

This large order contains about 1300 species classified in 18 families with a worldwide distribution. Neotropical Chiroptera from two families are known to be infested by several tick species from all genera of Ixodidae except *Haemaphysalis*, but chiropteran hosts are only important for two species of *Ixodes*, as detailed below.

Phyllostomidae: This family contains about 200 species, all of them found in the Americas. In the Neotropics, 12 of these species have occasionally been found infested with ixodids in the genera *Amblyomma*, *Ixodes*, and *Rhipicephalus*, but one phyllostomid, *Anoura geoffroyi*, is the only known host for the larva of *Ixodes downsi*.

Additionally, an adult of *Ixodes pararicinus* was collected from *Desmodus rotundus*, and larvae of *Rhipicephalus microplus* have been taken from *Sturnira lilium*. Different stages of 11 *Amblyomma* species have occasionally been found on nine species of Neotropical Phyllostomidae.

Vespertilionidae: This is a large family containing more than 430 species distributed worldwide. At least two species of Neotropical Vespertilionidae have on rare occasions been found infested with ticks of the genera *Amblyomma*, *Dermacentor*, and *Ixodes*, but one vespertilionid, *Myotis chiloensis*, is the only known host for the female and nymph of *Ixodes chilensis*.

Additionally, one nymph of *Amblyomma triste* was reported from *Myotis albescens*, and one nymph of *Dermacentor panamensis* was collected from *Myotis* sp.

5.1.4.4 Cingulata

This order contains 20 species found in the Americas, and in the Neotropical Region 16 of these have been found infested with ticks of the genera *Amblyomma*, *Dermacentor*, *Haemaphysalis*, and *Ixodes*. Cingulata are important hosts for four *Amblyomma* species.

Chlamyphoridae: This family contains 13 species, all of them occurring in the Neotropical Region, where 11 species have been found infested with ticks of the genera *Amblyomma* and *Haemaphysalis*. Chlamyphoridae are of importance in the life cycles of two species of *Amblyomma*. Thus, males, females, nymphs, and larvae of *Amblyomma auricularium* have been found on *Euphractes sexcinctus* and *Tolypeutes matacus*, while males, females, and nymphs of *Amblyomma auricularium* have been taken from *Chaetophractus vellerosus*, *Chaetophractus villosus*, and *Tolypeutes tricinctus*. Males and females of *Amblyomma auricularium* and *Amblyomma pseudoconcolor* have been collected from *Cabassous tatouay*, and only males and females of *Amblyomma auricularium* have been recovered from *Cabassous centralis*. Males and females of *Amblyomma pseudoconcolor* have been found on *Chaetophractus vellerosus*, *Chaetophractus villosus*, *Euphractes sexcinctus*, *Priodontes maximus*, and *Tolypeutes matacus*.

Chlamyphorids help sustain populations of several *Amblyomma* species that do not otherwise exhibit a clear pattern of host preference because adults and/or immature stages feed on several orders of mammals. Thus, adults and nymphs of *Amblyomma sculptum* have been taken from *Cabassous unicinctus* and *Priodontes maximus*, while adults of *Amblyomma parvum* have been recovered from *Chaetophractus vellerosus*, *Euphractes sexcinctus*, *Priodontes maximus*, and *Tolypeutes matacus*, and adults of *Amblyomma cajennense* have been found on *Cabassous unicinctus*.

Additionally, males and females of *Amblyomma pseudoparvum* have been found on *Chaetophractus villosus* and *Tolypeutes matacus*, the latter species also hosting males of *Haemaphysalis juxtakochi*.

Dasypodidae: This family contains seven species that chiefly occur in the Neotropics, although one species reaches the southern Nearctic. In the Neotropical Region, five species of Dasypodidae have been found infested with specimens of the genus *Amblyomma*, and, to a lesser extent, with members of the genera *Ixodes* and *Dermacentor*. Dasypodids are important hosts for four species of *Amblyomma*. Thus, all parasitic stages and males and females of *Amblyomma auricularium* have been collected from *Dasypus novemcinctus* and *Dasypus sabanicola*, respectively, and males and females of *Amblyomma fuscum* have been taken from *Dasypus novemcinctus* and *Dasypus septemcinctus*. The nine-banded armadillo, *Dasypus novemcinctus*, is the only known host for the male and female of *Amblyomma beaurepairei*, and this host has also been found parasitized by males and females of *Amblyomma pseudoconcolor*.

Dasypodids help sustain populations of several *Amblyomma* species that do not otherwise exhibit a clear pattern of host preference because adults and/or immature stages feed on several orders of mammals. Thus, adults, nymphs and larvae of *Amblyomma mixtum* and *Amblyomma sculptum* have been recovered from *Dasypus novemcinctus*, while adults and nymphs of *Amblyomma cajennense*, males and females of *Amblyomma parvum* and nymphs of *Amblyomma humerale* have also been collected from this host.

Additionally, *Dasypus novemcinctus* has been recorded as hosting the female of *Ixodes loricatus*, males and females of *Amblyomma ovale* and *Dermacentor dispar*, the female of *Amblyomma rotundatum*, and the nymph of *Amblyomma coelebs.*

5.1.4.5 Didelphimorphia

This order contains one family of about 100 species that are mostly Neotropical, although at least one species is also found in the Nearctic Region. In the Neotropics, several members of this order have been found infested with ixodids from the genera *Amblyomma*, *Haemaphysalis*, and *Ixodes*, as discussed below.

Didelphidae: This family contains at least 34 species that are important hosts for six species of *Ixodes*, three *Amblyomma* and one species of *Haemaphysalis.* Thus, *Didelphis albiventris* is a principal host for males, females, nymphs, and larvae of *Ixodes loricatus* and *Ixodes luciae*, as well as for larvae and nymphs of *Amblyomma coelebs*. Similarly, *Didelphis aurita* hosts the same stages of *Ixodes loricatus* and *Amblyomma coelebs*, but it is a principal host for nymphs and larvae of *Amblyomma fuscum* and *Amblyomma yucumense.* The common opossum, *Didelphis marsupialis*, is an important host for males, females, nymphs, and larvae of *Ixodes loricatus* and *Ixodes luciae*, but it is the principal host for females of *Ixodes tropicalis*, while *Lutreolina crassicaudata* is another important host for males, females, nymphs, and larvae of *Ixodes loricatus.* Males, females, nymphs, and larvae of *Ixodes luciae* have been collected from *Marmosa robinsoni*, while *Monodelphis brevicaudata* is an important host for nymphs and larvae of *Ixodes luciae* and *Ixodes venezuelensis.* All parasitic stages of *Ixodes luciae* as well as males and females of *Ixodes loricatus* have been collected from *Philander opossum.* The principal host for the female of *Ixodes amarali* is *Monodelphis domestica*, while *Marmosa demerarae* is a hypothetical principal host for the female of *Ixodes catarinensis.*

Didelphids help sustain populations of *Haemaphysalis juxtakochi* and several *Amblyomma* species that do not otherwise exhibit a clear pattern of host preference because adults and/or immature stages feed on several orders of mammals. Thus, females, nymphs, and larvae of *Amblyomma sculptum* have been recovered from *Didelphis albiventris*, while adults and nymphs of *Amblyomma oblongoguttatum* have been found on *Chironectes minimus.* Males and females of *Amblyomma parvum* have been taken from *Didelphis* sp., nymphs and larvae of *Amblyomma brasiliense* have been collected from *Didelphis aurita*, and nymphs and larvae of *Amblyomma mixtum* have been recovered from *Didelphis marsupialis.* Nymphs and larvae of *Haemaphysalis juxtakochi* have been found on *Didelphis aurita* and *Didelphis marsupialis*, and nymphs of *Amblyomma cajennense* have been collected from *Didelphis imperfecta.* Nymphs of *Amblyomma humerale* have been taken from *Didelphis marsupialis*, *Marmosa constantiae*, *Metachirus nudicaudatus*, and *Philander opossum.*

Additionally, nymphs of *Haemaphysalis leporispalustris* have been collected from *Metachirus nudicaudatus*. Different stages of six species of *Ixodes* and 17 species of *Amblyomma* have been found on several didelphids.

5.1.4.6 Eulipotyphla

This worldwide order contains about 490 species classified in five families. Only one species of Eulipotyphla has been found infested with Ixodidae in the Neotropical Region.

Soricidae: This family is large, containing about 400 species globally. However, the only Neotropical record of tick parasitism is that of a female of *Ixodes guatemalensis* taken from *Cryptotis goodwini*, and this host is not considered important in the life cycle of this tick.

5.1.4.7 Lagomorpha

This worldwide order contains about 100 species classified into two families. Several Neotropical members of one family have been found infested with ixodids representing all genera in this region except *Dermacentor*, lagomorphs being especially important in the life cycles of *Ixodes* and *Haemaphysalis* species.

Leporidae: This family comprises about 65 species worldwide, and at least seven leporids in the Neotropical Region are parasitized by ticks of the genera *Amblyomma*, *Haemaphysalis*, *Ixodes*, and *Rhipicephalus*. Leporids are signifcant hosts for three species of *Ixodes* and one species of *Haemaphysalis*. Thus, *Romerolagus diazi* is the only known host of *Ixodes spinipalpis* in the Neotropics. *Sylvilagus brasiliensis* and *Sylvilagus floridanus* are the principal hosts for males and females of *Ixodes pomerantzi*, and for males, females, nymphs, and larvae of *Haemaphysalis leporispalustris*, while *Sylvilagus dicei* is the only known host of the male and female of *Ixodes dicei*.

Leporids also help sustain populations of *Haemaphysalis juxtakochi* and two species of *Amblyomma* that do not otherwise exhibit a clear pattern of host preference because adults and/or immature stages feed on several orders of mammals. Thus, males and females of *Amblyomma parvum* have been taken from *Sylvilagus floridanus*, and nymphs of *Amblyomma sculptum* have been collected from *Sylvilagus brasiliensis*. Nymphs and larvae of *Haemaphysalis juxtakochi* have been taken from *Sylvilagus brasiliensis*.

Additionally, males and females of *Amblyomma inornatum* have been reported from a "hare," and nymphs and larvae of *Amblyomma dubitatum* have been taken from *Lepus europaeus*, while nymphs of *Amblyomma coelebs* have been found on *Sylvilagus brasiliensis*. Females of *Haemaphysalis juxtakochi* have been found on *Sylvilagus floridanus*. Males, females, nymphs, and larvae of *Rhipicephalus sanguineus sensu lato* have been collected from domestic rabbit raised in the

Neotropics, while nymphs and larvae of *Rhipicephalus microplus* have been collected from *Lepus europaeus*.

5.1.4.8 Microbiotheria

This order of marsupials is Neotropical and of great evolutionary significance, because it is regarded as the ancestor of Australasian marsupials. Microbiotheria contains one monotypic family.

Microbiotheriidae: This family consists of one extant species, *Dromiciops gliroides*, whose geographic distribution encompasses a small territory in southern South America straddling the borders of Argentina and Chile. *Dromiciops gliroides* is the only known host of the female, nymph, and larva of *Ixodes neuquenensis*.

5.1.4.9 Paucituberculata

This order contains seven species in one Neotropical family; one of these species has been found infested with ticks of the genus *Ixodes*.

Caenolestidae: *Caenolestes fuliginosus*, the only member of this family, was found infested with larvae of *Ixodes jonesae* on one occasion, but this host is not considered important in the life cycle of this tick.

5.1.4.10 Perissodactyla

This order contains 16 species classified in three families that are currently found in the Afrotropical, Neotropical and Oriental Zoogeographic Regions, but domesticated perissodactyls are now present all over the world. Members of two of these families have been found infested by all ixodid genera present in the Neotropics, and several perissodactyls are importanct as hosts for seven species of *Amblyomma*, three *Dermacentor* and one species of *Ixodes*.

Equidae: This family contains seven species established in Africa and Asia, but domestic equids are known Neotropical hosts for ixodids of all genera found in the region, being important hosts for two species of *Dermacentor* and one species of *Amblyomma*. Thus, horses are the principal hosts for males, females, nymphs, and larvae of *Dermacentor dissimilis* and *Dermacentor nitens*, and for males and females of *Amblyomma tenellum*.

Equids help sustain populations of several *Amblyomma* species that do not otherwise exhibit a clear pattern of host preference because adults and/or immature stages feed on several orders of mammals. Thus, males, females, nymphs, and larvae of *Amblyomma mixtum* and *Amblyomma sculptum*, and males, females, and nymphs of *Amblyomma cajennense* and *Amblyomma oblongoguttatum* have been found on horses. Males and females of *Amblyomma triste* have also been collected from horses, while adults of *Amblyomma parvum* have been taken from donkeys and horses.

Additionally, all parasitic stages of *Rhipicephalus microplus* and males, females, and nymphs of *Haemaphysalis juxtakochi* have been collected from horses, while adults of *Rhipicephalus sanguineus sensu lato* have been taken from donkeys and horses. Equids have been found infested with different stages of five species of *Ixodes* and 14 species of *Amblyomma.*

Tapiridae: This family contains four species found in a wide area of the Neotropical Region and in a small portion of the Oriental Region. Specimens of all genera of Ixodidae have been found on the three species of Neotropical Tapiridae, and these hosts are important in the life cycle of six species of *Amblyomma* and one species each of *Dermacentor* and *Ixodes.* Thus, *Tapirus bairdii* is the principal host for males, females, and nymphs of *Amblyomma tapirellum*, and males and females of *Dermacentor latus* and *Ixodes tapirus.* The lowland tapir, *Tapirus terrestris*, is the principal host for males and females of *Amblyomma coelebs*, *Amblyomma incisum*, *Amblyomma latepunctatum*, and *Amblyomma scalpturatum*, while *Tapirus pinchaque* is the principal host for males and females of *Amblyomma multipunctum.*

Tapirids help sustain populations of several *Amblyomma* species that do not otherwise exhibit a clear pattern of host preference because adults and/or immature stages feed on several orders of mammals. Thus, males, females, nymphs, and larvae of *Amblyomma brasiliense*, *Amblyomma cajennense* and *Amblyomma sculptum*, males and females of *Amblyomma parvum* and *Amblyomma triste*, and nymphs of *Amblyomma naponense* have been found on *Tapirus terrestris*. Males and females of *Amblyomma mixtum* have been collected from *Tapirus bairdii.*

Additionally, males, females, and nymphs of *Haemaphysalis juxtakochi* have been found on *Tapirus bairdii*, and males, females, and nymphs of *Rhipicephalus microplus* have been collected from *Tapirus terrestris.* Males and females of *Dermacentor latus* and *Dermacentor nitens*, and females of *Ixodes boliviensis* have been taken from *Tapirus bairdii.* Females of *Ixodes bocatorensis* have been recovered from *Tapirus* sp., and females of *Rhipicephalus sanguineus sensu lato* have been found on *Tapirus terrestris.* Different stages of nine *Amblyomma* species have been collected from Tapiridae.

5.1.4.11 Pilosa

This order contains ten species classified into four families that are exclusively Neotropical in distribution. Members of these families have been found parasitized by ticks of the genera *Amblyomma*, *Haemaphysalis*, and *Rhipicephalus*, pilosans being of importance as hosts for six *Amblyomma* species.

Bradypodidae: This family contains four species, three of which have been found infested by ticks of the genera *Amblyomma* and *Rhipicephalus*, but two species of Bradypodidae are principal hosts for two *Amblyomma* species. Thus, both *Bradypus* *tridactylus* and *Bradypus variegatus* are principal hosts for males, females, and nymphs of *Amblyomma geayi* and males and females of *Amblyomma varium.*

Species of Bradypodidae also help sustain two *Amblyomma* species that do not otherwise exhibit a clear pattern of host preference because adults and/or immature stages feed on several orders of mammals. Thus, males, females, and nymphs of *Amblyomma cajennense* have been collected from *Bradypus tridactylus*, and females of *Amblyomma sculptum* have been found on *Bradypus torquatus* and *Bradypus variegatus.*

Additionally, one female of *Rhipicephalus sanguineus sensu lato* was found on a *Bradypus tridactylus*, and different stages of seven *Amblyomma* species have been collected from three species of Bradypodidae.

Choloepodidae: This family contains two species that have been found infested with ticks of the genus *Amblyomma*, but choloepodids are not considered important hosts for these ticks. Males and females of *Amblyomma varium* have been found on *Choloepus didactylus* and *Choloepus hoffmanni*, while males, females and nymphs of *Amblyomma geayi* have been collected from *Choloepus didactylus*, and males and females of *Amblyomma geayi* have also been taken from *Choloepus hoffmanni.* Males, females, and nymphs of *Amblyomma dissimile* have been recovered from *Choloepus didactylus*, females of *Amblyomma calcaratum* have been found on *Choloepus hoffmanni*, and there is a nymphal record of *Amblyomma oblongoguttatum* from the latter host.

Cyclopedidae: This family contains a single species, *Cyclopes didactylus*, that has been found infested with males and females of *Ambyomma calcaratum*, females of *Ixodes bocatorensis*, nymphs of *Amblyomma geayi* and a nymph of *Amblyomma humerale*, but this host is not considered important in maintaining these ticks in nature.

Myrmecophagidae: This family contains three species that have been found infested with ticks of the genera *Amblyomma*, *Haemaphysalis*, and *Rhipicephalus*, myrmecophagids being of importance as hosts for four species of *Amblyomma.* Thus, *Myrmecophaga tridactyla* is the principal host for males and females of *Amblyomma pictum*, while *Tamandua tetradactyla* is the principal host for males and females of *Amblyomma calcaratum* and *Amblyomma goeldii.* Males and females of *Amblyomma nodosum* are usually collected from *Tamandua mexicana* and *Tamandua tetradactyla.*

Myrmecophagids help sustain populations of several *Amblyomma* species that do not otherwise exhibit a clear pattern of host preference because adults and/or immature stages feed on several orders of mammas. Thus, males, females, nymphs, and larvae of *Amblyomma mixtum* have been collected from *Tamandua mexicana*, and males, females, nymphs, and larvae of *Amblyomma sculptum* have been found on *Myrmecophaga tridactyla*, while males, females, and nymphs of *Amblyomma brasiliense* have been taken from the latter host. Males and females of *Amblyomma oblongoguttatum* have been collected from *Myrmecophaga tridactyla* and *Tamandua mexicana*, while males and females of *Amblyomma parvum* have been recovered from *Myrmecophaga tridactyla* and *Tamandua tetradactyla.* Males and females of *Amblyomma cajennense* and *Amblyomma triste* have been found on *Myrmecophaga tridactyla.* Nymphs of *Amblyomma naponense* have been collected from *Myrmecophaga tridactyla, Tamandua mexicana,* and *Tamandua tetradactyla,* while nymphs of *Amblyomma humerale* have been taken from the last-named host.

Additionally, all parasitic stages of *Rhipicephalus microplus* have been found on *Myrmecophaga tridactyla,* and adults of *Rhipicephalus sanguineus sensu lato* have been collected from *Tamandua mexicana* and *Tamandua tetradactyla,* while nymphs of *Haemaphysalis juxtakochi* have been taken from *Tamandua tetradactyla.* Different stages of 15 *Amblyomma* species have been found on Myrmecophagidae.

5.1.4.12 Primates

This order contains about 465 species classified in 16 families, with most nonhuman primates found in the Afrotropical, Neotropical, and Oriental Zoogeographic Regions. Primates belonging to four Neotropical families have been found infested with ticks of the genera *Amblyomma, Haemaphysalis,* and *Rhipicephalus,* primates being important hosts for two species of *Amblyomma.*

Atelidae: This family contains 28 Neotropical species, five of which have been found infested with ticks of the genera *Amblyomma* and *Rhipicephalus,* atelids being important hosts for one species of *Amblyomma.* Thus, *Alouatta guariba* is one of the principal hosts for the nymph of *Amblyomma parkeri.*

Also, *Alouatta guariba* has been found infested with males, females, and nymphs of *Amblyomma sculptum,* all parasitic stages of which are known to feed on various orders and families of hosts.

Additionally, a female of *Rhipicephalus microplus* was collected from *Alouatta seniculus.* Different tick stages of five *Amblyomma* species have been found on Atelidae.

Callitrichidae: This family comprises about 45 Neotropical species, three of which have occasionally been found infested with ticks of the genera *Amblyomma* and *Rhipicephalus,* although these primates are not important hosts for these ticks. Adults of *Rhipicephalus sanguineus sensu lato* have been collected from *Mico melanurus* and *Saguinus bicolor,* while larvae of the same tick have been taken from *Saguinus geoffroyi.* Nymphs of *Amblyomma geayi* have been recovered from *Saguinus bicolor.*

Cebidae: This family contains about 20 Neotropical species, four of which have occasionally been found infested with ticks of genera *Amblyomma* and *Haemaphysalis.* There is a record of adults, nymphs, and larvae of *Amblyomma oblongoguttatum* from *Cebus capucinus,* while one female each of *Amblyomma ovale* and *Amblyomma parvum* was collected from *Sapajus apella.* A female of *Amblyomma coelebs* was taken from a *Sapajus microcephalus,* and a male of *Heamaphysalis juxtakochi* was found on a *Cebus olivaceus.*

Pitheciidae: This family contains about 55 Neotropical species, one of which, *Callicebus nigrifrons*, is the only known host of the males, females, and nymphs of *Amblyomma romarioi*.

Additionally, *Callicebus nigrifrons* has been found infested with nymphs of *Amblyomma parkeri*.

5.1.4.13 Rodentia

This order, the largest in the class Mammalia, consists of over 2300 species classified in about 33 families and is worldwide in distribution. Numerous species of Rodentia representing 15 families have been found infested with ticks from all genera of Ixodidae present in the Neotropical Region. Rodentia are important hosts for 26 *Ixodes* species, 10 *Amblyomma*, and one species of *Dermacentor*, but they are of particular importance as hosts of *Ixodes* species.

Abrocomidae: This Neotropical family contains ten species, one of which, *Abrocoma bennettii*, is regarded as one of the principal hosts of adults of *Ixodes abrocomae*. Additionally, *Abrocoma bennettii* has been found infested with nymphs of *Ixodes sigelos*.

Capromyidae: This family contains about 15 Neotropical species, one of which, *Capromys pilorides*, is the only known host of *Ixodes capromydis*.

Caviidae: This Neotropical family includes about 20 species, nine of which have been found infested with ticks representing all genera of Ixodidae present in the Neotropical Region, with the exception of the genus *Haemaphysalis*. Caviids are of importance as hosts of five species of *Amblyomma* and one species of *Ixodes*. Thus, *Cavia aperea* and *Cavia tschudii* are regarded as principal hosts for females of *Ixodes longiscutatus*, while *Hydrochoerus hydrochaeris* is the principal host for males, females, nymphs, and larvae of *Amblyomma dubitatum* as well as for the males, females, and nymphs of *Amblyomma romitii*. Males and females of *Amblyomma pseudoparvum* are usually collected from *Dolichotis salinicola*. Nymphs and larvae of *Amblyomma parvum* and nymphs of *Amblyomma tigrinum* are frequently taken from *Galea musteloides* within the southern range of these ticks.

Species of Caviidae also help sustain populations of *Amblyomma* species that do not otherwise exhibit a clear pattern of host preference because adults and/or immature stages feed on several orders of mammals. Thus, males, females, nymphs, and larvae of *Amblyomma mixtum* have been collected from *Hydrochoerus isthmius*, and males, females, nymphs, and larvae of *Amblyomma sculptum* have been found on *Hydrochoerus hydrochaeris*, while adults and nymphs of *Amblyomma brasiliense* have been taken from the latter host. Males and females of *Amblyomma oblongoguttatum* have been collected from *Hydrochoerus hydrochaeris* and *Hydrochoerus isthmius*, while adults of *Amblyomma parvum* have been recovered from *Hydrochoerus hydrochaeris* and *Kerodon rupestris*. Males and females of *Amblyomma cajennense* and *Amblyomma triste* have been found on *Hydrochoerus hydrochaeris*, and nymphs of *Amblyomma humerale* and *Amblyomma naponense* have also been collected from this host.

Additionally, males, females, and nymphs of *Rhipicephalus microplus* have been collected from *Hydrochoerus hydrochaeris*, while females and nymphs of *Rhipicephalus sanguineus sensu lato* have been taken from the same host, and a male was found on a *Dolichotis patagonum.* Males and females of *Dermacentor nitens* have been recovered from *Hydrochoerus hydrochoeris*, and females of *Ixodes loricatus* have been collected from *Cavia* sp. Different stages of 11 *Amblyomma* species have been taken from Caviidae.

Chinchillidae: This family consists of seven Neotropical species, two of which have been found infested with ticks of the genera *Ixodes* and *Amblyomma*, chinchillids being of importance as hosts of one species of *Ixodes.* Thus, *Lagidium viscacia* is the only known host for males, females, nymphs, and larvae of *Ixodes nuttalli.*

Lagostomus maximus has been found infested with males and females of *Amblyomma parvum* and larvae of *Amblyomma sculptum*, species that are known to parasitize various orders of mammalian hosts.

Additionally, males, females, and nymphs of *Amblyomma auricularium*, males and females of *Amblyomma pseudoparvum*, and nymphs of *Amblyomma dubitatum* have been found on *Lagostomus maximus*, but this host is not important in the life cycles of these ticks.

Cricetidae: This is a large family with about 700 species worldwide, although cricetids are absent in Africa and Australasia. In the Neotropical Region, at least 94 species of Cricetidae have been found infested with ticks from all genera of Ixodidae with the exception of genus *Rhipicephalus*, cricetids being important hosts for 18 *Ixodes* species, three *Amblyomma*, and one species of *Dermacentor.* Thus, *Thomasomys laniger* is the principal host for males, females, nymphs, and larvae of *Ixodes jonesae*; *Phyllotis* sp. and *Nectomys squamipes* are the preferred hosts for females, nymphs, and larvae of *Ixodes andinus* and *Ixode schulzei*, respectively, and females, nymphs, and larvae of *Ixodes sigelos* are usually found on *Abrothrix olivaceus* and *Phyllotis xanthopygus.* The long-haired akodont, *Abrothrix longipilis*, and *Abrothrix olivaceus* are considered principal hosts of *Ixodes abrocomae*, and *Phyllotis xanthopygus* is the only known host for the female of *Ixodes abrocomae.* Females and nymphs of *Ixodes galapagoensis* are only known to parasitize *Aegialomys galapagoensis*, and *Ixodes nectomys* is known only from a male found on *Nectomys apicalis.* Cricetid species in the genera *Peromyscus* and *Reithrodontomys* are considered the principal hosts of *Ixodes tancitarius,* while *Cerradomys subflavus* and *Oligoryzomys nigripes* are regarded as principal hosts for larvae and nymphs of *Ixodes amarali*, and *Oxymycterus nasutus* and *Scapteromys tumidus* are principal hosts for nymphs and larvae of *Ixodes longiscutatus.* The large vesper mouse, *Calomys callosus*, is the preferred host for larvae and nymphs of *Ixodes pararicinus*, and *Handleyomys alfaroi* is the usual host for *Ixodes sinaloa*, while *Nephelomys albigularis* is the principal host for the immature stages of *Ixodes tropicalis.* Larvae and nymphs of *Ixodes loricatus* have been found on *Akodon azarae*, *Calomys* sp., *Holochilus brasiliensis*, *Hylaeamys megacephalus*, *Oligoryzomys flavescens*, *Oligoryzomys nigripes*, *Rhipidomys* sp., *Scapteromys aquaticus*, and *Scapteromys tumidus*, while *Hylaeamys perenensis* and *Hylaeamys yunganus* are principal hosts for nymphs of *Ixodes luciae*, but only in the case of Peruvian populations of this tick. Nymphs of Neotropical populations of *Ixodes spinipalpis* are considered prone to feed on *Peromyscus* sp., while *Sooretamys angouya* is a hypothetical principal host for the female of *Ixodes catarinensis*. The only known Neotropical female of *Ixodes woodi* was collected on *Neotoma* sp.

Nymphs and larvae of *Amblyomma ovale* and *Amblyomma triste* have been found on *Oligoryzomys flavescens* and *Oligoryzomys nigripes*, while larvae and nymphs of *Amblyomma ovale* have been collected from *Akodon* sp., *Euryoryzomys russatus*, *Transandinomys bolivaris*, and *Zygodontomys brevicauda*; and larvae and nymphs of *Amblyomma triste* have been taken from *Akodon azarae*, *Calomys callosus*, *Holochilus brasiliensis*, *Holochilus sciureus*, *Necromys lasiurus*, *Oxymycterus nasutus*, *Oxymycterus rufus*, *Rhipidomys* sp., *Scapteromys aquaticus*, and *Scapteromys tumidus*. Larvae of *Amblyomma tigrinum* have been found on *Akodon dolores*, *Akodon molinae*, *Akodon spegazzini*, *Calomys musculinus*, *Calomys venustus*, *Graomys chacoensis*, *Necromys benefactus*, *Oligoryzomys longicaudatus*, and *Phyllotis xanthopygus*. Larvae and nymphs of *Dermacentor panamensis* have been collected from *Isthmomys flavidus*, *Peromyscus aztecus*, *Peromyscus mexicanus*, *Reithrodontomys creper*, and *Reithrodontomys mexicanus*.

Species of Cricetidae help sustain populations of *Haemaphysalis juxtakochi* and several *Amblyomma* species that do not otherwise exhibit a clear pattern of host preference because adults and/or immature stages feed on several orders of mammals. Thus, males and females of *Amblyomma parvum* have been found on *Sigmodon* sp., females of *Amblyomma sculptum* have been collected from *Holochilus* sp., and nymphs of the same ticks have been recovered from *Calomys callosus*, *Holochilus sciureus*, *Rhipidomys macrurus*, and *Sooretamys angouya*. Nymphs and larvae of *Amblyomma mixtum* have been found on *Transandinomys bolivaris*. Nymphs of *Amblyomma naponense* and *Amblyomma oblongoguttatum* have been collected from *Oryzomys* sp., while *Euryoryzomys* sp. and *Neacomys spinosus* have been found infested with nymphs of *Amblyomma humerale*. Nymphs of *Haemaphysalis juxtakochi* have been found on *Tylomys nudicaudus*, and larvae of this tick have been collected from *Akodon azarae* and *Oxymycterus rufus*.

Additionally, larvae of *Haemaphysalis leporispalustris* have been found on *Peromyscus* sp. Different stages of five *Ixodes* species and seven species of *Amblyomma* have been taken from several species of Cricetidae.

Ctenomyidae: This family contains about 60 species established in South America. One undetermined species, *Ctenomys* sp., is the only known host of *Ixodes cooleyi*. Additionally, a few adults and several larvae of *Amblyomma aureolatum* have been collected from *Ctenomys* sp.

Cuniculidae: This family contains two Neotropical species, one of which has been found infested with ticks of the genera *Ixodes*, *Amblyomma*, and *Dermacentor*, this cuniculid being an important host for two species of *Ixodes* and one *Amblyomma*. Thus, *Cuniculus paca* is considered to be the principal host for the female of *Ixodes lasallei* and one of the principal hosts for the female of *Ixodes bocatorensis*, as well as the preferred host for males and females of *Amblyomma pacae*.

Cuniculus paca helps sustain populations of several *Amblyomma* species that do not otherwise exhibit a clear pattern of host preference because adults and/or immature stages feed on several orders of mammals. Thus, adults and nymphs of *Amblyomma oblongoguttatum*, males and females of *Amblyomma cajennense*, adults of *Amblyomma mixtum*, and nymphs of *Amblyomma brasiliense* and *Amblyomma naponense* have been found on *Cuniculus paca*.

Additionally, females of *Ixodes loricatus*, *Ixodes luciae*, and *Ixodes spinosus*, males and females of *Dermacentor nitens*, adults and nymphs of *Amblyomma coelebs* and *Amblyomma scalpturatum*, and adults of *Amblyomma ovale* have been collected from *Cuniculus paca*, but this host is not important in the life cycles of these ticks.

Dasyproctidae: This family contains about 15 Neotropical species, at least six of which have been found infested with ticks from the genera *Amblyomma*, *Haemaphysalis*, and *Ixodes*, dasyproctids being important hosts for two *Ixodes* species. Thus, *Dasyprocta* sp. is the principal host for females, nymphs, and larvae of *Ixodes spinosus*, and *Dasyprocta punctata* is one of the preferred hosts for females of *Ixodes bocatorensis*.

Species of Dasyproctidae help sustain populations of *Haemaphysalis juxtakochi* and several *Amblyomma* species that do not otherwise show a clear pattern of host preference because adults and/or immature stages feed on several orders of mammals. Thus, males, females, and nymphs of *Amblyomma sculptum* have been found on *Dasyprocta azarae*, and females and nymphs of *Amblyomma cajennense* have been collected from *Dasyprocta* sp. and *Dasyprocta azarae*; the latter host has also been found infested with nymphs and larvae of *Amblyomma brasiliense*. Nymphs of *Amblyomma naponense* have been taken from *Dasyprocta azarae*, *Dasyprocta fuliginosa*, *Dasyprocta leporina*, and *Dasyprocta punctata*, while nymphs of *Amblyomma oblongoguttatum* have been collected from *Dasyprocta* sp., *Dasyprocta fuliginosa*, and *Dasyprocta punctata*, and nymphs of *Amblyomma humerale* have been found on *Myoprocta acouchy*. Nymphs and larvae of *Haemaphysalis juxtakochi* have been recovered from *Dasyprocta azarae*, *Dasyprocta fuliginosa*, and *Dasyprocta punctata*.

Additionally, females of *Ixodes lasallei* have been found on *Dasyprocta fuliginosa* and *Dasyprocta leporina*, while immature stages of *Ixodes venezuelensis* have been collected from the latter host but also from *Myoprocta acouchy* and *Myoprocta pratti*. Males of *Haemaphysalis leporispalustris*, and females and larvae of this tick have been taken from *Dasyprocta* sp. and *Dasyprocta punctata*, respectively. Different stages of ten *Amblyomma* species have been found on several species of Dasyproctidae.

Echimyidae: This family comprises about 90 species, all found only in the Neotropical Region. At least 15 of these have been found infested with ticks of the genera *Ixodes*, *Amblyomma*, and *Haemaphyalis*, echimyids being important hosts for one *Amblyomma* species. Thus, Echimyidae are important hosts for the immature stages of *Amblyomma ovale*, especially the nymph. *Hoplomys gymnurus*, *Proechimys canicollis*, and *Proechimys semispinosus* have also been found infested with nymphs and larvae of *Amblyomma ovale*, while *Proechimys guyannensis*, *Proechimys quadriplicatus*, *Thrichomys apereoides*, *Thrichomys pachyurus*, and *Trinomys setosus* are known to be parasitized by nymphs of this tick.

Species of Echimyidae help sustain populations of *Haemaphysalis juxtakochi* and several *Amblyomma* species that do not otherwise exhibit a clear pattern of host preference because adults and/or immature stages feed on several orders of mammals. Thus, nymphs of *Amblyomma humerale* have been found on *Proechimys* sp., nymphs of *Amblyomma mixtum* have been taken from *Hoplomys gymnurus* and *Proechimys semispinosus*, nymphs of *Amblyomma naponense* and *Amblyomma oblongoguttatum* have been collected from *Proechimys semispinosus*, and nymphs of *Amblyomma sculptum* have been taken from *Thrichomys pachyurus*. Nymphs and larvae of *Haemaphysalis juxtakochi* have been recovered from *Hoplomys gymnurus* and *Proechimys semispinosus*.

Additionally, females of *Ixodes tropicalis* have been found on *Dactylomys boliviensis* and *Dasyprocta leporina*, while immature stages of *Ixodes venezuelensis* have been collected from *Proechimys* sp., *Proechimys guyannensis* and *Proechimys semispinosus*, and larvae of *Ixodes amarali* have been taken from *Trinomys setosus*. Nymphs of *Haemaphysalis leporispalustris* have been recovered from *Proechimys guyannensis*. Different stages of 12 *Amblyomma* species have been found on several species of Echimyidae.

Erethizontidae: This family contains about 20 species in the Americas. In the Neotropical Region, at least nine of these species have been found infested with ticks of the genera *Ixodes*, *Amblyomma*, and *Dermacentor*, erethizontids being important as hosts for two *Amblyomma* species and one *Dermacentor*. Thus, *Coendou prehensilis*, *Coendou quichua*, and *Coendou spinosus* are regarded as significant hosts for males and females of *Amblyomma longirostre*, while *Coendou prehensilis* and *Coendou spinosus* are principal hosts for males and females of *Amblyomma parkeri*, and *Coendou mexicanus* is the preferred host for adults of *Dermacentor panamensis*.

Species of Erethizontidae help sustain populations of *Haemaphysalis juxtakochi* and several *Amblyomma* species that do not otherwise exhibit a clear pattern of host preference because adults and/or immature stages feed on several orders of mammals. Thus, males, females, and nymphs of *Amblyomma sculptum* have been found on *Coendou prehensilis* and *Coendou spinosus*, while nymphs of *Haemaphysalis juxtakochi* have been collected from *Coendou quichua*.

Additionally, several females of *Ixodes boliviensis* have been taken from *Coendou mexicanus*. Different stages of 12 *Amblyomma* species have been found on several species of Erethizontidae. However, these hosts are not considered important in the life cycles of these ticks.

Geomyidae: This American family contains about 35 species in North and Central America. One of these, *Orthogeomys grandis*, is the only known host of *Ixodes tecpanensis*, and this is also the only instance of tick parasitism of Geomyidae reported from the Neotropical Region.

Heteromyidae: This American family consists of about 65 species. In the Neotropical Region, at least five of these species have been found infested with ticks of the genera *Ixodes*, *Amblyomma*, and *Dermacentor*, heteromyids being important hosts for two *Ixodes* species. Thus, *Heteromys anomalus* and *Heteromys desmarestianus* are principal hosts for the females of *Ixodes venezuelensis*, while *Heteromys pictus* is the principal host for the female of *Ixodes sinaloa*.

Additionally, nymphs of *Amblyomma ovale* have been taken from *Heteromys anomalus*, and nymphs and larvae of *Dermacentor panamensis* have been collected from *Heteromys desmarestianus* and *Heteromys adspersus*, respectively.

Muridae: This large family comprises more than 750 species worldwide. At least three of these species have been found infested with ticks of the genera *Ixodes*, *Amblyomma*, and *Rhipicephalus*, but Muridae are not significant hosts for any species of Neotropical Ixodidae. A *Rattus* sp. was found parasitized by larvae of *Amblyomma mixtum*, a species known to also parasitize various orders of hosts.

Additionally, adults, nymphs, and larvae of *Amblyomma ovale* have been found on *Rattus rattus*, while nymphs of *Amblyomma parvum* and larvae of *Amblyomma variegatum* have been collected from *Mus musculus*, and larvae of the latter species as well as larvae of *Amblyomma tigrinum* have been taken from *Rattus rattus*. Adults, larvae, and nymphs of *Rhipicephalus sanguineus sensu lato* have been found on *Rattus norvegicus* and *Rattus rattus*. Different stages of six *Ixodes* species have been collected from Muridae in the Neotropical Region.

Myocastoridae: This family has one species of Neotropical origin that is now found in many parts of the world. Females of *Ixodes rubidus* and nymphs of *Amblyomma dubitatum* have been collected from *Myocastor coypus*, but this is not an important host in the survival of these ticks.

Octodontidae: This family consists of about 15 Neotropical species, two of which have been found infested with *Ixodes sigelos*. Thus, *Aconaemys fuscus* has been found parasitized by females, nymphs, and larvae of *Ixodes sigelos*, while *Octodon degus* has been found infested with larvae. Octodontids are not regarded as principal hosts of *Ixodes sigelos*, but their role in the maintenance of populations of this tick may be more important than currently recognized.

Sciuridae: This family contains about 285 species distributed throughout the world. At least four of these species have been found infested with ticks from all genera of Ixodidae present in the Neotropics, with the exception of *Rhipicephalus*, sciurids being important hosts for two *Ixodes* species. Thus, *Sciurus aureogaster* is the principal host for females of *Ixodes guatemalensis*, and *Sciurus granatensis* is the only known host for the adults, nymphs, and larvae of *Ixodes tiptoni*.

Species of Sciuridae help sustain populations of *Amblyomma mixtum*, whose active stages feed on several orders of mammals, as well as immature stages of *Haemaphysalis juxtakochi*, whose nymphs and larvae also parasitize several orders of mammals. Thus, males of *Amblyomma mixtum* have been found on *Sciurus aestuans*, and nymphs and larvae of *Haemaphysalis juxtakochi* have been collected from *Sciurus aestuans* and *Sciurus igniventris*.

Additionally, females of *Ixodes boliviensis* and nymphs of *Ixodes spinipalpis* have been taken from *Sciurus* sp., while nymphs of *Ixodes venezuelensis* have been recovered from *Sciurus igniventris*. Females of *Amblyomma aureolatum* and nymphs of *Amblyomma fuscum* have been found on *Sciurus aestuans*, while nymphs of *Amblyomma longirostre*, and nymphs and larvae of *Dermacentor panamensis* have been collected from *Sciurus granatensis*.

5.1.4.14 Synopsis

A total of 13 orders of mammals have been found infested with ixodid ticks in the Neotropics, but two orders, Eulipotyphla and Paucituberculata, contain no species that are principal hosts for any species of Ixodidae. Mammals are the only principal hosts for Neotropical *Dermacentor* and *Rhipicephalus*, but they are also the most important hosts for other Neotropical genera, sometimes in association with other vertebrate classes.

One species of *Haemaphysalis*, three *Ixodes*, and eight *Amblyomma* species utilize both mammals and birds as principal hosts. Seven species of *Ixodes* and two *Amblyomma* preferentially parasitize two orders of mammals. However, 12 species of *Amblyomma* exhibit a more complex pattern of parasitism, with nine species principally feeding on at least three orders of mammals as main hosts, two species feeding on mammals and reptiles, and one combining Mammalia, Aves, and Reptilia as principal hosts. These host–ixodid associations are summarized in Tables 1.29, 2.53, 3.6, 5.1, 5.2, 5.3, and 5.6, but see also Table 5.5, which contrasts mammal host utilization between the two most abundant Neotropical genera, *Ixodes* and *Amblyomma*.

The genus *Ixodes* is unique in containing the only Neotropical ixodid found exclusively on Chiroptera (*Ixodes chilensis*), while *Ixodes downsi* has been found on Chiroptera and Aves, and *Ixodes neuquenensis* is the only tick that parasitizes Microbiotheria.

Amblyomma is peculiar in that it includes species that feed only on Pilosa, such as *Amblyomma goeldii*, *Amblyomma pictum*, and *Amblyomma varium*, although hosts for the nymphs and larvae of these ticks are uncertain, while Pilosa and birds are principal hosts of *Amblyomma calcaratum*, *Amblyomma geayi*, and *Amblyomma nodosum*. Cingulata are essential hosts for *Amblyomma auricularium*, *Amblyomma beaurepairei*, and *Amblyomma pseudoconcolor*, but *Amblyomma fuscum* principally parasitizes Cingulata, Didelphimorphia, and Squamata. Aves and Primates are important hosts for *Amblyomma romarioi*, while *Amblyomma parkeri* preferentially parasitizes Primates, Rodentia, and Aves.

Table 5.5 Numbers of Neotropical species of *Ixodes* and *Amblyomma* utilizing only Mammalia as principal hosts

Order of Mammalia	*Ixodes*	*Amblyomma*
1. Artiodactyla	4	8
2. Carnivora	4	1
3. Chiroptera	1	0
4. Cingulata	0	3
5. Didelphimorphia	0	1
6. Lagomorpha	2	0
7. Microbiotheria	1	0
8. Perissodactyla	1	5
9. Pilosa	0	3
10. Rodentia	20	4
11. Artiodactyla + Perissodactyla	0	1
12. Artiodactyla + several orders	0	1
13. Carnivora + Rodentia	0	1
14. Didelphimorphia + Rodentia	6	0
15. Lagomorpha + Rodentia	1	0
16. Carnivora + Didelphimorphia + Perissodactyla	0	1
17. Several orders	0	5
18. Rodentia + several orders	0	2
Total	40	36
Number of species found in the Neotropics	55	67

Artiodactyla are important hosts for all genera of Neotropical Ixodidae. Eight *Amblyomma* species feed only on Artiodactyla; *Amblyomma tenellum* and *Amblyomma naponense* principally parasitize artiodactyls and other mammals; and *Amblyomma parvitarsum* feeds on Artiodactyla and Squamata. Three *Ixodes* (*Ixodes montoyanus*, *Ixodes stilesi*, and *Ixodes taglei*) feed mostly on artiodactlyls, while Artiodactyla, other mammals and birds are the principal hosts of *Ixodes pararicinus*, a pattern that may also apply to *Ixodes fuscipes*, although this remains unconfirmed. *Dermacentor dispar*, *Dermacentor imitans*, and *Rhipicephalus microplus* feed on artiodactyls, while *Haemaphysalis juxtakochi* parasitizes Artiodactyla, other mammal orders and Aves.

Rodentia are principal hosts for species of Neotropical *Ixodes*, *Dermacentor*, and *Amblyomma*, but *Ixodes* is characterized by its singularly wide utilization of rodent hosts, with 20 species feeding only on Rodentia, and six species parasitizing both Rodentia and Didelphimorphia. Rodentia and Lagomorpha are the hosts of *Ixodes spinipalpis* in the Neotropics, and *Ixodes pararicinus* utilizes Rodentia as one of its principal hosts for immature stages. Four *Amblyomma* (*Amblyomma dubitatum*, *Amblyomma pacae, Amblyomma pseudoparvum*, and *Amblyomma romitii*) principally parasitize Rodentia. *Amblyomma ovale*, *Amblyomma parvum*, and *Amblyomma triste* feed on Rodentia and other mammals. *Amblyomma longirostre* preferentially parasitizes Rodentia and Aves, while two *Amblyomma* (*Amblyomma tigrinum* and

Table 5.6 Numbers of species per Neotropical ixodid genus utilizing classes of vertebrate hosts, singly and in combination

Hosts	*Ixodes*	*Amblyomma*	*Dermacentor*	*Haemaphysalis*	*Rhipicephalus*
Aves	9	0	0	0	0
Mammalia	40	36	6	1	3
Reptilia	0	16	0	0	0
Amphibia + Reptilia	0	2	0	0	0
Aves + Mammalia	3	8	0	1	0
Mammalia + Reptilia	0	2	0	0	0
Aves + Mammalia + Reptilia	0	1	0	0	0

Amblyomma parkeri) utilize Rodentia, other mammals, and Aves as their principal hosts. *Dermacentor panamensis* is exclusively found on Rodentia.

Four *Ixodes* (*Ixodes affinis*, *Ixodes boliviensis*, *Ixodes rubidus*, and *Ixodes texanus*) and adults of four *Amblyomma* (*Amblyomma aureolatum*, *Amblyomma maculatum*, *Amblyomma ovale*, and *Amblyomma tigrinum*) utilize Carnivora as principal hosts, and immature stages of *Amblyomma coelebs* feed mostly on Carnivora and Didelphimorphia. *Rhipicephalus sanguineus sensu stricto* and *sensu lato* utilize only Carnivora as their principal hosts.

Perissodactyla are principal hosts for Neotropical *Ixodes*, *Amblyomma*, and *Dermacentor. Ixodes tapirus* principally parasitizes Perissodactyla, while adults of five *Amblyomma* species (*Amblyomma incisum*, *Amblyomma latepunctatum*, *Amblyomma multipunctum*, *Amblyomma scalpturatum*, and *Amblyomma tapirellum*) feed mostly on Persissodactyla, although hosts of their immature stages are uncertain, and Perissodactyla and other mammals are principal hosts of *Amblyomma coelebs* and *Amblyomma tenellum*. Three species of *Dermacentor* (*Dermacentor dissimilis*, *Dermacentor latus*, and *Dermacentor nitens*) chiefly parasitize only Perissodactyla.

Didelphimorphia are important hosts for Neotropical *Ixodes* and *Amblyomma*. Six *Ixodes* species principally parasitize Didelphimorphia and Rodentia. The immature stages of *Amblyomma yucumense* have only been found on Didelphimorphia, but hosts for the adults of this tick are unknown, while *Amblyomma coelebs* utilizes Didelphimorphia as one of its principal orders of mammalian hosts, and *Amblyomma fuscum* principally parasitizes Didelphimorphia, Cingulata, and Squamata.

Neotropical Lagomorpha are the only preferred hosts for *Ixodes dicei* and *Ixodes pomerantzi*, while *Ixodes spinicoxalis* feeds on both Lagomorpha and Rodentia. Lagomorphs are the only principal hosts of *Haemaphysalis leporispalustris*.

Several species of *Amblyomma* are able to utilize different orders of mammals, sometimes in conjunction with parasitism of other vertebrate classes. Thus, five *Amblyomma* (*Amblyomma brasiliense*, *Amblyomma cajennense*, *Amblyomma mixtum*, *Amblyomma oblongoguttatum*, and *Amblyomma sculptum*) rely on several orders of mammals as hosts for their various life history stages. However, other species utilize several mammals combined with avian or reptilian hosts. And *Amblyomma humerale* is peculiar because nymphs are frequently found on several orders of mammals as well as avian and squamatan hosts, while tortoises are the principal hosts for adult ticks.

See the appendix for a full list of mammal host species and the ticks that parasitize them in the Neotropical Region.

5.2 Host Profiles of Neotropical Genera of Ixodidae

In this section, defining themes of host utilization at the levels of vertebrate classes and orders are analyzed with respect to the prostriate and metastriate genera found in the Neotropics. As emphasized earlier, most ixodid tick genera are associated with particular classes of vertebrates—Amphibia, Reptilia, Aves, or Mammalia—but 17 ixodid species utilize a combination of vertebrate classes as their principal hosts. This situation is summarized in Table 5.6, constructed with data from Tables 1.29, 2.53, 3.6, 5.1, 5.2, and 5.3, for the ixodid species whose hosts are known: *Ixodes* (51 species), *Amblyomma* (65 species), *Dermacentor* (6 species), *Haemaphysalis* (2 species), and *Rhipicephalus* (3 species). This section also discusses some working hypotheses concerning the origins of the Neotropical Ixodidae, including evolutionary theories related to ticks found in this region, particularly with respect to the genus *Amblyomma*, the paradigmatic Neotropical genus.

5.2.1 Host Profile of Prostriata: Genus Ixodes

In the Neotropical Region, there are no records of *Ixodes* parasitizing amphibians, and only one record of one species feeding on Reptilia: Squamata, which relates to several nymphs and larvae of *Ixodes affinis* collected from *Ameiva* sp. (Teiidae). The role of this type of host in the life cycle of *Ixodes affinis* is unclear, but the Nearctic *Ixodes pacificus* and the Palearctic *Ixodes inopinatus*, among a few other *Ixodes* species, chiefly parasitize Squamata as immatures, and additional studies may therefore clarify the role of Teiidae as hosts of Neotropical *Ixodes*. Otherwise, throughout the Neotropics, the principal hosts of *Ixodes* are members of the classes Aves, Mammalia, or the combination Aves + Mammalia.

Ixodes is the only Neotropical genus containing species that exclusively utilize Aves as their principal hosts, but this observation is not peculiar to the Neotropical Zoogeographic Region. Guglielmone et al. (2014) and Guglielmone and Nava (2017) stressed the importance of Aves as exclusive hosts of significant numbers of *Ixodes* species worldwide. This pattern of parasitism is only reinforced by the nine Neotropical species of *Ixodes* that exclusively utilize five orders of Aves as principal hosts: Caprimulgiformes (*Ixodes cuernavacensis* and *Ixodes paranaensis*), Galliformes (*Ixodes cornuae*), Passeriformes (*Ixodes auritulus*, *Ixodes copei*, *Ixodes mexicanus*, and *Ixodes silvanus*), Sphenisciformes (*Ixodes uriae*), and Trogoniformes (*Ixodes bequaerti*).

The usage of avian hosts by Neotropical *Ixodes* (Prostriata) reveals a parasitic strategy unlike that seen in the Metastriata, where only the genera *Amblyomma* and *Haemaphysalis* contain species known to partially utilize birds in the order Passeriformes. The differences in Avian usage between Neotropical *Ixodes* and *Amblyomma* are clearly shown in Table 5.4, where 14 combinations have been found for Aves alone or Aves together with other classes of hosts, but only one combination (Passeriformes + Rodentia) is shared by both *Ixodes* and *Amblyomma*. The unique relationship between *Ixodes* and Aves most probably developed well before the evolution of the order Passeriformes. One *Ixodes* species known from both the Neotropics and other parts of the world, the marine bird parasite *Ixodes uriae*, is thought to be of evolutionary importance, its ancestors having originated before the Cretaceous-Paleogene mass extinction, as hypothesized by Dietrich et al. (2014) and discussed by Guglielmone and Nava (2017). It has been proposed that *Ixodes uriae* and its ancestors are survivors of the mass extinction, having dispersed from southern lands adjoining the Antarctic, and reaching the Arctic thereafter.

Despite the role of Aves as hosts for Neotropical *Ixodes*, it is clear that Mammalia is the chief vertebrate class parasitized by this genus because 40 species exclusively utilize mammals as their principal hosts. The relationships between *Ixodes* and some orders of Mammalia are unique among Neotropical Ixodidae; thus, *Ixodes chilensis* is exclusively a parasite of Chiroptera, while *Ixodes neuquenensis* feeds only on the monotypic order Microbiotheria, a precursor of Australian marsupials. Additionally, Artiodactyla are known to be the exclusive principal hosts of four species of *Ixodes* because the adults of *Ixodes fuscipes* and *Ixodes taglei*, and adults and immature stages of *Ixodes montoyanus* and *Ixodes stilesi* are tied to artiodactyls. Also, four species have been found only on Carnivora: *Ixodes affinis* (adult ticks), *Ixodes boliviensis* (adults), *Ixodes rubidus* (females and nymphs), and *Ixodes texanus* (females and nymphs). As well, two species, *Ixodes dicei* (adults) and *Ixodes pomerantzi* (adults and nymphs), utilize only Lagomorpha as principal hosts, and the adults of one species, *Ixodes tapirus*, are known to feed exclusively on Perissodactyla. However, as will be seen, the most important parasitic characteristics of Neotropical *Ixodes* lie in the relationship between this genus and the order Rodentia. Overall, mammal parasitization by *Ixodes* differs markedly from that observed in the genus *Amblyomma*. As evidenced in Table 5.5, of 18 combinations of mammalian orders that are exclusive principal hosts for Neotropical *Ixodes* and *Amblyomma*, only four combinations are shared by both genera.

The preponderance of rodents as hosts of Neotropical *Ixodes* is attested by a total of 29 species that utilize Rodentia as principal hosts during at least one parasitic stage, a number that represents 53% of *Ixodes* species found in the Neotropics (see Table 5.1). Twenty *Ixodes* species are exclusively found on Rodentia, six species principally parasitize rodents and mammals of other orders, while *Ixodes minor* feeds on Rodentia and Aves, its principal hosts for adults and larvae, respectively, and *Ixodes pararicinus* adults feed mainly on Artiodactyla, while larvae and nymphs parasitize Rodentia and Aves. Although several families of Rodentia, such as Abrocomidae, Capromyidae, Chinchillidae, Ctenomyidae, Cuniculidae, Dasyproctidae, Geomydae, Heteromyidae, and Sciuridae, are known as principal hosts for Neotropical *Ixodes*, members of the major family Cricetidae are numerically significant because 18 species of *Ixodes* are found on this family of rodents. Thus, eight species of *Ixodes* feed only on Cricetidae, three parasitize Cricetidae and other families of rodents, six are found on Cricetidae and mammals from other orders, and *Ixodes pararicinus* feeds on Cricetidae, Artiodactyla, and Aves. The main role of Cricetidae when combined with other types of principal hosts is to sustain the immature stages of a given species.

It is hypothesized that Neotropical *Ixodes* are more prone to utilize rodents as principal hosts than *Ixodes* found in other zoogeographic regions. This hypothesis was preliminarily tested by broadly comparing the host profiles of the species of *Ixodes* found in all zoogeographic regions using these criteria: (1) species in which at least one parasitic stage feeds principally on rodents, (2) species known only from rodents, and (3) species collected from rodents plus another type of host. Figures in Guglielmone et al. (2014) and data published thereafter until the end of 2019 (not shown) were used in this analysis. It was found that 11% of the Australasian *Ixodes* species, 28% of Afrotropical, 25% of Palearctic, and 30% of Oriental species met the above criteria, but this percentage was 51% for Nearctic *Ixodes* and 52% for Neotropical species. Thus, utilization of rodents as hosts is not peculiar to Neotropical *Ixodes* but a characteristic shared with Nearctic *Ixodes*, indeed the *Ixodes* fauna of the entire Western Hemisphere.

The predominance of rodents as hosts for American *Ixodes* is probably a consequence of the ability of members of this genus to colonize new niches following the natural invasion by Nearctic non-hystricomorph rodents, basically Cricetidae, into southern Mexico, Central America, and, finally, South America, together with the reverse route taken by hystricomorph rodents. These invasions occurred before the Great American Interchange that accompanied the merging of South and Central America during the Pliocene Epoch of the Neogene Period (Cenozoic Era), but the Great American Interchange event surely favored an even more intense interchange of Nearctic and Neotropical Rodentia. This is particularly evident in the case of the family Cricetidae that hypothetically invaded northern South America along with its parasites (ticks of the genus *Ixodes* included) and reached the southern tip of the continent in Argentina and Chile, providing *Ixodes* species the opportunity to colonize new lands and diversify together with their rodent hosts.

The South American invasion by non-hystricomorphs resulted in a host–*Ixodes* association peculiar to the Neotropics, where the presence of ancient South American hosts (Didelphimorphia) and newcomers (cricetid rodents) led to the evolution of five species: *Ixodes amarali*, *Ixodes catarinensis*, *Ixodes loricatus*, *Ixodes luciae*, and *Ixodes tropicalis*. This assemblage later reached northern Neotropical lands. Thus, while *Ixodes amarali*, *Ixodes catarinensis*, *Ixodes loricatus*, and *Ixodes tropicalis* are present only in South America, *Ixodes luciae* occurs from Argentina to southern Mexico. The singular case of *Ixodes luciae* is hypothetically considered to be a reversal of the rodent invasion of South America, aided by the colonization of Central America and southern Mexico by South American marsupials during the Great American Interchange, which propelled the relationship between marsupials, rodents, and *Ixodes* beyond the southern Neotropics.

Three species of *Ixodes* are known to preferentially parasitize both mammals and birds. Thus, *Ixodes downsi* is unique in utilizing Caprimulgiformes and Chiroptera as provisional principal hosts of females and larvae, respectively, while *Ixodes minor* principally parasitizes Rodentia as adult ticks and Passeriformes as larvae in the Neotropics. And as noted above, Artiodactyla, Rodentia, and Passeriformes are principal hosts of *Ixodes pararicinus*.

5.2.2 *Host Profile of Metastriata*

Genus *Amblyomma*

The genus *Amblyomma* is characterized by its ability to adapt to different classes of vertebrates, being the only Neotropical ixodid genus that utilizes species of Amphibia, Reptilia, Aves, and Mammalia as principal hosts (Tables 2.53 and 5.2). Consequently, *Amblyomma* also is the genus with the highest number of instances where combinations of vertebrate classes are parasitized (Table 5.6). This wide host utilization by *Amblyomma* is not restricted to the Neotropics, but is seen worldwide (Guglielmone et al. 2014). In the eyes of most researchers, *Amblyomma* is the paradigmatic Neotropical genus, although, in fact, its diversity is based on the numerous taxa found on the South American mainland, with 28 endemic species there (Table 4.2), which would appear to justify a South American origin for Neotropical *Amblyomma*.

Neotropical *Amblyomma* are unique in having members that principally parasitize the class Amphibia. Thus, *Amblyomma dissimile* and *Amblyomma rotundatum* are commonly found on Anura: Bufonidae, although other principal hosts of these ticks include reptiles in the order Squamata. It is unclear whether this host–parasite relationship is related to the South American origin of the Bufonidae during Late Cretaceous and is basal for Neotropical *Amblyomma*, or whether it is a consequence of the dramatic Cenozoic radiation of both vertebrates and *Amblyomma*, providing these ticks the opportunity to colonize new niches, including those occupied by some amphibians and squamatans. In any event, species of *Amblyomma* that feed on Anura and Squamata are thought to have evolved in the Neotropical Region (probably South America), with *Amblyomma dissimile* and *Amblyomma rotundatum* later reaching the Nearctic, and the latter species has already been collected from a bufonid introduced on a Pacific Island.

Although the genus *Amblyomma* possesses the evolutionary plasticity to parasitize all classes of terrestrial vertebrates, the great majority of *Amblyomma* species, as well as those in other Neotropical genera of Ixodidae, are exclusively parasites of Mammalia, with 36 of the 67 *Amblyomma* species utilizing members of this class as principal hosts. However, mammal host utilization of Neotropical *Amblyomma* differs from host usage recorded for *Ixodes*, the other large genus of Neotropical ixodids, with 54 species found in the region. This difference is apparent in Table 5.5, where of 18 combinations of mammalian orders that are exclusive principal hosts for Neotropical *Ixodes* and *Amblyomma*, only four combinations are shared by both genera.

Amblyomma species parasitize a wider range of mammal hosts than other genera of Neotropical ixodids, and five species—*Amblyomma brasiliense*, *Amblyomma cajennense*, *Amblyomma mixtum*, *Amblyomma oblongoguttatum*, and *Amblyomma sculptum*—have been collected from so many mammal species that it is impossible to single out a preferred host for any of them. Moreover, *Amblyomma auricularium*, *Amblyomma beaurepairei*, and *Amblyomma pseudoconcolor* are the only species known to parasitize the paradigmatic South American order Cingulata, but the immature stages of these ticks are known only for *Amblyomma auricularium*, a problematic species that is difficult to identify. *Amblyomma goeldii*, *Amblyomma pictum*, and *Amblyomma varium* parasitize Pilosa, another paradigmatic South American order whose members are the only principal hosts for adults of these three tick species, but hosts for the immature stages remain uncertain. Additionally, Rodentia are exclusive hosts for *Amblyomma dubitatum*, *Amblyomma pacae*, *Amblyomma pseudoparvum*, and *Amblyomma romitii*, but contrary to the genus *Ixodes*, several species of which feed on non-hystricomorph rodents, these four species of *Amblyomma* parasitize Hystricomorpha. Artiodactyla are well represented as principal hosts of Neotropical *Amblyomma*, with eight species exclusively found on them, and species such as *Amblyomma neumanni* and *Amblyomma tonelliae* have the capacity to use introduced hosts (cattle) to feed all parasitic stages. Perissodactyla are the principal hosts for adults of *Amblyomma incisum*, *Amblyomma latepunctatum*, *Amblyomma multipunctum* (also for nymphs), *Amblyomma scalpturatum*, and *Amblyomma tapirellum*. Only one Neotropical species in this genus, *Amblyomma maculatum*, has been found feeding chiefly on Carnivora, while immature stages of *Amblyomma yucumense* principally parasitize only Didelphimorphia, although the host ranges of these ticks should be regarded as provisional (see "Synopsis" in Sect. 5.1.4.14).

Another characteristic of Neotropical *Amblyomma* is the usage of more than one order of Mammalia as principal hosts. Thus, *Amblyomma tenellum* utilizes Artiodactyla and Perissodactyla as important hosts. *Amblyomma ovale* males and females are restricted to Carnivora, but immature stages parasitize Rodentia, while the adults of *Amblyomma naponense* principally parasitize Artiodactyla, but nymphs feed on mammals of several orders. Larvae and nymphs of *Amblyomma parvum* and

Amblyomma triste utilize Rodentia as principal hosts, but adults feed on several orders of mammals. And *Amblyomma coelebs* shares Carnivora and Didelphimorphia as principal hosts of nymphs and larvae, while Perissodactyla are the principal hosts for adult ticks. *Amblyomma* is also the only genus containing species that chiefly parasitize Primates; thus, *Amblyomma romarioi* has Primates as principal hosts of males, females, and nymphs, and larvae feed mostly on Aves (Passeriformes), while adults and nymphs of *Amblyomma parkeri* have Rodentia as principal hosts and larvae feed chiefly on Passeriformes. These combinations and others discussed below have not been found in other Neotropical Ixodidae and, as will be seen, have a bearing on hypotheses concerning the original hosts of Neotropical *Amblyomma* parasitic on mammals.

Reptilia are very important as hosts for *Amblyomma* in the Neotropical Zoogeographic Region and elsewhere, with 16 Neotropical *Amblyomma* species principally parasitizing members of this vertebrate class. Ten species are specific to Squamata, although only one of them, *Amblyomma fulvum*, is restricted to the South American mainland. Five of the specific parasites of Squamata—*Amblyomma albopictum*, *Amblyomma antillorum*, *Amblyomma cruciferum*, *Amblyomma quadricavum*, and *Amblyomma torrei*—are distributed in the Caribbean, but one of these species, *Amblyomma albopictum*, is also found on squamatans in Central America and northern South America. Six species of Neotropical *Amblyomma* parasitize Testudines as their only principal hosts; five of these, *Amblyomma argentinae* (South American mainland), *Amblyomma crassum* (Central America and South American mainland), *Amblyomma macfarlandi* (Galápagos Islands), *Amblyomma pilosum* (Galápagos Islands), and *Amblyomma usingeri* (Galápagos Islands), parasitize members of the family Testudinidae. A majority of the ticks reported from Testudines are established in the Galápagos Islands, whose tortoises' common ancestor is probably the South American *Chelonoidis chilensis*, which is the principal host of *Amblyomma argentinae*. Therefore, it is also possible that the common ancestor of ticks parasitizing Neotropical Testudinidae may be related to *Amblyomma argentinae*. Land tortoises may have invaded South America by rafting during the continent's 50 million years of isolation, i.e., the Eocene Epoch (Paleogene Period). Several species of *Amblyomma* are known to parasitize African Testudinidae, and it is also possible that *Amblyomma* species invaded South America along with their tortoise hosts. If that were the case, any rafting would have to have been of short duration, so that the final phase of feeding and host detachment occurred in South America in numbers sufficient to guarantee host finding and reproduction for the first generation of ticks in the new continent. Such a hypothesis should be tested through phylogenetic analysis of *Amblyomma* species currently infesting Testudinidae in the Afrotropical and Neotropical Zoogeographic Regions.

Aves are important hosts for nine species of *Amblyomma*, but none of them are exclusively parasites of birds, and all principal avian hosts of *Amblyomma* belong to the order Passeriformes. This utilization of Neotropical Aves as principal hosts of *Amblyomma* species differs from that seen in bird-infesting *Ixodes*, as shown in Table 5.4. Eight tick–host associations involve Passeriformes and Mammalia as principal hosts for immature stages and adult ticks, respectively, with three species,

Amblyomma calcaratum, *Amblyomma geayi*, and *Amblyomma nodosum*, utilizing the combination Passeriformes + Pilosa as principal hosts; while the principal hosts for *Amblyomma aureolatum* are Passeriformes + Carnivora, and as stated above, the combination for *Amblyomma romarioi* is Passeriformes + Primates. *Amblyomma longirostre* has mainly been found on Rodentia and Passeriformes, while *Amblyomma tigrinum* parasitizes the combination Passeriformes + Carnivora + Rodentia, and *Amblyomma parkeri* is found on the combination Passeriformes + Primates + Rodentia. Passeriformes is one of the principal host groups for the nymph of *Amblyomma humerale,* a tick whose adults chiefly parasitize Testudines, while nymphs have also been frequently collected from Squamata and members of several orders of mammals. All thcsc associations most likely evolved during the Paleogene Epoch (Cenozoic Era), favored by the major radiation of Passeriformes during that time, and it would further appear that many of these associations arose after the Great American Interchange (late Cenozoic) and the irruption of several orders of mammals into the South American mainland.

South America was isolated for about 50 million years during the Cenozoic Era, and this gave rise to a singular mammalian fauna that collapsed after the Great American Interchange, an event that occurred about 3 million years ago when a land bridge connected Central America to South America. This major geographic change facilitated the invasion of artiodactyls, carnivores, perissodactyls, and other mammals, from northern to southern lands, followed by a rapid displacement of the endemic fauna and mass extinctions, with few South American exceptions, such as marsupials (Didelphimorphia) and xenarthrans (Cingulata and Pilosa).

Amblyomma is the paradigmatic genus of Neotropical Ixodidae in general and South America in particular, and the historical events related above surely shaped the relationships between *Amblyomma* species and their mammal hosts. Two phylogenetic studies strongly suggest that several species of *Amblyomma* originated in South America before and after the Great American Interchange. Thus, Beati et al. (2013) found that the *Amblyomma cajennense* species complex evolved about 13 million years ago, during the Middle Miocene Epoch (Cenozoic Era) in northern South America and invaded northern Neotropical territories during the end of the Pliocene or early Pleistocene Epochs (Cenozoic). Later, Lado et al. (2018) found that *Amblyomma tigrinum* (South America) split from *Amblyomma maculatum* (Central America-southern Mexico + South America + Nearctic) and *Amblyomma triste* (South America + Nearctic), placing the most probable occurrence of this event in South America after the Great American Interchange. Therefore, South America seems to have been a suitable place for the genus *Amblyomma* to radiate, both during its period of isolation and after the Great American Interchange, resulting in an abundance of species not matched in any other regions of the world.

It can be hypothesized that the first *Amblyomma* species parasitizing Neotropical mammals were of African origin. Hystricomorph rodents invaded South America from Africa by rafting about 42 million years ago during the Middle Eocene (Cenozoic), giving rise to South American Caviodea, comprising the families Caviidae, exclusive hosts of *Amblyomma dubitatum*, *Amblyomma pseudoparvum*, and *Amblyomma romitii*, and Cuniculidae, only hosts of *Amblyomma pacae*.

However, no African *Amblyomma* principally parasitize rodents (Guglielmone et al. 2014), and an analysis of the 16S rDNA gene of Neotropical *Amblyomma* by Martins et al. (2019b) shows that those species linked to Caviodea are not closely related. Therefore, the hypothesis for an African origin of South American *Amblyomma* appears unsound.

Beati and Klompen (2019) hypothesized that *Amblyomma* most likely originated in southern South America, stating that the basal South American lineage consisted of *Amblyomma boeroi*, *Amblyomma neumanni*, and *Amblyomma parvitarsum*, which separated from Australasian *Amblyomma* about 60 million years ago (early Cenozoic Era), implying that the Australasian-South American ancestors evolved before that date (late Mezozoic Era?). However, the basal lineage of Beati and Klompen (2019) appears to be incongruent with their time of origin for South American *Amblyomma* because the proposed lineage contains species most probably derived from the Great American Interchange invaders (about 3 million years ago), rather than tick species that can be placed more convincingly as present in the late Mesozoic-early Cenozoic. Thus, *Amblyomma boeroi* is an exclusive parasite of *Catagonus wagneri* (Artiodactyla: Tayassuidae), while cattle (Artiodactyla: Bovidae) are the principal hosts of all parasitic stages of *Amblyomma neumanni*, and most other hosts of this species are descendants of the Great American Interchange invaders; also, adults of *Amblyomma parvitarsum* principally parasitize South American camelids (Artiodactyla: Camelidae). In short, none of these hosts were present in South America 60 million years ago, and the presence of *Amblyomma boeroi*, *Amblyomma neumanni* and *Amblyomma parvitarsum* at the time proposed by Beati and Klompen (2010) appears unlikely. It can be argued that such *Amblyomma* species, or their ancestors, are basal for the genus, but under that proposal the point of origin of the genus *Amblyomma* should be in northern lands along with their artiodactyl hosts, a hypothesis that is difficult to support.

Additional phylogenetic studies are needed to determine the origin of the *Amblyomma* species found in South America (and elsewhere), perhaps by focusing on species that still feed on ancestral South American mammals (Cingulata? Pilosa?), thereby shedding light on this unresolved historical event.

Genus *Dermacentor*

In the Neotropical Region, members of the genus *Dermacentor* characteristically parasitize mammals as their principal hosts (Tables 5.3 and 5.4), which is typical for this genus worldwide (Guglielmone et al. 2014). Thus, all Neotropical records of *Dermacentor* ticks are from Mammalia, with the sole exception of a case in which an amphibian was found infested by a female of *Dermacentor nitens*.

Definitive host information for the nine species of Neotropical *Dermacentor* is somewhat uncertain because of the scarcity of Neotropical data for three Nearctic–Neotropical species: *Dermacentor albipictus*, *Dermacentor halli*, and *Dermacentor variabilis*. Nevertheless, records from the Nearctic Region indicate that all parasitic stages of *Dermacentor albipictus* are usually found on Artiodactyla and Perissodactyla; *Dermacentor halli* has been found on a variety of mammals, though it is uncertain whether there are particularly preferred hosts for their adults, nymphs, and larvae; and adults of *Dermacentor variabilis* are frequent parasites of Artiodactyla, Perissodactyla, and Carnivora, although larvae and nymphs are commonly collected from Rodentia (several families) (Guglielmone et al. 2014). These host–parasite associations are probably also common in the Neotropics.

Three of the six Neotropical species of *Dermacentor* (*Dermacentor dissimilis*, *Dermacentor latus*, and *Dermacentor nitens*) principally parasitize Perissodactyla, highlighting the importance of this order in the region. Adults of *Dermacentor dispar* and *Dermacentor imitans* usually feed on Artiodactyla, while *Dermacentor panamensis* is chiefly a parasite of Rodentia. As noted earlier, Neotropical species of *Dermacentor* are restricted to southern Mexico and Central America, with only two records from northern South America (*Dermacentor nitens* excepted), and five species established in both the Nearctic and Neotropical Zoogeographic Regions. This limited Neotropical distribution indicates that the genus *Dermacentor* may have arrived relatively recently in the region, and it is hypothesized that their invasion of South America occurred well after the onset of the Great American Interchange.

Genus *Haemaphysalis*

Only three of the 176 species of *Haemaphysalis* worldwide have been detected in the Neotropics (Guglielmone et al. 2015). The host for one of them, *Haemaphysalis cinnabarina*, is unknown. The other two species utilize different hosts (Table 5.6), but both ticks are specific parasites of mammals. *Haemaphysalis leporispalustris* has been collected from several mammals and birds, but all active life history stages of this tick chiefly parasitize Lagomorpha, while *Haemaphysalis juxtakochi* depends on Cervidae in the adult stage, the larvae and nymphs feeding principally on several orders of mammals, although larvae have also been found frequently on passeriform birds.

One species, *Haemaphysalis cinnabarina*, is known from two specimens collected in the Neotropics, although several authors doubt the Neotropical origin of this species. *Haemaphysalis leporispalustris* and *Haemaphysalis juxtakochi* also occur in the Nearctic Region. The genus *Haemaphysalis* originated in the Oriental Zoogeographic Region, but few species reached the Nearctic Region, and those that were able to penetrate the Neotropical Region from the Nearctic probably arrived rather recently after the merging of Central and South America.

Genus *Rhipicephalus*

There are at least three species of *Rhipicephalus* in the Neotropics. *Rhipicephalus microplus* has rarely been found on Aves, and there are odd records of *Rhipicephalus sanguineus sensu lato* from Aves and Reptilia, but all species of *Rhipicephalus* in the Neotropics utilize mammals as their principal hosts. Thus, *Rhipicephalus sanguineus sensu stricto* and *Rhipicephalus sanguineus sensu lato* parasitize Carnivora, while *Rhipicephalus microplus* most often feeds on Artiodactyla.

None of the *Rhipicephalus* found in the Neotropics are exclusive to this region, all having been introduced as a result of human activities. *Rhipicephalus microplus* is thought to have arrived in the Neotropics with cattle brought by European settlers about 400 years ago. *Rhipicephalus sanguineus sensu stricto* most likely arrived with dogs infested in Europe, while the origin of Neotropical *Rhipicephalus sanguineus sensu lato* is unknown but, again, it is probable that this tick was introduced with dogs not longer than 400 years ago.

5.3 Human Parasitism

Forty-nine (36%) of the 137 Neotropical species of Ixodidae have been found biting people across the region (Table 5.7). These figures are higher than the number and percentage of tick species reported from humans up to 2003, when Guglielmone et al. (2003a) found 33 species, or 29% of the total of 114 Neotropical taxa known at the time, that had been collected from humans.

The great majority of cases of human parasitism in the Neotropics have involved ticks of the genus *Amblyomma*; 39 species, or 58% of the total for the genus (67 taxa), are known to feed on humans. These numbers represent a sharp increase in comparison with the picture presented by Guglielmone et al. (2003a), when 26 species, or 46% of the total, were documented as having been collected from humans. Seven new or reinstated species have contributed to the current figures: *Amblyomma hadanii*, *Amblyomma latepunctatum*, *Amblyomma mixtum*, *Amblyomma patinoi*, *Amblyomma romitii*, *Amblyomma sculptum*, and *Amblyomma tonelliae*. As well, eight species that were known to be valid before 2003 have since that year been found parasitizing humans: *Amblyomma auricularium*, *Amblyomma fuscum*, *Amblyomma humerale*, *Amblyomma parkeri*, *Amblyomma pecarium*, *Amblyomma pseudoconcolor*, *Amblyomma sabanerae*, and *Amblyomma varium*. Moreover, supposed cases of human parasitism by *Amblyomma extraoculatum* (now considered an Oriental species) were found to be caused by *Amblyomma romitii*, while cases attributed to *Amblyomma imitator* (now *Amblyomma tenellum*), which was treated as a human parasite in Guglielmone et al. (2003a), have since been recognized as restricted to the Nearctic Region.

Ixodes is the second largest genus of Neotropical ixodids, with 55 species, but only three (5% of the total) are human parasites: *Ixodes boliviensis*, *Ixodes pararicinus*, and *Ixodes tancitarius* (provisional). This profile also differs from that described in Guglielmone et al. (2003a), where 45 *Ixodes* species were reported from the region, with only *Ixodes boliviensis* and *Ixodes luciae* listed as human parasites. The inclusion of *Ixodes luciae* was subsequently treated as a probable diagnostic error by Guglielmone et al. (2011), while *Ixodes tancitarius* was not considered a Neotropical species in Guglielmone et al. (2003a).

As human parasites, Neotropical species of *Amblyomma* and *Ixodes* are a study in contrasts, since members of the former genus are commonly found on people, but the opposite is true for *Ixodes*. The many species of *Amblyomma* parasitize a broad range of hosts and are widely distributed in the Neotropical Region (Tables 2.53, 4.2, and 5.2), with cases of human parasitism known from 21 countries/territories (Table 5.8) and all parasitic stages of 11 species found biting people (Table 5.7). By contrast, the range of hosts of Neotropical *Ixodes* is narrower (Tables 1.29, 5.1, and 5.6), and many *Ixodes* species are restricted to rodent hosts (Tables 1.29 and 5.1), with the result that human parasitism by *Ixodes* is a very rare event that has so far been reported from only four countries (Argentina, Belize, southern Mexico and Panama) (Table 5.8).

Of course, humans are never principal hosts of ticks, either in the Neotropics or elsewhere. Rather, human parasitism is a consequence of people entering tick-infested habitats. Several species of *Amblyomma*, e.g., *Amblyomma cajennense*, *Amblyomma mixtum*, and *Amblyomma sculptum*, among others, are catholic feeders and attack persons eagerly if the opportunity arises, but no species of Neotropical *Ixodes* behaves in this manner. And unlike *Amblyomma*, all stages of several *Ixodes* species feed on rodents and birds in their hosts' nests or burrows (i.e., they are endophilous). However, there remains an element of speculation in these assessments because the life histories of many Neotropical ixodids are basically unknown, a situation that should encourage major near-term research.

Three species of *Dermacentor*, two *Haemaphysalis*, and two species of *Rhipicephalus* have been found feeding on people (Table 5.7). With the exception of *Rhipicephalus sanguineus sensu lato*, all these species are rare or sporadic parasites of humans. The case of *Rhipicephalus sanguineus sensu lato* is interesting because human parasitism is frequently reported under this taxonomically uncertain name, and such instances appear to be a consequence of interaction with infested dogs in kennels, on premises and in houses, circumstances that all but guarantee tick contact with humans.

Table 5.8 lists the species of Neotropical ixodids causing human parasitism on a country/territory basis. Although tick parasitism of people is common, not all countries and territories are represented. As expected, *Amblyomma* species are the chief human parasites in the majority of countries, but there are also several exceptions: no representatives of this genus have been reported attacking people in Bolivia, Chile, Curação, El Salvador, and Peru. Equally surprising, *Rhipicephalus sanguineus sensu lato* is the only tick that has been found causing human parasitism in Chile and Curação, and *Rhipicephalus microplus* is the only hard tick known to feed on people in El Salvador. Just two species, *Dermacentor nitens* and *Rhipicephalus microplus*, have been found attacking people in Bolivia, and two others, *Rhipicephalus microplus* and *Rhipicephalus sanguineus sensu lato*, have been collected from humans in Peru. The absence of cases of human parasitism by *Amblyomma* in Bolivia and Peru is unexpected because many *Amblyomma* species are found there, including some that are often collected from humans in other Neotropical countries. This situation is most probably due to the lack of published reports on tick infestations of humans in these countries, rather than actual absence of this phenomenon.

Table 5.7 Species of Ixodidae and tick stages causing human parasitism in countries/territories of the Neotropical Region. A = adult tick of undetermined sex, M = male, F = female, N = nymph, L = larva, U = undetermined stage

Ticks species	Tick stages	Countries-Territories
1. *Ixodes boliviensis*	MF	Belize, southern Mexico, Panama
2. *Ixodes pararicinus*	N	Argentina
3. *Ixodes tancitarius*	F*	Southern Mexico
1. *Amblyomma aureolatum*	MFN	Argentina, Brazil, French Guiana, Paraguay
2. *Amblyomma auricularium*	N	Brazil
3. *Amblyomma brasiliense*	MFNL	Argentina, Brazil, Paraguay
4. *Amblyomma cajennense*	MF,N*L*	Brazil, French Guiana, Guyana, Suriname
5. *Amblyomma calcaratum*	A	Trinidad and Tobago
6. *Amblyomma coelebs*	MFN, L*	Argentina, Belize, Brazil, Costa Rica, southern Mexico, Paraguay
7. *Amblyomma dissimile*	FN	Brazil, Colombia, southern Mexico, Panama, Venezuela
8. *Amblyomma dubitatum*	MFNL	Argentina, Brazil, Uruguay
9. *Amblyomma fuscum*	MF	Brazil
10. *Amblyomma hadanii*	F*NL	Argentina
11. *Amblyomma humerale*	L*	French Guiana
12. *Amblyomma incisum*	MFN, L*	Argentina, Brazil, Paraguay
13. *Amblyomma inornatum*	U*	Southern Mexico
14. *Amblyomma latepunctatum*	MFN, L*	Brazil, French Guiana
15. *Amblyomma longirostre*	MFN	Brazil, French Guiana, Venezuela
16. *Amblyomma maculatum*	F	Belize, southern Mexico
17. *Amblyomma mixtum*	MFNL	Belize, Colombia*, Costa Rica, Cuba, Guatemala, Honduras, Jamaica, southern Mexico, Nicaragua, Panama, Trinidad and Tobago
18. *Amblyomma naponense*	MFN, L*	Brazil, French Guiana, Panama
19. *Amblyomma neumanni*	MFNL	Argentina
20. *Amblyomma oblongoguttatum*	MFN	Brazil, Colombia, French Guiana, Guyana, Panama, Suriname, Venezuela
21. *Amblyomma ovale*	MF	Argentina, Brazil, Colombia, Costa Rica, Ecuador, French Guiana, southern Mexico, Panama, Paraguay, Suriname, Venezuela
22. *Amblyomma pacae*	M, L*	French Guiana, Suriname
23. *Amblyomma parkeri*	N	Brazil
24. *Amblyomma parvum*	MFN	Argentina, Brazil, southern Mexico*, Panama*
25. *Amblyomma patinoi*	MF	Colombia
26. *Amblyomma pecarium*	N*	Costa Rica, Panama
27. *Amblyomma pseudoconcolor*	A	Argentina
28. *Amblyomma pseudoparvum*	MF	Argentina
29. *Amblyomma romitii*	ML	Brazil
30. *Amblyomma rotundatum*	F	Brazil
31. *Amblyomma sabanerae*	AN	Colombia, Panama
32. *Amblyomma scalpturatum*	MFN	Brazil
33. *Amblyomma sculptum*	MFN, L*	Argentina, Brazil
34. *Amblyomma tapirellum*	A, N*	Panama
35. *Amblyomma tigrinum*	MFN	Argentina, Brazil, French Guiana, Paraguay, Uruguay
36. *Amblyomma tonelliae*	MFN, L*	Argentina, Paraguay
37. *Amblyomma triste*	MF	Argentina, Uruguay, Venezuela
38. *Amblyomma variegatum*	U	Unknown Caribbean Island
39. *Amblyomma varium*	NL	Costa Rica, French Guiana, Panama
1. *Dermacentor imitans*	F	Colombia, Panama
2. *Dermacentor latus*	MF	Panama
3. *Dermacentor nitens*	MFNL	Bolivia, Brazil, Colombia, Venezuela

(continued)

Table 5.7 (continued)

1. *Haemaphysalis juxtakochi*	MFNL	Argentina, Brazil, French Guiana, Panama, Uruguay, Venezuela
2. *Haemaphysalis leporispalustris*	U	Argentina
1. *Rhipicephalus microplus*	MFNL	Argentina, Bolivia, Brazil, Colombia, Costa Rica, El Salvador, southern Mexico, Peru, Venezuela
2. *Rhipicephalus sanguineus s.l.*	MFNL	Argentina, Brazil, Chile, Colombia, Curação, French Guiana, Guyana, southern Mexico, Panama, Peru, Uruguay, Venezuela

* Provisional diagnosis

Table 5.8 Ixodid species found on humans, arranged by country/territory

Country/territory	Tick species
Argentina	*Ixodes pararicinus*
	Amblyomma aureolatum, A. brasiliense, A. coelebs, A. dubitatum, A. hadanii, A. incisum, A. neumanni, A. ovale, A. parvum, A. pseudoconcolor, A. pseudoparvum, A. sculptum, A. tigrinum, A. tonelliae, A. triste
	Haemaphysalis juxtakochi, H. leporispalustris
	Rhipicephalus microplus, R. sanguineus s.l.
Belize	*Ixodes boliviensis*
	Amblyomma coelebs, A. maculatum, A. mixtum
Bolivia	*Dermacentor nitens*
	Rhipicephalus microplus
Brazil	*Amblyomma aureolatum, A. auricularium, A. brasiliense, A. cajennense, A. coelebs, A. dissimile, A. dubitatum, A. fuscum, A. incisum, A. latepunctatum, A. longirostre, A. naponense, A. oblongoguttatum, A. ovale, A. parkeri, A. parvum, A. romitii, A. rotundatum, A. scalpturatum, A. sculptum, A. tigrinum*
	Dermacentor nitens
	Haemaphysalis juxtakochi
	Rhipicephalus microplus, R. sanguineus s.l.
Chile	*Rhipicephalus sanguineus s.l.*
Colombia	*Amblyomma dissimile, A. mixtum, A. oblongoguttatum, A. ovale, A. patinoi, A. sabanerae*
	Dermacentor imitans, D. nitens
	Rhipicephalus microplus, R. sanguineus s.l.
Costa Rica	*Amblyomma coelebs, A. mixtum, A. ovale, A. pecarium, A. varium*
	Rhipicephalus microplus
Cuba	*Amblyomma mixtum*
Curação	*Rhipicephalus sanguineus s.l.*
Ecuador	*Amblyomma ovale*
El Salvador	*Rhipicephalus microplus*
French Guiana	*Amblyomma aureolatum, A. cajennense, A. humerale, A. latepunctatum, A. longirostre, A. naponense, A. oblongoguttatum, A. ovale, A. pacae, A. tigrinum, A. varium*
	Haemaphysalis juxtakochi

(continued)

Table 5.8 (continued)

	Rhipicephalus sanguineus s.l.
Guatemala	*Amblyomma mixtum*
Guyana	*Amblyomma cajennense, A. oblongoguttatum*
	Rhipicephalus sanguineus s.l.
Honduras	*Amblyomma mixtum*
Jamaica	*Amblyomma mixtum*
Mexico (southern)	*Ixodes boliviensis, I. tancitarius*
	Amblyomma coelebs, A. dissimile, A. inornatum, A. maculatum, A. mixtum, A. ovale, A. parvum
	Rhipicephalus microplus, R. sanguineus s.l.
Nicaragua	*Amblyomma mixtum*
Panama	*Ixodes boliviensis*
	Amblyomma dissimile, A. mixtum, A. naponense, A. oblongoguttatum, A. ovale, A. parvum, A. pecarium, A. sabanerae, A. tapirellum, A. varium
	Dermacentor imitans, D. latus
	Haemaphysalis juxtakochi
	Rhipicephalus sanguineus s.l.
Paraguay	*Amblyomma aureolatum, A. brasiliense, A. coelebs, A. incisum, A. ovale, A. tigrinum, A. tonelliae*
Peru	*Rhipicephalus microplus, R. sanguineus s.l.*
Suriname	*Amblyomma cajennense, A. oblongoguttatum, A. ovale, A. pacae*
Trinidad and Tobago	*Amblyomma calcaratum, A. mixtum*
Uruguay	*Amblyomma dubitatum, A. tigrinum, A. triste*
	Haemaphysalis juxtakochi
	Rhipicephalus sanguineus s.l.
Venezuela	*Amblyomma dissimile, A. longirostre, A. oblongoguttatum, A. ovale, A. triste*
	Dermacentor nitens
	Haemaphysalis juxtakochi
	Rhipicephalus microplus, R. sanguineus s.l.
Unknown Caribbean country	*Amblyomma variegatum*

Conclusions

Important new observations have been published since the last review of Neotropical hard ticks appeared in 2003, including several seminal changes in tick taxonomy and systematics. Thus, the former genus *Boophilus* has been relegated to a subgenus of *Rhipicephalus*, with the result that the economically important cattle tick formerly known as *Boophilus microplus* is now *Rhipicephalus* (*Boophilus*) *microplus*. Similarly, the horse tick previously classified by some workers as *Anocentor nitens* is currently included in the genus *Dermacentor*. During this 17-year period, conventional morphological diagnoses have increasingly yielded to molecular methods of identification, but the combination of both techniques has been helpful in resolving some taxonomic problems, while also shedding light on previously unrecognized issues and greatly improving descriptions of new or reinstated Neotropical tick taxa. Additionally, combined molecular and morphological methods have proved to be powerful tools for researchers interested in understanding the geographic distribution and host relationships of Neotropical ixodids. However, numerous morphological studies continue to be published, and these too have materially contributed to our understanding of tick taxonomy and ecology, as repeatedly demonstrated in the present analysis.

At this time, 137 species of Ixodidae are known to occur in the Neotropical Zoogeographic Region and, as expected, the numbers of species in all genera have increased since 2003. This is especially true for the genera *Amblyomma* and *Ixodes*, which currently comprise 67 and 55 species, respectively, and together represent 89% of all Neotropical Ixodidae, while *Dermacentor* is recognized as containing nine species in the Neotropics, *Haemaphysalis* three, and *Rhipicephalus* at least three species. Considerable difficulties attend the diagnosis of adults of several species that are morphologically close, and the correct morphological identification of immatures is sometimes all but impossible, but molecular taxonomy promises not only to facilitate diagnoses of nymphs and larvae, but to aid in identifying and separating members of species complexes.

A. A. Guglielmone et al., *Neotropical Hard Ticks (Acari: Ixodida: Ixodidae)*,
https://doi.org/10.1007/978-3-030-72353-8

In this regard, the paradigmatic Neotropical genus *Amblyomma* is of particular concern because the morphological definitions of several species are very often poor and sometimes confusing or contradictory (see *Amblyomma fulvum*). The case of *Amblyomma pictum* is unique because there are no illustrations of the female, but several tick species, such as *Amblyomma albopictum*, *Amblyomma auricularium*, *Amblyomma crassum*, *Amblyomma cruciferum*, *Amblyomma tapirellum*, and *Amblyomma torrei*, among others, require redescription because it appears that, in some cases, different workers classified dissimilar taxa under the same name. The cooperation of scientists at laboratories throughout the Neotropical Region and elsewhere may assist in resolving such problems through meticulous morphological and molecular analyses of specimens from diverse localities, thereby obtaining definitive results already seen in the case of the *Amblyomma cajennense* species complex (Beati et al. 2013; Nava et al. 2014a). Other questions may be answered by carefully redescribing type material, although, unfortunately, type specimens are no longer available for some problematic species, such as *Amblyomma auricularium*.

A significant number of species, including *Amblyomma incisum*, *Amblyomma oblongoguttatum*, *Amblyomma ovale*, *Amblyomma parvum*, *Ixodes affinis*, *Ixodes sigelos*, *Ixodes texanus*, *Ixodes uriae*, *Haemaphysalis juxtakochi*, and *Haemaphysalis leporispalustris* have been well defined morphologically, but molecular data, sometimes associated with morphological descriptions, indicate that more than one species may be included under these names. It is expected that ongoing taxonomic and biological studies will ultimately reveal such cryptic species, in which case some of the names listed here will become stand-ins for species complexes.

Rhipicephalis sanguineus sensu lato is a special case because under this umbrella name there is likely more than one species, found on dogs in the Neotropical Region and elsewhere. Resolution of this apparent species complex may prove to be particularly difficult because of the vast geographical distribution of the ticks involved and the inherent difficulties attending morphological identification of *Rhipicephalus* species, a challenge that will not be overcome without global cooperation among tick research laboratories. Walker et al. (2000) clearly demonstrated that the identification of *Rhipicephalus* species is not an easy task, particularly in Africa and South Asia, a point underscored by Horak et al. (2018) and further discussed in Guglielmone et al. (2020).

Neotropical Ixodidae have been systematically collected from southern Mexico to southern South America and on many islands of the region. A total of 101 species are exclusively found in the Neotropics, while 35 species occur in the Neotropical Region and elsewhere. Twenty-eight taxa are established in the Neotropical and Nearctic Regions. Seven species have different geographic distributions. Thus, *Amblyomma variegatum* is a Neotropical and Afrotropical species; *Ixodes auritulus* is found in the Neotropical, Nearctic, and Australasian Regions; *Amblyomma rotundatum* has been collected in the Neotropical and Nearctic Regions and on one Pacific island; *Rhipicephalus sanguineus sensu stricto* has been found in the Neotropical, Nearctic, and Palearctic Regions; *Rhipicephalus microplus* is a pantropical species; *Ixodes uriae* is regarded as a circumpolar species; and

Rhipicephalus sanguineus sensu lato has a worldwide range. Not many Neotropical species have an especially broad geographic distribution; in fact, only 26 taxa are found over the vastness of South and Central America, as well as in southern Mexico and on Caribbean Islands, and 16 of these widespread species occur in the Neotropical Region and elsewhere, representing almost half of the 35 species found in the Neotropics but also outside the region. Thus, the extensive Neotropical distribution of four invasive species—*Rhipicephalus microplus*, *Rhipicephalus sanguineus sensu stricto* (hypothetically treated as widely distributed in the region), *Rhipicephalus sanguineus sensu lato*, and *Dermacentor nitens*—is a matter of concern because all these ticks are of economic importance. Chile is outside the range of *Rhipicephalus microplus*, while neither that country nor Uruguay is infested with *Dermacentor nitens*. The cases of Chile and Uruguay are exceptional because of abiotic factors that are unfavorable to the nonparasitic periods in the life cycles of these ticks, but the absence of records of *Rhipicephalus microplus* and *Dermacentor nitens* in Neotropical territories north of Chile is probably a consequence of insufficient tick sampling rather than a reflection of these species' absence. In the case of *Rhipicephalus sanguineus sensu lato*, this last statement applies to the entire Neotropical Region, no part of which is unsuitable for the maintenance of dog tick populations, while *Rhipicephalus sanguineus sensu stricto* surely has a wider Neotropical range than currently recognized.

Another 12 ixodid species with ranges that extend beyond the Neotropical Region are also widespread within the Neotropics, namely *Ixodes affinis*, *Ixodes auritulus*, *Amblyomma auricularium*, *Amblyomma coelebs*, *Amblyomma dissimile*, *Amblyomma maculatum*, *Amblyomma mixtum*, *Amblyomma oblongoguttatum*, *Amblyomma ovale*, *Amblyomma rotundatum*, *Haemaphysalis juxtakochi*, and *Haemaphysalis leporispalustris*. The species of *Amblyomma* in this list are regarded as Neotropical intruders into the Nearctic Region, but the two species of *Haemaphysalis* are considered to have naturally entered the Neotropical Region from the Nearctic, while there is no evidence to support hypotheses concerning the origin of *Ixodes affinis* and *Ixodes auritulus*. The broad Neotropical distribution of these ticks embraces mainland South and Central America, and in most cases southern Mexico, but tick distribution in the Caribbean Islands is less uniform; thus, *Amblyomma dissimile* and *Amblyomma rotundatum* have been found on many islands, *Amblyomma mixtum* has been collected on four Caribbean Islands, *Haemaphysalis leporispalustris* has been detected on two islands, while Caribbean records for *Amblyomma auricularium*, *Amblyomma ovale*, and *Haemaphysalis juxtakochi* exist only for Trinidad and Tobago.

Only ten exclusively Neotropical ixodid species have especially broad ranges. Two of these species are *Ixodes* (*Ixodes boliviensis* and *Ixodes luciae*) and eight belong to *Amblyomma* (*Amblyomma calcaratum*, *Amblyomma geayi*, *Amblyomma longirostre*, *Amblyomma naponense*, *Amblyomma nodosum*, *Amblyomma pacae*, *Amblyomma parvum*, and *Amblyomma varium*), and most of them are thought to be South American ticks that later entered Central America and southern Mexico. The contribution of these wide-ranging species to the Caribbean ixodid fauna is rather poor, because while four species—*Ixodes luciae*, *Amblyomma auricularium*,

Amblyomma calcaratum, and *Amblyomma nodosum*—have been collected there, all occur only on Trinidad and Tobago, the Caribbean Islands closest to the South American mainland, which is their probable point of origin. With the exception of *Dermacentor nitens*, no species of *Dermacentor* is found throughout the Neotropical Region, and none have been found on Caribbean islands. Even so, the zoogeographic status of several of the 26 species currently thought to be widely distributed may change if, as presumed, some are eventually found to constitute more than one taxon.

Species of *Ixodes* and *Amblyomma* are widespread in the mainland Neotropics, but the highest numbers of endemic species for both genera are found in South America, which has the richest ixodid fauna not only within the Neotropical Region but throughout the Western Hemisphere. The genus *Ixodes* is not well represented on Neotropical islands, but insular *Amblyomma* includes five endemic species in the Caribbean and seven species found only in the small area encompassing the Galápagos Islands, a distribution situation that merits further investigation. The few Neotropical species of *Dermacentor* are concentrated in southern Mexico and Central America (*Dermacentor nitens* excepted), this being the only genus of Neotropical Ixodidae with a meager representation in South America.

The number of invasive ixodids that have been detected in the Neotropical Region is low compared with those in a Nearctic country like the USA (Burridge 2011) or Australasian New Zealand (Heath 2001; Heath and Hardwick 2011), just to cite two examples. Most cases of exotic ixodids in the Neotropics concern ticks collected on imported reptiles destined for the pet trade. However, there have also been recent reports of *Hyalomma* and *Rhipicephalus* species found on domestic animals imported to Brazil, underscoring the necessity for cooperation among Neotropical countries and territories to ensure consistent regional surveillance to impede and control these potentially harmful ticks.

The list of names incorrectly applied to ticks found or allegedly found in the Neotropical Region is long. Several names are currently categorized as *nomina dubia* or *incertae sedis*, and others are original names for Neotropical species whose taxonomic and nomenclatural status changed following scientific study. Many names are junior synonyms of valid species, but published data associated with them may be of value and should not be ignored during taxonomic revisions.

A more serious problem concerns published identifications of valid species that do not occur in the Neotropical Region. Some of these incorrect accounts resulted from a past lack of literature that would have enabled investigators to correctly separate species, e.g., *Ixodes angustus* (a Palearctic and Nearctic species), *Ixodes hexagonus* (Palearctic), and *Amblyomma americanum* (Nearctic), to cite a few cases. However, some recent uncritical reviews still list such species as established in the Neotropics, and these studies only serve to cloud our understanding of the regional tick fauna. Other cases claim to present evidence for the presence of ticks that are wholly unexpected in the Neotropical Region, such as *Dermacentor andersoni* (a Nearctic species), *Hyalomma* spp. (an Afrotropical, Oriental, and Palearctic genus), *Rhipicephalus turanicus* (a Palearctic species, but probably also established elsewhere), without providing taxonomic support for the authors' diagnoses, which frequently concern ticks of medical and veterinary importance. While it is obvious that any of these ticks might at some point be found in the Neotropics and perhaps become established in the region, it is also incumbent upon authors to provide convincing data for the occurrence of exotic ticks in the Neotropical Region. To cite another example, a supposed record from Colombia of *Amblyomma variegatum*, a tick species of great medical and economic importance that is generally believed to be restricted to some Caribbean Islands in the Neotropics, manifestly needs verification.

Mammals are the most important hosts for all but three extant genera of Ixodidae worldwide. The three exceptions are the Australasian genus *Archaeocroton* (monotypic), the Nearctic genus *Robertsicus* (monotypic), and the Afrotropical genus *Africaniella* (two species), which are parasites of Reptilia and important in tracing the natural history of the Ixodidae. The affinity of Ixodidae for mammals is a probable consequence of hard tick speciation in tandem with the radiation of mammals after the end of the Cretaceous period. The subfamily Rhipicephalinae is here considered to contain what are postulated to be the most recent genera (*Anomalohimalaya*, *Cosmiomma*, *Dermacentor*, *Hyalomma*, *Nosomma*, *Rhipicentor*, and *Rhipicephalus*), which arose during the Cenozoic Era, although this view is not accepted by all tick workers. With few exceptions, the species in these genera parasitize mammals as their principal hosts. Members of the Rhipicephalinae are not well represented in the Neotropical Region since only nine species of *Dermacentor* and three species of *Rhipicephalus* are currently found there, but all these species feed on mammals. Thus, Neotropical *Dermacentor* feed on Artiodactyla, Perissodactyla, and Rodentia as their principal hosts, while Artiodactyla and Carnivora are principal hosts for the species of *Rhipicephalus*. Consequently, Neotropical hosts other than mammals (Amphibia, Reptilia, and Aves) are parasitized by ticks belonging to the Prostriata group (*Ixodes*) and to the remaining non-rhipicephalid genera of the Metastriata (*Amblyomma* and *Haemaphysalis*).

But while Neotropical hosts other than mammals may be parasitized by species of *Ixodes*, *Amblyomma*, and/or *Haemaphysalis* (a genus with only two species whose hosts are known), it is important to stress that mammals remain the principal hosts of the majority of species in these genera. Thus, 39 of the 51 Neotropical species of *Ixodes* whose hosts are known parasitize mammals as their principal hosts, and these figures are 36 of 64 for ticks of the genus *Amblyomma*. However, the principal hosts for these two genera differ, with *Ixodes* characterized by many species feeding on Rodentia, and *Amblyomma* attacking a great variety of hosts belonging to various orders. In South America, where both genera prevail, most *Ixodes* and *Amblyomma* ticks feed on hosts that entered there just before or after the Great American Interchange, an indication that most tick species are relatively new here, when considered in evolutionary terms.

As emphasized in Chap. 5, Aves are important hosts for Neotropical *Ixodes*, this being the only ixodid genus containing species that feed exclusively on birds representing five different orders, an indication of a unique relationship with avian hosts that is not shared by other genera of Ixodidae in the Neotropics or elsewhere. It is speculated that this condition may reflect an ancient pre-Cenozoic origin for the Prostriata (*Ixodes*), whose ancestors fed on the predecessors of Aves—primeval birds. By contrast, Neotropical *Amblyomma* includes no species that feed exclusively on Aves, although this genus contains eight species whose adults feed on mammals but whose immature stages parasitize Passeriformes, a strategy probably related to host evolution during the Cenozoic.

It has already been noted several times that Amphibia (order Anura) and Reptilia (Squamata and Testudines) are the only significant hosts for several species of Neotropical *Amblyomma*, and that there are some unusual combinations in the associations of Mammalia + Reptilia (*Amblyomma fuscum* and *Amblyomma parvitarsum*), and Mammalia + Aves + Reptilia (*Amblyomma humerale*), indicating great plasticity in *Amblyomma* host relationships. This is one of the reasons for the success of this genus in colonizing Neotropical habitats in general, and South America in particular, *Amblyomma* being the only genus that is well represented on Neotropical islands. It has been postulated that *Amblyomma* may have arisen in southern South America, but the basal species for this genus remain unknown.

The tick-host relationships tabulated in this review yield some startling figures. Thus, a total of 1027 vertebrates have been found infested with ixodids in the Neotropics, representing 506 birds (49%), 319 mammals (31%), 179 reptiles (17%), and 23 anurans (2%). These figures seemingly indicate that birds are more important as hosts for Neotropical hard ticks than mammals—but they are not. The number of bird species found infested greatly surpasses the number of mammal species; moreover, 23 orders of Aves contain at least one species infested with Ixodidae. But only five avian orders contain principal hosts for Neotropical hard ticks, while 11 of the 13 orders of Neotropical mammals include species that are principal hosts of ixodids. And while many birds have been found infested with ixodids, most of those records are exceptional rather than usual infestations. Thus, mammals are confirmed as the most common hosts for Neotropical Ixodidae.

Human parasitism by ticks is assuming major importance as tick-borne diseases increase in number and distribution. Among Neotropical tick genera, *Ixodes*, *Dermacentor*, *Haemaphysalis*, and *Rhipicephalus* species appear to only occasionally parasitize humans. However, the picture differs markedly with respect to genus *Amblyomma*, because not only do many members of this genus attack people, but some eagerly feed on humans if provided the opportunity. Nevertheless, no cases of human parasitism by ticks have been reported in some Neotropical countries and territories, a situation that surely reflects a failure to publish medical information, rather than the actual absence of this global phenomenon.

Finally, the appendix to the present work provides a list of all species of tick hosts, arranged by class, order, and family, together with the tick species and parasitic stages found on them, the hope being that such a compilation will aid readers interested in tick-host specificity. Here too, however, users are urged to proceed cautiously, since the literature on vertebrate taxonomy is vast and dynamic, and some names included here may no longer be valid.

Appendix: List of Hosts for Neotropical Ixodidae with Tick Species and Parasitic Stages Found on Them

The hosts for the species of Neotropical Ixodidae are listed below by class (Amphibia, Reptilia, Aves, and Mammalia) followed by orders, families, and species within each class. Each host is accompanied by the hard tick species found on it, with tick stages abbreviated as follows: A = adult ticks of undetermined sex, M = male, F = female, N = nymph, and L = larva.

Amphibia

Anura

Bufonidae

Anaxyrus terrestris: *Amblyomma rotundatum* (L)
Peltophryne fustiger: *Amblyomma dissimile* (A)
Peltophryne peltocephala: *Amblyomma dissimile* (MFNL)
  Dermacentor nitens (F)
Rhaebo guttatus: *Amblyomma rotundatum* (FN)
Rhinella sp.: *Amblyomma argentinae* (MF), *A. goeldii* (M), *A. mixtum* (N)
Rhinella alata: *Amblyomma sabanerae* (N)
Rhinella arenarum: *Amblyomma fuscum* (MF)
Rhinella bergi: *Amblyomma dissimile* (MF), *A. rotundatum* (FN)
Rhinella crucifer: *Amblyomma rotundatum* (FN)
Rhinella diptycha: *Amblyomma rotundatum* (F)
Rhinella gildae: *Amblyomma rotundatum* (N)
Rhinella granulosa: *Amblyomma dissimile* (ANL), *A. rotundatum* (FN)
Rhinella icterica: *Amblyomma rotundatum* (FNL)
Rhinella jimi: *Amblyomma dissimile* (MF), *A. rotundatum* (FNL)

A. A. Guglielmone et al., *Neotropical Hard Ticks (Acari: Ixodida: Ixodidae)*,
https://doi.org/10.1007/978-3-030-72353-8

Rhinella major: *Amblyomma dissimile* (FN), *A. rotundatum* (F)
Rhinella margaritifera: *Amblyomma dissimile* (AN), *A. rotundatum* (FN)
Rhinella marina: *Amblyomma dissimile* (MFNL), *A. rotundatum* (FNL), *A. varium* (M)
Rhinella ornata: *Amblyomma rotundatum* (FN)
Rhinella pygmaea: *Amblyomma rotundatum* (F)
Rhinella schneideri: *Amblyomma dissimile* (MFN), *A. rotundatum* (FNL)

Leptodactylidae

Leptodactylus pentadactylus: *Amblyomma rotundatum* (F)
Physalaemus nattereri: *Amblyomma rotundatum* (F)

Pipidae

Pipa pipa: *Amblyomma rotundatum* (F)

Reptilia

Crocodilia

Alligatoridae

Caiman crocodilus: *Amblyomma dissimile* (MFN), *A. geayi* (M), *A. humerale* (M)
Rhipicephalus sanguineus s.l. (F)
Caiman latirostris: *Amblyomma fuscum* (M), *A. rotundatum* (F)
Paleosuchus palpebrosus: *Amblyomma rotundatum* (F)
Paleosuchus trigonatus: *Amblyomma dissimile* (MF), *A. humerale* (N), *A. rotundatum* (FN)

Crocodylidae

Crocodylus acutus: *Amblyomma dissimile* (F), *A. mixtum* (F)
Crocodylus moreletii: *Amblyomma dissimile* (F)

Squamata

Boidae

Boa constrictor: *Amblyomma argentinae* (MFN), *A. dissimile* (MFNL), *A. fuscum* (MF), *A. goeldii* (M), *A. nodosum* (MF), *A. rotundatum* (M-rarely, FNL), *A. varium* (MF)
Rhipicephalus sanguineus s.l. (F)

Boa nebulosa: *Amblyomma antillorum* (M)
Chilabothrus angulifer: *Amblyomma albopictum* (MFN), *A. quadricavum* (MF)
Chilabothrus striatus: *Amblyomma quadricavum* (M)
Chilabothrus strigilatus: *Amblyomma dissimile* (MNL)
Chilabothrus subflavus: *Amblyomma quadricavum* (F)
Corallus hortulanus: *Amblyomma dissimile* (MN), *A. rotundatum* (FN)
Corallus ruschenbergerii: *Amblyomma dissimile* (N)
Epicrates cenchria: *Amblyomma argentinae* (F), *A. dissimile* (F), *A. rotundatum* (FN)
Epicrates maurus: *Amblyomma dissimile* (MFN)
Eunectes murinus: *Amblyomma dissimile* (MFN), *A. fulvum* (M), *A. varium* (M)
Eunectes notaeus: *Amblyomma argentinae* (MF), *A. dissimile* (N)

Colubridae

Chironius carinatus: *Amblyomma dissimile* (A), *A. humerale* (N)
Chironius flavolineatus: *Amblyomma rotundatum* (N)
Chironius laevicollis: *Amblyomma dissimile* (MFN)
Chironius laurenti: *Amblyomma dissimile* (MF)
Chironius multiventris: *Amblyomma rotundatum* (FNL)
Chironius scurrulus: *Amblyomma dissimile* (MF), *A. rotundatum* (FN)
Drymarchon caudomaculatus: *Amblyomma dissimile* (N)
Drymarchon corais: *Amblyomma rotundatum* (N)
Leptophis ahaetulla: *Amblyomma rotundatum* (N)
Leptophis depressirostris: *Amblyomma sabanerae* (N)
Mastigodryas bifossatus: *Amblyomma dissimile* (M)
Mastigodryas boddaerti: *Amblyomma dissimile* (FN), *A. rotundatum* (N)
Oxybelis aeneus: *Amblyomma dissimile* (MNL)
Phrynonax poecilonotus: *Amblyomma dissimile* (MF)
Spilotes pullatus: *Amblyomma dissimile* (MFN), *A. geayi* (MF), *A. rotundatum* (FN)

Corytophanidae

Basiliscus basiliscus: *Amblyomma dissimile* (N)

Dactyloidae

Anolis auratus: *Amblyomma dissimile* (N)

Dipsadidae

Alsophis sibonius: *Amblyomma antillorum* (L)
Atractus guentheri: *Amblyomma rotundatum* (N)
Borikenophis portoricensis: *Amblyomma quadricavum* (M)
Clelia clelia: *Amblyomma fuscum* (F)
Cubophis cantherigerus: *Amblyomma quadricavum* (MF)
Dipsas indica: *Amblyomma dissimile* (FN)

Dipsas neuwiedi: *Amblyomma rotundatum* (FN)
Dipsas turgida: *Amblyomma rotundatum* (FN)
Erythrolamprus melanotus: *Amblyomma dissimile* (FL)
Erythrolamprus reginae: *Amblyomma dissimile* (N)
Erythrolamprus typhlus: *Amblyomma humerale* (N)
Helicops carinicaudus: *Amblyomma rotundatum* (F)
Helicops polylepis: *Amblyomma dissimile* (N)
Hydrodynastes gigas: *Amblyomma dissimile* (MFN), *A. rotundatum* (F)
Imantodes cenchoa: *Amblyomma dissimile* (N)
Leptodeira annulata: *Amblyomma rotundatum* (FN)
Leptodeira rhombifera: *Amblyomma dissimile* (N)
Lygophis dilepis: *Amblyomma rotundatum* (N)
Mussurana bicolor: *Amblyomma dissimile* (MN)
Oxyrhopus guibei: *Amblyomma rotundatum* (F)
Oxyrhopus melanogenys: *Amblyomma rotundatum* (N)
Oxyrhopus trigeminus: *Amblyomma rotundatum* (N)
Phalotris matogrossensis: *Amblyomma dissimile* (F)
Philodryas nattereri: *Amblyomma rotundatum* (N)
Philodryas olfersii: *Amblyomma rotundatum* (FN)
Phylodryas viridissima: *Amblyomma rotundatum* (N)
Phimophis guerini: *Amblyomma dissimile* (F)
Phimophis guianensis: *Amblyomma dissimile* (N)
Pseudoboa sp.: *Amblyomma dissimile* (N)
Pseudoboa neuwiedii: *Amblyomma dissimile* (F)
Pseudoboa nigra: *Amblyomma dissimile* (MFN), *A. rotundatum* (N)
Thamnodynastes gambotensis: *Amblyomma dissimile* (L)
Xenodon merremii: *Amblyomma dissimile* (MF), *A. rotundatum* (FN)
Xenodon severus: *Amblyomma dissimile* (A)
Xenodon werneri: *Amblyomma humerale* (N)
Xenopholis scalaris: *Amblyomma rotundatum* (N)

Elapidae

Micrurus sp.: *Amblyomma rotundatum* (FNL)
Micrurus averyi: *Amblyomma dissimile* (F)
Micrurus corallinus: *Amblyomma rotundatum* (N)
Micrurus ibiboboca: *Amblyomma rotundatum* (FN)
Micrurus lemniscatus: *Amblyomma dissimile* (MFN), *A. rotundatum* (N)

Gekkonidae

Hemidactylus mabouia: *Amblyomma dissimile* (N)

Iguanidae

Amblyrhynchus sp.: *Amblyomma williamsi* (MF)
Amblyrhynchus cristatus: *Amblyomma darwini* (MFNL)
Conolophus subcristatus: *Amblyomma boulengeri* (F), *A. usingeri* (MF), *A. williamsi* (MF)
Ctenosaura acanthura: *Amblyomma scutatum* (A), *A. varium* (F)
Ctenosaura bakeri: *Amblyomma dissimile* (MFN)
Ctenosaura pectinata: *Amblyomma dissimile* (M), *A. scutatum* (MFNL)
Ctenosaura similis: *Amblyomma dissimile* (F), *A. mixtum* (F), *A. scutatum* (MF)
Cyclura cornuta: *Amblyomma albopictum* (MF), *A. cruciferum* (MFN), *A. dissimile* (F)
Cyclura cychlura: *Amblyomma albopictum* (MFNL), *A. dissimile* (MFNL), *A. torrei* (MFNL)
Cyclura lewisi: *Amblyomma albopictum* (MFNL)
Cyclura nubila: *Amblyomma albopictum* (MF), *A. torrei* (MFN)
Cyclura pinguis: *Amblyomma antillorum* (MFNL)
Cyclura stejnegeri: *Amblyomma cruciferum* (MFN), *A. torrei* (M)
Iguana sp.: *Amblyomma auricularium* (F)
Iguana delicatissima: *Amblyomma antillorum* (MFNL)
Iguana iguana: *Amblyomma dissimile* (MFNL), *A. geayi* (F), *A. mixtum* (F), *A. rotundatum* (M-rarely, FN), *A. sabanerae* (MF), *A. sculptum* (F), *A. scutatum* (A)
Iguana rhinolopha: *Amblyomma dissimile* (MF)
Iguana tuberculata: *Amblyomma dissimile* (M), *A. scutatum* (MF)
"Land iguana": *Amblyomma usingeri* (MFNL)

Liolaemidae

Liolaemus alticolor: *Amblyomma parvitarsum* (L)
Liolaemus andinus: *Amblyomma parvitarsum* (L)
Liolaemus annectens: *Amblyomma parvitarsum* (NL)
Liolaemus copiapensis: *Amblyomma parvitarsum* (L)
Liolaemus eleodori: *Amblyomma parvitarsum* (NL)
Liolaemus jamesi: *Amblyomma parvitarsum* (L)
Liolaemus nigriceps: *Amblyomma parvitarsum* (L)
Liolaemus patriciaiturrae: *Amblyomma parvitarsum* (L)
Liolaemus pleopholis: *Amblyomma parvitarsum* (NL)
Liolaemus puritamensis: *Amblyomma parvitarsum* (N)

Phrynosomatidae

Phrynosoma sp.: *Amblyomma rotundatum* (FN)

Scincidae

Mabuya mabouya: *Amblyomma rotundatum* (NL)
Trachylepis atlantica: *Amblyomma rotundatum* (N)

Teiidae

Ameiva sp.: *Ixodes affinis* (NL)
Ameiva ameiva: *Amblyomma dissimile* (FN), *A. rotundatum* (F)
Ameiva bifrontata: *Amblyomma dissimile* (N)
Ameiva praesignis: *Amblyomma dissimile* (N)
Cnemidophorus gaigei: *Amblyomma dissimile* (N)
Kentropyx calcarata: *Amblyomma humerale* (N)
Kentropyx pelviceps: *Amblyomma humerale* (N)
Pholidoscelis fuscatus: *Amblyomma antillorum* (L)
Salvator sp.: *Amblyomma rotundatum* (L)
Salvator merianae: *Amblyomma fuscum* (M), *A. rotundatum* (N), *A. sculptum* (M)
Teius teyou: *Amblyomma argentinae* (NL)
Tupinambis teguixin: *Amblyomma dissimile* (MFNL), *A. fuscum* (M)

Tropiduridae

Leiocephalus carinatus: *Amblyomma albopictum* (NL), *A. dissimile* (N)
Leiocephalus macropus: *Amblyomma torrei* (M)
Microlophus sp.: *Amblyomma darwini* (A)
Microlophus albemarlensis: *Amblyomma darwini* (N)
Microlophus barringtonensis: *Amblyomma boulengeri* (F)
Microlophus delanonis: *Amblyomma boulengeri* (MF)
Plica plica: *Amblyomma humerale* (N), *A. rotundatum* (F)
Plica umbra: *Amblyomma humerale* (N)
Tropidurus sp.: *Amblyomma rotundatum* (M-rarely, FNL)
Tropidurus etheridgei: *Amblyomma argentinae* (N)
Tropidurus hispidus: *Amblyomma dissimile* (N), *A. rotundatum* (F)
Tropidurus spinulosus: *Amblyomma argentinae* (N)
Tropidurus torquatus: *Amblyomma rotundatum* (N)
Uranoscodon superciliosus: *Amblyomma humerale* (N), *A. rotundatum* (FN)

Viperidae

Bothrops alternatus: *Amblyomma rotundatum* (FN)
Bothrops asper: *Amblyomma dissimile* (MFNL), *A. rotundatum* (A)
Bothrops atrox: *Amblyomma dissimile* (MFNL), *A. rotundatum* (FN)
Bothrops erythromelas: *Amblyomma rotundatum* (N)
Bothrops insularis: *Amblyomma rotundatum* (FN)
Bothrops jararaca: *Amblyomma rotundatum* (FN)
Bothrops jararacussu: *Amblyomma rotundatum* (FN)
Bothrops lanceolatus: *Amblyomma rotundatum* (F)
Bothrops leucurus: *Amblyomma rotundatum* (FN)
Bothrops matogrossensis: *Amblyomma dissimile* (AL)
Bothrops moojeni: *Amblyomma rotundatum* (F)
Bothrops neuwiedi: *Amblyomma argentinae* (M), *A. rotundatum* (F)

Crotalus durissus: *Amblyomma argentinae* (MFN), *A. dissimile* (MFN), *A. rotundatum* (FN)
Lachesis muta: *Amblyomma dissimile* (MN), *A. goeldii* (MF), *A. rotundatum* (FNL)
Porthidium lansbergii: *Amblyomma dissimile* (M)

Testudines

Chelidae

Acanthochelys macrocephala: *Amblyomma dissimile* (MFN)
Hydromedusa tectifera: *Amblyomma rotundatum* (F)
Mesoclemmys tuberculata: *Amblyomma rotundatum* (F)
Mesoclemmys vanderhaegei: *Amblyomma rotundatum* (F)
Phrynops geoffroanus: *Amblyomma dissimile* (F)
Phrynops hilarii: *Amblyomma argentinae* (F)
Platemys platycephala: *Amblyomma rotundatum* (F)

Emydidae

Terrapene carolina: *Amblyomma mixtum* (F), *A. sabanerae* (MF)
Terrapene mexicana: *Amblyomma dissimile* (F), *A. sabanerae* (MF)
Trachemys callirostris: *Amblyomma dissimile* (MFN)
Trachemys dorbigni: *Amblyomma rotundatum* (F)
Trachemys scripta: *Amblyomma dissimile* (MFNL), *A. mixtum* (MF), *A. sabanerae* (M)

Geoemydidae

Heosemys annandalii: *Amblyomma rotundatum* (F)
Rhinoclemmys annulata: *Amblyomma crassum* (F), *A. sabanerae* (MFNL)
Rhinoclemmys areolata: *Amblyomma dissimile* (M), *A. mixtum* (F), *A. rotundatum* (F), *A. sabanerae* (MFNL)
Rhinoclemmys funerea: *Amblyomma sabanerae* (MFNL)
Rhinoclemmys melanosterna: *Amblyomma dissimile* (MFN), *A. sabanerae* (MFN)
Rhinoclemmys nasuta: *Amblyomma sabanerae* (MFNL)
Rhinoclemmys pulcherrima: *Amblyomma dissimile* (M), *A. sabanerae* (MFNL)
Rhinoclemmys punctularia: *Amblyomma sabanerae* (N)
Rhinoclemmys rubida: *Amblyomma dissimile* (M)

Kinosternidae

Kinosternon sp.: *Amblyomma sabanerae* (MFN)
Kinosternon acutum: *Amblyomma dissimile* (F)
Kinosternon leucostomum: *Amblyomma crassum* (F), *A. dissimile* (F)

Kinosternon scorpioides: *Amblyomma dissimile* (MFN), *A. rotundatum* (F), *A. sabanerae* (F)

Testudinidae

Chelonoidis sp.: *Amblyomma macfarlandi* (MF)
Chelonoidis becki: *Amblyomma usingeri* (MFNL)
Chelonoidis carbonarius: *Amblyomma crassum* (F), *A. dissimile* (F), *A. humerale* (MF), *A. rotundatum* (FN)
Chelonoidis chilensis: *Amblyomma argentinae* (MFNL)
Chelonoidis darwini: *Amblyomma pilosum* (MFNL)
Chelonoidis denticulata: *Amblyomma cajennense* (MF), *A. crassum* (F), *A. geayi* (MF), *A. humerale* (MFN), *Amblyomma rotundatum* (M-rarely, FNL)
Chelonoidis duncanensis: *Amblyomma pilosum* (MFNL)
Chelonoidis microphyes: *Amblyomma usingeri* (MFNL)
Chelonoidis porteri: *Amblyomma macfarlandi* (MFN)
Chelonoidis vanderburghi: *Amblyomma usingeri* (MFNL)
Chelonoidis vicina: *Amblyomma macfarlandi* (MFNL)

Aves

Accipitriformes

Accipitridae

Accipiter bicolor: *Amblyomma longirostre* (N)
Busarellus nigricollis: *Amblyomma sculptum* (A)
Buteo brachyurus: *Amblyomma sculptum* (N)
Buteo nitidus: *Amblyomma coelebs* (N)
Buteo platypterus: *Amblyomma longirostre* (L)
Buteogallus coronatus: *Amblyomma sculptum* (N)
Buteogallus lacernulatus: *Amblyomma coelebs* (N), *A. longirostre* (N), *A. nodosum* (N)
Buteogallus meridionalis: *Amblyomma sculptum* (AN)
Geranoaetus albicaudatus: *Rhipicephalus sanguineus s.l.* (N)
Harpia harpyja: *Amblyomma cajennense* (N)
Haemaphysalis juxtakochi (N)
Leptodon cayanensis: *Amblyomma parkeri* (N)
Parabuteo unicinctus: *Amblyomma sculptum* (N)
Rostrhamus sociabilis: *Amblyomma sculptum* (N)
Rupornis magnirostris: *Amblyomma aureolatum* (N), *A. calcaratum* (N), *A. longirostre* (NL), *A. mixtum* (A), *A. parkeri* (N), *A. sculptum* (N)
Haemaphysalis juxtakochi (N)
Rhipicephalus microplus (L)

Spizaetus melanoleucus: *Amblyomma brasiliense* (N), *A. longirostre* (N)
Spizaetus tyrannus: *Amblyomma parkeri* (N), *A. sculptum* (N)

Caprimulgiformes

Apodidae

Cypseloides fumigatus: *Ixodes paranaensis* (F)
Cypseloides senex: *Ixodes paranaensis* (FN)
Streptoprocne biscutata: *Ixodes paranaensis* (FNL)
Streptoprocne semicollaris: *Ixodes cuernavacensis* (F)
Streptoprocne zonaris: *Ixodes paranaensis* (FN)

Caprimulgidae

Nyctidromus albicollis: *Amblyomma longirostre* (MN)
Setopagis parvula: *Amblyomma sculptum* (N)

Steatornithidae

Steatornis caripensis: *Ixodes downsi* (F)

Trochilidae

Amazilia tzacatl: *Amblyomma longirostre* (N)
Aphantochroa cirrochloris: *Amblyomma longirostre* (N)
Campylopterus longipennis: *Amblyomma longirostre* (L)
Chlorostilbon lucidus: *Amblyomma longirostre* (N)
Clytolaema rubricauda: *Amblyomma longirostre* (N)
Eupetomena macroura: *Amblyomma longirostre* (N), *A. nodosum* (N)
Florisuga fusca: *Amblyomma longirostre* (N)
Leucochloris albicollis: *Amblyomma longirostre* (N)
Phaethornis eurynome: *Amblyomma calcaratum* (N)
Phaethornis guy: *Amblyomma longirostre* (L)
Phaethornis superciliosus: *Amblyomma calcaratum* (NL), *A. longirostre* (N)
Thalurania glaucopis: *Amblyomma longirostre* (NL), *A. parkeri* (L)

Cariamiformes

Cariamidae

Cariama cristata: *Amblyomma sculptum* (MFNL)

Cathartiformes

Cathartidae

Coragyps atratus: *Ixodes affinis* (F)
Amblyomma oblongoguttatum (N), *A. ovale* (A), *A. parvum* (MF), *A. sculptum* (N)

Charadriiformes

Haematopodidae

Haematopus ater: *Ixodes uriae* (N)

Laridae

Leucophaeus scoresbii: *Ixodes uriae* (F)

Stercorariidae

Stercorarius antarcticus: *Ixodes uriae* (F)

Columbiformes

Columbidae

Columba livia: *Rhipicephalus sanguineus s.l.* (F)
Columbina passerina: *Amblyomma variegatum* (L)
Columbina picui: *Amblyomma tigrinum* (L)
Columbina talpacoti: *Ixodes auritulus* (F)
Amblyomma longirostre (N), *A. nodosum* (N)
Geotrygon montana: *Amblyomma longirostre* (N), *A. nodosum* (N), *A. sabanerae* (L)
Leptotila cassinii: *Amblyomma geayi* (NL)
Leptotila rufaxilla: *Amblyomma longirostre* (N), *A. varium* (L)
Leptotila verreauxi: *Amblyomma coelebs* (L), *Amblyomma longirostre* (N), *A. nodosum (N)*, *A. parvum* (N)
Zenaida auriculata: *Amblyomma tigrinum* (NL)

Coraciiformes

Alcedinidae

Chloroceryle aenea: *Amblyomma longirostre* (N)

Momotidae

Baryphthengus martii: *Amblyomma calcaratum* (N), *A. geayi* (NL), *A. longirostre* (N), *A. naponense* (NL), *A. ovale* (NL)
Baryphthengus ruficapillus: *Amblyomma brasiliense* (N), *A. calcaratum* (N), *A. coelebs* (N), *A. longirostre* (NL), *A. nodosum* (N)
Momotus momota: *Amblyomma coelebs* (N), *A. humerale* (N), *A. longirostre* (L)

Cuculiformes

Cuculidae

Coccyzus merlini: *Amblyomma mixtum* (NL)
Crotophaga ani: *Amblyomma longirostre* (N)
Crotophaga major: *Amblyomma longirostre* (N), *A. nodosum* (N)
Guira guira: *Amblyomma tigrinum* (NL)
Piaya cayana: *Amblyomma longirostre* (N), *A. nodosum* (N)

Falconiformes

Falconidae

Caracara plancus: *Amblyomma sculptum* (AN)
Falco sparverius: *Amblyomma sculptum* (N)
Herpetotheres cachinans: *Amblyomma longirostre* (N)
Micrastur ruficollis: *Amblyomma longirostre* (N)
Milvago chimachina: *Amblyomma sculptum* (N)
Rhipicephalus microplus (A)
Phalcoboenus australis: *Ixodes auritulus* (FNL)

Galbuliformes

Bucconidae

Malacoptila panamensis: *Amblyomma calcaratum* (N), *A. coelebs* (N), *A. longirostre* (N), *A. varium* (N)
Malacoptila striata: *Amblyomma coelebs* (N), *A. longirostre* (NL)

Galbulidae

Galbula ruficauda: *Amblyomma longirostre* (L)

Galliformes

Cracidae

Crax blumenbachii: *Amblyomma sculptum* (F)
Crax fasciolata: *Amblyomma cajennense* (N), *A. sculptum* (NL)
Crax rubra: *Ixodes boliviensis* (FN)
Amblyomma oblongoguttatum (MN)
Mitu tuberosum: *Amblyomma cajennense* (NL), *A. naponense* (N), *A. ovale* (L)
Oreophasis derbianus: *Ixodes auritulus* (F)
Ortalis ruficauda: *Amblyomma mixtum* (N)
Penelope sp.: *Ixodes auritulus* (NL)
Penelope obscura: *Amblyomma brasiliense* (N), *A. longirostre* (F), *A. parkeri* (N), *A. sculptum* (N)
Haemaphysalis juxtakochi (L)
Penelope superciliaris: *Ixodes auritulus* (N)
Amblyomma brasiliense (N), *A. cajennense* (NL), *A. ovale* (A)

Odontophoridae

Callipepla californica: *Amblyomma tigrinum* (NL)
Odontophorus guttatus: *Ixodes auritulus* (F)

Phasianidae

"Caille": *Ixodes cornuae* (FN on tentative host)
Chicken: *Amblyomma mixtum* (NL)
Pheasant: *Ixodes boliviensis* (F)
Turkey: *Amblyomma mixtum* (N)

Gruiformes

Aramidae

Aramus guarauna: *Amblyomma sculptum* (N)

Rallidae

Anurolimnas viridis: *Amblyomma triste* (L)
Aramides saracura: *Amblyomma brasiliense* (N)
Laterallus albigularis: *Amblyomma longirostre* (L)
"Saracura": *Amblyomma ovale* (A)

Passeriformes

Cardinalidae

Cyanoloxia brissonii: *Amblyomma auricularium* (N), *A. longirostre* (N), *A. parvum* (NL), *A. romarioi* (L)
Cyanoloxia cyanoides: *Amblyomma calcaratum* (N), *A. geayi* (L), *A. longirostre* (NL), *A. nodosum* (N)
Habia fuscicauda: *Amblyomma longirostre* (N), *A. nodosum* (N), *A. varium* (N)
Habia gutturalis: *Amblyomma longirostre* (L)
Habia rubica: *Amblyomma longirostre* (NL), *A. nodosum* (N), *A. romarioi* (L)
Piranga flava: *Amblyomma longirostre* (NL)

Conopophagidae

Conopophaga lineata: *Ixodes auritulus* (NL)
Amblyomma aureolatum (NL), *A. calcaratum* (NL), *A. coelebs* (N), *A. longirostre* (NL), *A. nodosum* (NL), *A. ovale* (L), *A. parkeri* (L), *A. romarioi* (L)
Conopophaga melanops: *Amblyomma calcaratum* (NL), *A. longirostre* (L), *A. nodosum* (N), *A. parkeri* (L)

Corvidae

Cyanocorax chrysops: *Ixodes pararicinus* (NL)
Haemaphysalis juxtakochi (NL), *H. leporispalustris* (L)
Cyanocorax cristatellus: *Amblyomma longirostre* (N)

Cotingidae

Pyroderus scutatus: *Amblyomma longirostre* (N)
Querula purpurata: *Amblyomma longirostre* (N)
Rupicola rupicola: *Amblyomma oblongoguttatum* (F)

Formicariidae

Formicarius analis: *Amblyomma ovale* (N)

Fringillidae

Chlorophonia cyanea: *Amblyomma nodosum* (N)
Euphonia hirundinacea: *Amblyomma longirostre* (L)
Euphonia laniirostris: *Amblyomma geayi* (L), *A. longirostre* (NL)
Euphonia pectoralis: *Amblyomma longirostre* (NL)
Euphonia violacea: *Amblyomma longirostre* (N)
Euphonia xanthogaster: *Amblyomma longirostre* (N)

Furnariidae

Anabacerthia amaurotis: *Amblyomma longirostre* (N)
Anabacertia lichtensteini: *Amblyomma calcaratum* (N)
Anabazenops fuscus: *Amblyomma calcaratum* (NL), *A. longirostre* (NL), *A. nodosum* (NL), *A. parkeri* (L)
Aphrastura spinicauda: *Ixodes auritulus* (NL)
Asthenes baeri: *Amblyomma tigrinum* (L)
Asthenes fuliginosa: *Ixodes auritulus* (N)
Automolus leucophthalmus: *Amblyomma calcaratum* (N), *A. longirostre* (NL), *A. nodosum* (L), *A. romarioi* (L)
Automolus ochrolaemus: *Amblyomma longirostre* (L), *A. varium* (N)
Berlepschia rikeri: *Amblyomma longirostre* (N)
Campylorhamphus falcularius: *Amblyomma longirostre* (NL)
Campylorhampus trochilirostris: *Ixodes affinis* (NL) *Amblyomma longirostre* (N)
Certhiasomus stictolaemus: *Amblyomma longirostre* (L)
Certhiaxis cinnamomeus: *Amblyomma longirostre* (N)
Cinclodes antarcticus: *Ixodes auritulus* (F)
Cinclodes fuscus: *Ixodes auritulus* (FN)
Cinclodes patagonicus: *Ixodes auritulus* (N)
Clibanornis dendrocolaptoides: *Ixodes auritulus* (FNL)
Coryphistera alaudina: *Amblyomma tigrinum* (NL)
Cranioleuca obsoleta: *Amblyomma aureolatum* (NL)
Cranioleuca pallida: *Amblyomma aureolatum* (L)
Cranioleuca vulpina: *Amblyomma nodosum* (N)
Deconychura longicauda: *Amblyomma longirostre* (NL)
Dendrocincla anabatina: *Amblyomma longirostre* (N)

Dendrocincla fuliginosa: *Amblyomma longirostre* (NL), *A. naponense* (N), *A. varium* (NL)
Dendrocincla merula: *Amblyomma longirostre* (NL)
Dendrocincla turdina: *Amblyomma longirostre* (L), *A. parkeri* (L), *A. romarioi* (L), *A. sculptum* (L)
Dendrocolaptes certhia: *Amblyomma longirostre* (N)
Dendrocolaptes hoffmannsi: *Amblyomma geayi* (L), *A. humerale* (NL), *A. longirostre* (NL)
Dendrocolaptes picumnus: *Amblyomma longirostre* (N)
Dendrocolaptes platyrostris: *Amblyomma calcaratum* (N), *A. nodosum* (N)
Dendroplex picus: *Amblyomma longirostre* (NL), *A. nodosum* (NL)
Drymornis bridgesii: *Amblyomma tigrinum* (L)
Furnarius cristatus: *Amblyomma tigrinum* (NL)
Furnarius leucopus: *Amblyomma auricularium* (L), *A. triste* (L)
Furnarius rufus: *Ixodes auritulus* (F)
Amblyomma aureolatum (NL), *A. calcaratum* (N), *A. dubitatum* (L), *A. tigrinum* (NL), *A. triste* (NL)
Glyphorynchus spirurus: *Amblyomma calcaratum* (N), *A. geayi* (L), *A. longirostre* (NL), *A. nodosum* (N), *A. varium* (NL)
Hylexetastes perrotii: *Amblyomma calcaratum* (NL), *A. geayi* (L), *A. longirostre* (NL)
Lepidocolaptes angustirostris: *Amblyomma longirostre* (N), *A. nodosum* (N)
Lepidocolaptes falcinellus: *Amblyomma romarioi* (L)
Lepidocolaptes souleyetii: *Amblyomma longirostre* (NL)
Lepidocolaptes squamatus: *Amblyomma longirostre* (L), *A. nodosum* (L), *A. parkeri* (L)
Lochmias nematura: *Ixodes auritulus* (F)
Megaxenops parnaguae: *Amblyomma auricularium* (N)
Phacellodomus ruber: *Amblyomma triste* (NL)
Phacellodomus striaticollis: *Ixodes auritulus* (N), *I. fuscipes* (NL)
Philydor atricapillus: *Amblyomma longirostre* (N), *A. parkeri* (L)
Philydor rufum: *Amblyomma longirostre* (NL)
Pseudoseisura lophotes: *Amblyomma tigrinum* (NL)
Pygarrhichas albogularis: *Ixodes auritulus* (L)
Sclerurus sp.: *Amblyomma humerale* (N)
Sclerurus caudacutus: *Amblyomma humerale* (L)
Sclerurus guatemalensis: *Amblyomma naponense* (N)
Sclerurus scansor: *Amblyomma romarioi* (L)
Sittasomus griseicapillus: *Amblyomma humerale* (N), *A. longirostre* (NL), *A. ovale* (L), *A. parkeri* (L), *A. romarioi* (L)
Synallaxis albescens: *Amblyomma tigrinum* (NL)
Synallaxis cinerascens: *Amblyomma aureolatum* (NL)
Synallaxis frontalis: *Ixodes silvanus* (L)
Synallaxis ruficapilla: *Ixodes auritulus* (NL)
Amblyomma aureolatum (N), *A. longirostre* (NL)

Synallaxis spixi: *Ixodes auritulus* (NL)
Amblyomma longirostre (N)
Syndactyla rufosuperciliata: *Ixodes auritulus* (N), *I. fuscipes* (NL), *I. pararicinus* (NL), *I. silvanus* (N)
Amblyomma calcaratum (N), *A. longirostre* (N)
Tarphonomus certhioides: *Amblyomma tigrinum* (NL)
Xenops minutus: *Amblyomma longirostre* (NL), *A. nodosum* (N), *A. romarioi* (L)
Xiphocolaptes albicollis: *Amblyomma calcaratum* (N), *A. longirostre*(N)
Xiphorhynchus elegans: *Amblyomma longirostre* (NL)
Xiphorhynchus fuscus: *Amblyomma longirostre* (NL), *A. romarioi* (L)
Xiphorhynchus guttatus: *Amblyomma humerale* (N), *A. longirostre* (NL), *A. nodosum* (N), *A. varium* (N)
Xiphorhynchus lachrymosus: *Amblyomma longirostre* (N)
Xiphorhynchus ocellatus: *Amblyomma calcaratum* (N)
Xiphorhynchus pardalotus: *Amblyomma geayi* (L), *A. longirostre* (L)
Xiphorhynchus susurrans: *Amblyomma longirostre* (NL), *A. nodosum* (N)

Icteridae

Agelaioides badiusy
Cacicus cela: *Amblyomma longirostre* (N)
Cacicus haemorrhous: *Amblyomma longirostre* (N), *A. nodosum* (N)
Cacicus uropygialis: *Amblyomma longirostre* (NL)
Curaeus curaeus: *Ixodes auritulus* (F)
Gnorimopsar chopi: *Rhipicephalus sanguineus s.l.* (N)
Icterus chrysater: *Amblyomma longirostre* (N), *A. nodosum* (N)
Icterus icterus: *Amblyomma longirostre* (N)
Leistes loyca: *Ixodes auritulus* (F)
Molothrus bonariensis: *Amblyomma nodosum* (N), *A. ovale* (L), *A. tigrinum* (NL), *A. varium* (L)
Quiscalus lugubris: *Amblyomma variegatum* (NL)

Mimidae

Mimus polyglottos: *Ixodes copei* (F)
Mimus saturninus: *Amblyomma calcaratum* (N)

Mitrospingidae

Mitrospingus cassinii: *Amblyomma coelebs* (N)

Parulidae

Basileuterus culicivorus: *Ixodes affinis* (L), *I. auritulus* (NL)
Amblyomma calcaratum (N), *A. longirostre* (NL), *A. nodosum* (N), *A. romarioi* (L)
Basileuterus rufifrons: *Amblyomma longirostre* (FNL)

Cardellina canadensis: *Amblyomma longirostre* (L)
Geothlypis aequinoctialis: *Ixodes auritulus* (L)
Amblyomma calcaratum (N), *A. longirostre* (N)
Myioborus brunniceps: *Ixodes pararicinus* (L)
Myiothlypis bivittata: *Ixodes pararicinus* (L), *I. silvanus* (L)
Haemaphysalis leporispalustris (L)
Myiothlypis flaveola: *Amblyomma longirostre* (NL), *A. nodosum* (N), *A. parvum* (NL)
Myiothlypis leucoblephara: *Ixodes affinis* (L), *I. auritulus* (L), *I. fuscipes* (L)
Amblyomma aureolatum (NL), *A. longirostre* (N)
Oporornis agilis: *Amblyomma dissimile* (L)
Parkesia noveboracensis: *Amblyomma ovale* (N)
Seiurus aurocapilla: *Ixodes auritulus* (L)
Setophaga castanea: *Amblyomma longirostre* (L)

Passerellidae

Arremon flavirostris: *Ixodes affinis* (L), *I. pararicinus* (L)
Amblyomma calcaratum (N), *A. nodosum* (N), *A. ovale* (NL)
Haemaphysalis juxtakochi (L), *H. leporispalustris* (L)
Arremon taciturnus: *Amblyomma calcaratum* (N)
Arremon torquatus: *Ixodes pararicinus* (L)
Arremon semitorquatus: *Amblyomma aureolatum* (N)
Haemaphysalis leporispalustris (NL)
Arremonops conirostris: *Amblyomma geayi* (N), *A. longirostre* (N), *A. nodosum* (N), *A. ovale* (F)
Haemaphysalis leporispalustris (L)
Atlapetes citrinellus: *Ixodes pararicinus* (L), *I. silvanus* (N)
Atlapetes pallidinucha: *Ixodes auritulus* (NL)
Atlapetes schistaceus: *Ixodes auritulus* (L)
Junco phaeonotus: *Ixodes mexicanus* (F)
Junco vulcani: *Ixodes auritulus* (N)
Rhynchospiza strigiceps: *Amblyomma tigrinum* (NL)
Zonotrichia capensis: *Ixodes auritulus* (NL), *I. silvanus* (N)
Amblyomma aureolatum (N), *A. auricularium* (N), *A. brasiliense* (L), *A. dubitatum* (L), *A. parvum* (NL), *A. tigrinum* (NL), *A. triste* (NL)
Haemaphysalis leporispalustris (NL)

Pipridae

Antilophia galeata: *Amblyomma longirostre* (NL)
Ceratopipra erythrocephala: *Amblyomma longirostre* (NL)
Ceratopipra mentalis: *Amblyomma calcaratum* (NL), *A. coelebs* (N), *A. geayi* (NL), *A. longirostre* (NL), *A. naponense* (N), *A. nodosum* (N), *A. varium* (N)
Chiroxiphia caudata: *Amblyomma brasiliense* (N), *A. calcaratum* (N), *A. longirostre* (NL), *A. nodosum* (NL), *A. parkeri* (L), *A. romarioi* (L)

Chiroxiphia linearis: *Amblyomma coelebs* (N), *A. longirostre* (L)
Chiroxiphia pareola: *Amblyomma cajennense* (N), *A. geayi* (L), *A. longirostre* (NL)
Dixiphia pipra: *Amblyomma calcaratum* (N), *A. geayi* (NL), *A. longirostre* (NL), *A. nodosum* (NL)
Ilicura militaris: *Amblyomma longirostre* (NL)
Lepidothrix coronata: *Amblyomma calcaratum* (NL), *A. longirostre* (NL), *A. naponense* (NL), *A. nodosum* (N)
Machaeropterus pyrocephalus: *Amblyomma longirostre* (N)
Machaeropterus regulus: *Amblyomma longirostre* (N)
Manacus candei: *Amblyomma longirostre* (NL), *A. maculatum* (N), *A. sabanerae* (N)
Manacus manacus: *Amblyomma calcaratum* (NL), *A. geayi* (L), *A. longirostre* (NL), *A. nodosum* (N), *A. parkeri* (L)
Manacus vitellinus: *Amblyomma geayi* (NL), *A. nodosum* (N), *A. varium* (NL)
Neopelma pallescens: *Amblyomma auricularium* (N), *A. longirostre* (NL), *A. varium* (N)
Pipra aureola: *Amblyomma geayi* (L), *A. longirostre* (NL)
Pipra fasciicauda: *Amblyomma calcaratum* (N), *A. longirostre* (NL), *A. nodosum* (N)
Pipra filicauda: *Amblyomma geayi* (N)

Polioptilidae

Microbates cinereiventer: *Amblyomma nodosum* (N)
Polioptila dumicola: *Amblyomma tigrinum* (NL)

Rhinocryptidae

Rhinocrypta lanceolata: *Amblyomma tigrinum* (N)
Scytalopus sp.: *Ixodes auritulus* (L)

Thamnophilidae

Biatas nigropectus: *Amblyomma longirostre* (N)
Cercomacra tyrannina: *Amblyomma longirostre* (L)
Cymbilaimus lineatus: *Amblyomma calcaratum* (N), *A. longirostre* (N), *A. nodosum* (N)
Drymophila malura: *Amblyomma calcaratum* (N)
Drymophila ferruginea: *Amblyomma longirostre* (N)
Drymophila rubricollis: *Amblyomma ovale* (N)
Drymophila squamata: *Amblyomma nodosum* (N)
Dysithamnus mentalis: *Amblyomma aureolatum* (N), *A. calcaratum* (N), *A. longirostre* (NL), *A. nodosum* (L)
Dysithamnus stictothorax: *Amblyomma longirostre* (N), *A. nodosum* (N)
Epinecrophylla leucophthalma: *Amblyomma longirostre* (N)
Formicivora grisea: *Amblyomma nodosum* (N), *A. ovale* (L)

Formicivora melanogaster: *Amblyomma longirostre* (N)
Formicivora rufa: *Amblyomma nodosum* (N)
Gymnopithys leucaspis: *Amblyomma auricularium* (N), *A. calcaratum* (N), *A. geayi* (L), *A. longirostre* (NL), *A. nodosum* (N), *A. sabanerae* (N), *A. varium* (NL)
Hylophylax naevioides: *Amblyomma longirostre* (N), *A. nodosum* (N)
Hylophylax naevius: *Amblyomma calcaratum* (N), *A. humerale* (L), *A. longirostre* (L)
Haemaphysalis juxtakochi (NL)
Hypocnemoides maculicauda: *Amblyomma longirostre* (NL), *A. nodosum* (N)
Mackenziaena severa: *Amblyomma calcaratum* (N), *A. longirostre* (N), *A. ovale* (N), *A. parkeri* (N)
Myrmelastes hyperythrus: *Amblyomma nodosum* (N)
Myrmelastes leucostigma: *Amblyomma humerale* (L)
Myrmoderus ferrugineus: *Amblyomma longirostre* (L)
Myrmoderus squamosus: *Amblyomma aureolatum* (N)
Myrmorchilus strigilatus: *Amblyomma auricularium* (N), *A. parvum* (N)
Myrmotherula axillaris: *Amblyomma dissimile* (N), *A. humerale* (N), *A. longirostre* (NL)
Myrmotherula longipennis: *Amblyomma longirostre* (L)
Myrmotherula schisticolor: *Amblyomma nodosum* (N)
Percnostola rufifrons: *Amblyomma geayi* (L), *A. longirostre* (NL)
Phaenostictus macleannani: *Amblyomma longirostre* (NL), *A. nodosum* (N)
Phlegopsis nigromaculata: *Amblyomma coelebs* (L), *A. longirostre* (L), *A. nodosum* (N)
Poliocrania exsul: *Amblyomma nodosum* (N), *A. ovale* (NL)
Haemaphysalis leporispalustris (NL)
Pyriglena leuconota: *Amblyomma coelebs* (L), *A. humerale* (L), *A. longirostre* (NL)
Pyriglena leucoptera: *Amblyomma aureolatum* (NL), *A. brasiliense* (N), *A. calcaratum* (N), *A. longirostre* (NL), *A. naponense* (N), *A. romarioi* (L)
Sakesphorus cristatus: *Amblyomma auricularium* (N), *A. parvum* (N)
Sciaphylax hemimelaena: *Amblyomma nodosum* (N)
Taraba major: *Amblyomma longirostre* (NL), *A. nodosum* (N), *A. parvum* (N)
Terenura maculata: *Amblyomma calcaratum* (N)
Thamnomanes caesius: *Amblyomma geayi* (NL), *A. humerale* (N), *A. longirostre* (L)
Thamnomanes schistogynus: *Amblyomma humerale* (N), *A. nodosum* (N)
Thamnophilus aethiops: *Amblyomma calcaratum* (N), *A. longirostre* (N)
Thamnophilus atrinucha: *Amblyomma calcaratum* (N), *A. dissimile* (NL), *A. geayi* (NL), *A. longirostre* (NL), *A. nodosum* (N), *A. oblongoguttatum* (N), *A. sabanerae* (N)
Thamnophilus bridgesi: *Amblyomma longirostre* (N)
Thamnophilus caerulescens: *Ixodes auritulus* (FN)
Amblyomma aureolatum (NL), *A. calcaratum* (NL), *A. longirostre* (MNL), *A. nodosum* (N), *A. parkeri* (L)
Haemaphysalis leporispalustris (L)

Thamnophilus doliatus: *Amblyomma auricularium* (N), *A. geayi* (L), *A. longirostre* (NL), *A. nodosum* (N), *A. parvum* (N)
Thamnophilus multistriatus: *Amblyomma longirostre* (N)
Thamnophilus nigriceps: *Amblyomma longirostre* (N)
Thamnophilus pelzelni: *Amblyomma calcaratum* (N), *A. coelebs* (N), *A. longirostre* (NL), *A. naponense* (N), *A. nodosum* (NL), *A. ovale* (N), *A. parvum* (N)
Thamnophilus punctatus: *Amblyomma longirostre* (NL)
Thamnophilus ruficapillus: *Ixodes auritulus* (FNL)
    Amblyomma aureolatum (L)
Thamnophilus schistaceus: *Amblyomma nodosum* (N)
Thamnophilus torquatus: *Amblyomma nodosum* (N)
Willisornis poecilonotus: *Amblyomma calcaratum* (N), *A. geayi* (L), *A. humerale* (NL), *A. longirostre* (NL)
    Haemaphysalis juxtakochi (L)

Thraupidae

Anisognathus igniventris: *Ixodes auritulus* (F)
Anisognathus lacrymosus: *Ixodes auritulus* (L)
Coereba flaveola: *Amblyomma longirostre* (NL)
    Rhipicephalus sanguineus s.l. (M)
Coryphospingus cucullatus: *Ixodes pararicinus* (L)
    Amblyomma calcaratum (N), *A. nodosum* (N), *A. tigrinum* (NL)
Coryphospingus pileatus: *Amblyomma auricularium* (L), *A. parvum* (L)
Cyanerpes cyaneus: *Amblyomma nodosum* (N)
Dacnis cayana: *Amblyomma longirostre* (NL), *A. parvum* (N)
Diglossa albilatera: *Ixodes auritulus* (L)
Diglossa plumbea: *Ixodes auritulus* (N)
Eucometis penicillata: *Amblyomma geayi* (L), *A. longirostre* (NL), *A. nodosum* (N)
Haplospiza unicolor: *Ixodes auritulus* (F)
    Amblyomma aureolatum (N), *A. longirostre* (N)
Herpsilochmus atricapillus: *Amblyomma nodosum* (N)
Iridosornis rufivertex: *Ixodes auritulus* (L)
Lophospingus pusillus: *Amblyomma tigrinum* (NL)
Microspingus cabanisi: *Ixodes auritulus* (NL)
Microspingus lateralis: *Ixodes auritulus* (L)
    Amblyomma aureolatum (N)
Microspingus torquatus: *Amblyomma tigrinum* (L)
Nemosia pileata: *Amblyomma longirostre* (N)
Nephelornis oneilli: *Ixodes auritulus* (L)
Paroaria coronota: *Amblyomma tigrinum* (N)
Phrygilus fruticeti: *Ixodes auritulus* (NL)
Pipraeidea bonariensis: *Ixodes auritulus* (FN)
Pipraeidea melanonota: *Amblyomma longirostre* (N)
Poospiza nigrorufa: *Ixodes auritulus* (NL)

Amblyomma aureolatum (N), *A. calcaratum* (N), *A. triste* (NL)
Ramphocelus bresilius: *Amblyomma longirostre* (NL), *A. nodosum* (N), *A. romarioi* (L), *A. tigrinum* (N)
Ramphocelus carbo: *Amblyomma calcaratum* (N), *A. longirostre* (NL), *A. nodosum* (N), *A. parvum* (N)
Ramphocelus sanguinolentus: *Amblyomma longirostre* (N)
Saltator atriceps: *Amblyomma longirostre* (L)
Saltator aurantiirostris: *Ixodes auritulus* (NL)
Amblyomma parvum (N), *A. tigrinum* (NL)
Saltator coerulescens: *Amblyomma nodosum* (N)
Saltator maximus: *Amblyomma longirostre* (NL), *A. nodosum* (N), *A. parkeri* (L)
Saltator similis: *Amblyomma aureolatum* (NL), *A. auricularium* (N), *A. calcaratum* (N), *A. longirostre* (NL), *A. nodosum* (N), *A. parvum* (L)
Haemaphysalis leporispalustris (N)
Saltator striatipectus: *Amblyomma nodosum* (N)
Saltatricula multicolor: *Amblyomma parvum* (NL), *A. tigrinum* (NL)
Sicalis flaveola: *Amblyomma longirostre* (N), *A. parvum* (L), *A. tigrinum* (L)
Sicalis luteola: *Amblyomma tigrinum* (L)
Sicalis olivascens: *Ixodes auritulus* (FN)
Sporophila angolensis: *Amblyomma longirostre* (N), *A. nodosum* (N)
Sporophila caerulescens: *Ixodes auritulus* (L)
Amblyomma aureolatum (N), *A. longirostre* (N), *A. nodosum* (N), *A. parvum* (N), *A. triste* (L)
Sporophila corvina: *Amblyomma longirostre* (N)
Sporophila frontalis: *Amblyomma calcaratum* (N)
Sporophila funerea: *Amblyomma calcaratum* (N)
Sprophila intermedia: *Amblyomma nodosum* (N)
Sporophila leucoptera: *Amblyomma longirostre* (N)
Sporophila lineola: *Amblyomma longirostre* (N), *A. nodosum* (N)
Sporophila nigricollis: *Amblyomma longirostre* (NL), *A. ovale (MN)*, *A. parvum* (N)
Stephanophorus diadematus: *Ixodes auritulus* (NL)
Tachyphonus coronatus: *Amblyomma aureolatum* (L), *A. brasiliense* (N), *A. calcaratum* (N), *A. longirostre* (NL), *A. naponense* (L), *A. nodosum* (N), *A. ovale* (L), *A. parkeri* (N), *A. romarioi* (L), *A. sculptum* (L)
Tachyphonus cristatus: *Amblyomma longirostre* (NL), *A. nodosum* (N)
Tachyphonus luctuosus: *Amblyomma longirostre* (N), *A. nodosum* (N)
Tachyphonus phoenicius: *Amblyomma nodosum* (N)
Tachyphonus rufus: *Ixodes affinis* (N)
Amblyomma longirostre (NL), *A. nodosum* (N), *A. parvum* (N)
Tachyphonus surinamus: *Amblyomma humerale* (NL), *A. longirostre* (L)
Tangara cayana: *Amblyomma calcaratum* (N), *A. longirostre* (N), *A. nodosum* (N)
Tangara peruviana: *Amblyomma longirostre* (N)
Tangara seledon: *Amblyomma longirostre* (N)
Tangara vassorii: *Ixodes auritulus* (L)
Tersina viridis: *Amblyomma longirostre* (NL)

Thlypopsis pyrrhocoma: *Amblyomma calcaratum* (N), *A. longirostre* (N)
Haemaphysalis juxtakochi (F)
Thlypopsis sordida: *Ixodes affinis* (N), *I. silvanus* (N)
Amblyomma longirostre (N)
Thraupis cyanocephala: *Ixodes auritulus* (F)
Thraupis ornata: *Amblyomma longirostre* (N), *A. nodosum* (N)
Thraupis palmarum: *Amblyomma calcaratum* (N), *A. longirostre* (NL), *A. nodosum* (N)
Thraupis sayaca: *Amblyomma calcaratum* (N), *A. longirostre* (NL), *A. nodosum* (N)
Tiaris bicolor: *Amblyomma variegatum* (L)
Tiaris fuliginosus: *Amblyomma calcaratum* (N), *A. parkeri* (L)
Tiaris olivaceus: *Amblyomma longirostre* (N)
Trichothraupis melanops: *Ixodes auritulus* (N)
Amblyomma aureolatum (NL), *A. brasiliense* (L), *A. calcaratum* (NL), *A. longirostre* (NL), *A. nodosum* (NL), *A. ovale* (L), *A. parkeri* (NL), *A. romarioi* (L)
Haemaphysalis leporispalustris (L)
Volatinia jacarina: *Ixodes bequaerti* (F)
Amblyomma tigrinum (L)
Haemaphysalis leporispalustris (L)
Xenospingus concolor: *Ixodes auritulus* (L)

Tityridae

Pachyramphus castaneus: *Amblyomma longirostre* (L)
Pachyramphus polychopterus: *Amblyomma longirostre* (NL), *A. nodosum* (N)
Pachyramphus validus: *Ixodes silvanus* (F)
Schiffornis turdina: *Amblyomma longirostre* (NL), *A. ovale* (NL)
Schiffornis virescens: *Amblyomma calcaratum* (N), *A. longirostre* (NL), *A. romarioi* (L)

Troglodytidae

Campylorhynchus gularis: *Ixodes mexicanus* (F)
Cantorchilus longirostris: *Amblyomma auricularium* (L), *A. longirostre* (N), *A. parvum* (N)
Cantorchilus modestus: *Amblyomma calcaratum* (L), *A. longirostre* (L)
Cantorchilus nigricapillus: *Ixodes auritulus* (L)
Amblyomma longirostre (NL), *A. nodosum* (N)
Haemaphysalis leporispalustris (L)
Cyphorhinus phaeocephalus: *Amblyomma longirostre* (N), *A. naponense* (NL), *A. ovale* (L), *A. varium* (NL)
Henicorhina leucosticta: *Amblyomma longirostre* (L)
Microcerculus marginatus: *Amblyomma calcaratum* (N), *A. longirostre* (N)
Pheugopedius atrogularis: *Amblyomma longirostre* (L)
Pheugopedius genibarbis: *Ixodes auritulus* (L)
Amblyomma nodosum (N)

Haemaphysalis leporispalustris (L)
Thryophilus rufalbus: *Amblyomma longirostre* (L)
Thryorchilus browni: *Ixodes auritulus* (FNL)
Troglodytes sp.: *Ixodes auritulus* (FN)
Troglodytes aedon: *Ixodes affinis* (N), *I. auritulus* (FNL), *I. minor* (L), *I. pararicinus* (L)
Amblyomma aureolatum (NL), *A. auricularium* (L), *A. calcaratum* (L), *A. longirostre* (NL), *A. nodosum* (L), *A. ovale* (N), *A. parkeri* (L), *A. parvum* (N), *A. tigrinum* (NL)
Haemaphysalis leporispalustris (L)

Turdidae

Catharus sp.: *Ixodes bequaerti* (F)
Catharus aurantiirostris: *Amblyomma varium* (N)
Catharus dryas: *Haemaphysalis leporispalustris* (L)
Catharus frantzii: *Ixodes minor* (L)
Catharus gracilirostris: *Ixodes auritulus* (L)
Catharus ustulatus: *Amblyomma calcaratum* (L), *A. longirostre* (NL), *A. nodosum* (N)
Hylocichla mustelina: *Amblyomma calcaratum* (N)
Turdus albicollis: *Ixodes auritulus* (FNL), *I. fuscipes* (NL)
Amblyomma aureolatum (NL), *A. calcaratum* (N), *A. coelebs* (N), *A. geayi* (L), *A. longirostre* (NL), *A. nodosum* (N), *A. ovale* (NL), *A. parkeri* (L), *A. romarioi* (L)
Haemaphsyalis juxtakochi (NL), *H. leporispalustris* (NL)
Turdus amaurochalinus: *Ixodes affinis* (NL), *I. auritulus* (FNL), *I. pararicinus* (NL)
Amblyomma aureolatum (NL), *A. auricularium* (NL), *A. calcaratum* (N), *A. coelebs* (N), *A. longirostre* (NL), *A. nodosum* (N), *A. ovale* (N), *A. parvum* (NL), *A. romarioi* (L), *A. tigrinum* (L)
Turdus falcklandii: *Ixodes auritulus* (FNL)
Turdus flavipes: *Amblyomma longirostre* (N), *A. romarioi* (L)
Turdus fumigatus: *Amblyomma longirostre* (NL)
Turdus grayi: *Amblyomma auricularium* (N), *A. calcaratum* (N), *A. longirostre* (NL), *A. nodosum* (N)
Turdus hauxwelli: *Amblyomma longirostre* (N)
Turdus ignobilis: *Amblyomma geayi* (N), *A. longirostre* (N), *A. nodosum* (N)
Turdus jamaicensis: *Ixodes copei* (L)
Turdus leucomelas: *Amblyomma aureolatum* (N), *A. cajennense* (L), *A. calcaratum* (N), *A. coelebs* (N), *A. geayi* (L), *A. longirostre* (NL), *A. nodosum* (NL), *A. ovale* (N), *A. romarioi* (L), *A. sculptum* (N)
Haemaphysalis juxtakochi (L)
Turdus nigrescens: *Ixodes auritulus* (FL)
Turdus nigriceps: *Ixodes auritulus* (FL), *I. pararicinus* (N), *I. silvanus* (FN)
Amblyomma longirostre (N)

Turdus nudigensis: *Amblyomma longirostre* (N)
Turdus obsoletus: *Amblyomma longirostre* (L)
Turdus plebejus: *Ixodes auritulus (FL)*
Turdus rufiventris: *Ixodes affinis* (NL), *I. auritulus* (FNL), *I. fuscipes* (NL), *I. pararicinus* (NL), *I. silvanus* (FNL)
Amblyomma aureolatum (NL), *A. auricularium* (N), *A. calcaratum* (N), *A. coelebs* (N), *A. dubitatum* (L), *A. longirostre* (NL), *A. nodosum* (NL), *A. ovale* (NL), *A. parvum* (N), *A. romarioi* (L), *A. triste* (L)
Haemaphysalis juxtakochi (NL), *H. leporispalustris* (NL)
Turdus sanchezorum: *Amblyomma longirostre* (N)
Turdus subalaris: *Amblyomma calcaratum* (N), *A. longirostre* (NL)

Tyrannidae

Attila rufus: *Amblyomma calcaratum* (L), *A. longirostre* (N), *A. naponense* (N), *A. romarioi* (L)
Attila spadiceus: *Amblyomma geayi* (L), *A. longirostre* (N)
Camptostoma obsoletum: *Amblyomma longirostre* (N)
Casiornis rufus: *Amblyomma longirostre* (N), *A. nodosum* (N), *A. parvum* (N)
Cnemotriccus fuscatus: *Amblyomma auricularium* (L), *A. calcaratum* (N), *A. longirostre* (N), *A. nodosum* (N), *A. parvum* (N)
Cnipodectes subbrunneus: *Haemaphysalis juxtakochi* (N)
Corythopis delalandi: *Haemaphysalis leporispalustris* (L)
Elaenia albiceps: *Ixodes auritulus* (L)
Elaenia chiriquensis: *Amblyomma nodosum* (N)
Elaenia cristata: *Amblyomma longirostre* (MN), *A. nodosum* (N)
Elaenia flavogaster: *Amblyomma longirostre* (NL), *A. nodosum* (N)
Elaenia mesoleuca: *Amblyomma longirostre* (N), *A. nodosum* (N)
Elaenia obscura: *Amblyomma nodosum* (N)
Elaenia parvirostris: *Amblyomma longirostre* (N)
Empidonax alnorum: *Amblyomma longirostre* (L), *A. nodosum* (N)
Empidonax virescens: *Amblyomma longirostre* (L)
Haemaphysalis leporispalustris (L)
Fluvicola nengeta: *Haemaphysalis leporispalustris* (L)
Hemitriccus margaritaceiventer: *Amblyomma calcaratum* (N), *A. longirostre* (N), *A. nodosum* (NL)
Hemitriccus minor: *Amblyomma longirostre* (L)
Hemitriccus nidipendulus: *Amblyomma longirostre* (N)
Knipolegus nigerrimus: *Ixodes auritulus* (F)
Lathrotriccus euleri: *Amblyomma calcaratum* (N), *A. longirostre* (N), *A. nodosum* (N)
Leptopogon amaurocephalus: *Amblyomma calcaratum* (N), *A. geayi* (L), *A. longirostre* (NL), *A. nodosum* (N), *A. parkeri* (L), *A. romarioi* (L)
Leptopogon superciliaris: *Amblyomma longirostre* (L)
Machetornis rixosa: *Amblyomma longirostre* (N)

Megarynchus pitangua: *Amblyomma longirostre* (N)
Mionectes macconnelli: *Amblyomma longirostre* (L)
Mionectes oleagineus: *Amblyomma calcaratum* (N), *A. geayi* (L), *A. longirostre* (NL)
Mionectes olivaceus: *Amblyomma longirostre* (L)
Mionectes rufiventris: *Amblyomma calcaratum* (N), *A. longirostre* (NL)
Myiarchus ferox: *Amblyomma calcaratum* (N), *A. longirostre* (NL), *A. ovale* (N), *A. parkeri* (N)
Myiarchus tuberculifer: *Amblyomma longirostre* (N)
Myiarchus tyrannulus: *Amblyomma calcaratum* (N), *A. nodosum* (N), *A. parvum* (NL)
Haemaphysalis leporispalustris (L)
Myiobius barbatus: *Amblyomma longirostre* (L)
Myiodynastes maculatus: *Amblyomma longirostre* (N)
Myiopagis viridicata: *Amblyomma longirostre* (L)
Myiophobus fasciatus: *Amblyomma longirostre* (L)
Myiozetetes similis: *Amblyomma longirostre* (N), *A. nodosum* (N)
Oncostoma olivaceum: *Amblyomma longirostre* (NL)
Onychorhynchus coronatus: *Amblyomma longirostre* (L)
Phylloscartes kronei: *Amblyomma longirostre* (N)
Phylloscartes ventralis: *Ixodes auritulus* (N)
Amblyomma longirostre (N)
Pitangus sulphuratus: *Amblyomma longirostre* (N), *A. parvum* (N), *A. tigrinum* (N)
Platyrinchus cancrominus: *Amblyomma longirostre* (L)
Platyrinchus coronatus: *Amblyomma longirostre* (N)
Platyrinchus leucoryphus: *Amblyomma longirostre* (L)
Platyrinchus mystaceus: *Amblyomma aureolatum* (L), *A. calcaratum* (NL), *A. longirostre* (NL), *A. nodosum* (NL), *A. parkeri* (L), *A. romarioi* (L)
Poecilotriccus latirostris: *Amblyomma calcaratum* (N), *A. longirostre* (L), *A. nodosum* (N)
Poecilotriccus plumbeiceps: *Amblyomma calcaratum* (NL)
Poecilotriccus sylvia: *Amblyomma longirostre* (L)
Ramphotrigon megacephalum: *Amblyomma calcaratum* (N), *A. nodosum* (N)
Rhynchocyclus brevirostris: *Amblyomma longirostre* (N)
Rhynchocyclus olivaceus: *Amblyomma geayi* (L), *A. humerale* (N), *A. longirostre* (NL), *A. nodosum* (N)
Stigmatura napensis: *Amblyomma auricularium* (N), *A. parvum* (N)
Todirostrum cinereum: *Amblyomma longirostre* (L)
Todirostrum maculatum: *Amblyomma nodosum* (N)
Tolmomyias poliocephalus: *Amblyomma longirostre* (L), *A. parkeri* (L)
Tolmomyias sulphurescens: *Amblyomma longirostre* (FNL), *A. nodosum* (N)
Tyrannus melancholicus: *Amblyomma longirostre* (N), *A. ovale* (N)
Tyrannus savana: *Amblyomma longirostre* (N), *A. parvum* (N)
Xolmis cinereus: *Amblyomma longirostre* (N)

Xolmis velatus: *Amblyomma longirostre* (N), *A. parvum* (N)
Zimmerius chrysops: *Amblyomma varium* (L)

Vireonidae

Cyclarhis gujanensis: *Amblyomma auricularium* (L), *A. calcaratum* (N), *A. longirostre* (N)
Hylophilus poicilotis: *Amblyomma calcaratum* (N)
Vireo chivi: *Amblyomma longirostre* (L)

Pelecaniformes

Ardeidae

Bubulcus ibis: *Amblyomma variegatum* (NL)
Rhipicephalus microplus (L)
Cochlearius cochlearius: *Amblyomma dissimile* (M)
"Heron": *Amblyomma mixtum* (FN)
Nycticorax nycticorax: *Ixodes auritulus* (FN)

Pelecanidae

Pelecanus occidentalis: *Amblyomma variegatum* (M)

Threskiornithidae

Theristicus caudatus: *Amblyomma sculptum* (N)

Piciformes

Capitonidae

Capito auratus: *Amblyomma longirostre* (N), *A. nodosum* (N)

Picidae

Celeus flavescens: *Amblyomma longirostre* (N)
Colaptes rupícola: *Ixodes auritulus* (F)
Dryobates maculifrons: *Amblyomma longirostre* (N)
Picumnus cirratus: *Amblyomma longirostre* (N)
Picumnus exilis: *Amblyomma longirostre* (NL)
Picumnus temminckii: *Amblyomma longirostre* (N)

Ramphastidae

Pteroglossus castanotis: *Amblyomma longirostre* (N)
Ramphastos dicolorus: *Amblyomma longirostre* (N)
Ramphastos vitellinus: *Amblyomma geayi* (N), *A. longirostre* (N)

Procellariiformes

Procellariidae

Daption capense: *Ixodes uriae* (N)
Pelecanoides magellani: *Ixodes auritulus* (FNL)

Psittaciformes

Psittacidae

Amazona aestiva: *Amblyomma sculptum* (N)

Rheiformes

Rheidae

Rhea americana: *Amblyomma dubitatum* (N), *A. sculptum* (F)
Rhea pennata: *Amblyomma parvitarsum* (F)

Sphenisciformes

Spheniscidae

Eudyptes chrysocome: *Ixodes uriae* (FNL)
Pygoscelis adeliae: *Ixodes uriae* (FNL)
Pygoscelis antarcticus: *Ixodes uriae* (FNL)
Pygoscelis papua: *Ixodes uriae* (FNL)
Spheniscus magellanicus: *Ixodes uriae* (FNL)
Amblyomma parvitarsum (F)

Strigiformes

Strigidae

Asio clamator: *Amblyomma cajennense* (A), *A. longirostre* (NL), *A. sculptum* (N), *A. tigrinum* (N)
Rhipicephalus sanguineus s.l. (A)
Asio stygius: *Amblyomma longirostre* (N)
Athene cunicularia: *Amblyomma auricularium* (N)
Megascops choliba: *Amblyomma dubitatum* (A), *A. geayi* (NL), *A. longirostre* (N), *A. parkeri* (N), *A. sculptum* (AN)
Pulsatrix koeniswaldiana: *Amblyomma longirostre* (N)
Pulsatrix perspicillata: *Amblyomma calcaratum* (N), *A. nodosum* (N)

Tytonidae

Tyto alba: *Amblyomma aureolatum* (N), *A. sculptum* (N)

Suliformes

Phalacrocoracidae

Phalacrocorax atriceps: *Ixodes uriae* (NL)

Tinamiformes

Tinamidae

Crypturellus tataupa: *Haemaphysalis leporispalustris* (N)
Eudromia elegans: *Amblyomma tigrinum* (L)
Nothoprocta pentlandii: *Ixodes auritulus* (FN)
Nothoprocta perdicaria: *Amblyomma tigrinum* (NL)
Nothura boraquira: *Amblyomma pseudoconcolor* (MNL)
Nothura maculosa: *Amblyomma pseudoconcolor* (MNL), *A. tigrinum* (N)
Rhynchotus rufescens: *Amblyomma pseudoconcolor* (MNL), *A. tigrinum* (N)

Trogoniformes

Trogonidae

Pharomachrus mocinno: *Ixodes bequaerti* (F)
Trogon collaris: *Ixodes bequaerti* (F)
Trogon viridis: *Amblyomma longirostre* (N), *A. nodosum* (N)

Mammalia

Artiodactyla

Bovidae

Buffalo (domestic): *Amblyomma dubitatum* (N), *A. mixtum* (MF), *A. parvum* (MF), *A. pseudoconcolor* (F), *A. sculptum* (MFNL)
Rhipicephalus microplus (MFN)
Cattle: *Ixodes affinis* (MF), *I. boliviensis* (MF), *I. fuscipes* (MF), *I. longiscutatus* (F), *I. pararicinus* (MF)
Amblyomma aureolatum (MF), *A. cajennense* (MF), *A. dissimile* (AN), *A. dubitatum* (N), *A. hadanii* (MFN), *A. maculatum* (MF), *A. mixtum* (MFNL), *A. neumanni* (MFNL), *A. oblongoguttatum* (MF), *A. parvitarsum* (A), *A. parvum* (MFN), *A. patinoi* (MF), *A. pseudoconcolor* (F), *A. pseudoparvum* (MFN), *A. sculptum* (MFNL), *A. tapirellum* (MF), *A. tenellum* (MF), *A. tigrinum* (MF), *A. tonelliae* (MFNL), *A. triste* (MF), *A. variegatum* (MFNL)
Dermacentor albipictus (F), *D. dissimilis* (ANL), *D. nitens* (MFNL)
Haemaphysalis juxtakochi (MFNL)
Rhipicephalus microplus (MFNL), *R. sanguineus s.l.* (MFN)
Goat: *Amblyomma mixtum* (M), *A. neumanni* (MFN), *A. oblongoguttatum* (A), *A. parvitarsum* (MF), *A. parvum* (MF), *A. pseudoparvum* (F), *A. tigrinum* (F), *A. triste* (MF), *A. variegatum* (MFNL)
Dermacentor nitens (F)
Haemaphysalis juxtakochi (MF)
Rhipicephalus microplus (MFNL), *R. sanguineus s.l.* (MFNL)
Sheep: *Amblyomma dissimile* (A), *A. maculatum* (MFNL), *A. mixtum* (MFN), *A. neumanni* (MN), *A. parvitarsum* (A), *A. parvum* (MF), *A. sculptum* (MF), *A. tenellum* (MF), *A. tigrinum* (MF), *A. variegatum* (MFNL)
Haemaphysalis juxtakochi (N)
Rhipicephalus microplus (MFNL)
Syncerus caffer: *Amblyomma cajennense* (MF)

Camelidae

Alpaca: *Amblyomma parvitarsum* (MF)

Haemaphysalis juxtakochi (F)
Lama guanicoe: *Amblyomma parvitarsum* (A)
Llama: *Amblyomma parvitarsum* (MF)
Vicugna vicugna: *Amblyomma parvitarsum* (MF)

Cervidae

Axis axis: *Amblyomma dubitatum* (MNL)
Haemaphysalis juxtakochi (FN)
Rhipicephalus microplus (FN)
Blastocerus dichotomus: *Amblyomma longirostre* (F), *A. ovale* (M), *A. sculptum* (MF), *A. tigrinum* (MF), *A. triste* (MF)
Dermacentor nitens (MF)
Rhipicephalus microplus (MFN)
Cervus elaphus: *Amblyomma mixtum* (MFNL)
Rhipicephalus microplus (ANL)
"Deer": *Amblyomma interandinum* (MF), *A. maculatum* (F)
Dermacentor dispar (MF)
Rhipicephalus sanguineus s.l. (MF)
Mazama sp.: *Ixodes montoyanus* (F)
Amblyomma tonelliae (M), *A. triste* (M)
Dermacentor imitans (M)
Mazama americana: *Amblyomma brasiliense* (NL), *A. coelebs* (L), *A. mixtum* (F), *A. neumanni* (MFNL), *A. oblongoguttatum* (F), *A. ovale* (MF), *A. sculptum* (MF)
Haemaphysalis juxtakochi (MFNL)
Rhipicephalus microplus (MF)
Mazama bororo: *Amblyomma incisum* (N)
Haemaphysalis juxtakochi (MFNL)
Mazama gouazoubira: *Ixodes affinis* (MF), *I. fuscipes* (MF)
Amblyomma aureolatum (MF), *A. brasiliense* (N), *A. cajennense* (N), *A. dubitatum* (FL), *A. incisum* (N), *A. naponense* (N), *A. neumanni* (MFNL), *A. ovale* (F), *A. parvum* (MFN), *A. pseudoconcolor* (A), *A. pseudoparvum* (MF), *A. sculptum* (MFN), *A. tigrinum* (M)
Dermacentor nitens (MFNL)
Haemaphysalis juxtakochi (MFNL)
Rhipicephalus microplus (MFNL)
Mazama nana: *Haemaphysalis juxtakochi* (MF)
Mazama rufina: *Ixodes fuscipes* (M)
Mazama temama: *Ixodes affinis* (MF)
Amblyomma calcaratum (M), *A. oblongoguttatum* (MFN), *A. pecarium* (A)
Haemaphysalis juxtakochi (M)
Odocoileus virginianus: *Ixodes affinis* (MF), *I. boliviensis* (F)
Amblyomma mixtum (M), *A. naponense* (A), *A. oblongoguttatum* (MFNL), *A. parvum* (MF), *A. tapirellum* (M)
Dermacentor nitens (MFNL)

Haemaphysalis juxtakochi (MFN)
Rhipicephalus microplus (MFNL)
Ozotoceros bezoarticus: *Amblyomma triste* (MF)
Haemaphysalis juxtakochi (F)
Rhipicephalus microplus (MFNL)
Pudu sp.: *Ixodes montoyanus* (MFNL)
Pudu mephistophiles: *Ixodes montoyanus* (FNL)
Pudu puda: *Ixodes stilesi* (MFNL), *I. taglei* (MF)

Suidae

Domestic pig: *Amblyomma cajennense* (MF), *A. coelebs* (A), *A. dubitatum* (MFNL), *A. maculatum* (MF), *A. mixtum* (AL), *A. naponense* (A), *A. neumanni* (MFN), *A. oblongoguttatum* (MFN), *A. ovale* (FN), *A. parvum* (AN), *A. rotundatum* (F), *A. scalpturatum* (FN), *A. sculptum* (MFNL), *A. tapirellum* (A), *A. triste* (MF), *A. variegatum* (MFNL)
Dermacentor nitens (A)
Haemaphysalis juxtakochi (F)
Rhipicephalus microplus (F), *R. sanguineus s.l.* (M)
Feral pig: *Amblyomma brasiliense* (MFN), *A. ovale* (MF), *A. pacae* (M), *A. parvum* (MFN), *A. sculptum* (MFN), *A. tigrinum* (MF)
Haemaphysalis juxtakochi (MFN)
Rhipicephalus microplus (F)

Tayassuidae

Catagonus wagneri: *Amblyomma boeroi* (MFNL), *A. neumanni* (F), *A. parvum* (MF), *A. tonelliae* (MF)
Rhipicephalus microplus (MF)
Pecari tajacu: *Amblyomma auricularium* (N), *A. brasiliense* (MF), *A. cajennense* (MFN), *A. coelebs* (A), *A. mixtum* (MFL), *A. naponense* (MFN), *A. neumanni* (MFN), *A. oblongoguttatum* (MFNL), *A. ovale* (A), *A. parvum* (A), *A. pecarium* (MFN), *A. scalpturatum* (AN), *A. sculptum* (MFN), *A. tapirellum* (MFN), *A. tonelliae* (MF)
Dermacentor imitans (MF)
Haemaphysalis juxtakochi (MFNL)
Rhipicephalus microplus (MFN)
"Peccary": *Amblyomma pacae* (MF), *A. tigrinum* (F)
Tayassu sp.: *Amblyomma varium* (M)
Tayassu pecari: *Amblyomma brasiliense* (MFNL), *A. cajennense* (MFN), *A. coelebs* (MFN), *A. dubitatum* (F), *A. latepunctatum* (N), *A. mixtum* (F), *A. naponense* (MFN), *A. oblongoguttatum* (MFN), *A. ovale* (A), *A. pecarium* (MF), *A. sculptum* (MFN), *A. tapirellum* (MF), *A. tonelliae* (MF)
Dermacentor dispar (MF), *D. imitans* (MF)
Haemaphysalis juxtakochi (F)
Rhipicephalus microplus (F)

Carnivora

Canidae

Canis latrans: *Amblyomma mixtum* (MFN), *A. oblongoguttatum* (MF), *A. ovale* (MFN), *A. parvum* (MF)
Haemaphysalis juxtakochi (N)
Rhipicephalus sanguineus s.l. (A)

Cerdocyon thous: *Amblyomma aureolatum* (MFN), *A. auricularium* (M), *A. brasiliense* (N), *A. cajennense* (N), *A. coelebs* (N), *A. dubitatum* (ANL), *A. fuscum* (F), *A. incisum* (N), *A. longirostre* (A), *A. maculatum* (F), *A. naponense* (A), *A. neumanni* (MF), *A. oblongoguttatum* (MN), *A. ovale* (MFN), *A. parvum* (MFN), *A. sculptum* (MFNL), *A. tigrinum* (MFN), *A. triste* (FN)
Dermacentor nitens (FN)
Haemaphysalis juxtakochi (N)
Rhipicephalus microplus (MFNL), *R. sanguineus s.l.* (MFN)

Chrysocyon brachyurus: *Amblyomma aureolatum* (MF), *A. auricularium* (MF), *A. brasiliense* (N), *A. dubitatum* (N), *A. ovale* (MF), *A. parvum* (A), *A. sculptum* (MFNL), *A. tigrinum* (MF), *A. triste* (MF)
Haemaphysalis juxtakochi (N)
Rhipicephalus microplus (M), *R. sanguineus s.l.* (A)

Domestic dog: *Ixodes affinis* (MF), *I. boliviensis* (MFN), *I. fuscipes* (F), *I. luciae* (MF), *I. pararicinus* (N)
Amblyomma aureolatum (MFN), *A. auricularium* (MNL), *A. brasiliense* (FNL), *A. cajennense* (MFN), *A. calcaratum* (F), *A. dubitatum* (MN), *A. fuscum* (L), *A. goeldii* (F), *A. hadanii* (N), *A. inornatum* (MF), *A. longirostre* (MF), *A. maculatum* (MFN), *A. mixtum* (MFNL), *A. naponense* (FN), *A. neumanni* (MFNL), *A. nodosum* (A), *A. oblongoguttatum* (MFNL), *A. ovale* (MFL), *A. pacae* (A), *A. parvum* (MFNL), *A. pictum* (M), *A. pseudoconcolor* (MF), *A. pseudoparvum* (F), *A. sabanerae* (M), *A. scalpturatum* (AN), *A. sculptum* (MFNL), *A. tapirellum* (A), *A. tenellum* (MF), *A. tigrinum* (MFN), *A. tonelliae* (M), *A. triste* (MFN), *A. varium* (A)
Dermacentor dissimilis (F), *D. latus* (M), *D. nitens* (MFN)
Haemaphysalis juxtakochi (MN)
Rhipicephalus microplus (MFNL), *R. sanguineus s.l.* (MFNL), *R. sanguineus s.s* (MFNL)

Feral dog: *Amblyomma variegatum* (MFNL)

Lycalopex culpaeus: *Amblyomma tigrinum* (MF)

Lycalopex griseus: *Amblyomma tigrinum* (MF)

Lycalopex gymnocercus: *Amblyomma aureolatum* (MF), *A. neumanni* (MFNL), *A. ovale* (MF), *A. parvum* (MF), *A. tigrinum* (MF)
Rhipicephalus microplus (N)

Lycalopex vetulus: *Amblyomma auricularium* (N), *A. sculptum* (N), *A. tigrinum* (MF), *A. triste* (MF)

Dermacentor nitens (MFN)
Rhipicephalus microplus (MFNL), *R. sanguineus s.l.* (FL)
Speothos venaticus: *Ixodes boliviensis* (MFN)
Amblyomma ovale (MF), *A. sculptum* (MF)
Urocyon cinereoargenteus: *Ixodes rubidus* (FN)
Amblyomma auricularium (N), *A. brasiliense* (N), *A. mixtum* (N), *A. oblongoguttatum* (MF), *A. ovale* (MN), *A. parvum* (M)
Rhipicephalus sanguineus s.l. (M)

Felidae

Domestic cat: *Ixodes affinis* (MF), *I. boliviensis* (F), *I. catarinensis* (F)
Amblyomma aureolatum (MF), *A. auricularium* (N), *A. mixtum* (MF), *A. oblongoguttatum* (MF), *A. ovale* (MFN), *A. parvum* (MF), *A. sculptum* (MF), *A. tapirellum* (MF), *A. tigrinum* (MF), *A. triste* (M)
Dermacentor nitens (MF)
Rhipicephalus microplus (A), *R. sanguineus s.l.* (MF)
Feral cat: *Amblyomma variegatum* (N)
Herpailurus yagouaroundi: *Amblyomma aureolatum* (A), *A. coelebs* (N), *A. oblongoguttatum* (AL), *A. ovale* (MF), *A. parvum* (M), *A. triste* (A)
Rhipicephalus sanguineus s.l. (F)
Leopardus colocolo: *Amblyomma triste* (N)
Rhipicephalus microplus (AN)
Leopardus geoffroyi: *Amblyomma aureolatum* (MF), *A. ovale* (MF), *A. parvum* (MF)
Leopardus guuttulus: *Amblyomma aureolatum* (F)
Leopardus pardalis: *Ixodes affinis* (MF), *I. boliviensis* (MF), *I. lasallei* (F)
Amblyomma aureolatum (MF), *A. ovale* (MFN), *A. parvum* (MF), *A. sculptum* (MFNL), *A. tigrinum* (F)
Rhipicephalus microplus (A)
Leopardus tigrinus: *Amblyomma aureolatum* (A), *A. longirostre* (N), *A. ovale* (F)
Dermacentor nitens (A)
Rhipicephalus microplus (A)
Leopardus wiedii: *Amblyomma aureolatum* (MF), *A. ovale* (F)
Rhipicephalus sanguineus s.l. (A)
Panthera onca: *Ixodes affinis* (MF), *I. boliviensis* (MF)
Amblyomma brasiliense (N), *A. cajennense* (FN), *A. coelebs* (MNL), *A. incisum* (N), *A. mixtum* (M), *A. oblongoguttatum* (F), *A. ovale* (MFN), *A. parvum* (MF), *A. scalpturatum* (N), *A. sculptum* (MFN), *A. tigrinum* (MFL), *A. triste* (MFL)
Dermacentor nitens (AN)
Rhipicephalus microplus (FN), *R. sanguineus s.l.* (F)
Puma concolor: *Ixodes affinis* (MF)
Amblyomma aureolatum (MFN), *A. brasiliense* (N), *A. coelebs* (ANL), A. dubitatum (N), *A. incisum* (N), *A. oblongoguttatum* (MFL), A. ovale (MFN),

A. parvum (MF), *A. sculptum* (MFN), *A. tigrinum* (FL), *A. tonelliae* (F), *A. triste* (ML)
Dermacentor nitens (MFN)
Haemaphysalis juxtakochi (F)
Rhipicephalus microplus (MFNL), *R. sanguineus s.l.* (M)

Herpestidae

Herpestes auropunctatus: *Amblyomma variegatum* (NL)
Rhipicephalus microplus (L)

Mephitidae

Conepatus semistriatus: *Ixodes rubidus* (FN)
Amblyomma auricularium (MFNL), *A. ovale* (F), *A. sculptum* (N)

Mustelidae

Eira barbara: *Ixodes rubidus* (F)
Amblyomma coelebs (NL), *A. longirostre* (M), *A. oblongoguttatum* (A), *A. ovale* (MFN)
Galictis sp.: *Amblyomma naponense* (N), *A. pseudoconcolor* (MN)
Galictis cuja: *Amblyomma aureolatum* (M), *A. auricularium* (MF), *A. ovale* (MF)
Rhipicephalus sanguineus s.l. (A)
Galictis vittata: *Amblyomma aureolatum* (A), *A. auricularium* (M), *A. ovale* (MF), *A. parvum* (MF)
Lontra longicaudis: *Amblyomma ovale* (MF)
Mustela frenata: *Ixodes guatemalensis* (F), *I. rubidus* (F)
Pteronura brasiliensis: *Amblyomma cajennense* (N)

Procyonidae

Bassaricyon gabbii: *Ixodes rubidus* (FN)
Bassariscus sp.: *Ixodes texanus* (FN)
Nasua narica: *Ixodes affinis* (MF), *I. rubidus* (F)
Amblyomma auricularium (M), *A. coelebs* (N), *A. dissimile* (F), *A. mixtum* (MF), *A. naponense* (F), *A. oblongoguttatum* (MFN), *A. ovale* (MFN), *A. parvum* (M), *A. pecarium* (N), *A. tapirellum* (MN), *A. varium* (F)
Haemaphysalis juxtakochi (MNL)
Nasua nasua: *Ixodes loricatus* (F)
Amblyomma aureolatum (M), *A. auricularium* (N), *A. brasiliense* (ANL), *A. cajennense* (NL), *A. coelebs* (NL), *A. dubitatum* (N), *A. humerale* (N), *A. incisum* (N), *A. naponense* (N), *A. oblongoguttatum* (N), *A. ovale* (MFNL), *A. parkeri* (F), *A. parvum* (MFN), *A. rotundatum* (F), *A. sculptum* (MFN), *A. tigrinum* (A)
Haemaphysalis juxtakochi (NL)
Rhipicephalus microplus (F), *R. sanguineus s.l.* (F)
Nasuella olivacea: *Ixodes montoyanus* (F)

Procyon sp.: *Amblyomma mixtum* (N), *A. naponense* (N), *A. oblongoguttatum* (N), *A. parvum* (MF)
Procyon cancrivorus: *Amblyomma aureolatum* (MF), *A. calcaratum* (A), *A. dubitatum* (N), *A. naponense* (F), *A. oblongoguttatum* (A), *A. ovale* (MFN), *A. parvum* (A), *A. sabanerae* (N), *A. sculptum* (MFN)
Procyon lotor: *Ixodes boliviensis* (MFNL), *I. rubidus* (FNL)
Amblyomma auricularium (MN), *A. oblongoguttatum* (F), *A. ovale* (MFN), *A. tapirellum* (FN)

Ursidae

Tremarctos ornatus: *Ixodes boliviensis* (F)

Chiroptera

Phyllostomidae

Anoura geoffroyi: *Ixodes downsi* (L)
Artibeus jamaicensis: *Amblyomma mixtum* (L)
Artibeus lituratus: *Amblyomma geayi* (L), *A. longirostre* (N)
Carollia perspicillata: *Amblyomma oblongoguttatum* (M), *A. tapirellum* (M)
Carollia subrufa: *Amblyomma parvum* (F)
Choeroniscus minor: *Amblyomma cajennense* (F), *A. rotundatum* (F)
Chrotopterus auritus: *Amblyomma humerale* (N)
Desmodus rotundus: *Ixodes pararicinus* (A)
Glossophaga soricina: *Amblyomma dubitatum* (F)
Macrophyllum macrophyllum: *Amblyomma calcaratum* (M)
Platyrrhinus umbratus: *Amblyomma calcaratum* (M)
Sturnira lilium: *Rhipicephalus microplus* (L)

Vespertilionidae

Myotis sp.: *Dermacentor panamensis* (N)
Myotis albescens: *Amblyomma triste* (N)
Myotis chiloensis: *Ixodes chilensis* (FN)

Cingulata

Chlamyphoridae

Cabassous centralis: *Amblyomma auricularium* (MF)
Cabassous tatouay: *Amblyomma auricularium* (MF), *A. pseudoconcolor* (MF)

Cabassous unicinctus: *Amblyomma cajennense* (M), *A. pseudoconcolor* (F), *A. sculptum* (FN)
Calyptophractus retusus: *Amblyomma pseudoconcolor* (MN)
Chaetophractus vellerosus: *Amblyomma auricularium* (MFN), *A. parvum* (A), *A. pseudoconcolor* (MFL)
Chaetophractus villosus: *Amblyomma auricularium* (MFN), *A. pseudoconcolor* (MF), *A. pseudoparvum* (MF)
Euphractus sexcinctus: *Amblyomma auricularium* (MFNL), *A. nodosum* (MF), *A. parvum* (MFN), *A. pseudoconcolor* (MF), *A. sculptum* (F)
Priodontes maximus: *Amblyomma parvum* (MF), *A. pseudoconcolor* (MF), *A. sculptum* (MFN)
Tolypeutes sp.: *Amblyomma pseudoconcolor* (MFL)
Tolypeutes matacus: *Amblyomma auricularium* (MFNL), *A. parvum* (MF), *A. pseudoconcolor* (MF), *A. pseudoparvum* (MF)
Haemaphysalis juxtakochi (M)
Tolypeutes tricinctus: *Amblyomma auricularium* (MFN), *A. sculptum* (MF)
Zaedyus pichiy: *Amblyomma auricularium* (MN), *A. pseudoconcolor* (MNL)

Dasypodidae

Dasypus hybridus: *Amblyomma auricularium* (M), *A. pseudoconcolor* (F)
Dasypus kappleri: *Amblyomma auricularium* (A), *A. parvum* (M)
Dasypus novemcinctus: *Ixodes loricatus* (F)
Amblyomma auricularium (MFNL), *A. beaurepairei* (MF), *A. cajennense* (MFN), *A. coelebs* (N), *A. fuscum* (MF), *A. humerale* (N), *A. mixtum* (FNL), *A. ovale* (MF), *A. parvum* (MF), *A. pseudoconcolor* (MF), *A. rotundatum* (F), *A. sculptum* (MFN)
Dermacentor dispar (MF)
Dasypus sabanicola: *Amblyomma auricularium* (MF)
Dasypus septemcinctus: *Amblyomma auricularium* (MFN), *A. fuscum* (MF)

Didelphimorphia

Didelphidae

Caluromys lanatus: *Ixodes luciae* (F)
Amblyomma geayi (N)
Caluromys philander: *Ixodes luciae* (L)
Amblyomma cajennense (N)
Chironectes minimus: *Ixodes loricatus* (F)
Amblyomma oblongoguttatum (ANL)
Didelphis sp.: *Ixodes rubidus* (F)
Amblyomma parvum (MF)
Didelphis albiventris: *Ixodes amarali* (F), *I. loricatus* (MFNL), *I. luciae* (MFNL)

Amblyomma aureolatum (M), *A. auricularium* (MFNL), *A. cajennense* (N), *A. coelebs* (NL), *A. dubitatum* (NL), *A. fuscum* (N), *A. ovale* (FNL), *A. parkeri* (N), *A. parvum* (NL), *A. pseudoconcolor* (N), *A. sculptum* (FNL)

Didelphis aurita: *Ixodes loricatus* (MFNL), *I. luciae* (MF)

Amblyomma aureolatum (MFN), *A. brasiliense* (NL), *A. coelebs* (NL), *A. dubitatum* (N), *A. fuscum* (NL), *A. incisum* (L), *A. ovale* (NL), *A. sculptum* (FN), *A. yucumense* (NL)

Haemaphysalis juxtakochi (NL)

Didelphis imperfecta: *Amblyomma cajennense* (N)

Didelphis marsupialis: *Ixodes affinis* (MF), *I. boliviensis* (F, N-near I. boliviensis), *I. loricatus* (MFL), *I. luciae* (MFNL), *I. tropicalis* (FNL), *I. venezuelensis* (NL)

Amblyomma auricularium (MFNL), *A. cajennense* (NL), *A. coelebs* (NL), *A. dissimile* (NL), *A. dubitatum* (N), *A. geayi* (MFNL), *A. humerale* (NL), *A. latepunctatum* (N), *A. longirostre* (L), *A. mixtum* (NL), *A. naponense* (NL), *A. oblongoguttatum* (N), *A. ovale* (NL), *A. pacae* (NL), *A. parkeri* (N), *A. romitii* (M), *A. sabanerae* (NL), *A. scalpturatum* (NL), *A. tapirellum* (N), *A. triste* (A), *A. varium* (FNL)

Haemaphysalis juxtakochi (NL)

Didelphis virginiana: *Amblyomma mixtum* (MFNL), *A. parvum* (MFNL)

Gracilinanus agilis: *Ixodes loricatus* (N)

Amblyomma auricularium (NL), *A. parvum* (N), *A. sculptum* (NL), *A. triste* (L)

Lutreolina crassicaudata: *Ixodes loricatus* (MFNL), *I. luciae* (F)

Amblyomma aureolatum (M), *A. coelebs* (N), *A. dubitatum* (N)

Marmosa sp.: *Ixodes loricatus* (FN), *I. venezuelensis* (L)

Amblyomma humerale (N)

Haemaphysalis juxtakochi (L)

Marmosa constantiae: *Ixodes luciae* (N)

Amblyomma humerale (N)

Marmosa demerarae: *Ixodes catarinensis* (F), *I. lasallei* (MF), *I. loricatus* (MFNL), *I. luciae* (FN)

Amblyomma dissimile (NL), *A. fuscum* (NL)

Marmosa mexicana: *Ixodes luciae* (N), *I. venezuelensis* (N)

Marmosa murina: *Ixodes loricatus* (NL), *I. luciae* (FNL)

Amblyomma dissimile (L)

Marmosa robinsoni: *Ixodes affinis* (L), *I. luciae* (MFNL)

Amblyomma geayi (L), *A. naponense* (N), *A. ovale* (NL), *A. sabanerae* (L)

Marmosops sp.: *Amblyomma humerale* (L)

Marmosops impavidus: *Ixodes luciae* (F)

Metachirus nudicaudatus: *Ixodes loricatus* (A), *I. luciae* (F)

Amblyomma cajennense (N), *A. coelebs* (L), *A. fuscum* (NL), *A. humerale* (NL), *A. mixtum* (L), *A. pacae* (L), *A. sabanerae* (NL)

Haemaphysalis juxtakochi (N), *H. leporispalustris* (N)

Monodelphis sp.: *Ixodes luciae* (A)

Monodelphis adusta: *Ixodes venezuelensis* (F)

Monodelphis americana: *Ixodes fuscipes* (NL), *I. loricatus* (L)
Monodelphis brevicaudata: *Ixodes amarali* (F), *I. luciae* (NL), *I. venezuelensis* (FNL)
Monodelphis dimidiata: *Ixodes fuscipes* (N), *I. loricatus* (NL)
Amblyomma dubitatum (NL), *A. triste* (NL)
Monodelphis domestica: *Ixodes amarali* (FNL)
Amblyomma auricularium (NL), *A. fuscum* (N), *A. parvum* (NL), *A. triste* (L)
Monodelphis glirina: *Ixodes amarali* (F)
Monodelphis kunsi: *Ixodes amarali* (FNL)
Monodelphis scalops: *Ixodes loricatus* (F)
Philander andersoni: *Ixodes luciae* (F)
Philander frenatus: *Ixodes loricatus* (FNL)
Amblyomma sculptum (N)
Philander opossum: *Ixodes loricatus* (MF), *I. luciae* (MFNL), *I. venezuelensis* (NL)
Amblyomma auricularium (MFNL), *A. dissimile* (N), *A. geayi* (N), *A. humerale* (NL), *A. oblongoguttatum* (N), *A. ovale* (N), *A. pacae* (NL), *A. sabanerae* (NL)
Thylamys cinderella: *Ixodes luciae* (N)
Thylamys karimii: *Amblyomma parvum* (L)
Thylamys macrurus: *Amblyomma parvum* (N)
Thylamys venustus: *Ixodes luciae* (N), *I. pararicinus* (L)
Tlacuatzin canescens: *Ixodes luciae* (N), *I. sinaloa* (N)

Eulipotyphla

Soricidae

Cryptotis goodwini: *Ixodes guatemalensis* (F)

Lagomorpha

Leporidae

Domestic rabbit: *Haemaphysalis leporispalustris* (MFN)
Rhipicephalus sanguineus s.l. (MFNL)
"Hare": *Amblyomma inornatum* (MF)
Lepus europaeus: *Amblyomma dubitatum* (NL)
Haemaphysalis leporispalustris (MF)
Rhipicephalus microplus (NL)
Romerolagus diazi: *Ixodes spinipalpis* (F)
Sylvilagus brasiliensis: *Ixodes pomerantzi* (MFN)
Amblyomma coelebs (N), *A. sculptum* (N)

Haemaphysalis juxtakochi (NL), *H. leporispalustris* (MFNL)
Sylvilagus dicei: *Ixodes dicei* (MF), *I. pomerantzi* (MFN)
Sylvilagus floridanus: *Ixodes pomerantzi* (M)
Amblyomma parvum (MF)
Haemaphysalis juxtakochi (F), *H. leporispalustris* (MFNL)
Sylvilagus gabbi: *Haemaphysalis leporispalustris* (FNL)

Microbiotheria

Microbiotheriidae

Dromiciops gliroides: *Ixodes neuquenensis* (FNL)

Paucituberculata

Caenolestidae

Caenolestes fuliginosus: *Ixodes jonesae* (L)

Perissodactyla

Equidae

Donkey: *Amblyomma cajennense* (MF), *A. maculatum* (M), *A. mixtum* (MFN), *A. neumanni* (MFN), *A. ovale* (M), *A. parvum* (F), *A. sculptum* (MF), *A. tenellum* (M)
Dermacentor nitens (MFNL)
Rhipicephalus microplus (F), *R. sanguineus s.l.* (A)
Horse: *Ixodes affinis* (MF), *I. boliviensis* (F), *I. fuscipes* (M), *I. longiscutatus* (F), *I. pararicinus* (MF)
Amblyomma auricularium (F), *A. cajennense* (MFN), *A. coelebs* (F), *A. dissimile* (L), *A. dubitatum* (MF), *A. hadanii* (FN), *A. incisum* (M), *A. longirostre* (A), *A. maculatum* (MF), *A. mixtum* (MFNL), *A. neumanni* (MFNL), *A. oblongoguttatum* (MFN), *A. ovale* (MF), *A. parvum* (MF), *A. sculptum* (MFNL), *A. tapirellum* (MF), *A. tenellum* (MF), *A. tigrinum* (MF), *A. tonelliae* (MFNL), *A. triste* (MF), *A. variegatum* (MFNL)
Dermacentor dissimilis (MFNL), *D. nitens* (MFNL)
Haemaphysalis juxtakochi (MFN)
Rhipicephalus microplus (MFNL), *R. sanguineus s.l.* (MF)

Mule: *Ixodes boliviensis* (F)
Amblyomma mixtum (MFN), *A. neumanni* (MFN), *A. ovale* (M), *A. sculptum* (MF)
Dermacentor nitens (MFNL)
Haemaphysalis juxtakochi (F)

Tapiridae

Tapirus sp.: *Ixodes bocatorensis* (F)
Tapirus bairdii: *Ixodes boliviensis* (F), *I. tapirus* (MF)
Amblyomma coelebs (MF), *A. mixtum* (MF), *A. oblongoguttatum* (FN), *A. ovale* (MF), *A. pacae* (F), *A. tapirellum* (MFN)
Dermacentor halli (MF), *D. latus* (MF), *D. nitens* (MF)
Haemaphysalis juxtakochi (MFN)
Tapirus pinchaque: *Ixodes tapirus* (F)
Amblyomma multipunctum (MFN)
Tapirus terrestris: *Amblyomma brasiliense* (MFN), *A. cajennense* (MFN), *A. calcaratum* (FNL), *A. coelebs* (MFN), *A. dubitatum* (F), *A. hadanii* (M), *A. incisum* (MFN), *A. latepunctatum* (MF), *A. naponense* (MFN), *A. oblongoguttatum* (MF), *A. ovale* (MF), *A. parvum* (MF), *A. pecarium* (M), *A. pseudoconcolor* (MF), *A. scalpturatum* (MF), *A. sculptum* (MFN), *A. triste* (MF)
Dermacentor nitens (F)
Haemaphysalis juxtakochi (MF)
Rhipicephalus microplus (MFN), *R. sanguineus s.l.* (F)

Pilosa

Bradypodidae

Bradypus torquatus: *Amblyomma longirostre* (F), *A. sculptum* (F), *A. varium* (MFN)
Bradypus tridactylus: *Ixodes bocatorensis* (F)
Amblyomma aureolatum (N), *A. cajennense* (MFN), *A. dissimile* (MFN), *A. geayi* (MFN), *A. goeldii* (MF), *A. humerale* (MF), *A. longirostre* (N), *A. varium* (MFN)
Rhipicephalus sanguineus s.l. (F)
Bradypus variegatus: *Amblyomma aureolatum* (F), *A. geayi* (MFN), *A. nodosum* (F), *A. rotundatum* (F), *A. sculptum* (F), *A. varium* (MF)

Choloepodidae

Choloepus didactylus: *Amblyomma dissimile* (MFN), *A. geayi* (MFN), *A. varium* (MF)
Choloepus hoffmanni: *Amblyomma calcaratum* (F), *A. geayi* (MF), *A. oblongoguttatum* (N), *A. varium* (MF)

Cyclopedidae

Cyclopes didactylus: *Ixodes bocatorensis* (F)
Amblyomma calcaratum (MF), *A. geayi* (N), *A. humerale* (N)

Myrmecophagidae

Myrmecophaga tridactyla: *Amblyomma auricularium* (MN), *A. brasiliense* (MFN), *A. cajennense* (MF), *A. calcaratum* (MFN), *A. dubitatum* (MFNL), *A. naponense* (MFN), *A. neumanni* (MN), *A. nodosum* (MF), *A. oblongoguttatum* (MF), *A. ovale* (F), *A. parvum* (MFN), *A. pictum* (MF), *A. pseudoconcolor* (MF), *A. scalpturatum* (MF), *A. sculptum* (MFNL), *A. tapirellum* (MF), *A. tonelliae* (MF), *A. triste* (MF)
Rhipicephalus microplus (MFNL)
Tamandua sp.: *Amblyomma pictum* (M)
Tamandua mexicana: *Amblyomma auricularium* (MF), *A. calcaratum* (MFN), *A. geayi* (F), *A. mixtum* (MFNL), *A. naponense* (FNL), *A. nodosum* (MF), *A. oblongoguttatum* (MFL), *A. parvum* (F)
Haemaphysalis juxtakochi (N)
Rhipicephalus sanguineus s.l. (F)
Tamandua tetradactyla: *Ixodes bocatorensis* (F)
Amblyomma aureolatum (F), *A. auricularium* (MFNL), *A. brasiliense* (N), *A. cajennense* (F), *A. calcaratum* (MFN), *A. coelebs* (N), *A. dubitatum* (NL), *A. geayi* (MF), *A. goeldii* (MF), *A. humerale* (MFN), *A. naponense* (N), *A. nodosum* (MF), *A. pacae* (F), *A. parvum* (MFN), *A. pseudoconcolor* (F), *A. rotundatum* (F), *A. sculptum* (MFN), *A. tonelliae* (F), *A. varium* (M)
Rhipicephalus sanguineus s.l. (MF)

Primates

Atelidae

Alouatta sp.: *Amblyomma parkeri* (F)
Alouatta caraya: *Amblyomma dubitatum* (N)
Alouatta guariba: *Amblyomma aureolatum* (MF), *A. dubitatum* (N), *A. ovale* (A), *A. parkeri* (N), *A. sculptum* (MFN)
Alouatta nigerrima: *Amblyomma geayi* (L)
Alouatta palliata: *Amblyomma geayi* (N)
Alouatta seniculus: *Rhipicephalus microplus* (F)

Callitrichidae

Mico melanurusk: Rhipicephalus sanguineus s.l. (M)
Saguinus bicolor: *Amblyomma geayi* (N)
Rhipicephalus sanguineus s.l. (F)

Saguinus geoffroyi: *Rhipicephalus sanguineus s.l.* (L)

Cebidae

Cebus capucinus: *Amblyomma oblongoguttatum* (ANL)
Cebus olivaceus: *Haemaphysalis juxtakochi* (M)
Sapajus apella: *Amblyomma ovale* (F), *A. parvum* (F)
Sapajus macrocephalus: *Amblyomma coelebs* (F)

Pitheciidae

Callicebus nigrifrons: *Amblyomma parkeri* (N), *A. romarioi* (MFN)

Rodentia

Abrocomidae

Abrocoma bennettii: *Ixodes abrocomae* (M), *I. sigelos* (N)

Capromyidae

Capromys pilorides: *Ixodes capromydis* (MFNL)

Caviidae

Cavia sp.: *Ixodes loricatus* (F)
Cavia aperea: *Ixodes longiscutatus* (FNL)
Amblyomma dubitatum (NL), *A. triste* (NL)
Rhipicephalus sanguineus s. l. (F)
Cavia tschudii: *Ixodes longiscutatus* (FNL)
Dolichotis patagonum: *Rhipicephalus sanguineus s.l.* (M)
Dolichotis salinicola: *Amblyomma auricularium* (FN), *A. pseudoparvum* (MFN)
Galea musteloides: *Amblyomma parvum* (NL), *A. tigrinum* (NL)
Galea spixii: *Amblyomma auricularium* (N), *A. parvum* (NL), *A. tigrinum* (N)
Hydrochoerus hydrochaeris: *Amblyomma aureolatum* (F), *A. brasiliense* (FN), *A. cajennense* (MF), *A. coelebs* (MFN), *A. dissimile* (MN), *A. dubitatum* (MFNL), *A. humerale* (N), *A. incisum* (M), *A. latepunctatum* (N), *A. naponense* (MN), *A. oblongoguttatum* (MF), *A. ovale* (MF), *A. pacae* (FN), *A. parvum* (F), *A. romitii* (MFN), *A. rotundatum* (N), *A. scalpturatum* (MN), *A. sculptum* (MFNL), *A. tigrinum* (MF), *A. tonelliae* (MF), *A. triste* (MFN)
Dermacentor nitens (MF)
Rhipicephalus microplus (MFN), *R. sanguineus s.l.* (FN)
Hydrochoerus isthmius: *A. auricularium* (M), *A. mixtum* (MFNL), *A. oblongoguttatum* (MF), *A. tapirellum* (M)
Kerodon rupestris: *Amblyomma parvum* (MF)

Chinchillidae

Lagidium viscacia: *Ixodes nuttalli* (MFNL)
Lagostomus maximus: *Amblyomma auricularium* (MFN), *A. dubitatum* (N), *A. parvum* (MF), *A. pseudoparvum* (MF), *A. sculptum* (L)

Cricetidae

Abrothrix longipilis: *Ixodes abrocomae* (M), *I. sigelos* (NL)
Abrothrix olivaceus: *Ixodes abrocomae* (M), *I. sigelos* (FNL)
Aegialomys galapagoensis: *Ixodes galapagoensis* (FN)
Aepeomys lugens: *Ixodes auritulus* (L)
Akodon sp.: *Ixodes fuscipes* (L)
Amblyomma fuscum (L), *A. ovale* (NL)
Akodon azarae: *Ixodes fuscipes* (N), *I. loricatus* (NL)
Amblyomm dubitatum (NL), *A. triste* (NL)
Haemaphysalis juxtakochi (L)
Akodon dolores: *Amblyomma tigrinum* (NL)
Akodon lutescens: *Ixodes longiscutatus* (L)
Akodon molinae: *Amblyomma tigrinum* (NL)
Akodon montensis: *Ixodes amarali* (N), *I. schulzei* (NL)
Amblyomma fuscum (N), *A. ovale* (N), *A. yucumense* (N)
Akodon spegazzini: *Ixodes sigelos* (N)
Amblyomma tigrinum (NL)
Bruceipattersonius sp.: *Ixodes fuscipes* (L)
Calomys sp.: *Ixodes loricatus* (NL)
Calomys callosus: *Ixodes luciae* (N), *I. pararicinus* (NL)
Amblyomma sculptum (N), *A. triste* (NL)
Calomys fecundus: *Ixodes pararicinus* (NL)
Calomys laucha: *Amblyomma tigrinum* (N)
Calomys musculinus: *Ixodes sigelos* (N)
Amblyomma tigrinum (L)
Calomys tener: *Amblyomma triste* (L)
Calomys sorellus: *Ixodes andinus* (NL)
Calomys venustus: *Amblyomma tigrinum* (NL)
Cerradomys maracajuensis: *Amblyomma dubitatum* (N)
Cerradomys marinhus: *Amblyomma parvum* (L), *A. triste* (L)
Cerradomys subflavus: *Ixodes amarali* (FNL), *I. loricatus* (F)
Amblyomma fuscum (N)
Chelemys macronyx: *Ixodes sigelos* (L)
Eligmodontia morgani: *Ixodes sigelos* (L)
Euneomys sp.: *Ixodes sigelos* (N)
Euneomys chinchilloides: *Ixodes sigelos* (L)
Euryoryzomys sp.: *Amblyomma humerale* (N)
Euryoryzomys russatus: *Ixodes loricatus* (N), *I. schulzei* (N)
Amblyomma fuscum (L), *A. ovale* (NL), *A. yucumense* (N)

Graomys chacoensis: *Amblyomma parvum* (L), *A. tigrinum* (NL)
Handleyomys alfaroi: *Ixodes sinaloa* (NL)
Holochilus sp.: *Amblyomma sculptum* (F)
Holochilus brasiliensis: *Ixodes loricatus* (NL)
Amblyomma triste (NL)
Holochilus sciureus: *Amblyomma ovale* (N), *A. sculptum* (N), *A. triste* (NL)
Hylaeamys megacephalus: *Ixodes amarali* (L), *I. loricatus* (NL), *I. luciae* (N)
Amblyomma dubitatum (N), *A. longirostre* (N), *A. ovale* (NL)
Hylaeamys perenensis: *Ixodes luciae* (N)
Hylaeamys yunganus: *Ixodes luciae* (N)
Ichthyomys pittieri: *Ixodes luciae* (F), *I. venezuelensis* (F)
Isthmomys flavidus: *Dermacentor panamensis* (NL)
Juliomys pictipes: *Ixodes fuscipes* (L)
Loxodontomys micropus: *Ixodes sigelos* (N)
Lundomys molitor: *Amblyomma dubitatum* (NL)
Melanomys caliginosus: *Ixodes venezuelensis* (F)
Amblyomma ovale (N)
Microryzomys minutus: *Ixodes auritulus* (L), *I. jonesae* (FL), *I. venezuelensis* (NL)
Neacomys spinosus: *Amblyomma coelebs* (N), *A. humerale* (N)
Necromys benefactus: *Amblyomma tigrinum* (NL)
Necromys lasiurus: *Ixodes amarali* (FN)
Amblyomma dubitatum (N), *A. ovale* (N), *A. triste* (NL)
Necromys obscurus: *Ixodes longiscutatus* (L)
Amblyomma triste (N)
Nectomys apicalis: *Ixodes nectomys* (M)
Nectomys rattus: *Amblyomma dubitatum* (NL), *A. fuscum* (N)
Nectomys squamipes: *Ixodes schulzei* (FNL)
Amblyomma ovale (L)
Neotoma sp.: *Ixodes woodi* (F)
Nephelomys albigularis: *Ixodes auritulus* (L), *I. tropicalis* (NL), *I. venezuelensis* (F)
Nephelomys devius: *Dermacentor panamensis* (N)
Nyctomys sumichrasti: *Ixodes luciae* (N)
Dermacentor panamensis (L)
Oecomys sp.: *Ixodes luciae* (A)
Oecomys auyantepui: *Ixodes luciae* (N)
Oecomys bicolor: *Ixodes luciae* (L)
Oecomys concolor: *Ixodes luciae* (NL)
Oecomys mamorae: *Amblyomma parvum* (N)
Oligoryzomys sp.: *Amblyomma dubitatum* (N)
Dermacentor panamensis (N)
Oligoryzomys delticola: *Ixodes fuscipes* (N)
Oligoryzomys destructor: *Ixodes pararicinus* (L)
Oligoryzomys eliurus: *Ixodes amarali* (N)
Oligoryzomys flavescens: *Ixodes longiscutatus* (L), *I. loricatus* (NL), *I. schulzei* (N)
Amblyomma dubitatum (L), *A. ovale* (NL), *A. triste* (NL)

Oligoryzomys fulvescens: *Amblyomma tigrinum* (AN)
Oligoryzomys longicaudatus: *Ixodes sigelos* (FL), *I. stilesi* (N)
Amblyomma tigrinum (L)
Oligoryzomys microtis: *Ixodes amarali* (N), *I. luciae* (N)
Oligoryzomys nigripes: *Ixodes amarali* (NL), *Ixodes fuscipes* (NL), *I. longiscutatus* (N), *I. loricatus* (NL), *I. schulzei* (N)
Amblyomma ovale (NL), *A. triste* (NL)
Oryzomys sp.: *Ixodes luciae* (N), *I. venezuelensis* (FL)
Amblyomma naponense (NL), *A. oblongoguttatum* (N)
Dermacentor panamensis (L)
Oryzomys couesi: *Ixodes sinaloa* (F)
Oxymycterus sp.: *Amblyomma fuscum* (L)
Oxymycterus delator: *Amblyomma auricularium* (N), *A. dubitatum* (N), *A. triste* (N)
Oxymycterus nasutus: *Ixodes longiscutatus* (NL), *I. loricatus* (L)
Amblyomma triste (NL)
Oxymycterus rufus: *Ixodes loricatus* (NL)
Amblyomma ovale (N), *A. triste* (NL), *A. yucumense* (N)
Haemaphysalis juxtakochi (L)
Peromyscus sp.: *Ixodes spinipalpis* (N), *I. tancitarius* (F)
Haemaphysalis leporispalustris (L)
Peromyscus aztecus: *Dermacentor panamensis* (NL)
Peromyscus grandis: *Ixodes guatemalensis* (F)
Peromyscus mexicanus: *Ixodes sinaloa* (NL), *I. venezuelensis* (N)
Dermacentor panamensis (NL)
Phyllotis darwini: *Ixodes andinus* (FNL, unconfirmed host, treated as *Phyllotis* sp.)
Phyllotis xanthopygus: *Ixodes abrocomae* (F), *I. sigelos* (FNL)
Amblyomma tigrinum (L)
Reithrodon auritus: *Ixodes sigelos* (F)
Reithrodontomys sp.: *Ixodes tancitarius* (F)
Reithrodontomys creper: *Dermacentor panamensis* (NL)
Reithrodontomys mexicanus: *Dermacentor panamensis* (NL)
Reithrodontomys sumichrasti: *Ixodes sinaloa* (N)
Rhipidomys sp.: *Amblyomma dubitatum* (N), *A. triste* (NL)
Rhipidomys cariri: *Ixodes loricatus* (NL)
Rhipidomys fulviventer: *Ixodes venezuelensis* (L)
Rhipidomys macrurus: *Amblyomma coelebs* (N), *A. sculptum* (N)
Rhipidomys mastacalis: *Ixodes amarali* (L)
Scapteromys aquaticus: *Ixodes loricatus* (NL)
Amblyomma dubitatum (L), *Amblyomma triste* (NL)
Scapteromys tumidus: *Ixodes longiscutatus* (NL), *I. loricatus* (NL)
Amblyomma dubitatum (NL), *A. triste* (NL)
Scotinomys teguina: *Dermacentor panamensis* (N)
Sigmodon sp.: *Ixodes luciae* (NL)

Amblyomma auricularium (N), *A. inornatum* (A), *A. maculatum* (FN), *A. ovale* (MN), *A. parvum* (MF)

Sigmodon peruanus: *Ixodes affinis* (F)

Sigmodontomys alfari: *Ixodes venezuelensis* (F)

Sooretamys angouya: *Ixodes catarinensis* (F)

Amblyomma ovale (N), *A. sculptum* (N), *A. yucumense* (N)

Tanyuromys aphrastus: *Ixodes sinaloa* (NL)

Thaptomys nigrita: *Ixodes amarali* (L)

Amblyomma ovale (N)

Thomasomys aureus: *Ixodes tropicalis* (F)

Thomasomys laniger: *Ixodes jonesae* (MFNL)

Transandinomys bolivaris: *Ixodes affinis* (N)

Amblyomma mixtum (NL), *A. ovale* (NL)

Transandinomys talamancae: *Ixodes venezuelensis* (F)

Amblyomma ovale (MN), *A. sabanerae* (NL)

Tylomys nudicaudus: *Dermacentor panamensis* (N)

Haemaphysalis juxtakochi (N)

Tylomys panamensis: *Ixodes venezuelensis* (F)

Zygodontomys sp.: *Ixodes loricatus* (F)

Zygodontomys brevicauda: *Ixodes luciae* (N), *I. venezuelensis* (F)

Amblyomma ovale (MNL)

Ctenomyidae

Ctenomys sp.: *Ixodes cooleyi* (F)

Amblyomma aureolatum (MFL)

Cuniculidae

Cuniculus paca: *Ixodes bocatorensis* (MF), *I. lasallei* (MF), *I. loricatus* (F), *I. luciae* (F), *I. spinosus* (F)

Amblyomma brasiliense (N), *A. cajennense* (MF), *A. coelebs* (FN), *A. mixtum* (A), *A. naponense* (MN), *A. oblongoguttatum* (AN), *A. ovale* (A), *A. pacae* (MFN), *A. scalpturatum* (AN)

Dermacentor nitens (MF)

Dasyproctidae

Dasyprocta sp.: *Ixodes spinosus* (FNL)

Amblyomma cajennense (F), *A. coelebs* (A), *A. naponense* (A), *A. ovale* (A)

Dasyprocta azarae: *Amblyomma auricularium* (L), *A. brasiliense* (NL), *A. cajennense* (N), *A. coelebs* (NL), *A. naponense* (N), *A. oblongoguttatum* (N), *A. ovale* (L), *A. parvum* (N), *A. scalpturatum* (N), *A. sculptum* (MFN)

Haemaphysalis juxtakochi (NL), *H. leporispalustris* (M)

Dasyprocta fuliginosa: *Ixodes lasallei* (F)

Amblyomma latepunctatum (N), *A. naponense* (N), *A. oblongoguttatum* (N), *A. pacae* (A), *A. scalpturatum* (N)
Haemaphysalis juxtakochi (NL)
Dasyprocta leporina: *Ixodes bocatorensis* (MF), *I. lasallei* (F), *I. venezuelensis* (NL)
Amblyomma dissimile (M), *A. naponense* (N), *A. varium* (M)
Dasyprocta punctata: *Ixodes bocatorensis* (F)
Amblyomma dissimile (M), *A. naponense* (N), *A. oblongoguttatum* (N), *A. ovale* (N), *A. pacae* (MF), *A. tapirellum* (N)
Haemaphysalis juxtakochi (N), *H. leporispalustris* (FL)
Dasyprocta variegata: *Ixodes lasallei* (F)
Myoprocta acouchy: *Ixodes venezuelensis* (N)
Amblyomma humerale (N), *A. naponense* (MFN)
Myoprocta pratti: *Ixodes spinosus* (F), *I. venezuelensis* (N)

Echimyidae

Clyomys laticeps: *Amblyomma auricularium* (N), *A. parvum* (N)
Dactylomys boliviensis: *Ixodes tropicalis* (F)
Euryzygomatomys spinosusy
Hoplomys gymnurus: *Amblyomma auricularium* (NL), *A. mixtum* (N), *A. naponense* (L), *A. ovale* (NL), *A. pacae* (L), *A. sabanerae* (NL)
Haemaphysalis juxtakochi (NL)
Proechimys sp.: *Ixodes venezuelensis* (NL)
Amblyomma geayi (N), *A. humerale* (N)
Proechimys canicollis: *Amblyomma ovale* (NL)
Proechimys guyannensis: *Ixodes venezuelensis* (N)
Amblyomma longirostre (N), *A. ovale* (N)
Haemaphysalis leporispalustris (N)
Proechimys quadruplicatus: *Amblyomma ovale* (N)
Proechimys roberti: *Amblyomma humerale* (L)
Proechimys semispinosus: *Ixodes venezuelensis* (N)
Amblyomma auricularium (NL), *A. dissimilee* (MNL), *A. geayi* (L), *A. mixtum* (N), *A. naponense* (NL), *A. oblongoguttatum* (NL), *A. ovale* (FNL), *A. pacae* (NL), *A. sabanerae* (NL), *A. varium* (NL)
Haemaphysalis juxtakochi (NL)
Thrichomys apereoides: *Amblyomma auricularium* (NL), *A. ovale* (N), *A. parvum* (NL), *A. triste* (NL)
Thrichomys inermis: *Amblyomma auricularium* (NL), *A. parvum* (N)
Thrichomys laurentius: *Amblyomma auricularium* (FN), A. fuscum (N)
Thrichomys pachyurus: *Amblyomma coelebs* (N), *A. dubitatum* (N), *A. ovale* (N), *A. sculptum* (N)
Trinomys dimidiatus: *Amblyomma longirostre* (F)
Trinomys setosus: *Ixodes amarali* (L)
Amblyomma ovale (N)

Erethizontidae

Chaetomys subspinosus: *Amblyomma longirostre* (MF)
Coendou sp.: *Amblyomma dissimile* (F), *A. parkeri* (L)
Coendou bicolor: *Amblyomma longirostre* (MF)
Coendou insidiosus: *Amblyomma aureolatum* (F), *A. longirostre* (MF)
Coendou mexicanus: *Ixodes boliviensis* (F)
Amblyomma auricularium (N), *A. longirostre* (MF)
Dermacentor panamensis (MFN)
Coendou nycthemera: *Amblyomma longirostre* (A)
Coendou prehensilis: *Amblyomma geayi* (M), *A. longirostre* (MFN), *A. parkeri* (MF), *A. sculptum* (F)
Coendou quichua: *Amblyomma longirostre* (MFN)
Haemaphysalis juxtakochi (AN)
Coendou spinosus: *Amblyomma dubitatum* (N), *A. longirostre* (MFN), *A. ovale* (M), *A. parkeri* (MFN), *A. sculptum* (FN)
Coendou vestitus: *Amblyomma longirostre* (M)

Geomyidae

Orthogeomys grandis: *Ixodes tecpanensis* (F)

Heteromyidae

Heteromys adspersus: *Dermacentor panamensis* (L)
Heteromys anomalus: *Ixodes venezuelensis* (FN)
Amblyomma ovale (N)
Heteromys desmarestianus: *Ixodes venezuelensis* (F)
Dermacentor panamensis (N)
Heteromys pictus: *Ixodes sinaloa* (F)
Heteromys salvini: *Ixodes sinaloa* (FN)

Muridae

Mus sp.: *Ixodes loricatus* (MF)
Mus musculus: *Ixodes venezuelensis* (F)
Amblyomma parvum (N), *A. variegatum* (L)
"Rat": *Ixodes luciae* (MF), *I. minor* (MF)
Rattus sp.: *Amblyomma mixtum* (L)
Rattus norvegicus: *Ixodes loricatus* (MFL), *I. sigelos* (N)
Rhipicephalus sanguineus s.l. (MNL)
Rattus rattus: *Ixodes amarali* (N), *I. loricatus* (N)
Amblyomma ovale (ANL), *A. tigrinum* (L), *A. variegatum* (L)
Rhipicephalus sanguineus s.l. (ANL)

Myocastoridae

Myocastor coypus: *Ixodes rubidus* (F)
Amblyomma dubitatum (N)

Octodontidae

Aconaemys fuscus: *Ixodes sigelos* (FNL)
Octodon degus: *Ixodes sigelos* (L)

Sciuridae

Sciurus sp.: *Ixodes boliviensis* (F), *I. spinipalpis* (N)
Sciurus aestuans: *Amblyomma aureolatum* (F), *A. fuscum* (N)
Haemaphysalis juxtakochi (N)
Sciurus aureogaster: *Ixodes guatemalensis* (F)
Amblyomma mixtum (M)
Sciurus granatensis: *Ixodes tiptoni* (MFNL)
Amblyomma longirostre (N)
Dermacentor panamensis (NL)
Sciurus igniventris: *Ixodes venezuelensis* (N)
Haemaphysalis juxtakochi (L)

References

Abarca K, López J, Acosta-Jamett G, aLepe P, Soares JF, Labruna MB (2012) A third *Amblyomma* species and the first tick-borne rickettsia in Chile. J Med Entomol 49:219–222

Abarca K, Gárate D, López J, Acosta-Jamett G (2016) Flea and tick species from dogs in urban and rural areas in four districts in Chile. Arch Med Vet 48:109–115

Acero EJ, Calixto OJ, Prieto AC (2011) Garrapatas (Acari: Ixodidae) prevalentes en caninos no migrantes [sic] del noroccidente de Bogotá, Colombia. Nova 9(15):158–165

Acevedo-Gutiérrez LY, Paternina LE, Pérez-Pérez JC, Londoño AF, López G, Rodas JD (2020) Garrapatas duras (Acari: Ixodidae) de Colombia, una revisión a su conocimiento en el país. Acta Biol Colomb 25:126–139

Acosta ICL, Martins TF, Marcili A, Soares HS, Krawczak FS, Vieira FT, Labruna MB (2016) Ticks (Acari: Ixodidae, Argasidae) from humans, domestic and wild animals in the State of Espírito Santo, Brazil, with notes on rickettsial infection. Vet Parasitol Reg Stud Rep 3–4:66–69

Acosta ICL, Martins TF, Nóbrega YC, Mantovani TC, Silva-Soares T, Santos MRD, Tadokoro CE, Labruna MB (2019) First record of *Amblyomma rotundatum* Koch, 1844 (Acari: Ixodidae) parasitizing wild *Caiman latirostris* (Reptilia: Crocodylidae) in the Atlantic rainforest biome, southeastern Brazil. Herpetol News 12:9–11

Aguilar-Domínguez M, Romero-Salas D, Sánchez-Montes S, Barradas-Piña F, Rosas-Saito G, Cruz-Romero A, Ibarra-Priego N, Becker I, Lohmeyer KH, Pérez de León A (2018) Occurrence of *Amblyomma mixtum* on the water buffalo (*Bubalus bubalis*) in Mexico. Int J Parasitol Parasites Wildl 7:405–408

Aguirre AAR, Rodrigues VS, Costa IN, García MV, Csordas BG, Andreotti R, Medeiros JF (2019) *Amblyomma scalpturatum* Neumann, 1906 (Acari: Ixodidae): confirmation in Acre State, Brazil, and description of parasitism in a human. Braz J Vet Parasitol 28:473–478

Aitken THG, Worth CB, Tikasingh ES (1968) Arbovirus studies in Bush Bush Forest, Trinidad, W. I., September 1959 – December 1964. III. Entomological studies. Am J Trop Med Hyg 17:253–268

Alanazi AD, Al-Mohammed HI, Alyousif MS, Said AE, Salim B, Abdel-Shafy S, Shaapan RM (2019) Species diversity and seasonal distribution of hard ticks (Acari: Ixodidae) infesting mammalian hosts in various districts of Riyadh Province, Saudi Arabia. J Med Entomol 56:1027–1032

Alcántara EP, Ferreira-Silva C, Ávila RW, Pacheco RC, Martins TF, Muñoz-Leal S, Morais DH (2018) Ticks (Acari: Argasidae and Ixodidae) infesting amphibians and reptiles in northeastern Brazil. Syst Appl Acarol 23:1497–1508

Aléssio FM, Dantas-Torres F, Siqueira DB, Lizée MH, Marvulo MFV, Martins TF, Labruna MB, Silva JCR, Mauffrey JC (2012) Ecological implications on the aggregation of *Amblyomma*

A. A. Guglielmone et al., *Neotropical Hard Ticks (Acari: Ixodida: Ixodidae)*,
https://doi.org/10.1007/978-3-030-72353-8

fuscum (Acari: Ixodidae) on *Thrichomys laurentius* (Rodentia: Echimyidae), in northeastern Brazil. Exp Appl Acarol 57:83–90

Allerdice MEJ, Snellgrove AN, Hecht JA, Hartzer K, Jones ES, Biggerstaff BJ, Ford SL, Karpathy SE, Delgado-de la Mora J, Delgado-de la Mora D, Licona-Enríquez JD, Goddard J, Levin ML, Paddock CD (2020) Reproductive incompatibility between *Amblyomma maculatum* (Acari: Ixodidae) group ticks from two disjuunct geographical regions within the USA. Exp Appl Acarol 82:543–557

Allred DM, Beck DE, White LD (1960) Ticks of the genus *Ixodes* in Utah. Brigham Young Univ Sci Bull Biol Ser 1(4):42 pp

Álvarez VC, Hernández VF, Hernández JG (2005) Catálogo de garrapatas suaves (Acari: Argasidae) y duras (Acari: Ixodidae) de Costa Rica. Brenesia 63–64:81–88

Álvarez DO, Corona-González B, Rodríguez-Mallón A, Rodríguez-González I, Alfonso P, Ramos AAN, Díaz-Sánchez AA, Navarrete MG, Fernández RR, Mellor LM, Catanese HN, Peláez M, Gainza YA, Marrero-Perera R, Roblejo-Arias L, Lobo-Rivero E, Silva CB, Fonseca AH, López ER, Cabezas-Cruz A (2020) Ticks and tick-borne diseases in Cuba, half a century of scientific research. Pathogens 9(article 616):24 pp

Alvarez-Robles E, Fuentes-Rousselin H, Meoño-Sánchez E, Recinos-Donis R, Figueroa L, Guerra-Centeno D (2018) Aproximación al estudio de los parásitos externos del venado cola blanca (*Odocoileus virginianus*) en la reserva natural de la Brigada Paracaidistas de San José, Escuintla, Guatemala. REDVET Rev Electrón 19 (artículo 5):11 pp

Amorim M, Serra-Freire NM (1994a) Descrição morfológico do estádio de larva de carrapato (Acari: Ixodidae). 4. *Amblyomma dissimile* Koch, 1844. Bol Mus Para Emilio Goeldi 10:273–288

Amorim M, Serra-Freire NM (1994b) *Amblyomma nodosum* Neumann, 1899 descrição morfológico do estádio de larva. Rev Bras Parasitol Vet 3:131–142

Amorim M, Serra-Freire NM (1995) Descrição morfológica do estádio de larva de carrapato (Acari: Ixodidae) 1. *Amblyomma rotundatum* Koch, 1844. Parasitol Día 19:9–19

Amorim M, Serra-Freire NM (1996) Morphological description of tick larval stage (Acari: Ixodidae). 3. *Amblyomma varius* [sic] Kock [sic], 1844. Entomol Vectores 3:67–80

Amorim M, Serra-Freire NM (1999a) Descrição morfológica do estádio de larva de carrapato (Acari: Ixodidae) 6. *Amblyomma cooperi* Nuttall & Warburton, 1907 [sic]. Entomol Vectores 6:26–155

Amorim M, Serra-Freire NM (1999b) Descrição morfológica do estádio de larva de carrapato (Acari: Ixodidae) 6. *Amblyomma geayi* Neumann, 1899. Entomol Vectores 6:517–542

Amorim M, Serra-Freire NM (2000a) Redescrição morfológica do estádio de larva de carrapato (Acari: Ixodidae) 9. *Amblyomma parvum* Aragão, 1908. Caderno Técnico-Científico da Escola de Medicina Veterinária 1(1):51–80

Amorim M, Serra-Freire NM (2000b) Morphological description of tick larval stage (Acari: Ixodidae) 7. *Amblyomma auricularium* (Conil, 1878). Entomol Vectores 7:297–309

Amorim M, Gazeta GS, Guerim L, Serra-Freire NM (1997) Morphological description of tick larval stage (Acari: Ixodidae). 5. *Anocentor nitens*. Braz J Vet Parasitol 6:143–156

Amorim M, Gazeta GS, Peralta AS, Teixeira RHF, Serra-Freire NM (1999) Ixodofauna de quelônios do Brasil. Rev Univ Rural Ser Ciência Vida 20:31–35

Amorim M, Gazeta GS, Bossi DEP, Linhares AX (2003) Carrapatos *Ixodes* (*Haemixodes*) *serrafreirei* sp. n. em roedores silvestres dos estados do Rio de Janeiro e de São Paulo. Entomol Vectores 10:407–410

Amorim M, Serra-Freire NM, Pedro MJV, Teixeira RHF, Gazêta GS (2013) Caso índice de *Amblyomma crassum* Robinson, 1926 (Acari: Ixodidae) em sucuri (Reptilia: Ophidia) no Brasil. Rev Uniabeu 6(14):355–362

Anastos G, Smith CN (1957) The male, nymph, and larva of *Ixodes brunneus* Koch (Acarina: Ixodidae). J Parasitol 43:535–541

Anderson JF, Magnarelli LA, Keirans JE (1981) *Aponomma quadricavum* (Acari: Ixodidae) collected from an imported boa, *Epicrates striatus*, in Connecticut. J Med Entomol 18:123–125

Andreotti R, García MV, Barros JC, Koller WW, Rodrigues VS (2018) Museu do carrapato da Embrapa Gado de Corte: espécimes de carrapatos descritos no Brasil e depositados na coleção até a momento. Documento 251, Embrapa, Brasilia, 54 pp

Antonucci AM, Oda FB, Signorelli L, Santana NF, Mendes MC (2012) Parasitismo de *Amblyomma rotundatum* (Koch, 1844) [sic] (Acari: Ixodidae) em *Rhinella schneideri* (Wener, 1894) (Anura: Bufonidae) no estado do Paraná, Brasil. Nat Line 10:5–6

Apanaskevich DA, Bermúdez SE (2013) Description of a new *Dermacentor* (Acari: Ixodidae) species, a parasite of wild mammals in Central America. J Med Entomol 50:1190–1201

Apanaskevich DA, Bermúdez SE (2017) Description of a new species of *Ixodes* Latreille, 1795 (Acari: Ixodidae) and redescription of *I. lasallei* Méndez & Ortiz, 1958, parasites of agoutis and pacas (Rodentia: Dasyproctidae, Cuniculidae) in Central and South America. Syst Parasitol 94:463–475

Apanaskevich DA, Domínguez LG, Torres SS, Bernal JA, Montenegro V, Bermúdez SE (2017) First description of the male and redescription of the female of *Ixodes tapirus* Kohls, 1956 (Acari: Ixodidae), a parasite of tapirs (Perissodactyla: Tapiridae) from the mountains of Colombia, Costa Rica and Panama. Syst Parasitol 94:413–422

Aragão HB (1908a) Algunas novas especies de carrapatos brazileiros. Braz Méd 22:111–115

Aragão HB (1908b) Mais uma nova especie de carrapato brazileiro. Braz Méd 22:251–252

Aragão HB (1908c) Mais um novo carrapato brazileiro. Braz Méd 22:431–432

Aragão HB (1911) Notas sobre ixódidas brazileiros. Mem Inst Oswaldo Cruz 3:145–195

Aragão HB (1912a) Nota sobre os Ixodidas collecionados durante a expedição do Sr. Coronel Rondon nos estados de Goyaz e Matto Grosso pelo Snr. Dr. Murillo de Campos, medico da expedição. Nova especie de *Amblyomma*: *A. conspicuum*. Braz Méd 26:429–430

Aragão HB (1912b) Contribução para a sistematica e biolojia dos ixódidas. Partenojeneze em carrapatos. Mem Inst Oswaldo Cruz 4:96–119

Aragão HB (1913) Notas sobre algumas coleções de carrapatos brazileiros. Mem Inst Oswaldo Cruz 5:263–271

Aragão HB (1918) Notas ixodológicas. Rev Mus Paulista 10:375–417

Aragão HB (1935) Observações sobre os ixodideos da República Argentina. Mem Inst Oswaldo Cruz 30:519–533

Aragão HB (1936) Ixodidas brasileiros e de algunos paízes limitrophes. Mem Inst Oswaldo Cruz 31:759–843

Aragão HB (1938) Nota sobre os ixodideos da Republica Argentina. Mem Inst Oswaldo Cruz 33:319–327

Aragão HB, Fonseca F (1951a) Notas de ixodología. III. Confirmação de *Ixodes aragãoi* [sic], de *Ixodes amarali* e lista das espécies do género *Ixodes* que ocorrem no Brasil. Ciênc Cult 3:300

Aragão HB, Fonseca F (1951b) Notas de ixodologia. I. Duas novas espécies do género *Ixodes* e um novo nomen para *Haemaphysalis kochi* Aragão, 1908 (Acari: Ixodidae). Mem Inst Oswaldo Cruz 49:567–574

Aragão HB, Fonseca F (1952) Notas de ixodología. III. Confirmação de *Ixodes aragãoi* [sic], de *Ixodes amarali* e lista das espécies do género *Ixodes* que ocorrem no Brasil. (Acari, Ixodidae). Mem Inst Oswaldo Cruz 50:727–728

Aragão HB, Fonseca F (1953a) Notas de ixodologia. V. A propósito da validade de algumas espécies do genero *Amblyomma* do continente Americano (Acari: Ixodidae). Mem Inst Oswaldo Cruz 51:485–492

Aragão HB, Fonseca F (1953b) Nota de ixodologia. 7. *Otocentor nitens* Neumann, 1897 versus *Anocentor columbianus* Schulze, 1937 e comentários sôbre a rápida disseminação desse ixodideo no Brasil (Acari: Ixodidae). Mem Inst Oswaldo Cruz 51:499–501

Aragão HB, Fonseca F (1961a) Nota de ixodologia. VIII. Lista e chave para os representantes da fauna ixodológica brasileira. Mem Inst Oswaldo Cruz 59:115–129

Aragão HB, Fonseca F (1961b) Nota de ixodologia. 9. O complexo *ovale* do género *Amblyomma*. Mem Inst Oswaldo Cruz 59:131–148

Arana-Guardia R, Baak-Baak CM, Cigarroa-Toledo N, Reyes-Solís GC, Ojeda-Robertos NF, García-Rejón JE, Zarza H, Ceballos G, Machain-Williams C (2015) Ticks (Acari: Ixodidae) from wild mammals in fragmented environments in the south of Yucatán Peninsula, Mexico. Southwest Entomol 40:657–660

Araújo IM, Magalhães-Matos PC, Cordeiro MD, Pacheco A, Rocha BS, Feitoza LFBB, Sanavria A, Fonseca AH (2019) Ticks in wild animals at a resort and the first documentation of *Amblyomma sculptum* infesting Emu (*Dromaius novaehollandiae*). Braz J Vet Parasitol 28:177–179

Arias JJ, Betancourt A, Valencia CE, Sardi H (1991) Identificación de garrapatas de bovinos en el municipio de Tumaco (Colombia). Rev ICA 26:145–151

Arnold KA (1970) Notes on avian ectoparasites from Costa Rica. I. Acarina and Diptera. Rev Biol Trop 16:259–265

Arthur DR (1960a) Ticks. A monograph of the Ixodoidea. Part V. On the genera Dermacentor, Anocentor, Cosmiomma, Boophilus and Margaropus. Cambridge University Press, London, 251 pp

Arthur DR (1960b) A review of some ticks (Acarina: Ixodidae) of sea birds. Part II The taxonomic problems associated with the *Ixodes auritulus – percavatus* group of species. Parasitology 50:199–226

Arthur DR (1963) British ticks. Butterworths, London, 213 pp

Arzua M, Barros-Battesti D (1999) Parasitism of *Ixodes* (*Multidentatus*) *auritulus* Neumann (Acari: Ixodidae) on birds from the city of Curitiba, State of Paraná, southern Brazil. Mem Inst Oswaldo Cruz 94:597–603

Arzua M, Barros DM, Linardi PM, Botelho JM (1994) Noteworthy records of *Ixodes auritulus* Neumann, 1904 (Acari, Ixodida) on birds from Paraná, southern Brazil. Mem Inst Oswaldo Cruz 89:129

Arzua M, Navarro da Silva M, Famadas KM, Beati L, Barros-Battesti DM (2003) *Amblyomma aureolatum* and *Ixodes auritulus* (Acari: Ixodidae) on birds in southern Brazil, with notes on their ecology. Exp Appl Acarol 31:283–286

Arzua M, Onofrio VC, Barros-Battesti DM (2005) Catalogue of the tick collection (Acari: Ixodida) of the Museu de História Natural Capao da Imbuia, Curitiba, Paraná, Brazil. Rev Bras Zool 22:623–632

Audebad G, Courmes E (1962) Ixodides de Guadeloupe. Présence d'*Amblyomma dissimile*. *Archives de l'Institut Pasteur Guadeloupe*. Rapp Tech 1961:60–61

Ault CN (1943) Descripción de nuevos ejemplares de *Ixodes auritulus* Neumann 1899, hallados en la República Argentina. Rev Med Vet (Buenos Aires) 25:167–174

Ault CN (1946) Control of *Boophilus australis* in the Argentine by the gamma isomer of hexachlorociclohexane ("Gammexane"). Nature 157:699–700

Autino AG, Nava S, Venzal JM, Mangold AJ, Guglielmone AA (2006) La presencia de *Ixodes luciae* Sénevet en el noroeste argentino y nuevos huéspedes para *Ixodes pararicinus* y algunas especies de *Amblyomma* (Acari: Ixodidae). Rev Soc Entomol Argent 65:27–32

Bakkes DK, Chitimia-Dobler L, Matloa D, Oosthuysen M, Mumcuoglu KY, Mans BJ, Matthee CA (2020) Integrative taxonomy and species delimitation of *Rhipicephalus turanicus* (Acari: Ixodida: Ixodidae). Int J Parasitol 50:577–594

Banks N (1907) A catalogue of the Acarina, or mites, of the United States. Proc US Natl Mus 32:595–625

Banks N (1908) A revision of the Ixodoidea, or ticks, of the United States. US Dept Agric Bur Entomol Tech Ser 15:61 pp

Banks N (1909) Three new ticks from the United States. Proc Entomol Soc Washington 10:170–173

Banks N (1924) Arachnida of the Williams Galapagos Expedition. Zoologica (New York) 5:93–99

Barbará B, Dios RL (1918) Contribución al estudio de la sistemática y la biología de los Ixodidae de la República Argentina y de algunos países vecinos. Rev Inst Bacteriol Dept Nacional Higiene 1:285–322

Barber CA (1895) The tick pest in the tropics. Nature 52:197–200

Barbieri FS, Brito LG, Labruna MB, Barros-Battesti DM, Camargo LMA, Famadas KM (2008a) Description of the larva of *Amblyomma ovale* Koch, 1844 (Acari: Ixodidae) by light and scanning electron microscopy. Syst Appl Acarol 13:109–119

Barbieri FS, Brito LG, Labruna MB, Barros-Battesti DM, Camargo LMA, Famadas KM (2008b) Description of the larva of *Amblyomma pacae* Aragão, 1911 (Acari: Ixodidae) by light and scanning electron microscopy. Syst Appl Acarol 13:195–203

Barbieri ARM, Romero L, Labruna MB (2012a) *Rickettsia bellii* infecting *Amblyomma sabanerae* ticks in El Salvador. Pathog Glob Health 106:188–190

Barbieri FS, Brito LG, Barros-Battesti DM, Famadas KF, Labruna MB, Camargo LMA (2012b) Redescription of the larva of *Amblyomma oblongoguttatum* Koch, 1844 (Acari: Ixodidae) by light and scanning electron microscopy. Zootaxa 3579:80–88

Barbieri FS, Brito LG, Labruna MB, Barros-Battesti DM, Camargo LMA, Famadas KM (2013) Description of the larva of *Amblyomma calcaratum* Neumann, 1899 (Acari: Ixodidae) by light and scanning electron microscopy. Ticks Tick-Borne Dis 4:531–536

Barbieri ARM, Szabó MPJ, Costa FB, Martins TF, Soares HS, Pascoli G, Torga K, Saraiva DG, Ramos VN, Osava C, Gerardi M, Dias RA, Moraes EA, Ferreira F, Castro MB, Labruna MB (2019) Species richness and seasonal dynamics of ticks with notes on rickettsial infection in a natural park of the Cerrado biome in Brazil. Ticks Tick-Borne Dis 10:442–453

Barker SC, Burger TD (2018) Two new genera of hard ticks, *Robertsicus* n. gen., and *Archaeocroton* n. gen., and the solution of the mystery of Hoogstraal's and Kaufman's "primitive" tick from the Carpathian Mountains. Zootaxa 4500:543–552

Barker SC, Murrell A (2004) Systematics and evolution of ticks with a list of valid genus and species names. Parasitology 129:S15–S36

Barker SC, Murrell A (2008) Systematics and evolution of ticks with a list of valid genus and species names. In: Bowman AS, Nuttall P (eds) Ticks: biology, disease and control. Cambridge University Press, Cambridge, pp 1–39

Barrantes-González AV, Jiménez-Rocha AE, Romero-Zuñiga JJ, Dolz G (2016) Serology, molecular detection and risk factors of *Ehrlichia canis* infection in dogs in Costa Rica. Ticks Tick-Borne Dis 7:1245–1251

Barré N, Garris GI, Borel G, Camus E (1988) Hosts and population dynamics of *Amblyomma variegatum* (Acari: Ixodidae) on Guadeloupe, French West Indies. J Med Entomol 25:111–115

Barré N, Garris G, Camus E (1995) Propagation of the tick *Amblyomma variegatum* in the Caribbean. Sci Tech Rev World Org Anim Health 14:841–855

Barros DM, Baggio D (1992) Ectoparasites Ixodida Leach, 1817 on wild animals in the State of Paraná, Brazil. Mem Inst Oswaldo Cruz 87:291–296

Barros-Battesti DM, Knysak I (1999) Catalogue of the Brazilian *Ixodes* (Acari: Ixodidae) material in the mite collection of Instituto Butantan, São Paulo, Brazil. Pap Avulsos Zool 41:49–57

Barros-Battesti DM, Labruna MB (2005) First record of *Haemaphysalis leporispalustris* (Acari: Ixodidae) on *Lepus europaeus* (Lagomorpha: Leporidae), an introduced host into the New World. Entomol News 116:266–267

Barros-Battesti DM, Arzua M, Pichorim M, Keirans JE (2003) *Ixodes* (*Multidentatus*) *paranaensis* n. sp. (Acari: Ixodidae) a parasite of *Streptoprocne biscutata* (Sclater, 1865) (Apodiformes: Apodidae) birds in Brazil. Mem Inst Oswaldo Cruz 98:93–102

Barros-Battesti DM, Arzua M, Rebello VMM, Barbieri FS, Famadas KM (2005a) Description of the larva of *Amblyomma longirostre* (Koch, 1844) (Acari: Ixodidae) by light and scanning electron microscopy. Braz J Vet Parasitol 14(2):51–57

Barros-Battesti DM, Onofrio VC, Labruna MB, Martins JR, Guglielmone AA (2005b) Redescription of *Amblyomma fuscum* Neumann, 1907 (Acari: Ixodidae), a rare South America tick confirmed in Brazil. Syst Parasitol 61:85–92

Barros-Battesti DM, Arzua M, Onofrio VC, Labruna MB (2007a) Validation and redescription of *Amblyomma romitii* Tonelli Rondelli, 1939 (Acari: Ixodidae). Syst Parasitol 68:79–86

Barros-Battesti DM, Onofrio VC, Faccini JLH, Labruna MB, Arruda-Santos AD, Giacomin FG (2007b) Description of the immature stages and redescription of the female of *Ixodes schulzei*

Aragão & Fonseca, 1951 (Acari: Ixodidae), an endemic tick species of Brazil. Syst Parasitol 68:157–166

Barros-Battesti DM, Onofrio VC, Arzua M, Labruna MB (2008) Comments on the validity of *Haemaphysalis cinnabarina* Koch, 1844 (Acari: Ixodidae), a taxon known solely by the specimens from northern Brazil. Braz J Vet Parasitol 17(1):53–55

Barros-Battesti DM, Reyes Hernández M, Famadas KM, Onofrio VC, Beati L, Guglielmone AA (2009) The ixodid ticks (Acari: Ixodidae) of Cuba. Syst Appl Acarol 14:101–128

Barros-Battesti DM, Ramírez DG, Sampaio JS, Famadas KM, Faccini JLH, Nunes PH, Martins TF, Ogrzewalska M, Labruna MB, Marcili A, Barbieri FS (2012) Description of larva of *Amblyomma romitii* (Acari: Ixodidae) by optical and scanning electron microscopy, including porotaxy and phylogenetic analysis. Exp Appl Acarol 60:271–280

Barroso SM (1922) O que todo devemos saber os parasitas vegetaes e animaes que se implantam no nosso corpo e os males que nos causam. Meios de os evitar, Bahia, 230 pp

Bastos TSA, Madrid DMC, Faria AM, Freitas TMS, Linhares GFC (2016) Carrapatos em animais silvestres do bioma Cerrado triados pelos CETAS, IBAMA, Goiás. Ciênc Anim Bras 17:296–302

Basu AK, Charles RA (2017) Ticks of Trinidad and Tobago – an overview. Elsevier-Academic, London, 89 pp

Basu AK, Basu M, Adesiyun AA (2012) A review on ticks (Acari: Ixodoidea, Argasidae), associated pathogens and diseases of Trinidad and Tobago. Acarologia 52:5–16

Beati L, Klompen H (2019) Phylogeography of ticks (Acari: Ixodida). Annu Rev Entomol 64:379–397

Beati L, Nava S, Burkman EJ, Barros-Battesti D, Labruna MB, Guglielmone AA, Cáceres AG, Guzman Cornejo C, Léon R, Durden LA, Faccini JLH (2013) *Amblyomma cajennense* (Fabricius, 1787) (Acari: Ixodidae), the Cayenne tick: phylogeography and evidence for allopatric speciation. BMC Evol Biol 13 (article 267):20 pp

Beatty HA (1944) The Arachnida of St. Croix, V. I. J Agric Univ P R 28:111–113

Bechara GH, Szabó MPJ, Almeida WV, Bechara JN, Pereira RJG, García JE, Pereira MC (2002) Ticks associated with armadillo (*Euphractus sexcinctus*) and anteater (*Myrmecophaga tridactyla*) of Emas National Park, State of Goiás, Brazil. Ann NY Acad Sci 969:290–293

Becker GK, Silva Filho AL, Sinkoc AL, Brum JGW (1997) *Amblyomma parvitarsum* Neumann, 1901 (Acari: Ixodidae) em pingüim de Magalhães *Spheniscus magellanicus* (Spheniscidae) na Praia do Cassino, Rio Grande do Sul, Brasil. Arq Inst Biol 64:81–82

Becklund WW (1959) A note on the distribution of the tick *Amblyomma imitator* Kohls, 1958. J Parasitol 45:430

Becklund WW (1968) Ticks of veterinary significance found on imports in the United States. J Parasitol 54:622–628

Bedford GAH (1920) Ticks found on man and his domestic animals and poultry in South Africa. J Dept Agric 1:317–340

Bedford GAH (1934) South African ticks. Part I Onderstepoort. J Vet Sci Anim Ind 2:49–99

Beldoménico PM, Baldi JC, Orcellet V, Peralta JL, Venzal JM, Mangold AJ, Guglielmone AA (2003a) Ecological aspects of *Ixodes pararicinus* (Acari: Ixodidae) and other tick species parasitizing sigmodontin [sic] mice (Rodentia: Muridae) in the northwestern Argentina. Acarologia 54:15–21

Beldoménico PM, Baldi JC, Antoniazzi LR, Orduna GM, Mastropaolo M, Macedo AC, Ruiz MF, Orcellet V, Peralta JL, Venzal JM, Mangold AJ, Guglielmone AA (2003b) Ixodid ticks (Acari: Ixodidae) present at Parque Nacional El Rey, Argentina. Neotrop Entomol 32:273–277

Benavides-Montaño JA, Jaramillo-Cruz CJ, Mesa-Cobo NC (2018) Garrapatas Ixodidae (Acari) en el Valle del Cauca, Colombia. Boletín Científico Mus Hist Nat Caldas 22:131–150

Benítez-Ibalo AP, Aguiar LD, Di Benedetto IMD, Mangold AJ, Milano F, Debárbora VN (2020) Ectoparasites associated with rodents (Rodentia) and marsupials (Didelphimorphia) from northeastern Argentina: new host and locality records. Rev Mex Biodivers 91 (article e913161):8 pp

Bentley JW, Rodríguez G (2001) Honduran folk entomology. Curr Anthropol 42:285–300

Bequaert J (1925) Report of an entomological trip to the Truxillo Division, Honduras, to investigate the sand fly problem. In: 13th Report of the United Fruit Company Medical Department, pp 193–206

Bequaert J (1926) Medical report of the Hamilton Rice seventh expedition to the Amazon in conjunction with the department of tropical medicine of Harvard University, 1924-1925. Contrib Harv Inst Trop Biol Med 4:155–257

Bequaert J (1932a) *Amblyomma dissimile* Koch, a tick indigenous to the United States (Acarina: Ixodidae). Psyche 39:45–47

Bequaert JC (1932b) On the ornate nymphs of the tick genus *Amblyomma* (Acarina: Ixodidae). Z Parasitenkd 4:776–783

Bequaert JC (1938) Notes on the arthropods of medical importance in Guatemala. In: Medical survey of the Republic of Guatemala. Publications of the Carnegie Institution of Washington (499), pp 223–228

Bequaert JC (1945) The ticks, or Ixodoidea, of the northeastern United States and eastern Canada. Entomol Am 25:121–184

Berlese A (1888) Acari austro-americani. Boll Soc Entomol Ital 20:171–242

Berlese A, Trouessart EL (1889) Diagnoses d'acariens nouveaux ou peu connus. Bull Bibl Sci l'Ouest 2:121–143

Bermúdez SE, Miranda RJ (2011a) De mascotas exóticas y turistas: nuevas oportunidades para la introducción de ectoparásitos en Panamá. Bol Soc Entomol Aragonesa 48:491–492

Bermúdez SE, Miranda RJ (2011b) Distribution of ectoparasites of *Canis lupus familiaris* L. (Carnivora: Canidae) from Panama. Rev MVZ Córdoba 16:2274–2282

Bermúdez SE, Miranda RJ, Smith D (2010a) Tick species (Ixodida) in the Summit Municipal Park and adjacent areas, Panama City, Panama. Exp Appl Acarol 52:439–448

Bermúdez SE, Miranda RJ, Zaldívar YL, Page K (2010b) *Dermacentor variabilis* (Ixodida: Ixodidae) in Panama: report associated with tourism. J Vector Ecol 35:208–209

Bermúdez SE, Zaldívar AY, Spolidorio MG, Moraes-Filho J, Miranda RJ, Caballero CM, Mendoza Y, Labruna MB (2011) Rickettsial infection in domestic mammals and their ectoparasites in El Valle de Antón, Coclé, Panamá. Vet Parasitol 177:134–138

Bermúdez SE, Miranda R, Zaldívar Y, González P, Berguido G, Trejos D, Pascale JM, Labruna MB (2012a) Detección de *Rickettsia* spp. en ectoparásitos de animales domésticos y silvestres de la Reserva Natural Privada Cerro Chucantí y comunidades aledañas, Panamá, 2007-2010. Biomedica 32:189–195

Bermúdez SE, Castro A, Esser M, Liefting Y, García G, Miranda RJ (2012b) Ticks (Ixodida) on humans from central Panama, Panama. Exp Appl Acarol 58:81–88

Bermúdez SE, González D, García G (2013) Ticks (Acari: Ixodidae, Argasidae) of coyotes in Panama. Syst Appl Acarol 18:112–115

Bermúdez SE, Esser M, Miranda RJ, Moreno RS (2015a) Wild carnivores (Mammalia) as hosts for ticks (Ixodida) in Panama. Syst Appl Acarol 20:13–19

Bermúdez SE, Mejía L, Hernández L, Apanaskevich DA (2015b) First records of *Ixodes boliviensis* Neumann, 1904 and *Dermacentor dissimilis* Cooley, 1947 (Ixodida: Ixodidae) as parasites of domestic mammals in Nicaragua. Syst Appl Acarol 20:462–464

Bermúdez SE, Torres S, Aguirre Y, Domínguez L, Vega JAB (2015c) A review of *Ixodes* (Acari: Ixodidae) parasitizing wild birds in Panama, with the first records of *Ixodes auritulus* and *Ixodes bequaerti*. Syst Appl Acarol 20:847–853

Bermúdez S, Apanaskevich DA, Domínguez L (2018a) Garrapatas Ixodidae de Panamá. Secretaría Nacional de Ciencia, Tecnología e Innovación, Panamá, 129 pp

Bermúdez SE, Meyer N, Moreno R, Artavia A (2018b) Notas sobre *Pecari tajacu* (L., 1758) y *Tayassu pecari* (Link, 1795) (Artiodactyla: Tayassuidae) como hospederos de garrapatas duras (Acari: Ixodidae) en Panamá. Tecnociencia 20:61–70

Bermúdez SE, Domínguez L, Ochoa I, Oliveira JG, Lemos ERS, Castillo B, Smith D, Herrera J, Ogrzewalska M (2020) Molecular detection of rickettsial agents of hard ticks (Acari: Ixodidae) collected from wild birds in Panama. Syst Appl Acarol 25:622–632

Bernasconi MV, Casati S, Péter O, Piffaretti JC (2002) *Rhipicephalus* ticks infected with *Rickettsia* and *Coxiella* in southern Switzerland (Canton Ticino). Infect Genet Evol 2:111–120

Betancourt JA, García O, Roqueme L, Navarretes M (1992) Distribución y niveles de infestación por garrapatas en bovinos de Córdoba, noroeste de Sucre y noreste de Antioquía. Rev ICA 27:63–76

Billeter SA, Cáceres AG, Gonzales-Hidalgo J, Luna-Caypo D, Kosoy MY (2011) Molecular detection of *Bartonella* species in ticks from Peru. J Med Entomol 48:1257–1260

Binetruy F, Chevillon C, Thoisy B, Garnier S, Duron O (2019) Survey of ticks in French Guiana. Ticks Tick-Borne Dis 10:77–85

Binetruy F, Buysse M, Barosi R, Duron O (2020a) Novel *Rickettsia* genotypes in ticks in French Guiana, South America. Sci Rep 10(article 2537):11 pp

Binetruy F, Garnier S, Boulanger N, Talagrand-Reboul E, Loire E, Faivre B, Noël V, Buysse M, Duron O (2020b) A novel *Borrelia* species, intermediate between Lyme disease and relapsing fever groups, in Neotropical passerine-associated ticks. Sci Rep 10(article 10596):10 pp

Bishopp FC (1911) Some new North American Ixodidae with notes on other species. Proc Biol Soc Washington 24:197–208

Bishopp FC (1912) A new species of *Dermacentor* and notes on other North American Ixodidae. Proc Biol Soc Washington 25:29–37

Bishopp FC, Trembley HL (1945) Distribution and hosts of certain North American ticks. J Parasitol 31:1–54

Bitencourth K, Teixeira RF, Amorim M, Gazêta GS, Serra-Freire NM (2007) Análise do número de cerdas internas do órgão de Haller de larvas de *Amblyomma nodosum* Neumann (Acari: Ixodidae). Rev Bras Entomol 51:58–61

Bitencourth K, Amorim M, Voloch CM, De Oliveira SV, Caetano RL, Gazeta GS (2017) *Amblyomma sculptum*: genetic diversity and rickettsias in the Brazilian Cerrado biome. Med Vet Entomol 31:427–437

Bitencourth K, Amorim M, De Oliveira SV, Voloch CM, Gazeta GS (2019) Genetic diversity, population structure and rickettsias in *Amblyomma ovale* in areas of epidemiological interest for spotted fever in Brazil. Med Vet Entomol 33:256–268

Blair PJ, Jiang J, Schoeler GB, Moron C, Anaya E, Céspedes M, Cruz C, Felices V, Guevara C, Mendoza L, Villaseca P, Summer JW, Richards AL, Olson JG (2004) Characterization of spotted fever group rickettsiae in flea and tick specimens from northern Peru. J Clin Microbiol 42:4961–4967

Blanchard R (1909) L'insecte et l'infection, premier fascicule: Acariens. Librairie Scientifique et Littéraire, Paris, 160 pp

Blanco CM, Teixeira BR, Silva AG, Oliveira RC, Strecht L, Ogrzewalska M, Lemos EBS (2017) Microorganisms in ticks (Acari: Ixodidae) collected on marsupials and rodents from Santa Catarina, Paraná and Mato Grosso do Sul states, Brazil. Ticks Tick-Borne Dis 8:90–98

Bodkin G (1918) The biology of *Amblyomma dissimile* Koch. With an account of its power of reproducing parthenogenetically. Parasitology 1:10–17

Boero JJ (1944a) Notas ixodológicas. (I). *Ixodes longiscutatum* [sic], nueva especie. Rev Asoc Med Argent 58:353–354

Boero JJ (1944b) Notas ixodológicas. (II). Nueva lista de los ixodoideos argentinos y sus huéspedes. Rev Asoc Med Argent 58:354–355

Boero JJ (1944c) Los ixodideos de la República Argentina. Boletín Técnico de la Dirección de de Ganadería, Ministerio de Agricultura y Ganadería de la República Argentina, 68 pp

Boero JJ (1945) *Ixodes brunneus* Koch 1844. Nueva especie para la acarofauna argentina. Rev Med Vet (Buenos Aires) 27:128–130

Boero JJ (1947) *Amblyomma pseudoconcolor* Aragão, nuevo huésped en la República Argentina. Rev Med Vet (Buenos Aires) 29:900–904

Boero JJ (1955) Los ixodoideos de la República Argentina y sus huéspedes. Rev Fac Agron Vet (Buenos Aires) 13:505–514

Boero JJ (1957) Las garrapatas de la República Argentina (Acarina: Ixodoidea). Departamento Editorial de la Universidad de Buenos Aires, Buenos Aires, 113 pp

Boero JJ, Delpietro H (1970) *Amblyomma longirostre* Koch 1844 (Acarina-Ixodidae) Primera comprobación en la Argentina. Rev Med Vet (Buenos Aires) 51:335–337

Boero JJ, Delpietro H (1971) *Amblyomma calcaratum* Neumann 1899 (Acarina-Ixodidae) Nueva especie para la fauna argentina. Rev Med Vet (Buenos Aires) 52:339–341

Boero JJ, Prosen (1955) Ixodideos de "anta". Mis Estud Patol Reg Norte Argent 26:47–55

Boero JJ, Prosen (1959) Ixodideos de "anta" II. *Amblyomma multipunctum* Neumann 1899. Anal Inst Med Reg 5:95–97

Boero JJ, Prosen (1960) Ixodideos de "anta." III. *Amblyomma scalpturatum* Neumann 1906. Descripción del alotipo macho y redescripción de la hembra. Anal Inst Med Reg 6:115–119

Boffy ACM, Almeida-Santos P (2010) *Rhinella pygmaea* (Rio Parahyba toad). Ectoparasitism. Herpetol Rev 41:342–343

Bohmfalk GT, Price MA (1979) Chaetotactic and morphologic comparison in larval *Boophilus annulatus* (Say) and *Boophilus microplus* (Canestrini) (Acarina: Ixodidae). Southwest Entomol 4:102–116

Bokma BH, Shaw JL (1993) Eradication of a new focus of *Amblyomma variegatum* in Puerto Rico. Rev Élevage Méd Vét Pays Tropicaux 46:355–358

Bool PH, Sutmöller P (1957) *Ehrlichia canis* infection in dogs on Aruba (Netherlands Antilles). J Am Vet Med Assoc 130:418–420

Borght-Elbl A (1977) Ixodid ticks (Acarina: Ixodidae) of Central Africa. Vol. V. The larval and nymphal stages of the more important species of the genus *Amblyomma* Koch, 1844. Ann Mus R I 'Afrique Centrale, Série 8, Sci Zool (222):158 pp

Borsoi ABP, Bitencourth K, Oliveira SV, Amorim M, Gazêta GS (2019) Human parasitism by *Amblyomma parkeri* ticks infected with *Candidatus* Rickettsia paranaensis, Brazil. Emerg Infect Dis 25:2339–2341

Botelho MCN, Leite LMRM, Neto IPB, Silva LAM, Campello MLCB, Aguiar MCA, Serra-Freire NM, Oliveira JB (2002) *Amblyomma dissimile* Kock [sic], 1844 (Acari, Ixodidae) em mamíferos silvestres no Estado de Pernambuco, Brasil. Entomol Vectores 9:71–78

Bothelo JR, Linardi PM, Encarnação CD (1989) Interrelações entre Acari Ixodidae e hospedeiros Edentata da da Serra da Canastra, Minas Gerais, Brazil. Mem Inst Oswaldo Cruz 84:61–64

Brinton EP, Beck DE, Allred DM (1965) Identification of the adults, nymphs and larvae of ticks of the genus *Dermacentor* Koch (Ixodidae) in the western United States. Brigham Young Univ Sci Bull Biol Ser 5(4):44 pp

Brum JGW, Rickes EM (2003) *Amblyomma dissimile* Koch, 1844 (Acari: Ixodidae) em serpente sucuri (*Eunectes murinus*) (Reptilia: Boidae) no Parque Zoológico do Rio Grande do Sul. Arq Inst Biol 70:215–216

Brum JGW, Ribeiro PB, Costa PRP, Oliveira CMB (1987) Artrópodos parasitos dos animais domésticos da zona sul do estado do Rio Grande do Sul, Brasil. Arq Bras Med Vet Zoot 39:533–537

Buitendijk AM (1945) Voorloopige catalogus van de Acari in de Collectie-Oudemans. Zool Meded 24:281–391

Burridge MJ (2011) Non-native and invasive ticks. Threats to human and animal health in the United States. University of Florida Press, Gainesville, 448 pp

Burridge MJ, Simmons LA (2003) Exotic ticks introduced into the United States on imported reptiles from 1962 to 2001 and their potential roles in international dissemination of diseases. Vet Parasitol 113:289–320

Cáceres AG, Beati L, Keirans JE (2002) First evidence of the ocurrence of *Amblyomma calcaratum* Neumann, 1899 in Peru. Rev Peruana Entomol 9:116–117

Calzada V (1936) Comprobación de *Ixodes ricinus* (nueva sub-especie) en el Uruguay. Boletín Mensual de la Dirección de Ganadería del Uruguay 20:103–109

Camacho Escobar MA, Pérez-Lara E (2009) Ectoparásitos de iguana verde (*Iguana iguana*) y negra (*Ctenosaura pectinata*) en condiciones de crianza intensiva en la costa de Oaxaca, México. Ciencia y Mar 13:15–22

Cameron AE, Fulton JS (1927) A local outbreak of the winter or moose tick, *Dermacentor albipictus*, Pack. [sic] (Ixodoidea) in Saskatchewan. Bull Entomol Res 17:249–257

Camicas JL (1970) Contribution à l'étude des tiques du Sénégal (Acarina, Ixodoidea). I Les larves d'*Amblyomma* Koch et de *Hyalomma* Koch. Acarologia 12:71–102

Camicas JL, Hervy JP, Adam F, Morel PC (1998) Les tiques du monde. Nomenclature, stades décrits, hôtes, répartition (Acarida, Ixodida). Orstom, Paris, 233 pp

Campos Pereira M, Szabó MJP, Bechara GH, Matushima ER, Duarte JMB, Rechav Y, Fielden L, Keirans JE (2000) Ticks (Acari: Ixodidae) associated with wild animals in the Pantanal region of Brazil. J Med Entomol 37:979–983

Camus E, Barre N (1995) Vector situation of tick-borne diseases in the Caribbean islands. Vet Parasitol 57:167–176

Cançado PHD, Faccini JLH, Mourao GM, Piranda EM, Onofrio VC, Barros-Battesti DM (2017) Current status of ticks and tick-host relationship in domestic and wild animals from Pantanal wetlands in the State of Mato Grosso do Sul, Brazil. Iheringia Ser Zool 107:1–8

Candanedo-Guerra RMSN, Serra-Freire NM (1999) *Amblyomma oblongoguttatum* Koch, 1844 (Acari: Ixodidae) in *Carollia perspicillata* (L.) (Chiroptera: Phyllostomidae): report and ecological reflexion [sic]. Entomol Vectores 6:63–73

Canestrini G (1888) Intorno da alcuni Acari ed Opilonidi dell'America. Atti Soc Veneto-Trentina Sci Nat Residente Padova 11:100–109

Canestrini G (1890) Prospetto dell'acarofauna italiana. IV. Famiglie: Tetranychini, Ixodini, Argasini. Edizione Prosperini, Padova, pp 427–540

Canestrini G, Fanzago F (1878) Intorno agli acari italiani. Atti Reale Istit Veneto Sci Lett Arti, Ser 5 (4):69–208

Cansi ER, Bonorino R, Mustafa VS, Guedes KMR (2012) Multiple parasitism in wild maned wolf (*Chrysocyon brachyurus*, Mammalia: Canidae) in central Brazil. Comp Clin Pathol 21:489–493

Canto-Osorio JM, Cuxim-Coyoc A, Ruiz-Piña HA, Morales-Malacara JB, Reyes-Novelo E (2020) Ectoparasites of *Didelphis virginiana* from Yucatán, Mexico. J Med Entomol 57:1821–1829

Capriles JM, Gaud SM (1977) The ticks of Puerto Rico (Arachnida: Acarina). J Agric Univ P R 61:402–404

Cardona-Romero M, Martínez-Sánchez ET, Londoño JA, Tobón-Escobar WD, Ossa- López PA, Pérez-Cárdenas JE, Ramírez-Chaves HE, Blandón-Marín G, Cuervo LA Castaño-Villa GJ, Rivera-Pñez FA (2020) *Rickettsia parkeri* strain Atlantic rainforest in ticks (Acari: Ixodidae) of wild birds in Arauca, Orinoquia region of Colombia. Int J Parasitol Parasites Wildl 13:106–113

Cardozo H (1996) Situación de la resistencia del *Boophilus microplus* en Uruguay. Medidas para controlarla. Veterinaria (Montevideo) 32:15–18

Carrascal J, Oviedo T, Monsalve S, Torres A (2009) *Amblyomma dissimile* (Acari: Ixodidae) parásito de *Boa constrictor* en Colombia. Rev MVZ Córdoba 14:1745–1749

Carreno RA, Durden LA, Brooks DR, Abrams A, Hoberg EP (2001) *Parelaphostrongylus tenuis* (Nematoda: Protostrongylidae) and other parasites of white-tailed deer (*Odocoileus virginianus*) in Costa Rica. Comp Parasitol 68:177–184

Carvalho LA, Armúa-Fernández MT, Sosa N, Félix ML, Venzal JM (2017) *Anaplasma platys* in Uruguay. Ticks Tick-Borne Dis 8:241–245

Carvalho LA, Maya L, Armúa-Fernández MT, Félix ML, Bazzano V, Barbieri AM, González EM, Lado P, Colina R, Díaz P, Labruna MB, Nava S, Venzal AJ (2020) *Borrelia burgdorferi* sensu lato infecting *Ixodes auritulus* ticks in Uruguay. Exp Appl Acarol 80:109–125

Castillo GN, Acosta JC, Rodríguez-Muñoz M, Martínez T, González-Rivas CJ, Nava S (2019) New association between *Amblyomma parvitarsum* (Acari: Ixodidae) and the endemic lizard *Liolaemus eleodori* (Iguania: Liolaemidae) in Argentina. Ann Parasitol 65:27–33

Castrezana SJ (2010) Artrópodos terrestres no hexápodos. In: Molina-Freaner FE, Vandevender TR (eds) Diversidad biológica de Sonora. UNAM, México, pp 293–314

Cerdeña J, Quequejana V, Rivera ALL (2019) First record of the hard tick *Amblyomma parvitarsum* (Acari: Ixodidae) on the *Liolaemus* (Reptilia: Squamata) lizards from Peru. Int J Acarol 45:106–107

Černý V (1966a) Nueva especie de garrapata del género *Ixodes* Latreille (Ixodoidea, Ixodidae) en la jutía conga de la isla de Pinos. Poeyana Ser A 24:9 pp

Černý V (1966b) Nuevas garrapatas (Ixodoidea) en aves y reptiles de Cuba. Poeyana Ser A 26:10 pp

Černý V (1969a) The tick fauna of Cuba. Folia Parasitol 16:279–284

Černý V (1969b) Nuevos conocimientos sobre la ixodofauna cubana. Torreia (21):12 pp

Černý V (1969c) Las garrapatas ectoparásitas del ganado vacuno. Los ectoparásitos de las gallinas. Acad Ciencias Cuba, Ser Ganadería (2):26 pp

Černý V, Sixl W, Waltinger H (1975) Scanning electron microscopic studies of the morphology of *Ixodes capromydis* nymph and adults. Folia Parasitol 22:141–143

Cervantes M, Masgo D, Ramírez L, Álvarez G, Li O, Ydrogo AV, Puerta LA, Hoyos L (2020) Identificación morfológica y molecular de garrapatas colectadas de perros (*Canis lupus familiaris*) con ehrlichiosis en Chiclayo, Perú. Rev Investig Vet Perú 31(2) (article e17820):10 pp

Charles R, Basu A, Sanford B, King-Cenac A, Melville-Edwin S, Pow-Brown P, Sant C, Georges K (2020) Survey of ticks of domestic dogs and cattle in three Caribbean islands. Transbound Emerg Dis 67:129–134

Charruau P, Pérez-Flores J, Cedeño-Vázquez JR, González-Solís D, González-Desales GA, Monroy-Vilchis O, Desales-Lara MA (2016) Occurrrence of *Amblyomma dissimile* on wild crocodylians in southern Mexico. Dis Aquat Org 121:167–171

Chavarría C (1943) La anaplasmosis en los bovinos de Costa Rica. Rev Agric 15:343–346

Chitimia-Dobler L, Pfeffer T, Dunlop JA (2018) *Haemaphysalis cretacea* a nymph of a new species of hard tick in Burmese amber. Parasitology 145:1440–1451

Chitimia-Dobler L, Schaper S, Mansfeld P, Gonschorrek J, Bröker M, Nava S (2020) Detection of *Amblyomma mixtum* (Acari: Ixodidae) in Germany on a human traveler returning from Cuba. J Med Entomol 57:962–964

Cicuttin GL, Sassaroli JC, Ardiles MI, Zotter AC, Guglielmone AA, Nava S (2013) Presencia de dos especies de garrapatas (Acari: Ixodidae) con importancia médica en la ciudad de Buenos Aires. Medicina (Buenos Aires) 73:389–390

Cicuttin GL, De Salvo MN, Nava S (2017) Especies de garrapatas duras en un área urbana protegida de la Ciudad Autónoma de Buenos Aires. Rev Argent Salud Públ 8:7–12

Cicuttin GL, Wiemeyer G, Guglielmone AA, Nava S (2019a) Detección de la garrapata *Amblyomma varanense* (Acari: Ixodidae) en especímenes importados de *Varanus salvator* (Squamata: Varanidae) en la Argentina. InVet 21:1–9

Cicuttin GL, De Salvo MN, Venzal JM, Nava S (2019b) *Borrelia* spp. in ticks and birds from a protected urban area in Buenos Aires city, Argentina. Ticks Tick-Borne Dis 10 (6) (article 101282):8 pp

Clarkson MJ, Owen LN (1959) The parasites of domestic animals in the Bahama islands. Ann Trop Med Parasitol 53:341–346

Clavijo JJ, Vásquez CL, Valera NC (2009) Garrapatas duras de los géneros *Amblyomma* Koch y *Rhipicephalus* Koch (Acari: Ixodidae) presentes en la colección de zoología agrícola, decanato de agronomía, Universidad Centro Occidental Lisandro Alvarado, Venezuela. Entomotropica 24:41–50

Clements JF, Schulenberg TS, Iliff MJ, Roberson D, Fredericks TA, Sullivan BL, Wood CL (2018) The eBird/Clements checklist of birds of the world: v 2018. http://wwwbirdscornelledu/clementschecklist/download/. Last accession 31 July 2020

Clifford CM, Hoogstraal H (1980) A new species of *Ixodes* parasitizing the rice rat in the Galapagos (Ixodoidea: Ixodidae). Proc Entomol Soc Washington 82:378–383

Clifford CM, Kohls GM (1962) Description of the female of *Dermacentor latus* Cooley and of *Amblyomma albopictum* Neumann (Acarina-Ixodidae). J Parasitol 48:486–489

Clifford CM, Anastos G, Elbl A (1961) The larval ixodid ticks of the eastern United States (Acarina – Ixodidae). Misc Publ Entomol Soc Am 5:214–237

Clifford CM, Sonenshine DE, Keirans JE, Kohls GM (1973) Systematics of the subfamily Ixodinae (Acarina: Ixodidae). 1. The subgenera of *Ixodes*. Ann Entomol Soc Am 66:489–500

Coelho MC, Ramos VN, Limongi JE, Lemos ERS, Guterres A, Neto SFC, Rozental T, Bonvicino CR, D'Andrea PS, Moraes-Filho J, Labruna MB, Szabó MPJ (2016) Serologic evidence of the exposure of small mammals to spotted-fever *Rickettsia* and *Rickettsia belli* in Minas Gerais, Brazil. J Infect Dev Ctries 10:275–282

Coimbra-Dores MJ, Jaarsma RI, Carmo AO, Maia-Silva M, Fonville M, Costa DFF, Brandão RML, Azevedo F, Casero M, Oliveira AC, Afonso SMS, Sprong H, Rosa F, Dias D (2020) Mitochondrial sequences of *Rhipicephalus* and *Coxiella* endosymbiont reveal evidence of lineages co-cladogenensis. FEMS 96(6) (article fiaa072):18 pp

Coley K (2015) Identification guide to larval stages of ticks of medical importance in the USA. Georgia South Univ Stud Res Pap 110:34 pp

Colle AC, Mendonça FB, Maia MO, Freitas LC, Martins TF, Muñoz-Leal S, Aguiar DM, Labruna MB, Rossi RV, Pacheco RC (2020) Rickettsial survey and ticks infesting small mammals from the Amazon forest in midwestern Brazil. Syst Appl Acarol 25:78–91

Colombo VC, Lareschi M, Monje LD, Nava S, Antoniazzi LR, Beldoménico PM, Guglielmone AA (2013) Garrapatas (Ixodida) y ácaros (Mesostigmata) parásitos de roedores sigmodontinos del Delta del Paraná, Argentina. Rev FAVE Sección Ciencias Vet 12:39–50

Colombo VC, Antoniazzi LR, Fasano AA, Beldoménico PM, Nava S (2016a) *Amblyomma triste* en simpatría con *Amblyomma tigrinum* (Acari: Ixodidae) en la provincia de Santa Fe, Argentina. Medicina (Buenos Aires) 76:304–306

Colombo VC, Pereira JA, Nava S, Beldoménico PM (2016b) *Amblyomma aureolatum* (Acari: Ixodidae): nuevos registros en las provincias de Buenos Aires y Santa Fe, Argentina. Rev FAVE Sección Ciencias Vet 15:1–4

Conil PA (1877) Description d'une nouvelle espèce d'ixode, *Ixodes testudinis*. Acta Acad Nacional Ciencias Exactas (Argentina) 3:25–37

Conil PA (1878) Description d'une nouvelle espèce d'ixode, *Ixodes auricularius*. Acta Acad Nacional Ciencias Exactas (Argentina) 3:99–110

Cooley RA (1937) Two new dermacentors [sic] from Central America. J Parasitol 23:259–264

Cooley RA (1938) The genera *Dermacentor* and *Otocentor* (Ixodidae) in the United States, with studies in variation. Natl Inst Health Bull 171:89 pp

Cooley RA (1944a) *Ixodes montoyanus* (Ixodidae) a new tick from Colombia. Bol Oficina Sanit Panam 23:804–806

Cooley RA (1944b) *Ixodes neotomae*, a new species from California (Acarina, Ixodidae). Pan Pac Entomol 20:7–12

Cooley RA (1945) *Ixodes tovari*, a new species from Mexico (Ixodidae). Pan Pac Entomol 21:144–148

Cooley RA (1946) The genera *Boophilus*, *Rhipicephalus*, and *Haemaphysalis* (Ixodoidea) of the New World. Natl Inst Health Bull 187:54 pp

Cooley RA (1947) *Dermacentor dissimilis*, a new species of tick from southern Mexico. Am Mus Novit 1332:1–3

Cooley RA, Kohls GM (1939) *Amblyomma philipi* – a new tick from Texas and Mexico, with a key to known species of *Amblyomma* in the United States. Publ Health Rep 54:44–47

Cooley RA, Kohls GM (1942a) *Antricola* new genus, *Amblyomma gertschi* new species, and notes on *Ixodes spinipalpis* (Acarina: Ixodidae). Publ Health Rep 57:1733–1736

Cooley RA, Kohls GM (1942b) *Ixodes mexicanus* n. sp. et *Ixodes tancitarius* n. sp. dos nuevas garrapatas mexicanas. Rev Soc Mexi Hist Nat 3:149–154

Cooley RA, Kohls GM (1943) *Ixodes californicus* Banks, 1904, *Ixodes pacificus* n. sp., and *Ixodes conepati* n. sp. Pan Pac Entomol 19:139–147

Cooley RA, Kohls GM (1944) The genus *Amblyomma* (Ixodidae) in the United States. J Parasitol 30:77–111

Cooley RA, Kohls GM (1945) *The* genus *Ixodes* in North America. Natl Inst Health Bull 184:246 pp

Cooper WF, Robinson LE (1908) On six new species of Ixodidae including a second species of the new genus *Rhipicentor* N. and W. Proc Camb Philos Soc 14:457–470

Corn JL, Kavanaugh DM, Creekmore TE, Robinson JL (1994) Wildlife as hosts for ticks (Acari) in Antigua, West Indies. J Med Entomol 31:57–61

Corn JL, Berger P, Mertins JW (2009) Survey for ectoparasites on wildlife associated with *Amblyomma variegatum* (Acari: Ixodidae)-infested livestock in St. Croix, U.S. Virgin Islands. J Med Entomol 46:1483–1489

Coronel-Benedett KC, Ojeda-Robertos NF, González-Garduño R, Martínez-Ibañez F, Rodríguez-VIvas RI (2018) Prevalence, intensity and population dynamics of hard ticks (Acari: Ixodidae) on sheep in the humid tropics of Mexico. Exp Appl Acarol 74:99–105

Correa O (1955) Carrapatos determinados no R. G. do Sul. Biología, patología e controle. Arq Inst Pesquisas Vet Desiderio Finamor 1:35–50

Corrier DE, Vizcaíno O, Terry M, Betancourt A, Kuttler KL, Carson CA, Treviño G, Ristic M (1979) Mortality, weight loss and anaemia in *Bos taurus* calves exposed to *Boophilus microplus* ticks in the tropics of Colombia. Trop Anim Health Prod 11:215–221

Cortés M, Pantoja A, Robles W, Pantoja J (2005) Importancia del control de garrapatas en Puerto Rico. J Agric Univ P R 89:59–66

Costa FB, Costa AP, Moraes-Filho J, Martins TF, Soares HS, Ramírez DG, Dias RA, Labruna MB (2017) *Rickettsia amblyommatis* infecting ticks and exposure of domestic dogs to *Rickettsia* spp. in an Amazon-Cerrado transition region of northeastern Brazil. PLoS One 12(6) (article e0179163):17 pp

Costa FB, Martins TF, Muñoz-Leal S, Serpa MCA, Ogrzewalska M, Luz HR, Barros-Battesti DM, Mesquita ETKC, Costa AP, Labruna MB (2020) Retrospective and new records of ticks (Acari: Argasidae, Ixodidae) from the State of Maranhão, an Amazon-Cerrado transition area of Brazil. Vet Parasitol Reg Stud Rep 21(article 100413):9 pp

Cotes-Perdomo A, Santodomingo A, Castro LR (2018) Hemogregarine and rickettsial infection in ticks of toads from northwestern Colombia. Int J Parasitol Parasites Wildl 7:237–242

Cotes-Perdomo AP, Oviedo A, Castro LR (2020) Molecular detection of pathogens in ticks associated with domestic animals from the Colombian Caribbean region. Exp Appl Acarol 82:137–150

Cox CB, Moore PD (2005) Biogeography. An ecological and evolutionary approach, 7th edn. Blackwell, Oxford, 428 pp

Crom RL (1992) Eradication of *Boophilus microplus* (Acari: Ixodidae) in Puerto Rico. Ann N Y Acad Sci 653:64–71

Crosbie PR, Boyce WM, Rodwell TC (1998) DNA sequence variation in *Dermacentor hunteri* and estimated phylogenies of *Dermacentor* spp. (Acari: Ixodidae) in the New World. J Med Entomol 35:277–288

Cruz-Aldán E, Torres IL, Güiris-Andrade DM, Osorio-Sarabia D, Quintero MT (2006) Parásitos del tapir centroamericano *Tapirus bairdii* (Perissodactyla: Tapiridae) en Chiapas, México. Rev Biol Trop 54:445–450

Cunha NC, Lemos ERS, Rozental T, Teixeira RC, Cordeiro MD, Lisboa RS, Favacho AR, Barreira JD, Rezende J, Fonseca AH (2014) Rickettsiae of the spotted fever group in dogs, horses and ticks: an epidemiological study in an endemic region of the State of Rio de Janeiro. Rev Bras Med Vet 36:294–300

Cuthbert R, Ryan PG, Cooper J, Hilton G (2003) Demographic and population trends of the Atlantic yellow-nosed albatross. Condor 105:439–452

Daciuk J, Cicchino AC, Mauri R, Capri JJ (1981) Notas faunísticas y bioecológicas de Península Valdés y Patagonia. XXIV. Artrópodos ectoparásitos de mamíferos y aves colectadas en la

Península Valdés y alrededores (provincia de Chubut, Argentina). Physis, Sección C 39 (97):41–48

Dall'Agnol B, Michel T, Weck B, Souza U, Webster A, Leal BF, Klafke G, Martins JR, Ott R, Venzal JM, Ferreira CAS, Reck J (2017) *Borrelia burgdorferi* sensu lato in *Ixodes longiscutatus* ticks from Brazilian pampa. Ticks Tick-Borne Dis 8:928–932

Danielová V, Marhoul Z, Dusbábek F, Ryba J, Fernández A, De La Cruz J, Abreu R, Herrera M, Rodríguez P, Cantelar N (1982) Isolation of Hughes virus from ticks in Cuba. Acta Virol 26:186–189

Dantas-Torres F, Oliveira EF, Soares FAM, Souza BOF, Valenca RBP, SÁ, F.B. (2008) Ticks infesting amphibians and reptiles in Pernambuco, northeastern Brazil. Rev Bras Parasitol Vet 17:218–221

Dantas-Torres F, Onofrio VC, Barros-Battesti DM (2009a) The ticks (Acari: Argasidae, Ixodidae) of Brazil. Syst Appl Acarol 14:30–46

Dantas-Torres F, Bernardi LF, Souza-Silva M, Ferreira RL, Onofrio VC, Barros-Battesti DM, Labruna MB (2009b) New records of *Ixodes paranaensis* (Acari: Ixodidae) from Minas Gerais, southeastern Brazil. Syst Appl Acarol 14:213–215

Dantas-Torres F, Ferreira DRA, Melo LM, Lima PACP, Siqueira DB, Albuquerque LCR, Melo AV, Ramos JAC (2010a) Ticks on captive and free-living wild animals in northeastern Brazil. Exp Appl Acarol 50:181–189

Dantas-Torres F, Siqueira DB, Rameh-de-Albuquerque LC, Silva e Souza DD, Zanotti AP, Ferreira DRA, Martins TF, Senna MB, Wagner PGC, Silva MA, Marvulo MFV, Labruna MB (2010b) Ticks infesting wildlife species in northeastern Brazil with new host and locality records. J Med Entomol 47:1243–1246

Dantas-Torres F, Aléssio FP, Siqueira DB, Mauffrey JF, Marvulo MFV, Martins TF, Moraes-Filho J, Camargo MCG, D'Auria SRN, Labruna MB, Silva JCR (2012) Exposure of small mammals to ticks and rickettsiae in Atlantic Forest patches in the metropolitan area of Recife, north-eastern Brazil. Parasitology 139:83–91

Dantas-Torres F, Mascarenhas-Junior PB, Anjos HR, Santos EM, Correia JMS (2019a) Tick infestation on caimans: a casual tick-host association in the Atlantic rainforest biome? Exp Appl Acarol 79:411–420

Dantas-Torres F, Martins TF, Muñoz-Leal S, Onofrio VC, Barros-Battesti DM (2019b) Ticks (Ixodida: Argasidae, Ixodidae) of Brazil: updated species checklist and taxonomic keys. Ticks Tick-Borne Dis 10(6)(article 101252):45 pp

De La Cruz J (1995) Nota sobre algunas garrapatas (Acarina: Ixodidae) de Nicaragua. Rev Nicaragüense Entomol 31:35–36

De La Cruz J (2001) Patterns in the biogeography of West Indian ticks. In: Woods CA, Sergile FE (eds) Biogeography of the West Indies: pattern and perspectives, 2nd edn. CRC, Boca Raton, pp 85–106

Debárbora VN, Oscherov EB, Guglielmone AA, Nava S (2011) Garrapatas (Acari: Ixodidae) asociadas a perros en diferentes ambientes de la provincia de Corrientes, Argentina. InVet 13:45–51

Debárbora VN, Nava S, Cirignoli S, Guglielmone AA, Poi ASG (2012) Ticks (Acari: Ixodidae) parasitizing endemic and exotic wild mammals in the Esteros del Iberá wetlands, Argentina. Syst Appl Acarol 17:243–250

Debárbora VN, Acosta JL, Mangold AJ (2015) Primer asociación parásito hospedador de *Amblyomma argentinae* Neumann, 1904 (Acari: Ixodidae) con tres especies de saurios del chaco semiárido argentino. Rev Argent Parasitol 3(3):6–7

Delabra-Vaca G, Fragoso-Sánchez H, Franco-Bello R, Martínez-Ibañez F, Ortiz-Estrada M, Ortiz-Nájera A, Osorio-Miranda J, Santamaría-Vargas M, Soberanes-Céspedes N (1996) Manual de identificación de las especies de garrapatas de importancia en México. Secretaría de Agricultura Ganadería y Desarollo Rural; Instituto Interamericano de Cooperación para la Agricultura, 89 pp

Di Benedetto IMD, Nava S, Oscherov EB (2013) Primer registro de *Didelphis albiventris* Lund, 1841 (Didelphimorphia: Didelphidae) como hospedador para adultos y ninfas de *Amblyomma ovale* Koch, 1844 (Acari: Ixodidae) en Argentina. Rev Argent Parasitol 1(3):45–48

Díaz MM, Nava S, Venzal JM, Sánchez N, Guglielmone AA (2007) Tick collections from the Peruvian Amazon, with new host records for species of *Ixodes* Latreille, 1795 (Acari: Ixodidae) and *Ornithodoros* Koch, 1844 (Acari: Argasidae). Syst Appl Acarol 12:127–133

Díaz NM, Méndez GS, Grijalva CJ, Walden HS, Cruz M, Aragón E, Hernández JA (2016) Dog overpopulation and burden of exposure to canine distemper virus and other pathogens on Santa Cruz Island, Galapagos. Prev Vet Med 123:128–137

Díaz-Sánchez AA, Chilton NB, Roblejo-Arias L, Fonseca-Rodríguez O, Marrero-Perera R, Diyes CP, Yunik MEM, Lobo-Rivero E, Corona-González B (2021) Molecular detection and identification of spotted fever group rickettsiae in ticks collected from horses in Cuba. Med Vet Entomol 35:207–212

Díaz Ungría C (1957) Nota sobre las especies de Acarina de Venezuela. Rev Sanid Asist Soc 22:457–467

Dietrich M, Kempf F, Boulinier T, McCoy KD (2014) Tracing the colonization and diversification of the worldwide seabird ectoparasite *Ixodes uriae*. Mol Ecol 23:3292–3305

Diogo AAR, Guerim L, Pires JR, Cuoto ALG, Serra-Freire NM (2003) Parasitismo por *Rhipicephalus sanguineus* Latreille, 1806 em *Columba livia* Linnaeus na cidade do Rio de Janeiro, Brasil. Entomol Vectores 10:277–280

Dios R (1917) Sistemática y biología de los ixodídeos argentinos. Contribución para su estudio. Anal Soc Rural Argent 51:249–251

Dios RL, Knopoff R (1930) Sobre Ixodoidea de la República Argentina. Rev Soc Argent Biol 6:593–627

Dios RL, Knopoff R (1934) Sobre Ixodoidea de la República Argentina. Rev Inst Bacteriol Dept Nacional Hyg 6:359–412

Dolz G, Castro R, Jiménez-Rocha AE, Retamosa M, Alberti A (2019) Strain diversity of *Rickettsia amblyommatis* in ticks infesting birds in the North Huetar conservation area of Costa Rica. Ticks Tick-Borne Dis 10:1109–1112

Domínguez L, Hernández A, Estrada E, Correa N, Cleghorn J, Bermúdez SE (2016) Inusual parasitismo de *Amblyomma mixtum* (Ixodida: Ixodidae) en *Trachemys scripta* (Testudines: Emydae [sic]) en Panamá. Rev Ibérica Acarol 28:137–139

Domínguez LG, Miranda RJ, Torres S, Moreno R, Ortega J, Bermúdez SE (2019) Hard tick (Acari: Ixodidae) survey of Oleoducto trail, Soberanía National Park, Panama. Ticks Tick-Borne Dis 10:830–837

Domínguez LG, Lezcano JJ, Rodríguez I, Miranda RJ, Bermúdez SC (2020a) Is *Geckobiella stamii* (Acari: Pterygosomatidae) a hyperparasite or phoretic on *Amblyomma dissimile* (Acari: Ixodidae) associated with *Iguana iguana* from Panama? Acarologia 60:40–44

Domínguez LG, Pérez E, Bermúdez S (2020b) First report of abnormal morphology in the tick *Dermacentor dissimilis* (Acari: Ixodidae) and evidence of molt nymph-adult on hosts from Nicaragua. Acarol Stud 2:46–50

Domínguez LG, Ortega J, Moreno R, Bermúdez S (2020c) Free-living *Panthera onca* (Carnivora: Felidae) as host of *Amblyomma mixtum* and *Rhipicephalus microplus* (Acari: Ixodidae) in Darién, Panamá. Acarol Stud 2:51–53

Dönitz W (1909) Über das Zeckengenus *Amblyomma*. Sber Ges Naturf Freunde Berl 8:440–482

Dönitz W (1910) Die Zecken Südafrikas. Denkschriften Medizinisch-Nat Ges Jena 16:397–494

Donoso R (1953) Ixodoidea de Chile. Rev Chil Entomol 3:132–134

Doss MA, Anastos G (1977) Index – catalogue of medical and veterinary zoology. Special publication N°. 3. Ticks and tickborne diseases. III. Checklist of families, genera, species and subspecies of ticks. United States Government Printing Office, Washington, 97 pp

Doss MA, Farr MM, Roach KF, Anastos G (1974a) Index – catalogue of medical and veterinary zoology. Special publication 3. Ticks and tick borne diseases. I. Genera and species of ticks. Part

1. Genera A – G. United States Department of Agriculture, United States Government Printing Office, Washington, 429 pp

Doss MA, Farr MM, Roach KF, Anastos G (1974b) Index – catalogue of medical and veterinary zoology. Special publication 3. Ticks and tick borne diseases. I. Genera and species of ticks. Part 2. Genera H – N. United States Department of Agriculture, United States Government Printing Office, Washington, 593 pp

Doss MA, Farr MM, Roach KF, Anastos G (1978) Index – catalogue of medical and veterinary zoology. Special publication 3. Ticks and tick borne diseases. IV. Geographical distribution of ticks. United States Department of Agriculture, United States Government Printing Office, Washington, 648 pp

Drake MC, Zieger U, Groszkowski A, Gallardo B, Sages P, Reavis R, Faircloth L, Jacobson K, Lonce N, Pinckney R, Cole RA (2014) Survey of helminths, ectoparasites and chytrid fungus of an introduced population of cane toads, *Rhinella marina* (Anura: Bufonidae), from Grenada, West Indies. J Parasitol 100:608–615

Drummond RO, Cadogan BL, Wilson TM (1981) Ticks on livestock in Barbados. Vet Parasitol 8:253–259

Dubie TR, Grantham R, Coburn L, Noden BH (2017) Pictorial key for identification of immature stages of common ixodid ticks found on pastures in Oklahoma. Southwest Entomol 42:1–14

Dugès AAD (1888) Description d'un nouvel ixodidé. Bull Soc Zool Fr 13:129–133

Dugès AAD (1902) *Trichodectes geomydis*, Osborn var. *expansus*, Alf. Dug. Mem Soc Científica Antonio Alzate 18:185–195

Dumbleton LJ (1953) The ticks (Ixodoidea) of the New Zealand subregion. Cape Expedition Ser Bull 14:5–35

Dumbleton LJ (1961) The ticks (Acarina: Ixodoidea) of sea birds in New Zealand waters. N Z J Sci 4:760–769

Dunn LH (1918) Studies on the iguana tick, *Amblyomma dissimile*, in Panama. J Parasitol 5:1–10

Dunn LH (1929) Notes on some insects and other arthropods affecting man and animals in Colombia. Am J Trop Med 9:493–508

Dunn LH (1933) Two new species of ticks from Panama (*Amblyomma tapirellum* and *A. pecarium*). Parasitology 25:353–358

Dunn LH (1934) Entomological investigations in the Chiriqui region of Panama. Psyche 41:166–183

Durden LA, Keirans JE (1994) Description of the larva, diagnosis of the nymph and female based on scanning electron microscopy, hosts, and distribution of *Ixodes* (*Ixodes*) *venezuelensis*. Med Vet Entomol 8:310–316

Durden LA, Keirans JE (1996) Nymphs of the genus *Ixodes* (Acari: Ixodidae) of the United States: taxonomy, identification key, distribution, hosts, and medical/ veterinary importance. Thomas Say Publ Entomol 9:95 pp

Durden LA, Knapp CR (2005) Ticks parasitizing reptiles in the Bahamas. Med Vet Entomol 19:326–328

Durden LA, Knapp CR, Beati L, Dold S (2015) Reptile-associated ticks from Dominica and the Bahamas with notes on hyperparasitic erythraeid mites. J Parasitol 101:24–27

Durden LA, Beckmen KB, Gerlach RF (2016) New records of ticks (Acari: Ixodidae) from dogs, cats, humans, and some wild vertebrates in Alaska: invasion potential. J Med Entomol 53:1391–1395

Düttmann C, Flores B, Kadoch N, Bermúdez S (2016) Hard ticks (Acari: Ixodidae) of livestock in Nicaragua, with notes about distribution. Exp Appl Acarol 70:125–135

Eads RB, Borom M (1975) Host and distributional records for the tick *Amblyomma inornatum* (Banks) (Acarina: Ixodidae), with description of the immature stages. J Med Entomol 12:493–496

Eberhardt AT, Fernández C, Zurvera DA, Monje LD (2016) First report of *Amblyomma dubitatum* (Acari: Ixodidae) parasitizing the crab-eating racoon, *Procyon cancrivorus* (Carnivora: Procyonidae). Rev FAVE Sección Ciencias Vet 14:20–21

Egizi AM, Robbins RG, Beati L, Nava S, Evans CR, Occi JL, Fonseca DM (2019) A pictorial key to differentiate the recently detected exotic *Haemaphysalis longicornis* Neumann, 1901 (Acari, Ixodidae) from native congeners in North America. ZooKeys 818:117–128

Elbl A, Anastos G (1966) Ixodid ticks (Acarina, Ixodidae) of Central Africa. Volume I. General introduction. Genus *Amblyomma* Koch, 1844. Ann Mus R l'Afrique Centrale Sér 8, Sci Zool 145:275 pp

Encinosa PEG, Martínez YG, Cruz RL, Ramos DA, Afonso YF, Soto YB, De La Guardia C, Alfaro YG, García AD, Estrada MP, Rodríguez-Mallon A (2021) Morphological and molecular characterization supporting *Amblyomma mixtum* presence in Cuba. Ticks Tick-Borne Dis 12 (1)(article101602):8 pp

Enríquez S, Guerrero R, Arrivillaga-Henríquez J, Araujo P, Villacrés E, Enríquez A, Benítez-Ortiz W (2020) New records of ticks of the genus *Amblyomma* Koch, 1844 (Acari: Ixodidae) for Ecuador. Acta Parasitol 65:430–440

Ernst CH, Ernst EM (1977) Ectoparasites associated with Neotropical turtles of the genus *Callopsis* (Testudines, Emydidae, Batagurinae). Biotropica 9:139–142

Ernst SE, Gladney WJ (1975) *Dermacentor albipictus*: hybridization of the two forms of the winter tick. Ann Entomol Soc Am 68:63–67

Escalante JA (1981) Ectoparásitos de animales domésticos en el Cusco. Rev Peruana Entomol 24:123–125

Espinoza-Gómez F, Newton-Sánchez O, Flores-Cazares G, De La Cruz-Ruiz M, Melnikov V, Austria-Tejeda J, Rojas-Lario F (2011) Tick paralysis caused by *Amblyomma maculatum* on the Mexican Pacific coast. Vector-Borne Zoon Dis 11:945–946

Esser HJ, Foley JE, Bongers F, Herre EA, Miller MJ, Prins HHT, Jansen PA (2016a) Host body size and the diversity of tick assemblages on Neotropical vertebrates. Int J Parasitol Parasites Wildl 5:295–304

Esser HJ, Herre EA, Blüthgen N, Loaiza JR, Bermúdez SE, Jansen PA (2016b) Host specificity in a diverse Neotropical tick community: an assessment using quantitative network analysis and host phylogeny. Parasites Vec 9(article 372):14 pp

Estébanes-González ML, Cervantes FA (2005) Mites and ticks associated with some small mammals in Mexico. Int J Acarol 31:23–37

Estrada-Peña A, Castellá J (1994) Check-list of the species of ticks (Ixodoidea). Part II. Genera *Amblyomma, Anomalohimalaya, Aponomma, Boophilus, Hyalomma, Margaropus, Palpoboophilus, Rhipicentor and Uroboophilus. Documents de Travail de l'Institut Royal des Sciences Naturelles de Belgique* (76):117 pp

Estrada-Peña A, Guglielmone AA, Mangold AJ, Castellá J (1993a) Patterns of cuticular hydrocarbon variation and genetic similarity between natural populations of *Amblyomma cajennense* (Acari: Ixodidae). Acta Trop 55:61–78

Estrada-Peña A, Guglielmone AA, Mangold AJ, Castellá J (1993b) A description of *Amblyomma tigrinum* Koch, *A. neumanni* Ribaga, and *A. testudinis* (Conil) immatures (Acarina: Ixodidae). Folia Parasitol 40:147–153

Estrada-Peña A, Venzal JM, Guglielmone AA (2002) *Amblyomma dubitatum* Neumann: description of nymph and redescription of adults, together with the description of the immature stages of *A. triste* Koch. Acarologia 42:323–333

Estrada-Peña A, Venzal JM, Mangold AJ, Cafrune MM, Guglielmone AA (2005) The *Amblyomma maculatum* Koch, 1844 (Acari: Ixodidae: Amblyomminae) tick group: diagnostic characters, description of the larva of *A. parvitarsum* Neumann, 1901, 16S rDNA sequences, distribution and hosts. Syst Parasitol 60:99–112

Estrada-Peña A, Venzal AJ, Nava S, Mangold A, Guglielmone AA, Labruna MB, de la Fuente J (2012) Reinstatement of *Rhipicephalus* (*Boophilus*) *australis* Fuller, the Australian cattle tick (Acari: Ixodidae) with redescription of the adult and larval stages. J Med Entomol 49:794–802

Estrada-Peña A, Mihalca AD, Petney TN (2017) Ticks of Europe and North Afrrica. Springer, Cham, 404 pp

Estrada-Peña A, Nava S, Tarragona E, de la Fuente J, Guglielmone AA (2020) A community approach to the Neotropical ticks-hosts interactions. Sci Rep 10(article 9269):9 pp

Evans HE (1947) Notes on Panamanian reptiles and amphibians. Copeia 1947:166–170

Evans DE, Martins JR, Guglielmone AA (2000) A review of the ticks (Acari, Ixodida) of Brazil, their hosts and geographical distribution – 1. The State of Rio Grande do Sul, southern Brazil. Mem Inst Oswaldo Cruz 95:453–470

Everard COR, Tikansingh ES (1973) Ecology of the rodents, *Proechimys guyannensis trinitatis* and *Oryzomys capito velutinus*, on Trinidad. J Mammal 54:875–886

Eyo JE, Ekeh FN, Ivoke N, Atama CI, Onah IE, Ezenwaji NE, Ikele CB (2014) Survey of tick infestation of cattle at four selected grazing sites in the tropics. Glob Vet 12:479–486

Fabricius JC (1775) Systema Entomologiae sistens insectorum classes, ordines, genera, species, adiectis synonymis, locis, descriptionibus, observationibus. Flensburg and Leipzig, 832 pp

Fabricius JC (1787) Mantisa insectorum sistens species nuper detectas adiectis synonymis, observationibus, descriptionibus, emendationibus. Hafniae 2:382 pp

Fabricius JC (1794) Entomologia systematica emendata et aucta, secundum classes, ordines, genera, species adjectis synonimis, loci, observationibus, descriptionibus. Hafniae 4:472 pp

Fabricius JC (1798) Supplementum Entomologiae Systematicae. Hafniae, 572 pp

Fabricius JC (1805) Systema Antliatorum secundum ordines, genera, species adiectis synonymis, locis, observationibus, descriptionibus. Carolum Reichard, Brunsvigae, 372 pp

Faccini JLH, Prata MCA, Daemon E, Barros-Battesti DM (1999) Características biológicas da fase nao parasitária do *Ixodes amarali* (Acari: Ixodidae) em gambá (*Didelphis* sp.) no estado do Rio de Janeiro. Arq Bras Med Vet Zoot 51:267–270

Faccini-Martínez AA, Ramírez-Hernández A, Barreto C, Forero-Becerra E, Millán D, Valbuena E, Sánchez-Alfonso AC, Imbacuán-Pantoja WO, Cortés-Vecino JA, Polo-Terán LJ, Yaya-Lancheros N, Jácome J, Palomar AM, Santibañez S, Portillo A, Oteo JA, Hidalgo M (2017) Epidemiology of spotted fever group rickettsioses and acute undifferentiated febrile illness in Villeta, Colombia. Am J Trop Med Hyg 97:782–788

Faccini-Martínez AA, Martins TF, Silveira I, Labruna MB (2021) Detection of *Dermacentor andersoni* (Acari: Ixodidae) in Brazil on a human traveler returning from the United States. J Med Entomol 58:947–949

Fairchild GB (1943) An annotated list of the bloodsucking insects, ticks and mites known from Panama. Am J Trop Med Hyg 23:569–591

Fairchild GB, Kohls GM, Tipton VJ (1966) The ticks of Panama (Acarina: Ixodoidea). In: Wenzel WR, Tipton VJ (eds) Ectoparasites of Panama. Field Museum of Natural History, Chicago, pp 167–219

Famadas KM, Faccini JLH (1989) Variação morfológica de *Boophilus microplus* (Canestrini, 1887) (Acari: Ixodidae) no Brasil. Arq Univ Federa Rural Rio de Janeiro 12:73–81

Famadas KM, Serra-Freire NM, Lanfredi RM (1997) Redescription of the larva of *Amblyomma cajennense* (Fabricius) (Acari: Ixodidae) using optical and scanning electron microscopy. Acarologia 38:101–109

Fantozzi MC, Tarragona EL, Dayenoff P, Cuervo PF, Mera y Sierra R, Nava S (2018) Report of *Amblyomma parvitarsum* infesting goats in midwestern Argentina. Vet Parasitol Reg Stud Rep 14:99–102

Fecchio A, Martins TF, Bell JA, De La Torre GM, Pinho JB, Weckstein JD, Tkach VV, Labruna MB, Dias RI (2020) Low host specificity and lack of parasite avoidance by immature ticks in Brazilian birds. Parasitol Res 119:2039–2045

Fernandes MMP, Medeiros FER, Carvalho YK, Ribeiro VMF, Souza SF (2018) Ectoparasitas de cães domiciliados e errantes do município de Rio Branco-Acre. Enciclopedia Biosfera 15:442–450

Ferreira CGT, Rego IG, Ahid SMM (2008) Parasitismo em humano por *Amblyomma parvum* Aragão, 1908 (Acari: Ixodidae) em Paus dos Ferros, Rio Grande do Norte, Brasil. Pubvet 3(1) (article 478):6 pp

Ferreira CGT, Bezerra ACDS, Carvalho OV, Almeida MR, Mafra C (2013) First occurrence of *Amblyomma ovale* in the State of Rio Grande do Norte, Brazil. Braz J Vet Parasitol 22:167–170

Ferrell AM, Brinkerhoff RJ, Bernal J, Bermúdez SE (2017) Ticks and tick-borne pathogens of dogs along an elevational and land-use gradient in Chiriquí Province, Panama. Exp Appl Acarol 71:371–385

Fiasson R (1949) Contribución al estudio de los ácaros de Venezuela. Rev Grancolombina Zoot Higiene Med Vet 3:567–588

Figueiredo LTM, Badra SJ, Pereira LE, Szabó MPJ (1999) Report on ticks collected in the southeast and mid-west regions of Brazil: analyzing the potential transmission of tick-borne pathogens to man. Rev Soc Bras Med Trop 32:613–619

Figueiredo MAP, Santos ACG, Guerra RMSNC (2010) Ectoparasitos de animais silvestres no Maranhão. Pesquisa Vet Bras 30:988–990

Filippova NA (1977) Ixodid ticks (Ixodinae). Fauna USSR New Series 4(4), Nauka, Moscow, Leningrad, 396 pp [in Russian]

Filippova NA (2008) Type specimens of argasid and ixodid ticks (Ixodoidea: Argasidae, Ixodidae) in the collection of the Zoological Institute, Russian Academy of Sciences (St. Petersburg). Entomol Rev 88:1002–1011

Floch H, Abonnenc E (1940) Ixodidés de la Guyane Française, I. Arch l'Inst Pasteur Guyane Française l'Inini 3:46 pp

Floch H, Abonnenc E (1941) Ixodidés de la Guyane Française II. Arch l'Inst Pasteur Guyane Française l'Inini 4:31 pp

Floch H, Abonnenc E (1942) Ixodidés de la Guyane Française, III. Arch l'Inst Pasteur Guyane Française l'Inini 32:16 pp

Floch H, Fauran P (1958) Ixodidés de la Guyane et des Antilles Françaises. Arch l'Inst Pasteur Guyane l'Inini 446:94 pp

Floch H, Fauran P (1959a) Les Ixodidae du genre *Amblyomma* en Guyane et aux Antilles Françaises. Acarologia 1:216–227

Floch H, Fauran P (1959b) Les ixodidés, autres que ceux du genre *Amblyomma* en Guyane et aux Antilles Françaises. Acarologia 1:393–407

Floch H, Fauran P, Alexandre A (1955) Ixodidés de la Guyane Française, V Le genre *Ixodes*. Arch l'Inst Pasteur Guyane Française l'Inini 364:6 pp

Flores FS, Nava S, Batallán G, Tauro LB, Contigiani MS, Díaz LA, Guglielmone AA (2014) Ticks (Acari: Ixodidae) on wild birds in north-central Argentina. Ticks Tick-Borne Dis 5:715–721

Flores FS, Saracho-Bottero MN, Sebastian PS, Venzal JM, Mangold AJ, Nava S (2020) *Borrelia* genospecies in *Ixodes* sp. cf. *Ixodes affinis* (Acari: Ixodidae) from Argentina. Ticks Tick-Borne Dis 11(6)(article 101546) 6 pp

Fonseca F (1935a) Notas de acareologia. XV. Ocorrência de subspecies de *Ixodes ricinus* (L., 1758) no estado de S. Paulo (Acarina, Ixodidae). Mem Inst Butantan 9:131–135

Fonseca F (1935b) Notas de acareologia. XVI. *Ixodes amarali*, sp. n. (Acarina: Ixodidae). Mem Inst Butantan 9:137–143

Fonseca F (1937) Occurrence of a new subspecies of *Ixodes ricinus* (L.) in the State of S. Paulo, Brazil. Comptes Rendus 12th. Congrès International de Zoologie, Lisbon 1935, pp 1587–1592

Fonseca F (1958a) Notas de acarologia XLIV. Inquérito sobre a fauna acarológica de parasitas no nordeste do Brasil. Mem Inst Butantan 28:99–186

Fonseca F (1958b) Notas de acarología. XL Ácaros do estado do Maranhão. Pap Avulsos Zool 13:133–139

Fonseca F (1959) Notas de acarología. XLVI Acarofauna zooparasita na Bolivia. Mem Inst Butantan 29:89–141

Fonseca F (1960) Notes d'acarologie. XLV Enquête acarologique au Pérou. Acarologia 2:1–34

Fonseca F, Aragão HB (1952a) Notas de ixodología. II. Uma nova espécie do gênero *Amblyomma* e uma nova espécie do gênero *Ixodes* (Acari, Ixodidae). Mem Inst Oswaldo Cruz 50:713–726

Fonseca F, Aragão HB (1952b) Notas de ixodología. IV. Considerações sobre a nomenclatura de algumas espécies do género *Amblyomma* do Brasil e países limítrofes (Acari: Ixodidae). Mem Inst Oswaldo Cruz 50:729–731

Fonseca F, Aragão HB (1953) Notas de ixodología. IV. Descrição da femea de *Amblyomma multipunctum* Neumann 1899 e redescrição de macho (Acari: Ixodidae). Mem Inst Oswaldo Cruz 51:493–498

Fonseca F, Trindade G (1957) Notas de acarología. XLIII Fauna acarológica de Ouro Preto. Mem Inst Butantan 28:59–66

Fonseca MS, Bahiense TC, Silva AAB, Onofrio VC, Barral TD, Souza BMP, Lira da Silva RM, Biondi I, Meyer R, Portela RW (2020) Ticks and associated pathogens from rescued wild animals in rainforest fragments of northeastern Brazil. Front Vet Sci 7(article 117):11 pp

Fontalvo MC, Favacho ARM, Araujo AC, Santos NM, Oliveira GMB, Aguiar DM, Lemos ERS, Horta MC (2017) *Bartonella* species pathogenic for humans infect pets, free-ranging wild mammals and their ectoparasites in the Caatinga biome, northeastern Brazil: a serological and molecular study. Braz J Infect Dis 21:290–296

Forlano M, Mujica F, Coronado A, Meléndez RD, Linardi PM, Botelho JR, Bellosta P, Barrios N (2008) Especies de *Amblyomma* (Acari: Ixodidae) parasitando perros (*Canis familiaris*) en áreas rurales de los estados de Lara, Yaracuy, Carabobo y Falcón, Venezuela. Rev Científica FCV-LUZ 18:662–666

Forlano M, Mujica F, Gallardo J, Rodríguez J (2013) Identificación de ectoparásitos en oso melero (*Tamandua tetradactyla*), estado Lara, Venezuela. Rev MVZ 18(Suplemento):3738–3742

Fournier GFSR, Pinter A, Santiago R, Muñoz-Leal S, Martins TF, Lopes MG, McCoy KD, Toty C, Horta MC, Labruna MB, Dias RA (2019) A high gene flow in populations of *Amblyomma ovale* ticks found in distinct fragments of Brazilian Atlantic rainforest. Exp Appl Acarol 77:215–228

Fox I (1950) The brown dog tick on rats in Puerto Rico. J Econ Entomol 43:722–723

Fox I, León D (1983) The tropical horse tick and the southern cattle tick in Puerto Rico. J Agric Univ P R 67:57–59

Freire JJ (1972) Revisão das espécies da familia Ixodidae. Rev Med Vet (São Paulo) 8:1–16

Fujita H, Takada N (2007) Identification of immature ticks in Japan. In: Organization Committee of SADI (ed) Acari and emerging/reemerging diseases. Kyoiku Kyokai, Tokyo, pp 53–68. [in Japanese]

Fuller C (1899) Notes on the Queensland cattle tick and its relationship to the Texas fever and the blue tick of Cape Colony, (South Africa). Queensl Agric J 4:389–394

Furman DP, Loomis EC (1984) The ticks of California (Acari: Ixodida). Bull Calif Insect Surv 25:239 pp

Gadsen H, Guerra G (1991) Los ácaros ectoparásitos de *Amblyrhynchus cristatus* Bell (Sauria: Iguanidae) como trazadores zoogeográficos, en las islas Galápagos, Ecuador. Folia Entomol Mexic 83:183–197

Galli-Valerio G (1909) Notes de parasitologie et de technique parasitologique. Zentralblatt Bakteriol Parasitenkd Infekt Hyg 51:538–545

Garcés-Restrepo MF, Giraldo A, Carr JL, Brown LD (2013) Turtle ectoparasites from the Pacific coastal region of Colombia. Biota Neotrop 13:74–79

García GG, Castro AM, Rodríguez I, Bermúdez SC (2014) Ixodid ticks of *Hydrochoerus isthmius* Goldman, 1912 (Rodentia: Caviidae) in Panama. Syst Appl Acarol 19:404–408

Garnham PCC, Lewis DJ (1959) Parasites of British Honduras with special reference to leishmaniasis. Trans R Soc Trop Med Hyg 53:12–40

Garris GI (1987) *Amblyomma variegatum* (Acari: Ixodidae): population dynamics and hosts used during an eradication program in Puerto Rico. J Med Entomol 24:82–86

Garris GI, George JE (1983) Efficacy of crotoxyphos sprays for the control of the southern cattle tick in Puerto Rico. J Agric Univ P R 67:256–261

Garris GI, George JE (1985) Field evaluation of amitraz applied to cattle as sprays for control of *Boophilus microplus* (Acari: Ixodidae) in the eradication program in Puerto Rico. Prev Vet Med 3:363–369

Garris GI, Scotland K (1985) Ticks on livestock in St. Lucia. Vet Parasitol 18:367–373

Garris GI, Bokma BH, Strickland RK, Combs GP (1989) Evaluation of the eradication program for *Amblyomma variegatum* (Acari: Ixodidae) on Puerto Rico. Exp Appl Acarol 6:67–76

Gerardi M, Martins MM, Nava S, Szabó MPJ (2013) Comparing feeding and reproductive parameters of *Amblyomma parvum* tick populations (Acari: Ixodidae) from Brazil and Argentina on various host species. Vet Parasitol 197:312–317

Gerstäcker A (1873) Gliederthiere (Insekten, Arachniden, Myriapoden und Isopoden). In: Kersten O (ed) Baron Carl Claus von der Decken's Reisen in Ost Africa in den Jahre 1859 bis 1865, vol 3. Wissenschaftlicher Theil, C.J. Winter'sche Verlaghandlung, Leipzig, 542 pp

Gervais P (1842) Quinzaine d'espèces d'insectes aptères. Ann Soc Entomol France, Bull Entomol 11:45–48

Gervais P (1844) Acarides. In: Walckenear CA (ed) Histoire naturelle des insects. Libraire Encyclopédique de Roret, Paris, pp 132–288

Gervais P (1849) Arácnidos. In: Gay C (ed) Historia física y política de Chile, vol 4, Paris, pp 5–52

Gianizella SL, Junior JM, Nascimento CAR, Martins TF (2018a) Primeiro registro de machos de *Amblyomma rotundatum* (Acari: Ixodidae) em jabuti-tinga (*Cheolonoidis denticulatus*) no estado do Amazonas, Amazonia brasileira: relato de caso. Arq Bras Med Vet Zoot 70:195–198

Gianizella SL, Martins TF, Onofrio VC, Aguiar NO, Gravena W, Nascimento CAR, Neto LC, Faria DL, Lima NAS, Solorio MR, Maranhão L, Lima IJ, Cobra IVD, Santos T, Lopes GP, Ramalho EE, Luz HR, Labruna MB (2018b) Ticks (Acari: Ixodidae) of the State of Amazonas, Brazil. Exp Appl Acarol 74:177–183

Gittens RA, Almanza A, Bennett KL, Mejía LC, Sanchez-Galan JE, Merchan F, Kern J, Miller MJ, Esser HJ, Hwang R, Dong M, De León LF, Álvarez E, Loaiza JR (2020) Proteomic fingerprinting of Neotropical hard tick species (Acari: Ixodidae) using a self-curated mass spectra reference library. PLoS Negl Trop Dis 14(10)(article e0008849):18 pp

Gladney WJ, Dawkins CC, Price MA (1977) *Amblyomma inornatum* (Acarina: Ixodidae): natural hosts and laboratory biology. J Med Entomol 14:85–88

Glenny M, Mendoza L, Falconi E (2004) Detección de anticuerpos contra *Borrelia burgdorferi* e identificación de garrapatas ixodidas en Piura y Amazonas, Peru. Rev Peru Med Exp Salud Publ 21:23–27

Gomes SN, Pesenti TC, Müller G (2015) Parasitismo por *Amblyomma ovale* e *Amblyomma fuscum* (Acari: Ixodidae) em *Dasypus novemcinctus* (Xenarthra: Dasypodidae) no Brasil. Arq Inst Biol 82:1–4

González IHL, Labruna MB, Chagas CRF, Salgado PAB, Monticelli C, Morais LH, Moraes AA, Antunes TC, Ramos PL, Martins TF (2017) Ticks infesting captive and free-roaming wild animal species at the São Paulo Zoo, São Paulo, Brazil. Braz J Vet Parasitol 26:496–499

González-Acuña D, Guglielmone AA (2005) Ticks (Acari: Ixodoidea: Argasidae, Ixodidae) of Chile. Exp Appl Acarol 35:147–163

González-Acuña D, Venzal JM, Guglielmone AA (2003a) Primer registro de *Rhipicephalus sanguineus* (Latreille, 1806) (Acari: Ixodidae) en *Rattus norvegicus* (Mammalia: Rodentia) en Chile. Gayana 67:121–122

González-Acuña D, Venzal JM, Keirans JE, Guglielmone AA (2003b) The genus *Amblyomma* Koch, 1844 (Acari: Ixodidae) in Chile, with new records of *A. argentinae* Neumann, 1904 and *A. tigrinum* Koch, 1844. Syst Appl Acarol 8:85–88

González-Acuña D, Venzal JM, Guglielmone AA (2004a) Nuevos hospedadores y localidades de *Ixodes sigelos* e *Ixodes auritulus* (Acari: Ixodidae) en Chile. Gayana 68:113–116

González-Acuña D, Venzal JM, Fabry M, Guglielmone AA (2004b) *Liolaemus jamesi* (Boulanger, 1891) (Reptilia: Tropiduridae), a host for the larva of *Amblyomma parvitarsum* Neumann, 1901 (Acari: Ixodidae). Syst Appl Acarol 9:33–36

González-Acuña D, Beldoménico PM, Venzal JM, Fabry M, Keirans JE, Guglielmone AA (2005a) Reptile trade and the risk of exotic tick introductions into southern South American countries. Exp Appl Acarol 35:335–339

González-Acuña D, Venzal JM, Keirans JE, Robbins RG, Ippi S, Guglielmone AA (2005b) New host and locality records for the *Ixodes auritulus* (Acari: Ixodidae) species group, with a review of host relationships and distribution in the Neotropical Zoogeographic Region. Exp Appl Acarol 37:147–156

González-Acuña D, Nava S, Mangold AJ, Guglielmone AA (2008) *Ixodes downsi* Kohls, 1957 in Venezuela. Syst Appl Acarol 13:39–42

González-Acuña D, Mangold AJ, Robbins RG, Guglielmone AA (2009) New host and locality records for the *Ixodes auritulus* Neumann, 1904 (Acari: Ixodidae) species group in northern Chile. Syst Appl Acarol 14:47–50

González-Acuña DA, Saracho-Bottero MN, Ossa G, Guglielmone AA, Nava S (2018) *Ixodes chilensis* Kohls, 1956 (Acari, Ixodida: Ixodidae): re-description of the female, description of the nymph, and phylogenetic position inferred from mitochondrial DNA sequences of the 16S rRNA gene. Syst Parasitol 95:959–967

González-Martín del Campo F, Navarrete-Gutiérrez DA, Enríquez PL, Gordillo-Pérez MG, Cabrera-Romo S (2018) Ticks of wild birds at sites with different land uses at Campeche, Mexico. Southwest Entomol 43:677–681

Gorham JR (1995) First report of *Amblyomma coelebs* from Honduras, with notes on other arthropods. Rev Nicaragüense Entomol 33:7–12

Gothe R (1967) Ticks in the South African zoological survey collection. Part XII – the genera *Boophilus* Curtice, 1891 and *Margaropus* Karsch, 1879. Onderstepoort J Vet Sci 34:81–108

Graham OH, Gladney WJ, Beltrán LG (1975) Comparación de la distribución e importancia económica de *Amblyomma maculata* [sic] Koch (Acarina: Ixodidae) en México y los Estados Unidos. Folia Entomol Mexic 33:66–67

Gregson JD (1956) The Ixodoidea of Canada. Can Dept Agric Sci Service Entomol Div Publ (930):92 pp

Gruhn KD, Ogrzewalska M, Rozental T, Farikoski IO, Blanco C, Freitas LS, Lemos ERS, Ribeiro VMF (2019) Evaluation of rickettsial infection in free-range capybaras (*Hydrochoerus hydrochaeris* Linnaeus, 1766 [sic]) (Rodentia: Caviidae) and ticks (Acari: Ixodidae) in the Western Amazon, Brazil. Ticks Tick-Borne Dis 10:981–986

Guérin-Méneville F-E (1844) Iconographie du règne animal de G. Cuvier (1829–1843), ou représentation d'après nature de l'une des espèces les plus remarquables et souvent non encore figurées de chaque genre d'animaux (4). Arachnides, Paris, 20 pp

Guerra RMSNC, Brito DRB (2004) Ixodofauna de mamíferos domésticos da Ilha de Sño Luis, estado do Maranhão, Brasil. Entomol Vectores 11:435–444

Guerrero RP (1931) Nota preliminar sobre las garrapatas de los animales domésticos en Colombia. Rev Med Vet (Bogotá) 3(17):45–54

Guerrero R (1996) Las garrapatas de Venezuela (Acarina: Ixodoidea). Listado de especies y claves para su identificación. Bol Dirección Malariología Saneamiento Ambiental 36:1–24

Guerrero R, Rodríguez-Acosta A (2014) Primer reporte de infestación de la serpiente reinita *Erythrolamprus melanotus* (Shaw, 1802) (Serpentes: Dipsadidae) por el ácaro *Amblyomma dissimile* Koch, 1844 (Acarina: Ixodidae) en Venezuela. Cuadernos Herpetol 28(2):141–143

Guglielmone AA (1995) Epidemiology of babesiosis and anaplasmosis in South and Central America. Vet Parasitol 57:109–119

Guglielmone AA, Hadani H (1980) Hallazgos de *Amblyomma parvum* Aragão, 1908, en Catamarca y Salta. Rev Med Vet (Buenos Aires) 61:121–129

Guglielmone AA, Hadani H (1981) La distribución geográfica de *Amblyomma neumanni* Ribaga, 1902, en la Argentina. Gaceta Vet 42:754–760

Guglielmone AA, Hadani H (1985) *Amblyomma neumanni* Ribaga, 1902: su nomenclatura y taxonomía. Vet Argent 2:721–728

Guglielmone AA, Mangold AJ (1987) Áreas del noroeste argentina conideradas libres de *Boophilus microplus* (Can. 1888). Anal INTA Salta 1:27–30

Guglielmone AA, Nava S (2005) Las garrapatas de la familia Argasidae y de los géneros *Dermacentor, Haemaphysalis, Ixodes* y *Rhipicephalus* (Ixodidae) de la Argentina: distribución y hospedadores. Rev Invest Agropecuarias 34(2):123–141

Guglielmone AA, Nava S (2006) Las garrapatas argentinas del género *Amblyomma* (Acari: Ixodidae): distribución y hospedadores. Rev Invest Agropecuarias 35(3):135–155

Guglielmone AA, Nava S (2010) Hosts of *Amblyomma dissimile* Koch, 1844 and *Amblyomma rotundatum* Koch, 1844 (Acari: Ixodidae). Zootaxa 2541:27–49

Guglielmone AA, Nava S (2011) Rodents of the subfamily Sigmodontinae (Myomorpha: Cricetidae) as hosts for South American hard ticks (Acari: Ixodidae) with hypothesis of life history. Zootaxa 2904:45–65

Guglielmone AA, Nava S (2014) Names for Ixodidae (Acari: Ixodoidea): valid, synonyms, incertae sedis, nomina dubia, nomina nuda, lapsus, incorrect and suppressed names – with notes on confusions and misidentifications. Zootaxa 3767:1–256

Guglielmone AA, Nava S (2017) Birds and hard ticks (Acari: Ixodidae), with discussions about hypotheses on tick evolution. Rev FAVE Sección Ciencias Vet 16:13–29

Guglielmone AA, Robbins RG (2018) Hard ticks (Acari: Ixodida: Ixodidae) parasitizing humans. A global overview. Springer, Cham, 314 pp

Guglielmone AA, Viñabal AE (1994) Claves morfológicas dicotómicas e información ecológica para la identificación de las garrapatas del género *Amblyomma* Koch, 1844 en la Argentina. Rev Invest Agropecuarias 25(1):39–67

Guglielmone AA, Mangold AJ, Hadani A (1982) *Amblyomma tigrinum* Koch, 1844 en la Argentina. Su diagnóstico erróneo como *Amblyomma maculatum* y su distribución geográfica. Gaceta Vet 44:57–63

Guglielmone AA, Viñabal AE, Mangold AJ, Aguirre DH, Gaido AB (1989) Un estudio epizootiológico sobre garrapatas del grupo *Rhipicephalus sanguineus* en el Valle de Lerma, Salta, Argentina. Rev Med Vet (Buenos Aires) 70:230–237

Guglielmone AA, Mangold AJ, Keirans JE (1990) Redescription of the male and female of *Amblyomma parvum* Aragão, 1908, and description of the nymph and larva, and description of all stages of *Amblyomma pseudoparvum* sp. n. (Acari: Ixodida: Ixodidae). Acarologia 31:143–159

Guglielmone AA, Mangold AJ, Aufranc CR (1992a) *Haemaphysalis juxtakochi, Ixodes pararicinus* (Ixodidae) and *Otobius megnini* (Argasidae) in relation to the phytogeography of Argentina. Ann Parasitol Hum Comp 67:91–93

Guglielmone AA, Mangold AJ, Oyola BC (1992b) Ciclo de vida del *Amblyomma cajennense* (Fabricius, 1787) (Acari: Ixodidae) en condiciones de laboratorio. Rev Med Vet (Buenos Aires) 73:184–187

Guglielmone AA, Castellá J, Mangold AJ, Estrada-Peña A, Viñabal E (1999) Phenotypical anomalies in specimens of a Neotropical tick (Ixodidae) collection. Acarologia 40:127–132

Guglielmone AA, Luciani CA, Mangold AJ (2001) Aspects of the ecology of *Amblyomma argentinae* Neumann, 1904 [= *Amblyomma testudinis* (Conil, 1877)] (Acari: Ixodidae). Syst Appl Acarol 8:1–12

Guglielmone AA, Estrada-Peña A, Keirans JE, Robbins RG (2003a) Ticks (Acari:Ixodida) of the Neotropical Zoogeographic Region. Special Publication of the Integrated Consortium on Ticks Tick-Borne Dis-2, Atalanta, Houten, 173 pp

Guglielmone AA, Estrada-Peña A, Luciani CA, Mangold AJ, Keirans JE (2003b) Hosts and distribution of *Amblyomma auricularium* (Conil 1878) and *Amblyomma pseudoconcolor* Aragão, 1908 (Acari: Ixodidae). Exp Appl Acarol 29:131–139

Guglielmone AA, Estrada-Peña A, Mangold AJ, Barros-Battesti DM, Labruna MB, Martins JR, Venzal JM, Arzua M, Keirans JE (2003c) *Amblyomma aureolatum* (Pallas, 1772) and *Amblyomma ovale* Koch, 1844 (Acari: Ixodidae): hosts, distribution and 16S rDNA sequences. Vet Parasitol 113:273–288

Guglielmone AA, Venzal JM, Amico G, Mangold AJ, Keirans JE (2004) Description of the larva and redescription of the nymph and female of *Ixodes neuquenensis* Ringuelet, 1947 (Acari:

Ixodidae), a parasite of the endangered species *Dromiciops gliroides* Thomas (Microbiotheria: Microbiotheriidae). Syst Parasitol 57:211–219

Guglielmone AA, Romero J, Venzal JM, Nava S, Mangold AJ, Villavicencio J (2005a) First record of *Haemaphysalis juxtakochi* Cooley, 1946 (Acari: Ixodidae) from Peru. Syst Appl Acarol 10:33–35

Guglielmone AA, González-Acuña D, Autino AG, Venzal JM, Nava S, Mangold AJ (2005b) *Ixodes sigelos* Keirans, Clifford & Corwin, 1976 (Acari: Ixodidae) in Argentina and southern Chile. Syst Appl Acarol 10:37–40

Guglielmone AA, Beati L, Barros-Battesti DM, Labruna MB, Nava S, Venzal JM, Mangold AJ, Szabó MPJ, Martins JR, González-Acuña D, Estrada-Peña A (2006a) Ticks (Ixodoidea) on humans in South America. Exp Appl Acarol 40:83–100

Guglielmone AA, Venzal JM, González-Acuña D, Nava S, Mangold AJ (2006b) The phylogenetic position of *Ixodes stilesi* Neumann, 1911 (Acari: Ixodidae): morphological and preliminary molecular evidences from 16S rDNA sequences. Syst Parasitol 65:1–11

Guglielmone AA, Nava S, Mangold AJ, Caparrós JA (2007a) Las garrapatas del norte de Córdoba, Argentina. Rev Med Vet (Buenos Aires) 88:57–61

Guglielmone AA, Nava S, González-Acuña D, Mangold AJ, Robbins RG (2007b) Additional observations on the morphology and hosts of *Ixodes stilesi* Neumann, 1911 (Acari: Ixodidae). Syst Appl Acarol 12:135–139

Guglielmone AA, Robbins RG, Apanaskevich DA, Petney TN, Estrada-Peña A, Horak IG (2009) Comments on controversial tick (Acari: Ixodida) species names and species described or resurrected from 2003 to 2008. Exp Appl Acarol 48:311–327

Guglielmone AA, Nava S, Bazán-León EA, Mangold AJ (2010a) Redescription of the male and description of the female of *Ixodes abrocomae* Lahille, 1916 (Acari: Ixodidae). Syst Parasitol 77:153–160

Guglielmone AA, Robbins RG, Apanaskevich DA, Petney TN, Estrada-Peña A, Horak IG, Shao R, Barker SC (2010b) The Argasidae, Ixodidae and Nuttalliellidae (Acari: Ixodida) of the world: a list of valid species names. Zootaxa 2528:1–28

Guglielmone AA, Nava S, Díaz MM (2011) Relationships of South American marsupials (Didelphimorphia, Microbiotheria and Paucituberculata) and hard ticks (Acari: Ixodidae) with distribution of four species of *Ixodes*. Zootaxa 3086:1–30

Guglielmone AA, Nava S, Mastropaolo M, Mangold AJ (2013) Distribution and genetic variation of *Amblyomma triste* (Acari: Ixodidae) in Argentina. Ticks Tick-Borne Dis 4:386–390

Guglielmone AA, Robbins RG, Apanaskevich DA, Petney TN, Estrada-Peña A, Horak IG (2014) The hard ticks of the world (Acari: Ixodida: Ixodidae). Springer, Dordrecht, 738 pp

Guglielmone AA, Sánchez ME, Franco LG, Nava S, Rueda LM, Robbins RG (2015, continuously updated) Hard ticks (Acari: Ixodida: Ixodidae): a non-profit open-access web portal for original descriptions of tick species (valid and invalid), dubious and uncertain names, and selected *nomina nuda*. http://rafaela.inta.gob.ar/nombresgarrapatas/. Last accession January 20, 2021

Guglielmone AA, Petney TN, Mastropaolo M, Robbins RG (2017) Genera, subgenera, species and subspecies of hard ticks (Acari: Ixodidae) described, named, renamed or given new rank by Paul Schulze (1887-1949) and their current status. Zootaxa 4325:1–66

Guglielmone AA, Petney TN, Robbins RG (2020) Ixodidae (Acari: Ixodoidea): descriptions and redescriptions of all known species from 1758 to December 31, 2019. Zootaxa 4871:1–322

Guimarães JH, Tucci ED, Barros-Battesti DM (2001) Ectoparasitos de importancia veterinária. Pleiade-FAPESP, São Paulo, 218 pp

Guzmán-Cornejo C, Robbins RG (2010) The genus *Ixodes* (Acari: Ixodidae) in Mexico: adult identification keys, diagnoses, hosts and distribution. Rev Mex Biodivers 81:289–298

Guzmán-Cornejo C, Robbins RG, Pérez TM (2007) The *Ixodes* (Acari: Ixodidae) of Mexico: parasite-host and host-parasite checklists. Zootaxa 1553:47–58

Guzmán-Cornejo C, Robbins RG, Guglielmone AA, Montiel-Parra G, Pérez TM (2011) The *Amblyomma* (Acari: Ixodida: Ixodidae) of Mexico: identification keys, distribution and hosts. Zootaxa 2998:16–38

Guzmán-Cornejo C, García-Prieto L, Acosta-Gutiérrez R, Falcón-Ordaz J, León-Paniagua L (2012) Metazoarios parásitos de *Tlacuatzin canescens* y *Marmosa mexicana* (Mammalia: Didelphimorphia) de México. Rev Mex Biodivers 83:557–561

Guzmán-Cornejo C, Robbins RG, Guglielmone AA, Montiel-Parra G, Rivas G, Pérez TM (2016) The *Dermacentor* (Acari, Ixodida, Ixodidae) of Mexico: hosts, geographical distribution and new records. ZooKeys 569:22 pp

Guzmán-Cornejo C, Herrera-Mares A, Ugalde-Medina A, López-Pérez AM, Castillo-Martínez L, Acosta-Gutiérrez R, Cabrera-Garrido M, Morales-Malacara JB (2020) Arthropods associated with mammals. Their importance as part of the richness in a biosphere reserve in Mexico. J Med Entomol 57:780–787

Hadrill DJ, Walker AR (1994) Effect of acaricide control of *Amblyomma variegatum* ticks on bovine dermatophilosis on Nevis. Trop Anim Health Prod 26:28–34

Hall AV (1951) Short history of tick eradication in the British Virgin Islands. J Am Vet Med Assoc 118:108–110

Harrison BA, Rayburn WH, Toliver M, Pouwell EE, Engber BR, Durden LA, Robbins RG, Prendergast BF, Whitt PB (2010) Recent discovery of widespread *Ixodes affinis* (Acari: Ixodidae) distribution in North Carolina with implications for Lyme disease studies. J Vector Ecol 35:174–179

Hayman D (2006) Evaluating tick-peccary interaction for peccary pelt certification. In: Bodmer R (ed) Reports on the diversity and abundance of wildlife in the greater Yavarí Valley, Peru, pp 25–101

Hays KL, Lawrence WH (1957) *Ixodes banksi* Bishopp (Acarina: Ixodidae) I. Description of the male and larva. J Parasitol 43:213–216

Heath ACG (2001) Exotic tick interceptions 1980-2000. Surveillance 28(4):13–15

Heath ACG, Hardwick S (2011) The role of humans in the importation of ticks to New Zealand: a threat to public health and biosecurity. N Z Med J 124:67–82

Heller EL, Wright CL, Nadolny RM, Hynes WL, Gaff HD, Walters EL (2016) New records of *Ixodes affinis* (Acari: Ixodidae) parasitizing avian hosts in southeastern Virginia. J Med Entomol 52:441–445

Hermans P, Dwinger RH, Buening GM, Herrero MV (1994) Seasonal incidence and hemoparasite infection rates of ixodid ticks (Acari: Ixodidae) detached from cattle in Costa Rica. Rev Biol Trop 42:623–632

Hernández SM, Peters VE, Weygandt PL, Jiménez C, Villegas P, O'Connor B, Yabsley MJ, García M, Riblet SM, Carroll CR (2013) Do shade-grown coffee plantations pose a disease risk for wild birds? EcoHealth 10:145–148

Hernández-Divers SM, Aguilar R, Loria DL, Foerster CR (2005) Health evaluation of a radiocollared population of free-ranging Baird's tapirs (*Tapirus bairdii*) in Costa Rica. J Zoo Wildl Med 36:176–187

Heyne E (1986) Differentiation of *Boophilus decoloratus* and *Boophilus microplus*. J S Afr Vet Assoc 57:251–252

Hirst S, Hirst LF (1910) Description of five new species of ticks (Ixodidae). Ann Magaz Nat Hist Ser *8*(6):299–308

Hoffmann A (1946) *Amblyomma boneti* n. sp. de México. Rev Inst Salubr Enferm Trop 7:207–220

Hoffmann A (1950) Presencia en México de *Amblyomma curruca* Schulze 1936. Anal Escuela Nacional Ciencia Biol 6:77–80

Hoffmann A (1962) Monografía de los Ixodoidea de México. I parte. Rev Soc Mex Hist Nat 23:191–307

Hoffmann A (1969) Un caso de parálisis por picadura de garrapata. Rev Latinoam Microbiol Parasitol 11:75–77

Hoffmann A, López-Campos G (2000) Biodiversidad de los ácaros en México. Jiménez Editores e Impresores, México, 230 pp

Homsher PJ, Robbins RG, Keirans JE (1991) Scanning electron microscopy of Haller's organ for subgeneric systematic studies in the genus *Ixodes*. In: Dusbábek F, Bukva V (eds) Modern Acarology. Academia, Prague and SPB Academic Publishing, The Hague, vol 2, pp 335–342

Hoogstraal H (1956) African Ixodoidea. I. Ticks of the Sudan (with special reference to Equatoria Province and with preliminary reviews of the genera Boophilus, Margaropus and Hyalomma). Research report NM 005 050.29.07, Department of the Navy, Bureau of Medicine and Surgery, Washington, DC, 1101 pp

Hoogstraal H (1973) Redescription of the type material of *Haemaphysalis* (*Aboimisalis*) *cinnabarina* (revalidated) and its junior synonym *H.* (*A.*) *sanguinolenta* described by Koch in 1844 from Brazil (Ixodoidea: Ixodidae). J Parasitol 59:379–383

Hooker WA (1909) The geographical distribution of American ticks. J Econ Entomol 2:403–428

Hooker WA, Bishopp FC, Wood HP (1912) The life history and bionomics of some North American ticks. Bull Bur Entomol US Dept Agric (106):214 pp

Horak IG, Camicas JL, Keirans JE (2002) The Argasidae, Ixodidae and Nuttalliellidae (Acari: Ixodida): a world list of valid tick names. Exp Appl Acarol 28:27–54

Horak IG, Heyne H, Williams R, Gallivan GJ, Spickett A, Bezuidenhout JD, Estrada-Peña A (2018) The ixodid ticks (Acari: Ixodidae) of southern Africa. Springer, Cham, 676 pp

Horta MC, Saraiva DG, Oliveira GMB, Martins TF, Labruna MB (2015) *Rickettsia belli* in *Amblyomma rotundatum* ticks parasitizing *Rhinella jimi* from northeastern Brazil. Microbes Infect 17:856–858

Hubbard HG (1894) The insect guests of the Florida land tortoise. Insect Life 6:302–315

Hubbard CA (1941) Ectoparasites of western Lagomorpha. Part I. The ectoparasites of 100 western conies (*Ochotona*) Part II. The ectoparasites of western rabbits (*Lepus-Sylvilagus-Brachylagus*). Pac Univ Bull 37:1–8

Huerto-Medina E, Dámaso-Mata B (2015) Factores asociados a la infección por *Ehrlichia canis* en perros infestados con garrapatas en la ciudad de Huánuco, Perú. Rev Peru Med Exp Salud Publica 32:756–760

Hun L, Herrero L, Fuentes L, Vargas M (1991) Tres nuevos casos de fiebre manchada de las montañas Rocosas en Costa Rica. Rev Costarric Cienc Med 12:51–56

Hunter WD, Hooker WA (1907) Information concerning the North American fever tick with notes on other species. Bull Bur Entomol US Dept Agric 72:87 pp

Hutson LR (1944) Observations on anaplasmosis in Antigua, B.W.I. Can J Comp Med 8:214–215

Instituto Interamericano para la Cooperación en Agricultura (1984) Proyecto de fomento de la producción bovina y sanidad animal. Honduras. Capítulo II. Marco de referencia. IICA, 168 pp

Instituto Interamericano para la Cooperación en Agricultura (1985) Proyecto de control de garrapata y tórsalo en Honduras, 19 pp

Instituto Interamericano para la Cooperación en Agricultura (1988) Estudio ecológico y epidemiológico de garrapatas en Guatemala, 43 pp

International Union for Conservation of Nature (2018) The IUCN red list of threatened species. Version 2018-1. http://www.iucntredlist.org. Last accesed 31 May 2020

Isaac C, Ohiolei JA, Igbinosa IB, Nmorsi OPG (2019) The prevalence of ectoparasites of livestock and dogs in Edo State (Sout-South), Nigeria. Niger Vet J 40:62–72

Ito FH, Vasconcellos SA, Bernardi F, Nascimento AA, Labruna MB, Arantes IG (1998) Evidencia sorologica de brucelose e leptospirose e parasitismo por ixodídeos em animais silvestres do Pantanal-Sul-Mato-Grossense. ARS Vet 14:302–310

Ivancovich JC (1973) Nuevas subspecies de garrapatas del género *Amblyomma* (Ixodidae). Rev Invest Agropecuarias Ser 4 Patol Anim 10:55–64

Ivancovich JC, Luciani CA (1992) Las garrapatas de Argentina. Monografía de la Asociación Argentina de Parasitología Veterinaria, Buenos Aires, 95 pp

Ivanova LB, Tomova A, González-Acuña D, Murúa R, Moreno CX, Hernández C, Cabello J, Daniels TJ, Godfrey HP, Cabello FC (2014) *Borrelia chilensis*, a new member of the *Borrelia burgdorferi* sensu lato complex that extends the range of this genospecies in the Southern Hemisphere. Environ Entomol 16:1069–1080

Jakowska S (1972) Lesions produced by ticks, *Amblyomma dissimile*, in *Bufo marinus* toads from the Dominican Republic. Am Zool 12:731

Jarquín-Díaz VH, Barbachano-Guerrero A, Maldonado-Rodríguez R, Vásquez-Aguilar AA, Aguilar-Faissal JL (2016) First molecular evidence of *Hepatozoon canis* in domestic dogs and ticks in fragmented rainforest areas in Mexico. Vet Parasitol Reg Stud Rep 6:4–8

Jiménez AE, Castro R, Solórzano A, Montenegro V, Bermúdez S, Viquez C, Dolz G (2015) First report of *Amblyomma tapirellum* Dunn, 1933 (Ixodida: Ixodidae) in Costa Rica. Syst Appl Acarol 20:471–477

Joan T (1920) Sobre el género *Amblyomma* y dos de sus especies en la Argentina. Dirección de Laboratorios e Investigaciones Agrícolo-Ganaderas, Ministerio de Agricultura de la Nación, 20 pp

Joan T (1928) Nota sobre el *Amblyomma neumanni, A. testudinis* y *A. auriculare*. Actas 4ª Reunión de la Sociedad Argentina de Patología Regional del Norte, mayo 1928, Santiago del Estero, Argentina, pp 665–667

Joan T (1930) El amblyomma (sic) de Cooper y demás garrapatas de los carpinchos. 5ª Reunión de la Sociedad Argentina de Patologia Regiona del Norte, Jujuy, Argentina, octubre 1929, 2, pp 1168–1179

Jones EK, Clifford CM, Keirans JE, Kohls GM (1972) The ticks of Venezuela (Acarina: Ixodoidea) with a key to the species of *Amblyomma* in the Western Hemisphere. Brigham Young Univ Sci Bull Biol Ser 17(4):40 pp

Kadarsan S (1971) Larval ixodid ticks of Indonesia (Acarina: Ixodidae). Ph.D. Dissertation, University of Maryland, 182 pp

Kamilar V, Brundage A, Mohammed RS (2020) An evaluation of hard ticks (Ixodidae) infesting cane toads *Rhinella marina* (Bufonidae) in northwestern Trinidad, W.I. Living World, J Trinidad Tobago Field Naturalists' Club, 1–6

Karsch F (1879) Zwei neue Arachniden des Berliner Museums. Mitteilungen Münchener Entomol Vereins 3:95–96

Karsch F (1880) Vier neue Ixodidae des Berliner Museums. Mitteilungen Münchener Entomol Vereins 4:141–142

Kaufman TS (1972) A revision of the genus *Aponomma* Neumann, 1899 (Acarina: Ixodidae). Ph.D. Dissertation, University of Maryland, 390 pp

Keegan HL, Toshioka S (1957) Ixodid ticks of Japan, Korea and the Ryukyu Islands. Medical General Laboratory, Camp Zama, 37 pp

Keirans JE (1972) Redescription of *Amblyomma fulvum* Neumann, 1899 (Acarina: Ixodidae), a parasite of the giant anaconda in northern South America. J Med Entomol 9:138–139

Keirans JE (1973) *Ixodes* (*I.*) *montoyanus* Cooley (Acarina: Ixodidae): first description of the male and immature stages, with records from deer in Colombia and Venezuela. J Med Entomol 10:249–254

Keirans JE (1976) *Dermacentor venustus* Marx Ms. in Neumann, 1897: proposed suppression under the plenary powers so as to conserve *Dermacentor andersoni* Stiles, 1908 (Acarina: Ixodidae). Bull Zool Nomencl 32:261–264

Keirans JE (1982) The tick collection (Acarina: Ixodidae) of the Hon. Nathaniel Charles Rothschild deposited in the Nuttall and general collections of the British Museum (Natural History). Bull Br Mus Natl Hist Zool Ser 42:1–36

Keirans JE (1985a) *Amblyomma antillorum* Kohls, 1969 (Acari: Ixodidae): description of the immature stages from the rock iguana, *Iguana pinguis* (Sauria: Iguanidae) in the British Virgin Islands. Proc Entomol Soc Washington 87:821–825

Keirans JE (1985b) George Henry Falkiner Nuttall and the Nuttall tick catalogue. US Dept Agric Agric Res Service, Miscellaneous Publication (1438), 1785 pp

Keirans JE (1992) Systematics of the Ixodida (Argasidae, Ixodidae, Nuttalliellidae): an overview and some problems. In: Fivaz B, Petney T, Horak I (eds) Tick vector biology. Medical and veterinary aspects. Springer, Berlin, pp 1–21

Keirans JE, Ajohda S (2003) *Ixodes* (*Ixodes*) *dicei*, a new tick species (Acari: Ixodida: Ixodidae) from a lagomorph in a restricted high-altitude habitat in Talamanca, Costa Rica. Int J Acarol 29:63–67

Keirans JE, Brewster BE (1981) The Nuttall and British Museum (Natural History) tick collections: lectotype designations for ticks (Acarina: Ixodoidea) described by Nuttall, Warburton, Cooper and Robinson. Bull Br Mus Nat Hist (Zoology) 41:153–178

Keirans JE, Clifford CM (1978) The genus *Ixodes* in the United States: a scanning electron microscope study and key to adults. J Med Entomol (Suppl 2):149 pp

Keirans JE, Clifford CM (1984) A checklist of types of Ixodoidea (Acari) in the collection of the Rocky Mountain Laboratories. J Med Entomol 21:310–320

Keirans JE, Durden LA (1998) Illustrated keys to nymphs of the tick genus *Amblyomma* (Acari: Ixodidae) found in the United States. J Med Entomol 35:489–495

Keirans JE, Durden LA (2001) Invasion: exotic ticks (Acari: Argasidae, Ixodidae) imported into the United States. A review and new records. J Med Entomol 38:850–861

Keirans JE, Eckerlin RP (2005) Description of the male and redescription of the female of *Ixodes* (*Ixodes*) *guatemalensis* and description of the male, nymph and larva of *Ixodes* (*Afrixodes*) *moreli* and redescription of the female (Acari: Ixodida: Ixodidae). Int J Acarol 31:259–268

Keirans JE, Garris GI (1986) *Amblyomma arianae* n. sp. (Acari: Ixodidae), a parasite of *Alsophis portoricensis* (Reptilia: Colubridae) in Puerto Rico. J Med Entomol 6:622–625

Keirans JE, Hillyard PD (2001) A catalogue of the type specimens of Ixodida (Acari: Argasidae, Ixodidae) deposited in The Natural History Museum, London. Occasional Pap Syst Entomol (13) 74 pp

Keirans JE, Jones EK (1972) Description of the immature stages of *Ixodes* (*I.*) *sinaloa* Kohls and Clifford (Acarina: Ixodidae) from rodents in Mexico and Nicaragua. Acarologia 13:471–475

Keirans JE, Klompen JSH (1996) *Amblyomma quadricavum* (Schulze) (new combination), and *Amblyomma arianae* Keirans and Garris, a new junior synonym of *Amblyomma quadricavum* (Acari: Ixodidae). Proc Entomol Soc Washington 98:164–165

Keirans JE, Litwak TR (1989) Pictorial key to the adults of hard ticks, family Ixodidae (Ixodida: Ixodoidea), east of the Mississippi River. J Med Entomol 26:435–448

Keirans JE, Oliver JH (1993) First description of the male and redescription of immature stages of *Amblyomma rotundatum* (Acari: Ixodidae), a recently discovered tick in the U.S.A. J Parasitol 79:860–865

Keirans JE, Restifo RA (1993) *Haemaphysalis juxtakochi* Cooley (Acari: Ixodidae), a neotropical tick species found in Ohio. J Med Entomol 30:1074–1075

Keirans JE, Hoogstraal H, Clifford CM (1973) The *Amblyomma* (Acarina: Ixodidae) parasitic on giant tortoises (Reptilia: Testudinidae) of the Galapagos Islands. Ann Entomol Soc Am 66:673–688

Keirans JE, Clifford CM, Corwin D (1976) *Ixodes sigelos*, n. sp. (Acarina: Ixodidae), a parasite of rodents in Chile, with a method for preparing ticks for examination by scanning electron microscopy. Acarologia 18:217–225

Keirans JE, Clifford CM, Guglielmone AA, Mangold AJ (1985) *Ixodes* (*Ixodes*) *pararicinus* n. sp. (Acari: Ixodoidea: Ixodidae), a South American cattle tick long confused with *Ixodes ricinus*. J Med Entomol 22:401–407

Keirans JE, Hutcheson HJ, Durden LA, Klompen JSH (1996a) *Ixodes* (*Ixodes*) *scapularis* (Acari: Ixodidae): redescription of all active stages, distribution, hosts, geographical variation, and medical and veterinary importance. J Med Entomol 33:297–318

Keirans JE, Brown RN, Lane RS (1996b) *Ixodes* (*Ixodes*) *jellisoni* and *I.* (*I.*) *neotomae* (Acari: Ixodidae): descriptions of the immature stages from California. J Med Entomol 33:319–327

Keirans JE, Needham GR, Oliver JH (1999) The *Ixodes ricinus* complex worldwide: diagnosis of the species in the complex, hosts and distribution. In: Needham GR, Horn DJ, Welbourn WC (eds) Symposia, Acarology 9, vol 2, pp 341–347

Kelehear C, Keightley SJ, Ritzi C (2017a) First record of *Amblyomma dissimile* infesting the south American common toad (*Rhinella margaritifera*) in French Guiana. Comp Parasitol 84:163–164

Kelehear C, Hudson CM, Mertins JW, Shine R (2017b) First report of exotic ticks (*Amblyomma rotundatum*) on the island of Hawai'i. Ticks Tick-Borne Dis 8:330–333

Khattak RM, Ali S, Jahangir M, Khan MN, Rasul A, Iqbal F (2012) Prevalence of ectoparasites in wild and domesticated grey (*Francolinus pondicerianus*) and black partridge (*Francolinus francolinus*) from Khyber Pakhtoonkhwa Province of Pakistan. Pak J Zool 44:1239–1244

Kishida K (1930) Notes on Japanese ticks of the genus *Ixodes* Latreille, 1795. Lansania 2(11):1–5. [in Japanese]

Kleinjan JE, Lane RS (2008) Larval keys to the genera of Ixodidae (Acari) and species of *Ixodes* (Latreille) [sic] ticks established in California. Pan Pac Entomol 84:121–142

Klompen JSH, Dobson SJ, Barker SC (2002) A new subfamily, Bothriocrotoninae n. subfam., for the genus *Bothriocroton* Keirans, King & Sharrad, 1994 status amend. (Ixodida: Ixodidae), and the synonymy of *Aponomma* Neumann, 1899 with *Amblyomma* Koch, 1844. Syst Parasitol 53:101–107

Kluyber D, Martins TF, Desbiez AJL, Amorim M, Massocato G, Labruna MB (2016) Ticks (Acari: Ixodidae) infesting armadillos (Cingulata: Dasypodidae) in the Pantanal wetland, Mato Grosso do Sul, Brazil. Syst Appl Acarol 21:1087–1091

Kmetiuk LB, Krawczak FS, Machado FP, Paploski IAD, Martins TF, Tieder-Junior PI, Serpa MCA, Barbieri ARM, Bach RVW, Barros-Filho IR, Lipinski LC, Santos AP, Labruna MB, Biondo AW (2019) Ticks and serosurvey of anti-*Rickettsia* spp. antibodies in wild boars (*Sus scrofa*), hunting dogs and hunters of Brazil. PLoS Negl Trop Dis 13(5)(article e0007405):14 pp

Ko RC (1973) The larva of *Ixodes cookei* (Acarina: Ixodidae). Can Entomol 105:245–248

Koch CL (1844) Systematische Übersicht über die Ordnung der Zecken. Arch Nat 10:217–239

Koch CL (1847) Übersicht des Arachnidensystems. Nürnberg 4:136 pp

Koch L (1867) Beschreibungen neuer Arachniden und Myriapoden. Verhandlungen Kaiserlich-Königlichen Zool-Bot Ges Wien 17:173–250

Kocher C, Morrison AC, Legula M, Loyola S, Castillo RM, Gálvez HA, Astete H, Flores-Mendoza C, Ampuero JS, Bausch DG, Halsey ES, Céspedes M, Zevallos K, Jiang J, Richards AL (2016) Rickettsial disease in the Peruvian Amazon basin. PLoS One Negl Trop Dis 10 (7) (article e04843):13 pp

Kohls GM (1950) Description of the male of *Ixodes woodi* Bishopp (Acarina: Ixodidae). J Parasitol 36:65–66

Kohls GM (1953) *Ixodes venezuelensis*, a new species of tick from Venezuela, with notes on *Ixodes minor* Neumann, 1902 (Acarina: Ixodidae). J Parasitol 39:300–303

Kohls GM (1956a) Eight new species of *Ixodes* from Central and South America (Acarina: Ixodidae). J Parasitol 42:636–649

Kohls GM (1956b) Concerning the identity of *Amblyomma maculatum*, *A. tigrinum*, *A. triste*, and *A. ovatum* of Koch, 1844. Proc Entomol Soc Washington 58:143–147

Kohls GM (1956c) The identity of *Ixodes boliviensis* Neumann, 1904 and *Ixodes bicornis* Neumann, 1906. Proc Entomol Soc Washington 58:232–233

Kohls GM (1957a) Insects of Micronesia. Acarina Ixodoidea. Insects Micronesia 3:85–104

Kohls GM (1957b) *Ixodes downsi*, a new species of tick from a cave in Trinidad, British West Indies (Acarina – Ixodidae). Proc Entomol Soc Washington 59:257–264

Kohls GM (1958) *Amblyomma imitator*, a new species of tick from Texas and Mexico, and remarks on the synonymy of *A. cajennense* (Fabricius) (Acarina – Ixodidae). J Parasitol 44:430–433

Kohls GM (1960) Records and new synonymy of New World *Haemaphysalis* ticks, with descriptions of the nymph and larva of *H. juxtakochi* Cooley. J Parasitol 46:355–361

Kohls GM (1969a) *Ixodes taglei* n. sp. (Acarina: Ixodidae) a parasite of the deer, *Pudu pudu* [sic] (Wol.), in Chile. J Med Entomol 6:280–283

Kohls GM (1969b) A new species of *Amblyomma* from iguanas in the Caribbean (Acarina: Ixodidae). J Med Entomol 6:439–442

Kohls GM (1969c) New records of ticks from the Lesser Antilles. Stud Fauna Curaçao Other Caribbean Islands 28:126–134

Kohls GM, Clifford CM (1962) *Ixodes tiptoni*, a new species of tick from Panama (Acarina – Ixodidae). J Parasitol 48:182–184

Kohls GM, Clifford CM (1964) *Ixodes eadsi*, a new species of tick from rodents in southern Texas (Acarina: Ixodidae). J Parasitol 50:466–470

Kohls GM, Clifford CM (1966) Three new species of *Ixodes* from Mexico and description of the male of *I. auritulus auritulus* Neumann, *I. conepati* Cooley and Kohls, and *I. lasallei* Méndez and Ortiz (Acarina: Ixodidae). J Parasitol 52:810–820

Kohls GM, Clifford CM (1967) *Ixodes* (*Haemixodes*) *uruguayensis*, new subgenus, new species (Acarina: Ixodidae) from small rodents in Uruguay. Ann Entomol Soc Am 60:391–394

Kohls GM, Dalmat HT (1952) The male of *Dermacentor dissimilis* Cooley (Acarina: Ixodidae). J Parasitol 38:140–142

Kohls GM, Sonenshine DE, Clifford CM (1969) *Ixodes* (*Exopalpiger*) *jonesae* sp. n. (Acarina: Ixodidae), a parasite of rodents in Venezuela. J Parasitol 55:447–452

Kolonin GV (2008) Birds as hosts of ixodid ticks (Acarina: Ixodidae). Entomol Rev 88:1012–1015

Kolonin GV (2009) Fauna of ixodid ticks of the world. https://archive.is/CtZk. Last accession 26 Feb 2020

Kramer P, Neuman CJ (1883) Acariden wahrend der Vega-Expedition eingesammelt. In: Vega-Expeditiones Vetenskapliga Jakttagelser, E. Nordensköld, Stockholm 3:519–529

Krawczak FS, Martins TF, Oliveira CS, Binder LC, Costa FB, Nunes PH, Gregori F, Labruna MB (2015) *Amblyomma yucumense* n. sp. (Acari: Ixodidae), a parasite of wild mammals in southern Brazil. J Med Entomol 52:28–37

L'Hostis M, Rose-Rosette F, Thomas N, Fourgead P (1998) Tick infestation on feral dogs in Martinique. Ann NY Acad Sci 849:395–397

Labruna MB, Homem VSF, Heinemann MB, Neto JSF (2000) Ticks (Acari: Ixodidae) associated with rural dogs in Uruará, eastern Amazon, Brazil. J Med Entomol 37:774–776

Labruna MB, Pinter A, Roncati NV, Pereira MC, Faccini JLH (2001) Exotic ticks on horses imported from Portugal to Brazil. Arq Bras Med Vet Zoot 53:179–182

Labruna MB, Marrelli LT, Heinemann MB, Fava AB, Cortez A, Soares RM, Sakamoto SM, Richtzenhain LJ, Marinotti O, Schumaker TTS (2002a) Taxonomic status of *Ixodes didelphidis* (Acari: Ixodidae). J Med Entomol 39:135–142

Labruna MB, Camargo LMA, Schumaker TTS, Camargo EP (2002b) Parasitism of domestic swine by *Amblyomma* ticks (Acari: Ixodidae) on a farm at Monte Negro, western Amazon, Brazil. J Med Entomol 39:241–243

Labruna MB, Camargo LMA, Terrassini FA, Schumaker TTS, Camargo EP (2002c) Notes on parasitism by *Amblyomma humerale* (Acari: Ixodidae) in the State of Rondônia, western Amazon, Brazil. J Med Entomol 39:814–817

Labruna MB, Paula CD, Lima TF, Sana DA (2002d) Ticks (Acari: Ixodidae) on wild animals from the Porto-Primavera hydroelectric power station area, Brazil. Mem Inst Oswaldo Cruz 97:1133–1136

Labruna MB, Silva MNJ, Oliveira MF, Barros-Battesti DM, Keirans JE (2003) New records and laboratory-rearing data for *Ixodes schulzei* (Acari: Ixodidae) in Brazil. J Med Entomol 40:116–118

Labruna MB, Jorge RSP, Sana DA, Jácomo ATA, Kashivakura CK, Furtado MM, Ferro C, Pérez SA, Silveira L, Santos TS, Marques SR, Morato RG, Nava A, Adania CH, Teixeira RHF, Gomes AAB, Conforti VA, Azevedo FCC, Prada CS, Silva JCR, Batista AF, Marvulo MFV, Morato RLG, Alho CJR, Pinter A, Ferreira PM, Ferreira F, Barros-Battesti DM (2005a) Ticks (Acari: Ixodida) on wild carnivores in Brazil. Exp Appl Acarol 36:149–163

Labruna MB, Terrassini FA, Camargo LMA (2005b) First report of the male of *Amblyomma rotundatum* (Acari: Ixodidae) from a field-collected host. J Med Entomol 42:945–947

Labruna MB, Keirans JE, Camargo LMA, Ribeiro AF, Soares RM, Camargo EP (2005c) *Amblyomma latepunctatum*, a valid tick species (Acari: Ixodidae) long misidentified with both *Amblyomma incisum* and *Amblyomma scalpturatum*. J Parasitol 91:527–541

Labruna MB, Camargo LM, Terrassini FA, Ferreira F, Schumaker TTS, Camargo EP (2005d) Ticks (Acari: Ixodidae) from the State of Rondônia, western Amazon, Brazil. Syst Appl Acarol 10:17–32

Labruna MB, Sanfilippo LF, Demetrio C, Menezes AC, Pinter A, Guglielmone AA, Silveira LF (2007a) Ticks collected on birds in the State of São Paulo, Brazil. Exp Appl Acarol 43:147–160

Labruna MB, Ahid SMM, Soares HS, Suassuna ACD (2007b) Hyperparasitism in *Amblyomma rotundatum* (Acari : Ixodidae). J Parasitol 93:1531–1532

Labruna MB, Onofrio VC, Beati L, Arzua M, Bertola PB, Ribeiro AF, Barros-Battesti DM (2009) Redescription of the female, description of the male and several new records of *Amblyomma parkeri* (Acari: Ixodidae), a South American tick species. Exp Appl Acarol 49:243–260

Labruna MB, Barbieri FS, Martins TF, Brito LG, Ribeiro FDS (2010a) New tick records in Rondônia, western Brazilian Amazon. Braz J Vet Parasitol 19:192–194

Labruna MB, Romero M, Martins TF, Tobler M, Ferreira F (2010b) Ticks of the genus *Amblyomma* (Acari: Ixodidae) infesting tapirs (*Tapirus terrestris*) and peccaries (*Tayassu pecari*) in Peru. Syst Appl Acarol 15:109–112

Labruna MB, Martins TF, Nunes PB, Borges FC, Portero F, Venzal JM (2013) New records of *Amblyomma multipunctum* and *Amblyomma naponense* from Ecuador, with the description of the nymph of *A. multipunctum*. J Parasitol 99:973–977

Labruna MB, Onofrio VC, Barros-Battesti DM, Gianizella SL, Venzal JM, Guglielmone AA (2020a) Synonymy of *Ixodes aragaoi* with *Ixodes fuscipes*, and reinstatement of *Ixodes spinosus* (Acari : Ixodidae). Ticks Tick-Borne Dis 11(2) (article 101349):9 pp

Labruna MB, Muñoz-Leal S, Vanstreels RET, Acosta ICL, Repenning M, Martins TF, Bugoni L (2020b) Records of *Ixodes percavatus* sensu lato on Atlantic yellow-nosed albatrosses (*Thalassarche chlororhynchos*) on the Brazilian coast and offshore waters. Syst Appl Acarol 25:957–962

Lado P, Nava S, Labruna MB, Szabó MPJ, Durden LA, Bermúdez S, Montagna M, Quirós ACS, Beati L (2016) *Amblyomma parvum* Aragão, 1908 (Acari: Ixodidae): phylogeography and systematic considerations. Ticks Tick-Borne Dis 7:817–827

Lado, P., Nava, S., Mendoz-Uribe, L., Cáceres, A.G., Delgado-de la Mora J, Licona-Enríquez JD, Delgado-de la Mora D, Labruna MB, Durden LA, Allerdice MEJ, Paddock CD, Szabó MPJ, Venzal JM, Guglielmone AA, Beati L (2018) The *Amblyomma maculatum* Koch, 1844 (Acari: Ixodidae) group of ticks: phenotypic plasticity or incipient speciation? Parasites Vect 11 (article 610):22 pp

Lahille F (1905) Contribution à l'étude des ixodidés de la République Argentine. Anal Ministerio Agric Repúbl Argent 2:166 pp

Lahille F (1913) Sobre dos *Ixodes* de la República Argentina y la medición de las garrapatas. Bol Ministerio Agric Repúbl Argent 16:278–289

Lahille F (1914) Nota sobre la presencia accidental en Buenos Aires de unas garrapatas del camello. Bol Ministerio Agric 17:289–293

Lahille F (1916) Descripción de un nuevo ixódido chileno. Rev Chil Hist Nat 20:107–108

Lahille F (1920) Enumeración sistemática de los pedicúlidos, malófagos, pulícidos, linguatúlidos y ácaros (1ª parte) encontrados en la República Argentina con una nota sobre una especie de piojo. Ministerio de Agricultura, Buenos Aires, 41 pp

Lamattina D, Tarragona EL, Costa SA, Guglielmone AA, Nava S (2014) Ticks (Acari: Ixodidae) of northern Misiones Province, Argentina. Syst Appl Acarol 19:393–398

Lamattina D, Venzal JM, Guglielmone AA, Nava S (2016) *Ixodes schulzei* Aragao & Fonseca, 1951 in Argentina. Syst Appl Acarol 21:865–867

Lamattina D, Venzal JM, Costa SA, Arrabal JP, Flores S, Berrozpe PE, González-Acuña D, Guglielmone AA, Nava S (2018a) Ecological characterization of a tick community across a landscape gradient exhibiting differential anthropogenic disturbance in the Atlantic Forest ecoregion in Argentina. Med Vet Entomol 32:271–281

Lamattina D, Venzal JM, Tarragona EL, Mangold AJ, Nava S (2018b) *Amblyomma yucumense* Krawczak, Martins & Labruna, 2015 (Acari: Ixodidae) in Argentina. Syst Appl Acarol 23:799–802

Lamattina D, Costa SA, Arrabal JP, Venzal JM, Guglielmone AA, Nava S (2018c) Factors associated with hard tick (Acari: Ixodidae) parasitism in medium-sized mammals in the Atlantic Rainforest region of Argentina. Ticks Tick-Borne Dis 9:1451–1458

Lampo M, Rangel Y, Mata A (1997) Genetic markers for the identification of two tick species, *Amblyomma dissimile* and *Amblyomma rotundatum*. J Parasitol 83:382–386

Lampo M, Rangel Y, Mata A (1998) Population genetic structure of a three-host tick, *Amblyomma dissimile*, in eastern Venezuela. *J Parasitol* 84:1137–1142

Lancaster JL (1973) A guide to the ticks of Arkansas. Bull Agric Exp Station Univ Arkansas 779:39 pp

Landívar S, Ratti F, Farías R (1985) Garrapatas (Ixodides) de *Canis familiari* [sic] en la ciudad de Guayaquil. Rev Ecuat Hig Med Trop 35:35–47

Lane RS, Poinar GO (1986) First fossil tick (Acari: Ixodidae) in New World amber. Int J Acarol 12:75–78

Lareschi M, Savchenko E, Urdapilleta M (2019) Ectoparasites associated with sigmodontine rodents from northeastern Argentina. Therya 10:103–108

Latreille PA (1806) Genera crustaceorum et insectorum secundum ordinem naturalem in familia disposita, iconibus exemplisque plurimis explicata, vol 1, Paris et Argentorati, 302 pp

Lavina MS, Souza AP, Souza JC, Bellato V, Sartor AA, Moura AB (2011) Ocorrencia de *Amblyomma aureolatum* (Pallas, 1772) e *A. ovale* (Koch, 1844) [sic] (Acari: Ixodidae) parasitando *Alouatta clamitans* Cabrera, 1940 (Primates: Atelidae) na região norte do estado de Santa Catarina. Arq Bras Med Vet Zoot 63:266–269

Lavina MS, Souza AP, Bellato V, Sartor AA, Moura AB, Famadas KM (2014) Ixodídeos coletados em equinos e caninos no estado de Santa Catarina. Rev Bras Med Vet 36:79–84

Lavina MS, Souza AP, Sartor AA, Moura AB (2015) Ixodids in wild animals of the mountainous plateau region of Santa Catarina State, Brazil. Semina Ciencias Agrarias 36:3173–3180

Leach WE (1815) A tabular view of the external characters of four classes of animals, which Linné arranged under Insecta: with the distribution of the genera composing three of these classes into orders, and descriptions of several new genera and species. Trans Linn Soc Lond 11:306–400

Leite RC, Faccini JLH, Granger SA (2001) Ocorrência de carrapatos em bovinos no estado de Roraima. Braz J Vet Parasitol 10:49–50

Leo SST, Pybus MJ, Sperling FAH (2010) Deep mitochondrial DNA lineage divergences within Alberta populations of *Dermacentor albipictus* (Acari: Ixodidae) do not indicate distinct species. J Med Entomol 47:565–574

León D (1986) Efficacy of ivermectin administered subcutaneously against cattle tick, *Boophilus microplus* (Canestrini). J Agric Univ P R 7:315–318

León D, Ruiz T, Quiñones R (1987) Ectrin spray in combination with ectrin eartags and the control of *Boophilus microplus* (Canestrini) in dairy cattle. J Agric Univ P R 71:241–247

Leydig F (1855) Zum feineren Bau der Arthropoden. In: Müller J (ed) Archiv für Anatomie, Physiologie und Wissenschaftliche Medicin. Veit et Comp, pp 376–480

Light JA, Eckerlin RP, Durden LA (2019) Checklist of ectoparasites of Canidae and Felidae in Mexico. Therya 10:109–119

Light JA, Durden LA, O'Connor BM, Preisser WC, Acosta R, Eckerlin RP (2020) Checklist of ectoparasites of cricetid and heteromyid rodents in Mexico. Therya 11:79–136

Lima MA, Martins TF, Muñoz-Leal S, Guilherme E, Ogrzewalska M, Labruna MB (2018) Ticks and tick-associated spotted fever group *Rickettsia* from birds in the southwestern Brazilian Amazon. Rev Colomb Ciencias Pecuarias 31:26–35

Linardi PM, Botelho JR, Neves DP, Cunha HC (1984) Sobre alguns ectoparasitos de roedores silvestres de Belo Horizonte, MG. Rev Bras Biol 44:215–219

Linardi PM, Botelho JR, Ximenez A, Padovani CR (1991) Notes on ectoparasites of some small mammals from Santa Catarina State, Brazil. J Med Entomol 28:183–185
Lindquist EE, Galloway TD, Artsob H, Lindsay LR, Drebot M, Wood H, Robbins RG (2016) A Handbook to the Ticks of Canada (Ixodida: Ixodidae, Argasidae). Biol Surv Can Monogr Ser 7:317 pp
Linnaeus C (1758) Systema naturae per regna tria naturae, secundum classes, ordines, genera, species cum characteribus, differentiis, synonymis, locis. Editio decima, reformata. Holmiae, vol 1, 823 pp
Linthicum KJ, Logan TM, Bailey CL, Gordon SW, Peters CJ, Monath TP, Osorio J, Francy DB, McLean RG, Leduc JW, Graham RR, Jahrling PB, Moulton JR, Dohm DJ (1991) Venezuelan equine encephalomyelitis virus infection in and transmission by the tick *Amblyomma cajennense* (Arachnida: Ixodidae). J Med Entomol 28:405–409
Lira-Amaya JJ, Ojeda-Robertos NF, Álvarez-Martínez JA, Rojas-Martínez C, Bautista-Garfias CR, Figueroa-Millán JV (2015) Identificación de garrapatas en una explotación de ovinos. Entomol Mex 2:721–726
Llanos-Soto S, Muñoz B, Moreno L, Landaeta-Aqueveque C, Kinsella JM, Mironov S, Cicchino A, Barrientos C, Torres-Fuentes G, González-Acuña D (2017) External and gastrointestinal parasites of the rufous-collared sparrow *Zonotrichia capensis* (Passeriformes, Emberizidae) in Chile. Braz J Vet Parasitol 28:314–322
Londoño AF, Díaz FJ, Valbuena G, Gazi M, Labruna MB, Hidalgo M, Mattar S, Contreras V, Rodas JD (2014) Infection of *Amblyomma ovale* by *Rickettsia* sp. strain Atlantic rainforest, Colombia. Ticks Tick-Borne Dis 5:672–675
Londt JGH, Arthur D (1975) The structure and parasite life cycle of *Boophilus microplus* (Canestrini, 1888) in South Africa (Acarina: Ixodidae). J Entomol Soc South Afr 38:321–340
Lopes SG, Andrade GV, Costa-Júnior LM (2010) A first record of *Amblyomma dissimile* (Acari: Ixodidae) parasitizing the lizard *Ameiva ameiva* (Teiidae) in Brazil. Braz J Vet Parasitol 19:262–264
Lopes MG, Junior JM, Foster RJ, Harmsen BJ, Sánchez E, Martins TF, Quigley H, Marcilli A, Labruna MB (2016) Ticks and rickettsiae from wildlife in Belize, Central America. Parasites Vec 9(article 62):7 pp
Lopes MG, Muñoz-Leal S, Lima JTR, Fournier GFSR, Acosta ICL, Martins TF, Ramírez DG, Gennari SM, Labruna MB (2018) Ticks, rickettsial and ehrlichial infection in small mammals from Atlantic forest remnants in northeastern Brazil. Int J Parasitol Parasites Wildl 7:380–385
Lopes MG, Krawczak FS, Lima JTR, Fournier GFSR, Acosta ICL, Ramírez DG, Marcili A, Labruna MB, Gennari SM (2019) Occurrence of *Ehrlichia canis* and *Hepatozoon canis* and probable exposure to *Rickettsia amblyommatis* in dogs and cats in Natal, RN. Braz J Vet Parasitol 28:151–156
López G (2017) Garrapatas (Acari: Ixodidae y Argasidae) de importancia médica y veterinaria, procedentes de Norte, Centro y Suramérica. Editorial Universidad de Antioquia, Medellín, 110 pp
López G, Parra D (1985) *Amblyomma neumanni,* Ribaga 1902. Primera comprobación en Colombia y claves para las especies de *Amblyomma*. Rev Inst Colomb Agropecuario 20:152–162
Lourenço EC, Alves-da-Silva AF, Gaspar É, Bergallo HG (2020) Host-related factors affecting the occurrence of tick on dogs in Ilha Grande, Rio de Janeiro State, Brazil. Int J Acarol 46:424–430
Lucas H (1846) Sur quelques espèces nouvelles d'*Ixodes* qui vivent parasites sur les serpens (*Boa constrictor* et *Python sebae*) et sur l'ornythorhynque (*Ornithorhynchus paradoxus*, Blum). Annal Soc Entomol France 4:53–64
Lucas H (1847) *Ixodes gervaisii* et *Ixodes fusco-lineatus* nn. spp. Bull Soc Entomol France Sér 2(5):99–101
Lucas H (1852) *Ixodes variegatus*, Lucas. Bull Soc Entomol France Sér 2(7):88–89
Lucas H (1873) *Ixodes fuscomaculatus*. Bull Soc Entomol France Ser 5(3):31–32

Lugarini C, Martins TF, Ogrzewalska M, Vasconcelos NCT, Ellis VA, Oliveira JB, Pinter A, Labruna MB, Silva JCR (2015) Rickettsial agents in avian ixodid ticks in northeast Brazil. Ticks Tick-Borne Dis 6:364–375

Luque Forero G (1948) *Amblyomma nodosum* en *Tamandua tetradactyla*. Rev Facult Med Vet Zoot 17:119–120

Luque Forero A (1949) *Amblyomma oblongoguttatum* sobre la piel de un campesino en la región de selvática de Barrancabermeja. Rev Facult Med Vet Zoot 18:213–214

Luz HR, Faccini JLH, Landulfo GA, Berto BP, Ferreira I (2012) Bird ticks in an area of the Cerrado of Minas, southeast Brazil. Exp Appl Acarol 58:89–99

Luz HR, Faccini JLH, Pires MS, Silva HR, Barros-Battesti MB (2013a) Life cycle and behavior of *Amblyomma rotundatum* (Acari: Ixodidae) under laboratory conditions and remarks on parasitism of toads in Brazil. Exp Appl Acarol 60:55–62

Luz HR, Faccini JLH, Landulfo GA, Sampaio JS, Neto SFC, Famadas KM, Onofrio VC, Barros-Battesti DM (2013b) New host records of *Ixodes luciae* (Acari: Ixodidae) in the State of Pará. Braz J Vet Parasitol 22:152–154

Luz HR, Mathias C, Faccini JLH (2014) Carrapatos parasitando caes em uma área insular do Rio de Janeiro, RJ, Brasil. Rev Bras Med Vet 36:437–442

Luz HR, Faccini JLH, Silva HR (2015) Patterns of parasitism by *Amblyomma rotundatum* (Ixodidae) on *Rhinella schneideri* (Bufonidae) in islands of São Francisco River, Minas Gerais, Brazil. Pesquisa Vet Bras 35:579–582

Luz HR, Faccini JLH, Landulfo GA, Neto SFC, Famadas KM (2016a) New records for *Amblyomma sculptum* (Ixodidae) on non-passerine birds in Brazil. Braz J Vet Parasitol 25:124–126

Luz HR, Faccini JLH, Landulfo GA, Berto BP (2016b) New host records of ticks (Ixodidae) infesting birds in an Atlantic Forest fragment in southeastern Brazil. Syst Appl Acarol 21:1107–1115

Luz HR, Morgana H, Martins TF, Pichorim M, Labruna MB, Faccini JLH (2017a) Additional information on ticks (Ixodidae) infesting birds in Atlantic Forest fragments in State of Paraná, Brazil. Syst Appl Acarol 22:1813–1821

Luz HR, Faccini JLH, McIntosh D (2017b) Molecular analyses reveal an abundant diversity of ticks and rickettsial agents associated with wild birds in two regions of primary Brazilian Atlantic rainforest. Ticks Tick-Borne Dis 8:657–665

Luz HR, Neto SFC, Weksler M, Gentile R, Faccini JLH (2018a) Ticks parasitizing wild mammals in Atlantic Forest areas in the State of Rio de Janeiro, Brazil. Braz J Vet Parasitol 27:409–414

Luz HR, Silva-Santos E, Costa-Campos CE, Acosta I, Martins TF, Muñoz-Leal S, McIntosh D, Faccini JLH, Labruna MB (2018b) Detection of *Rickettsia* spp. in ticks parasitizing toads (*Rhinella marina*) in the northern Brazilian Amazon. Exp Appl Acarol 75:309–318

Machado RZ, Machado CR, Toledo CZP, Ferreira FA, Rocha UF (1985) Ecología em carrapatos. XIII. *Boophilus microplus* (Canestrini, 1887) em infestações naturais de veados (*Ozotocerus bezoarticus bezoarticus*, Linnaeus, 1762) dos estados de São Paulo e Mato Grosso do Sul. ARS Vet 1:47–50

Machado-Ferreira E, Vizzoni VF, Balsemão-Pires E, Moerbeck L, Gazeta GS, Piesman J, Voloch CM, Soares CAG (2016) *Coxiella* symbionts are widespread into hard ticks. Parasitol Res 115:4691–4699

Maes JM, Palacios Vargas JG, Jiménez ML (1989) Catálogo de los artrópodos terrestres no insectos. Rev Nicaragüense Entomol 7:1–55

Maia MO, Koppe VC, Muñoz-Leal S, Martins TF, Marcili A, Labruna MB, Pacheco RC (2018) Detection of *Rickettsia* spp. in ticks associated to wild mammals in northeastern Brazil, with notes on an undetermined *Ornithodoros* sp. collected from marsupials. Exp Appl Acarol 76:52–535

Mallo O (1990) Control de garrapatas en caninos mediante el uso de ivermectina. Vet Argent 6:631–633

Manilla G (1984) Su alcune zecche (Acari: Ixodoidea) fortuitamente raccoltein paesi extraeuropei e considerazioni sulle “malattie emergente”. Riv Parassitol 45:105–124

Manzanilla J, García ME, Moissant E, García FA, Tortolero E (2002) Dos especies de garrapatas del género *Amblyomma* (Acari: Ixodidae) en perros del estado Aragua, Venezuela. Entomotropica 17:177–180

Marín-Vial P, González-Acuña D, Cellis-Diez JL, Cattan PE, Guglielmone AA (2007) Presence of *Ixodes neuquenensis* Ringuelet, 1947 (Acari: Ixodidae) on the endangered Neotropical marsupial monito del monte (*Dromiciops gliroides*, Thomas, 1894, Microbiotheria: Microbiotheriidae) at Chiloé island, Chile. Eur J Wildl Res 53:73–75

Marques S, Barros-Battesti DM, Faccini JLH, Onofrio VC (2002) Brazilian distribution of *Amblyomma varium* Koch, 1844 (Acari: Ixodidae), a common parasite of sloths (Mammalia: Xenarthra). Mem Inst Oswaldo Cruz 97:1141–1146

Marques S, Barros-Battesti DM, Onofrio VC, Famadas KM, Faccini JLH, Keirans JE (2004) Redescription of larva, nymph and adults of *Ixodes* (*I.*) *loricatus* Neumann, 1899 (Acari: Ixodidae) based on light and scanning electron microscopy. Syst Parasitol 59:135–146

Martínez-Sánchez ET, Cardona-Romero M, Ortiz-Giraldo M Tobón-Escobar WD, López DM, Ossa-López PA, Pérez-Cárdenas JE, Labruna MB, Martins TF, Rivera-Páez FA, Castaño-Villa GJ (2020) Associations between wild birds and hard ticks (Acari: Ixodidae) in Colombia. Ticks Tick-Borne Dis 11(6)(article 101534):8 pp

Martins JR, Salomão EL, Doyle RL, Onofrio V, Barros-Battesti DM, Guglielmone AA (2007) *Haemaphysalis juxtakochi* Cooley, 1946 (Acari: Ixodidae) parasitando *Mazama nana* (Hemsel, 1872) (Artiodactyla: Cervidae) no estado do Rio Grande do Sul. Braz J Vet Parasitol 16:171–173

Martins TF, Spolidorio MG, Batista TCA, Oliveira IAS, Yoshinari NH, Labruna MB (2009a) Ocorrência de carrapatos (Acari: Ixodidae) no município de Goiatins, Tocantins. Braz J Vet Parasitol 18:52–54

Martins TF, Dantas-Torres F, Nieri-Bastos FA, Marcili A, Siquiera DB, Aléssio F, Mauffrey JF, Marvulo MFV, Silva JCR, Labruna MB (2009b) Host records for the immature stages of the South American tick *Amblyomma fuscum* (Acari: Ixodidae). Entomol News 120:370–374

Martins TF, Onofrio VC, Barros-Battesti DM, Labruna MB (2010) Nymphs of the genus *Amblyomma* (Acari: Ixodidae) of Brazil: descriptions, redescriptions, and identification key. Ticks Tick-Borne Dis 1:75–99

Martins TF, Furtado MM, Jácomo AT, Silveira L, Sollman R, Torres NM, Labruna MB (2011) Ticks on free-living wild mammals in the region of Emas National Park, Goiás State, central Brazil. Syst Appl Acarol 16:201–207

Martins TF, Scofield A, Oliveira WBL, Nunes PH, Ramírez DG, Barros-Battesti DM, Sá LRM, Ampuero F, Souza JC, Labruna MB (2013) Morphological description of the nymphal stage of *Amblyomma geayi*, and new nymphal records of *Amblyomma parkeri*. Ticks Tick-Borne Dis 4:181–184

Martins TF, Venzal JM, Terassini FA, Costa FB, Marcili A, Camargo LMA, Barros-Battesti DM, Labruna MB (2014a) New tick records from the state of Rondônia, western Amazon, Brazil. Exp Appl Acarol 62:121–128

Martins TF, Fecchio A, Labruna MB (2014b) Ticks of the genus *Amblyomma* (Acari: Ixodidae) on wild birds in the Brazilian Amazon. Syst Appl Acarol 19:385–392

Martins TF, Labruna MB, Mangold AJ, Cafrune MM, Guglielmone AA, Nava S (2014c) Taxonomic key to nymphs of the genus *Amblyomma* (Acari: Ixodidae) in Argentina, with description and redescription of the nymphal stage of four *Amblyomma* species. Ticks Tick-Borne Dis 5:753–770

Martins TF, Lado P, Labruna MB, Venzal JM (2014d) El género *Amblyomma* (Acari: Ixodidae) en Uruguay: especies, distribución, hospedadores, importancia sanitaria y claves para la determinación de adultos y ninfas. Veterinaria (Montevideo) 50:26–41

Martins TF, Teixeira RH, Labruna MB (2015a) Ocorrencia de carrapatos em animais silvestres recebidos e atendidos pelo Parque Zoológico Municipa Quinzinho de Barros, Sorocaba, São Paulo, Brasil. Braz J Vet Res Anim Sci 52:319–324

Martins TF, Arrais RC, Rocha FL, Santos JP, May Júnior JA, Azevedo FC, Paula RC, Morato RG, Rodrigues FHG, Labruna MB (2015b) Carrapatos (Acari: Ixodidae) em mamíferos silvestres do Parque Nacional da Serra da Canastra e arredores, Minas Gerais, Brasil. Ciência Rural 45:288–291

Martins TF, Gianizella SL, Nunes PH, Faria DCLO, Nascimento CAR, Abrahão CR, Miranda FR, Teixeira RHF, Ramirez DG, Barros-Battesti DM, Labruna MB (2015c) New records of *Amblyomma goeldii* (Acari: Ixodidae) and description of the nymphal stage. Zootaxa 3949:439–444

Martins TF, Peres MG, Costa FB, Bacchiega TS, Appolinario CM, Antunes JMAP, Vicente AF, Megid J, Labruna MB (2016a) Ticks infesting wild small rodents in three areas of the State of São Paulo, Brazil. Ciência Rural 46:871–875

Martins TF, Barbieri ARM, Costa FB, Terassini FA, Camargo LMA, Peterka CRL, Pacheco RC, Dias RA, Nunes PH, Marcili A, Scofield A, Campos AK, Horta MC, Guilloux AGA, Benatti HR, Ramírez DG, Barros-Battesti DM, Labruna MB (2016b) Geographical distribution of *Amblyomma cajennense* (sensu lato) ticks (Parasitiformes: Ixodidae) in Brazil, with description of the nymph of *A. cajennense* (sensu stricto). Parasites Vec 9(article 186):14 pp

Martins TF, Milanelo L, Krawczak FS, Furuya HR, Fitorra LS, Dores FT, Pedro VS, Hippolito AG, Labruna MB (2017) Diversity of ticks in the wildlife screening center of São Paulo city, Brazil. Ciência Rural 47:1–6

Martins TF, Bernardi LFO, Muñoz-Leal S, Marcili A, Labruna MB (2019a) New records of *Ixodes amarali* (Acari: Ixodidae) in the Amazon biome, with description of the male. Syst Appl Acarol 24:2552–2558

Martins TF, Luz HR, Muñoz-Leal S, Ramírez DG, Milanelo L, Marques S, Sanches TC, Onofrio VC, Acosta ICL, Benatti HR, Maturano R, Oliveira PB, Albuquerque GR, Marcili A, Flausino W, Silveira LF, McIntosh D, Faccini JLH, Labruna MB (2019b) A new species of *Amblyomma* (Acari: Ixodidae) associated with monkeys and passerines of the Atlantic rainforest biome, southeastern Brazil. Ticks Tick-Borne Dis 10(6)(article 101259):10 pp

Martins TF, Miranda FR, Lopes AMC, Portela S, Labruna MB (2020a) Ticks parasitizing free-ranging armadillos in the Caatinga biome, Brazil. Edentata 21:29–33

Martins TF, Reis JL, Viana EB, Luz HR, Oda FH, Dantas SP, Labruna MB (2020b) Ticks (Acari: Ixodidae) on captive and free-ranging wild animals in Tocantins State, a Cerrado-Amazon transition region of northern Brazil. Int J Acarol 46:254–257

Martins TF, Teixeira RHF, Benatti HR, Minervino AHH, Soares HS, Soares JF, Labruna MB (2020c) Life cycle of the tick *Amblyomma humerale* (Parasitiformes: Ixodida) in the laboratory. Int J Acarol 46(46):351–356

Massard CL, Fonseca AH, Souza AC, Linares FFM, Silva GVO, Faccini JLH, Souza JCP, Famadas KM, Riera MDF, Flausino W (2002) Diagnóstico de *Hyalomma dromedarii* Koch, 1844 (Acari: Ixodidae) em *Camelus dromedarius* importados das Ilhas Canarias – Espanha, para o estado do Rio Grande do Norte – Brasil. Hora Vet 21:58–60

Massi Pallarés R, Benítez Usher CA (1982) De la distribución de Ixodina (Vander Hammen, 1968) en el Paraguay. Rev Paraguaya Microbiol 17:49–52

Mastropaolo M, Orcellet V, Guglielmone AA, Mangold AJ (2008) *Ixodes pararicinus* Keirans & Clifford 1985 y *Amblyomma tigrinum* Koch, 1884 [sic] (Acari: Ixodidae): nuevos registros para Argentina y Chile. Rev FAVE Sección Ciencias Vet 7:67–70

Mastropaolo M, Nava S, Venzal JM, Mangold AJ, Guglielmone AA (2010) Estudio de una colección de garrapatas (Acari: Ixodidae) de pecaríes (Artiodactyla: Tayassuidae) de Salta, Argentina. Rev FAVE Sección Ciencias Vet 9:27–30

Mastropaolo M, Nava S, Schmidt E, Guglielmone AA, Mangold AJ (2012) First report of *Amblyomma boeroi* Nava, Mangold, Mastropaolo, Venzal, Oscherov and Guglielmone

(Acari: Ixodidae) from the chacoan peccary, *Catagonus wagneri* (Artiodactyla: Tayassuidae), in Paraguay. Syst Appl Acarol 17:7–9

Mastropaolo M, Martín GM, Nava S (2013) Nuevos registros de garrapatas (Acari: Ixodidae) asociadas a roedores sigmodontinos en Chubut, Argentina. Rev Argent Parasitol 1:119–123

Mastropaolo M, Beltrán-Saavedra LF, Guglielmone AA (2014) The ticks (Acari: Ixodida: Argasidae, Ixodidae) of Bolivia. Ticks Tick-Borne Dis 5:186–194

Mattar S, López Valencia G (1998) Searching for Lyme disease in Colombia: a preliminary study on the vector. J Med Entomol 35:324–326

Matthysse JG, Colbo MH (1987) The ixodid ticks of Uganda. Entomological Society of America, College Park, 426 pp

Maturano R, Faccini JLH, Daemon E, Fazza POC, Bastos RR (2015) Additional information about tick parasitism in Passeriformes birds in an Atlantic Forest in southeastern Brazil. Parasitol Res 114:4181–4193

Maya-Delgado A, Madder M, Benítez-Ortiz W, Saegerman C, Berkvens D, Ron-Garrido L (2020) Molecular screening of cattle ticks, tick-borne pathogens and amitraz resistance in ticks of Santo Domingo de los Tsáchilas Province in Ecuador. Ticks Tick-Borne Dis 11(5)(article 101492):8 pp

Mazioli R, Szabó M, Mafra C (2012) *Amblyomma nodosum* (Acari: Ixodidae) parasitizing a domestic dog in Colatina, Espírito Santo, Brazil. Braz J Vet Parasitol 21:428–429

McCain CM, Timm RM, Weksler M (2007) Redescription of the enigmatic long-tailed rat *Sigmodontomys aphrastus* (Cricetidae: Sigmodontinae) with comments on taxonomy and natural history. Proc Biol Soc Washington 120:117–136

McChesney TC, Jones LP (1965) Equine piroplasmosis in Puerto Rico. J Am Vet Med Assoc 146:23

McIntosh A (1931) A new species of tick from the Texas peccary. J Parasitol 18:124

McIntosh A (1932) Description of a tick, *Dermacentor halli*, from the Texas peccary, with a key to the North American species of *Dermacentor*. Proc US Natl Mus 82(article 4):6 pp

Medeiros AP, Souza AP, Moura AB, Lavina MS, Bellato V, Sartor AA, Nieri-Bastos FA, Richtzenhain LJ, Labruna MB (2011) Spotted fever group *Rickettsia* infecting ticks (Acari: Ixodidae) in the State of Santa Catarina, Brazil. Mem Inst Oswaldo Cruz 106:926–930

Medem F (1956) Informe sobre reptiles colombianos (I). Noticias sobre el primer hallazgo de la tortuga *Geoemyida annulata* (Gray) en Colombia. Caldasia 7(34):317–325

Medlin JS, Cohen JI, Beck DL (2015) Vector potential and population dynamics for *Amblyomma inornatum*. Ticks Tick-Borne Dis 6:463–472

Medri IM, Martins JR, Doyle RL, Mourao G, Marinho-Filho J (2010) Ticks (Acari: Ixodidae) from yellow armadillo, *Euphractus sexcinctus* (Cingulata: Dasypodidae), in Brazil's Pantanal wetlands. Neotrop Entomol 39:823–825

Méndez Arocha, M. (1956) *Amblyomma margaritae*, n. sp. (Acarina: Ixodidae) de la isla Margarita (Venezuela). Novedades Científicas Contribuciones Ocasionales del Museo de Historia Natural La Salle, Serie Zoología (17):8 pp

Méndez Arocha M, Ortiz I (1957) Descripción del macho y redescripción de la hembra de *Amblyomma crassum* (Acarina: Ixodidae). Mem Soc Ciencias Nat La Salle 17:190–199

Méndez Arocha M, Ortiz I (1958) Revisión de las garrapatas venezolanas del género *Ixodes* Latreille, 1795 y estudio de un nuevo *Amblyomma* (Acarina: Ixodidae). Mem Soc Ciencias Nat Salle 18:196–208

Mendoza-Roldán J, Ribeiro SR, Castilho-Onofrio V, Grazziotin FG, Rocha B, Ferreto-Fiorillo B, Pereira JS, Benelli G, Otranto D, Barros-Battesti DM (2020) Mites and ticks of reptiles and amphibians in Brazil. Acta Trop (article 105515):16 pp

Mendoza-Uribe L, Chávez-Chorocco J (2004) Ampliación geográfica de siete especies de *Amblyomma* (Acari: Ixodidae) y primer reporte de *A. oblongoguttatum* Koch, 1844 para Perú. Rev Peruana Entomol 44:69–72

Merten HA, Durden LA (2000) A state-by-state survey of ticks recorded from humans in the United States. J Vector Ecol 25:102–113

Mertins JW, Vigil SL, Corn JL (2017) *Amblyomma auricularium* (Ixodida: Ixodidae) in Florida: new hosts and distributional records. J Med Entomol 54:132–141
Michael AD (1893) On the genus *Ixodes*. Proc Zool Soc Lond (12):177 [presented by Flower, W. H.]
Michel T, Souza U, Dall'Agnol B, Webster A, Peters F, Christoff A, Luza AL, Kasper N, Becker M, Fiorentin G, Klafke G, Venzal J, Martins JR, Jardim MMA, Ott R, Reck J (2017) *Ixodes* spp. (Acari: Ixodida) ticks in Rio Grande do Sul State, Brazil. Syst Appl Acarol 12:2057–2067
Millán J, Travaini A, Zanet S, López-Bao JV, Trisciuoglio A, Ferroglio E, Rodríguez A (2016) Detection of *Leishmania* DNA in wild foxes and associated ticks in Patagonia, Argentina, 2000 km south of its known distribution area. Parasites Vec 9(article 241):7 pp
Miller MJ, Esser HJ, Loaiza JR, Herre EA, Aguilar C, Quintero D, Alvarez E, Bermingham E (2016) Molecular ecological insights into neotropical bird-tick interaction. PLoS One 11(5) (article e0155989):17 pp
Minervino AHH, Lima JTR, Soares HS, Malheiros AF, Marcili A, Krawczak FS, Lopes MG, Martins TF, Moreira TR, Ribeiro MFB, Labruna MB, Gennari SM (2015) Seroprevalence of tick-borne pathogens and tick infestation in dogs from Tapirapé and Karajá indigenous communities, Brazil. Vect-Borne Zoonot Dis 15:412–418
Minning W (1934) Beiträge zur Systematik und Morphologie der Zeckengattung *Boophilus* Curtice. Z Parasitenkd 7:1–43
Minoprio JL, Minoprio JE, Jörg ME (1994) Babesiosis, caso humano con sindrome hemocitofágico observado en Argentina. Prensa Med Argent 84:43–48
Miranda J, Mattar S (2015) Molecular detection of *Anaplasma* sp. and *Ehrlichia* sp. in ticks collected in domesticated animals, Colombia. Trop Biomed 32:726–735
Miranda J, Violet-Lozano L, Barrera S, Mattar S, Monsalve-Buriticá S, Rodas J, Contreras V (2020) *Candidatus* Rickettsia colombianensi in ticks from reptiles in Córdoba, Colombia. Vet World 13:1764–1770
Moerbeck L, Vizzoni VF, Machado-Ferreira E, Cavalcante RC, Oliveira SV, Soares CAG, Amorim M, Gazeta GS (2016) *Rickettsia* (Rickettsiales: Rickettsiaceae) vector biodiversity in high altitude Atlantic Forest fragments within a semiarid climate: a new endemic area of spotted-fever in Brazil. J Med Entomol 53:1458–1466
Mohler RF (1939) Report of the chief of the Bureau of Animal Industry (1938–1939). United States Department of Agriculture, Washington, DC, 82 pp
Molaei G, Karpathy SE, Andreadis TG (2019) First report of the introduction of an exotic tick, *Amblyomma coelebs* (Acari: Ixodidae), feeding on a human traveler returning to the United States from Central America. J Parasitol 105:571–575
Molaei G, Mertins JW, Stafford KC (2020) Enduring challenge of invasive ticks: introduction of *Amblyomma oblongoguttatum* (Acari: Ixodidae) into the United States on a human traveler returning from Central America. J Parasitol 106:670–674
Monroy Lefebre C, Cajas González JV (1988) Diversidad de garrapatas descritas en la República de Guatemala. Bol Lab Central Diagnóstico Sanidad Anim 2:42–43
Monroy C, Solís S (1988) Informe del hallazgo de *Ixodes scapularis* (Say) (Acari: Ixodidae) en Guatemala. Bol Lab Central Diagnóstico Sanidad Anim 2:59–60
Montiel-Parra G, Fuentes-Moreno H, Vargas M (2007) Primer registro de *Ixodes cookei* (Acari: Ixodidae) para México. Rev Mex Biodivers 78:205–206
Moraes J, Pinter A, Pacheco RC, Gutmann TB, Barbosa SO, Gonzales ARM, Muraro MA, Cecílio SRM, Labruna MB (2009) New epidemiological data on Brazilian spotted fever in an endemic area of the State of São Paulo, Brazil. Vect-Borne Zoonot Dis 9:73–78
Moraes-Filho J, Marcili A, Nieri-Bastos FA, Richrzenhain LJ, Labruna MB (2011) Genetic analysis of ticks belonging to the *Rhipicephalus sanguineus* group in Latin America. Acta Trop 117:51–55
Morales-Soto M, Cruz-Vázquez C (1998) Fluctuaciones poblacionales de *Rhipicephalus sanguineus*, garrapata parásita de perros, en el valle de Cuernavaca, Morelos, México. Estudio preliminar. Vet Méx 29:299–301

Mora-Rivera C, Suárez-Páez F, Pacheco-Sierra G, Vargas-Cueva L, Padilla-Barreto M (2020) Tick infection of *Caiman crocodilus fuscus* at the Hidroprado Hydroelectric Dam in Colombia: new records, parasite prevalence, and blood loss rate. South Am J Herpetol 16:42–49

Moreira-Soto A, Carranza MV, Taylor L, Calderón-Arguedas H, L. & Troyo, A. (2016) Exposure of dogs to spotted fever group rickettsioses in urban sites associated with human rickettsioses in Costa Rica. Ticks Tick-Borne Dis 7:748–753

Morel PC (1966) Étude sur les tiques du bétail en Guadeloupe et Martinique. I. Les tiques et leur distribution (Acariens, Ixodoidea). Rev l'Élevage Méd Vét Pays Trop 19:307–321

Morel PC (1967) Les tiques des animaux sauvages des Antilles (Acariens, Ixodoidea). Acarologia 9:341–352

Morel PC, Pérez C (1973) Morphologie des stades préimaginaux des Ixodidae *s. str.* d'Europe occidentale. II. – Les larves des *Pholeoixodes* Schulze, 1942. Cahiers l'Office Recherche Sci Tech Outremer, Série Entomol Med Parasitol 11:275–284

Morel PC, Pérez C (1977) Morphologie des stases préimaginales des Ixodidae *s. str.* d'Europe occidentale. IV. Généralité sur le sous-genre *Ixodes* (*Ixodes*). Acarologia 19:201–208

Moritz M, Fischer SC (1981) Die Typen der Arachniden-Sammlung des Zoologischen Museums Berlin. Mitteil Zool Mus Berl 57:341–364

Morshed MG, Scott JD, Fernando K, Beati L, Mazerolle DF, Geddes G, Durden LA (2005) Migratory songbirds disperse ticks across Canada, and first isolation of the Lyme disease spirochete, *Borrelia burgdorferi*, from the avian tick, *Ixodes auritulus*. J Parasitol 91:780–790

Mullins MC, Lazzarini SM, Picando MCL, Serra-Freire NM (2004) *Amblyomma parvum* a parasite of *Dasypus kappleri* in the State of Amazonas, Brazil. Rev Ciências Agrárias 42:287–291

Muñoz LE, Casanueva ME (2001) Estado actual del conocimiento de las garrapatas (Acari: Ixodidae) asociada a *Canis familiaris* L. Gayana 65:193–210

Muñoz-García CI, Guzmán-Cornejo C, Rendón-Franco E, Villanueva-García C, Sánchez-Montes S, Acosta-Gutiérrez R, Romero-Callejas E, Díaz-López H, Martínez-Carrasco C, Berriatua E (2019) Epidemiological study of ticks collected from the northern tamandua (*Tamandua mexicana*) and a literature review of ticks of Myrmecophagidae anteaters. Ticks Tick-Borne Dis 10:1146–1156

Muñoz-Leal S, González-Acuña D (2015) The tick *Ixodes uriae* (Acari: Ixodidae): hosts, geographical distribution, and vector roles. Ticks Tick-Borne Dis 6:843–868

Muñoz-Leal S, González-Acuña D, Beltrán-Saavedra LF, Limachi JM, Guglielmone AA (2014) *Amblyomma parvitarsum* (Acari: Ixodidae): localities, hosts and host-parasite ecology. Exp Appl Acarol 62:91–104

Muñoz-Leal S, Tarragona EL, Martins TF, Martín CM, Burgos-Gallardo F, Nava S, Labruna MB, González-Acuña D (2016) *Liolaemus* lizards (Squamata: Liolaemidae) as hosts for the nymph of *Amblyomma parvitarsum* (Acari: Ixodidae), with notes on *Rickettsia* infection. Exp Appl Acarol 70:253–259

Muñoz-Leal S, Lopes MG, Marcili A, Martins TF, González-Acuña D, Labruna MB (2019a) Anaplasmataceae, *Borrelia* and *Hepatozoon* agents in ticks (Acari: Argasidae, Ixodidae) from Chile. Acta Trop 192:91–103

Muñoz-Leal S, Marcili A, Fuentes-Castillo D, Ayala M, Labruna MB (2019b) A relapsing fever *Borrelia* and spotted fever *Rickettsia* in ticks from an Andean valley, central Chile. Exp Appl Acarol 78:403–420

Muñoz-Leal S, Ramírez DG, Luz HR, Faccini JLH, Labruna MB (2020) "*Candidatus* Borrelia ibitipoquensis," a *Borrelia valaisiana*-related genospecies characterized from *Ixodes paranaensis* in Brazil. Microb Ecol 80:682–689

Murgas IL, Castro AM, Bermúdez SE (2013) Current status of *Amblyomma ovale* (Acari: Ixodidae) in Panama. Ticks Tick-Borne Dis 4:164–166

Murrell A, Barker SC (2003) Synonymy of *Boophilus* Curtice, 1891 with *Rhipicephalus* Koch, 1844 (Acari: Ixodidae). Syst Parasitol 56:169–172

Myers JG (1935) Second report of an investigation into the biological control of West Indian insect pests. Bull Entomol Res 26:181–252

Nadolny RM, Gaff HD (2018) Natural history of *Ixodes affinis* in Virginia. Ticks Tick-Borne Dis 9:109–119

Nakatsudi K (1942) One new tick from North Manchukou [sic]. J Agric Sci (Tokyo) 1:329–332. [in Japanese, Spanish translation by Terashima, E.]

Nascimento KKG, Veríssimo SMM, Raia VA, Guimarães RCS, Seade GCC, Azevedo ACP, Matos SP, Oliveira JM, Bezerra IA, Martins TF (2017) Tick fauna of wild animals received and attended at the Santarém Zoological Park, western Pará State, Brazil. Ciência Rural 47(10) (article e20170159):5 pp

Nava S, Guglielmone AA (2013) A meta-analysis of host specificity in Neotropical hard ticks (Acari: Ixodidae). Bull Entomol Res 103:216–224

Nava S, Lareschi M, Beldoménico PM, Zerpa C, Venzal JM, Mangold AJ, Guglielmone AA (2004) Sigmodontinae rodents as hosts for larvae and nymphs of *Ixodes loricatus* Neumann, 1899 (Acari: Ixodidae). Parasite 11:411–414

Nava S, Mangold AJ, Guglielmone AA (2006a) The natural hosts for larvae and nymphs of *Amblyomma neumanni* and *Amblyomma parvum* (Acari: Ixodidae). Exp Appl Acarol 40:123–131

Nava S, Mangold AJ, Guglielmone AA (2006b) The natural hosts of larvae and nymphs of *Amblyomma tigrinum* Koch, 1844 (Acari: Ixodidae). Vet Parasitol 140:124–132

Nava S, Lareschi M, Rebollo C, Benítez Usher C, Beati L, Robbins RG, Durden LA, Mangold AJ, Guglielmone AA (2007) The ticks (Acari: Ixodida: Argasidae, Ixodidae) of Paraguay. Ann Trop Med Parasitol 101:255–270

Nava S, Szabó MPJ, Mangold AJ, Guglielmone AA (2008) Distribution, hosts, 16S rDNA sequences and phylogenetic position of the Neotropical tick *Amblyomma parvum* (Acari: Ixodidae). Ann Trop Med Parasitol 102:409–425

Nava S, Estrada-Peña A, Mangold AJ, Guglielmone AA (2009a) Ecology of *Amblyomma neumanni* (Acari: Ixodidae). Acta Trop 111:226–236

Nava S, Mangold AJ, Mastropaolo M, Venzal JM, Oscherov EB, Guglielmone AA (2009b) *Amblyomma boeroi* n. sp. (Acari: Ixodidae), a parasite of the Chacoan peccary *Catagonus wagneri* (Rusconi) (Artiodactyla: Tayassuidae) in Argentina. Syst Parasitol 73:161–174

Nava S, Venzal JM, Labruna MB, Mastropaolo M, González EM, Mangold AJ, Guglielmone AA (2010a) Hosts, distribution and genetic divergence (16S rDNA) of *Amblyomma dubitatum* (Acari: Ixodidae). Exp Appl Acarol 51:335–351

Nava S, Velazco PM, Guglielmone AA (2010b) First record of *Amblyomma longirostre* (Koch, 1844) (Acari: Ixodidae) from Peru, with a review of this tick's host relationships. Syst Appl Acarol 15:21–30

Nava S, Mangold AJ, Mastropaolo M, Venzal JM, Fracassi N, Guglielmone AA (2011) Seasonal dynamics and hosts of *Amblyomma triste* (Acari: Ixodidae) in Argentina. Vet Parasitol 181:301–308

Nava S, Mastropaolo M, Mangold AJ, Venzal JM, Guglielmone AA (2012) Mitochondrial DNA analysis of *Rhipicephalus sanguineus* sensu lato (Acari: Ixodidae) in the Southern Cone of South America. Vet Parasitol 190:547–555

Nava S, Martínez JG, Arreguez GA, Guglielmone AA (2013) Ticks (Acari: Argasidae, Ixodidae) from Middle and pre-hispanic Late Holocene associated with human activities in northwestern Argentina. Ticks Tick-Borne Dis 4:167–169

Nava S, Beati L, Labruna MB, Cáceres AG, Mangold AJ, Guglielmone AA (2014a) Reassessment of the taxonomic status of *Amblyomma cajennense* (Fabricius, 1787) with the description of three new species, *Amblyomma tonelliae* n. sp., *Amblyomma interandinum* n. sp. and *Amblyomma patinoi* n. sp., and reinstatement of *Amblyomma mixtum* Koch, 1844 and *Amblyomma sculptum* Berlese, 1888 (Ixodida: Ixodidae). Ticks Tick-Borne Dis 5:252–276

Nava S, Mastropaolo M, Mangold AJ, Martins TF, Venzal JM, Guglielmone AA (2014b) *Amblyomma hadanii* n. sp. (Acari: Ixodidae), a tick from northwestern Argentina previously confused with *Amblyomma coelebs* Neumann, 1899. Syst Parasitol 88:261–272

Nava S, Beati L, Dunlop J, Guglielmone AA (2014c) Reestablishment of *Amblyomma tenellum* Koch, 1844 (Acari: Ixodidae). Ticks Tick-Borne Dis 5:620–624

Nava S, Estrada-Peña A, Petney T, Beati L, Labruna MB, Szabó MPJ, Venzal JM, Mastropaolo M, Mangold AJ, Guglielmone AA (2015) The taxonomic status of *Rhipicephalus sanguineus* (Latreille, 1806). Vet Parasitol 208:2–8

Nava S, Gerardi M, Szabó MP, Mastropaolo M, Martins TF, Labruna MB, Beati L, Estrada-Peña A, Guglielmone AA (2016) Different lines of evidence used to delimit species in ticks: a study of the South American populations of *Amblyomma parvum* (Acari: Ixodidae). Ticks Tick-Borne Dis 7:1168–1179

Nava S, Venzal JM, González-Acuña D, Martins TF, Guglielmone AA (2017) Ticks of the Southern Cone of America. Elsevier Academic, London, 348 pp

Nava S, Beati L, Venzal JM, Labruna MB, Szabó MPJ, Petney T, Saracho-Bottero MN, Tarragona EL, Santos Silva MM, Mangold AJ, Guglielmone AA, Estrada-Peña A (2018) *Rhipicephalus sanguineus* (Latreille, 1806): neotype designation, morphological re-description of all parasitic stages and molecular characterization. Ticks Tick-Borne Dis 9:1573–1585

Navarrete Abarca LR, Rodríguez Romero EA, Valle Martínez CA, Vargas Artiga MJ, Pérez Romero LE (2014) Principales especies de garrapatas (Ixodidae) en El Salvador. Universidad de El Salvador, Ministerio de Agricultura y Ganadería, Red de Laboratorios Veterinarios, 66 pp

Need JT, Dale WE, Keirans JM, Dasch GA (1991) Annotated list of ticks (Acari: Ixodidae: Argasidae) reported in Peru: distribution, hosts, and bibliography. J Med Entomol 28:590–597

Nelson SL, Durden LA, Reuter JD (2017) *Rhipicephalus microplus* and *Dermacentor nitens* (Acari: Ixodidae) coparasitize white-tailed deer on St. John, U.S. Virgin Islands. J Med Entomol 54:1440–1443

Neumann LG (1897) Révision de la famille des ixodidés (2e mémoire). Mém Soc Zool France 10:324–420

Neumann LG (1899) Révision de la famille des ixodidés (3e mémoire). Mém Soc Zool France 12:107–294

Neumann LG (1901) Révision de la famille des ixodidés (4e mémoire). Mém Soc Zool France 14:249–372

Neumann LG (1902) Notes sur les ixodidés. Arch Parasitol 6:109–128

Neumann LG (1904) Notes sur les ixodidés. II. Arch Parasitol 8:444–464

Neumann LG (1905) Notes sur les ixodidés. III. Arch Parasitol 9:225–241

Neumann LG (1906) Notes sur les Ixodidés. IV. Arch Parasitol 10:195–219

Neumann LG (1907) Quatre espèces nouvelles d'ixodidés. Notes Leyden Mus 29:88–100

Neumann LG (1908) Notes sur les ixodidés. XI. Notes Leyden Mus 30:73–91

Neumann LG (1910) Sur quelques espèces d'Ixodidae nouvelles ou insuffisamment connues. Annal Sci Nat Zool Sér *9*(12):161–176

Neumann LG (1911a) Ixodidae. Das Tierreich 26:169 pp

Neumann LG (1911b) Note rectificative à propos de deux espèces d'Ixodinae. Arch Parasitol 14:415

Neumann LG (1913) Un nouveau sous-genre et deux nouvelles espèces d'ixodidés. Bull Soc Zool Fr 38:147–151

Newstead R (1909) Reports of the twenty first expedition of the Liverpool School of Tropical Medicine Jamaica, 1908 – 1909. Section I Medical and economic entomology. Ann Trop Med Parasitol 3:421–468

Noda AA, Rodríguez I, Miranda J, Contreras V, Mattar S (2015) First molecular evidence of *Coxiella burnetti* infecting ticks in Cuba. Ticks Tick-Borne Dis 7:68–70

Nogueira BCF, Bezerra-Santos MA, Yamatogi RS, Campos AC (2020) *Amblyomma* ticks (Acari: Ixodidae) parasitizing feral pigs in the State of Minas Gerais, Brazil. Int J Acarol 46:171–173

Norris DE, Klompen JHS, Keirans JE, Lane RS, Piesman J, Black WC (1997) Taxonomic status of *Ixodes neotomae* and *I. spinipalpis* (Acari: Ixodidae) based on mitochondrial DNA evidence. J Med Entomol 34:696–703

Norris DE, Klompen JHS, Black WC (1999) Comparison of the mitochondrial 12S and 16S ribosomal DNA genes in resolving phylogenetic relationships among hard ticks (Acari: Ixodidae). Ann Entomol Soc Am 92:117–129

Nossa DN, Dias LCS, Acosta ICL, Martins TF, Srbek-Araujo AC (2019) Parasitism by *Amblyomma rotundatum* (Acari: Ixodidae) on *Rhinella diptycha* (Anura: Bufonidae) in the Atlantic Forest: geographic expansion of species interaction. Herpetol Notes 12:981–985

Novakova M, Literak I, Chévez L, Martins TF, Ogrzewalska M, Labruna MB (2015) Rickettsial infections in ticks from reptiles, birds and humans in Honduras. Ticks Tick-Borne Dis 6:637–642

Nowak M (2010) The characteristics of tick species (Acari: Ixodida) transferred on exotic reptiles to Poland. Wiadomości Parazytol 56:29–42

Nuñez JL, Cobeñas MEM, Moltedo HL (1982) Boophilus microplus*: la garrapata común del ganado vacuno.* Hemisferio Sur, Buenos Aires, 184 pp

Nuttall GHF (1910) New species of ticks (*Ixodes*, *Amblyomma*, *Rhipicephalus*). Parasitology 3:408–416

Nuttall GHF (1911) On the adaptation of ticks to the habits of their hosts. Parasitology 4:46–67

Nuttall GHF (1912) Notes on ticks. II. (1) New species (*Amblyomma, Haemaphysalis*) (2) *Ixodes putus*: description of the hitherto unknown larval stage. Parasitology 5:50–60

Nuttall GHF (1916) Notes on ticks. IV. Relating to the genus *Ixodes* and including description of three new species and two varieties. Parasitology 8:294–337

Nuttall GHF, Warburton C (1908) On a new genus of Ixodoidea together with a description of eleven new species. Proc Camb Philos Soc Math Phys Sci 14:392–416

Nuttall GHF, Warburton C (1911) Ticks. A monograph of the Ixodoidea. Part II. The Ixodidae. Section II Genus I. Ixodes Latreille 1795. Cambridge University Press, London, pp 133–293

Nuttall GHF, Warburton C (1915) Ticks. A monograph of the Ixodoidea. Part III. The genus Haemaphysalis. Cambridge University Press, London, pp 349–550

Obregón D, Corona-González B, Díaz-Sánchez AA, Armas Y, Roque E, Oliveira MCS, Cabezas-Cruz A (2020) Efficient transovarial transmission of *Babesia* spp. in *Rhipicephalus microplus* ticks fed on water buffalo (*Bubalus bubalis*). Pathogens 9(article 280):10 pp

Oda FH, Kitagawa C, Noronha JC, Rodrigues DJ, Martins TF, Valadao MC, Carvalho LM, Campos AK (2018) *Amblyomma* species infesting amphibians and reptiles in the seasonally dry Amazon forest, with new host records for *Amblyomma rotundatum* (Acari: Ixodidae). Syst Appl Acarol 23:387–392

Ogrzewalska M, Bermúdez S (2019) Detección molecular de *Rickettisia bellii* en *Amblyomma rotundatum* Koch, 1884 [sic] (Ixodida: Ixodidae) recolectada de *Rhinella marina* L. 1758 [sic] (Anura: Bufonidae) en Panamá. Tecnociencia 21:49–56

Ogrzewalska M, Pacheco RC, Uezu A, Ferreira F, Labruna MB (2008) Ticks (Acari: Ixodidae) infesting birds in an Atlantic forest area in the State of São Paulo, Brazil, with isolation of *Rickettsia* from the tick *Amblyomma longirostre*. J Med Entomol 45:770–774

Ogrzewalska M, Pacheco RC, Uezu A, Richtzenhain LJ, Ferreira F, Labruna MB (2009) Ticks (Acari: Ixodidae) infesting birds in an Atlantic rain forest region in Brazil. J Med Entomol 46:1225–1229

Ogrzewalska M, Uezu A, Labruna MB (2010) Ticks (Acari: Ixodidae) infesting wild birds in the eastern Amazon, northern Brazil, with notes on rickettsial infection in ticks. Parasitol Res 106:809–816

Ogrzewalska M, Uezu A, Labruna MB (2011) Ticks (Acari: Ixodidae) infesting wild birds in the Atlantic Forest in northeastern Brazil, with notes on rickettsial infection in ticks. Parasitol Res 108:665–670

Ogrzewalska M, Saraiva DG, Moraes-Filho J, Martins TF, Costa FB, Pinter A, Labruna MB (2012a) Epidemiology of Brazilian spotted fever in the Atlantic Forest, State of São Paulo, Brazil. Parasitology 139:1283–1300

Ogrzewalska M, Cárdenas-Callirgos JM, Capek M, Labruna MB (2012b) *Rickettsia bellii* in ticks *Amblyomma varium* Koch, 1844, from birds in Peru. Ticks Tick-Borne Dis 3:254–256

Ogrzewalska M, Martins T, Capek M, Literak I, Labruna MB (2013) A *Rickettsia parkeri*-like agent infecting *Amblyomma calcaratum* nymphs from wild birds in Mato Grosso do Sul, Brazil. Ticks Tick-Borne Dis 4:145–147

Ogrzewalska M, Literák I, Martins TF, Labruna MB (2014) Rickettsial infections in ticks from wild birds in Paraguay. Ticks Tick-Borne Dis 5:83–89

Ogrzewalska M, Literák I, Capek M, Sychra O, Álvarez Calderón V, Calvo Rodríguez B, Prudencio C, Martins TF, Labruna MB (2015) Bacteria of the genus *Rickettsia* in ticks (Acari: Ixodidae) collected from birds in Costa Rica. Ticks Tick-Borne Dis 6:478–482

Ogrzewalska M, Machado C, Rozental T, Forneas D, Cunha LE, De Lemos ERS (2019) Microorganisms in the ticks *Amblyomma dissimile* Koch 1844 and *Amblyomma rotundatum* Koch 1844 collected from snakes in Brazil. Med Vet Entomol 33:154–161

Ojeda-Chi MM, Rodríguez-Vivas RI, Esteve-Gasent MD, Pérez de León A, ModarellI JJ, Villegas-Pérez SL (2019a) Molecular detection of rickettsial tick-borne agents in white-tailed deer (*Odocoileus virginianus yucatanensis*), mazama deer (*Mazama temama*), and the ticks they host in Yucatán, Mexico. Ticks Tick-Borne Dis 10:365–370

Ojeda-Chi MM, Rodríguez-Vivas RI, Esteve-Gasent MD, Pérez de León A, Modarell JJ, Villegas-Pérez SL (2019b) Ticks infesting dogs in rural communities of Yucatán, Mexico and molecular diagnosis of rickettsial infection. Transbound Emerg Dis 66:102–110

Oliveira PR, Bechara GH, Denardi SE, Saito CS, Nunes ET, Szabó MPJ, Mathias MI (2005) Comparison of the external morphology of *Rhipicephalus sanguineus* (Latreille, 1806) (Acari: Ixodidae) ticks from Brazil and Argentina. Vet Parasitol 129:139–147

Oliveira HH, Gomes V, Amorim M, Gazeta GS, Serra-Freire NM, Quinelato IPF, Morelli-Amaral VF, Almeida AB, Carvalho RW, Carvalho AG (2014) Diversidade de ixodida em roedores e marsupiais capturados no Parque Estadual da Pedra Branca, Rio de Janeiro, Brasil. Arq Bras Med Vet Zoot 66:1097–1104

Oliveira GMB, Martins TF, Pereira LCM, Nicola PA, Horta MC (2017) Ocorrencia de carrapatos em *Tamandua tetradactyla* (Linnaeus, 1758) na Caatinga, nordeste do Brasil. Arq Bras Med Vet Zoot 69:865–869

Oliveira GMB, Silva IWG, Evaristo AMCF, Serpa MCA, Campos ANS, Dutra V, Nakasato L, Aguiar DM, Labruna MB, Horta MC (2020) Tick-borne pathogens in dogs, wild small mammals and their ectoparasites in the semi-arid Caatinga biome, northeastern Brazil. Ticks Tick-Borne Dis 11(4)(article 101409):7 pp

Oliver JH (1989) Biology and systematics of ticks (Acari: Ixodidae). Annu Rev Ecol Syst 20:397–430

Oliver JH, Keirans JE, Lavender DR, Hutcheson HJ (1987) *Ixodes affinis* Neumann (Acari: Ixodidae): new host and distribution records, description of immatures, seasonal activities in Georgia, and laboratory rearing. J Parasitol 73:646–652

Ondrejicka DA, Morey KC, Hanner RH (2016) DNA barcodes identify medically important tick species in Canada. Genome 64:74–84

Oniki Y (1999) Avian parasites and notes on habits of lice from Mato Grosso, Brazil. Iheringia Sér Zool 86:183–190

Oniki-Willis Y, Willis EO (2018) Tick (Acarina) diversity from South American birds and mammals. Atualidades Ornitol 206:33–46

Onofre-Barragán B (2015) Estudio preliminar de artrópodos de importancia veterinaria en El Carmen, Municipio de Xochistlahuaca, Guerrero, México. Bol Soc Mex Entomol Nueva Ser 1:118–121

Onofrio VC (2007) Revisâo do gênero *Amblyomma* Koch, 1844 (Acari: Ixodidae) no Brasil. Tese de Doutorado, Universidade Federal Rural do Rio de Janeiro, Instituto de Veterinária, 174 pp

Onofrio VC, Labruna MB, Barros-Battesti DM (2006a) Coméntarios e chaves para as espécies do gênero *Ixodes*. In: Barros-Battesti DM, Arzua M, Bechara GH (ed) Carrapatos de importância médico-veterinária da região neotropical: um guia ilustrada para identificação de especies. São Paulo/Vox/ICTTD-3/Butantan, pp 41–51

Onofrio VC, Labruna MB, Pinter A, Giacomin FG, Barros-Battesti DM (2006b) Comêntarios e chaves para as espécies do gênero *Amblyomma*. In: Barros-Battesti DM, Arzua M, Bechara GH (eds) Carrapatos de importância médico-veterinária da região neotropical: um guia ilustrada para identificação de especies. São Paulo/Vox/ICTTD-3/Butantan, pp 53–113

Onofrio VC, Barros-Battesti DM, Marques S, Faccini JLH, Labruna MB, Beati L, Guglielmone AA (2008) Redescription of *Amblyomma varium* Koch, 1844 (Acari: Ixodidae) based on light and scanning electron microscopy. Syst Parasitol 69:137–144

Onofrio VC, Barros-Battesti DM, Labruna MB, Faccini JLH (2009) Diagnoses of and illustrated key to the species of *Ixodes* Latreille, 1795 (Acari: Ixodidae) from Brazil. Syst Parasitol 72:143–157

Onofrio VC, Arzua M, Labruna MB, Faccini JLH, Barros-Battesti DM (2010a) First record of *Amblyomma scalpturatum* Neumann (Acari: Ixodidae) in the States of Paraná and Roraima, Brazil. Neotrop Entomol 39:451–453

Onofrio VC, Labruna MB, Faccini JLH, Barros-Battesti DM (2010b) Description of immature stages and redescription of adults of *Ixodes luciae* Sénevet (Acari: Ixodidae). Zootaxa 2495:53–64

Onofrio VC, Nieri-Bastos FA, Sampaio JS, Soares JF, Silva MJJ, Barros-Battesti DM (2013) Noteworthy records of *Ixodes schulzei* (Acari: Ixodidae) on rodents from the State of Paraná, southern Brazil. Braz J Vet Parasitol 22:159–161

Onofrio VC, Ramírez DG, Giovanni DNS, Marcili A, Mangold AJ, Venzal JM, Mendonça RZ, Labruna MB, Barros-Battesti DM (2014) Validation of the taxon *Ixodes aragaoi* Fonseca (Acari: Ixodidae) based on morphological and molecular data. Zootaxa 3860:361–370

Onofrio VC, Guglielmone AA, Barros-Battesti DM, Gianizella SL, Marcili A, Quadros RM, Marques S, Labruna MB (2020) Description of a new species of *Ixodes* (Acari: Ixodidae) and first report of *Ixodes lasellei* and *Ixodes bocatorensis* in Brazil. Ticks Tick-Borne Dis 11(4) (article 101423):9 pp

Ortiz F, Nava S, Guglielmone AA (2011) Análisis de una colección de garrapatas (Acari: Argasidae, Ixodidae) del norte argentino. Rev FAVE Sección Ciencias Vet 10:49–55

Ortiz CAR, Ríos MA, Soler-Tovar D (2020) Conocimiento de la distribución geográfica y ciclo de vida del género *Amblyomma* (Acari: Ixodidae) en Colombia. In: Posada-Arias J, Cabrera-Jaramillo A, Monsalve-Buriticá S (eds) Enfermedades rickettsiales en Latinoamérica. Fondo Editorial Biogénesis de la Universidad de Antioquía, Medellín, pp 123–150

Osava CF, Ramos VN, Rodrigues AC, Neto HVR, Martins MM, Pascoal JO, Yokosawa J, Szabó MPJ (2016) *Amblyomma sculptum* (*Amblyomma cajennense* complex) tick population maintained solely by domestic pig. Vet Parasitol Reg Stud Rep 6:9–13

Osorio G (2001) Búsqueda de la espiroqueta *Borrelia burgdorferi sensu lato* mediante PCR en garrapatas ixoideas [sic] chilenas silvestres. Rev Med Chil 129:270–276

Osorno Mesa E (1942) Las garrapatas de la República de Colombia. Rev Facult Nacional Agric Med 5:57–103

Oudemans AC (1906) Acarologische Aanteekeningen XXIV. Entomol Berichten 2(30):96–101

Oudemans AC (1936) Kritisch Historich overzicht der acarologie. Derde gedeelte: 1805–1850, vol B. E.J. Brill, Leiden, pp 430–797

Oudemans AC (1939) Neue Funde auf dem Gebiete der Systematik und der Nomenklatur der Acari. VI. Zool Anz 127:75–80

Pacheco RC, Arzua M, Nieri-Bastos FA, Moraes-Filho J, Marcili A, Richtzenhain LJ, Barros-Battesti DM, Labruna MB (2012) Rickettsial infection in ticks (Acari: Ixodidae) collected on birds in southern Brazil. J Med Entomol 49:710–716

Pacheco A, Cordeiro MD, Cepeda MB, Luz HR, Cardozo SV, Berto BP, Guterres A, Fonseca AH (2019) Hemoparasites in ticks of wild birds of Serra dos Órgãos National Park, State of Rio de Janeiro, Brazil. Braz J Vet Parasitol 28:238–244

Packard AS (1869) Report of the curator of Articulata. First Annual Report of the Trustees of the Peabody Academy of Sciences, pp 56–69

Pagenstecher HA (1861) Beiträge zur Anatomie der Milben. Heft II. *Ixodes ricinus*. W. Engelmann, Leipzig, 45 pp

Pallas PS (1772) Spicilegia Zoologica, vol 40, fascicule 9, 87 pp

Panzer GWF (1798) Faunae Insectorum Germanicae Initia, part 59, Nürnberg

Paredes-León R, García-Prieto L, Guzmán-Cornejo C, León-Regagnon V, Pérez TM (2008) Metazoan parasites of Mexican amphibians and reptiles. Zootaxa 1904:1–66

Pascoal JO, Amorim MP, Martins MM, Melo C, Silva Júnior EL, Ogrzewalska M, Labruna MB, Szabó MPJ (2013) Ticks on birds in a savanna (Cerrado) reserve on the outskirts of Uberlandia, Minas Gerais, Brazil. Braz J Vet Parasitol 22:46–52

Paternina LE, Díaz-Olmos Y, Paternina-Gómez M, Bejarano EE (2009) *Canis familiaris*, un nuevo hospedero de *Ornithodoros* (*A.*) *puertoricensis* Fox, 1947 (Acari: Ixodida) en Colombia. Acta Biol Colomb 14:153–160

Paulian P (1953) Pinnipèdes, cétacés, oiseaux des îles Kerguelen et Amsterdam. Mission Kerguelen 1951. Mém l'Inst Sci Madagascar Sér A Biol Anim 8:111–234

Payne RC, Scott JM (1982) Anaplasmosis and babesiosis in El Salvador. Trop Anim Health Prod 14:75–80

Payno Balazans JM (1978) Identificación de parásitos de importancia veterinaria en los animales de Bolivia. Misión Británica en Agricultura Tropical, Santa Cruz de la Sierra, 23 pp

Pearse AS (1936) Parasites of Yucatán. Publ Carnegie Inst Washington 457:45–59

Pedroso-Santos F, Santos ES, Sanches PR, Costa-Campos CE, Luz HR, Faccini JLH (2020) First record of *Amblyomma dissimile* (Acari: Ixodidae) infesting the granular toad *Rhinella major* (Anura: Bufonidae) in the eastern Amazon region. Herpetol Notes 13:385–387

Pegram RG, Clifford CM, Walker JB, Keirans JE (1987) Clarification of the *Rhipicephalus sanguineus* group (Acari, Ixodoidea, Ixodidae). I. *R. sulcatus* Neumann, 1908 and *Rhipicephalus turanicus* Pomerantsev, 1936. Syst Parasitol 10:3–26

Peniche-Lara G, Jiménez-Delgadillo B Munoz-Sansi C, Cárdenas-Marrufo M, Pérez-Osorio C, Arias-León J (2018) Presence of *Rickettsia* species in a marginalized area of Yucatán, Mexico. J Trop Med 2018 (article 7675828):4 pp

Pereira JS, Martins TF, Muñoz-Leal S, Lopes MG, Labruna MB, Paiva KAR, Oliveira MF, Ahid SMM (2017) Infestação por carrapatos Argasidae e Ixodidae em pequenos mamíferos silvestres da Estação Experimental Rafael Fernandes, Mossoró/RN. Pesquisa Vet Bras 37:741–748

Pérez Vigueras I (1934) On the ticks of Cuba, with description of a new species, *Amblyomma torrei* from *Cyclura macleayi* Gray. Psyche 41:13–18

Pesquera C, Portillo A, Palomar AM, Oteo JA (2015) Investigation of tick-borne bacteria (*Rickettsia* spp., *Ehrlichia* spp. and *Borrelia* spp.) in ticks collected from Andean tapirs, cattle and vegetation from a protected area in Ecuador. Parasites Vec 8(article 46):10 pp

Petraccia CA, Cardozo H, Nari A, Solari MA (1983) Estudios de resistencia a garrapaticidas organofosforados (OF) en *Boophilus microplus*. Resultados obtenidos en cepas de campo llegadas al C.I. Vet. "Miguel C. Rubino" durante los años 1981-1982. Veterinaria (Montevideo) 19:5–9

Petters J, Cirne LB, Gatica-Colima A, Martínez-Calderas JM, Movia J (2020) Reporte de *Amblyomma nodosum* Neumann (Acari: Ixodidae) en *Tamandua tetradactyla* (Linnaeus), departamento Cordillera, Paraguay. Bol Mus Nacional Hist Nat Paraguay 24:61–66

Pickard-Cambridge OP (1876) On a new order and some new genera of Arachnida from Kerguelen's Land. Proc Zool Soc Lond (17):258–265

Pietzsch M, Quest R, Hillyard PD, Medlock JM, Leach S (2006) Importation of exotic ticks into the United Kingdom via the international trade in reptiles. Exp Appl Acarol 38:59–65

Pinheiro MC, Lourenço EC, Patrício PMP, Sá-Hungaro IJB, Famadas KM (2014) Free-living ixodid ticks in an urban Atlantic Forest fragment, State of Rio de Janeiro, Brazil. Braz J Vet Parasitol 23:264–268

Pinter A (2007) Aspectos ecológicos da febre maculosa brasileira em um foco endémico no Estado de São Paulo. Tese de doutorado. Universidade de São Paulo, Faculdade de Medicina Veterinária e Zootecnia, 86 pp

Pinto C (1938) Zoo-parásitos de interesse medico e veterinario. Editora Científica, Rio de Janeiro, 376 pp

Pinto C (1945) Zoo-parásitos de interesse medico e veterinario, 2nd edn. Editora Científica, Rio de Janeiro, 461 pp

Pinto C, Lins de Almeida J (1939) Sobre o *Amblyomma neumanni* Rib., 1902, parasito do homen na Argentina. An Acad Bras Cienc 11:113–124

Polo G, Luz HR, Regolin AL, Martins TF, Winck GR, da Silva HR, Onofrio VC, Labruna MB, Faccini JL (2021) Distribution modeling of *Amblyomma rotundatum* and *Amblyomma dissimile* in Brazil: estimates of environmental suitability. Parasitol Res 120:797–806

Polsomboon S, Hoel DF, Murphy JR, Linton YM, Motoki M, Robbins RG, Bautista K, Briceño I, Achee NL, Grieco JP, Ching WM, Chao CC (2017) Molecular detection and identification of *Rickettsia* species in ticks (Acari: Ixodidae) collected from Belize, Central America. J Med Entomol 54:1718–1726

Pomerantzev BI, Matikashvili NV, Lotozky BV (1940) An ecological and faunistic outline of Ixodidae ticks occurring in Transcaucasia. Parazitol Sbornik 7:100–133. [in Russian, NAMRU-3 translation 51]

Power LA, Silvestri R, Chacón JC (1988) Nuevas observaciones sobre la presencia de *Boophilus microplus* y *Amblyomma cajennense* en ganado bovino de los estados Barinas, Lara y Guárico. Rev Facult Ciencias Vet 35:61–67

Price MA (1959) A new species of tick from the trans-Pecos region of Texas. J Parasitol 44:649–651

Prieto R (1974) Distribución geográfica, influencia ecológica y selectividad hospedera de *Amblyomma cajennense* (Fabricius, 1787) en Cuba. Rev Cubana Ciencias Vet 5:47–51

Quintero JC, Mignone J, Osorio L, Cienfuegos-Gallet AV, Rojas C (2021) Housing conditions linked to tick (Ixodida: Ixodidae) infestation in rural area of Colombia: a potential risk for rickettsial transmission. J Med Entomol 58:439–449

Quirino TF, Alcantara EP, Silva CF, Avila RW, Morais DH, Silva LAF (2016) *Tropidurus hispidus* (Peter's lava lizard). Ectoparasite. Herpetol Rev 47:305–306

Rainwater TR, Platt SG, Robbins RG, McMurry ST (2001) Ticks from a Morelet's crocodile in Belize. J Wildl Dis 37:836–839

Ramírez DG, Luz HR, Muñoz-Leal S, Flausino W, Acosta ICL, Martins TF, Peckle M, Santos HF, Furusawa GP, Labruna MB, Faccini JLH (2020) Immature ticks on wild birds and the molecular detection of a novel *Rickettsia* strain in the Ibitipoca State Park, south-eastern Brazil. Exp Appl Acarol 81:457–467

Ramírez-Barrios R, Chacín E, Barboza G, Fernández G, Valera Z, Villalobos A, Angulo F (2008) Garrapatas (Acari: Ixodidae) recolectadas de caninos bajo asistencia veterinaria en Maracaibo, Venezuela. Rev Científica FCV-LUZ 18:267–270

Ramos DG, Melo ALT, Martins TF, Alves AS, Pacheco TA, Pinto LB, Pinho JB, Labruna MB, Dutra V, Aguiar DM, Pacheco RC (2015) Rickettsial infection in ticks from wild birds from Cerrado and the Pantanal region of Mato Grosso, midwestern Brazil. Ticks Tick-Borne Dis 6:836–842

Ramos VN, Piovezan U, Franco AHA, Osava CF, Herrera HM, Szabó MPJ (2014) Feral pigs as hosts for *Amblyomma sculptum* (Acari: Ixodidae) populations in the Pantanal, Mato Grosso do Sul, Brazil. Exp Appl Acarol 64:393–406

Ramos VN, Lemos FG, Azevedo FC, Arrais RC, Lima CFM, Candeias IZ, Martins MM, Sandrin ACLG, Siqueira SM, Szabó MPJ (2020) Wild carnivores, domestic dogs and ticks: shared parasitism in the Brazilian Cerrado. Parasitology 147:689–698

Rawlins SC (1979) Seasonal variation in the population density of larvae of *Boophilus microplus* (Canestrini) (Acari: Ixodoidea) in Jamaica pastures. Bull Entomol Res 69:87–91

Rawlins SC, Mansingh A (1977) Relative susceptibility of Jamaican strains of *Boophilus microplus*, *Amblyomma cajennense* and *Anocentor nitens* to various acaricides. J Econ Entomol 70:697–698

Rawlins SC, Mansingh A (1978) Acaricidal susceptibility of five strains of *Boophilus microplus* from four Caribbean countries. J Econ Entomol 71:142–144

Rawlins SC, Mahadeo S, Martínez R (1993) A list of ticks affecting man and animals in the Caribbean. CARAPHIN News 6:8–9

Reck J, Souza U, Souza G, Kieling E, Dall'Agnol B, Webster A, Michel T, Doyle R, Martins TF, Labruna MB, Marks F, Ott R, Martins RJ (2018) Records of ticks on humans in Rio Grande do Sul State, Brazil. Ticks Tick-Borne Dis 9:1296–1301

Redell JR, Veni G (1996) Biology of the Chiquibul cave system, Belize and Guatemala. J Cave Karst Stud 58:131–138

Reyes RV (1938) Parásitos de los animales domésticos en Colombia. Rev Med Vet (Bogotá) 8:17–29

Reyne A (1923) Verslag van den entomoloog. Department van Landbouw, Nijverheid en Handel in Suriname 1922:32–39

Ribaga C (1902) Acari sudamericani. Zool Anz 25:502–508

Ribeiro AL, Faccini JLH, Daemon E (1996) Estudo das variações morfológicas de *Rhipicephalus sanguineus* (Latreille, 1806) (Acari: Ixodidae) no Brasil. Rev Ciência Vida 18:25–33

Ringuelet R (1947) La supuesta presencia de *Ixodes brunneus* Koch en la Argentina y descripción de una nueva garrapata *Ixodes neuquenensis* nov. sp. Notas Mus Plata 12:207–216

Ringuelet R (1948) Zooparásitos de interés veterinario. Su distribución en la Argentina según comprobaciones de la Dirección de Patología Animal 1935-1945. Publicación Miscelánea de la Dirección General de Ganadería, Ministerio de Agricultura de la Nación (281):54 pp

Rivas JA (1919) Contribución al estudio de los Ixodidae de Venezuela. Anal Dirección Sanidad Nacional 1:90–96

Rivas JA (2008) Ticks (*Amblyomma* spp.) on black iguanas (*Ctenosaura similis*) in Costa Rica. Iguana 15:25–27

Rivas G, Guzmán-Cornejo C (2016) Ácaros zooparásitos de la Sierra Madre del Sur. In: Luna-Veja I, Espinosa D, Contreras R (eds) Biodiversidad de la Sierra Madre del Sur, Universidad Nacional Autónoma de México, Ciudad de México, pp 325–333

Rivera-Páez FA, Labruna MB, Martins TF, Sampieri BR, Camargo-Mathias MI (2016) *Amblyomma mixtum* Koch, 1844 (Acari: Ixodidae): first record confirmation in Colombia using morphological and molecular analyses. Ticks Tick-Borne Dis 7:842–848

Rivera-Páez FA, Labruna MB, Martins TF, Pérez JE, Castaño-Villa GJ, Ossa-López PA, Gil CA, Sampieri BR, Aricapa-Giraldo HJ, Camargo-Mathias MI (2018) Contribution to the knowledge of hard ticks (Acari: Ixodidae) in Colombia. Ticks Tick-Borne Dis 9:57–66

Rivero R, Minoli P, Armúa-Fernández MC, Gianneechini E, Carvalho L, Venzal JM (2017) Descripción de un foco de rangeliosis en el litoral noroeste del Uruguay. Veterinaria (Montevideo) 208:15–22

Robayo-Sánchez LN, Ramírez-Hernández S, Cortés-Vecino JA (2020) *Amblyomma nodosum* Neumann, 1889 [sic] on collared anteaters (*Tamandua tetradactyla*) from the Andean region of Colombia. Braz J Vet Parasitol 29(4)(article e007120):8 pp

Robbins RG, Deem SL (2002) Ticks of the genus *Amblyomma* (Acari: Ixodida: Ixodidae) from a maned wolf, *Chrysocyon brachyurus*, with the first report of *A. ovale* from this endangered canid. Entomol News 113:25–28

Robbins RG, Keirans JE (1987) *Ixodes* (*Ixodiopsis*) *woodi* (Acari: Ixodidae): description of the larva and redescription of the nymph. J Med Entomol 24:310–314

Robbins GR, Karesh WB, Painter LE, Rosenberg S (1998) Ticks of the genus *Amblyomma* (Acari: Ixodida: Ixodidae) from white-lipped peccaries, *Tayassu pecari*, in northeastern Bolivia, with comments on host specificity. Entomol News 109:172–176

Robbins RG, Platt SG, Rainwater T, Weisman W (2001a) Statistical measures of association between *Amblyomma sabanerae* Stoll (Acari: Ixodida: Ixodidae) and the furrowed wood turtle, *Rhinoclemmys areolata* (Duméril and Bibron) (Testudines: Emydidae) in northern Belize. Proc Entomol Soc Washington 103:54–59

Robbins RG, Sorkin LN, Vuilleumier F (2001b) First report of *Ixodes auritulus* Neumann (Acari: Ixodida: Ixodidae) from the blacklish cinclodes, *Cinclodes antarcticus* (Garnot) (Aves: Passeriformes: Furnariidae), with additional records of parasitism of *Cinclodes* spp. by this tick species. Proc Entomol Soc Washington 103:265–268

Robbins RG, Deem SL, Occi JL (2003a) First report of *Amblyomma humerale* Koch (Acari: Ixodida: Ixodidae) from Bolivia, with a synopsis of collections of this tick from the South American yellow-footed tortoise, *Geochelone denticulata* (L.) (Reptilia: Testudines: Testudinae). Proc Entomol Soc Washington 105:502–505

Robbins RG, Deem SL, Noss AJ, Greco V (2003b) First report of *Amblyomma pseudoconcolor* Aragão (Acari: Ixodida: Ixodidae) from Bolivia, with a new record of this tick from the grey brocket deer, *Mazama gouazoupira* (G. Fischer) (Mammalia: Artiodactyla: Cervidae). Proc Entomol Soc Washington 105:1053–1055

Roberts FHS (1960) A systematic study of the Australian species of the genus *Ixodes* (Acarina: Ixodidae). Aust J Zool 8:392–485

Roberts FHS (1969) The larvae of Australian Ixodidae (Acarina: Ixodoidea). J Aust Entomol Soc 8:37–78

Roberts FHS (1970) Australian ticks. CSIRO, Melbourne, 267 pp

Robinson LE (1912) New species of ticks (*Haemaphysalis*, *Amblyomma*). Parasitology 4:478–484

Robinson LE (1926) Ticks. A monograph of the Ixodoidea. Part IV. The genus *Amblyomma*. Cambridge University Press, London, 302 pp

Rocha VJ, Sekiama ML, Gonçalves DD, Sampieri BR, Barbosa GP, Dias TC, Rossi HR, Souza PFP (2017) Capivaras (*Hydrochoerus hydrochaeris*) e a presença do carrapato (*Amblyomma sculptum*) no campus da UFSCAR-Araras, São Paulo. Ciência Anim Bras 18(article e44671):15 pp

Rodrigues DS, Maciel R, Cunha LM, Leite RC, Oliveira PR (2010) *Amblyomma rotundatum* (Koch, 1844) [sic] (Acari: Ixodidae) two-host life cycle on Viperidae snakes. Braz J Vet Parasitol 19:174–178

Rodríguez R, Flores A (2003) Identificación y prevalencia de garrapatas (Acari: Ixodidae) en caprinos del semi-árido falconiano. Vet Trop 28:169–175

Rodríguez González M, Lujambio L (1954) Los ixodidos del perro en el Uruguay. Anal Facult Vet 6:107–111

Rodríguez EV, Mollericona JL, Nallar R (2019) Garrapatas del género *Amblyomma* (Acari: Ixodidae) parasitando a tayasuidos silvestres (*Tayassu pecari* y *Pecari tajacu*) en la Reserva de la Biósfera Pilón Lajas, Beni-Bolivia. Ecol Bolivia 54:96–107

Rodríguez-Hidalgo R, Pérez-Otáñez X, Garcés-Carreras S, Vanwambeke O, Madder M, Benítez-Ortiz W. (2017) The current status of resistance to alpha-cypermethrin, ivermectin, and amitraz of the cattle tick (*Rhipicephalus microplus*) in Ecuador. PLoS One 12(4)(article e0174652):15 pp

Rodríguez-Peraza JL, Forlano MD, Meléndez RD, Carrero AA, Sánchez H, Sira E, Mujica FF (2014) Identificación morfológica de *Amblyomma longirostre* (Koch, 1844) (Acari: Ixodidae) en Venezuela. Gaceta Ciencias Vet 19(2):40–45

Rodríguez-Vivas RI, Ojeda-Chi MM, Rosado-Aguilar JA, Trinidad-Martínez IC, Torres-Acosta JFJ, Ticante-Pérez V, Castro-Marín JM, Tapia-Moo CA, Vázquez-Gómez G (2013) Red deer (*Cervus elaphus*) as a host for the cattle tick *Rhipicephalus microplus* (Acari: Ixodidae) in Yucatán, Mexico. Exp Appl Acarol 60:543–552

Rodríguez-Vivas RI, Apanaskevich DA, Ojeda-Chi MM, Trinidad-Martínez I, Reyes-Novelo E, Esteve-Gassent MD, Pérez de León AA (2016) Ticks collected from humans, domestic animals, and wildlife in Yucatán. Mexico. Vet Parasitol 215:106–113

Rohr CJ (1909) Estudos sobre ixódidas do Brasil. Gómes Irmao, Rio de Janeiro, 220 pp

Rojas G, Chávez G, Alvis R, Pino J, Shiga B (2010) Ocurrencia de ectoparasitismo de *Amblyomma rotundatum* Koch, 1844 (Acari: Ixodidae) en *Bufo marinus* Linnaeus, 1758. Científica 7:131–137

Román EM, Klober R, Manzanilla J (2002) *Amblyomma cajennense* (Fabricius, 1787) (Acari: Ixodidae) en los estados de Aragua y Cojedes. Rev Científica FCV-LUZ 12:94–96

Romero-Castañon S, Ferguson BG, Güiris D, González D, López S, Paredes A, Webber M (2008) Comparative parasitology of wild and domestic ungulates in the Selva Lacandona, Chiapas, México. Comp Parasitol 75:115–126

Rosas FCW, Cruz TMS, Gianizella SL, Ramalheira CS, Martins TF (2016) A first record of ticks in free-ranging giant otter (*Pteronura brasiliensis*) in the Brazilian Amazon. IUCN Otter Specialist Group Bull 33(1):3–7

Roveda R, Ringuelet R (1947) Lista de los parásitos de los animales domésticos en la Argentina. Gaceta Vet 9:67–78

Roy BC, Estrada-Peña A, Krücken J, Rehman A, Nihoj AM (2018) Morphological and phylogenetic analyses of *Rhipicephalus microplus* ticks from Bangladesh, Pakistan and Myanmar. Ticks Tick-Borne Dis 9:1069–1079

Salas LM, González-Acuña D (2015) *Rhipicephalus sanguineus* sensu lato en Chile. Parasitol Latinoam 64:68–75

Salvador CH, Carvalho-Pinto C, Carvalho R, Graipel ME, Simões-Lopes PC (2007) Interação parasito-hospedeiro entre ectoparasitos (Ixodida & Siphonaptera) e gambás *Didelphis aurita* Wied-Neuwied, 1826 (Mammalia: Didelphimorphia), no continente e em ilhas do litoral de Santa Catarina, sul do Brasil. Biotemas 20:81–90

Sameroff S, Tokarz R, Charles RA, Jain K, Oleynik A, Che X, Georges K, Carrington C, Lipkin WI, Oura C (2019) Viral diversity of tick species parasitizing cattle and dogs in Trinidad and Tobago. Sci Rep 9(article 1042):10 pp

Sanches GS, Bechara GH, Camargo-Mathias MI (2009) Morphological description of *Amblyomma brasiliense* Aragão, 1908 (Acari: Ixodidae) larvae [sic] and nymphs [sic]. Braz J Vet Parasitol 18 (3):15–21

Sanches GS, Martins TF, Lopes IT, Costa LFS, Nunes PH, Camargo-Mathias MI, Labruna MB (2013) Ticks infesting birds in Atlantic Forest fragments in Rio Claro, State of São Paulo, Brazil. Braz J Vet Parasitol 22:6–12

Sanches GS, Évora PM, Mangold AJ, Jittapalapong S, Rodríguez-Mallon A, Guzmán PEE, Bechara GH, Camargo-Mathias MI (2016) Molecular, biological, and morphometric comparisons between different geographical populations of *Rhipicephalus sanguineus* sensu lato (Acari: Ixodidae). Vet Parasitol 215:78–87

Sánchez JP, Nava S, Lareschi M, Udrizar Sauthier DE, Mangold AJ, Guglielmone AA (2010) Host range and geographical distribution of *Ixodes sigelos* (Acari: Ixodidae). Exp Appl Acarol 52:199–205

Sánchez-Montes S, Isaak-Delgado AB, Guzmán-Cornejo C, Rendón-Franco E, Muñoz-García CI, Bermúdez S, Morales-Díaz J, Cruz-Romero A, Romero-Sala D, Dzul-Rosado K, Lugo-Caballero C, Colunga-Salas P, Becker I (2019) *Rickettsia* species in ticks that parasitize amphibians and reptiles: novel report from Mexico and review of the worldwide record. Ticks Tick-Borne Dis 10:987–994

Sánchez-Montes S, Fernández-Figueroa E, González-Guzmán S, Cervantes VP, Ballados-González GG, RangeL-Escareño C, Cárdenas-Ovando RA, Becker I (2020) New records of *Haemaphysalis leporispalustris* in the Transmexican Volcanic Belt province of Mexico with detection of rickettsial infection. Parasitol Res 119:1969–1973

San-Martín-Órdenes J, Muñoz-Leal S, Garín CF, González-Acuña D (2019) A systematic review of parasites and micropredators of non-avian reptiles (Reptilia-Sauropsida) in Chile. Zootaxa 4543:301–340

Santodomingo A, Cotes-Perdomo A, Foley J, Castro LR (2018) Rickettsial infection in ticks (Acari: Ixodidae) from reptiles in the Colombian Caribbean. Ticks Tick-Borne Dis 9:623–628

Santodomingo A, Sierra-Orozco K, Cotes-Perdomo A, Castro LR (2019) Molecular detection of *Rickettsia* spp., *Anaplasma platys* and *Theileria equi* in ticks collected from horses in Tayrona National Park, Colombia. Exp Appl Acarol 77:411–423

Santos CA, Suzin A, Vogliotti A, Nunes PH, Barbieri ARM, Labruna MB, Szabó MPJ, Yokosawa J (2020) Molecular detection of a *Borrelia* sp. in nymphs of *Amblyomma brasiliense* ticks (Acari: Ixodidae) from Iguaçu National Park, Brazil, genetically related to *Borrelia* from Ethiopia and Côte d'Ivoire. Ticks Tick-Borne Dis 11(6)(article 101519):7 pp

Santos NJ, Delabie JHC, Saraiva DG, Nieri-Bastos FA, Martins TF, Labruna MB, Carneiro IO, Franke CR (2018) Effect of forest cover on ectoparasites of small mammals in the Brazilian Atlantic Forest. Oecol Australis 22:279–290

Santos Dias JAT (1953) Sobre a posição sistemática de algumas espécies africanas do género *Amblyomma* C.L. Koch (Acarina – Ixodoidea). Moçambique (73):119–139

Santos Dias JAT (1954) Um novo nome para o *Ixodes loricatus spinosus* Nuttall, 1910 (*nomen bis lectum*). Moçambique 79:79–81

Santos Dias JAT (1955) Identidade e sinonímia da espécie *Amblyomma extraoculatum* Neumann, 1899 (Acarina, Ixodoidea). Mem Estudos Mus Zool Univ Coimbra 229:6 pp

Santos Dias JAT (1958a) Notes on various ticks (Acarina – Ixodidae) in collection at some entomological institutes in Paris and London. Anais Inst Med Trop 15:459–563

Santos Dias JAT (1958b) Notas ixodológicas. IV Estudo de alguns espécimes-tipos de Schulze em colecção no Museu de Hamburgo. Mem Estudos Mus Zool Univ Coimbra 250:15 pp

Santos Dias JAT (1958c) Notas ixodológicas. V Acerca de alguns ixodídeos do Museu de Hamburgo. Mem Estudos Mus Zool Univ Coimbra 253:32 pp

Santos Dias JAT (1961) Reabilitação de duas espécies do género *Amblyomma* Koch, 1844 (Acarina – Ixodoidea). Anais Serviços Vet Moçambique 7:237–243

Santos Dias JAT (1989a) Ixodídeos (Acarina-Ixodoidea) em no Museu Zoológico de Amsterdão. García Orta Sér Zool 13:75–87

Santos Dias JAT (1989b) Notas ixodológicas. X – Considerações acerca de algumas espécies do género *Amblyomma* Koch, 1844, baseadas no estudo dos respectivos holótipos. García Orta Sér Zool 13:97–106

Santos Dias JAT (1993) Contribuição para o estudo da sistemática e taxonomia das espécies do género *Aponomma* Neumann, 1899 (Acarina-Ixodoidea). Estudos, Ensaios Documentos 157:1–204

Saracho-Bottero MN, Sebastian PS, Carvalho LA, Claps LG, Mastropaolo M, Mangold AJ, Venzal JM, Nava S (2017) Presence of *Borrelia* in different populations of *Ixodes pararicinus* from northwestern Argentina. Ticks Tick-Borne Dis 8:488–493

Saracho-Bottero MN, Tarragona EL, Sebastian PS, Venzal JM, Mangold AJ, Guglielmone AA, Nava S (2018) Ticks infesting cattle and humans in the Yungas Biogeographic Province of Argentina, with notes on the presence of tick-borne bacteria. Exp Appl Acarol 74:107–116

Saracho-Bottero MN, Venzal JM, Tarragona EL, Thompson CT, Mangold AJ, Beati L, Guglielmone AA, Nava S (2020) The *Ixodes ricinus* complex (Acari: Ixodidae) in the Southern Cone of America: *Ixodes pararicinus*, *Ixodes aragaoi* and *Ixodes* cf. *I. affinis*. Parasitol Res 119:43–54

Saracho-Bottero, MN, Beati L, Venzal JM, Claps LG, Thompson CS, Mangold AJ, Guglielmone AA, Nava S (2021) *Ixodes silvanus* n. sp. (Acari: Ixodidae), a new member of the subgenus *Trichotoixodes* Reznik, 1961, from northwestern Argentina. Ticks Tickborne Dis 12(1)(article 101572):11 pp

Saraiva DG, Fournier GFSR, Martins TF, Leal KPG, Vieira FN, Câmara MVC, Costa CG, Onofrio VC, Barros-Battesti DM, Guglielmone AA, Labruna MB (2012) Ticks (Acari: Ixodidae) associated with small terrestrial mammals in the State of Minas Gerais, southeastern Brazil. Exp Appl Acarol 58:159–166

Saunders PT (1914) Notes on some parasites of live-stock in the West Indies. West Indies Bull 14:132–138

Sauter A, Eiffert H, Bommer W (1999) Lyme-Borreliose: Erregersuche in Zecken aus der Region um Manaus (Brasilien). Mitteil Österreichischen Ges Tropenmedizin Parasitol 21:39–44

Savani EAMM, Costa FB, Silva EA, Couto ACF, Gutjahr M, Alves JNMO, Santos FCP, Labruna MB (2019) Fatal Brazilian spotted fever associated with dogs and *Amblyomma aureolatum* ticks, Brazil, 2013. Emerg Infect Dis 25:2322–2323

Say T (1821) An account of the arachnides of the United States. J Acad Natl Sci Phila 2:59–82

Schulze P (1932) Neue und wenig bekannte Arten der Zeckengattungen *Amblyomma* und *Aponomma*. Z Parasitenkd 4:459–476

Schulze P (1933a) Über Zeckengynander. Z Morphol Okol Tiere 26:427–436

Schulze P (1933b) Neue Ixodiden aus den Gattungen *Amblyomma* und *Aponomma*. Zool Anz 104:317–323

Schulze P (1936a) The Norwegian Zoological expedition to the Galapagos Islands 1925, conducted by Alf Wollebaek. XIII. Die Ixodoidea der Galapagos-Inseln. Meddelelser Zool Mus (45):155–162

Schulze P (1936b) Neue und wenig bekannte Amblyommen und Aponommen aus Afrika, Südamerika, Indien, Borneo und Australien. (Ixodidae). Z Parasitenkd 8:619–637

Schulze P (1937a) Beiträge zur Kenntnis der Zeckengattung *Amblyomma*. Z Parasitenkd 9:690–694

Schulze P (1937b) *Anocentor columbianus* n. g. n. sp. (Ixod.). Zool Anz 120:24–27

Schulze P (1941a) Ein neues *Amblyomma* und ein neues *Aponomma* mit Augenrudimenten aus Haiti. Zool Anz 133:225–229

Schulze P (1941b) Neues über brasilianische Amblyommen. Zool Anz 134:93–104

Schulze P (1941c) Über das Zeichnungsmuster von *Aponomma* und über einige bemerkenswerte geographische Rasen von *Aponomma* und *Amblyomma* aus Beludschistan, Australien und Argentinien. Zool Anz 134:179–187

Schulze P (1941d) Über Krötenzecken aus Cuba. Z Parasitenkd 12:133–138

Schulze P (1950) *Amblyomma diemeniae*, eine neue Schlangenzecke aus Australien nebst Bemerkungen über abweichende Clavalbildungen in der Gattung *Amblyomma*. Z Parasitenkd 14:432–441

Scott JD, Durden LA (2015a) First record of *Amblyomma rotundatum* tick (Acari: Ixodidae) parasitizing a bird collected in Canada. Syst Appl Acarol 20:155–161

Scott JD, Durden LA (2015b) *Amblyomma dissimile* Koch (Acari: Ixodidae) parasitizes bird captured in Canada. Syst Appl Acarol 20:854–860

Scott JD, Fernando K, Banerjee SN, Durden LA, Byrne SK, Banerjee M, Mann RB, Morshed MG (2001) Birds disperse ixodid (Acari: Ixodidae) and *Borrelia burgdorferi*-infected ticks in Canada. J Med Entomol 38:493–500

Scott JD, Lee MK, Fernando K, Durden LA, Jorgensen DR, Mak S, Morshed MG (2010) Detection of Lyme disease spirochete, *Borrelia burgdorferi* sensu lato, including three novel genotypes in ticks (Acari: Ixodidae) collected from songbirds (Passeriformes) across Canada. J Vector Ecol 35:124–139

Scott JD, Clark KL, Foley JE, Durden LA, Manord JM, Smith ML (2016) First record of *Ixodes affinis* tick (Acari: Ixodidae) infected with *Borrelia burgdorferi* sensu lato collected from a migratory songbird in Canada. J Bacteriol Parasitol 7(3)(article 1000281):10 pp

Scott JD, Clark KL, Foley JE, Anderson JF, Bierman BC, Durden LA (2018) Extensive distribution of the Lyme disease bacterium, *Borrelia burgdorferi* sensu lato, in multiple tick species parasitizing avian and mammalian hosts across Canada. Healthcare 6(3)(article 131):26 pp

Scott JD, Clark KL, Durden L (2019) Presence of *Babesia odocoilei* and *Borrelia burgdorferi* sensu stricto in a tick and dual parasitism of *Amblyomma inornatum* and *Ixodes scapularis* on a bird in Canada. Healthcare 7(article 46):12 pp

Sebastian PS, Saracho Bottero MN, Carvalho L, Mackenstedt U, Lareschi M, Venzal JM, Nava S (2016) *Borrelia burgdorferi* sensu lato in *Ixodes cf. neuquenensis* and *Ixodes sigelos* ticks from the Patagonian region of Argentina. Acta Trop 162:218–221

Sebastian PS, Flores FS, Saracho-Bottero MN, Tarragona EL, Venzal JM, Nava S (2020) Molecular detection of rickettsial bacteria in ticks of the genus *Ixodes* from the Southern Cone of America. Acta Tropica 210(article 105588):5 pp

Seifert H (1959) Bekämpfung der Ekto-und Endoparasiten bei Rindern und Schafen in Perú. Vet-Med Nachrichten Marburg 3:118–135

Sénevet G (1940) Quelques ixodidés de la Guyane française. Espèces nouvelles d'*Ixodes* et d'*Amblyomma*. 6° Congreso Internacional de Entomología, Madrid, España, setiembre 1935, pp 891–898

Sénevet G, Ripert C (1967) Les larves des espèces du genre *Ixodes*. Essai de revue d'ensemble. Ann Parasitol Humaine Comp 42:79–121

Serra-Freire NM (2010) Occurrence of ticks (Acari: Ixodidae) on human hosts, in three municipalities in the State of Pará, Brazil. Braz J Vet Parasitol 19(3):141–147

Serra-Freire NM (2014) Joined species of hard ticks on Amerindians, Quilombolas and domestic animals in the State of Paraíba. Rev UNIABEU 7(17):61–68

Serra-Freire NM, Leal A (2009) Vertical dispersion and transference strategy of *Amblyomma triste* Koch, 1844 (Acari: Ixodidae) in nature from vegetation to human host. Rev Bras Med Vet 31:118–124

Serra-Freire NM, Mello RP (1993) Encontro do *Amblyomma parvum* Aragão, 1911 no parque nacional de Iguaçú, Paraná, Brasil. Braz J Vet Parasitol 2(Suppl 1) 2

Serra-Freire NM, Mello RP (2006) Entomologia e acarologia na medicina veterinária. L.F. Livros, Rio de Janeiro, 199 pp

Serra-Freire NM, Peralta ASL, Teixeira RHF, Gazeta GS, Amorim M (1995) *Amblyomma rotundatum* parasitando *Homo sapiens* no parque zoobotánico do MPEG e em Itaboraí. Arq Soc Zool Bras 16:20

Serra-Freire NM, Amorim M, Gazêta GS, Guerim L, Desidério MHG (1996) Ixodofauna de cervídeos no Brasil. Rev Bras Ciências Vet 3:51–54

Serra-Freire NM, Sena LMM, Borsoi ABP (2011) Parasitismo humano por carrapatos na Mata Atlántica, Rio de Janeiro, Brasil. EntomoBrasilis 4:67–72

Serra-Freire NM, Batalla LA, Julca-Lozano R (2012) Malformação em gonandro de *Amblyomma (Cernyomma) extraoculatum* recolhido em *Geochelone denticulata* (Linnaeus, 1766) de ambiente silvestre no Peru – relato de caso. Rev Bras Med Vet 34:31–34

Serra-Freire NM, Amorim M, Gazêta GS (2015) Diagnostic of *Ixodes woodi* Bishopp, 1911 (Acari: Ixodidae) in Brazil: immigration on *Homo sapiens* – case report. Rev Bras Med Vet 37:97–99

Sevá AP, Martins TF, Muñoz-Leal S, Rodrigues AC, Pinter A, Luz HR, Angerami RN, Labruna MB (2019) A human case of spotted fever caused by *Rickettsia parkeri* strain Atlantic rainforest and its association to the tick *Amblyomma ovale*. Parasites Vec 12(article 471):5 pp

Silva LHM, Rocha e Silva R, Serra-Freire NM (1999) *Amblyomma longirostre* Koch, 1844 [sic], parasita de *Sphiggurus insidiosus* Cuvier, 1825. Entomol Vectores 6:455–459

Silva AS, Silva MK, Monteiro SG (2007) Parasitismo por *Amblyomma triste* em gato doméstico. Braz J Vet Parasitol 16:108–109

Silva JB, Fonseca AH, Barbosa JD (2015) Molecular characterization of *Anaplasma marginale* in ticks naturally feeding on buffaloes. Infect Genet Evol 35:38–41

Silva CB, Santos HA, Navarrete MG, Ribeiro CCDU, González BC, Zaldívar MF, Pires MS, Peckle M, Costa RL, Vitari GLV, Massard CL (2016a) Molecular detection and characterization of *Anaplasma platys* in dogs and ticks in Cuba. Ticks Tick-Borne Dis 7:938–944

Silva TKS, Blanco CM, Lemos ERS, Ogrzewalska M (2016b) Notes on parasitism and screening for microorganism of ticks *Amblyomma* (Acari: Ixodidae), Amazon, Brazil. Virus Rev Res 21:41–44

Silva AB, Cardoso KM, Oliveira SV, Costa RMF, Oliveira G, Amorim M, Alves LC, Monteiro MFM, Gazeta GS (2018a) *Rickettsia amblyommatis* infecting *Amblyomma pseudoconcolor* in area of new focus of spotted fever in northeast Brazil. Acta Trop 182:305–308

Silva MRL, Fornazari F, Martins TF, Hippólito AG, Rolim LS, Bisca JM, Teixeira CR, O'Dwyer, L.H. (2018b) A survey of hemoparasites and ectoparasites in *Nasua nasua* Linnaeus, 1766 with a redescription of *Hepatozoon procyonis* Richards, 1961 based on morphological and molecular data. Parasitol Res 117:2159–2169

Silva-Goytia R, Elizondo A (1952) Estudio sobre fiebre manchada en México. II. Parásitos hematófagos encontrados naturalmente infectados. Rev Med Mex 32:278–282

Silveira JAG, Oliveira PA, Curi NHA, Barata RSL, Chiarello AG, Ribeiro MFB (2008) Ocorrência de *Amblyomma longirostre* (Koch, 1844) em *Chaetomys subspinosus* (Olfers, 1818) oriundos da Mata Atlântica. Arq Bras Med Vet Zoot 60:772–774

Silveira I, Martins TF, Olegário MM, Peterka C, Guedes E, Ferreira F, Labruna MB (2015) Rickettsial infection in animals, humans and ticks in Paulicéia, Brazil. Zoonoses Public Health 62:525–533

Sinkoc AL, Brum JGW (1997) Ocorrência de *Amblyomma fuscum* Neumann, 1899 e *Amblyomma humerale* Koch, 1844 (Acari: Ixodoidea) em *Bufo arenalis* no estado de São Paulo, Brasil. Ciência Rural 27:339–340

Sinkoc AL, Brum JGW, Moraes W, Crawshaw P (1998) Ixodidae parasitos de animais silvestres na região de Foz do Iguaçú, Brasil e Argentina. Arq Inst Biol 65:29–33

Siuda K, Nowak M, Karbowiak G, Solarz G (2005) Transfer of exotic ticks (Acari, Ixodida) on hosts to Poland. Vestnik Zool (Suppl 19):336–338

Smith MW (1974) A survey of the distribution of the ixodid ticks *Boophilus microplus* (Canestrini, 1888) and *Amblyomma cajennense* (Fabricius, 1787) in Trinidad and Tobago and the possible influence of the survey results on planned livestock development. Trop Agric (Trinidad) 51:559–567

Smith CN, Gouck HK (1947) *Ixodes bishoppi*, a new species from Georgia (Acarina: Ixodidae). Ann Entomol Soc Am 40:75–81

Smith RL, Schnack JA, Schaefer EF, Kehr AI (2008) Ticks, *Amblyomma rotundatum* (Acari: Ixodidae), on toads, *Chaunus schneideri* and *Chaunus granulosus* (Anura: Bufonidae), in northern Argentina. J Parasitol 94:560–562

Soares HS, Barbieri ARM, Martins TF, Minervino AHH, Lima JTR, Marcili A, Gennari SM, Labruna MB (2015) Ticks and rickettsial infection in the wildlife of two regions of the Brazilian Amazon. Exp Appl Acarol 65:125–140

Solís-Hernández A, Rodríguez-Vivas RI, Pérez-Barrera MA, Esteve-Gasent MA, Apanaskevich DA (2015) *Ixodes affinis* (Acari: Ixodidae) in dogs from rural localities of Yucatán, Mexico: prevalence, abundance and associated factors. Vet Méx OA 21(article 3):9 pp

Sonenshine DE (1979) Ticks of Virginia (Acari: Metastigmata). Insects Virginia 13:44 pp

Sosa-Gutiérrez CG, Vargas-Sandoval M, Torres J, Gordillo-Pérez G (2016) Tick-borne rickettsial pathogens in questing ticks, removed from humans and animals in Mexico. J Vet Sci 17:353–360

Sousa KCM, Fernandes MP, Herrera HM, Benevenute JL, Santos FM, Rocha FL, Barreto WTG, Macedo GC, Campos JB, Martins TF, Pinto PCEA, Barros Battesti DM, Piranda EM, Cançado PHD, Machado RZ, André MR (2017) Molecular detection of *Hepatozoon* spp. in domestic dogs and wild mammals in Southern Pantanal, Brazil with implications in the transmission route. Vet Parasitol 237:37–46

Souza U, Dall'Agnol B, Webster A, Weck B, Doyle R, Kasper CB, Soares J, Martins JR, Trigo TC, Ott R, Jardim MMA, Reck J (2018) Molecular survey of *Rickettsia* spp. in the Neotropical deer tick *Haemaphysalis juxtakochi* from Brazilian Pampa. Parasitol Res 117:3293–3298

Souza VL, Martins TF, Guilherme E, Santos FGA (2020) New records of ticks (Acari: Ixodidae) infesting wild birds in a forest fragment in Acre, Brazilian Amazon. Sem Ciências Agrárias 41 (Supl 1):2163–2170

Sponchiado J, Melo GL, Martins TF, Krawczak FS, Labruna MB, Caceres NC (2015) Association patterns of ticks (Acari: Ixodida: Ixodidae, Argasidae) of small mammals in Cerrado fragments, western Brazil. Exp Appl Acarol 66:369–381

Springer A, Montenegro VM, Schicht S, Wolfel S, Schaper SR, Chitimia-Dobler L, Siebert S, Strube C (2018) Detection of *Rickettsia monacensis* and *Rickettsia amblyommatis* in ticks collected from dogs in Costa Rica and Nicaragua. Ticks Tick-Borne Dis 9:1565–1572

Squire FA (1972) Entomological problems in Bolivia. PANS 18:249–268

Stebbins RC, Hendrickson JR (1959) Field studies of amphibians in Colombia, South America. Univ Calif Publ Zool 56:497–540

Stiles CW (1908) The common tick (*Dermacentor andersoni*) of the Bitter Root Valley. Publ Health Rep US Marine Hosp Serv 23:949

Stiles CW (1910) The taxonomic value of the microscopic structure of the stigmal plate in the genus *Dermacentor*. Bull Hyg Lab US Publ Health Mar-Hosp Serv 62:67 pp

Stoll O (1894) *Arachnida Acaridae*. Biol Centrali-Am Ser Zool 1–55

Suárez M, Rubi J, Pérez D, Córdova V, Salazar Y, Vielma A, Barrios F, Gil CA, Segura N, Carrillo Y, Cartaya R, Palacios M, Rubio E, Escalona C, Ramírez RC, Baker RB, Machado H, Sordo Y, Bermudes J, Vargas M, Montero C, Cruz A, Puente P, Rodríguez JL, Mantilla E, Oliva O, Smith E, Castillo A, Ramos B, Ramírez Y, Abad Z, Morales A, González EM, Hernández A, Ceballo Y, Callard D, Cardoso A, Navarro M, González JL, Pina R, Cueto M, Borotto C, Pimentel C, Carpio Y, Estrada MP (2016) High impact and effectiveness of Gavac vaccine in the national program for control of bovine ticks *Rhipicephalus microplus* in Venezuela. Livest Sci 187:48–52

Suárez-González VM (2004) Veterinary hygienic situation of ruminant farming with special attention to endo-and ectoparasites in the humid tropics of South America – using the example of the Sucumbíos Province in Ecuador. J Agric Rural Dev Trop Subtrop (Suppl 79):140 pp

Superina M, Guglielmone AA, Mangold AJ, Nava S, Lareschi M (2004) New distributional and host records for *Amblyomma pseudoconcolor* Aragão, 1908 (Acari: Ixodidae: Amblyomminae). Syst Appl Acarol 9:41–43

Supino F (1897) Nuovi *Ixodes* della Birmania (nota preventiva). Atti Soc Veneto-Trentina Sci Nat Residente Padova, Ser 2(3):230–238

Szabó MPJ, Labruna MB, Vogliotti A, Duarte JMB (2006) Ticks (Acari: Ixodidae) on small red brocket deer (*Mazama bororo* Duarte) along deer trails in the Atlantic Rain Forest of south-eastern Brazil. Syst Appl Acarol 11:41–45

Szabó MPJ, Olegário MMM, Santos ALQ (2007) Tick fauna from two locations in the Brazilian savannah. Exp Appl Acarol 43:73–84

Szabó MPJ, Pascoli GVT, Marçal O, Franchin AG, Torga K (2008) Brown dog tick *Rhipicephalus sanguineus* parasitizing the bird *Coereba flaveola* in the Brazilian Cerrado. Ciência Rural 38:543–545

Szabó MPJ, Nieri-Bastos FA, Spolidorio MG, Martins TF, Barbieri AM, Labruna MB (2013) *In vitro* isolation from *Amblyomma ovale* (Acari: Ixodidae) and ecological aspects of the Atlantic rainforest *Rickettsia*, the causative agent of a novel spotted fever rickettsiosis in Brazil. Parasitology 140:719–728

Szabó MPJ, Martins MM, Castro MB, Pacheco RC, Tolesano-Pascoli GV, Santos KT, Martins TF, Souza LGA, May-Junior JA, Yokosawa J, Labruna MB (2018) Ticks (Acari: Ixodidae) in Serra da Canastra National Park in Minas Gerais, Brazil: species, abundance, ecological and seasonal aspects with notes on rickettsial infection. Exp Appl Acarol 76:381–397

Szabó MPJ, Pascoal JO, Martins MM, Ramos VN, Osava CF, Santos ALQ, Yokosawa J, Rezende LM, Tolesano-Pascoli GV, Torga K, Castro MB, Suzin A, Barbieri ARM, Werther K, Silva JMM, Labruna MB (2019) Ticks and *Rickettsia* in anteaters from south-east and central-west Brazil. Ticks Tick-Borne Dis 10:540–545

Szabó MPJ, Martins TF, Barbieri ARM, Costa FB, Soares HS, Tolesano-Pascoli GV, Torga K, Saraiva DG, Ramos VN, Osava CF, Castro MB, Labruna MB (2020) Ticks biting humans in the Brazilian savannah: attachment sites and exposure risk in relation to species, life stage and season. Ticks Tick-Borne Dis 11(2)(article 101328):6 pep

Tagle I (1971) Ixodoidea en Chile. Bol Chil Parasitol 26:46–49

Tarragona EL, Eberhardt MA, Zurvera D, Beldoménico PM, Mastropaolo M (2012) Primer registro de *Amblyomma aureolatum* (Pallas, 1772) y *Amblyomma ovale* Koch, 1844 (Acari: Ixodidae) en la provincia de Santa Fe, Argentina. Rev FAVE Sección Ciencias Vet 11:59–63

Tarragona EL, Mangold AJ, Mastropaolo M, Guglielmone AA, Nava S (2015) Ecology and genetic variation of *Amblyomma tonelliae* in Argentina. Med Vet Entomol 29:297–304

Tarragona EL, Sebastian PS, Bottero MNS, Martínez MI, Debárbora VN, Mangold AJ, Guglielmone AA, Nava S (2018) Seasonal dynamics, geographical range size, hosts, genetic diversity and phylogeography of *Amblyomma sculptum* in Argentina. Ticks Tick-Borne Dis 9:1264–1274

Tate HD (1941) The biology of the tropical cattle tick and other species of ticks in Puerto Rico, with notes on the effects on ticks of arsenical dips. J Agric Univ P R 25:1–24

Teixeira RHF, Amorim M, Gazeta GS, Serra-Freire NM (2003) Ixodofauna de répteis cativos no zoológico de Sorocaba, São Paulo, Brasil. Entomol Vectores 10:319–329

Teixeira RHF, Ferreira I, Amorim A, Gazeta GS, Serra-Freire NM (2008) Carrapatos em aves selvagens no Zoológico de Sorocaba – São Paulo, Brasil. Arq Bras Med Vet Zoot 60:1277–1280

Teixeira RHF, Miranda FR, Kluyber D, Lopes AMC, Ferreira VL, Guedes NMR, Bittencourt KG, Gazeta GS, Serra-Freire NM, Amorim M (2013) Ixodid fauna of Edentata (Mammalia: Xenarthra) in Brazil. Rev Saúde Física Mental-UNIABEU 3(2):59–73

Teixeira RHF, Martins TF, Labruna MB (2017) Ixodídeos coletados em animais selvagens no Zoológico de Sorocaba, São Paulo, Brasil. Rev Educação Continuada Med Vet Zoot CRMV-SP 15:8–14

Teixeira RHF, Luz HR, Pacheco RC, Onofrio VC, Amorim M, Gazêta GS, Silva PJ, Bitencourth K, Marques S, Mattos MO, Hernandes LSI, Milanelo L, Furuya HR, Silva VP, Petri B, Fitorra LS, Dores FT, Sanches TC, Joppert AMJ, Navas-Soares PE, Fagundes-Moreira R, Soares JF, Costa ALM, Galassi GG, Spina MA, Horta MC, Faccini JLH, Labruna MB, Martins TF (2020) Ticks (Acari: Ixodidae) on wild raptors in Brazil. Int J Acarol 46:357–363

Tendeiro J (1952) Ixodídeos da Guiné Portuguesa. Chaves para a identificação das carraças guineenses Lista de hospedeiros e de espécies. Anais Inst Med Trop 9:195–263

Teng KF, Jiang ZJ (1991) Economic insect fauna of China. Fascicule 39, Acari: Ixodidae. Science Press, Beijing, 355 pp [in Chinese]

Theiler G (1943) Notes on the ticks of domestic stock from Portuguese East Africa. Estação Antimalária, Lourenço Marques, 55 pp

Theiler G, Salisbury LE (1959) Ticks in the South African Zoological Survey Collection – Part IX – "The *Amblyomma marmoreum* group". Onderstepoort J Vet Res 28:47–124

Thompson GB (1950) Ticks in Jamaica, B.W.I. Records and notes (including a summary of the distribution of the West Indian species). Ann Mag Nat Hist Ser 12(3):220–229

Thompson AT, Domínguez K, Cleveland CA, Dergoussof SJ, Doi K, Falco RC, Greay T, Irwin P, Lindsay LR, Liu J, Mather TN, Oskam CL, Rodríguez-Vivas RI, Ruder MG, Shaw D, Vigil SL, White S, Yabsley MJ (2020) Molecular characterization of *Haemaphysalis* species and a molecular genetic key for the identification of *Haemaphysalis* of North America. Front Vet Sci 7(article 141):11 pp

Tolesano-Pascoli GV, Torga K, Franchin AG, Ogrzewalska M, Gerardi M, Olegário MMM, Labruna MB, Szabó MPJ, Marçal O (2010) Ticks on birds in a forest fragment of Brazilian Cerrado (savanna) in the Municipality of Uberlandia, State of Minas Gerais, Brazil. Braz J Vet Parasitol 19:244–248

Tolesano-Pascoli G, García FI, Gomes CRG, Diniz KC, Onofrio VC, Venzal JM, Szabó MPJ (2014) Ticks (Acari: Ixodidae) on swifts (Apodiformes: Apodidae) in Minas Gerais, southeastern Brazil. Exp Appl Acarol 64:259–263

Tomassone L, Nuñez P, Ceballos LA, Gürtler RE, Kitron UD, Farber M (2010) Detection of "*Candidatus* Rickettsia sp. strain Argentina" in *Amblyomma* ticks (Acari: Ixodidae) from northern Argentina. Exp Appl Acarol 52:93–100

Tonelli Rondelli M (1931) Ixodoidea del museo di Torino. Boll Mus Zool Anatom Comp 41:1–10

Tonelli Rondelli M (1937) Ixodoidea. Parte I. *Amblyomma ovale* Koch, *Amblyomma cajennense* Fabricius [sic] e le specie a loro affini nuove o poco note. Riv Parassitol 1:273–298

Tonelli Rondelli M (1939) Ixodoidea. Parte II. Contributo alla conoscenza della fauna ixodologica Sud-americana. Riv Parassitol 3:39–55

Tonn RJ, Kohls GM, Arnold K (1963) Ectoparasites of birds and mammals of Costa Rica. 2. Ticks. Rev Biol Trop 11:217–220

Torga K, Tolesano-Pascoli G, Vásquez JB, Silva Júnior EL, Labruna MB, Martins TF, Ogrzewalska M, Szabó MPJ (2013) Ticks on birds from Cerrado forest patches along the Uberlandia River in the Triangulo Mineiro region of Minas Gerais, Brazil. Ciência Rural 43:1852–1857

Torregiani G (1914) I zooparassiti osservati nella fauna Boliviana. Nouvo Ercolani 19:417–425

Torres AC, Minervino AHH, Santos AP, Sarturi C, Martins TF, Vale WG, Morini AC, Marcili A, Portela JC, Gennari SM (2018) *Amblyomma* ticks infesting amphibians and Squamata reptiles from the lower Amazon region, Brazil. Exp Appl Acarol 75:399–407

Torres-Mejía AM, de la Fuente J (2006) Risks associated with ectoparasites of wild mammals in the Department of Quindío, Colombia. Int J Appl Res Vet Med 4:187–192

Tovar RM (1944) Existencia del *Dermacentor parumapertus* Neumann, *Amblyomma inornatum* Banks y *Amblyomma maculatum* Koch en México. Rev Inst Salubr Enferm Trop 5:293–295

Trapido H, Sanmartín C (1971) Pichindé virus. A new virus of the Tacaribe group from Colombia. Am J Trop Med Hyg 20:631–641

Troyo A, Moreira-Soto A, Carranza M, Calderón-Arguedas O, Hun L, Taylor L (2014) Detection of an undescribed *Rickettsia* sp. in *Ixodes boliviensis* from Costa Rica. Ticks Tick-Borne Dis 5:883–886

Troyo A, Moreira-Soto A, Calderón-Arguedas O, Mata-Somarribas C, Ortiz-Tello J, Barbieri ARM, Avendaño A, Vargas-Castro LE, Labruna MB, Hun L, Taylor L (2016) Detection of rickettsiae in fleas and ticks from areas of Costa Rica with history of spotted fever group rickettsioses. Ticks Tick-Borne Dis 7:1128–1134

Turk FA (1948) On two new species of ticks. Parasitology 38:243–246

Uilenberg G, Hoogstraal H, Klein JM (1979) Les tiques (Ixodoidea) de Madagascar et leur rôle vecteur. Archives de l'Institut Pasteur de Madagascar, Numéro Spécial, 153 pp

Uilenberg G, Barré N, Camus E, Burridge MJ, Garris GI (1984) Heartwater in the Caribbean. Prev Vet Med 2:255–267

Ulloa-García A, Dzul-Rosado K, Bermúdez-Castillero SE, López-López N, Torres-Monazón JA (2020) Detección de *Rickettsia typhi* en *Rhipicephalus sanguineus s.l.* y *Amblyomma mixtum* en el sur de México. Salud Públ Méx 62:358–363

Uribe JE, Nava S, Murphy KR, Tarragona E, Castro LR (2020) Characterization of the complete mitochondrial genome of *Amblyomma ovale*, comparative analyses and phylogenetic considerations. Exp Appl Acarol 81:421–439

Urushiyama F, Matsubara K, Doi K, Taira M, Komiya T, Tokiwa T (2020) First record of infestation of a pet sloth in Japan with the exotic tick *Amblyomma geayi* (Acari: Ixodidae). Parasitol Int 78(article 102157):3 pp

Valencia CE, Appel VA, Ruiz GA, Salazar H (2017) Identification and prevalence of ticks (Acari, Ixodidae) in bovines in five municipalities of the Popayán Plateau (Colombia). Rev Colomb Ciencias Pecuarias 30:239–237

Valente JDM, Silva PW, Arzua M, Barros-Battesti DM, Martins TF, Silva AM, Labruna MB, Vieira RFC (2020) Records of ticks (Acari: Ixodidae) on humans and distribution of spotted-fever cases and its tick vectors in Paraná State, southern Brazil. Ticks Tick-Borne Dis 11(6) (article 101510):11 pp

Valle S, Caicedo W, Masapanta L (2020) Eficacia de un aislamiento nativo de *Metarhizium anisopliae* (T16301) para el control de garrapata (*Rhipicephalus microplus*) adultas en condiciones *in vitro*. Livestock Res Rural Dev 32(7)(article #114):10 pp

Vallejos-Rayos JD (2011) Diversidad de los géneros y especies de garrapatas en animales domésticos del municipio de Ciudad Darío, Matagalpa, en el período de enero a abril de 2010. Escuela de Medicina Veterinaria, Universidad Nacional de Nicaragua-León, 19 pp

Valle-Mendoza J, Rojas-Jaime J, Vásquez-Achaya F, Aguilar-Luis MA, Correa-Nuñez G, Silva-Caso W, Lescano AG, Song X, Liu Q, Li D (2018) Molecular identification of *Bartonella bacilliformis* in ticks collected from two species of wild mammals in Madre de Dios: Peru. BMC Res Notes 11(article 405):5 pp

Van Volkenberg HL (1939) An annotated check list of the parasites of animals in Puerto Rico. US Dept Agric Circ Puerto Rico Exp Station 22:12 pp

Vargas L (1955) Relación del papel patógeno de las garrapatas y listas de las especies mexicanas. Gac Med Mex 85:489–502

Vargas L, Hoffmann A (1952) *Amblyomma castañedai* [sic] n. sp. de México (Ixodidae). Prensa Méd Mex 17:128–130

Vargas-Sandoval M, Priego-Santander AG, Larrazábal A, Sosa-Gutiérrez CG, Lara-Chávez B, Ávila-Val T (2014) Potential species distribution and richness of Ixodidae ticks associated with wild vertebrates from Michoacán, México. J Geogr Inf Syst 6:467–477

Varma RGM (1973) Ticks (Ixodidae) of British Honduras. Trans R Soc Trop Med Hyg 67:92–102

Vásquez CL, Muro JJ, Clavijo JJ (2011) Garrapatas del género *Ixodes* Latreille, 1795 y *Rhipicephalus* (*Boophilus*) Koch, 1844 (Acari: Ixodidae) presentes en la colección de zoología agrícola, decanato de agronomía, UCLA, Lara, Venezuela. Entomotropica 26:89–97

Vázquez L (1946) *Ixodes scuticrenatus*, una nueva especie de garrapata de México (Acarina: Ixodidae). An Inst Bio 17:237–245

Vecino JAC, Echeverri JAB, Cárdenas JA, Herrera LAP (2010) Distribución de garrapatas *Rhipicephalus (Boophilus) microplus* en bovinos y fincas del Altiplano cundiboyacense (Colombia). Corpoica Ciencia Tecnol Agropecuaria 11:73–84

Venzal JM, Castro O, Cabrera P, Souza C, Fregueiro G, Barros-Battesti DM, Keirans JE (2001) *Ixodes* (*Haemixodes*) *longiscutatum* [sic] Boero (new status) and *I.* (*H.*) *uruguayensis* Kohls & Cilfford, a new synonym of *I.* (*H.*) *longiscutatum* [sic] (Acari: Ixodidae). Mem Inst Oswaldo Cruz 96:1121–1122

Venzal JM, Castro O, Guglielmone AA, Keirans JE (2002) First records of *Amblyomma auricularium* (Conil, 1878) and *Amblyomma pseudoconcolor* Aragão, 1908 (Acari: Ixodidae) from Uruguay. Syst Appl Acarol 7:109–111

Venzal JM, Catro O, Claramunt S, Guglielmone AA (2003a) Primer registro de *Amblyomma longirostre* (Acari: Ixodidae) en Uruguay. Parasitol Latinoam 58:72–74

Venzal JM, González EM, Capellino D, Estrada-Peña A, Guglielmone AA (2003b) First record of *Amblyomma triste* Koch, 1844 (Acari: Ixodidae) and new records of *Ornithodoros mimon* Kohls, Clifford & Jones, 1969 (Acari: Argasidae) from Neotropical bats. Syst Appl Acarol 8:93–96

Venzal JM, Castro O, Cabrera PA, Souza CG, Guglielmone AA (2003c) Las garrapatas de Uruguay: especies, hospedadores, distribución e importancia sanitaria. Veterinaria (Montevideo) 38:17–28

Venzal JM, Félix ML, Olmos A, Mangold AJ, Guglielmone AA (2005a) A collection of ticks (Ixodidae) from wild birds in Uruguay. Exp Appl Acarol 36:325–331

Venzal JM, Estrada-Peña A, Barros-Battesti DM, Onofrio VC, Beldoménico PM (2005b) *Ixodes* (*Ixodes*) *pararicinus* Keirans & Clifford, 1985 (Acari: Ixodidae): description of the immature stages, distribution, hosts and medical/veterinary importance. Syst Parasitol 60:225–234

Venzal JM, Estrada-Peña A, Castro O, De Souza CG, Portillo A, Oteo JA (2007a) Study on seasonal activity in dogs and ehrlichial infection in *Rhipicephalus sanguineus* (Latreille, 1806) (Acari: Ixodidae) from southern Uruguay. Parasitol Latinoam 62:23–26

Venzal JM, Nava S, Guglielmone AA (2007b) Garrapatas exóticas: hallazgo de *Amblyomma latum* Koch, 1844 (Acari: Ixodidae) parasitando *Pythons regius* (Serpentes: Boidae) en Uruguay. Veterinaria (Montevideo) 42(168):15–17

Venzal JM, Nava S, Beldoménico PM, Barros-Battesti DM, Estrada-Peña A, Guglielmone AA (2008) Hosts and distribution of *Ixodes longiscutatus* Boero, 1944. Syst Appl Acarol 13:102–108

Verbel-Vergara DE, Bejarano EE, Paternina LE (2015) First report of *Amblyomma dissimile* (Acari: Ixodidae) on *Spilotes pullatus* (Squamata: Colubridae) from Colombia. Rev Invest Med Trop 1:23–25

Vercammen-Grandjean PH (1966) Evolutionary problems of mite-host specificity and its relevance to studies on Galapagos organisms. In: Bowman RI (ed) The Galapagos. University of California Press, Berkeley, pp 236–239

Verdugo C, Jiménez O, Hernández C, Álvarez P, Espinoza A, González-Acuña D (2017) Infection with *Borrelia chilensis* in *Ixodes stilesi* ticks collected from *Pudu puda* deer. Ticks Tick-Borne Dis 8:733–740

Verea C, Aponte O (2004) Algunas aves del norte de Venezuela como hospederos de ectoparásitos ixodoideos. Ornitol Neotrop 15:247–251

Vergani F (1956) Datos epizootiológicos sobre ixódidos de Venezuela. Bol Inst Invest Vet 8:46–54

Vigueras IP (1934) Descripción de la hembra de *Amblyomma albopictum* Neumann 1899. Rev Univ Habana 3:128–129

Vogel H, Foley J, Fiorello CV (2018) *Rickettsia africae* and novel rickettsial strain in *Amblyomma* spp. ticks, Nicaragua, 2013. Emerg Infect Dis 24:385–387

Vogelsang EG (1928) Garrapatas (Ixodidae) del Uruguay. Bol Inst Clín Quirúrgica 4:668–670

Vogelsang EG, Cordero EH (1939) Sobre una pequeña colección de garrapatas (Ixodidae) de la Argentina, Paraguay y Uruguay. Rev Med Vet Parasitol 1:115–116

Vogelsang EG, Cordero EH (1940) Las garrapatas (Ixodidae) de Venezuela. Rev Med Vet Parasitol 2:71–76

Vogelsang EG, Santos Dias JAT (1953a) Contribución al estudio de la fauna ixodológica en Venezuela. Rev Med Vet Parasitol 12:3–62

Vogelsang EG, Santos Dias JAT (1953b) Nueva contribución al estudio de la fauna ixodológica en Venezuela. Rev Med Vet Parasitol 12:63–89

Voltzit OV (2007) A review of Neotropical *Amblyomma* species (Acari: Ixodidae). Acarina 15:3–134

Voltzit OV, Keirans JE (2003) A review of African *Amblyomma* species (Acari, Ixodida, Ixodidae). Acarina 11:135–214

Walker JB, Olwage A (1987) The tick vectors of *Cowdria ruminantium* (Ixodoidea, Ixodidae, genus *Amblyomma*) and their distribution. Onderstepoort J Vet Res 54:353–379

Walker JB, Keirans JE, Horak IG (2000) The genus *Rhipicephalus* (Acari: Ixodidae): a guide to the brown ticks of the world. Cambridge University Press, Cambridge, 643 pp

Walker AR, Bouattour A, Camicas JL, Estrada-Peña A, Horak IG, Latif AA, Pegram RG, Preston PM (2003) Ticks of domestic animals in Africa: a guide to identification of species. Biosciences Report, Edinburgh, 221 pp

Warburton C (1933) On five new species of ticks (Arachnida Ixodoidea). *Ixodes petauristae*, *I. ampullaceus*, *Dermacentor imitans*, *Amblyomma laticaudae* and *Aponomma draconis*, with notes on three previously described species, *Ornithodorus* [sic] *franchinii* Tonelli Rondelli, *Haemaphysalis cooleyi* Bedford and *Rhipicephalus maculatus* Neumann. Parasitology 24:558–568

Webb JP, Bennett SG, Challet GL (1990) The larval ticks of the genus *Ixodes* Latreille (Acari: Ixodidae) of California. Bull Soc Vec Ecol 15:73–124

Wells EA, D'Alessandro A, Morales GA, Ángel D (1981) Mammalian wildlife diseases as hazards to man and livestock in an area of the Llanos Orientales of Colombia. J Wildl Dis 17:153–162

Welschen NM, Tarragona EL, Nava S, Guglielmone AA (2012) Confirmación de la presencia de *Amblyomma brasiliense* Aragão, 1908 (Acari: Ixodidae) en la Argentina. Rev FAVE Sección Ciencias Vet 11:65–69

White A (1852) Insects and Aptera. In: Journal of the voyage in Baffin's Bay and Barrow Straits, in the years 1850 – 1851, vol 2 (appendix), pp 208–211

Whittick RJ (1939) On some ticks belonging to the genera *Aponomma* and *Amblyomma*. Parasitology 31:434–441

Wikelski M (1999) Influence of parasites and thermoregulation on grouping tendencies in marine iguanas. Behav Ecol 10:22–29

Wilson N (1967) Mesostigmata: Rhinonyssidae, Halarachnidae (nasal mites); Metastigmata: Ixodidae (ticks). Antarctic Res Ser 10:41–49

Wilson N (1970) Acarina: Metastigmata: Ixodidae of South Georgia, Heard and Kerguelen. Pac Insects Monogr 23:78–88
Wilson N (1980) Records of ticks from Jamaica with the description of *Ixodes* (*Ixodes*) *copei* n. sp. from the mockingbird, *Mimus polyglottos* (Acari: Ixodidae). Int J Acarol 6:157–162
Wilson N, Baker WW (1988) *Ixodes downsi* (Acari: Ixodidae) from Peru. Proc Entomol Soc Wash 91:54
Wilson DE, Reeder DM (2005) Mammal species of the world. The John Hopkins University Press, Baltimore, 2142 pp
Witter R, Martins TF, Campos AK, Melo ALT, Correa SHR, Morgado TO, Wolf RW, May-Júnior JA, Sinkoc AL, Strussmann C, Aguiar DM, Rossi RV, Semedo TBF, Campos Z, Desbiez ALJ, Labruna MB, Pacheco RC (2016) Rickettsial infection in ticks (Acari: Ixodidae) of wild animals in midwestern Brazil. Ticks Tick-Borne Dis 7:415–423
Woke PA (1947) Arthropods of sanitary importance in the Republic of Nicaragua, Central America. Am J Trop Med 27:357–375
Wolffhügel K (1916) Garrapatas (Ixodidae) del Uruguay. Rev Med Vet (Montevideo) 1:106–107
Woodham CB, González-Origel A, López-León A, Gereña-Morales R (1983) Progress in the eradication of *Boophilus* ticks in Mexico 1960-80. World Anim Rev 48:18–24
Xiroudakis N (1935) Lymphangite épizootique (farcin du bouef). Recueil Méd Vét Exotique 8:112–117
Xu G, Pearson P, Dykstra E, Andrews ES, Rich SM (2019) Human-biting *Ixodes* ticks and pathogen prevalence from California, Oregon, and Washington. Vect-Borne Zoonot Dis 19:106–114
Yabsley MJ, McKibben J, Macpherson CN, Cattan PF, Cherry NA, Hegarty BC, Breitschwerdt EB, O'Connor T, Chandrashekar R, Paterson T, Perea ML, Ball G, Friesen S, Goedde J, Henderson B, Sylvester W (2008) Prevalence of *Ehrlichia canis*, *Anaplasma platys*, *Babesia canis vogeli*, *Hepatozoon canis*, *Bartonella vinsonii berkhoffii*, and *Rickettsia* spp. in dogs from Grenada. Vet Parasitol 151:279–285
Yamaguti N, Kitaoka S (1980) Family Ixodidae. In: Ebara S (ed) Illustrations of the mites and ticks of Japan. Zenkoku Kyokai, Tokyo, pp 146–161. [in Japanese]
Yamaguti N, Tipton VJ, Keegan HL, Toshioka S (1971) Ticks of Japan, Korea, and the Ryukyu Islands. Brigham Young Univ Sci Bull Biol Ser 15(1), 226 pp
Yamauchi T (2010) On the author's names in ixodid ticks named by Mr. Kodi Nakatsudi. J Acarol Soc Jpn 19:23–25. [in Japanese]
Yoshizawa MAC, Souza JL, Bredt A, Baggio D (1996) Ectoparasitos de *Rattus norvegicus* no Distrito Federal, Brasil. Braz J Vet Parasitol 5:39–42
Yunker CE, Brennan JM, Hughes LE, Philip CB, Clifford CM, Peralta PH, Vogel J (1975) Isolation of viral and rickettsial agents from Panamian Acarina. J Med Entomol 12:250–255
Yunker CE, Keirans JE, Clifford CM, Easton ER (1986) *Dermacentor* ticks (Acari: Ixodoidea: Ixodidae) of the New World: a scanning electron microscope atlas. Proc Entomol Soc Washington 88:609–627
Yust HR, Cevallos MA (1955) Lista preliminar de plagas de la agricultura del Ecuador. Rev Ecuatoriana Entomol Parasitol 2:425–442
Zepeda N, Rimbaud E, Soto JL, Pineda M, Morales X, Rivera G (2006) Ectoparásitos encontrados en *Iguana iguana* y *Boa constrictor* de zoocriaderos en Nicaragua. *20° Congreso. Panamericano de Ciencias Veterinarias*, noviembre 2006, Santiago, Chile, 1 pp. CD version
Zeringóta V, Maturano R, Santolin IDAC, McIntosh D, Famadas KM, Daemon E, Faccini JLH (2016) New host records of *Haemaphysalis leporispalustris* (Acari: Ixodidae) on birds in Brazil. Parasitol Res 115:1107–1110
Zeringóta V, Maturano R, Luz HR, Senra TOS, Daemon E, Faccini JLH, McIntosh D (2017) Molecular detection of *Rickettsia rhipicephali* and other spotted fever group *Rickettsia* species in *Amblyomma* ticks infesting wild birds in the State of Minas Gerais, Brazil. Ticks Tick-Borne Dis 8:81–89

Zerpa C, Keirans JE, Mangold AJ, Guglielmone AA (2003a) Confirmation of the presence of *Amblyomma ovale* Koch 1844 and first records of *Amblyomma scalpturatum* Neumann 1906 (Acari: Ixodida: Ixodidae) in the Amazonian region of Ecuador. Proc Entomol Soc Washington 105:783–785

Zerpa C, Venzal JM, López N, Mangold AJ, Guglielmone AA (2003b) Garrapatas de Catamarca y Tucumán: estudio de una colección de hospedadores silvestres y domésticos. Rev FAVE Sección Ciencias Vet 2:167–171

Zimmerman RH, Garris GI (1985) Use of Taylor's power law in examining the spatial distribution of free-living *Boophilus microplus* (Acari: Ixodidae) larvae in pastures in Puerto Rico. Exp Appl Acarol 3:207–211

Zimmermann NP, Aguirre AAR, Rodrigues VS, García MV, Medeiros JF, Blecha IMZ, Duarte PO, Cruz BC, Cunha RC, Martins TF, Andreotti R (2018) Wildlife species, ixodid fauna and new host records for ticks in an Amazon forest area, Rondônia, Brazil. Braz J Vet Parasitol 27:177–182